Ethology and Behavioral Ecology of Marine Mammals

Series Editor

Bernd Würsig, Department of Marine Biology, Texas A&M University at Galveston, Galveston, TX, USA

The aim of this series is to provide the latest ethological information on the major groupings of marine mammals, in six separate books roughly organized in similar manner. These groupings are the 1) toothed whales and dolphins, 2) baleen whales, 3) eared seals and walrus, 4) true seals, 5) sea otter, marine otter and polar bear, and 6) manatees and dugong, the sirens. The scope shall present 1) general patterns of ethological ways of animals in their natural environments, with a strong bent towards modern behavioral ecology; and 2) examples of particularly well-studied species and species groups for which we have enough data. The scope shall be in the form of general and specific reviews for concepts and species, with an emphasis especially on data gathered in the past 15 years or so. The editors and authors are all established scientists in their fields, even though some of them are quite young.

More information about this series at https://link.springer.com/bookseries/15983

Daniel P. Costa · Elizabeth A. McHuron
Editors

Ethology and Behavioral Ecology of Phocids

Springer

Editors
Daniel P. Costa
Center for Ocean Health
University of California, Santa Cruz
Santa Cruz, CA, USA

Elizabeth A. McHuron
Cooperative Institute for Climate, Ocean, and Ecosystem Studies
University of Washington
Seattle, WA, USA

ISSN 2523-7500 ISSN 2523-7519 (electronic)
Ethology and Behavioral Ecology of Marine Mammals
ISBN 978-3-030-88922-7 ISBN 978-3-030-88923-4 (eBook)
https://doi.org/10.1007/978-3-030-88923-4

This Springer imprint is published by the registered company Springer Nature Switzerland AG
The registered company address is: Gewerbestrasse 11, 6330 Cham, Switzerland

This book is dedicated to the memory of Bernie McConnell, a co-author of Chapter 10 "The Harbor Seal: The Most Ubiquitous Phocid in the Northern Hemisphere," who passed away unexpectedly while this book was in press.

Remembering and Celebrating Bernie McConnell by Mike Fedak

As the oldest extant member of the Sea Mammal Research Unit (SMRU) by some margin, I certainly never expected to be writing a remembrance of my friend and SMRU colleague, Bernie McConnell. Odds were that the shoe would be on the other foot. His unexpected death was a terrible shock to everyone who knew him, especially of course to his family, but also to a remarkably large number of others in our research community, both locally and internationally. At least Bernie did more to generate material for such a remembrance than anyone I know. I am not a gambling man, but I am willing to bet that even those of you who knew him only briefly will have stories to tell, so imagine the memories I have collected over the 40 years that Bernie and I have worked together.

I first met Bernie when he joined SMRU in 1980, only a year after I did. We soon began working together to provide the means to study marine mammals at sea using technology. This followed another tragedy that SMRU suffered in 1983, a helicopter crash that resulted in the deaths of two of SMRU's founder members, Mike Curry and Bill Vaughan. Mike was involved in efforts to develop telemetry approaches. After the crash, Bernie, another new recruit to SMRU, Dave Thompson, and I took on the role of continuing those early efforts, working on Argos satellite telemetry and real-time tracking with VHF and acoustic devices. It was an exciting if, sometimes, frustrating time but also great fun. And Bernie always made sure that it was that. At that stage, the work involved lots of fieldwork in remote places, working in small boats and aircraft, using fragile equipment and a lot of glue. If you knew Bernie at all, you will immediately see the potential for a wealth of stories and memorable events.

In the 40 years since we began working together, our ability to learn about the lives of marine mammals while at sea has been transformed. Bernie's role in this transformation is recognized by us all and continued right up to his untimely death. While there is still much more, we need to learn about marine animals to understand their biology and be good stewards of their environment. Our ignorance of their life at sea back in 1980 is almost impossible to imagine today.

The preface of Gentry and Kooyman's 1986 book[1] on fur seals summed that situation up. In their introduction, they spoke of Kooyman's pioneering work using the first timed depth recorders (TDRs) on Weddell seals held "captive" in ice holes drilled in the sea ice off McMurdo Station in the Weddell Sea, and Gentry's idea to use TDRs on the "behaviorally" captive Antarctic fur seal mothers who had to return to feed pups. Gentry and Kooyman went on to say, "We knew nothing of how to capture, immobilize, or harness fur seals, and our first attempts must have been amusing to the uninvolved. Our elation at the return of the first instrument carrying data could not have been greater had the animal just returned from the moon." And remember that excitement was about recording strings of TDR profiles, free of geo-referencing or any other info. Bernie and I were at least as excited when, at last in 1990, we deployed our first Argos satellite relayed data loggers on elephant seals. We tracked a female elephant seal from South Georgia as she swam and dived on a 2000 km trip down to the Antarctic Peninsula. I vividly remember Bernie and I excitedly rolling 30 m of computer line printer paper with the dive and geolocation data down the hallway at the British Antarctic Survey, with me yelling out latitude and longitude data and Bernie running along it with Admiralty Charts in hand to relate her diving to the bathymetry. The joy of being able to do that was made all the sweeter by the frustrating struggle that preceded it, trying to force the technology of the time to deliver the data over the limited bandwidth within the energy constraints imposed by the hardware of the time.

Now we take for granted that we can sit in an airport lounge while traveling to a meeting, and in near real-time, use our laptop to see where a tagged animal foraging in the Southern Ocean or the high Arctic is, what it is doing, and what its immediate surroundings are like. Biologists like Bernie, working directly with engineers, have taken advantage of the rapid consumer-driven advances in low power microelectronics to transform our ability to observe marine mammals and monitor their environment. In particular, Bernie worked to develop GSM cell phone tags for use on seals, which use the cellular network to send data stored at sea via phone calls when the animals return to land. The phoned-in data stored in online databases can be accessed wherever there is Wi-Fi. Increasingly, data are collected by a sophisticated range of sensors that can provide more detailed behavioral data. Today, we can even estimate how successful animals acquire the resources they need to survive and reproduce. Bringing in additional information from remote sensing and modeling, we can visualize their surroundings in detail, almost as if we were observing it directly. Bernie is one of many who have created the means by which these advances have been accomplished, and we recognize the importance of his role. His contributions were crucial from the start of this revolution in developing tools to remotely observe animals freely ranging in nature.

But his remembrance is not intended to be so much about the science but rather to recall and celebrate the joy and fun that Bernie brought to and derived from the work. Anyone who has worked with Bernie in the field or interacted with him at

[1] *Fur Seals: Maternal Strategies on Land and at Sea* (*Princeton* Legacy Library) by Gentry, Roger L., Kooyman, Gerald L. (ISBN: 9780691084008)

meetings will know exactly what I mean. Amazingly, Bernie maintained this output of memorable events, unabated, throughout the entirety of his >40 years at SMRU. Most were great fun, others a bit scary. In my case, arguably, only one of these was acutely life-threatening, but a few others came close. Usually, they were just a great laugh. It is so sad to have these interactions terminated by his unexpected and premature death. Bernie said he aspired to write a book of collected experiences from the marine mammal community entitled "The real materials and methods." I bet he would eventually have done this, and it would have been great. My only fear is that such an account could not have been published in the current, somewhat fearful institutional climate. I can only hope that readers will all help to maintain the memories of Bernie by recounting their own stories to those coming along who will not have had the joy of interacting with him.

Preface

Phocid (or earless or true) seals are perhaps not the first taxonomic group that come to mind when thinking about the ethology and behavioral ecology of marine mammals. Not only are they less gregarious than their otariid cousins, but many phocids, though numbering in the millions, inhabit areas that are largely inaccessible and uninhabited by humans. Elephant seals are rare exceptions; video footage of bloody fights between males during the breeding season has made its way into countless nature documentaries. While underrepresented, phocids certainly have a lot to contribute to the field of behavioral ecology, particularly in providing insight into the underlying mechanisms that drive the expression of behavior. They are ecologically diverse, occupying habitats from the tropics to the poles in marine and freshwater and feeding on anything from small zooplankton to other marine mammals. The smallest species (ringed seal) is more than 20 times smaller than the largest (southern elephant seal), with marked sexual size dimorphism present in some species. As the most derived of the pinnipeds, phocids are intimately linked with aquatic habitats. Their behavior (among other things) strongly reflects the environments under which they evolved. While we hesitate to make the claim that this is the first book solely dedicated to phocid behavior, it is certainly one of the first and we hope it will not be the last.

The chapters in the book largely fall within one of two broad categories, general patterns of phocid behavior (Part I) and species-specific descriptions of behavior (Part II). There are only 18 (or 19 if the extinct Caribbean monk seal is considered) extant phocid species. Of those, little is known about the basic biology let alone the behavior of some species. Most of the patterns described in Part I are gleaned from a few species, primarily those represented in Part II. We have taken a holistic approach in this book, focusing not only on the behaviors themselves but also on the factors that constrain the expression of behavior and the proximate mechanisms driving behavior. For example, how can we hope to understand communication (Chap. 3) without understanding the information a seal can perceive from its environment (Chap. 2)? Similarly, body size plays a large role in inter- and intraspecific variation in life-history strategies and behavior, which is primarily driven by physiological capabilities (Chaps. 4 and 8).

The book starts out where all things begin and end, with evolution. Chapter 1 (Berta et al.) describes our current understanding of the taxonomic relationships among phocids, while also highlighting the evolution of key character traits that influence behavior, such as sensory capabilities, locomotion, and body size. In describing where we have been, the authors provide some context into what might be in store for phocids regarding these key character traits. Chapters 2 and 3 focus on sensory systems and communication, respectively. In Chap. 2, Hanke and Reichmuth describe sensory capabilities, focusing on comparisons with their terrestrial counterparts and how the need to balance a semi-aquatic lifestyle has influenced sensory adaptations. In Chap. 3, Charrier and Casey focus on communication during the breeding season, a time when most social interactions occur. Chapters 4 and 8 explore physiology's role in the expression of behavior and how physiological adaptations allow phocids to do what they do. In Chap. 4, Crocker and McDonald highlight the physiological challenges faced by an aquatic mammal, such as extreme pressure, heat loss, and the need to return to the surface to breathe, and how those challenges influence phocid behavior. In Chap. 8, Costa and Maresh examine the energetics of reproduction, with particular attention paid to how some phocids are able to exclusively rely on body stores to support lactation and the influence of the environment on the evolution of reproductive strategies. Chapters 5 (Hückstädt and Reisinger), 6 (Bowen and Jonsen), and 7 (Robinson and Pomeroy) comprise the bulk of available information on phocid behavior, including movements and associations with ocean features (Chap. 5), foraging ecology (Chap. 6), and reproductive behavior (Chap. 7). Chapters 5 and 6 include sections on the statistical approaches used to detect and describe behavior, not only because behavior is rarely observed at sea, but also because the statistical approach influences the conclusions that can be drawn about such behavior.

The second part of the book describes the behavior of individual species, primarily (although not exclusively) those that have been the best studied. This section starts with chapters on two widely recognized phocids, the gray seal (Chap. 9) and the harbor seal (Chap. 10). In Chap. 9, Twiss et al. explore on-land interactions during the breeding season, highlighting interactions with conspecifics and how the physical environmental alters those interactions. This chapter also touches on the topic of animal personality, which is an emerging topic in the behavioral literature and for marine mammals, one that has been almost exclusively studied in gray seals. In Chap. 10, Harvey et al. examine the adaptable (yet easily disturbed) harbor seal, which occupies the greatest range of phocid habitats, from sub-tropical to Arctic waters and even freshwater lakes. Chapters 11 (Beltran et al.) and 12 (Le Boeuf and Naito) are devoted to elephant seals, two species where many of the technologies and techniques that have been crucial to the study of phocid behavior were pioneered. As some of the most sexually size dimorphic of all mammals, northern and southern elephant seals provide unique insights into how body size drives behavioral variation. Chapters 13 (Shero and Burns), 14 (Hammill and Stenson), and 15 (Kelly) focus on pagophilic (ice-loving) species, Weddell, harp, and ringed seals. Together, these three chapters illustrate the intimate association between phocids and their environment, in this case sea ice, and how the threat of terrestrial predation (or lack thereof) has shaped behavior. In Chap. 16, Robinson et al. explore how behavioral research has been integrated into efforts to save Hawaiian monk seals, one of the most endangered marine mammals. This is a fitting conclusion to the book since the motivation for much of the ongoing behavioral research in phocids is related to conservation or management efforts.

A common theme throughout the book is the importance of technological developments, mainly instruments that can be attached to animals (biologgers), in understanding phocid behavior. The chapters in Part II remind us that there is still a place for more traditional approaches in contributing to behavioral studies, such as observations, long-term monitoring, and even Traditional Knowledge of Indigenous communities.

Editing this book has indeed been a pleasure, and a very intellectually stimulating endeavor for us, although not without its challenges. We thank the authors for not only agreeing to contribute to the book, but for producing such fantastic chapters (and putting up with one very Type A editor). Not only were we all struggling to cope with much-altered work environments due to the ongoing pandemic, but several authors were also affected by the wildfires that raged in California this past year. It is a testament to the hard work of the authors (and a few deadline extensions) that we were able to persevere in the face of such adversity. A special thanks to Charles Littnan for helping to make sure we had a chapter on monk seals. The book was much improved by Bernd Würsig and the 19 reviewers who took the time to provide feedback on each chapter; this was no small feat given the scope of some chapters. We also thank the editors and advisors of Springer International, in particular Éva Lőrinczi. We hope you will enjoy reading the book as much we enjoyed producing it. If you learn even a small portion of what we have learned throughout editing this book, we will have succeeded in opening a window into true seals' fascinating behavior.

Santa Cruz, USA Daniel P. Costa
Seattle, USA Elizabeth A. McHuron

Introduction to the Series

We—multiple topic editors and authors—are pleased to provide a series on ethology and behavioral ecology of marine mammals. We define ethology as "the science of animal behavior," and behavioral ecology as "the science of the evolutionary basis for animal behavior due to ecological pressures." Those ecological pressures include us, the humans.

We determine, somewhat arbitrarily but with some background, that "marine mammals" habitually feed in the sea, but also include several mammals that went from saltwater oceans back into rivers, as seen in the chapter by Sutaria et al., in the first book on Odontocetes. Polar bears represent a somewhat outlier "marine mammal," as they are quite at home in the sea, but can also feed on terrestrial mammals, birds, berries, lichens, and mosses.

In six books, we include toothed whales (the odontocetes); baleen whales (the mysticetes); sea lions and fur seals (the otariids) as well as the walrus; true seals (the phocids); the special cases of the sea otter and polar bear; and manatees and the dugong (the sirens). Each of our chosen editors and their chapter authors have their own schedules, so the series will not arrive in the order given above, but within the 5 years of 2019 through 2023, all six marine mammal books on Ethology and Behavioral Ecology of Marine Mammals will see the light of day, and you, the readers, will be able to ascertain their worth and their promise, as to current knowledge and to accumulating data while our fields of science advance.

Since the first book on odontocetes came out in 2019, we added a seventh final book, on The Human Factor, with chapters on past assaults on marine mammals, continuing assaults on the marine and other environments, dawning of awareness of assaults, and perhaps ways that we humans can and must do better. Several of us simply felt that to detail modern science of marine mammal ethology and behavioral ecology was not enough—we need to be aware of the amazingly destructive Anthropocene epoch in which we live, and try to improve, for all of nature (and therefore also for us). While topics of human influence run throughout each of the first six books, a concentration on human actions and potential solutions is needed.

Not all mammals that occur in marine waters are represented, nor all that have gone back to freshwater. Thus, there is nary a mention of marine-feeding bats, marine-feeding river otters, those aspects of beluga whales that foray way up into major rivers, seals living in land-locked lakes at times thousands of kilometers from the ocean, and other species that occasionally make the marine environment or—as generally accepted marine mammals—adjacent freshwater systems their home. Such are the ways of a summary, and we apologize that we do not fully encompass all.

As series editor, I have been a science partner to all major taxonomic entities of this series, but to this only because I have been in the marine mammal field for about 50 years now, with over 65 graduate students who—in aggregate—have conducted research on all seven continents. In no manner do I pretend to have kept up with all aspects of diverse fields of modern enquiry. It is a special privilege (and delight) to have multiple up-to-date editors and their fine authors involved in this modern compilation and am extremely grateful (and humbled) for this. Still learning, and ever so.

Each chapter is reviewed by the book editors, peer reviewed by other scientists as chosen by the editors, and perused and commented on by me. If you learned something new and imparted that to your colleagues, students, or your own mentors, then the series and sections of it shall have been worthwhile.

With respect and best wishes

Atawhai, Kaikoura, New Zealand
March 2021

Bernd Würsig

Contents

Part I
Patterns of Phocid Ethology and Behavioral Ecology

Chapter 1
The Origin of Phocid Seals and Evolution of Key Behavioral Character Traits

Annalisa Berta, Morgan Churchill, and Robert W. Boessenecker

Contents

Abstract The earliest phocid seals evolved 30–24 million years ago (Ma) in the western North Atlantic. There remains uncertainty in evolutionary relationships, with molecular data positioning phocids as the sister group to all other pinnipeds (fur seals, sea lions, and walruses), while morphology places them as sister to the extinct Desmatophocidae. The two major subfamilies of extant seals, Monachinae (southern seals) and Phocinae (northern seals) diverged 18 Ma and diversified in the North Atlantic and Paratethyan region. Both clades colonized the southern hemisphere, although southern hemisphere phocines quickly went extinct, while monachine seals declined in diversity in the North Atlantic. Monachines and phocines exhibited opposing body size trends with monachines increasing and phocines decreasing in size. Phocids are the only pinnipeds to display four feeding strategies: pierce, grip and tear, suction, and filter-feeding. Phocids were ancestrally shallow water divers with increased diving capabilities evolving among some large-bodied taxa. In terms of mating systems, ice breeding appears to have independently evolved in both clades. Although faunal dynamics are inadequately studied there is evidence during

A. Berta (✉)
Department of Biology, San Diego State University, San Diego, CA 92182, USA
e-mail: aberta@sdsu.edu

M. Churchill
Department of Biology, University of Wisconsin-Oshkosh, Oshkosh, WI 54901, USA

R. W. Boessenecker
Department of Geology and Environmental Geosciences, College of Charleston, Charleston, SC 29424, USA

D. P. Costa and E. McHuron (eds.), *Ethology and Behavioral Ecology of Phocids*, Ethology and Behavioral Ecology of Marine Mammals,
https://doi.org/10.1007/978-3-030-88923-4_1

the Plio-Pleistocene for the replacement of desmatophocids and walruses by otariids and phocids and for the replacement of otariids by phocids in South America.

Keywords Monachinae · Phocinae · Desmatophocidae · Pierce · Grip and tear and filter-feeding · Pelvic oscillation · Lateral undulation

Terminology

crown group—a monophyletic group containing all living members of a clade and all descendants of a common ancestor
diphyletic—having two separate origins
holotype—a single type specimen upon which a new species is based
junior synonym—a later name of a taxon given that the earliest published name is the senior synonym
monophyly—taxa that share a common ancestry
sister taxa—taxa derived from a common ancestral node
stem group—a group that includes some or all of the shared derived characters of a clade but falls outside the crown group
tribe—a taxonomic category that ranks above genus and below family

1.1 Introduction

The true or earless seals, family Phocidae, are the most morphologically diverse group of pinnipeds, known by 19 extant species and 47 fossil species. Two clades are widely recognized based on their current biogeography, the subfamilies, Monachinae (southern seals), and Phocinae (northern seals; Davis et al. 2004; Higdon et al. 2007; Amson and Muizon 2014). Within Monachinae, three tribes are recognized as successive sister taxa to one another, Monachini (monk seals *Neomonachus* and *Monachus*), Miroungini (elephant seals *Mirounga*), and Lobodontini (Antarctic seals, *Hydrurga, Lobodon, Ommatophoca*, and *Leptonychotes*). Extant Monachinae include the monk seals that occupy the Caribbean (now extinct), Hawaii, and Mediterranean; the elephant seals found along the eastern shores of the Pacific Ocean, western shores of the South Atlantic, and many Sub-Antarctic Islands; and the lobodontine seals that inhabit the Antarctic and sub-Antarctic waters. Within Phocinae, Erignathini (bearded seal *Erignathus*) and Cystophorini (hooded seal *Cystophora*) are successive sister taxa to Phocini (*Pagophilus, Histriophoca, Halichoerus, Phoca*, and *Pusa*) (Davis et al. 2004; Higdon et al. 2007; Fulton and Strobeck 2010). Extant Phocinae is primarily restricted to the North Atlantic, North Pacific, and Arctic Oceans.

Phocid seals are distinguished from otariids (fur seals and sea lions) and odobenids (walruses) based on several ear characters, including a thick, dense mastoid region of the temporal bone and greatly inflated auditory bulla (Wyss 1987). With regard to postcranial characters, phocids are unable to turn their hind flippers forward because of the peculiar structure of their ankles (de Muizon 1982; King 1983). Compared to otariids, phocids inhabit higher latitudes and possess more extreme adaptations toward breeding on sea ice. They spend more time in the water, swim using lateral pelvic oscillation, and have a more diverse skull anatomy as well as feeding and life history strategies (Jones and Goswami 2010; Kienle and Berta 2016).

There is considerable taxonomic uncertainty regarding fossil phocid taxa since the majority are known from isolated or incomplete postcrania (Table 1.1, Fig. 1.1; see also Valenzuela-Toro and Pyenson 2019). Quantitative analysis of morphometric data from the humerus and femur of extant taxa reveals that some characters, while useful in separating modern taxa, are often more variable in fossils (Churchill and Uhen 2019).

In this chapter, we review the origin and affinities of phocids, past distribution, and their functional anatomy. We discuss these in terms of the origin of key character complexes such as feeding, locomotion, and body size that are related to behavioral traits.

1.2 Origin and Affinities

Pinnipeds were traditionally considered a diphyletic group, with phocids evolving independently from musteloid ancestors (i.e., weasels, otters, raccoons, and their kin), while otariids and odobenids evolved from ancestors closer to Ursidae (bears, e.g., Tedford 1976; Repenning et al. 1979; de Muizon 1982). This hypothesis has not been confirmed in any rigorous phylogenetic analysis of morphology (Berta and Ray 1990; Berta and Wyss 1994), and it conflicts with molecular data that support pinniped monophyly (Arnason et al. 2006; Agnarsson et al. 2010).

Phocids are the among the earliest diverging lineage of crown pinnipeds, and morphological data have recovered them as the sister group to the extinct Desmatophocidae (Deméré and Berta 2002; Dewaele et al. 2018b; Velez-Juarbe 2017; Boessenecker and Churchill 2018). Desmatophocids, are known from the early middle Miocene (23–9 million years ago, Ma) in the North Pacific and include the first large pinnipeds. *Allodesmus* reached over 2 m in length, though some (*Allodesmus sadoensis*) material from Japan suggests that they attained even larger body sizes (Barnes and Hirota 1995; Kohno et al. 2007; Churchill et al. 2014; Berta et al. 2018).

1.3 Evolution and Biogeography

In the absence of a robust phylogenetic analysis for phocids that includes extant and extinct taxa, we provide a composite tree (Fig. 1.2) based on molecular data (Fulton and Strobeck 2010) with the addition of fossil taxa based primarily on Amson and Muizon (2014) and Dewaele et al. (2018a, b). The oldest definitive fossil phocid (30–24 Ma) is the stem monachine *Noriphoca gaudini* from the uppermost

Table 1.1 Holotypes of fossil phocid taxa, with holotype location, basis of holotype, and referred material. Museum abbreviations as follows: **ANSP**: Academy of Natural Sciences of Philadelphia, Philadelphia USA; **HNHM**: Hungarian Natural History Museum, Budapest, Hungary; **HMZ**: Museum of Zoology, Paleontological Department, Helsinki, Finland; **IRSNB**: Royal Belgium Institute of the Natural Sciences, Brussels, Belgium; **IZUAN** Shmalhausen Institute of Zoology of the Academy of Sciences of Ukraine, Kiev, Ukraine; **MAB**: Prehistoric Museum of Groene Poort, Broxel, the Netherlands; **MNHN**: Museum of Natural History, Paris, France; **MPGI**: Paleontological Museum of the Mining Institute, Saint-Petersburg, Russia; **MSNUP**: Museum of Natural History, University of Pisa, Pisa, Italy; **NHMW**: Natural History Museum, Vienna, Austria; **NMNHU-P**: National Museum of Natural History at the National Academy of Science of Ukraine, Kiev, Ukraine; **GPDHGS**: Geological and Paleontological Department of the Hungarian Geological Survey, Budapest, Hungary; **OGUM**: Mechnikov Paleontological Museum, State University of Odessa, Odessa, Ukraine; **PIN**: Paleontological Institute of the Academy of Sciences of Russia, Moscow, Russia; **SAM**: South African Museum, Cape Town, South Africa; **SNM**: Slovak National Museum of Natural History; **USNM**: National Museum of Natural History, Washington D.C., USA

Taxon	Holotype location	Holotype	Referred material
Acrophoca longirostris	Peru	MNHN SAS 563 Partial skull	Complete skulls and skeletons from Peru and Chile
Afrophoca libyca	Libya	ELNRP 2Z131 Mandible	None
Auroraphoca atlantica	North Carolina	USNM 181419 Humerus	Partial scapula, partial left humerus, and left ulna
Australophoca changorum	Peru	USNM 438707 Partial skeleton	Additional postcranial material
Batavipusa neerlandica	Netherlands	MAB 3798 Humerus	Femora and innominate from Netherlands and Belgium
Callophoca obscura	Belgium	IRSNB 1198-M203 Humerus	Mandibles, cranial fragments, and postcrania from Florida, Massachusetts, and North Carolina, USA
Cryptophoca maeotica	Moldova	HMZ N1815 Femur	Femora, humeri, and mandibles from Italy, Malta, Romania, Turkey, and Ukraine
Devinophoca claytoni	Slovakia	SNM Z14523 Partial skull	None
Devinophoca emryi	Slovakia	USNM 553684 Incomplete skull	Mandibles, teeth, and postcrania
Frisiphoca aberratum	Belgium	IRSNB 1191-M266 partial humerus	Isolated humerus from Massachusetts
Frisiphoca affine	Belgium	IRSNB 1118-M260 Partial humerus	Ulna, astralagus, and calcaneum from Belgium and Virginia

(continued)

Table 1.1 (continued)

Taxon	Holotype location	Holotype	Referred material
Gryphoca nordica	The Netherlands	MAB 4602 Distal portion of humerus	Proximal portion of humerus
Gryphoca similis	Belgium	IRSNB 1081 (Humerus)	Scapulae, humeri, innominate, femora, tibia, and fibula from Belgium, the Netherlands, and North Carolina, USA
Hadrokirus martini	Peru	MNHN.F.SAS 1627 Skull, mandible, and partial skeleton	Additional cranial and mandible material
Histriophoca alekseevi	Romania	IZUAN 40-121 (Rostrum)	None
Homiphoca capensis	South Africa	SAM PQ L15695 Skull	Numerous skeletal elements from South Africa; Dentition, mandibles, maxillae, femora, and humeri from Florida and North Carolina, USA
Kawas benegasorum	Argentina	MEF-PV 601 Partial skeleton	None
Leptophoca proxima	Maryland, USA	USNM 5359 (Humerus)	Partial Crania and postcrania from Delaware, Maryland, and Virginia, USA
Lobodon vetus	New Jersey, USA	Isolated Postcanine (now lost)	None
Messiphoca mauretanica	Algeria	MNHN.F.ORNI Left humerus, ulna, radius, six vertebrae	Partial skull
Monachopsis pontica	Ukraine	MPGI 17-113 Neotype (Humerus)	Femora and rostral fragments from Ukraine, Moldova, Romania, Russia, and Ukraine
Monotherium delognii	Belgium	IRSNB 1153-M257a Partial sacrum	None
"Monotherium" wymani	Maryland, USA	MCZ 8741 Left and right temporal bone	None; see Dewaele et al. (2018)
Nannophoca vitulinoides	Belgium	IRSNB M2276a-q Partial skeleton	Vertebrae, pelvic girdle, femur, radius, rib
Noriphoca gaudini	Italy	MSNUN123 Cranium and mandible	None

(continued)

Table 1.1 (continued)

Taxon	Holotype location	Holotype	Referred material
Pachyphoca chapskii	Ukraine	NMNHU-P 64-706 (femur)	Ukraine
Pachyphoca ukrainica	Ukraine	NMNHU-P 64-701 (humerus)	Numerous isolated postcranial elements from the Ukraine
Palmidophoca callirhoe	France	MNHN FS 3160 (postcanine tooth)	None
Phoca moori	England	No number given (Humerus)	None
Phocanella pumila	Belgium	IRSNB 1080 Ct. M. 227 (Humerus)	Femora, humeri, tibia, fibula, radius, and sacrum from Florida and North Carolina, USA, and the UK and Netherlands
Piscophoca pacifica	Peru	MNHN SAS 564 (skull and partial skeleton)	Additional skeletal material from Peru and Chile
Platyphoca danica	Denmark	GM1 Distal portion of humerus	None
Platyphoca vulgaris	Belgium	IRSNB 1117 Ct. M. 210 (humerus)	Innominate, tibia, phalanges, and metatarsals from Belgium and North Carolina, USA
Pliophoca etrusca	Italy	MSNUP I-13993 (Cranium and partial skeleton)	Crania and postcrania from Italy, Spain, and France
Pontophoca jutlandica	Denmark	MSM 1788 Incomplete femur	None
Pontophoca sarmatica	Moldova	OGUM no number (femur)	Numerous isolated postcranial elements from Azerbaijan, Moldova, Romania, Russia, and Ukraine
Praepusa boeska	Belgium	MAB 4686 humerus	Additional humeri
Praepusa magyaricus	Hungary	GPDHGS 5394 (humerus)	Humeri, innominate, and femora from Hungary
Praepusa pannonica	Hungary	HGI AW n1 (mandible)	Partial tibia and fibula from Hungary
Praepusa tarchankutica	Ukraine	IZUAN 64-469 (partial cranium)	None

(continued)

Table 1.1 (continued)

Taxon	Holotype location	Holotype	Referred material
Praepusa vindobonensis	Austria	NHMW no number (femur)	Partial and complete crania and postcrania from Austria, Azerbaijan, Kazakstan, Moldova, Slovakia, Turkey, and Ukraine
Properiptychus argentinus	Argentina	MACN 11593 Maxilla with P^3	None
Prophoca rousseaui	Belgium	IRSNB 1149-M274 Humerus	Additional postcrania and vertebrae
Sarmatonectes sintsovi	Moldova	PIN 1713/140 (femur)	None
Virginiaphoca magurai	Virginia, USA	USNM 639748 Humerus	None

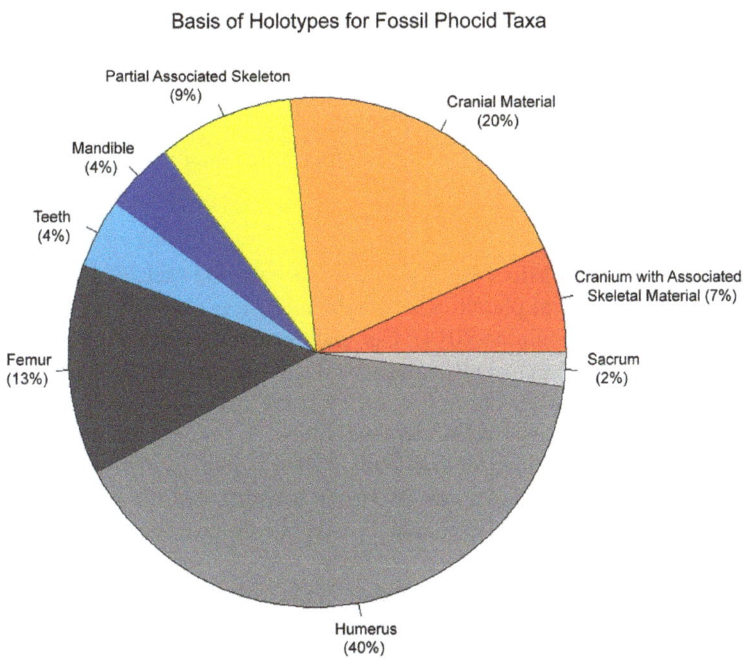

Fig. 1.1 Pie chart illustrating relative proportion of fossil phocid holotypes represented by specific skeletal material, based on data assembled in Table 1.1. Holotypes based on femora and humeri include species based on both complete limb elements and partial limb elements

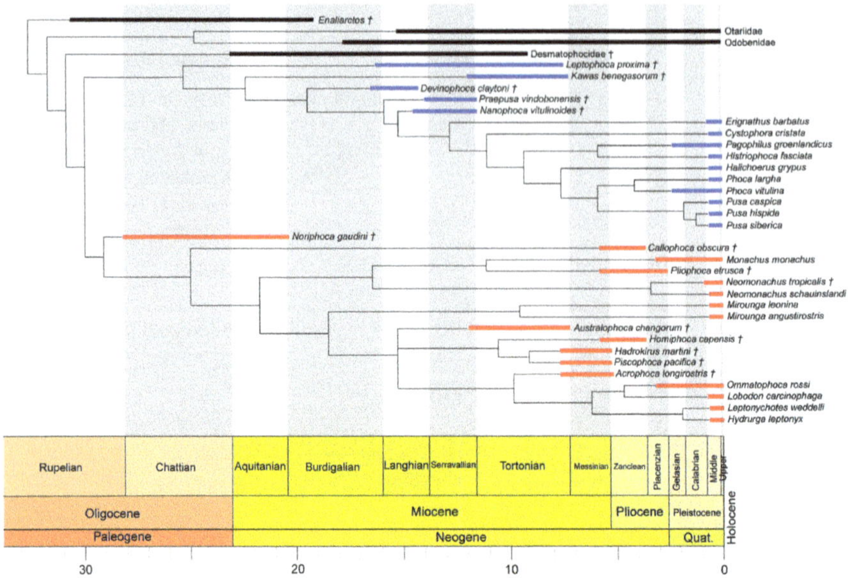

Fig. 1.2 Time-scaled phylogeny of earless seals, with geologic age on the bottom. Topology is a composite tree, based on phylogenies from Fulton and Strobeck (2010), Amson and Muizon (2014), Dewaele et al. (2017, 2018b). Bars represent the span of time that a given taxon is estimated to have lived, using first and last appearance data from Berta et al. (2018). Black bars represent other major pinniped clades, blue bars represent phocine seals

Oligocene or lowermost Miocene (Aquitanian) of Italy (Dewaele et al. 2018a). The previously reported oldest phocid was *Afrophoca libyca* from 19–14 Ma rocks in Africa (Koretsky and Domning 2014). Our understanding of the phylogenetic relationships of fossil seals has recently improved (Dewaele et al. 2018a, b, c). However, the taxonomy is unclear as the holotypes of many fossil phocid species consist of isolated, incomplete postcranial elements (Table 1.1, Fig. 1.1), many of which may not be diagnostic (Churchill and Uhen 2019). *Leptophoca proxima* (Dewaele et al. 2017a) from the eastern coast of North America and the North Sea is the oldest (18–13.8 Ma) known phocine seal. Fossil phocids have been found worldwide (Fig. 1.3).

The occurrence of monachine and phocine lineages during the middle Miocene on both sides of the North Atlantic (Figs. 1.3 and 1.4) suggests that they either evolved there or later dispersed to the North Atlantic after evolving in the North Pacific. The latter possibility supports a desmatophocid ancestry. A southern route of dispersal is more likely, as discussed by Costa (1993; Chap. 8), and Bininda-Emonds and Russell (1996) given that the Bering land bridge blocked access to the Arctic through much of the late Oligocene and early Miocene. It may also be that the colder climate through the Arctic Basin hindered early phocid dispersal along the northern route. The Central American Seaway (CAS) was open during this time and is consistent with the warm-water affinities of North Atlantic phocids at this time.

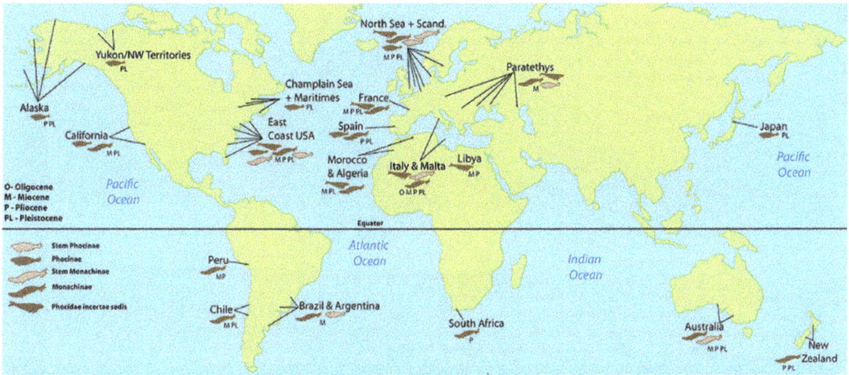

Fig. 1.3 Major localities for fossil phocids: stem Phocinae, stem Monachinae, crown Monachinae, crown Phocinae, Phocidae of uncertain affinities

1.3.1 Stem Monachines

Monachines had established a circum-Atlantic distribution by the late Miocene that included the Mediterranean and Paratethys Seas (Fig. 1.4a). There is a long history of monachines in the Mediterranean, as implied from the presence of *Noriphoca* from latest Oligocene or earliest Miocene of Italy (Dewaele et al. 2018b), an indeterminate monachine from the middle Miocene of Malta (Bianucci et al. 2011), and *Afrophoca libyca* (Koretsky and Domning 2014) from the lower middle Miocene of Libya. *Noriphoca* represents one of the best known monachine seals from this region, known from a partial skull. With fossil material known from both sides of the North Atlantic (Fig. 1.4a), *Callophoca* represents one of the largest known fossil monachines, having a potential body length of 3 m. This taxon was considered to be closely related to *Mirounga* (Koretsky and Ray 2008; Amson and Muizon 2014); however, this assertion largely rests on the potential presence of sexual dimorphism within this taxon, and other phylogenetic analyses have recovered this as a stem monachine (Berta et al. 2015b). The holotype of this taxon consists of an isolated humerus; more recently, Dewaele et al. (2018c) question referral of other isolated monachine skeletal elements to this taxon, and a more thorough taxonomic revision is needed.

The remaining Miocene stem monachine taxa or potential stem taxa from this region are represented by fragmentary or isolated material, preventing the determination of their precise phylogenetic affinities. These include *Auroraphoca*, *Messiphoca*, *Monotherium*, *Pontophoca*, and *Virginiaphoca*. Given the incompleteness and non-associated fossils assigned to the taxon *Terranectes* described by Rahmat et al. (2017) from the upper Miocene of Virginia and North Carolina, we follow Dewaele et al. (2018c) in identifying this taxon as a *nomen dubium* in recognition of its taxonomic uncertainty.

There is evidence that at least some dispersal of monachine seals into the North Pacific happened as early as 8.5 Ma (Fig. 1.4a; Velez-Juarbe and Valenzuela-Toro 2019). This is based on the description of two isolated teeth from the Monterey

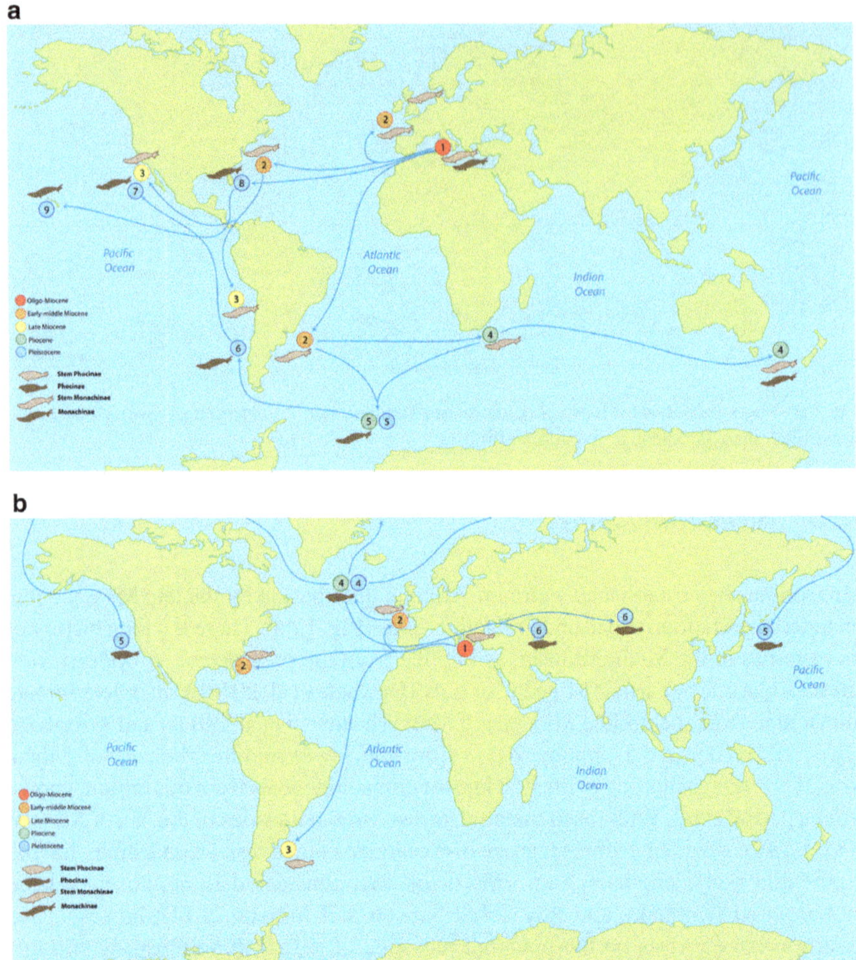

Fig. 1.4 **a** Monachine key biogeographic events: (1) origin in the Mediterranean and Paratethys Seas (Italy, Malta, Libya) during the Oligo-Miocene (2) dispersal of stem taxa to North and South Atlantic (Virginia, North Carolina) during the Early Middle Miocene, (3) dispersal of stem taxa east to west to North Pacific (California) through the Central American seaway during the late Miocene, (4) dispersal of stem taxa to South Pacific during the Pliocene, (5) radiation of lobodontines in Antarctic and Sub-Antarctic waters beginning in the Pliocene, (6) dispersal of crown taxa from South Pacific to North Pacific during the Pleistocene. **b** Phocinae key biogeographic events: (1) origin in the Mediterranean and Paratethys Seas during the Oligo-Miocene, (2) dispersal of stem taxa to North Sea and North Atlantic during the Miocene (3) dispersal of stem taxa to South Atlantic during the Late Miocene, (4) dispersal of crown taxa to higher latitudes of the North Atlantic during the Pliocene and Pleistocene, (5) dispersal of crown taxa to the North Pacific during the Pleistocene

Formation of California, and as such, a further taxonomic appraisal is not possible. Although too old to represent a potential ancestor to the Hawaiian monk seal *N. schauinslandi*, this at least suggests a possible route of dispersal through the CAS (Fig. 1.4a). However, given that these are isolated teeth and there is widespread convergence in dental morphology amongst marine tetrapods, caution is warranted, and definitive interpretation of these as phocid teeth awaits the discovery of more completely preserved specimens.

1.3.2 Monk Seals (Monachini) and Elephant Seals (Miroungini)

The fossil monk seal *Pliophoca etrusca* is known from a partial skeleton from the upper Pliocene (3.19–2.82 Ma) of Italy. *Pliophoca* is closely related to the extant Mediterranean monk seal *Monachus monachus* (Berta et al. 2015b; Dewaele et al. 2018b). Monachine seals were extinct in the Mediterranean region by the late Pleistocene, except for the extant *M. monachus*.

The study of morphologic and molecular data supports recognition of a new genus of living monk seals *Neomonachus* (Scheel et al. 2014). According to Scheel et al. (2014), the common ancestor of *Neomonachus* may have been occupied the CAS during the late Pliocene. Final closure of the CAS 2.5–2.0 Ma resulted in the evolution of two populations *Neomonachus tropicalis* in the Caribbean, first reported from the middle Pleistocene (1.7–1.05 Ma) of Florida (Berta 1995) that became extinct in the past century, and the extant Hawaiian species, *N. schauinslandi*.

In contrast to monk seals, there is little fossil material available for elephant seals. A fragmentary rostrum resembling *Mirounga* from the lower Pleistocene (2.6–2.4 Ma) of New Zealand suggests a southern hemisphere origin for Miroungini (Fig. 1.4a; Boessenecker and Churchill 2016). *Mirounga* sp. are reported from Chile (Valenzuela-Toro et al. 2015) during the Middle-late Pleistocene, and undescribed specimens of *Mirounga* sp., cf. *M. angustirostris* are also known from the upper Pleistocene of southern California (Miller 1971). A supposed Pliocene specimen of *Mirounga* from California (Barnes and Mitchell 1975) appears to be a short-tusked odobenine walrus (Boessenecker 2013; Boessenecker and Churchill 2016).

1.3.3 Antarctic Seals (Lobodontini) and Southern Hemisphere Monachines

Very early in their evolutionary history phocids reached the southern hemisphere, unlike southern otariids that arrived later (Fig. 1.4a; Churchill et al. 2014). The oldest record of a southern hemisphere seal, *Kawas*, is surprisingly a presumed phocine—though all extant southern seals are monachines. The oldest southern hemisphere monachine is *Properiptychus argentinus* from the upper Miocene Parana Formation,

Argentina (Muizon and Bond 1982; Perez et al. 2010). South American monachine seals include four closely related genera: *Acrophoca, Piscophoca, Hadrokirus,* and *Australophoca.* Recent phylogenetic analysis (e.g., Amson and Muizon 2014) place the South American fossil seals with the exception of *Australophoca* and the South African seal *Homiphoca* in a clade with lobodontine seals (Fig. 1.2).

Extant lobodontine seals, also known as Antarctic seals, include four monotypic genera: crabeater seal, *Lobodon carcinophaga*, leopard seal, *Hydrurga leptonyx*, Weddell seal, *Leptonychotes weddellii*, and Ross seal *Ommatophoca rossii* (Fig. 1.2). This clade diverged from elephant seals approximately 6.9 Ma. Lobodontines may have dispersed along the western South American coastline and spread eastwards around Antarctica beginning at least 3.4 Ma based on the occurrence of *Homiphoca* in South Africa. No extant genera are known from the pre-Holocene except for *Ommatophoca* from the Plio-Pleistocene of New Zealand (Fig. 1.4a; King 1973).

The fossil monachine, *Acrophoca longirostris*, is a medium-large sized seal (approximately 2.5 m in length) from upper Miocene to lower Pliocene of Peru and Chile (Muizon 1981; Walsh and Naish 2002; Valenzuela-Toro et al. 2013, 2015). *Acrophoca* is distinct in having a long skull, slender, flexible neck, and elongated body (Fig. 1.5). Likely it was a generalized pierce feeder with a fish diet. The anatomy of the pelvis and hind flippers suggest that it was a hind limb swimmer. The closely related and coeval *Piscophoca pacifica* (Muizon 1981) is, however, more similar to typical seals in cranial morphology and may have had a different feeding behavior.

Similar to extant lobodontine seals (Amson and Muizon 2014) the fossil *Hadrokirus* from the upper Miocene-lower Pliocene, 5.75 Ma Pisco Formation, Peru exhibits sexual dimorphism. The skull is similar to *Piscophoca* (Fig. 1.6). The strong masticatory and neck muscles and robust teeth, some showing breakage, and extensive tooth wear suggest a durophagous diet that included shellfish.

Valenzuela-Toro et al. (2015) described *Australophoca changorum* from the upper Miocene Bahia Inglesa Formation, northern Chile, and the Pisco Formation of southern Peru. This new seal is in the size range of the smallest phocines (e.g., the extant genus *Pusa* and even smaller fossil seal *Praepusa boeska*, Valenzuela-Toro et al. 2015). The dwarf seal *Australophoca* long-snouted *Acrophoc*a and the

Fig. 1.5 Fossil monachine *Acrophoca longirostris* on exhibit at the Smithsonian

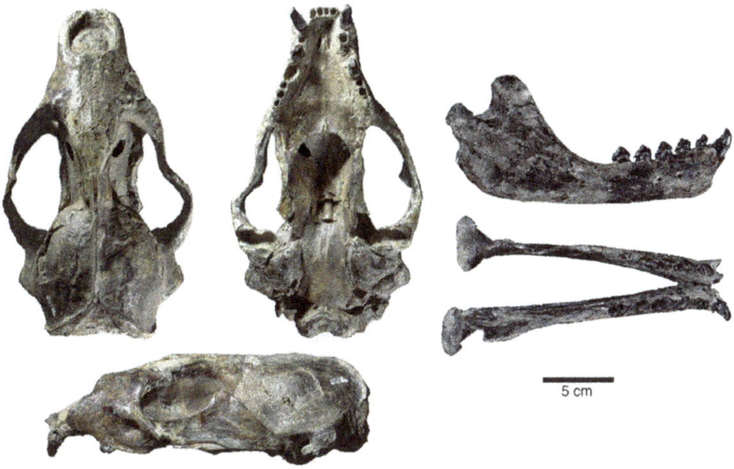

Fig. 1.6 Skull and mandibles of *Hadrokirus martini* (from Amson and Muizon 2014)

durophagous *Hadrokirus*, suggest an exceptional degree of ecological diversity in South American phocids. Additional material (i.e., skulls) of *Austrophoca* are currently under study, and its present phylogenetic position within Monachinae may change.

Phocids from Langebaanweg, South Africa, include *Homiphoca capensis* and as many as six additional seal taxa (Govender 2015, 2018). The remains of some pups and immature animals provide evidence that *Homiphoca was* breeding in this area. Anatomical evidence for the adaptation of *Homiphoca* to colder climates is suggested by countercurrent exchangers that may have been developed in the voluminous maxilloturbinates. An earlier record of *Homiphoca* from North America (Koretsky and Ray 2008) has been disputed (Dewaele et al. 2018c). However, a recent report of this taxon from Spain based on an incomplete innominate (Rahmat et al. 2020) suggests that this southern hemisphere fossil taxon may have had a geographic range that included Western Europe. In addition to the well-documented presence of lobodontines in South Africa and South America, monachines, including undescribed fossils, are known from the late Miocene-early Pliocene of Australia and early Pliocene of New Zealand (Fordyce and Flannery 1983; McKee 1994; Rule et al. 2020), which supports a broader distribution of this lineage in the southern hemisphere (Fig. 1.4a).

1.3.4 Fossil Phocines

Phocine seals are primarily a northern hemisphere radiation (Fig. 1.4b), and in contrast to monachines, most taxa are represented by isolated and fragmentary material. This has made it considerably more challenging to determine the phylogenetic relationships of fossil taxa to extant species. The earliest stem phocines are several taxa from the middle Miocene of Antwerp, Belgium: *Frisiphoca aberrata* and *F. affinis* (formerly *Monotherium aberratum* and *M. affine*; Dewaele et al. 2018a, b), *Leptophoca proxima* (formerly *Prophoca proxima* and *Leptophoca lenis*) and *Prophoca rousseaui* (Fig. 1.4b; Dewaele et al. 2017b). The phocine affinities of *Frisiphoca* spp. are based on a few weak characters and the genus might be a monachine (Dewaele et al. 2018a, b). *Leptophoca proxima* is also reported from Virginia and Maryland. A reexamination of the *L. lenis* holotype (Plum Point Member, Calvert Formation, Maryland) suggests that it may be the oldest phocine at approximately 18 Ma (Dewaele et al. 2017b). *Leptophoca amphiatlantica* from the lower upper Miocene in Maryland and the Netherlands (Koretsky 2001; Koretsky et al. 2012) has been considered a junior synonym of *L. proxima* with specimens reassigned to *Leptophoca* cf. *L. proxima* (Dewaele et al. 2017b). *Devinophoca* from the lower middle Miocene (16.26–14.89 Ma) of Slovakia is one of the best-represented stem phocines with two species known, *D. claytoni* and *D. emyri* (Koretsky and Holec 2002; Koretsky and Rahmat 2015), known from complete skulls and postcrania. *Devinophoca* was initially placed within the monotypic extinct subfamily Devinophocinae, sister clade to crown Phocidae (Koretsky and Holec 2002), but more recently has been recovered as a stem phocine in phylogenetic analysis (Dewaele et al. 2017a, 2018b).

Extant phocines are absent from the southern hemisphere today, but there is evidence of dispersal of phocines into the South Atlantic during the Miocene. *Kawas benegasorum* was described by Cozzuol (2001) based on an articulated partial skeleton from the middle Miocene (12–14 Ma) of Patagonia (Fig. 1.4b). Notable among fossil pinnipeds, the preserved gut contents of *Kawas* suggests a diet of bony fish. Closely related to crown group phocines are the fossil taxa *Praepusa* and *Nanophoca* (Dewaele et al. 2017, 2019). *Nanophoca vitulinoides* (formerly "*Phoca*" *vitulinoides*), described based on a partial skeleton from the lower upper Miocene of the North Sea, is among the smallest of pinnipeds and is positioned as sister taxa to crown Phocinae (Dewaele et al. 2019). *Praepusa*, another small phocine, is also just outside the crown group and is represented by five species described from the Miocene of Europe. Like *Nanophoca*, this species is a dwarf taxon, and unlike most fossil phocines, good cranial material exists for this genus but has yet to be described. Other extinct phocines include *Cryptophoca*, *Gryphoca*, *Monachopsis*, *Pachyphoca*, *Phocanella*, *Platyphoca*, and *Praepusa*; they are primarily represented by disarticulated, non-associated postcranial material from various Miocene and Pliocene localities in the Paratethys region (Fig. 1.4b; Koretsky 2001; Koretsky and Grigorescu 2002; Koretsky and Holec 2002; Koretsky and Rahmat 2013), although their phylogenetic affinities or even validity remain uncertain (Churchill and Uhen 2019).

1.3.5 Crown Phocines

At present, only phocine seals occupy the higher latitudes of the North Atlantic Ocean. Sister taxa to remaining extant phocines is the bearded seal, *Erignathus barbatus* (Erignathini), and the hooded seal, *Cystophora cristata* (Cystophorini) (Fig. 1.2). This deepest phocine split was dated to 11 Ma, followed by a divergence time of 5.5 Ma for the hooded seal lineage (Fulton and Strobeck 2010). Recent studies have found support for the next branch of the tree being the ribbon seal, *Histriophoca*, and harp seal *Pagophilus* as sister group to remaining taxa (e.g., Fulton and Strobeck 2010). The ribbon and harp seal diverged approximately 3.4 Ma, likely due to sea-level fluctuations that drove allopatric speciation at higher latitudes of the North Atlantic (Fig. 1.4b). There are disagreements about the phylogenetic relationships of the remaining species, harbor (*Phoca*), ringed (*Pusa*), and gray (*Halichoerus*) seals (e.g., Árnason et al. 2006). Several studies (e.g., Fulton and Strobeck 2010; Nyakatura and Bininda-Emonds 2012) position the circum-Atlantic gray seal (*Halichoerus grypus*), as the sister species of the land-locked Caspian seal (*Pusa capsica*). In other studies (e.g., Fulton and Strobeck 2010), *Halichoerus* clusters within the more broadly distributed *Phoca* species found in both the North Pacific and Atlantic. Among evolutionary trends in later diverging phocines (e.g., *Histriophoca-Pagophilus* clade and the *Phoca-Pusa* clade) is body size reduction (Churchill et al. 2014).

Phylogenetic relationships of fossil phocines are, at present, too uncertain for proper referral of extinct fossil species to crown Phocinae. The mid Miocene *Pachyphoca* has been suggested as a relative of *Cystophora* (Koretsky and Rahmat 2013), although using a definition of Cystophorinae that is polyphyletic (in the context of abundant molecular and morphological analyses) because it includes members of Monachinae (*Mirounga*) and Phocinae (*Cystophora*). Another taxon, the Miocene *Histriophoca alekseevi* known from a partial rostrum, has been considered closely related to the extant ribbon seal (Koretsky 2001), but more thorough phylogenetic analysis is still needed.

No extinct species can be confidently referred to crown Phocinae, despite molecular divergence dates indicating late Miocene and Pliocene origins for the clade and clades within it (e.g., Phocini, Phocina). Furthermore, in contrast with the fossil record of Pacific pinnipeds (e.g., Deméré et al. 2003; Boessenecker 2013), there are no pre-Pleistocene fossil occurrences assignable to extant phocine genera, and all are Pleistocene at the oldest (e.g., Deméré et al. 2003). In contrast, many crown monachines are known from the Miocene and Pliocene, owing to a fossil record with more completely preserved crania and skeletons. The discovery of more completely preserved craniomandibular specimens and skeletons of new or existing taxa will likely result in the recognition of crown phocines in the pre-Pleistocene fossil record.

1.4 Functional Anatomy

The origin and evolution of key character traits that relate to behavioral traits (i.e., body size and sexual dimorphism, sensory abilities, feeding, and locomotion) are next discussed. Several interesting patterns in the morphological evolution of phocids suggest that further integrative study of phylogeny and behavior is warranted to better understand the interplay of these variables.

1.4.1 Size and Sexual Dimorphism

The common ancestor of Phocidae is reconstructed to have had a relatively small body size (total length of 2.04 m), smaller than most extant monachines. Monachines and phocines exhibit very different evolutionary patterns related to body size (Fig. 1.7). Monachines showed a trend toward increasing body size from the middle Miocene onward. Stem monachines were smaller than extant taxa at 1.8–1.9 m in total length. The one exception is *Callophoca*, at nearly 3 m in total length (Churchill et al. 2014). Body size increased from stem monachines through crown taxa (i.e., *Monachus* and *Mirounga*). This is partly reversed within Lobodontini, with size increases and decreases occurring.

Phocines showed the opposite pattern, with the earliest diverging lineages being larger than later diverging taxa (Churchill et al. 2014; but see Dewaele et al. 2017a, b). Although Wyss (1994) identified only one trend toward decreasing body size in phocines, Churchill et al. (2014) identified two possibly separate reductions in body size, within the *Histriophoca–Pagophilus* clade and the *Phoca–Pusa* clade. This would, in part, support Wyss (1994), as members of these clades show the most pronounced ontogenetic changes in morphology, and reduction in body size may be a result of heterochronic processes. Interpretation of the evolution of body size in Phocidae remains tentative; many fossil phocids are known only from fragmentary and isolated postcranial material, making estimation of body size using the regression equations of Churchill et al. (2014) impossible.

We performed ancestral character state reconstructions (ACSR) for sexual dimorphism using a source phylogenetic tree for phocids and their close relatives that included well-preserved fossil and extant taxa (Dewaele et al. 2018a; Fulton and Strobeck 2010). Results indicate that sexual dimorphism is the ancestral condition seen in stem pinnipeds. Stem phocids are ambiguous with regard to sexual dimorphism with the derived state (monomorphic body in males and females) evolving in phocines (except *Cystophora*). Among monachines, a few show reversal to the ancestral condition (*Callophoca* and *Mirounga*; Fig. 1.7).

1.4.2 Sensory Abilities

Visual capabilities of modern-day pinnipeds have been assessed using behavioral and experimental studies (e.g., Kastelein et al. 1993; Welsch et al. 2001; Chap. 2). Bony orbit size is correlated with diving depth in phocids (Debey and Pyenson 2013). In monachines, the ancestral state of bony orbit size is small, comparable with other phocids, but shows an increase among deeper diving taxa (e.g., *Mirounga*

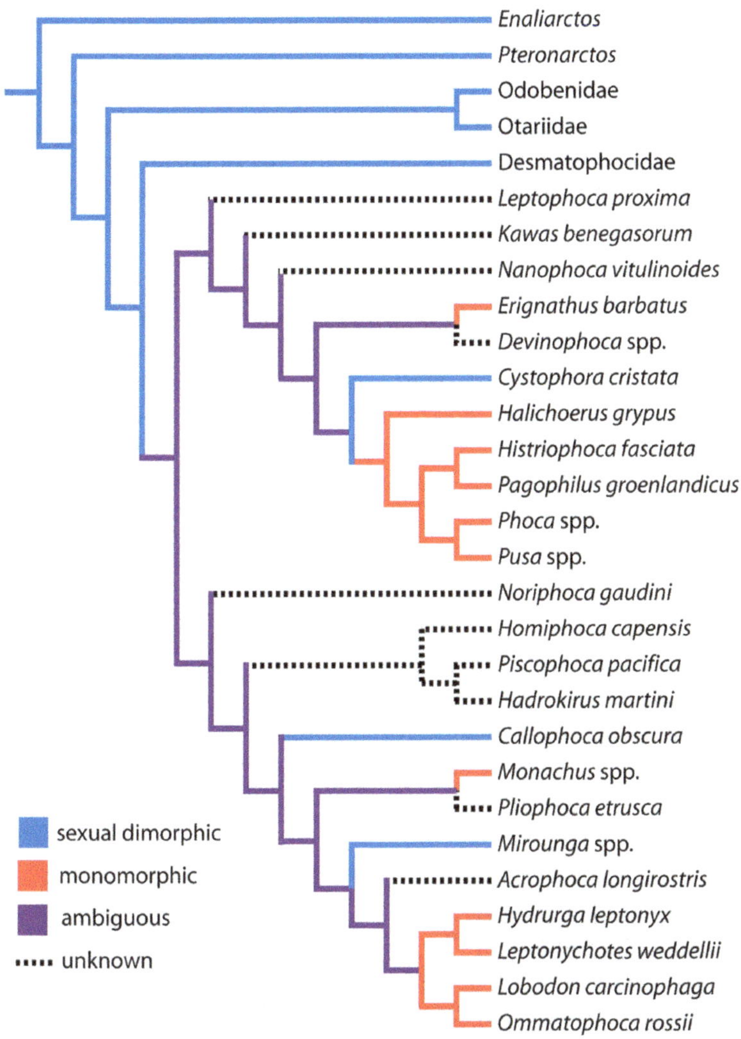

Fig. 1.7 Ancestral character state reconstructions for sexual dimorphism

spp.). Among phocines, *Leptophoca* exhibited the proportionally largest orbit size of any pinniped, living, or extinct, suggesting it may have been a deep diver (Debey and Pyenson 2013). Underwater hearing is also essential in pinniped prey detection (e.g., Kastak and Schusterman 1999; Schusterman et al. 2000). Several studies have examined the vestibular and ossicular anatomy of the ear of pinnipeds (Loza et al. 2017, 2018). The extremely aquatic (spending eight months exclusively at sea) lobodontine seal *Ommatophoca rossii* showed anatomical specializations (e.g., extreme size reduction of semicircular canals and parafloccular volume) hypothesized to be related to sensory function and increased time in the water (Loza et al. 2017).

1.4.3 Feeding

Analyses of skull shape and dentition have attempted to diagnose specific foraging behaviors based on morphology and to interpret their evolution (Jones and Goswami 2010; Jones et al. 2013; Kienle and Berta 2016, 2019). From these analyses, four pinniped feeding strategies have been identified: filter-feeding, suction feeding, grip, and tear feeding, pierce feeding (the latter two types are sometimes combined into biting) supported by cranial and mandibular characters, dietary information, and behavioral data when available (Adam and Berta 2002; Kienle and Berta 2016; Kienle et al. 2018). Phocids are the only pinniped family to employ all four feeding strategies. Most phocids are pierce feeders, using their teeth to capture and pull prey in the mouth, often swallowing the prey whole (Marshall et al. 2008, 2014; Kane and Marshall 2009; Hocking et al. 2013). Filter-feeding is a specialized strategy where multicusped teeth are used as sieves to trap prey and is seen in *Lobodon carcinophaga* (Fig. 1.8) and possibly the fossil phocid *Homiphoca capensis* (Govender 2018; Kienle and Berta 2019). In grip and tear feeding, seen in *Hydrurga leptonx,* the jaws, teeth, or forelimbs are used to hold prey that is then ripped into pieces and ingested (Adam and Berta 2002; Churchill and Clementz 2015; Kienle et al. 2018). Grip and tear feeding are considered the ancestral condition in pinnipeds retained in stem and most crown phocids (Adam and Berta 2002; Kienle and Berta 2019). *Hydrurga* also possesses specializations for suction feeding (Hocking et al. 2013). Specializations for durophagy have also been identified within *Monachus* and *Neomonachus* (Churchill and Clementz 2015), as well as the fossil phocid *Hadrokirus* (Amson and Muizon 2014). Morphological features linked to suction feeding (lengthened palate, transversely arched palate), where food is sucked into the oral cavity by the creation

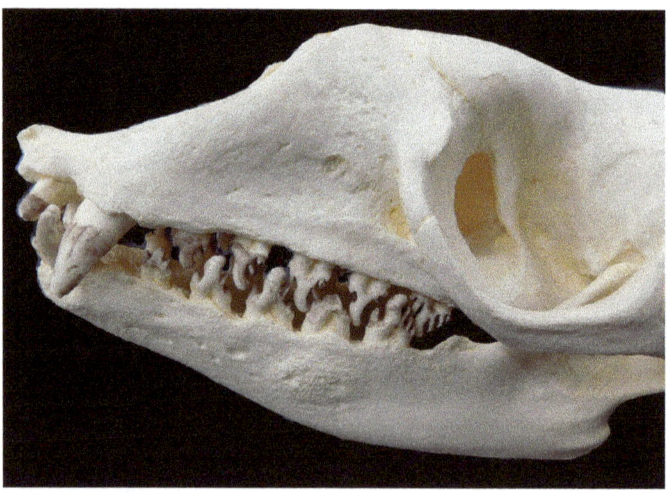

Fig. 1.8 Skull of crabeater seal, *Lobodon carcinophaga,* illustrating interdigitating multicusped teeth

of negative pressure with little use of the teeth, have been identified in *Erignathus*, *Cystophora*, and *Mirounga* (Adam and Berta 2002; Kienle and Berta 2016, 2019). Performance data indicate that some phocid species show behavioral flexibility, with individuals employing one or more feeding strategies depending on context (Hocking et al. 2013; Klages and Cockcroft 1990; Marshall et al. 2014; Kienle et al. 2018).

The ancestral strategy for stem pinnipeds and phocids is biting/crushing, as revealed by ACSRs for prey capture strategy employing the same source tree as in Fig. 1.7, with coding based on previous anatomical and behavioral data. Among phocines, biting/crushing/suction evolved a minimum of three times in *Erignathus, Cystophora*, and *Phoca* spp. Monachines exhibit more varied feeding strategies with biting, crushing, and filter-feeding, having evolved once among lobodontines, with *Lobodon* displaying an exclusive filter-feeding strategy. A combined biting, crushing, and suction feeding strategy evolved in *Mirounga* spp. (Fig. 1.9).

1.4.4 Locomotion and Diving Abilities

All three extant pinniped families have very different styles of locomotion associated with differences in the anatomy of their limbs and axial skeleton (Giffin 1992; King 1983; Pierce et al. 2011; Berta et al. 2015a). Reconstructions of the ancestral swimming mode in phocids are debated since studies disagree on whether ancestral pinnipeds were forelimb dominated or hind limb dominated swimmers (Berta and Adam 2001; Bebej 2009). Phocids rely upon pelvic oscillation, with thrust generated by undulations of the spine in conjunction with protraction and retraction of the hindlimbs while forelimbs provide direction (Fish et al. 1988). Skeletal features, including a rigid thoracic region attached to a highly flexible lumbar region, provide the mobility and leverage needed for pelvic oscillation (Pierce et al. 2011). Mirceta et al. (2013) employed body mass data with inferred concentrations of myoglobin to estimate the diving capabilities of ancestral phocids (Chap. 4). Their results suggest that ancestral phocids were shallow water divers initially, with later increases in their diving capabilities among some taxa (e.g., elephant seals, Weddell seals). Supporting evidence comes from the osteosclerotic-like (i.e., dense, compact bone) skeleton of *Nanophoca vitulinoides*. Increased bone density likely served as a "bone ballast," enabling this seal to forage benthically in the shallow, nearshore waters of the Miocene North Sea Basin (Dewaele et al. 2019).

Since phocids are incapable of turning their hind limbs forward, locomotion on land is accomplished by vertical undulations of the entire body, with limbs contributing little to movement (Fig. 1.10; Tarasoff et al. 1972). Phocines use their forelimbs to haul themselves forward when moving on uneven surfaces during terrestrial locomotion, dig lairs in the ice during the breeding season, and to flush buried fish out of sediments or to hold mammalian prey against the ground when feeding (see references cited in Hocking et al. 2018). Wild and captive phocine seals routinely use their forelimbs and sharp claws to assist with processing (tearing) large prey (e.g., salmonid fish) underwater (Fig. 1.11; Hocking et al. 2018). Anatomical features

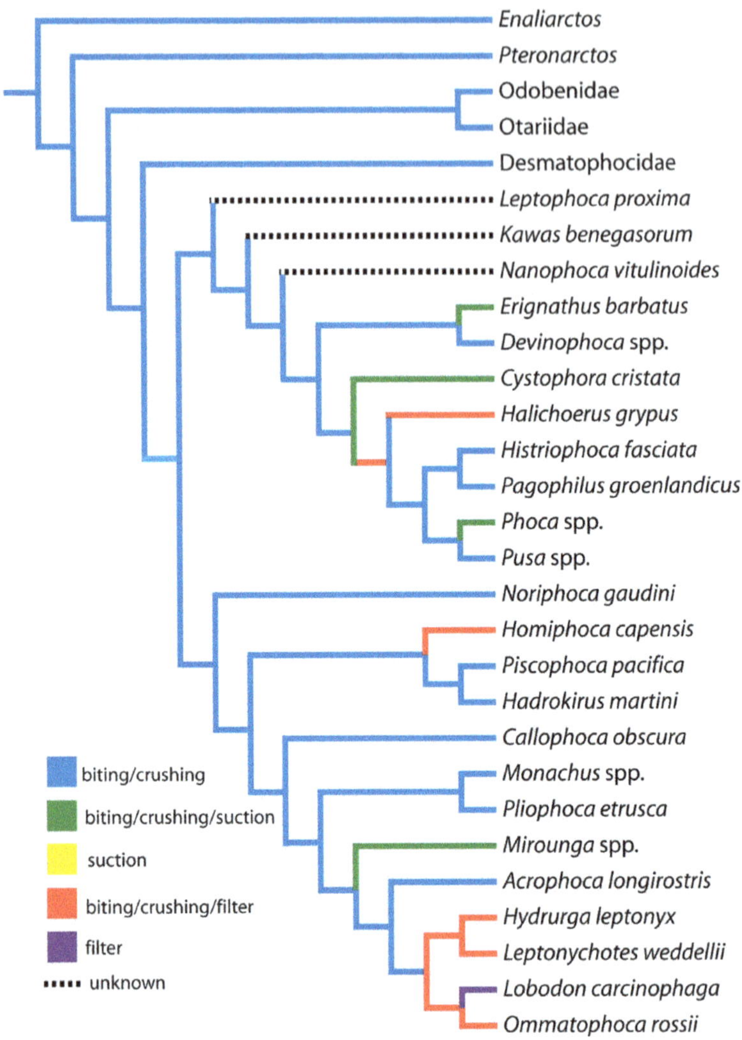

Fig. 1.9 Ancestral character state reconstructions for prey capture strategy

Fig. 1.10 Photo of harbor seal crawling along beach illustrating the use of forelimbs and sharp claws to haul themselves forward (from Berta et al. 2015b)

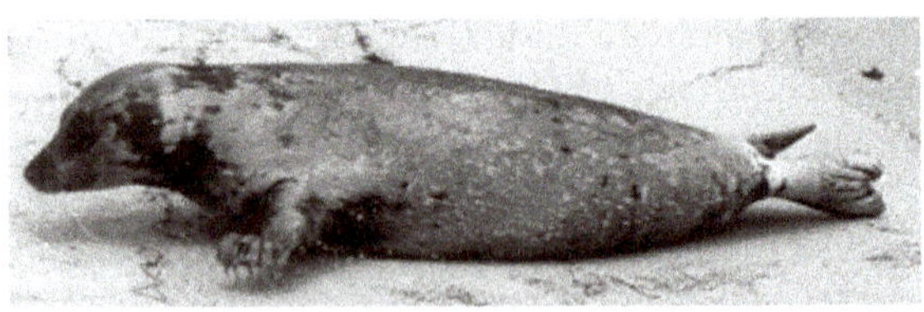

Fig. 1.11 Phocine seals use "hold and tear" prey processing to break large prey into pieces small enough to swallow (from Hocking et al. 2018 drawing courtesy of Kai Hagberg)

facilitating this "hold and tear" processing of prey include claws with trochleated phalanges and long ungual process, morphologies inherited from their terrestrial ancestors (Fig. 1.12; Hocking et al. 2018). By contrast, monachines show an otariid-like forelimb (i.e., reduced claws with cartilaginous extensions on the digits) and do not employ their forelimbs in prey processing (Hocking et al. 2018).

1.5 Diversity Changes Through Time

The decline and extinction of monachines in the North Atlantic in the late Pliocene, except for the extant Mediterranean monk seal, may be related to sea-level changes, trophic changes, or other environmental changes (Dewaele et al. 2018a). Today, phocine seals exclusively occupy the higher latitudes of the North Atlantic, and Plio-Pleistocene glaciations may be responsible for the survival and current distribution of North Atlantic phocids given that most are pagophilic (i.e., ice breeding), with a few temperate (*Phoca vitulina, Halichoerus grypus*) to tropical (*Neomonachus tropicalis*) exceptions. Colder climates likely adversely affected Pliocene monachines given their temperate distribution today. A curious dichotomy exists between the fossil record and the modern phocid fauna: today, only a minority of phocids are terrestrial (e.g., non-pagophilic) breeders on sandy or rocky coastlines (*Mirounga* spp., *Phoca vitulina, Pusa capsica, Halichoerus grypus, Monachus/Neomonachus* spp.), whereas the fossil record is dominated by low- and mid-latitude records of phocid seals far beyond the lowest latitudinal range of most extant pagophilic species. These records include temperate to warm-temperate Pliocene records of ancestral phocines as well as ancestral lobodontines in the northern and southern hemispheres, respectively (Koretsky and Ray 2008; Dewaele et al. 2017a, b; Muizon 1981; Amson and Muizon 2014), indicating a geologically recent and independent derivation of pagophily in each clade. Climate cooling and ice cap expansion during the Pleistocene are likely

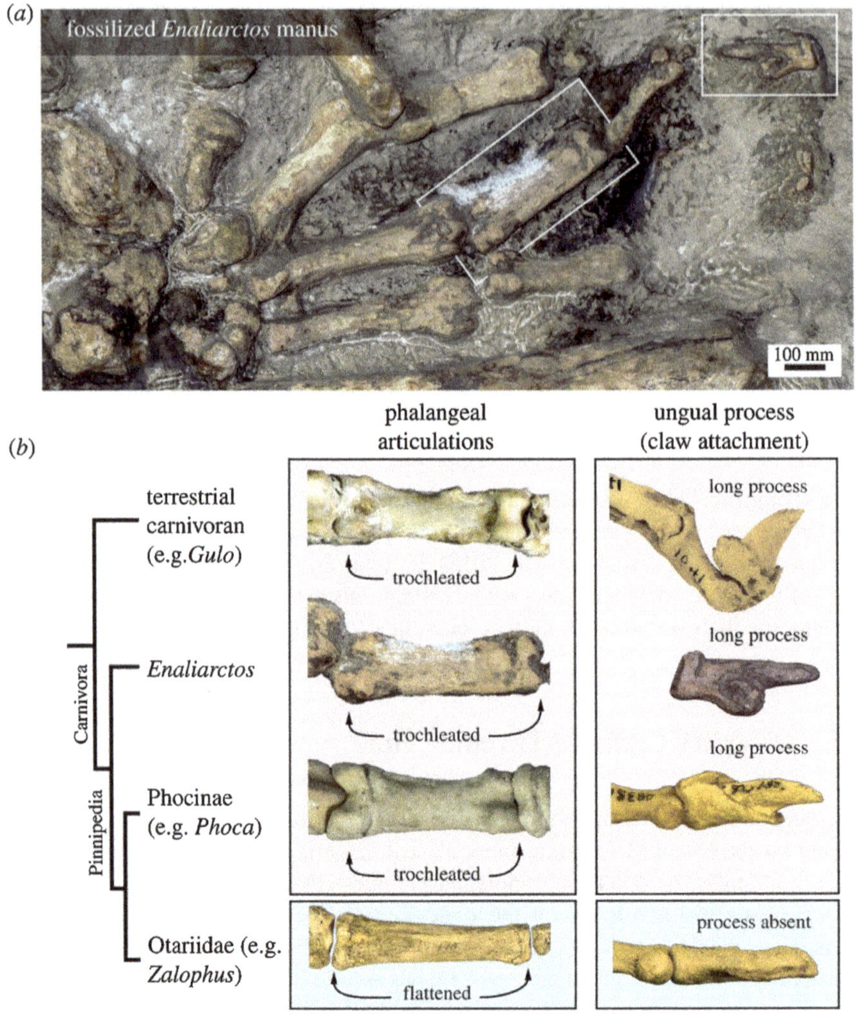

Fig. 1.12 Evolution of digit morphology in **a** ancestral pinniped *Enaliarctos mealsi*, **b** phylogeny illustrating that phocids (vs. otariids) retained the digit morphology (i.e., trochleated phalangeal articulations and long claw attachments) from their ancestors (from Hocking et al. 2018)

the major reasons for pagophily by phocids. Still, the loss of diverse temperate latitude (and presumed land breeders) that dominated the Miocene and Pliocene is more challenging to explain (Chap. 8).

Pinniped faunal turnover has been implicated in phocid changes in diversity during the Plio-Pleistocene. Phocids dominate the South American fossil record, whereas now otariids dominate the extant fauna in this region. Changing sea levels during the Pleistocene may have reduced the number of haul-out sites suitable for phocids and an increase in rocky islands surrounded by a deeper water environment that favored otariid seals (Valenzuela-Toro et al. 2013). This parallels the replacement of the Neogene desmatophocid and odobenid fauna by a Quaternary otariid/phocid fauna in the North Pacific (Boessenecker 2013). However, the South America transition appears to have been quite rapid, while the North Pacific transition seems to have had more overlap between pinniped groups, and both groups evolved in the same region in contrast to phocids and otariids. Regardless, the chronology along both continental margins is hampered by a dearth of well-sampled marine mammal fossil assemblages from the latest Pliocene and early Pleistocene (Boessenecker 2013; Valenzuela-Toro et al. 2013), and improved sampling is needed.

Pliocene assemblages from either side of the North Atlantic appear to contain extinct phocids (Koretsky and Ray 2008), contrasting sharply with middle and late Pleistocene assemblages that are dominated by extant genera and species (Deméré et al. 2003; Post 2005). Though incomplete and needing critical reassessment, these differences suggest significant faunal turnover, gross misinterpretation of Pliocene phocid taxonomy, or both. Reevaluation of Pliocene phocid assemblages and scattered Pleistocene fossils and/or improved field sampling of North Atlantic localities are necessary to draw further conclusions about a supposed Pliocene–Pleistocene extinction or period of faunal turnover (Pimiento et al. 2017).

In the eastern North Pacific, the late Miocene-Pliocene walrus dominated fauna disappeared during the Pliocene–Pleistocene transition. Modern sea lions and the harbor seal then invaded from the western North Pacific and through the Arctic from the North Atlantic, respectively, and by the middle Pleistocene (Boessenecker 2013 and references therein). The northward trans-equatorial dispersal of the northern elephant seal *Mirounga angustirostris* occurred at least by the late Pleistocene (Boessenecker and Churchill 2016), thereby completing the assembly of the modern pinniped fauna in the eastern North Pacific.

1.6 Conclusions

Phocid seals, the most diverse, species-rich pinniped lineage, first appear in the fossil record during the late Oligocene or earliest Miocene (30–24 Ma). Two distinct clades are recognized: the Monachine or "southern seals" (monk, elephant, and Antarctic seals) and the Phocinae or "northern seals" (e.g., bearded, ribbon, hooded, and harbor seals). Although the taxonomy of fossil phocids is unclear, both monachines and phocines appear to have evolved in the North Atlantic or dispersed thereafter evolving in the North Pacific, based on their shared ancestry with the extinct Desmatophocidae. Some dispersal of monachines into the North Pacific occurred as early as 8.5 Ma with a possible CAS dispersal. Monk seals originated in the North Atlantic with closure of the CAS creating two allopatric populations evolving separately in the Caribbean and Hawaii. Elephant seals had a more recent origin during the Pleistocene in the southern hemisphere. The speciose Antarctic seals diverged from elephant seals and may have dispersed along the western South American coast around Antarctica. Phocine seals radiated primarily in the northern hemisphere. There is evidence for dispersal of phocines into the southern hemisphere although they are absent from that region today, instead being the sole occupants of the high latitudes of the North Atlantic.

The mapping of behavioral traits onto a phocid phylogeny reveals that sexual dimorphism is the ancestral condition in stem pinnipeds, although phocids are ambiguous with regard to this trait. Among sensory traits, the importance of vision to locate food is indicated by bony orbit size, a proxy for diving ability. Ancestrally, phocids were shallow divers with later increases in their diving capabilities in both monachines and phocines. Most phocids are pierced feeders with a few taxa having developed specialized strategies (e.g., grip and tear feeding in leopard seals, filter-feeding in the crabeater seal and possibly the fossil phocid *Homiphoca*, and suction feeding in elephant, bearded, and hooded seals). Future work should include fossil taxa in rigorous phylogenetic analyses of phocids to establish the relationship among the behavioral characters discussed here and to elucidate the factors driving these patterns.

The fossil record includes temperate to warm temperate Pliocene records of species of both monachines and phocines, which suggests that ice breeding evolved independently in each lineage. There has been limited study, as yet, of pinniped faunal diversity changes through time. The South American fossil record of phocids during the Plio-Pleistocene suggests that changing sea levels reduced haul-out sites for phocids and an increase in rocky island habitats during that time favored otariids. Similarly, an environmental and habitat change might explain the replacement of Neogene desmatophocid and odobenid faunas by Quaternary otariid/phocid fauna in the North Pacific.

References

Adam PJ, Berta A (2002) Evolution of prey capture strategies and diet in the Pinnipedimorpha (Mammalia, Carnivora). Oryctos 4:83–107

Agnarsson I, Kuntner M, May-Collado LJ (2010) Dogs, cats, and kin: a molecular species-level phylogeny of Carnivora. Mol Phylog Evol 54:726–745

Amson E, de Muizon C (2014) A new durophagous phocid (Mammalia: Carnivora) from the late Neogene of Peru and considerations on monachine seal phylogeny. J Syst Paleo 12:523–548

Árnason U, Gullberg A, Janke A, Kullberg M, Lehman N et al (2006) Pinniped phylogeny and a new hypothesis for their origin and dispersal. Mol Phylog Evol 41:345–354

Barnes LG, Hirota K (1995) Miocene pinnipeds of the otariid subfamily Allodesminae in the North Pacific Ocean: systematics and relationships. Isl Arc 3:329–360

Barnes LG, Mitchell ED (1975) Late Cenozoic Northeast Pacific Phocidae. Conseil International pour l'Exploration de la Mer. Rapp Proc Verbaux Des Réunions 169:34–42

Bebej RM (2009) Swimming mode inferred from skeletal proportions in the fossil pinnipeds *Enaliarctos* and *Allodesmus* (Mammalia, Carnivora). J Mamm Evol 16:77–97

Berta A (1995) Fossil carnivores from Leisey Shell Pits. Bull Florida Mus Nat Hist 37:463–499

Berta A, Ray CE (1990) Skeletal morphology and locomotor capabilities of the archaic pinniped *Enaliarctos mealsi*. J Vert Paleo 10:141–157

Berta A, Wyss AR (1994) Pinniped phylogeny. Proc San Diego Soc Nat Hist 29:33–56

Berta A, Adam PJ (2001) Evolutionary biology of pinnipeds. In: Mazin JM, de Buffrenil V (eds) *Secondary adaptation of tetrapods to life in water*: proceedings of the international meeting, Poiters, 1996, pp 235–258. Verlag Dr Friedrich Pfeil, Munich
Berta A, Kienle S, Bianucci G, Sorbi S (2015a) A reevaluation of *Pliophoca etrusca* (Pinnipedia, Phocidae) from the Pliocene of Italy: phylogenetic and biogeographic implications. J Vert Paleo 35:e889144
Berta A, Sumich JL, Kovacs KM (2015b) Marine mammals: evolutionary biology, 3rd edn. Academic, San Diego, CA
Berta A, Churchill M, Boessenecker RW (2018) The origin and evolutionary biology of pinnipeds. Ann Rev Earth Planet Sci 46:203–228. https://doi.org/10.1146/annurev-erth-082517-010009
Bianucci G, Gatt M, Catanzariti R, Sorbi S, Bonavia CG et al (2011) Systematics, biostratigraphy and evolutionary pattern of the Oligo–Miocene marine mammals from the Maltese Islands. Geobios 44:549–85
Bininda-Emonds ORP, Russell AP (1996) A morphological perspective on the phylogenetic relationships of the extant phocid seals (Mammalia: Carnivora: Phocidae). Bonner Zool Monogr 41:1–256
Boessenecker RW (2013) A new marine vertebrate assemblage from the Late Neogene Purisima Formation in Central California. Part II. Pinnipeds and cetaceans. Geodiversitas 35:815–940
Boessenecker RW, Churchill M (2016) The origin of elephant seals: implications of a fragmentary late Pliocene seal (Phocidae: Miroungini) from New Zealand. NZ J Geol Geophys 59:544–550
Boessenecker RW, Churchill M (2018) The last of the desmatophocid seals: a new species of *Allodesmus* from the upper Miocene of Washington, USA. Zool J Linn Soc 184:211–235. https://doi.org/10.1093/zoolinnean/zlx098
Churchill M, Clementz MT (2015) Functional implications of variation in tooth spacing and crown size in Pinnipedimorpha (Mammalia: Carnivora). Anat Rec 298:878–902
Churchill M, Uhen MD (2019) Taxonomic implications of morphometric analysis of earless seal limb bones. Acta Palaeo Pol 64(2):213–230
Churchill M, Clementz MT, Kohno N (2014) Cope's rule and the evolution of body size in Pinnipedimorpha (Mammalia: Carnivora). Evolution 69:201–215
Costa DP (1993) The relationship between reproductive and foraging energetics and the evolution of the Pinnipedia. Symp Zool Soc Lond 66:293–314
Cozzuol MA (2001) A "northern" seal from the Miocene of Argentina: implications for phocid phylogeny and biogeography. J Vert Paleo 21:415–421
Davis CS, Delisle I, Stirling I, Siniff DB, Strobeck C (2004) A phylogeny of the extant Phocidae inferred from complete mitochondrial DNA coding regions. Mol Phylo Evol 33:363–373
de Muizon C (1981) Les Vertébrés Fossiles de la Formation Pisco (Pérou). Part 1. Recherche sur les Grandes Civilisations, Mem 6 Instituts Francais d'Études Andines, Paris
de Muizon C (1982) Phocid phylogeny and dispersal. Ann S Afr Mus 89:175–213
de Muizon C, Bond M (1982) Le Phocidae (Mammalia) de la Formation Parana (Entre Rios, Argentine). Bull Mus Nat d'Hist Natur Paris, 4 ser, sect C, 4:165–207
Debey LB, Pyenson ND (2013) Osteological correlates and phylogenetic analysis of deep diving in living and extinct pinnipeds: what good are big eyes? Mar Mamm Sci 29(1):48–83
Deméré TA, Berta A (2002) The Miocene Pinniped *Desmatophoca oregonensis* Condon, 1906 (Mammalia: Carnivora) from the Astoria Formation, Oregon. Smithsonian Contrib Paleobiol 93:113–148
Deméré TA, Berta A, Adam PJ (2003) Pinnipedimorph evolutionary biogeography. Bull Am Mus Nat Hist 279: 32–76
Dewaele L, Amson E, Lambert O, Louwye S (2017a) Reappraisal of the extinct seal "*Phoca*" *vitulinoides* from the Neogene of the North Sea Basin, with bearing on its phylogenetic affinities, and locomotion. PeerJ 5:e3316
Dewaele L, Lambert O, Louwye S (2017b) On *Prophoca* and *Leptophoca* (Pinnipedia, Phocidae) from the Miocene of the North Atlantic realm: redescription, phylogenetic affinities and paleobiogeographic implications. PeerJ 5:e3024

Dewaele L, Lambert O, Louwye S (2018a) A critical revision of the fossil record, stratigraphy and diversity of the Neogene seal genus *Monotherium* (Carnivora, Phocidae). R Soc Open Sci 5:171669. https://doi.org/10.1098/rsos.171669

Dewaele L, Lambert O, Louwye S (2018b) A late surviving Pliocene seal from high latitudes of the North Atlantic realm: the latest monachine seal on the southern margin of the North Sea. PeerJ 9:e5734. https://doi.org/10.7717.peerj.5734

Dewaele L, Peredo CM, Meyvisch P, Louwye S (2018c) Diversity of late Neogene Monachinae (Carnivora, Phocidae) from the North Atlantic, with the description of two new species. R Soc Open Sci 5:172437. https://doi.org/10.1098/rsos.172437

Dewaele L, Lambert O, Laurin M, deKock T, Louwye S, de Buffrenil V (2019) Generalized osteosclerotic condition in the skeleton of *Nanophoca vitulinoides*, a dwarf seal from the Miocene of Belgium. J Mamm Evol 26:517–543. https://doi.org/10.1007/s10914-018-9438-9

Fish FE, Innes S, Ronald K (1988) Kinematics and estimated thrust production of swimming harp and ringed seals. J Exp Biol 137:157–173

Fordyce RE, Flannery T (1983) Fossil phocid seals from the late Tertiary of Victoria. Proc R Soc Vict 95:99–100

Fulton TL, Strobeck C (2010) Multiple markers and multiple individuals refine true seal phylogeny and bring molecules and morphology back in line. Proc R Soc B 277:1065–1070

Giffin EB (1992) Functional implications of neural canal anatomy in recent and fossil marine carnivores. J Morphol 214:357–374

Govender R (2015) Preliminary phylogenetic and biogeographic history of the Pliocene seal, *Homiphoca capensis* from Langebaanweg, South Africa. Trans R Soc S Afr 70:25–39

Govender R (2018) Extension of the geographic range of *Homiphoca* into Northern Cape Province, South Africa, with comments on the possible feeding strategies used by *Homiphoca*. J Vert Paleo 38:e146536. https://doi.org/10.1080/02724634.2018.1463536

Higdon JW, Bininda-Emonds ORP, Beck RMD, Ferguson SH (2007) Phylogeny and divergence of the pinnipeds (Carnivora: Mammalia) assessed using a multigene dataset. BMC Evol Biol 7:216

Hocking DP, Evans AR, Fitzgerald EMG (2013) Leopard seals (*Hydrurga leptonyx*) use suction and filter feeding when hunting small prey underwater. Polar Biol 36:211–222

Hocking DP, Marx FG, Sattler R, Harris RN, Pollock TI, Sorrell KJ, Fitzgerald EMG, McCurry MR, Evans AR (2018) Clawed forelimbs allow northern seals to eat like their ancient ancestors. R Soc Open Sci 5:172393. https://doi.org/10.1098/rsos.172393

Jones KE, Goswami A (2010) Quantitative analysis of the influences of phylogeny and ecology on phocid and otariid pinniped (Mammalia; Carnivora) cranial morphology. J Zool 280:297–308. https://doi.org/10.1111/j.1469-7998.2009.00662.x

Jones KE, Ruff CR, Goswami A (2013) Morphology and biomechanics of the pinniped jaw: mandibular evolution without mastication. Anat Rec 296:1049–1063. https://doi.org/10.1002/ar.22710

Kane EA, Marshall CD (2009) Comparative feeding kinematics and performance of odontocetes: belugas, Pacific white-sided dolphins and long-finned pilot whales. J Exp Biol 212:3939–3950

Kastak D, Schusterman RJ (1999) In-air and underwater hearing sensitivity of a northern elephant seal (*Mirounga angustirostris*). Can J Zool 77:1751–1758

Kastelein RA, Zweypfenning RCVJ, Spekreijse H, Dubbeldam JL, Born EW (1993) The anatomy of the walrus head (*Odobenus rosmarus*). Part 3: The eyes and their function in walrus ecology. Aquat Mamm 19:61–92

Kienle SS, Berta A (2016) The better to eat you with: the comparative feeding morphology of phocid seals (Pinnipedia, Phocidae). J Anat 228:396–413

Kienle SS, Berta A (2019) The evolution of feeding strategies in phocid seals (Pinnipedia, Phocidae). J Vert Paleo 38:e1559172. https://doi.org/10.1080/02724634.2018.1559172

Kienle SS, Hermann-Sorensen HH, Costa DP, Reichmuth C, Mehta RS (2018) Comparative feeding strategies and kinematics in phocid seals: suction without specialized skull morphology. J Exp Biol 221: jeb179424

King JE (1973) Pleistocene Ross seal (*Ommatophoca rossi*) from New Zealand (Note). New Zealand J Mar Freshw Res 7:391–397
King JE (1983) Seals of the world. Cornell Univ Press, Ithaca, NY
Klages NTW, Cockcroft VG (1990) Feeding behavior of a captive crabeater seal. Polar Biol 10:403–404
Kohno N, Koike H, Narita K (2007) Outline of fossil marine mammals from the middle Miocene Bessho and Aoki Formations, Nagano Prefecture, Japan. Res Rep Shinshushinmachi Fossil Mus 10:1–45
Koretsky IA (2001) Morphology and systematics of Miocene Phocinae (Mammalia: Carnivora) from Paratethys and the North Atlantic Region. Geol Inst Hung, Budapest
Koretsky IA, Domning DP (2014) One of the oldest seals (Carnivora, Phocidae) from the Old World. J Vert Paleo 34:224–229
Koretsky IA, Grigorescu D (2002) The fossil monk seal *Pontophoca sarmatica* (Alekseev) (Mammalia: Phocidae: Monachinae) from the Miocene of eastern Europe. Smithsonian Contrib Paleobiol 93:149–162
Koretsky IA, Holec P (2002) A primitive seal (Mammalia: Phocidae) from the Badenian stage (early middle Miocene) of central Paratethys. Smithsonian Contrib Paleobiol 93:163–178
Koretsky IA, Rahmat S (2013) First record of fossil Cystophorinae (Carnivora, Phocidae): middle Miocene seals from the northern Paratethys. Riv Ital Paleo Strat 119:325–350
Koretsky IA, Rahmat SJ (2015) A new species of the subfamily Devinophocinae (Carnivora: Phocidae) from the central Paratethys. Riv Ital Paleo Strat 121(1):1–17
Koretsky IA, Ray CE (2008) Phocidae of the Pliocene of eastern USA. In: Ray CE, Bohaska D, Koretsky IA et al (eds) Geology and paleontology of the Lee Creek Mine, North Carolina, IV, vol 15. Vir Mus Nat Hist Sp Publ, pp 81–140
Koretsky IA, Ray CE, Peters N (2012) A new species of *Leptophoca* (Carnivora, Phocidae, Phocinae) from both sides of the North Atlantic Ocean (Miocene seals of the Netherlands, part D). *Deinsea* 15:1–2
Loza CM, Latimer AE, Sanchez-Villagra MR, Carlini AA (2017) Sensory anatomy of the most aquatic of carnivorans: the Antarctic Ross seal, and convergences with other mammals. Biol Lett 13:20170489. https://doi.org/10.1098/rsbl.2017.0489
Loza CM, Reutimann O, Sanchez-Villagra MR, Carlini AA, Aguirre-Fernandez G (2018) Evolutionary transformations of the malleus in pinnipeds, with emphasis on South Hemisphere tax. Contrib Zool 87(2):75–85
Marshall CD, Kovacs K, Lydersen C (2008) Feeding kinematics, suction and hydraulic jetting capabilities in bearded seals (*Erignathus barbatus*). J Exp Biol 211:699–708
Marshall CD, Wieskotten S, Hanke W, Hanke FD, Marsh A et al (2014) Feeding kinematics, suction, and hydraulic jetting performance of harbor seals (*Phoca vitulina*). PLOS ONE 9:e86710
McKee J (1994) Geology and vertebrate palaentology of the Tangahoe Formation, South Taranaki Coast, New Zealand. Geol Soc New Zealand Misc Publ B 80:63–91
Miller WE (1971) Pleistocene vertebrates of the Los Angeles Basin and vicinity (exclusive of Rancho La Brea). Bull Mus Nat Hist Los Angeles Cty 10, Los Angeles, CA
Mirceta S, Signore AV, Burns JM, Cossins AR, Campbell KL, Berenbrink M (2013) Evolution of mammalian diving capacity traced by myoglobin net surface charge. Science 340:1303
Nyakatura K, Bininda-Emonds ORP (2012) Updating the evolutionary history of Carnivora (Mammalia): a new species-level supertree complete with divergence time estimates. BMC Biol 10:12
Perez LM, Genta Iturreria SF, Griffin M (2010) Paleoecological and paleobiogeographic significance of two species of bivalves in the Parana Formation (late Miocene) of Entre Rios province, Argentina. Malacol 53:61–76
Pierce SE, Clack JA, Hutchinson JR (2011) Comparative axial morphology in pinnipeds and its correlation with aquatic locomotory behavior. J Anat 219:502–514

Pimiento C, Griffin JN, Clements CF, Silvestro D, Varela S, Uhen MD, Jaramillo C (2017) The Pliocene marine megafauna extinction and its impact on functional diversity. Nat Ecol Evol 1:1100–1106. https://doi.org/10.10.1038/s41559-017-0223-6

Post K (2005) A Weichselian marine mammal assemblage from the southern North Sea. Deinsea 11:21–27

Rahmat SJ, Koretsky IA, Osborne JE, Alford AA (2017) New Miocene Monachinae from the western shore of the Chesapeake Bay (Maryland, USA). Vestnik Zoologii 51:221–242. https://doi.org/10.1515/vzoo-2017-0029

Rahmat S, Muniz F, Toscano A, Esperante R, Koretsky I (2020) First European record of *Homiphoca* (Phocidae: Monachinae: Lobodontini) and its bearing on the paleobiogeography of the genus. Hist Biol 32:561–569. https://doi.org/10.1080/08912963.2018.1507030

Repenning CA, Ray CE, Grigorescu D (1979) Pinniped biogeography. In: Gray J, Boucot AJ (eds) Historical biogeography, plate tectonics, and the changing environment. Ore. State Univ. Press, Corvallis, pp 357–69

Rule JP, Adams JW, Fitzgerald EMG (2020) Colonization of the ancient southern oceans by small-sized Phocidae: new evidence from Australia. Zool J Linn Soc XX

Scheel DM, Slater GJ, Kolokotronis SO, Potter CW, Rotstein DS et al (2014) Biogeography and taxonomy of extinct and endangered monk seals illuminated by ancient DNA and skull morphology. ZooKeys 409:1–33

Schusterman RJ, Kastak D, Levenson DH, Reichmuth CJ, Southall BL (2000) Why pinnipeds don't echolocate. J Acoustic Soc Am 107:2256–2264

Tarasoff FJ, Bisaillon A, Pierard J, Whitt AP (1972) Locomotory patterns and external morphology of the river otter, sea otter, and harp seal (Mammalia). Can J Zool 50:915–929

Tedford RH (1976) Relationships of pinnipeds to other carnivores (Mammalia). Syst Zool 25:363–374

Valenzuela-Toro A, Pyenson ND (2019) What do we know about the fossil record of pinnipeds? A historiographical investigation. R Soc Open Sci 6:191394. https://doi.org/10.1098/rsos.191394

Valenzuela-Toro AM, Gutstein CS, Varas-Malca RM, Suarez ME, Pyenson ND (2013) Pinniped turnover in the South Pacific Ocean: new evidence from the Plio-Pleistocene of the Atacama Desert, Chile. J Vert Paleo 33:216–223

Valenzuela-Toro, AM, Gutstein CS, Suárez ME, Otero R, Pyenson ND (2015) Elephant seal (*Mirounga* sp.) from the Pleistocene of the Antofagasta region, northern Chile. J Vert Paleo 35(3):e918883

Velez Juarbe J (2017) *Eotaria citrica*, sp nov, a new otariid from the "Topanga" formation of southern California. *PeerJ* 5:e3022

Velez-Juarbe J, Valenzuela-Toro AM (2019) Oldest record of monk seals from the North Pacific and biogeographic implications. Biol Lett 15:20190108. https://doi.org/10.1098/rsbl.2019.0108

Walsh S, Naish D (2002) Fossil seals from late Neogene deposits in South America: a new pinniped (Carnivora, Mammalia) assemblage from Chile. Palaeo 45:821–842

Welsch U, Ramdohr S, Riedelsheimer B, Hebel R, Eisert R, Plötz J (2001) Microscopic anatomy of the eye of the deep-diving Antarctic Weddell seal (*Leptonychotes weddellii*). J Morphol 248:165–174

Wyss AR (1987) The walrus auditory region and monophyly of pinnipeds. Am Mus Novit 2871:1–31

Wyss AR (1994) The evolution of body size in phocids: some ontogenetic and phylogenetic observations. Proc San Diego Soc Nat Hist 29:69–75

Chapter 2
Phocid Sensory Systems and Cognition

Frederike D. Hanke and Colleen Reichmuth

Contents

Abstract Phocid seals experience disparate terrestrial and aquatic habitats through unique sensory filters. These filters enable them to gather information as they move through each environment, with sensory streams combining to form a salient and perceptually limited construct of the far more cluttered physical world. An understanding of how seals 'see' the world is fundamental to every aspect of their behavior. Such an understanding must be pieced together from available clues about their sensory biology and cognitive capabilities. These can be derived from anatomical, behavioral, physiological, and evolutionary studies—and consideration of the unique and sometimes opposing selective pressures operating on these perfectly amphibious carnivores. In this chapter, we summarize what is known about the sensory abilities of phocid seals and link sensory performance to behavior. We examine how vision, hearing, touch and hydrodynamic perception, chemoreception, and other senses operate effectively under the conflicting demands of marine and terrestrial living and consider how information from individual single sensory systems is integrated to form a multimodal representation of the environment.

F. D. Hanke (✉)
Institute for Biosciences, Neuroethology, Rostock, University of Rostock, Albert-Einstein-Str. 3, 18059 Rostock, Germany
e-mail: frederike.hanke@uni-rostock.de

C. Reichmuth
Long Marine Laboratory, Institute for Marine Sciences, University of California Santa Cruz, Santa Cruz, CA 95060, USA

D. P. Costa and E. McHuron (eds.), *Ethology and Behavioral Ecology of Phocids*, Ethology and Behavioral Ecology of Marine Mammals,
https://doi.org/10.1007/978-3-030-88923-4_2

Keywords Seals · Vision · Audition · Chemoreception · Olfaction · Gustation · Mechanoreception · Sensory adaptation · Cognition

2.1 Introduction

Behaviors and decisions made by animals largely depend on the perception and interpretation of information gathered from the environment. Different sensory systems provide the necessary information, which is transported via neural pathways to the brain. The brain collects, combines, and interprets the incoming data streams, which enable the selection of behavior appropriate for the specific situation. Sensory information is required to support the most important behavioral actions, including those related to orientation and navigation, predator avoidance, reproduction, and foraging. As a result, understanding the ethology of species involves consideration of their sensory biology in context to the surrounding world, that is, their 'sensory ecology'.

Sensory systems are tuned to the surrounding environment. In contrast to land-living carnivores and fully aquatic marine mammals, pinnipeds operate within dual environments due to their semi-aquatic or amphibious lifestyle. The continuous transition between air and water poses conflicting challenges for sensory systems. These media have drastically different physical properties, and these differences affect the stimuli that may be detected and perceived. Sound, for example, travels much faster through water than air, while light is quickly attenuated under water due to scattering and absorption. Moreover, the properties of the sensory systems themselves may vary with the surrounding medium. This can be illustrated in the visual domain, where the refractive index of the surrounding medium alters the cornea's refractive power; on land, a cornea is an optically effective focusing element as light passes from the air through the curved surface into the eye. However, the same structure is rendered useless in water due to the similar refractive power of seawater and the aqueous humor of the eye's anterior chamber. Thus, most sensory systems of semi-aquatic pinnipeds are adapted to gather information from terrestrial and aquatic habitats, the latter requiring innovative solutions to the previously mentioned challenges.

In this chapter, we summarize available information on the sensory adaptations and abilities of phocid seals that overall show more deviations from the terrestrial *Bauplan* compared to other pinnipeds, the otariids and odobenids. We consider how phocids cope with the challenges of an amphibious lifestyle and whether common themes can be discerned. At the same time, we examine species-specific adaptations that allow different seals to explore their dual environments, keeping in mind that sensory adaptations often reflect differences in habitat and life history. Phocids occupy habitats as diverse as tropical, temperate, and polar regions and some freshwater lakes and river inlets. Some seals are relatively shallow divers such as harbor seals (*Phoca vitulina*), a model species among the phocids (Fig. 2.1); others, such as hooded seals (*Cystophora cristata*) and elephant seals (*Mirounga* spp.), swim to depths beyond 1,000 m (Chaps. 4 and 11). Phocids also differ widely in their diets. Some species are generalists that consume a wide variety of pelagic and benthic prey

Fig. 2.1 The harbor seal (*Phoca vitulina*) is a model species among phocids regarding sensory and cognitive research. This image of harbor seal 'Sprouts' at Long Marine Laboratory illustrates several of the seal's sensory organs: large eyes, three types of vibrissae (mystacial, rhinal, and supraorbital), external openings of the ears, and the nose, which remains closed when the seal is diving or holding its breath. *Image* C. Reichmuth/NMFS18902

types. Others such as the crabeater seal (*Lobodon carcinophaga*) specialize in certain types of prey (Chap. 6). Consequently, the mode of foraging differs by species. In the social domain, reproductive tactics and social patterns in seals may deviate significantly from the 'typical' strategy of breeding aquatically and tending young on shore or ice (Chaps. 7 and 8). We attempt to explain how seals have responded to the unique selective pressures imposed upon their sensory systems by linking modifications in sensory structures and performance to variability in these and other life-history traits.

Ultimately, behavior related to orientation, navigation, foraging, predator avoidance, reproduction, and other essential activities can only be partially understood based on physical cues gathered from the surrounding environment. An understanding of cognitive abilities is also needed to infer how sensory information is used by seals for assessment and decision-making. Thus, in addition to reviewing their sensory abilities, we also describe current knowledge about phocid cognition and suggest possible avenues for future research.

2.2 The Visual System

One of the most notable features of seals is their eyes that are large relative to those of their terrestrial relatives (Fig. 2.1, 2.2 a,b). Nevertheless, it has often been presumed that diving marine mammals do not rely on visual cues because they often hunt in turbid waters and under low light conditions that occur at great depths or in the darkness of night. Consequently, some researchers have suggested a dependence on other sensory abilities (see, for example, Renouf and Davis 1982). This view is bolstered by the occasional observation of seemingly blind seals apparently thriving in wild conditions (see, for example, Newby et al. 1970; Hyvärinen 1989). However, we now have a greater understanding of how seals may see in the 'dark' and how their eyes' structures have evolved to enable acute, predatory vision on land and at sea.

2.2.1 Sensitivity—Adaptations to Dim Light Vision and Spectral Sensitivity

Phocids generally have large, round, and sometimes slightly elliptical eyes (Pardue et al. 1993). Reported equatorial diameters range from 38 mm in Caspian seals (*Pusa caspica*; Mass and Supin 2010) and bearded seals (*Erignathus barbatus*; Pardue et al. 1993) to 50 mm in northern elephant seals (*Mirounga angustirostris*; Smodlaka et al. 2016; Turner et al. 2017). Interestingly, orbit size measured from skulls—as an approximation for eye size—is correlated with diving depth in phocids (Debey and Pyenson 2013). This correlation suggests that a large eye can accommodate a large pupil for light gathering, which is required for vision in dim light conditions. While the pupil is circular when dilated, it constricts to a vertical slit as ambient light increases, and finally to a tiny pinhole (Hanke et al. 2009a; Levenson and Schusterman 1997). Among phocids, northern elephant seals have the largest pupil area; their pupil covers an area of 422 mm^2 in darkness whereas the ringed seal (*Pusa hispida*), harbor seal, and harp seal (*Pagophilus groenlandicus*) show much smaller pupillary areas of 124 mm^2, 162 mm^2, and 208 mm^2, respectively (Levenson and Schusterman 1997 and citations therein). The range of pupillary area—minimum to maximum size—is also largest in elephant seals at 469, compared to 71 in harbor seals. These differences in the pupil's dynamic range reflect that elephant seals are extraordinary divers that experience profound light changes when diving, which contrasts to shallower diving harbor seals that sometimes only submerge for a few meters (Lesage et al. 1999). When elephant seals dive from the brightly lit surface to depths exceeding 600 m, they experience drastic and rapid ambient light changes. To support this behavior, elephant seals are capable of full dark adaptation after only 6 min, close to the time it takes them to reach typical foraging depths (6.8 min; P. Robinson, personal communication). Conversely, the eyes of harbor seals take—while still much less time than terrestrial mammals—three times as long as that of the elephant seal to dark adapt (Levenson and Schusterman 1999). So far, only three species, northern elephant seals, California sea lions (*Zalophus californianus*), and harbor seals, have been examined regarding pupillary range and dark adaptation. Thus, further study is required to determine whether all deep-diving seals show enhanced dark adaptation abilities.

Besides the large aperture of the pupil, a well-developed light-reflecting layer also increases the eyes' sensitivity to light. This *tapetum lucidum* underlies the retina (Jamieson and Fisher 1971; Johnson 1901; Walls 1942). This structure reflects light not absorbed by the retinal photoreceptors on the first passage back through the photoreceptor layer, allowing absorption on the second passage. According to Walls (1942), seals possess the thickest *tapeta* among vertebrates, consisting of up to 35 cell layers; Smodlaka et al. (2016) even report an extraordinary 50 cell layers in the *tapetum* of northern elephant seals. The seal *tapetum* lines the eye fundus entirely, enabling light from any direction to be reflected back for another opportunity for phototransduction. In contrast, the *tapetum* in terrestrial carnivores can only be found behind the dorsal portion of the retina (Johnson 1901; Walls 1942). The pinniped tapetal arrangement reflects that these animals live in an environment where light is sparse and in which light can reach the eye from any direction as they change their body orientation continuously while swimming.

The retina of phocids is exquisitely sensitive to light because it is dominated by rod photoreceptors (Jamieson 1970; Jamieson and Fisher 1971; Landau and Dawson 1970; Nagy and Ronald 1970; Peichl et al. 2001; Peichl and Moutairou 1998). The rod pigment is blue-shifted in deep-diving species to match the wavelength of light penetrating into deep waters (Fasick and Robinson 2000; Lavigne and Ronald 1975b; Lythgoe and Dartnall 1970; Southall et al. 2002) and the wavelength of bioluminescence of their prey (Vacquie-Garcia et al. 2012). Fewer than 2% of photoreceptors are cones (Bernholz and Matthews 1974; Peichl and Moutairou 1998); cone photoreceptors mediate color vision in terrestrial species that mostly possess 2–4 different cone types. In phocids, the single cone type that is present maximally absorbs green light (Crognale et al. 1998; Fasick and Robinson 2000; Lavigne and Ronald 1975a; Levenson et al. 2006; Newman and Robinson 2005). Given their retinal photoreceptors, i.e., one rod and one cone type, seals should not have cone-based color vision. Early reports of color vision capability in seals (Wartzok and McCormick 1978) can probably be attributed to underestimating the seals' brightness discrimination abilities (Hanke et al. 2011; Scholtyssek et al. 2008). Color vision based on rods and cones could not be shown in a more rigorous behavioral experiment with harbor seals (Scholtyssek et al. 2015). Another experiment suggested that seals may have color vision (Oppermann et al. 2016); however, they did not control for ambient light. Thus, based on available assessments, it is apparent that seals are highly sensitive to small amounts of light, while their color vision remains unconfirmed.

Intriguingly, sensitivity to ultraviolet (UV) light (< 400 nm) has recently been suggested for hooded seals (Hogg et al. 2015) as their ocular media transmit, their *tapeta* reflect, and their rods respond to UV light. Apparent sensitivity to UV light in hooded seals is surprising, as UV sensitivity among mammals has only been documented in rodents and bats (Winter et al. 2003; Peichl 2005; Müller et al. 2007). However, the same authors report that Arctic reindeer (*Rangifer tarandus tarandus*) are also sensitive to UV light (Hogg et al. 2011), suggesting a possible environmental driver for this ability. Ultraviolet sensitivity might be beneficial for seeing at depth in water, where only light of short wavelengths penetrates, or in air to contrast the UV-absorbing fur of predators or conspecifics (Lavigne and Oritsland 1974; Reynolds and Lavigne 1981) against a UV-reflecting background of ice or snow. In general, the extent to which UV vision occurs in different mammals, and its relationship to color vision systems, remains poorly understood (Cronin and Bok 2016).

2.2.2 Resolution—Light-Focusing Apparatus and Retinal Adaptations

The dioptric, or light-focusing, apparatus consists of a cornea and a lens. The phocid cornea is generally very thick, protecting the eye during swimming (Pardue et al. 1993). Additionally, in many species, the cornea is notably flattened in the region overlying the pupil (Mass and Supin 2003; Pütter 1903; Sivak et al. 1989; Wilson 1970a, b). Building on studies that describe corneal flattening, corneal topography has been measured in detail for just one species, the harbor seal (Hanke et al. 2006a). The harbor seal's cornea is steeply curved in the periphery with radii of curvature of approximately 20 mm but possesses a central flat area with radii of curvature even up to 80 mm. The flattening is more pronounced in the vertical meridian, resulting in a high degree of astigmatism (Hanke et al. 2006a; Jamieson 1971). This corneal asymmetry has also been observed in a ringed seal (F. Hanke, unpublished data). A flattened cornea is beneficial for amphibious vision. In air, where bending of light occurs at the cornea and the lens, a flattened cornea contributes little to the eye's overall refractive power. This adaptation allows harbor seals to achieve a visual acuity of 5.5 min in air—comparable to the vision of terrestrial carnivores (Rahmann 1967) and superior to the fully aquatic cetaceans (e.g., Mass and Supin 1995; Mass et al. 2012a) and manatees (Mass et al. 2012b). This acuity is achieved only when the pupil is small in bright ambient light, and light passes through the flattened portion of the cornea to be received by the eye. By eliminating strong refraction at the cornea in air and allowing the lens to dominate, the seals' visual acuity in air is similar to their visual acuity under water when light is solely refracted by the lens (Hanke and Dehnhardt 2009; Jamieson and Fisher 1970; Schusterman and Balliet 1970; Weiffen et al. 2006). Whereas visual acuity in air is strongly affected by available light, visual acuity in water decreases with turbidity (Weiffen et al. 2006). Refractive measurements taken directly from the surface of the eye of numerous phocids (Hanke et al. 2006a; Johnson 1893, 1901; Piggins 1970; Sivak et al. 1989; Wilson 1970a, b) can indicate the possible refractive state of the eye, meaning whether the light-focusing apparatus can focus the light on, in front of or behind the retina. While such data are interesting from a comparative perspective, they cannot infer visual acuity in seals because other factors such as the shape of the cornea or the pupil's size have to be considered.

The lenses of phocid seals, like those of fully aquatic fish, are spherical for maximum refraction of light in water and possess a short focal length (e.g., Hanke et al. 2008b; Pardue et al. 1993; Sivak et al. 1989). There is experimental evidence that seals may be able to 'accommodate' (Hanke et al. 2006a; Piggins 1970; Sivak et al.

1989), meaning that the eye can bring objects at different distances into focus. Moreover, a multifocal lens has been documented in harbor seals (Hanke et al. 2008b). A multifocal lens possesses distinct zones within the round 'fish-eye' lens having different refractive indices, termed complex spherical aberration, which can create well-focused color images. The discovery of multifocal lenses in seals was surprising, as such lenses are usually considered an adaptation for color vision (Kröger et al. 1999). As seals are monochromats—they only possess one cone type and thus no cone-based color vision (see Sect. 2.2.1)—the multifocal lens alternatively might increase the depth of focus in dim ambient light, when the pupil is large and circularly dilated (Hanke et al. 2008b). In bright light, the pupil is constricted, and light rays only pass through the lens's central part; this optical situation is comparable to a pinhole camera. In this condition, the lens is more or less monofocal, meaning it only possesses one refractive index, so the highest possible acuity alongside a great depth of focus can be obtained. In contrast to the complex spherical aberration present in harbor seals, Sivak et al. (1989) described the hooded seal's lens as well corrected for spherical aberrations. Thus, future studies are required to obtain a more detailed understanding of the optics of phocid lenses.

The cornea and lens serve to focus light onto the rods and cones of the retina (see Sect. 2.2.1). These photoreceptors transmit the optic information via a dense network of amacrine, bipolar, and horizontal cells to the ganglion cells, the first neurons along the visual pathway. A peculiarity of the retinae of aquatic mammals in general as compared to terrestrial species is the presence of large ganglion cells reaching diameters of 40–80 μm in comparison to maximal diameters of 15–35 μm in terrestrial carnivores (Mass and Supin 2007). The distribution of ganglion cells within the retina has been described in some phocid species (Hanke et al. 2009b; Mass and Supin 2003, 2010; Smodlaka et al. 2016; Welsch et al. 2001). Whereas peripheral ganglion cell density is relatively low in all aquatic mammals, all phocids seem to have an *area centralis* with an increased ganglion cell density of 1,300–4,000 cells/mm^2, depending on species. The lowest ganglion cell density measured thus far in a seal has been for the deep-diving northern elephant seal, which might indicate a high degree of retinal convergence, meaning that the information gathered by many photoreceptors is converging on fewer ganglion cells to increase sensitivity (Smodlaka et al. 2016). Such increased sensitivity may be advantageous to deep-diving in contrast to shallow-diving species such as harbor seals. In addition to the *area centralis*, Hanke et al. (2009b) documented a visual 'streak' on the retina of harbor seals, with an increased ganglion cell density of approximately 1,000 cells/mm^2. A streak-like elongation of the area of highest ganglion cell density across the midline of the retina has also been found in Caspian seals (Mass and Supin 2010). The existence of such a region of heightened neural density might indicate the importance of a horizontal reference plane (Hughes 1977), such as the horizon or sea bottom, for seals.

2.2.3 Visual Neuroanatomy

Further along the visual pathway, optic information is conveyed to higher brain areas via the optic nerve with 170,000–750,000 axons (Pütter 1903; Turner et al. 2017; Wohlert et al. 2016). In comparison to phocids, river dolphins (*Inia geoffrensis*) that live in murky water have only 15,000–20,000 axons (Morgane and Jacobs 1972; Mass and Supin 1989), whereas the axon numbers of bottlenose dolphins (*Tursiops truncatus*; Morgane and Jacobs 1972; Dawson et al. 1982, 1983) and terrestrial carnivores (e.g., Hughes 1975; Bruesch and Arey 1942) fall into the range of axon numbers found in phocids. In contrast, humans and primates possess more than a million axons (e.g., Bruesch and Aurey 1942; Ogden and Miller 1966; Potts et al. 1972; Sanchez et al. 1986). Thus phocids apparently invest substantially in the transport of visual information, which is evidence of the importance of vision in these species. The discrepancy between the very high number of axons in the optic nerve and the very low number of ganglion cells that give rise to the optic nerve in the retina of northern elephant seals still needs to be resolved (Pütter 1903; Smodlaka et al. 2016).

The optic nerves of hooded seals (Dohmen et al. 2015) and likely harbor seals (F. Hanke, unpublished data) only partially cross at the optic chiasm at the base of the brain; this means that some fibers are originating from one eye project to the other (contralateral) side of the brain, whereas others project to the same (ipsilateral) side. Turner et al. (2017) described the structure of higher brain centers, including the superior colliculus, lateral geniculate nucleus, and the visual cortex in northern elephant seals; these brain centers show a number of similarities to the respective brain centers of terrestrial carnivores. Besides these few neuroanatomical studies, a detailed understanding of visual neuroanatomy and processing is still lacking.

2.2.4 Motion Vision and Other Functions of Vision

So far, we have only considered the visual adaptations of seals when the eye is static. However, in contrast to Walls' (1942) observations of limited eye movements in seals, phocids seem to be able to move their eyes considerably (Fig. 2.2b). Harbor seals can move their whole eyes at least 12 degrees to both sides and 64 degrees upwards, thereby substantially extending their visual field (Hanke et al. 2006b). Harbor seals also show reflexive eye movements to stabilize external motion (Hanke et al. 2008a) and appear to have excellent 'motion' vision. This species is furthermore very sensitive to motion patterns that fill large parts of the visual field (Weiffen et al. 2014) and are able to perceive 'optic flow' (Gläser et al. 2014), that is, the motion pattern of the surrounding environment that is elicited in the eye of the moving observer. Optic flow is induced when a seal moves over the seafloor, underneath the water surface, or through clouds of particles. Optic flow perception could mediate many functions, including estimation of distances to objects (Geurten et al. 2017), heading detection, determination of distance traveled, or collision avoidance. The

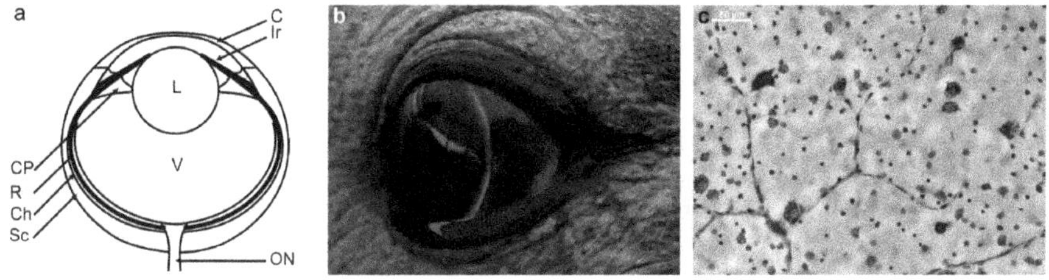

Fig. 2.2 **a** Schematic of a phocid eye. C cornea, Ch choroid, CP ciliary process, Ir iris, L lens, ON optic nerve, Sc sclera, V vitreous. **b** The right eye of a bearded seal, showing pronounced temporally oriented eye movements. **c** Giant ganglion cells in the periphery of the retina of a harbor seal. Scale 50 μm. *Images* C. Reichmuth/NMFS 18902 and F. D. Hanke

large ganglion cells found in the retinae of aquatic mammals (Fig. 2.2c; Mass and Supin 2007) may help to support motion vision (Dawson et al. 1982).

Given the many modifications of the phocid eye for function in water and low light, visual cues are likely involved in predator avoidance and foraging. Under good light conditions in relatively clear water, a seal can visually track and hunt its prey (Kilian et al. 2015). This ability is supported by forward-directed eyes with strongly overlapping visual fields (Hanke et al. 2006b) that most likely allow for excellent depth perception. Moreover, a seal might also detect a cryptic benthic fish from the coherent movement of the prey's body parts (Lui et al. 2012; Weiffen et al. 2014). Vision might especially come into play in close vicinity to prey (Hanke et al. 2011; Scholtyssek et al. 2008; Weiffen et al. 2006). Even the deepest diving phocids, such as elephant seals that operate in aphotic conditions, appear to use low levels of bioluminescent light within prey fields to exploit visual cues while foraging (Levenson and Schusterman 1999; Vacquie-Garcia et al. 2012, 2017; Warrant and Locket 2004).

Finally, vision can be used for orientation and navigation. According to Mauck et al. (2005), harbor seals can perceive many stars in the night sky and distinguish a single star from a full starry sky (Mauck et al. 2008). These abilities could be the basis for astronavigation, navigation using the stars. In contrast, the sky's polarization pattern around the sun or moon apparently cannot be used in this context as evidenced by the lack of polarized light perception in harbor seals (Hanke et al. 2013). Future studies of visual orientation could focus on the sun, the moon, or visual landmarks as possible orientation cues. The recent finding of optic flow perception in harbor seals also leads to interesting research questions regarding navigation. Future research could focus on whether seals are able to perform path integration (Etienne and Jeffrey 2004; Mittelstaedt and Mittelstaedt 1982) based on visual cues (e.g., optic flow) as they move through the environment, or solely based on idiothetic information (i.e., self-motion cues). Path integration has already been hypothesized, however not confirmed, for homing Weddell seals (*Leptonychotes weddellii;* Fuiman et al. 2020). Outside of the laboratory, evidence from free-ranging northern elephant seals supports the notion that phocids use visual cues and landmarks to navigate to their home colony, with individuals adjusting their bearing while at the surface (Matsumura et al. 2011).

2.3 Mechanoreception

The seals' body is covered by thick blubber that serves as insulation, energy storage, and structural material to sculpt the body into a streamlined shape. The skin that overlies the blubber layer contains sensory receptors for pressure and vibration detection and presumably sensory receptors for thermal perception. Not much is known about mechanoreception in the skin of phocids. In contrast, mechanoreception mediated by the whiskers, or vibrissae (Figs. 2.1 and 2.3b), has been well studied. Vibrissae are active in nearly all aspects of seal behavior, including exploratory, social, and foraging contexts. Investment in this sensory pathway is significant at many levels, and the diversity of vibrissae seen in different phocid species suggests significant specialization at the species level.

2.3.1 General Aspects of the Vibrissae

All phocid seals possess vibrissae at both sides of their snout, the so-called mystacial vibrissae. Additional vibrissae can be found on the nose, the rhinal vibrissae, and the supraorbital vibrissae above the eyes (Fig. 2.1). Some phocids also have vibrissae on the lower jaw, the submental vibrissae.

Most information about the morphology, innervation, and function of the vibrissae comes from the mystacial vibrissae, which are regularly arranged in columns and rows on the 'cheeks' of the seal. These vibrissae can be actively moved and protracted forward to form an erect array surrounding the muzzle. The number of mystacial vibrissae on each cheek pad ranges from less than 20–30 in Ross (*Ommatophoca rossii*) and leopard seals (*Hydrurga leptonyx*; King 1983; Ling 1977; Rogers et al. 2016) up to 120 in bearded seals (Fig. 2.3b; Marshall et al. 2006). Most phocids, however, possess 35–50 vibrissae per pad (Fig. 2.1; Adam 2005; Dehnhardt and Kaminski 1995; Ginter et al. 2010; Hyvärinen 1989; Hyvärinen et al. 2009; King 1969, 1983; Ling 1977; McGovern et al. 2015; McHuron et al. 2020). Bearded seals apparently have the longest vibrissae among phocids, reaching more than 20 cm (Marshall et al. 2006; McHuron et al. 2020), whereas in spotted seals (*Phoca largha*) and other small phocids, vibrissal length varies from a few to 15 cm (McHuron et al. 2016). These lengths indicate the visible part of the sensory hairs in a seal but do not include the small portion rooted within the follicle.

The vibrissae are composed of inert, keratinous tissue that is gradually deposited from the base as the whisker grows. In many species, the vibrissae are shed during or at the end of the annual molt, whereas in others the vibrissae seem to be shed and regrown without any seasonal trend (Beltran et al. 2015; Greaves et al. 2004;

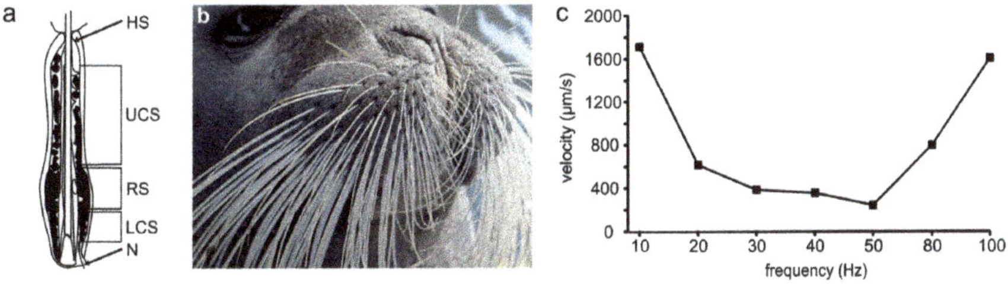

Fig. 2.3 **a** Follicle within which the vibrissal hair is inserted in the skin. HS hair shaft, LCS lower cavernous sinus, N nerve, RS ring sinus, UCS upper cavernous sinus. Figure adapted from Dehnhardt et al. (1998b) **b** Muzzle of a bearded seal illustrating the high density and smooth texture of the vibrissae. *Image* C. Reichmuth/NMFS18902 **c** Hydrodynamic thresholds of a harbor seal. Depicted is the seal's performance as a velocity threshold (in µm/s) as a function of the stimulus's frequency (in Hz). Figure adapted from Dehnhardt et al. (1998a)

Hall-Aspland et al. 2005; Hirons et al. 2001; Lübcker et al. 2016; McHuron et al. 2016, 2020; Newland et al. 2011; Rogers et al. 2016; Walters et al. 2014; Zhao and Schell 2004). The maximum lifespan of a seal's vibrissa seems to be approximately 2 years, much shorter than those of otariid pinnipeds. Even in those phocids that shed their vibrissae seasonally, the vibrissae are not lost at once. The advantage of not shedding all vibrissae simultaneously and quickly re-growing lost whiskers is that the seals can rely on a functional sensory system throughout the year. A putative reduced functionality during shedding might not adversely affect those species with seasonal shedding, as seals tend to spend more time on land and forage less during the annual molt. Vibrissal growth rates vary from 0.04 to 1.4 mm/day depending on species and time of year (Beltran et al. 2015; Greaves et al. 2004; Hirons et al. 2001; Lübcker et al. 2016; McHuron et al. 2016; Rogers et al. 2016). As each growing vibrissa encodes dietary information, these structures are often collected and used for isotopic analyses in ecological studies.

2.3.2 *Vibrissa Types*

The morphology of seal whiskers varies widely by species. The vibrissae of most phocids are flattened (oval) in cross-section and show a pronounced undulatory or 'beaded' pattern along the hair shaft (Fig. 2.1; Ginter et al. 2010, 2012; Hanke et al. 2010; Miersch et al. 2011; Murphy et al. 2013). When swimming, each vibrissa's thinnest blade faces into the oncoming water to reduce resistance (Murphy et al. 2013). The number of undulations per cm varies along the slender profile, with 1.1 bumps/cm in harp seals up to 3.4 bumps/cm in gray seals (*Halichoerus grypus*; Ginter et al. 2012). Smooth vibrissae are only present in monk, bearded (Fig. 2.3b), and possibly leopard seals (K. Morgenthal, unpublished data). The more typical undulation pattern seems to allow the vibrissae to remain relatively still even when the seal is swimming, as vortex-induced vibrations are reduced by the instability of the vortices shedding from the vibrissae (Hanke et al. 2010; Miersch et al. 2011; Morrison et al. 2016; Witte et al. 2012). Consequently, when an undulated vibrissa encounters a disturbance in the water, the 'signal' can be detected within the relatively low background 'noise' (Miersch et al. 2011; Murphy et al. 2017). In contrast, background noise appears higher for smooth vibrissae. Nevertheless, when placed in a flume, a smooth vibrissa can still respond to a disturbance in the flow: the noise signal of the vibrissa possesses a dominant frequency that is modulated by external events, and the frequency seems to correlate with flow velocity (Miersch et al. 2011). In contrast to Miersch et al. (2011), Murphy et al. (2013) could not identify differences in flow performance between the smooth and undulated vibrissal types, a discrepancy that merits further exploration. As the flume data were obtained with the smooth vibrissa of a California sea lion (Miersch et al. 2011), it also remains to be investigated if the smooth vibrissae of phocid seals perform comparably to those of sea lions.

Future research could clarify why undulated vibrissae evolved in phocids and whether the vibrissa type or characteristics can be related to life-history traits or specific foraging modes.

2.3.3 Follicle Sinus Complex

Each vibrissal hair is embedded in a follicle sinus complex in the skin that has several distinctive features (Fig. 2.3a). The follicles differ in many respects from the whisker follicles found in terrestrial animals (e.g., Ebara et al. 2002); these differences are considered adaptations to the aquatic medium (Hyvärinen 1989; Hyvärinen and Katajisto 1984; Hyvärinen et al. 2009).

First, seals' follicles are generally longer with a subdermal length of approximately 20 mm (Connolly Sadou et al. 2014; Hyvärinen 1989; Jones and Marshall 2019; Ling 1977; Marshall et al. 2006; McGovern et al. 2015). This increase in length can be explained by an additional blood sinus that surrounds the upper 60% of the hair shaft within the follicle. On the one hand, this arrangement causes additional blood to flow into the sensory organ as a form of regional thermoregulation (Erdsack et al. 2014; Mauck et al. 2000), which is supported by the presence of low-melting-point monoenoic fatty acids in the adipose tissue in the vicinity of the follicle (Käkelä and Hyvärinen 1993, 1996). On the other hand, this arrangement embeds the mechanoreceptors—including Merkel cell-neurite complexes, Lancelot, and lamellated endings—deeply into the skin. Both aspects enable the system to function independently of temperature changes in the surrounding medium (Dehnhardt et al. 1998b), a helpful feature as water temperature varies during diving and also seasonally or regionally. The increased blood supply may also play a role in the intentional movement or flexion of vibrissae, although this has not been explored in detail.

Second, the innervation density of the vibrissae of seals is extraordinary, more than ten times as high as the innervation density of the vibrissae of terrestrial mammals: 1,200–1,600 myelinated axons can be counted per seal vibrissa, resulting in an estimated number of 120,000–320,000 axons per vibrissal system in total (Hyvärinen et al. 2009; Jones and Marshall 2019; Marshall et al. 2006; Mattson and Marshall 2016; McGovern et al. 2015). These values may slightly overestimate innervation density, as innervation of the short (microvibrissae) and long (macrovibrissae) whiskers vary considerably (Mattson and Marshall 2016); regardless of such differences, it is apparent that the neural investment in the vibrissae is remarkable. In conclusion, seals' vibrissal follicles are quite specialized sensory structures that are surrounded by highly innervated tissue with a rich blood supply, suggesting an essential function of this sensory system.

2.3.4 Functions of the Vibrissal System

In phocids, the vibrissae serve many functions: they are part of facial expressions during greetings, agonistic behavior, and individual recognition, and may facilitate collision-free movement when visual cues are unavailable (Oliver 1978). The rhinal and supraorbital vibrissae likely provide information to the animal about the air–water boundary when surfacing (Ahl 1986). The vibrissae may aid seals in under-ice navigation as well as during the exploration of benthic substrates. It remains to be determined if vibrissae play a role during underwater orientation. External environmental cues such as tidal currents might provide seals with directional information, such as whether they are heading on- or off-shore.

The whisker array's primary function is to provide the animal with haptic information through active touch (direct contact) with surfaces and with hydrodynamic information gathered during underwater movement and foraging (Dehnhardt et al. 2014; Dehnhardt and Mauck 2008; Renouf 1980a). Most of the information regarding haptic and hydrodynamic performances in phocids summarized in the following sections has been obtained from harbor seals. Comparative data for phocids with smooth vibrissae are not yet available.

2.3.4.1 Haptics—The Sense of Touch

When their vibrissae are in direct contact with the surface of an object, trained harbor seals can discern fine details of features, including size, shape, and texture both in air and under water (Dehnhardt and Kaminski 1995; Dehnhardt et al. 1997a, b, 1998b). During haptic sensing, seals often touch the object's surface with their vibrissae while performing lateral head movements (Grant et al. 2013; Milne et al. 2020). For example, in behavioral size discrimination experiments, the seals' head movements center the vibrissae over the object. Thus, differences in deflection angle and/or the number of vibrissae in contact with the object can convey size information (Dehnhardt and Kaminski 1995; Grant et al. 2013). Other behavioral experiments demonstrate that harbor seals' vibrissae are most sensitive to stimuli vibrating with frequencies from 10 to 1,000 Hz; best sensitivity has been documented at 80 Hz to movement as small as 0.09 mm/s (Mills and Renouf 1986; Murphy et al. 2015; Renouf 1979).

While there are no direct data to describe the haptic performance of other phocids, observations suggest that there are probably species-level differences in the use of vibrissae for active touch. For example, bearded seals tend to brush objects and surfaces with forward 'sweeping' motions of their long, soft vibrissae, whether they are hauled out or exploring surfaces when submerged. In contrast, other species seem to use their stiff and erect vibrissae to identify surfaces by firmly 'pressing' against them, a behavior often seen by seal pups when attempting to nurse.

2.3.4.2 Hydrodynamics—Detection of Water Movements

Under water, trained seals can detect minute water movements caused by a dipole moving with frequencies from 10 to 100 Hz and can perceive this type of stimulus down to a water velocity of just 245 μm/s at a frequency of 50 Hz (Fig. 2.3c; Dehnhardt et al. 1998a). Such water movements can be found in the hydrodynamic trails left behind by conspecifics or prey (Hanke and Bleckmann 2004). Numerous experiments have revealed that harbor seals can detect and actively follow these hydrodynamic trails immediately after generation and up to 30 s later (Dehnhardt et al. 2001; Kilian 2010; Schulte-Pelkum et al. 2007; Wieskotten et al. 2010b). A seal's ability to detect and follow a hydrodynamic trail is mostly unaffected by whether the generator moves actively or glides through the water (Wieskotten et al. 2010b). The hydrodynamic trail left behind by a moving stimulus encodes information on the form, movement direction, and size of the trail generator; at least the latter two aspects can also be interpreted by seals from single vortices found within a complex trail (Krüger 2017; Krüger et al. 2018; Wieskotten et al. 2010a; Wieskotten et al. 2011). The ability to interpret single vortices is particularly interesting in the context of fleeing fish that perform 'C-starts' as they escape, fast evasive maneuvers that generate single vortex rings (Niesterok and Hanke 2013).

It has long remained speculative how seals might find stationary, benthic prey such as flatfish. Flatfish constitute a substantial proportion of harbor seal diets in some locations or during specific seasons (e.g., Härkönen 1987; Pierce et al. 1991; Tollit and Thompson 1996; Chap. 10). Niesterok et al. (2017a, b) hypothesized that seals might detect cryptic flatfish by sensing their subtle breathing currents while cruising over the sea bottom. The water measurements generated by respiration in flatfish are well within the hydrodynamic detection range of seal vibrissae (Bublitz 2010). Trained harbor seals can detect artificial breathing currents even when spaced out in a sizeable benthic area (Niesterok et al. 2017b). Seals can thus use hydrodynamic information to sense both moving and cryptic prey.

2.4 Audition

Early research on hearing in phocids was motivated in part by the idea that they might possess a specialized biosonar sense, as was demonstrated in dolphins in the 1950s (Kellogg 1958). We now know that seals and other pinnipeds do not echolocate (Schusterman et al. 2004). They do, however, rely on their auditory sense to observe the surrounding environment and receive biologically relevant acoustic cues. Hearing is thought to be of particular importance to seals, not simply because of the efficiency of sound transmission in water compared to that of other physical stimuli, but also because of the adaptations that seals have developed which enable acute hearing on land and at sea.

2.4.1 Ear Anatomy

The most noticeable feature of the phocid ear is the lack of a pinna, an external ear. The pinna was most likely reduced during evolution to streamline body shape and to minimize heat loss to the environment. In the absence of a pinna, only the ear canal's opening, the external auditory meatus, is discernable in phocids. This opening varies in size among species. The opening points to the back of the head. Beyond the opening, the long and narrow meatus runs along the outside of the skull for some distance before passing into the cranium (Boenninghaus 1903). The external auditory meatus can be closed by muscular control and is indeed closed when the animal is submerged (Møhl 1968b).

The ear canal itself and the lumen of the middle ear are lined with cavernous tissue (Møhl 1967; Repenning 1972). This tissue—which expands when engorged with blood—is thought to play a role in equilibrating pressure differences during diving. Consequently, it may protect the eardrum, or tympanic membrane, from rupture during deep-diving (Møhl 1967; Stenfors et al. 2001). The tympanic membrane is larger in phocids than in terrestrial carnivores, as are the middle ear's bony structures. The middle ear bones, or ossicles, are also heavier, denser, and more robust in phocids, conferring increased tolerance to hydrostatic pressure. Moreover, due to their increased mass and increased inertia (Nummela 2008), these modified ossicles likely support hearing via bone conduction in water.

The auditory bulla surrounding the ossicles and forming the middle ear cavity is enlarged and partially detached from the skull. This design feature is thought to improve directional hearing abilities in water by providing some degree of separation of each ear from the bones of the skull (Møhl 1968b). A remarkable anatomical specialization within the ears of phocids is the presence of sensory receptors in the cochlea of the inner ear that are capable of encoding sounds of very high frequency relative to most other carnivores (Fay 1988), including otariid and odobenid pinnipeds (Hemilä et al. 2006; Nummela 2008; Repenning 1972; Wartzok and Ketten 1999).

2.4.2 Sound Transmission

The 'prototypical' phocid ear is challenged to efficiently receive acoustic information both in air and under water. In air, sound is apparently received via the typical sound transmission pathway of mammals: pressure waves are transmitted through the opened meatus via the ear canal and amplification system of the middle ear to the sensorineural receptors (hair cells) within the cochlea of the inner ear. The cavernous tissue does not engorge with blood. The auditory pathway remains air-filled, enabling sound reception in the conventional mammalian manner.

When the seal is submerged, and the meatus is closed, the sound transmission pathway is less certain. Because bones, tissues, and seawater have similar acoustic impedances, sound waves can more readily penetrate the entire body when submerged. Phocids apparently lack the specialized 'acoustic fats' for ducting sound present in some other marine mammals (Nummela 2008). Most researchers favor the idea that underwater sounds are transmitted from the fluid environment through bone and tissue to the seals' inner ears, perhaps bypassing the outer and middle ears. This hypothesis is supported by experimental evidence that phocids remain sensitive to waterborne sounds even when the ear opening is above the water's surface (Kastelein et al. 2018). However, it is difficult to reconcile this proposed mechanism of bone and tissue conduction with good sound localization abilities documented for seals in water (see Sect. 2.4.3). It should be hard to gain information about sound directionality if sound waves are passed through the entire body to the inner ears. Furthermore, there are reasons to suspect that portions of the terrestrial sound transmission pathway (i.e., ducting sound along the external auditory meatus through the middle ear) may play a role in underwater hearing even if the cavernous tissue is inflated and air spaces are reduced (see Kastak and Schusterman 1998). Despite some compelling clues, the auditory pathway(s) used by submerged seals to receive sounds remain mysterious.

2.4.3 Sound Localization

Behavioral studies with trained individuals have provided insight into the ability of seals to localize sound sources in terrestrial and marine environments. The auditory performance of harbor seals and northern elephant seals shows that they can localize the origin of broadband noise in air in the horizontal plane with an accuracy—or minimum audible angle—of less than 5° (Holt et al. 2004; Terhune 1974). These seals can also localize the source of airborne tonal sounds to within 3°–15°, with both species performing best at lower frequencies (<4 kHz) and harbor seals also performing well at some higher frequencies (>16 kHz; Holt et al. 2005). Harbor seals can additionally localize the source of airborne sounds in the median (vertical) plane (J. Byl, unpublished data). These observations suggest that directional hearing in seals is not particularly limited by the absence of a sound-ducting pinna, a structure known to support sound localization in many terrestrial mammals (Blauert 1997; Muller and Bovet 1999; Wightman and Kistler 1997).

Under water, sound waves may pass directly into the body to stimulate the inner ears' neural receptors, as noted earlier. For this reason, human divers have difficulty in resolving the directionality of sounds when submerged (Hollien and Feinstein 1976). However, the behavior of trained harbor seals indicates that they can localize underwater sound sources in both the horizontal and the medial plane much better than humans, with best performance to sound frequencies less than 2 kHz (Bodson et al. 2006, 2007; Byl et al. 2016, 2019; Møhl 1964). While seals are capable of reasonable sound localization in water, the specific adaptations that enable this ability are not well understood. Bony features of their skull may help to acoustically isolate the inner ears; however, seals lack the extreme anatomical specializations and performance capabilities of odontocetes related to sound localization (Mooney et al. 2012).

2.4.4 Hearing Sensitivity

Hearing sensitivity in phocids is characterized by measuring the smallest amount of audible sound in air and water, either in behavioral experiments with trained animals (Fig. 2.4; e.g., Reichmuth et al. 2013) or through neurophysiological recordings of activity from the auditory brainstem or cortex (Lucke et al. 2016; Ridgway and Joyce 1975; Ruser et al. 2014; Wolski et al. 2003). Most hearing studies have focused on harbor seals (Cunningham and Reichmuth 2016; Kastak and Schusterman 1998; Kastelein et al. 2009a, b; Lucke et al. 2016; Møhl 1968a; Reichmuth et al. 2013; Terhune 1991; Turnbull and Terhune 1990; Wolski et al. 2003), which has allowed for a detailed description of this phocid's amphibious hearing abilities as a reference for comparison of hearing performance in other species (Fig. 2.4).

Harbor seals have a remarkably wide range of sensitive hearing in water that extends from 0.3 to 60 kHz, spanning more than eight octaves (Fig. 2.4a). Best hearing in water occurs in the region of 10–20 kHz with a maximum sensitivity of 54 dB re 1 μPa. Their ability to detect quiet sounds is comparable to the sensitivity of fully aquatic dolphins, albeit at much higher frequencies in the dolphin (Mooney et al. 2012). Although harbor seals' hearing ability declines at frequencies above 60 kHz, they are capable of detecting intense sounds up to at least 180 kHz (Cunningham and Reichmuth 2016; Møhl 1968a). Hastie et al. (2014) provide evidence that free-ranging gray seals show significant behavioral changes to intense sounds of even higher frequency. While seals can detect very high-pitched sounds (>80 kHz), they apparently cannot distinguish between sounds of different frequencies (Møhl 1968b); thus, their 'ultrasonic' hearing performance is less well developed than that of the odontocete cetaceans that rely on biosonar and also have good frequency resolution within this range.

Despite possessing an ear that is highly sensitive to underwater sounds, harbor seals also have acute hearing abilities in air. When out of the water, harbor seals display a narrower frequency range of hearing that extends to approximately 33 kHz

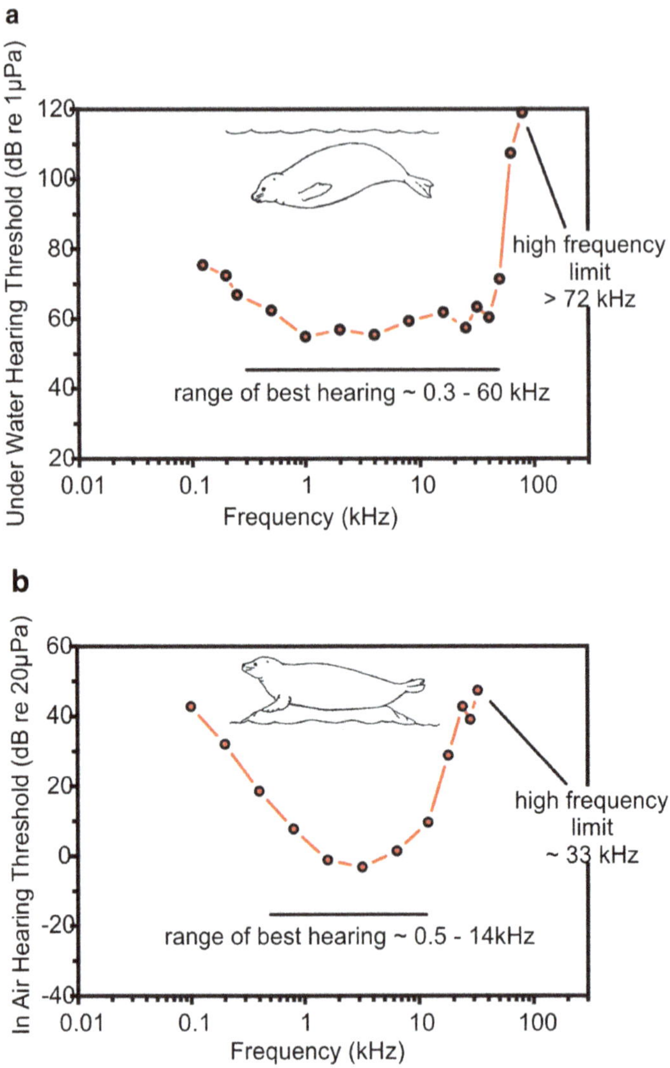

Fig. 2.4 Amphibious hearing profiles for harbor seals tested **a** in water (Kastelein et al. 2009b) and **b** in air (data taken from Reichmuth et al. 2013), highlighting acute hearing in air and a broad range of best hearing with an expanded upper-frequency hearing limit in water. More sensitive thresholds appear at the base of the plot. Figure drawing: R. Jones

(Fig. 2.4b), an upper-frequency limit of hearing similar to many other terrestrial carnivores. Their most sensitive hearing in quiet conditions extends to −4 dB re 20 μPa at 3.2 kHz, a hearing threshold better than that of human listeners and rivaling that of felid carnivores (Fay 1988). The harbor seal's ability to receive airborne sounds through an open, air-filled ear canal and through the conventional auditory pathway explains this high sensitivity to airborne sounds, despite their ability to constrict the air spaces within this pathway to protect the ear during diving.

With some notable exceptions, the audiograms of other phocid seals are similar to hearing profiles of harbor seals (see Erbe et al. 2016 for review). Spotted and ringed seals tested in similar environments with similar methods have a comparable hearing in air and water to harbor seals (Sills et al. 2014, 2015), as do bearded seals in water (Sills et al. 2020). Some differences in reported hearing abilities for other phocid species may be explained by differences in testing methodology or noise conditions (Reichmuth et al. 2013; Southall et al. 2019). However, hearing measurements for northern elephant seals (Kastak and Schusterman 1998; Reichmuth et al. 2013) and Hawaiian monk seals (*Neomonachus schauinslandi*; Ruscher et al. 2021; Sills et al. 2018, 2021; Thomas et al. 1990) suggest that there are biological differences in hearing between the two major clades of phocids.

Various experiments have shown that sound characteristics influence their detectability by seals. Hearing thresholds are lower for more complex stimuli that resemble naturally occurring sounds like vocalizations, such as sweeps, pulses, and harmonic signals (Cunningham et al. 2014a; Turnbull and Terhune 1993, 1994). Background noise is also relevant to hearing ability, regardless of whether seals are listening in air or water. Trained seals are particularly good at detecting signals in noisy conditions, as shown by very low critical ratios—the amount by which a sound must exceed the spectrum of surrounding noise to be heard (Fletcher 1940). Critical ratios have been measured for several species at different frequencies (Renouf 1980b; Sills et al. 2014, 2015, 2020; Southall et al. 2000, 2003; Terhune 1991; Terhune and Ronald 1971, 1975; Turnbull and Terhune 1990). These phocids seem to have lower critical ratios than terrestrial mammals, particularly at low frequencies (Erbe et al. 2016; Fay 1988). A good ability to detect signals in noise is likely advantageous for phocids living in noisy coastal and marine environments (Reichmuth 2014; Southall et al. 2000).

2.4.5 Function of Audition

Phocids may use their hearing during foraging, as the sounds emitted by many fish species (Kasumyan 2008; Wilson et al. 2004) could be used by seals to detect and localize potential prey. Seals can preferentially target fish marked with acoustic transmitters (Cunningham et al. 2014b; Stansbury et al. 2015), confirming they can learn the significance of different acoustic cues. Hearing is also used by seals to detect and recognize predation threats while on land and under water (e.g., Deecke et al. 2002). Their ultrasonic sensitivity enables them to detect the echolocation pulses of killer whales (*Orcinus orca*). Audition is required for the detection and interpretation of conspecific vocalizations which phocids use to delineate and claim territories, transmit dominance status, or attract females (e.g., Bartsh et al. 1992; Bjorgesaeter et al. 2004; Charrier et al. 2013; Hanggi and Schusterman 1994; Hayes et al. 2004). Perception of vocalizations is also a fundamental component of their maternal behavior and the ability to recognize conspecific individuals (Chap. 3).

The perception of soundscapes likely aids phocids in orientation, especially in the three-dimensional, frequently darkened expanse of the ocean. This idea has been proposed for whales (Norris 1967) and pinnipeds (Schusterman et al. 2004), as acoustic cues can indicate large topographic features. Along these lines, it has been suggested that pagophilic phocids use acoustic information to sense the location of breathing holes and breaks in the ice during polar winters (Elsner et al. 1989; Miksis-Olds and Madden 2014; Wartzok et al. 1992).

2.5 Chemoreception

Olfaction and gustation are the senses most neglected in marine mammal research and phocids in particular. Consequently, although the habitats occupied by seals contain a wealth of chemical information, little is known about their ability to receive and use chemical cues.

2.5.1 *Olfaction*

Olfaction is a sense that can be exclusively used in air; under water, the nose of phocids is closed, and the olfactory epithelium is unavailable for contact with airborne molecules. Given the absence of little direct measurement, researchers have attempted to estimate the significance of chemical odorants to seals and other marine mammals based on their anatomy—with the idea that if olfaction should be of only minor importance, then corresponding anatomical structures should be reduced relative to terrestrial mammals that are known to rely on chemical cues. However, whereas the olfactory bulbus of phocids is smaller than in terrestrial carnivores (Reep et al. 2006; Spitzka 1890), the ethmoid bones of the nasal cavity resemble those of terrestrial mammals (Philström et al. 2005). Nummela et al. (2013) concluded from the relative size of sensory organs and epithelia that aquatic carnivores are 'less olfactory oriented' than terrestrial carnivores. However, observations suggest that scent is important to phocids when out of the water and even when they are scanning the water's surface.

A single study has directly examined olfactory sensitivity in a phocid. Two harbor seals were trained to detect very low concentrations of dimethyl sulfide (DMS; Kowalewsky et al. 2006), an algal byproduct released into the air in highly productive coastal areas (Andreae et al. 1994; Bürgermeister et al. 1990). Such acute sensitivity may support foraging behavior. If seals can identify and follow a DMS gradient at the surface, they could systematically detect these high-productivity areas, as demonstrated for seabirds (Nevitt 2008; Nevitt and Bonadonna 2005). The potential role of olfaction is further suggested by an observation reported by Kim et al.

(2013), in which gray seals appeared downwind of an oil slick caused by conspecifics consuming prey at the surface, suggesting they were attracted by the fish oil scent.

Olfaction seems to play an important role in social behavior among phocids. Frequent nose–nose contact is common during the period of maternal dependence (Burton et al. 1975; Kovacs 1995), and female seals often sniff their dependent pups. For example, female harbor seals appear to distinguish their own pup among others by nasal investigation of the anal region (Renouf et al. 1983). Scent may also be important in other reproductive contexts—such as identifying estrous females—even from some distance, as suggested for Hawaiian monk seals (Kenyon and Rice 1959). Some phocids emit odors that may provide cues relevant to territorial behavior or female defense. For example, male spotted seals emit a musky odor during the breeding season (C. Reichmuth, unpublished data). Male ringed seals produce a chemical odorant from the facial skin described as smelling strongly of petrol (Hardy et al. 1991; Ryg et al. 1992), which might be used to mark breathing holes and birth lairs in the ice during the rut. While the production of chemical cues suggests correlated olfactory sensitivity, much remains to be learned about this sensory modality in phocids.

2.5.2 *Gustation*

Little is known about gustation in phocids. Sonntag (1923) attempted to describe the anatomy of the seal's tongue and concluded that, among the species investigated, the gustatory organs are best developed and show an arrangement comparable to other mammals in harbor seals and are least developed in gray seals. In the latter species, Sonntag was unable to find papillae on the tongue. Tuckerman (1890) had access to a tongue of a harbor seal and also concluded that 'the bulbs are far from numerous', suggesting reduced reliance on taste. Nevertheless, the sensory receptors in the tongue and mouth allowed trained harbor seals to resolve small differences ($\leq$4%) in salinity at natural concentrations (Sticken and Dehnhardt 2000). Thus, seals should be capable of chemosensory orientation such as locating foraging areas associated with salinity fronts (Bowman and Esaias 1978; Floodgate et al. 1981). Seals living in zoos and aquaria can distinguish prey species by taste and often exhibit food preferences or spit out novel prey types or fish containing vitamins or medications; anecdotally, a captive bearded seal preferentially selected capelin fish but spit out female capelin that were gravid based on taste alone (C. Reichmuth, unpublished data). Further studies from anatomical and functional perspectives are needed to determine how the shift toward marine living has influenced the sense of taste in phocids.

2.6 Other Senses

Consideration of sensory perception is focused on modalities with which scientists are most familiar and for which data are available. Besides these modalities, the vestibular and proprioceptive systems related to balance and movement are not well studied in marine mammals (Spoor and Thewissen 2008). However, two aspects of sensory performance that have been investigated to some extent in phocids include magnetoreception and the perception of time.

2.6.1 *Unconfirmed Magnetoreception*

Orientation using the information provided by the earth's magnetic field has been demonstrated for various species, including some birds, elasmobranchs, subterranean mammals, and marine reptiles (e.g., Nordmann et al. 2017). Thus far, attempts to train harbor seals to use magnetic cues have been unsuccessful (F. Hanke, unpublished data; Renouf 1991). It remains to be investigated whether this is a correct negative result that applies to all phocids. Some phocids, such as northern elephant seals that undertake complex annual migrations over large ocean expanses, could be guided by magnetoreception (Robinson et al. 2012). Magnetoreception has also been suggested as a possible mechanism to explain how migrating seals maintain a direct path while transiting under water for extended periods (Matsumura et al. 2011), as well as how some individuals exhibit extraordinary interannual route fidelity (Costa et al. 2012; Horton et al. 2017). Furthermore, navigation using geomagnetic cues has been suggested for Weddell seals that re-locate widely dispersed breathing holes in complete darkness (R. Davis, personal communication).

2.6.2 *Perception of Time*

It has been suggested that, besides the information provided by the classical senses, phocids might also benefit from a well-developed sense of time. In a behavioral experiment, trained harbor seals were presented with an optical stimulus visible for specific time intervals on a monitor. They learned to discriminate time intervals quickly and easily discriminated intervals from 100 ms to 30 s to determine difference thresholds (Heinrich et al. 2016, 2020). Testing revealed that the seals needed a difference of only 12–19% to discriminate time intervals. Thus, harbor seals apparently have good time resolution abilities, which might be advantageous for making decisions during foraging or estimating travel duration. Whilst time perception is not a sensory dimension per se, the ability to make temporal judgments over different time intervals provides information that can be integrated with sensory streams from other inputs for higher-level cognitive processing.

2.7 Cognition

Sensory information is gathered and integrated in real time, and requires higher-level processing by the brain to provide meaningful information. Following initial demonstrations of associative learning abilities in trained harbor seals (Hanggi and Schusterman 1995; Renouf and Gaborko 1988), only a few experimental studies have examined the cognitive abilities of phocid seals. These studies have had two experimental foci: reversal learning and concept formation. By their nature, such cognitive studies use arbitrary cues designed to reveal problem-solving abilities across species.

2.7.1 Reversal Learning

During reversal learning tasks, a subject learns to discriminate between two stimuli. One stimulus is defined as the positive stimulus, and the subject is rewarded for choosing this stimulus. The other is defined as the negative stimulus, and the choice of this stimulus is not reinforced. After successfully acquiring this simple discrimination, the signs of the stimuli are changed: the subject is now rewarded for choosing the formerly negative stimulus and therefore has to switch response behavior with respect to the choices presented. Usually, the signs of the stimuli are repeatedly changed, and the number of errors made to successfully achieve the reversal is documented. Most animals exhibit a decrease in the number of errors per reversal as the number of reversals they have experienced increases. Some species even achieve the ultimate performance, which is referred to as one-trial-learning; that is, they only need exposure to a single trial to realize that a reversal has occurred. Reversal learning studies go beyond discrimination studies, as they examine whether animals can flexibly respond to familiar stimuli. Such behavioral plasticity might be advantageous when conditions change rapidly in the environment.

Walsh et al. (2007) trained harp seals in a type of visual reversal learning experiment. In this case, the positive stimulus varied depending upon the experimental enclosure so that correct responding to visual shapes was conditional upon spatial cues. The harp seals learned to use this spatial cue to solve the reversal task. In contrast, it took two 'control' seals, which did not experience the context change, much longer to perform the same stimulus reversals. Harbor seals have also been successfully trained in visual and spatial reversal learning experiments (F. Hanke, unpublished data). When performance using spatial (position) cues was compared to performance using visual (shape) cues, the harbor seals solved the reversal task with more ease and with less inter-individual variability when spatial information was provided. Although the seals' performance progressively improved over successive reversals irrespective of the task, one-trial-learning was only achieved when spatial cues were provided. In a subsequent experiment, a harbor seal could use an auditory cue to predict the onset of a new reversal. Harp and harbor seals thus possess the cognitive abilities to flexibly respond to known stimuli and alter their behavior appropriately in response to changing conditions.

2.7.2 Concept Formation

Concept formation is demonstrated when an individual uses rule-based learning to solve novel problems. Studies of concept formation in phocid seals have focused on the identity or 'sameness' concept, which relies upon the rule of linking two or more stimuli that are alike. The ability to use this concept has been evaluated in trained harbor seals using 'matching-to-sample' procedures (Mauck and Dehnhardt 2005, 2007; Renouf and Gaborko 1988) and a 'same/different' procedure (Scholtyssek et al. 2013). In the matching-to-sample procedures, seals were required to select a comparison stimulus that was the same (the match) as a previously seen stimulus (the sample) while ignoring the alternative choice(s). In the same/different procedure, the seal compared two stimuli and determined whether they were the same or not. The training was conducted by rewarding correct responses to familiar stimuli with fish. Testing was conducted by measuring performance with unfamiliar problems.

Although harbor seals seem to have well-developed learning abilities, early studies failed to show identity concept formation with visual, tactile, and auditory stimuli (Renouf and Gaborko 1988). Harbor seals could apparently match familiar stimuli when they could use a spatial cue to solve the task, but the seals were not tested with novel problems. Mauck et al. (2005) revisited identity matching experiments and demonstrated successful concept formation in a harbor seal; following training, the seal could match novel objects in a task with two or four alternative choices. The matching performance was even better when the choice stimuli were spaced apart in the seal's enclosure; the seal could still find the correct match to the sample even after a 12 min delay. In contrast, much shorter delays impaired the seal's performance when choice stimuli were presented close together (Mauck and Dehnhardt 2007). In a related task with increased complexity, Gläser (2012) showed the ability of trained harbor seals to successfully match geometric shapes with their rotated or unrotated, mirrored matches. Finally, a concept of identity was demonstrated in a harbor seal using the same/different procedure (Scholtyssek et al. 2013); in this case, the seal applied the same/different rule to new stimulus dimensions such as pattern or brightness. Like some otariids, harbor seals can thus form a concept based on the 'sameness' of stimulus features. It is unknown whether they can achieve more advanced levels of concept formation using rules of logic. A question arising from these cognitive studies is whether phocids' habitat and lifestyle require the formation of concepts. Alternatively, phocids might solve difficult problems and form abstract concepts due to a well-developed information processing system, as speculated by Mauck and Dehnhardt (2005).

2.7.3 Concluding Remarks on Phocid Cognition

The few cognitive studies conducted with phocids demonstrate efficiency and flexibility in learning and decision-making that are adaptive in an ever-changing environment. Future studies on phocid seal cognition and neuroanatomy are required to determine if the seals' semi-aquatic lifestyle has led to specific cognitive adaptations. It is tempting to compare phocids' learning abilities to those of the better-studied otariids (for review see Cook et al. 2021; Schusterman and Kastak 2002). The much shorter maternal dependency period in phocids may require a higher level of innate rather than learned behavior. However, few data are available to consider phocids' cognitive abilities within a comparative context. A few field studies suggest that complex learning can be manifested by phocids in the social domain (Bishop et al. 2015; Casey et al. 2015) and with respect to vocal learning (see Reichmuth and Casey 2014 for review). However, the extent to which seals' cognitive abilities are reflected in ecological problem-solving remains to be explored.

2.8 Summary

While still incomplete, our knowledge of the sensory and cognitive abilities of phocids has increased substantially. The relationships between sensory abilities and various types of behavior are illustrated in Fig. 2.5. The systematic scientific work on harbor seals as a model species has, in particular, provided profound understanding of what a phocid senses and how it might perceive the world. This species' sensory systems are adapted for an amphibious lifestyle, allowing harbor seals to gather relevant information from terrestrial and marine habitats with vastly different physical constraints. However, even for harbor seals, future research is needed to fill the gaps in our understanding of sensory and cognitive abilities. For example, it would be desirable to understand how the photoreceptors are distributed over the retina or how sound is transmitted to the inner ear under water. Such insights would support future examination of the mechanisms that enable complex behaviors such as orientation and navigation. Research in this field will profit from close collaborations between experimental biologists working in controlled laboratory settings, neuroanatomists, and researchers studying phocids in their natural habitats.

The available information on the sensory and cognitive abilities of phocids indicates that adaptations to habitat and lifestyle observed in deep-diving species deviate substantially from adaptations found in more shallow-diving coastal species. Directed investigations of deep-diving phocids could reveal even more remarkable features of their sensory systems. In general, the field will benefit from comparative studies to determine if sensory adaptations can be traced back to a common ancestor or are determined by habitat or lifestyle.

Finally, research thus far has failed to investigate how the senses interact to provide a multimodal representation of the environment—one of the key questions

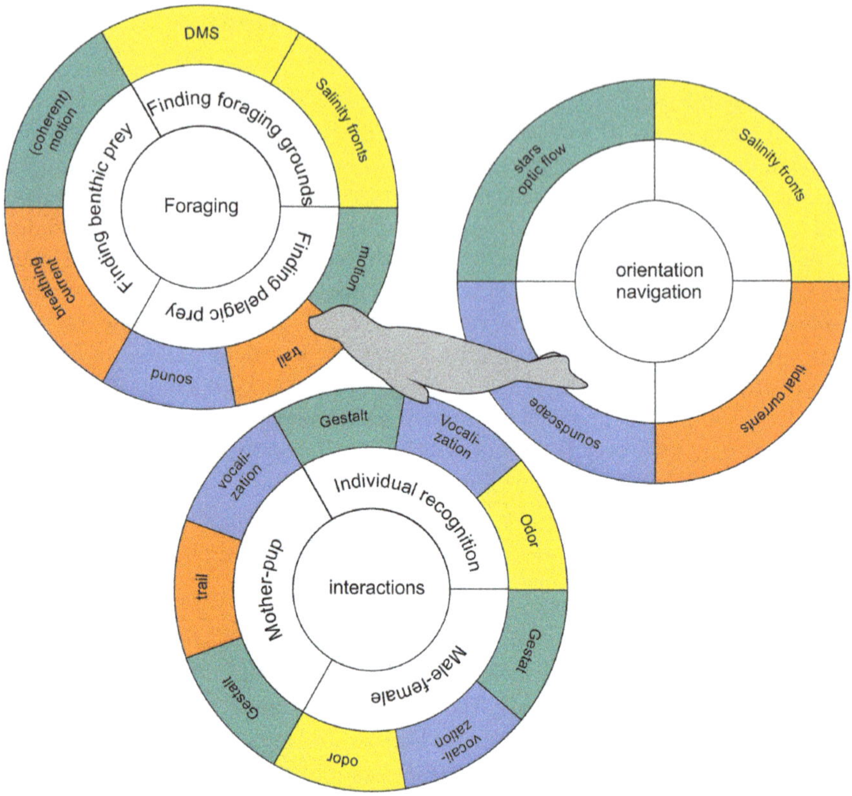

Fig. 2.5 Summary of the evidence, summarized in this chapter, of how sensory information relates or might relate to different types of behavior. Green sections refer to visual information used for the specific behavior, orange sections to haptic/hydrodynamic information, yellow sections to chemoreceptive information, blue sections to acoustic information

of sensory biology (Johnsen 2017). An integrative approach and improved knowledge of neuroanatomical structures will be required to fully understand how seals make sense of the surrounding world and successfully perform complex, multimodal behaviors such as foraging or navigating in the open ocean.

References

Adam PJ (2005) *Lobodon Carcinophaga.* Mammal Spec 772:1–14

Ahl AS (1986) The role of vibrissae in behavior: a status review. Vet Res Commun 10:245–268

Andreae TW, Andreae MO, Schebeske G (1994) Biogenic sulfur emissions and areosols over the tropical south Atlantic 1. Dimethylsulfide in seawater and the atmosphere boundary layer. J Geophys Res 99:22819–22829

Bartsh SS, Johnston SD, Siniff DB (1992) Territorial behavior and breeding frequency of male Weddell seal (*Leptonychotes weddelii*) in relation to age, size and concentrations of serum testosterone and cortisol. Can J Zool 70:680–692

Beltran RS, Sadou MC, Condit R, Peters SH, Reichmuth C, Costa DP (2015) Fine-scale whisker growth measurements can reveal temporal foraging patterns from stable isotope signatures. Mar Ecol Prog Ser 523:243–253

Bernholz CD, Matthews ML (1974) Critical flicker frequency in a harp seal: evidence for duplex retinal organization. Vis Res 15:733–737

Bishop AM, Pomeroy P, Twiss SD (2015) Variability in individual rates of aggression in wild gray seals: fine-scale analysis reveals importance of social and spatial stability. Behav Ecol Sociobiol 69:1663–1675

Bjorgesaeter A, Ugland KI, Bjorge A (2004) Geographic variation and acoustic structure of the underwater vocalization of harbor seals (*Phoca vitulina*) in Norway, Sweden and Scotland. J Acoust Soc Am 116:2459–2468

Blauert J (1997) Spatial hearing: the psychophysics of human sound localization. MIT Press, Cambridge

Bodson A, Miersch L, Mauck B, Dehnhardt G (2006) Underwater auditory localization by a swimming harbor seal (*Phoca vitulina*). J Acoust Soc Am 120:1550–1557

Bodson A, Miersch L, Dehnhardt G (2007) Underwater localization of pure tones by harbor seals (*Phoca vitulina*). J Acoust Soc Am 122:2263–2269

Boenninghaus G (1903) Das Ohr des Zahnwales, zugleich ein Beitrag zur Theorie der Schalleitung. Eine biologische Studie. Zoologische Jahrbuecher Abteilung Fuer Anatomie Und Ontogenie Der Tiere 19:1–175

Bowman MJ, Esaias WE (1978) Oceanic fronts in coastal processes. Springer, Berlin Heidelberg New York

Bruesch SR, Arey LB (1942) The number of myelinated and unmyelinaed fibers in the optic nerve of vertebrates. J Comp Neurol 77:61–656

Bublitz A (2010) Wasserbewegungen von stationären Fischen und ihre mögliche Bedeutung für fischfressende Tiere. Diploma thesis, University of Rostock

Bürgermeister S, Zimmermann RL, Georgii HW, Bingemer HG, Kirst GO, Janssen M, Ernst W (1990) On the biogenic origin of dimethylsulfide: relation between chlorophyll, ATP, organismic DMSP, phytoplankton species, and DMS distribution in Atlantic surface water and atmosphere. J Geophys Res 95:20607–20615

Burton RW, Anderson SS, Summers CF (1975) Perinatal activities in the grey seal (*Halichoerus grypus*). J Zool (Lond) 177:197–201

Byl JA, Miersch L, Wieskotten S, Dehnhardt G (2016) Underwater sound localization of pure tones in the median plane by harbor seals (*Phoca vitulina*). J Acoust Soc Am 140:4490

Byl JA, Miersch L, Wieskotten S, Dehnhard G (2019) Underwater sound localization abilities of harbor seals (*Phoca vitulina*) for high-frequency noise band stimuli in the median plane. J Acoust Soc Am 146:189–194

Casey C, Charrier I, Methevon N, Reichmuth C (2015) Rival assessment among northern elephant seals: evidence of associate learning during male-male contests. R Soc Open Sci 12:150228

Charrier I, Mathevon N, Aubin T (2013) Bearded sea males perceive geographic variations in their trills. Behav Ecol Sociobiol 67:1679–1689

Connolly Sadou M, Beltran RS, Reichmuth C (2014) A calibration procedure for measuring pinniped vibrissae using photogrammetry. Aquat Mamm 40:213–218

Cook P, Reichmuth C, Hanke FD (2021) The mind of a sea lion. In: Campagna C. Harcourt R (eds.) Ethology and Behavioral Ecology of Otariids and the Odobenid. Springer Nature Switzerland

Costa DP, Breed GA, Robinson PW (2012) New insights into pelagic migrations: implications for ecology and conservation. Annu Rev Ecol Evol Syst 4(1):73–96

Crognale MA, Levenson DH, Ponganis PJ, Deegan JF II, Jacobs GH (1998) Cone spectral sensitivity in the harbor seal (*Phoca vitulina*) and implications for color vision. Can J Zool 76:2114–2118

Cronin TW, Bok MJ (2016) Photoreception and vision in the ultraviolet. J Exp Biol 219:2790–2801

Cunningham KA, Reichmuth C (2016) High-frequency hearing in seals and sea lions. Hear Res 331:83–91
Cunningham DE, Southall BL, Reichmuth C (2014a) Auditory sensitivity of seals and sea lions in complex listening scenarios. J Acoust Soc Am 136:3410–3421
Cunningham KA, Hayes SA, Wargo Rub AM, Reichmuth C (2014b) Auditory detection of ultrasonic coded transmitters by seals and sea lions. J Acoust Soc Am 135:1978–1985
Dawson WW, Hawthorne MN, Jenkins RL, Goldston RT (1982) Ginat neural systems in the inner retina and optic nerve of small whales. J Comp Neurol 205:1–7
Dawson WW, Jope GM, Ulshafer RJ, Hawthrone MN, Jenkins RL (1983) Contents of the optic nerve of a small cetacean. Aquat Mamm 10(2):45–56
Debey LB, Pyenson ND (2013) Osteological correlates and phylogenetic analysis of deep diving in living and extinct pinnipeds: what good are big eyes? Mar Mamm Sci 29:48–83
Deecke VB, Slater PJB, Ford JKB (2002) Selective habituation shapes acoustic predator recognition in harbour seals. Nature 420:171–173
Dehnhardt G, Kaminski A (1995) Sensitivity of the mystacial vibrissae of harbour seals (*Phoca vitulina*) for size differences of actively touched objects. J Exp Biol 198:2317–2323
Dehnhardt G, Mauck B (2008) Mechanoreception in secondarily aquatic vertebrates. In: Thewissen JGM, Nummela S (eds) Sensory evolution on the threshold: adaptations in secondarily aquatic vertebrates. University of California Press, Berkeley, Los Angeles, pp 295–314
Dehnhardt G, Sinder M, Sachser N (1997a) Tactual discrimination of size by means of mystacial vibrissae in harbor seals: in air versus underwater. Zeitschrift Für Säugetierkunde 62:40–43
Dehnhardt G, Sinder M, Sachser N (1997b) Tactual discrimination of size by means of mystacial vibrissae in harbour seals: in air versus underwater. In: Proceedings of the 1st international symposium on physiology and ethology of wild and zoo animals, pp 40–43
Dehnhardt G, Mauck B, Bleckmann H (1998a) Seal whiskers detect water movements. Nature 394:235–236
Dehnhardt G, Mauck B, Hyvärinen H (1998b) Ambient temperature does not affect tactile sensitivity of mystacial vibrissae in harbour seals. J Exp Biol 201:3023–3029
Dehnhardt G, Mauck B, Hanke W, Bleckmann H (2001) Hydrodynamic trail-following in harbor seals (*Phoca vitulina*). Science 293:102–104
Dehnhardt G, Hanke W, Wieskotten S, Krüger Y, Miersch L (2014) Hydrodynamic perception in seals and sea lions. In: Bleckmann H (ed) Flow sensing in air and water. Springer, Berlin Heidelberg, pp 147–167
Dohmen M, Menzel M, Wiese H et al (2015) Understanding fiber mixture by simulation in 3D polarized light. Imaging. Neuroimage 111:464–475
Ebara S, Kumamoto K, Matsuura T, Mazurkiewicz JE, Rice F (2002) Similarities and differences in the innervation of mystacial vibrissa follicle-sinus complexes in the rat and cat: a confocal microscopic study. J Comp Neurol 449:103–119
Elsner R, Wartzok D, Sonafrank NB, Kelly BP (1989) Behavioral and physiological reactions of arctic seals during under-ice pilotage. Can J Zool 67:2506–2513
Erbe C, Reichmuth C, Cunningham K, Lucke K, Dooling R (2016) Communication masking in marine mammals: a review and research strategy. Mar Pollut Bull 103:15–38
Erdsack N, Dehnhardt G, Hanke W (2014) Thermoregulation of the vibrissal system in harbor seals (*Phoca vitulina*) and Cape fur seals (*Arctocephalus pusillus pusillus*). J Exp Mar Biol Ecol 452:111–118
Etienne AS, Jeffrey KJ (2004) Path integration in mammals. Hippocampus 14:180–192
Fasick JI, Robinson PR (2000) Spectral-tuning mechanisms of marine mammal rhodpsins and correlations with foraging depth. Vis Neurosci 17:781–788
Fay RR (1988) Hearing in vertebrates: a psychophysics databook. Hill-Fay, Winnetka
Fletcher H (1940) Auditory patterns. Rev Mod Phys 12:47–65
Floodgate GD, Fogg GE, Jones DA, Lochte K, Turley CM (1981) Microbiological and zooplankton activity at a front in Liverpool Bay. Nature 290:133–136

Fuiman LA, Williams TA, Davis RW (2020) Homing tactics of Weddell seals in the Antarctic fast-ice environment. Mar Biol 167(116). https://doi.org/10.1007/s00227-020-03730-w
Geurten B, Niesterok B, Dehnhardt G, Hanke FD (2017) Saccadic movement strategy in a semiaquatic species—the harbor seal (*Phoca vitulina*). J Exp Biol 220:1503–1508
Ginter CC, Fish FE, Marshall CD (2010) Morphological analysis of the bumpy profile of phocid vibrissae. Mar Mamm Sci 26:733–743
Ginter CC, DeWitt TJ, Fish FE, Marshall CD (2012) Fused traditional and geometric morphometrics demonstrate pinniped whisker diversity. PloS One 7:e34481
Gläser N, Mauck B, Kandil F, Lappe M, Dehnhardt G, Hanke FD (2014) Harbour seals (*Phoca vitulina*) can perceive optic flow underwater. PloS One 9:e103555
Gläser N (2012) Visual and hydrodynamic orientation abilities of pinnipeds. Dissertation, University of Rostock
Grant R, Wieskotten S, Wengst N, Prescott T, Dehnhardt G (2013) Vibrissal touch sensing in the harbor seal (*Phoca vitulina*): how do seals judge size? J Comp Physiol A 199:521–533
Greaves DK, Hammil MO, Eddington JD, Pettipas D, Schreer JF (2004) Growth rate and shedding of vibrissae in the gray seal, *Halichoerus grypus*: a cautionary note for stable isotope diet analysis. Mar Mamm Sci 20:296–304
Hall-Aspland SA, Rogers TL, Canfield RB (2005) Stable carbon and nitrogen isotope analysis reveals seasonal variation in the diet of leopard seals. Mar Ecol Prog Ser 305:249–259
Hanggi EB, Schusterman RJ (1995) Conditional discrimination learning in a male harbor seal (*Phoca vitulina*). In: Kastelein RA, Thomas JA, Nachtigall PE (eds) Sensory systems of aquatic mammals. DeSpil, Woerden, pp 543–559
Hanggi EB, Schusterman RJ (1994) Underwater acoustic displays and individual variation in male harbour seals, *Phoca vitulina*. Anim Behav 48:1275–1283
Hanke W, Bleckmann H (2004) The hydrodynamic trails of *Lepomis gibbosus* (Centrarchidae), *Colomesus psittacus* (Tetraodontidae) and *Thysochromis ansorgii* (Cichlidae) investigaed with scanning particle image velocimetry. J Exp Biol 207:1585–1596
Hanke FD, Dehnhardt G (2009) Aerial visual acuity in harbor seals (*Phoca vitulina*) as a function of luminance. J Comp Physiol A 195:643–650
Hanke FD, Dehnhardt G, Schaeffel F, Hanke W (2006a) Corneal topography, refractive state, and accommodation in harbor seals (*Phoca vitulina*). Vis Res 46:837–847
Hanke W, Römer R, Dehnhardt G (2006b) Visual fields and eye movements in a harbor seal (*Phoca vitulina*). Vis Res 46:2804–2814
Hanke FD, Hanke W, Hoffmann K-P, Dehnhardt G (2008a) Optokinetic nystagmus in harbor seals (*Phoca vitulina*). Vis Res 48:304–315
Hanke FD, Kröger RHH, Siebert U, Dehnhardt G (2008b) Multifocal lenses in a monochromat: the harbour seal. J Exp Biol 211:3315–3322
Hanke FD, Hanke W, Scholtyssek C, Dehnhardt G (2009a) Basic mechanisms in pinniped vision. Exp Brain Res 199:299–311
Hanke FD, Peichl L, Dehnhardt G (2009b) Retinal ganglion cell topography in juvenile harbor seals (*Phoca vitulina*). Brain Behav Evol 74:102–109
Hanke W, Witte M, Miersch L et al (2010) Harbor seal vibrissa morphology suppresses vortex-induced vibrations. J Exp Biol 213:2665–2672
Hanke FD, Scholtyssek C, Hanke W, Dehnhardt G (2011) Contrast sensitivity in a harbor seal (*Phoca vitulina*). J Comp Physiol A 197:203–210
Hanke FD, Miersch L, Warrant EJ, Mitschke FM, Dehnhardt G (2013) Are harbour seals (*Phoca vitulina*) able to perceive and use polarised light? J Comp Physiol A 199:509–519
Hardy MH, Roff E, Smith TG, Ryg M (1991) Facial skin glands of ringed and gray seals, and their possible function as odoriferous organs. Can J Zool 69:189–200
Härkönen T (1987) Seasonal and regional variations in the feeding habits of the harbour seal, *Phoca vitulina*, in the Skagerrak and the Kattegat. J Zool (Lond) 213:535–543
Hastie GD, Donovan C, Götz T, Janik VM (2014) Behavioral responses by grey seals (*Halichoerus grypus*) to high frequency sonar. Mar Pollut Bull 79:205–210

Hayes SA, Kumar A, Daniel PC, Mellinger DK, Harvey JT, Southall BL, LeBoeuf BJ (2004) Evaluating the function of the male harbor seal, *Phoca vitulina*, roar through playback experiments. Anim Behav 67:1133–1139
Heinrich T, Dehnhardt G, Hanke FD (2016) Harbor seals (*Phoca vitulina*) are able to time precisely. Anim Cogn 19:1133–1142
Heinrich T, Ravignani A, Hanke FD (2020) Visual timing abilities of a harbour seal (*Phoca vitulina*) and a South African fur seal (*Arctocephalus pusillus pusillus*) for sub- and supra-second time intervals. Anim Cogn 35(5):851–859
Hemilä S, Nummela S, Berta A, Reuter T (2006) High-frequency hearing in phocid and otariid pinnipeds: an interpretation based on inertial and cochlear constraints. J Acoust Soc Am 120:3463–3466
Hirons AC, Schell DM, Aubin DJS (2001) Growth rates of vibrissae of harbor seals (*Phoca vitulina*) and Steller sea lions (*Eumetopias jubatus*). Can J Zool 79:1053–1061
Hogg C, Neveu M, Stokkan K-A et al (2011) Artic reindeer extend their visual range into the ultraviolet. J Exp Biol 214:2014–2019
Hogg C, Neveu M, Folkow LP, Stokkan K-A, Kam JH, Douglas RH, Jeffrey G (2015) The eyes of the deep diving hooded seal (*Cystophora cristata*) enhance sensitivity to ultraviolet light. Biol Open 4:812–818
Hollien H, Feinstein S (1976) Hearing in divers. In: Drew EA (ed) Underwater research. Academic Press, Cambridge, pp 81–138
Holt MM, Schusterman RJ, Southall BL, Kastak D (2004) Localization of aerial broadband noise by pinnipeds. J Acoust Soc Am 115:2339–2345
Holt MM, Schusterman RJ, Kastak D, Southall BL (2005) Localization of aerial pure tones by pinnipeds. J Acoust Soc Am 118:3921–3926
Horton TW, Hauser N, Zerbini AN et al (2017) Route fidelity dring marine megafauna migration. Front Mar Sci 4:422
Hughes A (1975) A quantitative analysis of the cat retinal ganglion cell topography. J Comp Neurol 163:107–128
Hughes A (1977) Topography of vision in mammals. In: Crescitelli F (ed) Handbook of sensory physiology. Springer, Berlin Heidelberg, pp 613–756
Hyvärinen H (1989) Diving in darkness: whiskers as sense organs of the ringed seal (*Phoca hispida*). J Zool 218:663–678
Hyvärinen H, Katajisto H (1984) Functional structure of the vibrissae of the ringed seal (*Phoca hispida* Schr.). Acta Zoologica Fennica 171:27–30
Hyvärinen H, Palviainen A, Strandberg U, Holopainen IJ (2009) Aquatic environment and differentiation of vibrissae: comparison of sinus hair systems of ringed seal, otter and pole cat. Brain Behav Evol 74:268–279
Jamieson GS (1971) The functional significance of corneal distortion in marine mammals. Can J Zool 49:421–423
Jamieson GS, Fisher HD (1970) Visual discrimination in the harbour seal *Phoca vitulina*, above and below water. Vis Res 10:1175–1180
Jamieson GS, Fisher HD (1971) The retina of the harbour seal, *Phoca vitulina*. Can J Zool 49:19–23
Jamieson GS (1970) The eye of the harbour seal, *Phoca vitulina*. Dissertation, The University of British Columbia
Johnsen S (2017) Open questions: we don't really know anything, do we? Open questions in sensory biology. BMC Biol 15(1):43. https://doi.org/10.1186/s1291-017-0385-3
Johnson GL (1901) Contributions to the comparative anatomy of the mammalian eye, chiefly based on ophthalmoscopic examination. Philos Trans R Soc Biol Characters 194:1–30
Johnson GL (1893) Observations on the refraction and vision of the seal's eye. Proc R Soc Lond 719–723
Jones A, Marshall CD (2019) Does vibrissal innervation patterns and investment predict hydrodynamic trail following behavior of harbor seals (*Phoca vitulina*)? Anat Rec 302:1837–1845

Käkelä R, Hyvärinen H (1993) Fatty acid composition of fats around the mystacial and superciliary vibrissae differs from that of blubber in the Saimaa ringed seal (*Phoca hispida saimensis*). Comp Biochem Physiol 105:547–552

Käkelä R, Hyvärinen H (1996) Site-specific fatty acid composition in adipose tissues of several northern aquatic and terrestrial mammals. Comp Biochem Physiol 115:501–514

Kastak D, Schusterman RJ (1998) Low-frequency amphibious hearing in pinnipeds: methods, measurements, noise, and ecology. J Acoust Soc Am 103:2216–2228

Kastelein RA, Wensveen P, Hoek L (2009a) Underwater hearing sensitivity of harbor seals (*Phoca vitulina*) for narrow noise bands between 0.2 and 80 kHz. J Acoust Soc Am 126:476–483

Kastelein RA, Wensveen PJ, Hoek L, Verboom WC, Terhune JM (2009b) Underwater detection of tonal signals between 0.25 and 100 kHz by harbor seals (*Phoca vitulina*). J Acoust Soc Am 125:1222–1229

Kastelein RA, Helder-Hoek L, Terhune JM (2018) Hearing thresholds, for underwater sounds, of harbor seals (*Phoca vitulina*) at the water surface. J Acoust Soc Am 143:2554–2563

Kasumyan AO (2008) Sounds and sound production in fishes. J Ichthyol 48:981–1030

Kellog WN (1958) Echoranging in the porpoise. Science 128:982–988

Kenyon KW, Rice DW (1959) Life history of the Hawaiian monk seal. Pac Sci 13(3):215–252

Kilian M, Dehnhardt G, Hanke FD (2015) How harbor seals (*Phoca vitulina*) pursue schooling herring. Mamm Biol 80:385–389

Kilian M (2010) Hydrodynamische Spurverfolgung beim Seehund (*Phoca vitulina*). Diploma thesis, University of Rostock

Kim SK, Amundin M, Laska M (2013) Olfactory discrimination ability of South African fur seals (*Arctocephalus pusillus*) for enantiomers. J Comp Physiol A 199:535–544

King JE (1983) Seals of the world. Cambridge University Press, Cambridge

King JE (1969) Some aspects of the anatomy of the Ross seal, *Ommatophoca rossi* (Pinnipedia: Phocidae). British Antarctic Survey, Scientific reports, No. 63, London

Kovacs KM (1995) Mother-pup reunions in harp seals, *Phoca groenlandica*—cues for the relocation of pups. Can J Zool 73:843–849

Kowalewsky S, Dambach M, Mauck B, Dehnhardt G (2006) High olfactory sensitivity for dimethyl sulphide in harbor seals. Biol Lett 2:106–109

Kröger RHH, Campbell MCW, Fernald RD, Wagner H-J (1999) Multifocal lenses compensate for chromatic defocus in vertebrate eyes. J Comp Physiol A 184:361–369

Krüger Y, Hanke W, Miersch L, Dehnhardt G (2018) Detection and direction discrimination of single vortext rings by harbour seals (*Phoca vitulina*). J Exp Biol 221(8):jeb170753 https://doi.org/10.1242/jeb.170753

Krüger Y (2017) Perception of single vortex rings by harbour seals (*Phoca vitulina*). Dissertation, University of Rostock

Landau D, Dawson WW (1970) The histology of retinas from the pinnipedia. Vis Res 10:691–702

Lavigne DM, Oritsland NA (1974) Black polar bears. Nature 251:218–219

Lavigne DM, Ronald K (1975a) Evidence of duplicity in the retina of the California sea lion (*Zalophus californianus*). Comp Biochem Physiol 50A:65–70

Lavigne DM, Ronald K (1975b) Pinniped visual pigments. Comp Biochem Physiol 52:325–329

Lesage V, Hammil MO, Kovacs KM (1999) Functional classification of harbor seal (*Phoca vitulina*) dives using depth profiles, swimming velocity, and an index of foraging success. Can J Zool 77:74–87

Levenson DH, Schusterman RJ (1997) Pupillometry in seals and sea lions: ecological implications. Can J Zool 75:2050–2057

Levenson DH, Schusterman RJ (1999) Dark adaptation and visual sensitivity in shallow and deep-diving pinnipeds. Mar Mamm Sci 15:1303–1313

Levenson DH, Ponganis PJ, Crognale MA, Deegan JF II, Dizon A, Jacobs GH (2006) Visual pigments of marine carnivores: pinnipeds, polar bear, and sea otter. J Comp Physiol A 192:833–843

Ling JK (1977) Vibrissae of marine mammals. In: Harrison RJ (ed) Functional anatomy of marine mammals. Academic Press, London, pp 387–415

Lübcker N, Condit R, Beltran RS, de Bruyn N, Besler MN (2016) Vibrissal growth parameters of southern elephant seals (*Mirounga leonina*): obtaining fine-scale time-based stable isotope data. Mar Ecol Prog Ser 559:243–255

Lucke K, Hastie GD, Ternes K et al (2016) Aerial low-frequency hearing in captive and free-ranging harbour seals (*Phoca vitulina*) measured using auditory brainstem responses. J Comp Physiol A 202:859–868

Lui LL, Dobiecki AE, Bourne JA, Rosa MGP (2012) Breaking camouflage: responses of neurons in the middle temporal area to stimuli defined by coherent motion. Eur J Neurosci 36:2063–2076

Lythgoe JN, Dartnall HJA (1970) A 'deep sea rhodopsin' in a marine mammal. Nature 227:995–996

Marshall CD, Amin H, Kovacs KM, Lydersen C (2006) Microstructure and innervation of the vibrissal follicle-sinus complex in the bearded seal, *Erignathus barbatus* (Pinnipedia: Phocidae). Anat Rec 288A:13–25

Mass AM, Supin AY (1989) Distribution of ganglion cells in the retina of an amazon river dolphin *Inia geoffrensis*. Aquat Mamm 15(2):49–56

Mass AM, Supin AY (1995) Ganglion cell topography of the retina in the bottlenose dolphin, *Tursiops truncatus*. Brain Behav Evol 45:257–265

Mass AM, Supin AY (2003) Retinal topography of the harp seal *Pagophilus Groenlandicus*. Brain Behav Evol 62:212–222

Mass AM, Supin AY (2007) Adaptive features of aquatic mammals' eyes. Anat Rec 290:701–715

Mass AM, Supin AY (2010) Retinal ganglion cell layer of the Caspian seal *Pusa caspica*: topography and localization of the high-resolution area. Brain Behav Evol 76:144–153

Mass AM, Ketten DR, Odell DK, Supin AY (2012a) Ganglion cell distribution in the Florida manatee, *Trichechus manatus latirostris*. Anat Rec 296:177–186

Mass AM, Supin AY, Mukhametoc IM, Rozanova EI, Abramov AV (2012b) Morphological estimation of retinal resolution of a killer whale (*Orcinus orca*). Dokl Biol Sci 442:34–37

Matsumura M, Watanabe YY, Robinson PW, Miller PJO, Costa DP, Miyazaki N (2011) Underwater and surface behavior of homing juvenile northern elephant seals. J Exp Biol 214:629–636

Mattson EE, Marshall CD (2016) Follicle microstructure and innervation vary between pinniped micro- and macrovibrissae. Brain Behav Evol 88(1):43–58 https://doi.org/10.1159/000447551

Mauck B, Dehnhardt G (2005) Identity concept formation during visual multiple-choice matching in a harbor seal (*Phoca vitulina*). Learn Behav 33:428–436

Mauck B, Brown D, Schlosser W, Schaeffel F, Dehnhardt G (2005) How a harbour seal sees the night sky. Mar Mamm Sci 21(4):646–656

Mauck B, Dehnhardt G (2007) Spatial multiple-choice matching in a harbour seal (*Phoca vitulina*): differential encoding of landscape versus local feature information? Anim Cogn 10:397–405

Mauck B, Eysel U, Dehnhardt G (2000) Selective heating of vibrissal follicles in seals (*Phoca vitulina*) and dolphins (*Sotalia fluviatilis guianensis*). J Exp Biol 203:2125–2131

Mauck B, Gläser N, Schlosser W, Dehnhardt G (2008) Harbour seals (*Phoca vitulina*) can steer by the stars. Anim Cogn 11:715–718

McGovern KA, Marshall CD, Davis RW (2015) Are vibrissae viable sensory structures for prey capture in northern elephant seals, *Mirounga angustirostris*? Anat Rec 298:750–760

McHuron EA, Walcott SM, Skrovan S, Costa DP, Reichmuth C (2016) Whisker growth dynamics in two North Pacific pinnipeds: implications for determining foraging ecology from stable isotope analysis. Mar Ecol Prog Ser 554:213–224

McHuron EA, Williams T, Costa DP, Reichmuth C (2020) Contrasting whisker growth dynamics within the phocid lineage. Mar Ecol Prog Ser 634:231–236

Miersch L, Hanke W, Wieskotten S et al (2011) Flow sensing in pinniped whiskers. Philos Trans R Soc Lond B Biol Sci 366:3077–3084

Miksis-Olds JL, Madden LE (2014) Environmental predictors of ice seal presence in the Bering Sea. PloS One 9:e106998

Mills FHJ, Renouf D (1986) Determination of the vibration sensitivity of harbor seal *Phoca vitulina* (L.) vibrissae. J Exp Mar Biol Ecol 100:3–9
Milne AO, Smith C, Orton LD, Sullivan MS, Grant RA (2020) Pinnipeds orient and control their whiskers: a study on Pacific walrus, California sea lion and harbor seal. J Comp Physiol A 206:441–451
Mittelstaedt H, Mittelstaedt M-L (1982) Homing by path integration. In: Papi F, Wallraff HG (eds) Avian navigation—international symposium on avian navigation (ISAN) held at Tirrenia (Pisa), September 11–14, 1981. Springer, Berlin, pp 290–298
Møhl B (1964) Preliminary studies on hearing in seals. Videnskabelige Meddelelser Fra Dansk Naturhistorik Forening I Kjobenhaven 127:283–294
Møhl B (1967) Seal ears. Science 157:99
Møhl B (1968a) Auditory sensitivity of the common seal in air and water. J Audit Res 8:27–38
Møhl B (1968b) Hearing in seals. In: Harrison RJ, Hubbard RC, Peterson RS, Rise CE, Schusterman RJ (eds) The behavior and physiology of pinnipeds. Appleton Century Crofts, New York, pp 172–193
Mooney TA, Yamato M, Branstetter BK (2012) Hearing in cetaceans: from natural history to experimental biology. In: Lesser M (ed) Advances in marine biology. Academic Press, Cambridge, pp 197–246
Morgane PJ, Jacobs M (1972) Comparative anatomy of the cetacean nervous system. In: Harrison RJ (ed) Functional anatomy of marine mammals. Academic Press, New York, pp 117–244
Morrison HE, Brede M, Dehnhardt G, Leder A (2016) Simulating the flow and trail following capabilities of harbour seal vibrissae with the Lattice Boltzmann Method. J Comput Sci 17:394–402. https://doi.org/10.1016/j.jocs.2016.04.004
Müller B, Goodman SM, Peichl L (2007) Cone photorecepor diversity in the retinas of fruit bats (Megachiroptera). Brain Behav Evol 70:90–104
Muller BS, Bovet P (1999) Role of pinnae and head movements in localizing pure tones. Swiss J Psychol 58:170–179
Murphy CT, Reichmuth C, Mann D (2015) Vibrissal sensitvity in a harbor seal (*Phoca vitulina*). J Exp Biol 218:2463–2471
Murphy CT, Reichmuth C, Eberhardt WC, Calhoun BH, Mann DA (2017) Seal whiskers vibrate over broad frequencies during hydrodynamic tracking. Sci Rep 7:8350
Murphy CJ, Eberhardt WC, Calhoun BH, Mann KA, Mann DA (2013) Effect of angle on flow-induced vibrations of pinniped vibrissae. PloS One 8:e69872
Nagy AR, Ronald K (1970) The harp seal, *Pagophilus groenlandicus* (Erxleben, 1777). VI. Structure of retina. Can J Zool 48:367–370
Nevitt GA (2008) Sensory ecology on the high seas: the odor world of the procellariiforms seabirds—review. J Exp Biol 211:1706–1713
Nevitt GA, Bonadonna F (2005) Sensitivity to dimethyl sulphide suggests a mechanism for olfactory navigation by seabirds. Biol Lett 1:303–305
Newby TC, Hart FM, Arnold RA (1970) Weight and blindness of harbor seals. J Mammol 51(1):152
Newland C, Field IC, Cherel Y, Guinet C, Bradshaw CJA, McMahon CR, Hindell MA (2011) Diet of juvenile southern elephant seals reappraised by stable isotopes in whiskers. Mar Ecol Prog Ser 424:247–258
Newman LA, Robinson PR (2005) Cone visual pigments of aquatic mammals. Vis Neurosci 22:873–879
Niesterok B, Hanke W (2013) Hydrodynamic patterns from fast-starts in teleost fish and their possible relevance to predator-prey interactions. J Comp Physiol A 199:139–149
Niesterok B, Krüger Y, Wieskotten S, Dehnhard G, Hanke W (2017) Hydrodynamic detection and localization of artificial flatfish breathing currents by harbour seals (*Phoca vitulina*). J Exp Biol 220:174–185
Niesterok B, Dehnhard G, Hanke W (2017a) Hydrodynamic sensory threshold in harbour seals (*Phoca vitulina*) for artificial flatfish breathing currents. J Exp Biol 220:2364–2371

Nordmann GC, Hochstoeger T, Keays DA (2017b) Magnetoreception—a sense without a recepter. PLoS Biology 15(10):e2003234

Norris K (1967) Some observations on the migration and orientation of marine mammals. In: Storm RM (ed) Animal orientation and navigation. Oregon State University Press, Corvallis, OR, pp 101–125

Nummela S (2008) Hearing in aquatic mammals. In: Thewissen JGM, Nummela S (eds) Sensory evolution on the threshold: adaptations in secondarily aquatic vertebrates. University of California Press, Berkeley Los Angeles, London, pp 211–224

Nummela S, Pihlström H, Puolamäki K, Fortelius M, Hemilä S, Reuter T (2013) Exploring the mammalian sensory space: co-operations and trade-offs among senses. J Comp Physiol A 199:1077–1092

Ogden TE, Miller RF (1966) Studies of the optic nerve of the rhesus monkey: nerve fiber spectrum and physiological properties. Vis Res 6:485–506

Oliver GW (1978) Navigation in mazes by a grey seal, *Halichoerus grypus* (Fabricius). Behaviour 67:97–114

Oppermann D, Schramme J, Neumeyer C (2016) Rod-cone based color vision in seals under photopic conditions. Vis Res 125:30–40

Pardue MT, Sivak JG, Kovacs KM (1993) Corneal anatomy of marine mammals. Can J Zool 71:2282–2290

Peichl L (2005) Diversity of mammalian photoreceptor properties: adaptations to habitat and lifestyle? Anat Rec 287A:1001–1012

Peichl L, Moutairou K (1998) Absence of short-wavelength sensitive cones in the retinae of seals (*Carnivora*) and African giant rats (*Rodentia*). Eur J Neurosci 10:2586–2594

Peichl L, Behrmann G, Kröger RHH (2001) For whales and seals the ocean is not blue: a visual pigment loss in marine mammals. Eur J Neurosci 13:1520–1528

Philström H, Fortelius M, Hemilä S, Forsman R, Reuter T (2005) Scaling of mammalian ethmoid bones can predict olfactory organ size and performance. Proc R Soc B: Biol Sci 272:957–962

Pierce GJ, Thompson PM, Miller A, Diack JSW, Miller D, Boyle PR (1991) Seasonal variation in the diet of common seals (*Phoca vitulina*) in the Moray Firth area of Scotland. J Zool (Lond) 223:641–652

Piggins DJ (1970) Refraction of the harp seal, *Pagophilus groenlandicus* (Erxleben 1777). Nature 227:78–79

Potts AM, Hodges D, Shelman CB, Fritz KJ, Levy NS, Mangnall Y (1972) Morphology of the primate optic nerve—I. Method and total fiber count. Invest Ophthalmol 11(12):980–988

Pütter A (1903) Die Augen der Wassersäugethiere. Zoologische Jahrbuecher Abteilung Fuer Anatomie Und Ontogenie Der Tiere 17:99–402

Rahmann H (1967) Die Sehschärfe Bei Wirbeltieren. Naturwissenschaftliche Rundschau 1:10–14

Reep RL, Finlay BL, Darlington RB (2006) The limbic system in mammalian brain evolution. Brain Behav Evol 70:57–70

Reichmuth C (2014) Psychophysical studies of auditory masking in marine mammals: key concepts and new directions. In: Popper AN, Hawkins A (eds) The effects of noise on aquatic life. Springer, Berlin, pp 23–27

Reichmuth C, Casey C (2014) Vocal learning in seals, sea lions, and walruses. Curr Opin Neurobiol 28:66–71

Reichmuth C, Holt MM, Mulsow J, Sills JM, Southall BL (2013) Comparative assessment of amphibious hearing in pinnipeds. J Comp Physiol A 199:491–507

Renouf D (1979) Preliminary measurements of the sensitivity of the vibrissae of harbor seals (*Phoca vitulina*) to low frequency vibrations. J Zool (Lond) 188:443–450

Renouf D (1980a) Fishing in captive harbour seals (*Phoca vitulina concolor*): a possible role for vibrissae. Netherlands J Zool 30:504–509

Renouf D (1980b) Masked hearing thresholds of harbour seals (*Phoca vitulina*) in air. J Audit Res 20:263–269

Renouf D, Davis MB (1982) Evidence that seals may use echolocation. Nature 300(5893):635–637

Renouf D, Gaborko L (1988) Spatial matching to sample in harbour seals (*Phoca vitulina*). Biol Behav 13:73–81
Renouf D, Lawson J, Gabarko L (1983) Attachment between harbour seal (*Phoca vitulina*) mothers and pups. J Zool (Lond) 199:179–187
Renouf D (1991) Sensory reception and processing in Phocidae and Otariidae. In: Renouf D (ed) Behaviour in pinnipeds. Springer Science+Business Media, Dordrecht, pp 345–394
Repenning CA (1972) Underwater hearing in seals: functional morphology. In: Harrison RJ (ed) Functional anatomy of marine mammals. Academic Press, London, pp 307–331
Reynolds PS, Lavigne DM (1981) Visible and ultraviolet reflectance characteristics of Arctic homeotherms. Int J Biometeorol 25(4):299–308
Ridgway SH, Joyce PL (1975) Studies on seal brain by radiotelemetry. ICES J Mar Sci 169:81–91
Robinson PW, Costa DP, Crocker DE et al (2012) Foraging behavior and success of a mesopelagic predator in the Northeast Pacific Ocean: insights from a data-rich species, the northern elephant seal. PloS One 7:e36728
Rogers TL, Fung J, Slip D, Steindler L, O'Connell TC (2016) Calibrating the time span of longitudinal biomarkers in vertebrate tissues when fine-scale growth records are unavailable. Ecosphere 7:e01449
Ruscher B, Sills JM, Richter BP, Reichmuth C (2021) In-air hearing in Hawaiian monk seals: implications for understanding the auditory biology of Monachinae seals. J Comp Physiol A 1–13: https://doi.org/10.1007/s00359-021-01498-y
Ruser A, Dähne M, Sundermeyer J et al (2014) In-air evoked potential audiometry of grey seals (*Halichoerus grypus*) from the North and Baltic Sea. PloS One 9:e90824
Ryg M, Solberg Y, Lydersen C, Smith TG (1992) The scent of rutting male ringed seals (*Phoca hispida*). J Zool (Lond) 226:681–689
Sanchez RM, Dunkelberger GR, Quigley HA (1986) The number and diameter distribution of axons in the monkey optic nerve. Invest Ophthalmol vis Sci 27:1342–1350
Scholtyssek C, Kelber A, Dehnhardt G (2008) Brightness discrimination in the harbor seal (*Phoca vitulina*). Vis Res 48:96–103
Scholtyssek C, Kelber A, Hanke FD, Dehnhardt G (2013) Same different concept formation in a harbor seal (*Phoca vitulina*). Anim Cogn 16:915–925
Scholtyssek C, Kelber A, Dehnhardt G (2015) Why do seals have cones? Behavioral evidence for colorblindness in harbor seals. Anim Cogn 18:551–560
Schulte-Pelkum N, Wieskotten S, Hanke W, Dehnhardt G, Mauck B (2007) Tracking of biogenic hydrodynamic trails in harbour seals (*Phoca vitulina*). J Exp Biol 210:781–787
Schusterman RJ, Balliet RF (1970) Visual acuity of the harbour seal and the Stellar sea lion under water. Nature 226:563–564
Schusterman RJ, Kastak D (2002) Problem solving and memory. In: Hoelzel R (ed) Marine mammal biology–an evolutionary approach. Blackwell Publishing, Oxford, pp 371–387
Schusterman RJ, Kastak D, Levenson DH, Reichmuth Kastak C, Southall BL (2004) Pinniped sensory systems and the echolocation issue. In: Thomas JA, Moss C, Vater M (eds) Echolocation in bats and dolphins. University of Chicago Press, Chicago, pp 531–535
Sills JM, Southall BL, Reichmuth C (2014) Amphibious hearing in spotted seals (*Phoca largha*): underwater audiograms, aerial audiograms and critical ration measurements. J Exp Biol 217:726–734
Sills JM, Southall BL, Reichmuth C (2015) Amphibious hearing in ringed seals (*Pusa hispida*): underwater audiograms, aerial audiograms and critical ratio measurements. J Exp Biol 218:2250–2259
Sills J, Parnell K, Reichmuth C (2018) The production and reception of underwater sound in Hawaiian monk seals (*Neomonachus schauinslandi*). J Acoust Soc Am 144(3):1741–1741
Sills JM, Reichmuth C, Southall BL, Whiting A, Goodwin J (2020) Auditory biology of bearded seals (*Erignathus barbatus*). Polar Biol 43:1681–1691

Sills JM, Parnell KP, Lew C, Kendall T, Reichmuth C (2021) Underwater hearing and communication in the endangered Hawaiian monk seal, *Neomonachus schauinslandi*. Endanger Spec Res 44:61–78

Sivak JG, Howland HC, West J, Weerheim J (1989) The eye of the hooded seal, *Cristophora cristata*, in air and water. J Comp Physiol A 165:771–777

Smodlaka H, Khamas WA, Palmer L, Lui B, Borovac JA, Cohn BA, Schmitz L (2016) Eye histology and ganglion cell topography of northern elephant seals (*Mirounga angustirostris*). Anat Rec 299:798–805

Sonntag CF (1923) The comparative anatomy of the tongues of the mammalia. VIII, Carnivora. Proc Zool Soc Lond 9:129–153

Southall BL, Schusterman RJ, Kastak D (2000) Masking in three pinnipeds: underwater, low-frequency critical ratios. J Acoust Soc Am 108:1322–1326

Southall KD, Oliver GW, Lewis JW, Le Boeuf BJ, Levenson DH, Southall BL (2002) Visual pigment sensitivity in three deep diving marine mammals. Mar Mamm Sci 18:275–281

Southall BL, Schusterman RJ, Kastak D (2003) Auditory masking in three pinnipeds: aerial criticial ratios and direct critical bandwidth measurements. J Acoust Soc Am 114:1660–1666

Southall BL, Finneran JJ, Reichmuth C et al (2019) Marine mammal noise exposure criteria: updated scientific recommondations for residual hearing effects. Aquat Mamm 45:125–232

Spitzka EC (1890) Remarks on the brain of the seals. Am Nat 24:115–122

Spoor F, Thewissen JGM (2008) Comparative and functional anatomy of balance in aquatic mammals. In: Thewissen JGM, Nummela S (eds) Sensory evolution on the threshold: adaptations in secondarily aquatic vertebrates. University of California Press, Berkeley Los Angeles, London, pp 257–284

Stansbury AL, Götz T, Deecke VB, Janik VM (2015) Grey seals use anthopogenic signals from acoustic tags to locate fish: evidence from a simulated foraging task. Proc R Soc B: Biol Sci 282:20141595

Stenfors L-E, Sadé J, Hellstrom S, Anniko M (2001) How can the hooded seal dive to a depth of 1000 m without rupturing its tympanic membrane? A morphological and functional study. Acta Otolaryngol 121:689–695

Sticken J, Dehnhardt G (2000) Salinity discrimination in harbour seals: a sensory basis for spatial orientation in the marine environment? Naturwissenschaften 87:499–502

Terhune JM (1974) Directional hearing of a harbour seal in air and water. J Acoust Soc Am 56:1862–1865

Terhune JM (1991) Masked and unmasked pure tone detection thresholds of a harbour seal listening in air. Can J Zool 69:2059–2066

Terhune JM, Ronald K (1971) The harp seal, *Pagophilus groenlandicus* (Erxleben, 1777). X. The air audiogram. Can J Zool 49:385–390

Terhune JM, Ronald K (1975) Underwater hearing sensitivity of two ringed seals (*Pusa hispida*). Can J Zool 53:227–231

Thomas J, Moore P, Withrow R, Stoemer M (1990) Underwater audiogram of a Hawaiian monk seal (*Monachus schauinslandi*). J Acoust Soc Am 87:417–420

Tollit DJ, Thompson PM (1996) Seasonal and between-year variations in the diet of harbour seals in the Moray Firth, Scotland. Can J Zool 74:1110–1121

Tuckerman F (1890) On the gustatory organs of some of the mammalia. J Morphol 4:151–193

Turnbull SD, Terhune JM (1990) White noise and pure tone masking of pure tone thresholds of a harbour seal listening in air and underwater. Can J Zool 68:2090–2097

Turnbull SD, Terhune JM (1993) Repetition enhances hearing detection thresholds in a harbour seal (*Phoca vitulina*). Can J Zool 71:926–932

Turnbull SD, Terhune JM (1994) Descending frequency swept tones have lower thresholds than ascending frequency swept tones for a harbor seal (*Phoca vitulina*) and human listeners. J Acoust Soc Am 96(5):2631–2636

Turner EC, Sawyer EK, Kaas JH (2017) Optic nerve, superior colliculus, visual thalamus, and primary visual cortex of the northern elephant seal (*Mirounga angustirostris*) and California

sea lion (*Zalophus californianus*). J Comp Neurol 525:2109–2132. https://doi.org/10.1002/cne.24188
Vacquie-Garcia J, Mallefet J, Bailleul F, Picard B, Guinet C (2017) Marine bioluminescence: measurement by a classical light sensor and related foraging behvior of a deep diving predator. Photochem Photobiol 93:1312–1319
Vacquie-Garcia J, Royer F, Dragon A-C, Viviant M, Bailleul F, Guinet C (2012) Foraging in the darkness of the Southern ocean: influence of bioluminescence on a deep diving predator. PloS One 7:e43565
Walls GL (1942) The vertebrate eye and its adaptive radiation. Hafner Press, New York
Walsh SJ, Skinner DM, Martin GM (2007) Location serves as a conditional cue when harp seals (*Pagophilus groenlandicus*) solve object discrimination reversal problems. Can J Psychol 61:44–53
Walters A, Lea M-A, van den Hoff J et al (2014) Spatially explicit estimates of prey consumption reveal a new krill predator in the southern ocean. PloS One 9:e86452
Warrant EJ, Locket NA (2004) Vision in the deep sea. Biol Rev 79:671–712
Wartzok D, Ketten DR (1999) Marine mammal sensory systems. In: Reynolds JE, Rommel SA (eds) Biology of marine mammals. Smithsonian Institution Press, Washington London, pp 117–175
Wartzok D, McCormick MG (1978) Color discrimination by a Bering sea spotted seal, *Phoca largha*. Vis Res 18:781–784
Wartzok D, Elsner R, Stone H, Kelly BP, Davis RW (1992) Under-ice movements and the sensory basis of hole finding by ringed and Weddell seals. Can J Zool 70:1712–1722
Weiffen M, Möller B, Mauck B, Dehnhardt G (2006) Effect of water turbidity on the visual acuity of harbor seals (*Phoca vitulina*). Vis Res 46:1777–1783
Weiffen M, Mauck B, Dehnhardt G, Hanke FD (2014) Sensitivity of a harbor seal (*Phoca vitulina*) to coherent visual motion in random dot displays. Springerplus 3:688
Welsch U, Ramdohr S, Riedelsheimer B, Hebel R, Eisert R, Plötz J (2001) Microscopic anatomy of the eye of the deep-diving Antarctic Weddell seal (*Leptonychotes weddellii*). J Morphol 248:165–174
Wieskotten S, Dehnhardt G, Mauck B, Miersch L, Hanke W (2010a) Hydrodynamic determination of the moving direction of an artificial fin by a harbour seal (*Phoca vitulina*). J Exp Biol 213:2194–2200
Wieskotten S, Dehnhardt G, Mauck B, Miersch L, Hanke W (2010b) The impact of glide phases on the trackability of hydrodynamic trails in harbour seals (*Phoca vitulina*). J Exp Biol 213:3734–3740
Wieskotten S, Mauck B, Miersch L, Dehnhardt G, Hanke W (2011) Hydrodynamic discrimination of wakes caused by objects of different size or shape in a harbour seal (*Phoca vitulina*). J Exp Biol 214:1922–1930
Wightman FL, Kistler DJ (1997) Monaural sound localization revisited. J Acoust Soc Am 101:1050–1063
Wilson G (1970a) Some comments on the optical system of Pinnipedia as a result of observations on the Weddell seal (*Leptonychotes weddelli*). Br Antarct Surv Bull 23:57–63
Wilson G (1970b) Vision of the Weddell seal (*Leptonychotes weddelli*). In: Holgate MW (ed) Antarctic ecology. Academic Press, London New York, pp 490–494
Wilson B, Batty RS, Dill LM (2004) Pacific and Atlantic herring produce burst pulse sounds. Biol Lett 271:S95–S97
Winter Y, Lopez J, van Helversen O (2003) Ultraviolet vision in a bat. Nature 425:12–15
Witte M, Hanke W, Wieskotten S, Miersch L, Brede M, Dehnhardt G, Leder A (2012) On the wake flow dynamics behind harbor seal vibrissae—a fluid mechanical explanation for an extraordinary capability. In: Tropea C, Bleckmann H (eds) Nature-inspired fluid mechanics. Springer, Berlin, pp 241–260
Wohlert D, Kröger J, Witt M et al (2016) A comparative morphometric analysis of three cranial nerves in two phocids: the hooded seal (*Cystophora cristata*) and the harbor seal (*Phoca vitulina*). Anat Rec 299:370–378

Wolski LF, Anderson RC, Bowles AE, Yochem PK (2003) Measuring hearing in the harbor seal (*Phoca vitulina*): comparison of behavioral and auditory brainstem response techniques. J Acoust Soc Am 113:629–637
Zhao L, Schell DM (2004) Stable isotope ratios in harbor seal *Phoca vitulina* vibrissae: effects of growth patterns on ecological records. Mar Ecol Prog Ser 281:267–273

Chapter 3
Social Communication in Phocids

Isabelle Charrier and Caroline Casey

Contents

Abstract Communication plays a fundamental role in the social behavior of phocids, as individuals gain valuable information that aids in kin recognition, mate selection, and rival assessment. In this chapter, we provide an overview of the communication strategies used by phocid seals, which include acoustic, olfactory, and visual elements. We discuss the social contexts in which information exchange occurs, and how communication differs between females, pups, and adult males of different species during the breeding season. While significant effort has been directed toward characterizing the acoustic behavior of several seal species, much remains to be learned about how different phocids utilize available olfactory and visual information during interactions with conspecifics. We suggest that future work focuses on filling in these knowledge gaps while also experimentally testing the functional significance of different signaling strategies in the social lives of seals.

Keywords Animal communication · Vocalizations · Mother–young recognition · Male competition · Olfaction · Vision

I. Charrier (✉)
Institut des Neurosciences Paris-Saclay, Université Paris-Saclay, CNRS (UMR 9197), Université Paris-Sud, 151 route de la Rotonde, 91400 Saclay, France
e-mail: isabelle.charrier@u-psud.fr

C. Casey
Department of Ecology and Evolutionary Biology, University of California Santa Cruz, Santa Cruz 95060, USA
e-mail: cbcasey@ucsc.edu

D. P. Costa and E. McHuron (eds.), *Ethology and Behavioral Ecology of Phocids*, Ethology and Behavioral Ecology of Marine Mammals,
https://doi.org/10.1007/978-3-030-88923-4_3

3.1 Introduction

Animal communication is defined as the transfer of information from an emitter to a receiver (Shannon and Weaver 1949). Following the reception of information, the receiver will adapt its behavior according to the information encoded in the sensory signal or cue. In vertebrates, communication is involved in many vital functions related to breeding activities and social interactions such as territorial defense (food and breeding), mate selection, care of the young, avoidance of predators (i.e., alarm calls), and group coordination (Halliday 1983; Bradbury and Vehrencamp 2011). Several sensory modalities such as touch, olfaction, vision, and hearing can be integrated during communication, and animals use both signals and cues to extract information about the emitter. Cues can also provide information about the emitter; however, they have not evolved for that specific purpose (e.g., a tick using your body heat as an indication of where to move). Signals have evolved to convey information to the recipients and to impact the behavior of the receiver (Lehmann et al. 2014). Since the evolutionary origins of many sensory cues are often unknown and differentiating between cues and signals is sometimes challenging, we will use the term "cues" in this chapter. The structural properties of these sensory cues vary and, thus, show different patterns in their propagation (long- or short-range), persistence in the environment (lasting only during their production or long after), localization, and nocturnal use (Alcock 1998). For instance, acoustic cues can be transmitted over long distances; they can be used in the dark or an obstructed environment and are easily localizable yet do not persist in the environment for long periods of time.

In contrast, visual cues are not well perceived in a dark or obstructed environment, can be transmitted at short range, are easily localized, and can persist over time. Vocal, visual, and olfactory cues can change as a function of attenuation and transmission loss, as well as background noise during propagation through the environment. Thus, the detectability of a cue is primarily influenced by the perceptional abilities of the receiver.

It is important to consider the context in which these sensory cues are produced. Across different sensory modalities, a diversity of signals or cues can be emitted by a signaler to transmit information about itself (i.e., phenotype) and/or its motivation or behavioral state. While most vocalizations are produced in a specific behavioral context, visual cues (e.g., color, size, and shape of the feathers in birds, the color pattern of the fur in mammals), and some olfactory cues (e.g., body odors) persist irrespective of the behavioral state of the animal. As a result, both visual and olfactory cues can have functions other than communication (locomotion dynamics, anti-predator strategy) but can be used by conspecifics to gain information about individuals such as age class, sex, and body condition. Depending on biological function, active communication space, and the characteristics of the surrounding environment, animals preferentially utilize a specific sensory modality or combinations of different sensory modalities to secure the transmission of critical information to others.

Communication is a fundamental feature of all pinniped social systems and serves to support significant life-history events such as navigation, breeding, foraging, and parental care. In contrast to most otariids (sea lions and fur seals), breeding in all phocids is not necessarily tied to land or ice (Riedman 1990). In some phocid species, mating, courtship, and/or territory defense occur in water. Given that seals breed in a variety of aquatic and terrestrial environments, their social structure and mating systems show greater diversity than those of otariids (Chaps. 7 and 8). Breeding structures range from solitary to highly colonial, and their mating system ranges from serial monogamy (e.g., ringed seals *Pusa hispida*) to extreme polygyny (e.g., elephant seal *Mirounga* and gray seals, *Halichoerus grypus*). Within these polygynous breeding systems, phocids exhibit both resource (i.e., territory) and female defense polygyny. During social interactions, phocids use a diversity of sensory cues to transmit information. The nature of these interactions may range from affiliative to agonistic and may include vocal and non-vocal sounds (i.e., sneeze, puffing sounds, tail or body slap), olfactory cues, and body posturing. Fundamental to the evolution of phocid communication systems is the social environment in which information exchange occurs, and the sensory modalities involved in these social interactions. In this chapter, we explore the diversity of acoustic, olfactory, and visual cues used by phocids during social interactions, affiliative, and agonistic. We focus on the breeding season, where most social interactions occur and consequently where most of the available data have been presented. We describe the different sensory cues produced by females, pups, and adult males, and how these cues are involved in mother–pup recognition and male–male assessment.

3.2 Vocal Communication

A first step when studying the vocal communication of a species is to describe its vocal repertoire. Different call types define an animal's vocal repertoire, with each call characterized by its unique acoustic structure (often defined by three dimensions: time, frequency, and amplitude). The name of a call type is either related to its acoustic structure or sound (e.g., noisy call, whistle, trill, bell sound) or to its biological function (e.g., contact call, food call, alarm call, territorial call). Acoustic communication includes "vocal" sounds (i.e., produced by the vocal tract), and also non-vocal sounds produced by another part of the animal body such as wing flapping and beak drumming in birds, drumming in primates, and tail flapping in marine mammals (Bro-Jørgensen and Dabelsteen 2008; Remedios et al. 2009; Bradbury and Vehrencamp 2011). Depending on its biological function, its intended audience, and the communication range, different information may be encoded within such specialized sounds.

3.2.1 Females and Pups

3.2.1.1 Vocal Repertoire

Phocid females give birth to a single pup (on land or ice), and lactation is much shorter compared to otariids or odobenids, ranging from 4 days in hooded seals (*Cystophora cristata*; Bowen 1985) to up to 53 days in Weddell seals (*Leptonychotes weddelli*; Thomas and DeMaster 1983). One exception is the Mediterranean monk seal *Monachus monachus*, which exhibits a lactation duration of up to 4–5 months (Aguilar et al. 2007; Chap. 16).

Phocids have been described as capital breeders that fast during the entire lactation period (Riedman 1990, Chap. 8). In actuality, many exhibit a mixed capital–income breeding strategy (Wheatley et al. 2008), where females perform short foraging trips during lactation, as seen in harbor (*Phoca vitulina*), harp (*Pagophilus groenlandicus*), bearded (*Erignathus barbatus*), ringed, spotted (*Phoca largha*), Weddell, and Mediterranean monk seals (Hammill et al. 1991; Lydersen and Kovacs 1993; Bowen et al. 1994; Gazo and Aguilar 2005; Wheatley et al. 2008; Chaps. 7 and 8).

Ice-breeding phocids show different grouping strategies than land-breeding species (small to large groups versus clustering in dense colonies, respectively), as the stability of sea ice conditions influences their distribution. While most pagophilic seals are highly dispersed (e.g., bearded, leopard *Hydrurga leptonyx*, ringed seals), others can form large groups of individuals during molting and breeding (e.g., harp and spotted seals; Lydersen and Kovacs 1999). The instability of ice floes and severe weather conditions (i.e., strong wind and swell) can either lead to the separation of the mother and her pup or may complicate mother–pup reunions after mothers embark on short foraging events. In colonial species, the high density of animals, as well as the breeding activity of males (i.e., fighting, chasing, and harassing), may induce movements of females that result in temporary separations between mothers and their offspring. Consequently, in many phocid species, intentional and/or incidental mother–pup separations can occur, and individual vocal recognition helps to facilitate their reunion. While some female phocids are highly tolerant toward nonfilial pups and allonurse (e.g., Hawaiian monk seals, *Neomonachus schauinslandi*, Chap. 16), other species can be quite aggressive. They may bite and attack nonfilial pups that approach them (e.g., elephant and harp seals), which sometimes leads to the death of the pup (Leboeuf 1972; Kovacs 1987). Thus, the development of individual vocal recognition between mothers and pups is critical to offspring avoiding injuries.

The vocal interaction between a mother and her young may occur as early as birth (Kovacs 1995; Collins et al. 2006; Sauvé et al. 2015a) so that both individuals can recognize one another's respective voices. Vocalizations are exchanged throughout the entire lactation, even if they remain close to one another during this period. Pups often vocalize toward their mother not only to initiate suckling but also to initiate protection when other conspecifics get too close. These contact calls (also named "attraction calls" in the literature) have the same general structure across phocids and even across pinniped species. They are complex calls with a fundamental frequency and subsequent harmonics, and an inverted-U shape or chevron-like frequency modulation (FM) pattern (ascending FM at the beginning followed by a plateau, and ending with a descending FM; Fig. 3.1). Depending on species, these vocalizations can present a noisy broadband part (as shown in Fig. 3.1a, b) and a specific amplitude modulation pattern. Some phocid pups—including harbor and Weddell seals—produce contact calls in-air and underwater to reunite or to maintain contact with their mothers (Fig. 3.2).

In Weddell seals, adult females and pups of both sexes produce calls on ice, many of which are similar to the elaborate underwater vocalizations attributed to males (Terhune et al. 1994; Oetelaar et al. 2003). The biological function of these aerial calls is unknown, and it is unclear whether females and pups produce these calls underwater. Preliminary data using acoustic tags placed on adult lactating females indicate that females also produce complex calls underwater, which were previously attributed to males only (Charrier, unpublished data; Fig. 3.3).

Female phocids produce calls outside mother–pup social interactions, including aggressive vocalizations (during or outside the breeding season) emitted toward unwanted conspecifics when they approach too close. For example, elephant seal females emit protest calls when sub-adult males attempt to copulate with them, and these calls warn alpha males of the intrusion of a competitor into his harem (Cox and Le Boeuf 1977). Finally, adult females also produce non-vocal sounds in the air,

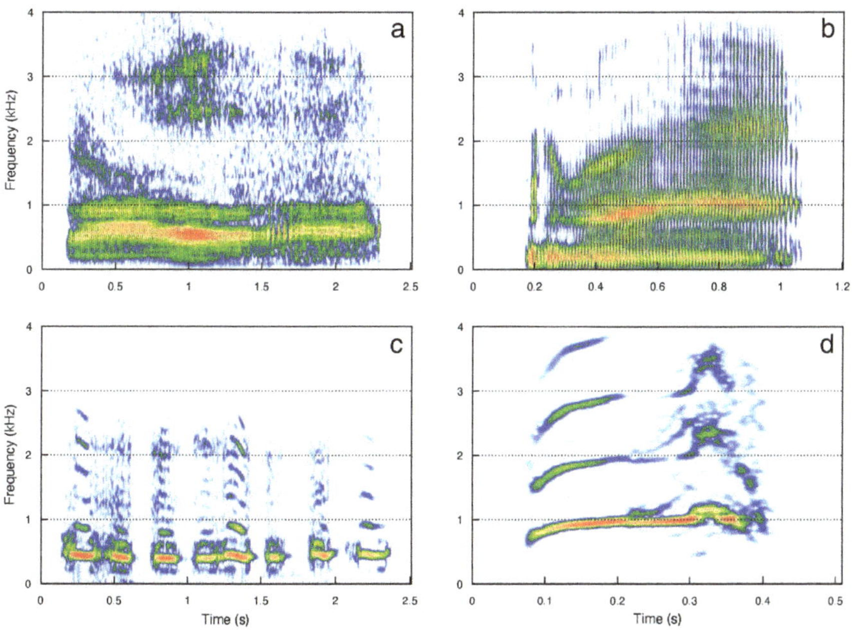

Fig. 3.1 Contact calls produced in-air by females and pups of two species of phocids, Hawaiian monk seal (**a** female, **b** pup) and harbor seal (**c** female, **d** pup). *Recordings Credits: I. Charrier*

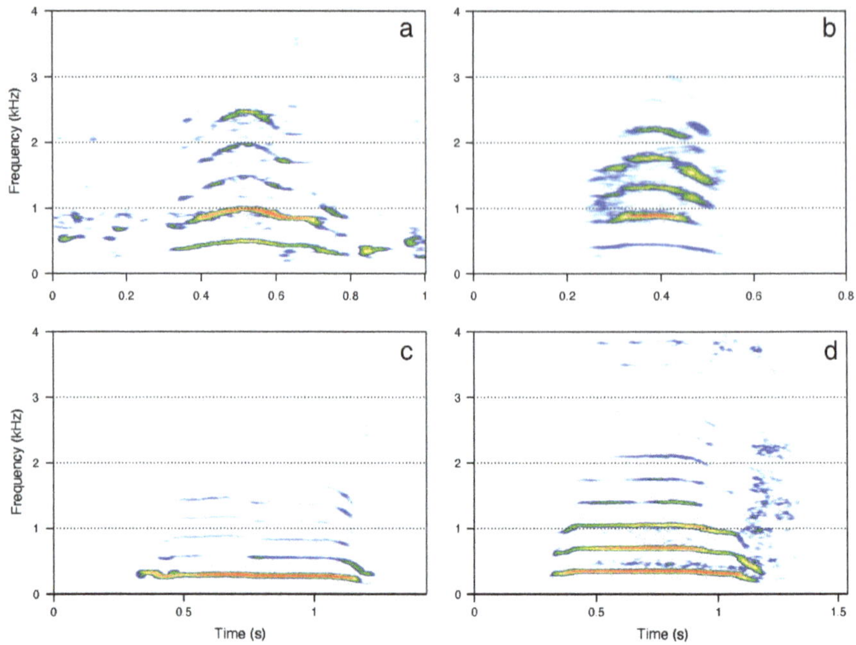

Fig. 3.2 Contact calls produced by harbor seal and Weddell seal pups under water (**a** and **c**, respectively) and in-air (**b** and **d**, respectively). Calls produced in the different media show a similar general structure, however, the spectral content differs between the two medium (calls produced under water have more energy at lower frequencies). Calls made under water are produced with a close mouth, which changes the resonance properties and thus the spectral content. *Recordings Credits: I. Charrier*

often during agonistic interactions with other conspecifics (adult of both sexes and pups) or toward humans. These include sneezes, snorts, puffs, and coughs that result from the expulsion of air through the mouth or the nostrils, as well as jaw snaps that include both acoustic and visual components.

3.2.1.2 Mother–Pup Vocal Recognition

In mammals, mother–pup vocal recognition allows females to direct nursing effort toward their offspring, to facilitate reunions after incidental or intentional separation, and enable pups to avoid risks of injury from non-related females. The selective pressures facilitating mother–pup vocal recognition depend on the surrounding social structure and mating system, which are highly diverse among phocids. Due to a striking difference in maternal traits, vocal recognition appears to be less developed in phocids than in otariids. Otariids exhibit longer maternal attendance, frequent and much more extended mother–pup separation periods during lactation, and high risk of confusion among individuals within the colony (see Chap. 3 of the Ethology of

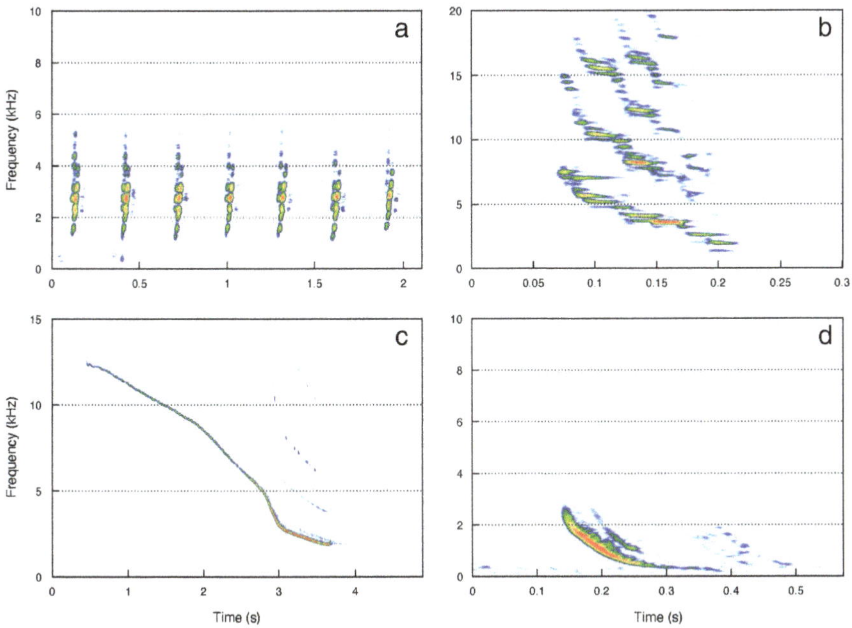

Fig. 3.3 Underwater calls produced by Weddell seal females during the breeding season at Durmont D'Urville French station. Calls were recorded using an acoustic tag (AUSOMS micro, Aquasound, Japan) attached to females. **a** Repetitive sound I3 (not previously described in the literature); **b** sdw (step descending whistle); **c** trill and **d** chirp. While most of these calls were previously described for males in the literature (Thomas and Kuechle 1982; Thomas and Stirling 1983; Terhune et al. 1994; Van Opzeeland et al. 2010), this is the first description of similar sound types being produced by females. *Recordings Credits: I. Charrier*

Otariids and Odobenids). As a result, the need for individual recognition is much higher in otariids than in most phocids.

Rates of allonursing and fostering behaviors are higher in some phocids species, but rare among otariids. In Hawaiian monk seals, 57–90% of females have been observed to nurse non-filial pups during lactation (Boness et al. 1998), and in some gray and northern elephant seal colonies, allonursing can reach rates of 28% and 18%, respectively (Riedman and Boeuf 1982; Perry et al. 1998). Several factors may contribute to allonursing behavior, including a lack of individual recognition, low cost for females' fitness, and benefits for females who had lost their pups (e.g., improving their maternal skills; Riedman and Boeuf 1982).

A prerequisite for individual vocal recognition is the presence of individual-specific vocalizations. Vocal stereotypy has been studied in some phocids species for mother and pup calls. Although individuality seems to be lower when comparing phocids to otariids or walrus (individual vocal stereotypy "IVS" ranging from 4.7 to 16.9 in otariids and walrus vs. 1.26–9.21 in phocids; Charrier 2020), vocal individuality occurs among different phocid species. However, variability in the individuality of calls in different phocids may be a result of the acoustic analysis performed and the variables measured (Table 3.2). A striking illustration is found in the pup calls produced by Hawaiian monk seals. Job et al. (1995) reported that these vocalizations are not very individual-specific (six acoustic variables measured), while Charrier and colleagues discovered that pups exhibit individual-specific vocalizations (10 acoustic variables measured, unpublished data—Table 3.2).

While individuality in calls is a prerequisite for individual vocal recognition, only playback experiments can determine the occurrence or absence of an animal's ability to recognize the calls of others. Experimental evidence of mother–pup vocal recognition is quite scarce among phocids, as only four species have been tested, and only the recognition of the pup by the mother was assessed. Gray seals (Sable Island, Canada; McCulloch and Boness 2000), northern elephant seals (Petrinovich 1974), and harbor seals (Sauvé et al. 2015a) appear able to discriminate the calls of their filial pups from those of nonfilial pups. This discrimination can occur soon after parturition; northern elephant and harbor seal mothers tested within 48 h of birth responded more strongly to the calls of their pup than those of non-filial pups (Linossier et al. unpublished data; Sauvé et al. 2015b). In contrast, Weddell seal mothers of pups 5- to 12-day old did not show vocal discrimination between filial and nonfilial offspring (Van Opzeeland et al. 2012). However, a lack of recognition could be explained by the fact that at such a young age, a pup's vocalization may not yet be individually distinct. Calls of Weddell seal pups become individual-specific when they are about 14-day old, matching the age at which they start entering the water with their mother (Collins et al. 2006). Consequently, further investigations are needed to confirm the lack of occurrence of individual vocal recognition in Weddell seals.

There is some evidence of intraspecific variation in vocal recognition, although the cause of such variation is unknown. Among gray seals, mothers from a Canadian colony showed vocal recognition of their pups (McCulloch and Boness 2000), but mothers from a Scottish colony did not (McCulloch et al. 1999). These populations of gray seals differ in their habitat, maternal attendance, and ancestral background (Chap. 9). At Sable Island (Canada), gray seals breed on sandy beaches (i.e., open area without prominent landmarks such as rocks) where there is a high density of conspecifics. This higher density increases the likelihood of misidentifying their offspring. At Isle of May (Scotland), gray seals breed on a rocky substrate with a high density of conspecifics leading to the risk of confusion. While females at the Isle of May perform short foraging trips at sea during lactation and show some allonursing events, gray seal females at Sable Island colony fast and stay near their pup and are rarely observed nursing nonfilial offspring. Despite these differences, both populations of gray seals present clear needs for individual vocal recognition (frequent mother–pup separations at the Isle of May, the high dense colony at Sable Island with no landmarks). An alternative explanation is that individual recognition found in the Canadian colony might be a residual behavior from an ice-breeding ancestry (McCulloch et al. 1999). Breeding in an unstable environment such as ice increases the chance of mother–pup separations, which may have led to the development of vocal recognition between mothers and pups in this region.

Some female phocids exhibit similar recognition abilities to those shown by otariid females (see Chap. 14 in Ethology of Otariids and Odobenids). Indeed, northern elephant seal females can recognize their pup's calls within 1–2 days after parturition (Linossier et al. 2021). Still, much remains to be investigated on the topic of mother–pup recognition among phocids. In contrast to otariids, the decoding process of individual identity—determining the specific acoustic features that are used by mothers and pups to identify each other—has not yet been examined. This will require a detailed analysis of the acoustic structure of pup vocalizations paired with playback experiments that test which specific features of the call help to convey individual identity. Further research is needed to compare the complexity of vocal signatures involved in mother–pup recognition to the ecological, physical, and social conditions in which communication occurs.

3.2.2 Adult Males

3.2.2.1 Vocal Repertoire

Phocid males produce sounds to defend their territory, guard females and attract potential mates (Figs. 3.4 and 3.5). Stirling and Thomas (2003) suggest that in aquatic-breeding phocids, there is a positive correlation between the size of the underwater vocal repertoire and the level of polygyny—and subsequent competition—in the species (Table 3.1). Harp and Weddell seals—which exhibit dense aggregations of females during the mating season—have the most extensive underwater vocal repertoire of any phocid (Stirling and Thomas 2003). Polygynous/gregarious species also show a high diversity (i.e., number of different call types) in vocal repertoire, and produce calls at high amplitude levels (up to 193 dB re 1 μPa peak at 1 m for Weddell seal, Thomas and Kuechle 1982; up to 126 dB re 20 μPa peak for elephant seals, Southall et al. 2019) and high rates. In contrast, most solitary species with low levels of polygyny have a small repertoire size (e.g., crabeater *Lobodon carcinophaga*, hooded, and Hawaiian monk seals; Table 3.1). Stirling and Thomas (2003) also suggest that the level of predation (polar bear, killer whales, sharks, and leopard seal) during the mating season may affect the size of the underwater vocal repertoire for some species (Table 3.2).

Terhune (2019) investigated vocal complexity of the repertoire of different seal species relative to their breeding systems and included both repertoire size as well as acoustic complexity (i.e., pure tone or harmonically structured call, the occurrence of amplitude and frequency modulation, broad or narrow band, mono or multi-element) in his analysis. He found that phocids exhibiting serial-monogyny (i.e., hooded, crabeater, harbor, spotted, and Hawaiian monk seals) showed the lowest levels of vocal complexity, produce low-frequency calls with burst pulses (i.e., short and low-frequency broadband sounds), and have few calls within their vocal repertoire (Terhune 2019). In contrast, phocids with the highest vocal complexity produce

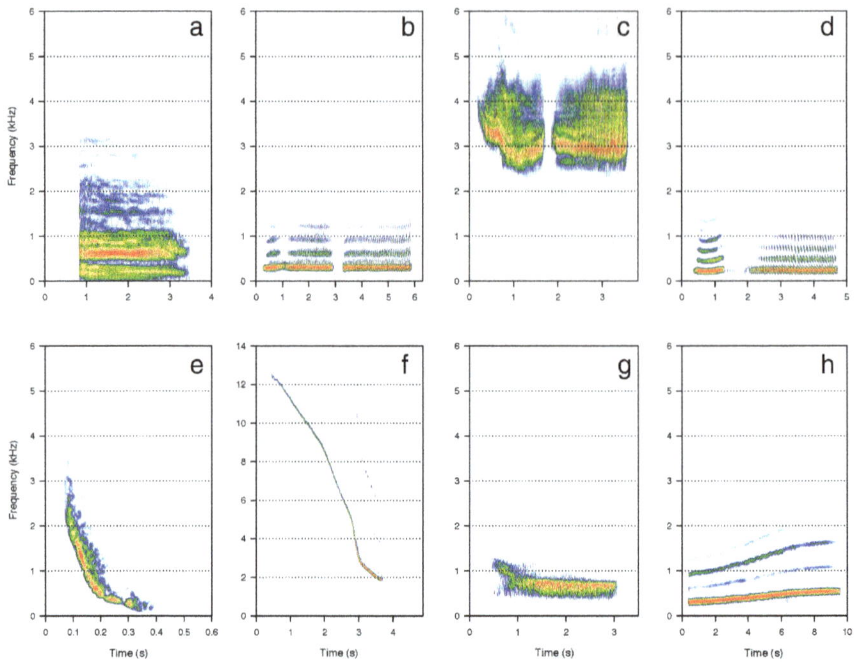

Fig. 3.4 Examples of underwater vocalizations produced by three species of Antarctic phocids: crabeater (**a**), leopard (**b–d**), and Weddell seals (**e–h**). **a** Moan, **b** double trill, **c** high double trill, **d** hoot single trill, **e** chirp, **f** trill, **g** falling tone, **h** rising tone. *Recordings Credits: I. Stirling (crabeater and leopard seal), I. Charrier (Weddell seals)*

calls with high diversity, have an extensive repertoire, and exhibit slight to high polygynous mating systems and form large breeding groups (Weddell and harp seals). Continued research is needed to describe the vocal behavior of male phocids more fully because we lack basic information on their vocal repertoire of some species (Hawaiian and Mediterranean monk seals), whereas, for others, the vocal repertoire is likely underestimated (Weddell and harp seals, Terhune 2019), or their mating system is still largely unknown (leopard, bearded, and Ross seals *Ommatophoca rossii*).

Land-breeding phocids—including gray seals and elephant seals—exhibit a limited airborne vocal repertoire. For instance, northern and southern elephant (*Mirounga leonina*) seal males produce only one call type (threat call; Bartholomew and Collias 1962; Sanvito and Galimberti 2000a), whereas gray seal males produce up to five in-air call types (kataro, gurgle, moan, growl, yodel; McCulloch 1999).

Similar to adult females, non-vocal sounds are also produced by aquatic- and land-breeding males during agonistic interactions (puff, cough, sneeze, and snort). Gray seals perform body slap behaviors during male–male competitive interactions, with vibrational cues on the substrate that give reliable size information to receivers (Bishop et al. 2015). This behavior creates a multimodal signal composed

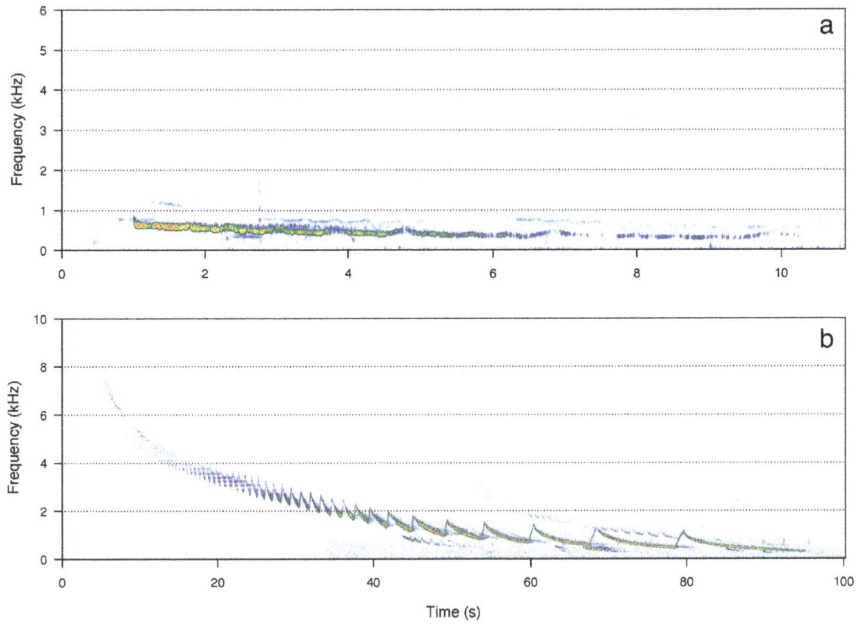

Fig. 3.5 Examples of underwater vocalizations produced by bearded seals. **a** Moan, **b** trill. *Recordings Credits: I. Charrier*

of three sensory components: acoustic, mechanic (substrate-vibration), and visual (body posture). Elephant seal males produce multimodal signals comprised of loud vocalizations, elevated visual posturing, and vibrational cues generated by slamming the chest against the ground (Shipley et al. 1992). Playback experiments on northern elephant seal males demonstrated that threat calls in combination with substrate vibrational cues elicited more robust responses than playback with threat calls alone (Shipley et al. 1992). Finally, harbor seal males perform flipper slapping during agonistic interactions and playback experiments in response to calls of territorial males (Hayes et al. 2004).

3.2.2.2 Seasonal and Diel Patterns

In aquatic-breeding phocids, the vocal behavior of males is linked to their breeding activity and plays a key role in mate attraction and rival competition. Year-round passive acoustic monitoring of Weddell and bearded seals have shown that the period of highest vocal activity corresponds to their respective breeding seasons (Green and Burton 1988; Van Parijs and Clark 2006; Van Opzeeland et al. 2010). A similar trend of high vocal activity during the breeding season has been observed in harbor seals (Van Parijs et al. 2000b), ringed seals (Stirling 1973), Ross, leopard, and crabeater

Table 3.1 Underwater and airborne vocalizations produced by males of aquatic-mating phocid species in relation to both social and mating systems. The number of known vocalizations for each medium is provided along with the names of each call type

	Species	Scientific name	Vocalization types: underwater (UW) and airborne (Air)	Social system	Mating system	References (acoustics)
Non-polar phocids	Gray seal (ice breeding)	*Halichoerus grypus*	– 7 UW: rup, rupe, growl, knocking, trot, click, and roar	Colonial	Moderate to high polygyny	– Asselin et al. (1993)
	Harbor seal	*Phoca vitulina*	– 5 UW: roar, bubbly growl, grunt, groan, creak – 6 air: bark, honk, grunt, growl, roar, moan	Small to large groups	Slight polygyny or promiscuity	– Hanggi and Schusterman (1994) – Van Parijs and Kovacs (2002)
	Hawaiian monk seal	*Neomonachus schauinslandi*	– 6 UW: croak, groan, growl, moan, rumble, whoop (from one captive male) – 4 air: bubble or gulp, roar, bark, growl	Solitary or small groups	Serial monogamy	– Parnell (2018), Miller and Job (1992), Charrier (unpublished data)
	Mediterranean monk seal	*Monachus monachus*	– UW unknown – 5 Air: bark, scream, short scream, chirp, grunt	Solitary or small groups	Slight polygyny	– Charrier et al. (2017)

(continued)

Table 3.1 (continued)

	Species	Scientific name	Vocalization types: underwater (UW) and airborne (Air)	Social system	Mating system	References (acoustics)
Arctic phocids	Harp seal	*Pagophilus groenlandicus*	– 26 UW: call type 1 to call type 26 – 1 Air: aerial call type 1	Large groups	Slight polygyny or promiscuity	– Terhune (1994), Serrano and Terhune (2002) – Serrano (2001)
	Hooded seal	*Cystophora cristata*	– 8 UW and 8 air: air exhalation, guttural growl, moaning growl, long growl, roar, FM growl, moaning-growling call, and hood/septum sounds	Solitary	Serial monogamy or slight polygyny	– Ballard and Kovacs (1995)
	Bearded seal	*Erignathus barbatus*	– 6 UW: trill with ascent, long trill, short trill, ascent, sweep and moan	Solitary	Serial monogamy suggested	Cleator et al. (1989), Van Parijs et al. (2001), Risch et al. (2007)
	Spotted seal	*Phoca largha*	– 6 UW: growl, drums, chirp, and bark – 8 Air: snort, creaky door, growl, drums, chirp, and bark	Solitary	Serial monogamy	Beier and Wartzok (1979)

(continued)

Table 3.1 (continued)

	Species	Scientific name	Vocalization types: underwater (UW) and airborne (Air)	Social system	Mating system	References (acoustics)
	Ribbon seal	*Histriophoca fasciata*	– 5 UW: downsweep, roar, grunt, yowl, hiss	Solitary	Slight polygyny	Watkins and Ray (1977), Miksis-Olds and Parks (2011), Jones et al. (2014)
	Ringed seal	*Pusa hispida*	– 4 UW: bark, growl, yelp, chirp. Pitch subcategories (low, medium or high) for each call	Solitary	Serial monogamy	Stirling (1973), Stirling et al. (1983)
Antarctic phocids	Weddell seal	*Leptonychotes weddellii*	– Up to 49 UW from 12 main types: trills, chirp, eeyoo, chug, click, teeth chatter, guttural glug, cricket call, knock, seitz, growl, mew – 12 Air: 2 whistles, rising chirp, falling chirp, descending tone, rising tone, grunt, whoop, contact call, jaw snap, clucking sound, growl	Colonial (moderate)	Moderate polygyny	– Terhune et al. (1994), Thomas and Kuechle (1982), Terhune (2019), Cziko et al. (2020)

(continued)

Table 3.1 (continued)

	Species	Scientific name	Vocalization types: underwater (UW) and airborne (Air)	Social system	Mating system	References (acoustics)
	Leopard seal	*Hydrurga leptonyx*	– 15 UW: low, middle, and high trills subdivided into single or double trills, sometimes paired with another type of call, the hoot + 3 ultrasonic underwater sounds (FM chirp, FM buzz, click train)	Solitary	Promiscuity suggested	– Stirling and Siniff (1979), Thomas et al. (1983), Rogers et al. (1995, 1996)
	Ross seal	*Ommatophoca rossii*	– 5 UW: high, mid and low siren calls, whoosh broadband component, whoosh tonal component	Solitary	Monogamy suggested	– Watkins and Wartzok (1985), Seibert (2007)
	Crabeater seal	*Lobodon carcinophagus*	– 2 UW: groan or low moan, high moan	Solitary	Serial monogamy	– Stirling and Siniff (1979), Klinck et al. (2010)

Table 3.2 Vocal individuality and occurrence of mother–pup vocal recognition with respect to reproductive behavior and ecological conditions. Individual vocal stereotypy (IVS) is the ratio between the correct classification rate and the chance of a call being attributed to the correct individual (defined as 100/total number of individuals, see Charrier 2020). Breeding density: high H, moderate M, low L; mating system: polygyny P, monogamy M; maternal absence during lactation: frequent and for several days Y, no absence N; Selective Pressures for Individual Recognition: high H, moderate Mod, low Low; Classification Rate in %, Vocal Recognition tested: Pup by Mom or Mom by Pup, Occurrence: Yes or No). * Average values

Phocid species	Breeding density	Mating system	Allo-suckling	Maternal absence	Sel. Pres. for IR	Mothers			Pups			Vocal recognition	References for classification rates/test on IR
						Cl. rate	n	IVS	Cl. rate	n	IVS		
N. elephant seal	H	High P	Common	No	Mod	54	8	4.32	64	8	5.12	Pup by Mom: Y	A/B
Gray seal	H-M	Moderate P	Common	No or Yes	Mod	–	–	–	31	20	6.2	Pup by Mom:	C/C, D
Weddell seal	M-L	Moderate P	Rare	Yes	Mod	56	9	5.04	44	10	4.4	Pup by Mom: N	E, F/G
Harp seal	M-L	Slight P	Unknown	Yes	Low	–	–	–	8–55	12–47	9.21*	Pup by Mom: N	H, I
Harbor seal	L	Slight P	Rare	Yes	Low	–	–	–	29	15	4.35	–	J
									42.6	15	6.39	Pup by Mom: Y	K/L
Haw. monk seal	L	Serial M	Common	No	Low	–	–	–	14	9	1.26	–	M
						53	5	2.65	54	10	5.4	–	N

A: Insley (1992); B: Petrinovich (1974); C: McCulloch et al. (1999); D: McCulloch and Boness (2000); E: Collins et al. (2005); F: Collins et al. (2006); G: Van Opzeeland et al. (2012); H:Van Opzeeland and Van Parijs (2004); I: Van Opzeeland et al. (2009); J: Khan et al. (2006); K: Sauvé et al. (2015b); L: Sauve et al. (2015a, b); M: Job et al. (1995); N: Charrier (unpub data) seals (Van Opzeeland et al. 2010), hooded seals (Ballard and Kovacs 1995), ice-breeding gray seals (Asselin et al. 1993), and bearded seals (Van Parijs et al. 2001). Weddell seal males exhibit high vocal activity in winter and spring (outside the breeding season), which suggests that males may maintain territorial defense year-round (Rouget et al. 2007; Van Opzeeland et al. 2010). These males might have higher breeding success than non-territorial or roaming males, as they already have an established territory with some females in attendance outside the breeding season and are familiar with competitive neighbors once the breeding season commences (Harcourt et al. 2008; Van Opzeeland et al. 2010).

In addition to seasonal patterns, the vocal activity of polar phocids also shows diel patterns. Harp seals are more vocal at night versus the day during the breeding season (Terhune and Ronald 1976; Serrano and Miller 2000). During winter (outside the breeding season), bearded seal males have a higher vocal production rate during late night/early morning in the Chukchi sea (Frouin-Mouy et al. 2016). Ice-breeding gray seal males show daily variations in underwater vocalizations, and while this does not seem related to time of day, clicks appear to be produced more often at night than during the day in this species (Asselin et al. 1993). Similarly, Weddell seals tend to be more vocal at night (Rouget et al. 2007), as well as leopard and crabeater seals that produce more vocalizations during nocturnal hours (1900 to 0600; Thomas and DeMaster 1982).

For land-breeding phocids, the majority of vocal activity occurs only during the breeding season; however, no study has focused on their vocal activity while at sea. One possible vocalization has been detected from an acoustic tag deployed on one juvenile northern elephant seal female (Burgess et al. 1998). Diel patterns occur in adult male northern elephant seals, which exhibit low vocal activity during the warmest hours of the day and a high vocal activity just after sunset (Shipley and Strecker 1986). As males fast during the entire breeding season, reducing their activity during the warmest period of the day may allow them to limit water loss (Lester and Costa 2006).

3.2.2.3 Geographic Variation

Phocid males exhibit similarities in vocal behavior specific to geographical regions or social groups that typically do not intersect. These population-level differences in vocal behavior have been documented for Weddell (Thomas and Stirling 1983; Abgrall et al. 2003), harp (Terhune 1994; Perry and Terhune 1999), leopard (Thomas and Golladay 1995), harbor (Van Parijs et al. 2000a; Bjørgesæter et al. 2004), elephant (Le Boeuf and Peterson 1969), and bearded seals (Cleator et al. 1989; Risch et al. 2007). Several social and ecological factors may lead to geographic differences in vocal behavior, including patterns of dispersal, breeding site fidelity, female selection, and the role of learning during vocal development. There is some evidence that these geographic differences are detectable (and potentially biologically meaningful) by the animals themselves. For example, a playback study on bearded seals showed that males were able to discriminate geographic variation by showing a stronger behavioral response to calls produced by local males than to calls from geographically distant males (Charrier et al. 2013). Such outcomes provide essential information on population structure and breeding site fidelity. These studies offer detailed characterizations of the vocal breeding behavior of males across multiple species and, thus, create a foundation for addressing more specific questions regarding the adaptive value (if any) of this variation.

3.2.2.4 Rival Assessment and Individual Recognition

Vocalizations produced by males are assumed to have two principal biological functions: an inter-sexual function to attract mates and an intra-sexual function to warn competitors of their presence or to establish dominance hierarchies. Experimental studies (i.e., playback tests) investigating the biological function of male underwater vocalizations are essential. Despite the critical insights they can provide, experimental work is surprisingly scarce among phocids. A pioneering study in 1968 led by Watkins and Schevill exposed Weddell seals to trills (produced by males) from breathing holes and showed that males responded vocally to the played-backed trills as if they came from a real seal (Watkins and Schevill 1968). Males checked the transducer intensively and were distracted from breathing upon arrival at the breathing hole.

Similarly, Hayes and colleagues (2004) experimentally tested the role of roaring among harbor seal males. In this species, younger and thus subordinate males produce shorter, higher-frequency roars compared with older and dominant males producing longer and lower-frequency roars (Nicholson 2000). The playback of both types of roars induced agonistic responses from territory-holding males, including flipper slapping at the surface (Hayes et al. 2004). Responses to short and high-frequency roars (younger and subordinate males) were more intense than those with long and low-frequency roars (older dominant males), suggesting that adult males can discriminate age, size, and social status based on roar characteristics.

Charrier and colleagues (2013) assessed the role of trills (the most common call types) in bearded seals. Bearded seal males produce trills mainly during the breeding season and appear to advertise male breeding status to females and to define underwater territories (Cleator et al. 1989; Van Parijs et al. 2001, 2003). Playback of trills during the breeding season elicited a significant decrease of vocal activity in comparison to the pre- and post-playback period, and surface behaviors of bearded seals were observed only after the playback had terminated. Such a response may represent the typical territorial defense of bearded seal males: males reduce their vocal activity to quietly patrol their territory when they detect a vocalizing intruder to fight or chase off the potential rival (Charrier et al. 2013).

The most comprehensive experimental work evaluating the function of male vocalizations in land-breeding colonial phocids has been conducted with northern elephant seals. In this species, reproduction is annually synchronous, and mature female seals aggregate each winter on breeding beaches. Males on shore establish dominance hierarchies that determine access to harems of up to 100 estrous females (Le Boeuf and Peterson 1969; Le Boeuf 1974). While the position of males within the dominance hierarchy is initially established through physical confrontations (Le Boeuf 1974; Haley et al. 1994), the majority of agonistic encounters between males are resolved through ritualized vocal displays.

The acoustic components of northern elephant seal threat displays appear to be of particular importance to con-specifics, as calling males can control the movement of rivals over long distances (Bartholomew and Collias 1962; Sandegren 1976), even under circumstances where visual cues are limited (Shipley and Strecker 1986). These calls appear to be used only during social exchanges between males, and there is no evidence to suggest that females exhibit mate choice or attend to the features of these calls. Throughout the breeding season, dominant individuals closely monitor the behavior of subordinate males on the rookery, who retreat from the vocal displays of higher-ranking males.

Fine-scale acoustic analyses of male elephant seal vocalization have been paired with close behavioral observations of known individuals to determine the potential functions of these acoustic signals (Casey et al. 2015). A series of sound analysis and playback experiments featuring natural and modified signals found that males do not use information about size or dominance status encoded in their vocalizations, but instead, learn to recognize individual acoustic signatures produced by their competitors (Casey et al. 2015). Responses to competitors' calls are modulated by relative position in the hierarchy, and males react differently to familiar challengers based on the outcome of previous interactions. Additional playback experiments revealed that males use two main acoustic features to individually identify their rivals, the pulse rate and the centroid of the spectrum (i.e., a spectral feature). Slight modifications of one of these two acoustic parameters disrupt individual vocal recognition among adult northern elephant seals (Mathevon et al. 2017). As males become more physically and socially mature, their threat calls are produced with less variability and show a stable and individual-specific vocal signature when fully mature (Casey et al. 2020). A series of playback experiments highlighted the fact that young males could extract age-related vocal cues from threat calls and responded in a different manner, according to the age class of the unfamiliar competitors (Casey et al. 2020).

Threat calls of southern elephant seal males are individually-specific (Sanvito and Galimberti 2000b) and contain age- and phenotype-related acoustic features (Sanvito et al. 2007). This suggests that these calls may convey honest information about size or dominance status to the receiver. No playback experiments have been performed on this species to assess whether males use phenotype-linked acoustic cues or/and individual vocal recognition to evaluate their competitors.

There is a clear need for experimental studies that investigate the biological functions of male phocid vocalizations in air and under water. Some species exhibit a highly diverse vocal repertoire; however, the function of these signals and/or the type of information (age, size, social status, or motivation) encoded in these vocalizations remained unresolved. The development of miniature electronic devices, including different sensors (e.g., hydrophone, video camera, 3D accelerometer, pressure) could bring new knowledge on the behavioral and social context of phocid underwater vocal production. Ongoing studies of adult and juvenile Hawaiian monk seals using such devices are evaluating their underwater vocal repertoire, which is currently undescribed. Additionally, passive acoustic monitoring is being used to study vocal behavior of the Mediterranean monk seal to assess whether males and/or females produce calls underwater during breeding activities. Continued development of new tag technology and the use of passive acoustics will provide further information on vocal repertoires of often cryptic phocid species.

3.3 Visual Communication

Intentional and non-intentional visual cues such as fur color pattern, shape, body size, and body postures can provide meaningful information from the emitter and thus can be involved in various social interactions. Color and shape/size can give information about the age, sex, identity (Fig. 3.6), and reproductive status (i.e., estrus, courtship, sexual maturity) of individuals, and these cues are available to receivers (Schaefer 2010) as long as individuals are in visual contact. Conversely, animals may also change body posture during discrete or limited periods to transmit information about the behavioral and emotional state of the individual (e.g., threat and submissive posturing).

Among pinnipeds, visual cues are involved in both affiliative (e.g., mother–pup reunions, mate choice) and agonistic interactions (e.g., male–male competition). Until recently, only observational studies have commented on the potential role that visual signals play during social interactions in phocids, and no experimental work has assessed whether seals utilize visual signaling during important life-history events.

Fig. 3.6 Color fur pattern in harbor seal females used for photo-identification. Orange circles indicate individual-specific visual features used to identify females. *Photos Credits: I. Charrier*

3.3.1 *Visual Acuity*

Visual acuity in phocids is quite good in air (under bright conditions; Piggins 1970; Lavigne and Ronald 1972) and underwater (Chap. 2), and as a consequence, they can likely detect differences in body size, scar and color fur pattern, and facial traits (Schusterman et al. 1970; Dehnhardt 2002; Mass and Supin 2007). Pinnipeds are myopic and astigmatismatic in air under low levels of light, as their visual acuity is generally better suited for underwater (Piggins 1970; Wartzok and Ketten 1999). While pinnipeds are considered color-blind from an anatomic perspective as they possess only one type of cone (Levenson et al. 2006), behavioral experiments have shown that some species can discriminate between colors (e.g., spotted seals; Wartzok and McCormick 1978). Additional research where brightness was controlled showed that harbor seals are indeed color blind (Scholtyssek et al. 2015). Therefore, it is likely that color perception in pinnipeds is based on brightness discrimination rather than color vision per se (Scholtyssek et al. 2008; Scholtyssek and Dehnhardt 2013; Chap. 2).

3.3.1.1 Mother–Pup Interactions

Visual cues such as topographic features or landmarks may facilitate mother–pup reunion when pairs have been separated intentionally (e.g., females that forage at sea during lactation) and/or accidentally (e.g., dramatic tidal changes, harsh weather conditions, aggression from conspecifics). In some populations of harbor seals, females perform short foraging trips and leave their pup alone at the breeding site. Some females show strong site fidelity during lactation, and thus topographic cues are likely used to facilitate reunion, especially since harbor seal females do not produce calls to relocate their pups (Godsell 1988; Sauvé et al. 2015b). In gray seals breeding at the Isle of May, Scotland, females may use spatial cues to reunite with their pups, as previous research indicates that they are not very good at recognizing their pups' vocalizations (McCulloch and Boness 2000; see Sect. 3.2.1.2). In ice-breeding harp seals, females begin short foraging trips 1–2 days after parturition, and pups are quite sedentary while their mothers are away (Kovacs 1995; Chap. 14). Behavioral observations indicate that females use a direct path on ice to relocate their pup, choosing the closest breathing hole or crack in the ice near the last location of her pup. Females quickly relocate their pups without vocalizations or the use of direct visual cues (i.e., pups can be hidden behind ice, and females can still locate them). Female spatial orientation appears critical to the successful reunion with their dependent pups in this species, which is quite surprising given that pack-ice is considered an unstable environment.

Studies are lacking on individual visual recognition between mothers and pups in phocids. Among all pinnipeds, there is no information describing whether females or pups use some physical characteristics (e.g., different fur color patterns, body size) to individually identify one another, or at least sort individuals into age/sex classes. The only exception is for Australian sea lions (*Neophoca cinerea*), where females sort pups by age class using visual cues (Wierucka et al. 2017). Variation in fur patterns has been used by researchers to identify individuals through photo-ID (Fig. 3.6) for the Mediterranean monk seal (Forcada and Aguilar 2000), harbor seal (Cunningham 2009), gray seal (Vincent et al. 2001), and Saimaa ringed seal *Pusa hispida saimensis* (Koivuniemi et al. 2016), which suggests that pups could also utilize these visual cues for individual identification. Since many terrestrial mammal species such as sheep (*Ovis aries*; Kendrick et al. 2001) and chimpanzee (*Pan troglodytes*; Parr and de Waal 1999) perform individual visual discrimination (non-kin and kin discrimination, respectively), it seems plausible that pinnipeds may have similar capabilities.

3.3.1.2 Adult Males and Females

Sexual dimorphism, secondary sexual trait, and color pattern differ between adult males and females. These visual traits may thus function to facilitate species- and sex identification. Additionally, visual cues such as changes in pelage color or scarring patterns may be useful in indicating age differences among individuals.

In land-breeding phocids, adult males are considerably larger than females (5–6 times more in elephant seals, two times in gray seals; Perrin et al. 2009) and develop additional secondary sexual traits during sexual maturity. Both species of elephant seal males develop a large proboscis that increases in size as they mature (Fig. 3.7; Le Boeuf 1974). Similarly, gray seal males develop a larger snout at sexual maturity, and color and scar patterns differ between males and females (Hall and Russell 2018). Hooded seal males have an oversized wide nose (hood), and while inflating it they can extrude their nasal septum from one of their nostrils, producing a large, red nasal balloon. This visual signal may either ward off potential competitors or attract females during the breeding season. In harp and ribbon seals, adult males have a darker fur pattern than females (Boveng and Lowry 2018; Lavigne 2018). In Mediterranean monk seals, adult males are much darker (nearly black/dark brown) than females and also have a pale area on their upper neck and a pale patch on their belly visible on both flank sides (Jefferson et al. 2011). In aquatic-breeding phocids, there are six species in which females are larger/longer than males: leopard, Weddell, crabeater, Ross, and both species of extant monk seals (Ralls 1976).

During the mating season, males exhibit ritualized visual cues such as upright body posturing, and open-mouth and snout displays during agonistic interactions. During competitive encounters, northern and southern elephant seal males show their large proboscis and position themselves in an upright posture to appear as tall as possible (Sandegren 1976; Sanvito et al. 2007). These visual displays are often produced in conjunction with their threat calls (Terhune and Ronald 1973; Ballard and Kovacs 1995). During agonistic interaction, male gray seals slap their body against the substrate (Bishop et al. 2015), producing vocalizations such as the yodel (Boness and James 1979).

Fig. 3.7 Proboscis size and shape change with age in male northern elephant seals. Development of secondary sexual characteristics associated with different age classes of northern elephant seals. **a** Sub-Adult 1 (SA1)—4-year old. Nose wide but not elongated, some chest wrinkles, no mid-nose indentation. **b** Sub-Adult 2 (SA2)—5-year old. Nose extends to mouth when resting, no mid-nose indentation, chest wrinkles no scaring. **c** Sub-Adult 3 (SA3)—6-year old. Nose begins to extend past the mouth when resting, mid-nose indentation begins to form, scaring begins to develop. **d** Sub-Adult 4 (SA4)—7-year old. Nose folds onto ground when resting, mid-nose indentation more pronounced, scaring more pronounced (but not pink. **e** Adult (AD)—8-year old. Nose extends to ground and fold under when resting, mid-nose indentation present, obvious calloused chest shield that often extends past the eye. *Photos Credits: Emma Levy*

3.4 Olfactory Communication

Chemical cues such as saliva, urine, feces, and glandular secretions can provide information about the emitter such as its species, sex, age class, individual identity, and emotional and reproductive state (Eisenberg and Kleiman 1972; Wyatt 2003). Additionally, olfaction is used by animals to extract information about their environment, to locate food sources, detect predators, and interact with conspecifics (Wyatt 2014). Pinnipeds—including phocids—are sensitive to natural odorant compounds in their environment. Harbor seals are extraordinarily sensitive to atmospheric dimethyl sulfide (DMS; Kowalewsky et al. 2006), a chemical indicator of primary production. Harbor seals could use such olfactory sensitivity to DMS to find profitable foraging areas.

Scent-marking is quite common in mammals to advertise male territory ownership, presence, and/or reproductive status. Typically, chemical cues are by-products of different sources (sebaceous glands, microbial activity, diet, environment) that generate a general body odor and can fluctuate with emotional state and reproductive status. Identifying the chemical compounds constituting body scent, and how conspecifics perceive them is quite challenging. Body odors are composed of a mixture of different chemical compounds (some at very low concentrations), and their identification can be difficult because odorous compounds are not well documented for all mammal species. These constraints explain why there is a gap in knowledge regarding the use of this sensorial modality in comparison to others (i.e., hearing and vision).

3.4.1 Mother–Pup Interactions

During reunions or close interactions, mothers often perform nose to nose or nose to body contacts with their pup (Riedman 1990, Fig. 3.8). Olfaction is likely a final check for individual recognition (Insley et al. 2003). Olfactory recognition has only been experimentally tested in two pinniped species, Australian sea lions and northern elephant seals. In a two-choice test, wild female Australian sea lions were tested with two pup models impregnated with the scent of their filial pup and those of non-filial pups, and mothers appeared to recognize their pup using these olfactory cues (Pitcher et al. 2011). A pilot study on northern elephant seal mothers (2019)—which also used a pup model impregnated with pup scent—revealed that females sniffed the model that was rubbed with the filial pup's scent significantly more often than the nonfilial

Fig. 3.8 Nose to nose contact between a harbor seal mother and her. While reuniting, mothers perform nose to nose contacts to definitively check the identity of their pup using olfactory cues. *Photos Credits: Emma Levy*

model (Linossier et al. unpublished data). These results indicate that some phocids are capable of olfactory discrimination. Further behavioral investigations paired with an evaluation of chemical compounds are needed in other phocids to understand the role of olfaction in individual recognition.

3.4.2 *Males*

During the breeding season, adult male-ringed seals develop a powerful scent (Hardy et al. 1991) that can be easily detectable by human observers. The Inuit refer to such seals as "tiggak". The strong and pungent odor of the ringed seal is a mixture of several organic nitrogen and sulfur compounds with at least two hydrocarbons (Ryg et al. 1992). The scent emanates from facial glands (Ryg et al. 1992), darkening the face of rutting ringed seal males (Hardy et al. 1991), and infuses the entire body and the snow around the breathing hole or the subnivean lairs. This scent production is primarily linked to breeding and territorial behavior (Ryg et al. 1992), thus may be used to advertise to both females and competitors the ownership of a territory (see Chap. 15 for further discussion of the mating system). Even if this strong smell is not intentionally produced by seals to repel terrestrial predators, it has been suggested that it could act as a repellent for predators such as polar bears (Smith 1980). This characteristic and robust odor also impregnates the meat and blubber of rutting seals and may explain why polar bears, humans, and dogs do not eat rutting males (Smith 1980). Such smell could also be used as a camouflage strategy by males to cover up the scent cues produced by females and their pups, and thus avoid polar bear predation on mother–pup pairs within their territory (Smith 1980). Such anti-predator effect would thus only be a by-product of the strong pungent smell of rutting males.

3.5 Conclusions

Sensory signals and cues play essential roles in the social life of phocids, as individuals can extract information that aids in kin recognition, mate selection, and rival assessment. Such information can be encoded in various modalities, leading to redundancy and thus enhancing the chances of being perceived by receivers. Communication is multimodal (Hebets and Papaj 2005), and further experimental investigations are needed to explore the synergistic roles of different sensory modalities. Such knowledge is crucial to better estimate the threats these marine species are exposed to in our ever-changing world. As anthropogenic disturbance and pollution (both noise and chemical) have dramatically increased, an understanding of how these disturbances impact phocid sensory perception, communication systems, and behavior is essential to propose mitigation measures that will aid in protecting these animals and their natural environment.

References

Abgrall P, Terhune JM, Burton HR (2003) Variation of Weddell seal (*Leptonychotes weddellii*) underwater vocalizations over mesogeographic ranges. Aquat Mamm 29:268–277. https://doi.org/10.1578/016754203101024202

Aguilar A, Cappozzo LH, Gazo M et al (2007) Lactation and mother-pup behaviour in the Mediterranean monk seal *Monachus monachus*: an unusual pattern for a phocid. J Mar Biol Assoc United Kingdom 87:93–99. https://doi.org/10.1017/S0025315407056147

Alcock J (1998) Animal behavior: an evolutionary approach. Sunderland, MA

Asselin S, Hammill MO, Barrette C (1993) Underwater vocalizations of ice breeding grey seals. Can J Zool 71:2211–2219. https://doi.org/10.1139/z93-310

Ballard K, Kovacs K (1995) The acoustic repertoire of hooded seals (*Cystophora cristata*). Can J Zool 73:1362–1374. https://doi.org/10.1139/z95-159

Bartholomew GA, Collias NE (1962) The role of vocalization in the social behaviour of the northern elephant seal. Anim Behav 10:7–8. https://doi.org/10.1016/0003-3472(62)90124-0

Beier JC, Wartzok D (1979) Mating behaviour of captive spotted seals (*Phoca largha*). Anim Behav 27:772–781. https://doi.org/10.1016/0003-3472(79)90013-7

Bishop AM, Denton P, Pomeroy P, Twiss S (2015) Good vibrations by the beach boys: magnitude of substrate vibrations is a reliable indicator of male grey seal size. Anim Behav 100:74–82. https://doi.org/10.1016/j.anbehav.2014.11.008

Bjørgesæter A, Ugland KI, Bjørge A (2004) Geographic variation and acoustic structure of the underwater vocalization of harbor seal (*Phoca vitulina*) in Norway, Sweden and Scotland. J Acoust Soc Am 116:2459. https://doi.org/10.1121/1.1782933

Boness DJ, James H (1979) Reproductive behaviour of the grey seal (*Halichoerus grypus*) on Sable Island, Nova Scotia. J Zool 188:477–500

Boness DDJ, Craig MPM, Honigman L, Austin S (1998) Fostering behavior and the effect of female density in Hawaiian monk seals, *Monachus Schauinslandi*. J Mammal 79:1060–1069. https://doi.org/10.2307/1383115

Boveng P, Lowry, L (2018) Ribbon seal: *Histriophoca fasciata*. In Wursig B, Thewissen JGM, Kovacs KM (eds) Encylopedia of marine mammals. Academic Press, London

Bowen W (1985) Birth to weaning in 4 days: remarkable growth in the hooded seal, *Cystophora cristata*. Can J Zool 63:2841–2846. https://doi.org/10.1139/z85-424

Bowen WD, Oftedal OT, Boness DJ, Iverson SJ (1994) The effect of maternal age and other factors on birth mass in the harbour seal. Can J Zool 72:8–14. https://doi.org/10.1139/z94-002

Bradbury JW, Vehrencamp SL (2011) Principles of animal communication. Sinauer Associates Inc., Sunderland, Massachusetts

Bro-Jørgensen J, Dabelsteen T (2008) Knee-clicks and visual traits indicate fighting ability in eland antelopes: multiple messages and back-up signals. BMC Biol 6:47. https://doi.org/10.1186/1741-7007-6-47

Burgess WC, Tyack PL, Le Boeuf BJ, Costa DP (1998) A programmable acoustic recording tag and first results from free-ranging northern elephant seals. Deep-Sea Res Pt II 45:1327–1351

Casey C, Charrier I, Mathevon N, Reichmuth C (2015) Rival assessment among northern elephant seals: Evidence of associative learning during male-male contests. R Soc Open Sci 2:150228. https://doi.org/10.1098/rsos.150228

Casey C, Charrier I, Mathevon N et al (2020) The genesis of giants: behavioural ontogeny of male northern elephant seals. Anim Behav 166:247–259. https://doi.org/10.1016/j.anbehav.2020.06.014

Charrier I (2020) Individual vocal recognition and social system in Pinnipeds. In: Aubin T, Mathevon N (eds) Coding strategies in vertebrate acoustic communication. Animal signals and communication, vol 7. Springer, Cham

Charrier I, Mathevon N, Aubin T (2013) Bearded seal males perceive geographic variation in their trills. Behav Ecol Sociobiol 67:1679–1689. https://doi.org/10.1007/s00265-013-1578-6

Charrier I, Marchesseau S, Dendrinos P, Tounta E, Karamanlidis AA (2017) Individual signatures in the vocal repertoire of the endangered Mediterranean monk seal: new perspectives for population monitoring. Endanger Species Res 32:459–470. https://doi.org/10.3354/esr00829
Cleator HJ, Stirling I, Smith TG (1989) Underwater vocalizations of the bearded seal (*Erignathus harhatus*). Can J Zool 67:1900–1910
Collins K, McGreevy PD, Wheatley KE, Harcourt RG (2005) Individual variation of in-air female "pup contact" calls in Weddell seals, *Leptonychotes weddellii*. Behaviour 142:167–189. https://doi.org/10.1163/1568539053627668
Collins KT, Terhune JM, Rogers TL et al (2006) Vocal individuality of in-air Weddell seal (*Leptonychotes weddellii*) pup "primary" calls. Mar Mamm Sci 22:933–951. https://doi.org/10.1111/j.1748-7692.2006.00074.x
Cox CR, Le Boeuf BJ (1977) Female incitation of male competition: a mechanism in sexual selection. Am Nat 111:317–335. https://doi.org/10.1086/283163
Cunningham L (2009) Using computer-assisted photo-identification and capture-recapture techniques to monitor the conservation status of harbour seals (*Phoca vitulina*). Aquat Mamm 35:319–329. https://doi.org/10.1578/AM.35.3.2009.319
Cziko PA, Munger LM, Santos NR, Terhune JM (2020) Weddell seals produce ultrasonic vocalizations. J Acoust Soc Am 148:3784–3796. https://doi.org/10.1121/10.0002867
Dehnhardt G (2002) Sensory systems. In: Hoelzel AR (ed) Marine mammal biology: an evolutionary approach. Blackwell Scientific Oxford, Oxford, pp 116–141
Eisenberg JF, Kleiman DG (1972) Olfactory communication in mammals. Annu Rev Ecol Syst 3:1–32. https://doi.org/10.1146/annurev.es.03.110172.000245
Forcada J, Aguilar A (2000) Use of photographic identification in capture-recapture studies of Mediterranean monk seals. Mar Mammal Sci 16:767–793. https://doi.org/10.1111/j.1748-7692.2000.tb00971.x
Frouin-Mouy H, Mouy X, Martin B, Hannay D (2016) Underwater acoustic behavior of bearded seals (*Erignathus barbatus*) in the northeastern Chukchi Sea, 2007–2010. Mar Mammal Sci 32:141–160. https://doi.org/10.1111/mms.12246
Gazo M, Aguilar A (2005) Maternal attendance and diving behavior of a lactating Mediterranean monk seal. Mar Mammal Sci 21:340–345. https://doi.org/10.1111/j.1748-7692.2005.tb01235.x
Godsell J (1988) Herd formation and haul-out behaviour in harbour seals (*Phoca vitulina*). J Zool Lond 215:83–98
Green K, Burton HR (1988) Annual and diurnal variations in the underwater vocalizations of Weddell seals. Polar Biol 8:161–164. https://doi.org/10.1007/BF00443447
Haley MP, Deutsch CJ, Le Boeuf BJ (1994) Size, dominance and copulatory success in male northern elephant seals, *Mirounga angustirostris*. Anim Behav 48:1249–1260
Hall AJ, Russell, DJF (2018) Gray seal: *Halichoerus grypus*. In: Wursig B, Thewissen JGM, Kovacs KM (eds) Encylopedia of Marine Mammals. Academic Press, London
Halliday TR (1983) Information and communication. In: Halliday TR, Slater PJB (eds) Animal behaviour: communication. Blackwell Scientific Oxford, Oxford, pp 43–81
Hammill MO, Lydersen C, Ryg M, Smith TG (1991) Lactation in the ringed seal (*Phoca hispida*). Can J Fish Aquat Sci 48:2471–2476. https://doi.org/10.1139/f91-288
Hanggi EB, Schusterman RJ (1994) Underwater acoustic displays and individual variation in male harbour seals, *Phoca vitulina*. Anim Behav 48:1275–1283. https://doi.org/10.1006/anbe.1994.1363
Harcourt RG, Kingston JJ, Waas JR, Hindell MA (2008) Foraging while breeding: alternative mating strategies by male Weddell seals? Aquat Conserv: Mar Freshw Ecosyst 17:S68–S78
Hardy MH, Roff E, Smith TG, Ryg M (1991) Facial skin glands of ringed and grey seals, and their possible function as odoriferous organs. Can J Zool 69:189–200. https://doi.org/10.1139/z91-029
Hayes SA, Kumar A, Costa DP et al (2004) Evaluating the function of the male harbour seal, *Phoca vitulina*, roar through playback experiments. Anim Behav 67:1133–1139. https://doi.org/10.1016/j.anbehav.2003.06.019

Hebets EA, Papaj DR (2005) Complex signal function: developing a framework of testable hypotheses. Behav Ecol Sociobiol 57:197–214. https://doi.org/10.1007/s00265-004-0865-7

Insley SJ (1992) Mother-offrsping separation and acoustic stereotypy: a comparison of call morphology in two species of pinnipeds. Behaviour 120:103–122. https://doi.org/10.1007/s13398-014-0173-7.2

Insley S, Phillips A, Charrier I (2003) A review of social recognition in pinnipeds. Aquat Mamm 29:181–201

Jefferson T, Webber M, Pitman R (2011) Marine mammals of the world: a comprehensive guide to their identification. Academic Press, San Diego

Job D, Boness D, Francis J (1995) Individual variation in nursing vocalizations of Hawaiian monk seal pups, *Monachus schauinslandi* (Phocidae, Pinnipedia), and lack of maternal recognition. Can J Zool 73:975–983

Jones JM, Thayre BJ, Roth EH et al (2014) Ringed, bearded, and ribbon seal vocalizations north of Barrow, Alaska: seasonal presence and relationship with sea ice. Arctic 67:203–222. https://doi.org/10.14430/arctic4388

Kendrick KM, da Costa AP, Leigh AE et al (2001) Sheep don't forget a face. Nature 414:165–166. https://doi.org/10.1038/35102669

Khan CBC, Markowitz H, McCowan B (2006) Vocal development in captive harbor seal pups, *Phoca vitulina richardii*: Age, sex, and individual differences. J Acoust Soc Am 120:1684. https://doi.org/10.1121/1.2226530

Klinck H, Mellinger DK, Klinck K, Hager J (2010) Long-range underwater vocalizations of the crabeater seal (*Lobodon carcinophaga*). J Acoust Soc Am 128:474–479. https://doi.org/10.1121/1.3442362

Koivuniemi M, Auttila M, Niemi M et al (2016) Photo-ID as a tool for studying and monitoring the endangered Saimaa ringed seal. Endanger Species Res 30:29–36. https://doi.org/10.3354/esr00723

Kovacs K (1987) Maternal behaviour and early behavioural ontogeny of harp seals, *Phoca groenlandica*. Anim Behav 35:844–855

Kovacs K (1995) Mother-pup reunions in harp seals, *Phoca groenlandica*: cues for the relocation of pups. Can J Zool 73:843–849

Kowalewsky S, Dambach M, Mauck B, Dehnhardt G (2006) High olfactory sensitivity for dimethyl sulphide in harbour seals. Biol Lett 2:106–109. https://doi.org/10.1098/rsbl.2005.0380

Lavigne DM, Ronald K (1972) The harp seal, *Pagophilus groenlandicus* (Erxleben 1777). XXIII. Spectral sensitivity. Can J Zool 1197–1206

Lavigne DM (2018) Harp seal: *Pagophilus groenlandicus*. In: Wursig B, Thewissen JGM, Kovacs KM (eds) Encylopedia of marine mammals. Academic Press, London

Le Boeuf BJ (1974) Male-male competition and reproductive success in elephant seals. Am Zool 14:163–176

Le Boeuf BJ, Peterson RS (1969) Dialects in elephant seals. Science 166:1654–1656. https://doi.org/10.1126/science.166.3913.1654

Leboeuf BJ (1972) Sexual behavior in the northern elephant seal *Mirounga angustirostris*. Behaviour 41:1–26

Lehmann KDS, Goldman BW, Dworkin I et al (2014) From cues to signals: Evolution of interspecific communication via aposematism and mimicry in a predator-prey system. PLoS ONE 9:e91783. https://doi.org/10.1371/journal.pone.0091783

Lester CW, Costa DP (2006) Water conservation in fasting northern elephant seals (*Mirounga angustirostris*). J Exp Biol 209:4283–4294. https://doi.org/10.1242/jeb.02503

Levenson DH, Ponganis PJ, Crognale MA et al (2006) Visual pigments of marine carnivores: pinnipeds, polar bear, and sea otter. J Comp Physiol A 192:833–843. https://doi.org/10.1007/s00359-006-0121-x

Linossier J, Casey C, Charrier I, Mathevon N, Reichmuth C (2021) Maternal responses to pup calls in a high-cost lactation species. Biol Lett 17:20210469. https://doi.org/10.1098/rsbl.2021.0469

Lydersen C, Kovacs KM (1993) Diving behaviour of lactating harp seal, *Phoca groenlandica*, females from the Gulf of St Lawrence, Canada. Anim Behav 46:1213–1221. https://doi.org/10.1006/anbe.1993.1312

Lydersen C, Kovacs KM (1999) Behaviour and energetics of ice-breeding, North Atlantic phocid seals during the lactation period. Mar Ecol Prog Ser 187:265–281

Mass AM, Supin AYA (2007) Adaptive features of aquatic mammals' eye. Anat Rec 290:701–715. https://doi.org/10.1002/ar.20529

Mathevon N, Casey C, Reichmuth C, Charrier I (2017) Northern elephant seals memorize the rhythm and timbre of their rivals' voices. Curr Biol 27:2352–2356.e2. https://doi.org/10.1016/J.CUB.2017.06.035

McCulloch S (1999) The vocal vehaviour of the grey seal (*Halichoerus grypus*). Dissertation, University of St. Andrews

McCulloch S, Boness D (2000) Mother–pup vocal recognition in the grey seal (*Halichoerus grypus*) of Sable Island, Nova Scotia, Canada. J Zool 251:449–455

McCulloch S, Pomeroy P, Slater P (1999) Individually distinctive pup vocalizations fail to prevent allo-suckling in grey seals. Can J Zool 77:716–723

Miksis-Olds JL, Parks SE (2011) Seasonal trends in acoustic detection of ribbon seal (*Histriophoca fasciata*) vocalizations in the Bering Sea. Aquat Mamm 37:464–471. https://doi.org/10.1578/AM.37.4.2011.464

Miller E, Job D (1992) Airborne acoustic communication in the Hawaiian monk seal, *Monachus schauinslandi*. In: Thomas JA, Kastelein RA, Supin AY (eds) Marine mammal sensory systems. Springer, Boston, pp 485–581

Nicholson T (2000) Social structure and underwater behavior of harbor seals in southern Monterey Bay. Thesis, San Francisco State University, California

Oetelaar ML, Burton HR, Terhune JM (2003) Can the sex of a Weddell seal (*Leptonychotes weddellii*) be identified by its surface call? Aquat Mamm 29:261–267. https://doi.org/10.1578/016754203101024194

Parnell K (2018) Underwater vocal repertoire of the endangered Hawaiian monk seal. Thesis, University of California Santa Cruz, Neomonachus schauinslandi

Parr LA, de Waal FB (1999) Visual kin recognition in chimpanzees. Nature 399:647–648. https://doi.org/10.1038/21345

Perrin WF, Wursig B, Thewissen JGM (eds) (2009) Encyclopedia of marine mammals. Academic Press, San Diego

Perry EA, Terhune JM (1999) Variation of harp seal (*Pagophilus groenlandicus*) underwater vocalizations among three breeding locations. Hydrobiologia 181–186

Perry E, Boness D, Fleischer R (1998) DNA fingerprinting evidence of nonfilial nursing in grey seals. Mol Ecol 7:81–85

Petrinovich L (1974) Individual recognition of pup vocalization by northern elephant seal mothers. Z Tierpsychol 34:308–312. https://doi.org/10.1111/j.1439-0310.1974.tb01803.x

Piggins DJ (1970) Refraction of the harp seal, *Pagophilus groenlandicus* (Erxleben 1777). Nature 78

Pitcher BJ, Harcourt RG, Schaal B, Charrier I (2011) Social olfaction in marine mammals: wild female Australian sea lions can identify their pup's scent. Biol Lett 7. https://doi.org/10.1098/rsbl.2010.0569

Ralls K (1976) Mammals in which females are larger than males. Q Rev Biol 51:245–276. https://doi.org/10.1086/409310

Remedios R, Logothetis NK, Kayser C (2009) Monkey drumming reveals common networks for perceiving vocal and nonvocal communication sounds. Proc Natl Acad Sci USA 106:18010–18015. https://doi.org/10.1073/pnas.0909756106

Riedman M (1990) The pinnipeds: seals, sea lions, and walruses. University of California Press, Berkeley

Riedman ML, Boeuf BJ (1982) Mother-pup separation and adoption in northern elephant seals. Behav Ecol Sociobiol 11:203–215. https://doi.org/10.1007/BF00300063

Risch D, Clark CCW, Corkeron PJP et al (2007) Vocalizations of male bearded seals, *Erignathus barbatus*: classification and geographical variation. Anim Behav 73:747–762. https://doi.org/10.1016/j.anbehav.2006.06.012
Rogers T, Cato D, Bryden MM (1996) Behavioral significance of underwater vocalizations of captive leopard seals, *Hydurga leptonyx*. Mar Mammal Sci 12:414–427. https://doi.org/10.1111/j.1748-7692.1996.tb00593.x
Rogers T, Cato DH, Bryden MM (1995) Underwater vocal repertoire of the leopard seal (*Hydrurga leptonyx*) in Prydz Bay, Antarctica. In: Kastelein R, Thomas JA, Nachtigall PE (eds) Sensory systems of aquatic mammals. De Spil, Woerden, pp 223–236
Rouget PA, Terhune JM, Burton HR (2007) Weddell seal underwater calling rates during the winter and spring near Mawson Station, Antarctica. Mar Mamm Sci 23:508–523. https://doi.org/10.1111/j.1748-7692.2007.00129.x
Ryg M, Solberg Y, Lydersen C, Smith TG (1992) The scent of rutting male ringed seals *Phoca hispida*. J Zool 226:681–689. https://doi.org/10.1111/j.1469-7998.1992.tb07509.x
Sandegren FE (1976) Agonistic behavior in the male northern elephant seal. Behaviour 57:136–157. https://doi.org/10.1163/156853976x00145
Sanvito S, Galimberti F (2000a) Bioacoustics of southern elephant seals. I. Acoustic structure of male aggressive vocalisations. Bioacoustics 10:259–285. https://doi.org/10.1080/09524622.2000.9753438
Sanvito S, Galimberti F (2000b) Bioacoustics of southern elephant seals. II. Individual and geographical variation in male aggressive vocalisations. Bioacoustics 10:259–285. https://doi.org/10.1080/09524622.2000.9753439
Sanvito S, Galimberti F, Miller EH (2007) Vocal signalling of male southern elephant seals is honest but imprecise. Anim Behav 73:287–299. https://doi.org/10.1016/j.anbehav.2006.08.005
Sauvé CC, Beauplet G, Hammill MO, Charrier I (2015a) Acoustic analysis of airborne, underwater, and amphibious mother attraction calls by wild harbor seal pups (*Phoca vitulina*). J Mammal 96:591–602. https://doi.org/10.1093/jmammal/gyv064
Sauvé CC, Beauplet G, Hammill MO, Charrier I (2015b) Mother-pup vocal recognition in harbour seals: Influence of maternal behaviour, pup voice and habitat sound properties. Anim Behav 105:109–120. https://doi.org/10.1016/j.anbehav.2015.04.011
Schaefer HM (2010) Visual communication: evolution, ecology, and functional mechanisms. Animal behaviour: evolution and mechanisms. Springer, Berlin Heidelberg, Berlin, Heidelberg, pp 3–28
Scholtyssek C, Dehnhardt G (2013) Brightness discrimination in the South African fur seal (*Arctocephalus pusillus*). Vis Res 84:26–32
Scholtyssek C, Kelber A, Dehnhardt G (2008) Brightness discrimination in the harbor seal (*Phoca vitulina*). Vis Res 48:96–103. https://doi.org/10.1016/j.visres.2007.10.012
Scholtyssek C, Kelber A, Dehnhardt G (2015) Why do seals have cones ? Behavioural evidence for colour-blindness in harbour seals. Animal Cogn 551–560. https://doi.org/10.1007/s10071-014-0823-3
Schusterman RJ, Balliet RF, ball, (1970) Conditioned vocalizations as a technique for determining visual acuity thresholds in sea lions. Science 169:498–501
Seibert A-M (2007) The Ross seal and its underwater vocalizations. Munich University, Thesis
Serrano A (2001) New underwater and aerial vocalizations of captive harp seals (*Pagophilus groenlandicus*). Can J Zool 79:75–81. https://doi.org/10.1139/cjz-79-1-75
Serrano A, Miller EH (2000) How vocal are harp seals (*Pagophilus groenlandicus*)? A captive study of seasonal and diel patterns. Aquat Mamm 26:253–259
Serrano A, Terhune J (2002) Stability of the underwater vocal repertoire of harp seals (*Pagophilus groenlandicus*). Aquat Mamm 28:93–101
Shannon C, Weaver W (1949) The mathematical theory of communication. University of Illinois Press, Urbana
Shipley C, Strecker G (1986) Day and night patterns of vocal activity of northern elephant seal bulls. J Mammal 67:775–778. https://doi.org/10.2307/1381150

Shipley C, Stewart BS, Bass J (1992) Seismic communication in northern elephant seals. In: Thomas JA, Kastelein RA, Supin AY (eds) Marine mammal sensory systems. Springer, Boston, pp 553–562
Smith TG (1980) Polar bear predation of ringed and bearded seals in the land-fast sea ice habitat. Can J Zool 58:2201–2209. https://doi.org/10.1139/z80-302
Southall BL, Casey C, Holt M et al (2019) High-amplitude vocalizations of male northern elephant seals and associated ambient noise on a breeding rookery. J Acoust Soc Am 146:4514–4524. https://doi.org/10.1121/1.5139422
Stirling I (1973) Vocalization in the ringed seal (*Phoca hispida*). J Fish Res Board Can 30:1592–1594. https://doi.org/10.1139/f73-253
Stirling I, Siniff DB (1979) Underwater vocalizations of leopard seals (*Hydrurga leptonyx*) and crabeater seals (*Lobodon carcinophagus*) near the South Shetland Islands, Antarctica. Can J Zool 57:1244–1248. https://doi.org/10.1139/z79-160
Stirling I, Thomas JA (2003) Relationships between underwater vocalizations and mating systems in phocid seals. Aquat Mamm 29:227–246. https://doi.org/10.1578/016754203101024176
Stirling I, Calvert W, Cleator H (1983) Underwater vocalizations as a tool for studying the distribution and relative abundance of wintering pinnipeds in the high Arctic. Arctic 38:262–274
Terhune J (1994) Geographical variation of harp seal underwater vocalizations. Can J Zool 72:892–897. https://doi.org/10.1139/z94-121
Terhune JM (2019) The underwater vocal complexity of seals (Phocidae) is not related to their phylogeny. Can J Zool 97:232–240. https://doi.org/10.1139/cjz-2018-0190
Terhune JM, Ronald K (1973) Some hooded seal (*Cystophora cristata*) sounds in March. Can J Zool 51:319–321. https://doi.org/10.1139/z73-045
Terhune JM, Ronald K (1976) Examining harp seal behavioural patterns via their underwater vocalizations. Appl Anim Ethol 2:261–264. https://doi.org/10.1016/0304-3762(76)90058-4
Terhune JM, Burton H, Green K (1994) Weddell seal in-air call sequences made with closed mouths. Polar Biol 14:117–122. https://doi.org/10.1007/BF00234973
Thomas JA, DeMaster DP (1982) An acoustic technique for determining diurnal activities in leopard (*Hydrurga leptonyx*) and crabeater (*Lobodon carcinophagus*) seal. Can J Zool 60:2028–2031. https://doi.org/10.1139/z82-260
Thomas JA, DeMaster DP (1983) Diel haul-out patterns of Weddell seal (*Leptonychotes weddelli*) females and their pups. Can J Zool 61:2084–2086. https://doi.org/10.1139/z83-273
Thomas JA, Kuechle VB (1982) Quantitative analysis of weddell seal (*Leptonychotes weddelli*) underwater vocalizations at McMurdo Sound, Antarctica. J Acoust Soc Am 72:1730–1738. https://doi.org/10.1121/1.388667
Thomas JA, Stirling I (1983) Geographic variation in the underwater vocalizations of Weddell seals (*Leptonychotes weddelli*) from Palmer Peninsula and McMurdo Sound, Antarctica. Can J Zool 61:2203–2212. https://doi.org/10.1139/z83-291
Thomas JA, Fisher SR, Evans WE, Awbrey FT (1983) Ultrasonic vocalizations of leopard seals (*Hydrurga leptonyx*). Antarct J 17:186
Thomas JA, Golladay CL (1995) Geographic variation in leopard seal (*Hydrurga leptonyx*) underwater vocalizations. In: Sensory systems of aquatic mammals, pp 201–221
Van Opzeeland IC, Van Parijs S (2004) Individuality in harp seal, *Phoca groenlandica*, pup vocalizations. Anim Behav 68:1115–1123
Van Parijs SM, Clark CW (2006) Long-term mating tactics in an aquatic-mating pinniped, the bearded seal, *Erignathus barbatus*. Anim Behav 72:1269–1277. https://doi.org/10.1016/j.anbehav.2006.03.026
Van Opzeeland IC, Corkeron PJ, Risch D et al (2009) Geographic variation in vocalizations of pups and mother-pup behavior of harp seals *Pagophilus groenlandicus*. Aquat Biol 6:109–120. https://doi.org/10.3354/ab00170
Van Opzeeland I, Van Parijs S, Bornemann H et al (2010) Acoustic ecology of antarctic pinnipeds. Mar Ecol Prog Ser 414:267–291. https://doi.org/10.3354/meps08683

Van Opzeeland IC, Van Parijs S, Frickenhaus S, Kreiss CM, Boebel O (2012) Individual variation in pup vocalizations and absence of behavioral signs of maternal vocal recognition in Weddell seals (*Leptonychotes weddellii*). Mar Mamm Sci 28:158–172. https://doi.org/10.1111/j.1748-7692.2011.00505.x

Van Parijs SM, Janik VM, Thompson PM (2000a) Display-area size, tenure length, and site fidelity in the aquatically mating male harbour seal, *Phoca vitulina*. Can J Zool 78:2209–2217. https://doi.org/10.1139/cjz-78-12-2209

Van Parijs SM, Hastie GD, Thompson PM (2000b) Individual and geographical variation in display behaviour of male harbour seals in Scotland. Anim Behav 59:559–568. https://doi.org/10.1006/anbe.1999.1307

Van Parijs SM, Kovacs KM, Lydersen C (2001) Spatial and temporal distribution of vocalising male bearded seals—implications for male mating strategies. Behaviour 138:905–922. https://doi.org/10.1163/156853901753172719

Van Parijs SM, Lydersen C, Kovacs KM (2003) Vocalizations and movements suggest alternative mating tactics in male bearded seals. Anim Behav 65:273–283. https://doi.org/10.1006/anbe.2003.2048

Vincent C, Meynier L, Ridoux V (2001) Photo-identification in grey seals: Legibility and stability of natural markings. Mammalia 65:363–372

Van Parijs SM, Kovacs KM (2002) In-air and underwater vocalizations of eastern Canadian harbour seals, Phoca vitulina. Can J Zool 80(7):1173–1179

Wartzok D, Ketten D (1999) Marine mammal sensory systems. In: Reynolds J, Rommel S (eds) Biology of marine mammals. Smithsonian Instituition Press, Washington DC, pp 117–175

Wartzok D, McCormick M (1978) Color discrimination by a Bering Sea spotted seal, *Phoca largha*. Vision Res 18:781–784

Watkins WA, Ray GC (1977) Underwater sounds from ribbon seal, *Phoca* (Histriophoca) *fasciata*. Fish Bull 75:450–453

Watkins WA, Schevill WE (1968) Underwater playback of their own sounds to Leptonychotes (Weddell seals). J Mammal 49:287–296

Watkins WA, Wartzok D (1985) Sensory biophysics of marine mammals. Mar Mamm Sci 1:219–260

Wheatley KE, Bradshaw CJA, Harcourt RG, Hindell MA (2008) Feast or famine: evidence for mixed capital–income breeding strategies in Weddell seals. Oecologia 155:11–20. https://doi.org/10.1007/s00442-007-0888-7

Wierucka K, Pitcher BJ, Harcourt R, Charrier I (2017) The role of visual cues in mother–pup reunions in a colonially breeding mammal. Biol Lett 13:20170444. https://doi.org/10.1098/rsbl.2017.0444

Wyatt TD (2003) Pheromones and animal behaviour: communication by smell and taste. Cambridge University Press, Cambridge

Wyatt TD (2014) Pheromones and animal behavior: chemical signals and signatures. Cambridge University Press, Cambridge

Chapter 4
Physiological Capacity and Constraint Impact Behavioral Phenotype in Phocid Seals

Daniel E. Crocker and Birgitte I. McDonald

Contents

Abstract The physiological challenges required for air-breathing phocids to forage and survive in the marine environment and then breed on land or ice have resulted in behavioral phenotypes that are highly constrained by physiological adaptation. Foraging behavior is enabled and constrained by species differences in the ability to withstand pressure, store oxygen, and manage oxygen consumption while swimming. Their reproductive behavior reflects a continuum of income to capital breeding strategies that include some of the highest sustained energy expenditure rates relative to body size found in nature while fasting. Metabolic adaptations for fasting while lactating, locomoting, and developing are critical components of most species' life-history strategies. Depending on the habitat, thermoregulatory adaptations that facilitate heat retention at depth may impose significant limitations on terrestrial behavior. Studies on metabolic and behavioral endocrinology suggest important activational roles for organizing behavior but are relatively unstudied in most species. For these reasons, phocids are an important taxon for comparative studies that examine the evolution of behavior and the role that performance capacity plays in selection for variation in life histories and behavioral strategies among closely related species.

Keywords Dive capacity · Physiological constraints · Dive response · Extended fasting

D. E. Crocker (✉)
Department of Biology, Sonoma State University, Rohnert Park, CA 94928, USA
e-mail: crocker@sonoma.edu

B. I. McDonald
Moss Landing Marine Laboratories, San Jose State University, Moss Landing, CA 95039, USA

D. P. Costa and E. McHuron (eds.), *Ethology and Behavioral Ecology of Phocids*,
Ethology and Behavioral Ecology of Marine Mammals,
https://doi.org/10.1007/978-3-030-88923-4_4

4.1 Physiological Capacities and Constraints

Analyses of the adaptive functions of animal behaviors frequently consider the relative fitness benefits and costs of variation in behavioral phenotypes exhibited by individuals. In contrast, studies of behavior in pinnipeds frequently focus on physiological constraints that limit behaviors and variations in performance characteristics necessary to enable behavioral phenotypes. For example, investigations of foraging behaviors in terrestrial mammals often focus on behaviors that optimize time/energy budgets or that minimize predator exposure. In contrast, studies on foraging in phocids might focus on the breath-hold abilities that constrain the structure of behavior (Carbone and Houston 1996; Costa et al. 2001; Mori 1999). These differences reflect the high degree of physiological adaptation required for air-breathing mammals to reinvade the aquatic environment and the unique amphibious life-histories of pinnipeds that combine marine foraging with terrestrial reproduction. The physiological capabilities of phocids have evolved as components of life-history strategies and align with reproductive and foraging strategies used by various species. This importance of physiological capability is particularly strong in capital breeding phocids. In some species, the more distinct separation of terrestrial reproduction and aquatic foraging has enabled life-history patterns that prioritize extreme fasting and enable foraging over wide spatial and temporal scales.

Strong selection for performance characteristics that enable air-breathing endotherms to be successful in the marine environment and then to forgo foraging during all or some component of reproduction has led to a tight association between physiological features and the diversity of behavioral strategies they support. For this reason, phocids are an excellent group to examine the evolution of behavior and the role that adaptations in physiology, biochemistry, and morphology play in selection for variation in life histories and behavioral strategies among closely related species. For simplicity, we lump physiology, biochemistry, and morphology together as 'physiology'. In doing so, we do not intend to ignore the distinctions between those components but emphasize the suite of organismal features that contribute to performance capabilities. In this view, interspecies and individual variation in physiology enables and constrains the behavior on which selection acts based on fitness gradients specific to the habitats and ecological variables encountered by populations (Fig. 4.1;

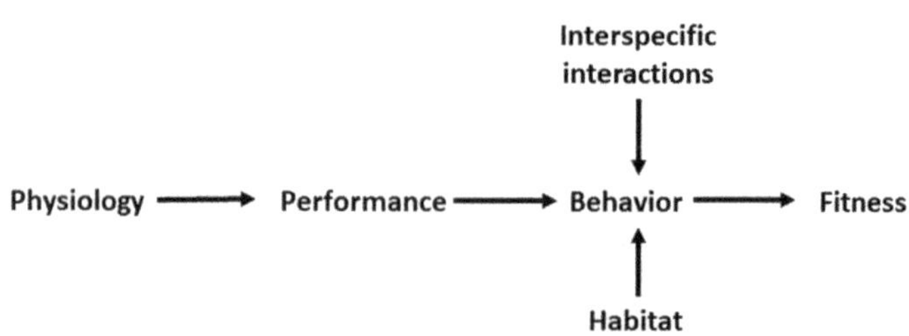

Fig. 4.1 Relationship between physiology, behavior, and fitness. Modified after: Garland and Losos (1994)

Garland and Losos 1994).

In this chapter, we consider several critical physiological challenges to phocids, to which the various taxa reflect differing levels of physiological adaptation and performance. Our focus is on the role that adaptation to these challenges has played in shaping behavioral strategies. Each of the challenges can be viewed as a set of conflicting physiological demands that constrain behavior. These challenges are: (1) the need to provide oxygen to support aerobic metabolism and avoid pressure-related injuries or illnesses during diving and foraging; (2) the need to endure variable periods fasting from food and water while undergoing energy-expensive behaviors like lactation and mating, and (3) the challenges of thermoregulating in the diverse thermal regimes required by the amphibious lifestyle of phocids. In the final part of this chapter, we review the sparse studies on the behavioral endocrinology of pinnipeds, focusing on reproductive and stress hormones, and identify critical areas where more research is needed.

4.2 Diving Physiology

Phocids, like all marine mammals, must hold their breath while diving to depth to find food. This presents two primary physiological challenges: (1) increased pressure at depth and (2) an oxygen-limiting environment. An animal's ability to deal with these challenges determines its dive capacity, which influences its foraging ecology and role in the ecosystem. Many of the pioneering diving physiology studies used phocids as model species. The initial studies forcefully submerged restrained seals while measuring a suite of physiological variables (Scholander et al. 1942; Scholander 1940). In the 1960s, Dr. Gerald Kooyman recognized a unique opportunity to study physiology during more natural diving of Weddell seals in Antarctica using the isolated dive hole protocol, which ensured that seals returned to a specific location for sampling and data logger retrieval (Kooyman 1965; Chap. 13). Studies with captive animals in the laboratory also provided opportunities to study physiology during trained dives (Williams and Kooyman 1985; Fedak et al. 1988; McKnight et al. 2019). With the advancement in data logging technologies over the last few decades, it is now possible to study the diving physiology of free-ranging wild phocids (Meir et al. 2009; Thompson and Fedak 1993). Research on phocids has provided foundational knowledge about the diving physiology of marine mammals.

Phocids exhibit a range of diving behaviors and capabilities. Most seals dive to depths that are less than 100 m and shorter than 10–15 min in duration but can dive deeper and longer when needed (summarized in Ponganis 2015; Chaps. 9–16). The largest phocids (i.e., Weddell (*Leptonychotes weddellii*), hooded (*Cystophora cristata*), northern elephant (*Mirounga angustirostris*), and southern elephant seals (*Mirounga leonina*) are the longest and deepest divers, with dives routinely exceeding depths of 400 m and 15 min in duration. The southern elephant seal holds the record for the deepest (2388 m) and the longest dive (120 min) for a pinniped (Costa et al. 2010; Hindell et al. 1991). What is even more impressive is that elephant seals and hooded seals spend 80–90% of their foraging migrations submerged. This contrasts with most other phocids that tend to perform bouts of dives with extended periods resting at the surface or on land or ice. These differences in behavior are likely to be associated with different physiological adaptations and mechanisms that allow them to dive to depth and manage oxygen (O_2) while transiting, hunting, and digesting at sea.

4.2.1 The Pressure Problem

The increased pressure associated with diving to depth leads to two challenges: (1) gas compression that can lead to tissue trauma and (2) increased solubility of gases in the blood at higher pressure, which can lead to elevated blood nitrogen (N_2) levels and increase the risk of decompression sickness. Compression effects are explained by Boyle's Law, which states that the product of volume and pressure on a fixed mass of gas is constant. For example, 1-L of air at the surface ($1 \text{ atm} \times 1 \text{ L} = 1 \text{ L}$) would be compressed to 0.5-L at 10 m of depth ($2 \text{ atm} \times 0.5 = 1 \text{ L}$). Elevated blood N_2 is explained by Henry's Law, which states that the amount of dissolved gas in a liquid is proportional to its partial pressure above the liquid (higher at depth).

4.2.1.1 Barotrauma

Marine mammals have anatomical adaptations that protect against barotrauma and tissue trauma caused by rapid or extreme ambient pressure changes that typically affect enclosed cavities such as the sinuses, middle ear, and lungs. Pinnipeds lack air-filled cranial sinuses present in most mammals. Additionally, the middle ear cavities are lined with venous plexuses that presumably engorge with blood at depth, eliminating air space (Odend'Hal and Poulter 1966; Stenfors et al. 2001).

Phocids also have anatomical modifications that prevent lung squeeze (also known as pulmonary barotrauma of descent). Lung squeeze occurs when the lung air volume is reduced within the incompressible rib cage, potentially leading to tissue damage. Phocids avoid lung squeeze by having highly compliant chest walls that provide little resistance to compression (Fahlman et al. 2014). Phocids also have modified airway structures, although they have the least modified airways (compared to terrestrial ancestors) among marine mammals. Some species have significant cartilaginous reinforcement of the trachea, and in all species, the bronchioles are reinforced with muscle. This reinforcement ensures that the alveoli (small air sacs where gas exchange occurs) collapse first and push air into the more rigid regions of the lung and respiratory system where gas exchange does not take place. This is important for avoiding some of the problems associated with increased solubility of gases at higher pressure, described below. There are differences among species, with deeper diving species tending to have more compressible airways (Moore et al. 2011). To facilitate rapid alveoli expansion as pressure decreases, marine mammals have modified lung surfactants with anti-adhesive properties (Miller et al. 2006).

4.2.1.2 Decompression Sickness, Nitrogen Narcosis, High-Pressure Nervous Syndrome

As the partial pressure of N_2 (P_{N2}) in the lung increases with depth, N_2 absorption into the blood and tissues increases. The elevated P_{N2} can lead to decompression sickness (DCS), which arises from dissolved gases coming out of the solution as bubbles in blood and tissues during rapid depressurization. Venous bubbles may be associated with circulatory obstruction, platelet aggregation, activation of coagulation, and tissue inflammation. Tissues that have high N_2 solubility but low perfusion are susceptible to localized tissue damage and hemorrhage.

Because breath-hold divers only take limited air to depth and N_2 absorption is minimized at depth due to alveoli collapse and decreased heart rate (see Sect. 4.2.2.2), it was thought marine mammals would have a low risk of DCS. However, studies found that seals and dolphins can accumulate high levels of N_2 after repeated dives (Kooyman et al. 1972; Ridgway and Howard 1979). More recently, signs of DCS, such as N_2 bubbles, have been documented in mass strandings of cetaceans associated with naval sonar and in bycaught animals (de Quirós et al. 2013; Jepson et al. 2003). There is little evidence of DCS in seals during natural diving, potentially because most species dive on exhalation and therefore minimize N_2 available for gas exchange on a dive. It is also possible that some species have behavioral mechanisms that reduce risk, such as decreasing the rate of ascent to facilitate offloading or spending time at the surface after a dive bout to offload N_2 before initiating another dive bout.

Other pressure-related illnesses that impact humans are nitrogen narcosis (anesthetic effect of N_2 at high pressure that can lead to drowsiness and loss of consciousness at deeper depths) and High-Pressure Nervous Syndrome (HPNS—tremors and convulsions caused by hydrostatic pressure). There is no evidence that phocids suffer either of these afflictions. The symptoms of nitrogen narcosis, usually associated with compressed air breathing in humans, start to appear at P_{N2} of 2400 mm Hg (~30 m depth); therefore, nitrogen narcosis is likely not a problem when breath-hold diving. In humans, HPNS symptoms appear at pressures exceeding 11 ATM (~100 m; Jain 1994). Many phocids exceed this depth for durations beyond those required for the onset of HPNS in humans, indicating that marine mammals could be susceptible to HPNS. It has been suggested that the elevated N_2 in the tissues at depth may mitigate symptoms of HPNS (Halsey 1982). In terrestrial mammals, anesthetic gases reduce symptoms of HPNS and increase pressure tolerance. Additionally, the critical P_{N2} for the prevention of HPNS was less than the P_{N2} that causes narcosis in humans (Halsey 1982). Although we are beginning to improve our understanding of how phocids tolerate pressure, there is still much we do not know.

4.2.2 The Oxygen Problem—Breath-Hold Diving

Phocids can perform long-duration dives, often only spending short periods at the surface between dives. The key adaptations to performing these long dives are (1) enhanced body O_2 stores, (2) reduced rate of O_2 consumption while on a dive, and (3) tolerance to hypoxia (low O_2).

4.2.2.1 Oxygen Stores

Phocids, like most marine mammals, have elevated body O_2 stores compared to terrestrial mammals (Fig. 4.2). These enhanced O_2 stores are distributed among the lungs, blood, and skeletal muscle. Compared to terrestrial and some marine mammals, the lung O_2 store of phocids is only a small percentage of total body O_2 stores despite having lung volumes similar to terrestrial mammals (Fig. 4.2). For example, about 42% of human total body O_2 stores are in the lungs compared with ~3–12% in seals (Fahlman et al. 2011; Kooyman 1973). This is because (1) most phocids exhale before a dive, so it is estimated they are diving at 50% total lung capacity, and (2) O_2 storage capacity in blood and muscle is elevated.

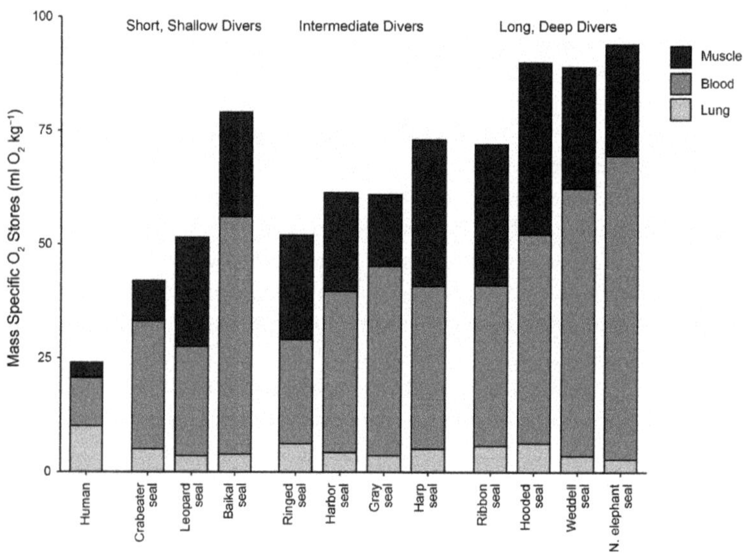

Fig. 4.2 Magnitude and distribution of O_2 stores in phocids. Deep, long divers tend to have greater O_2 stores, primarily located in the blood and muscle. Values obtained from: Burns et al. (2005), Fowler et al. (2007), Hassrick et al. (2010), Kuhn et al. (2006), Lenfant et al. (1970), Lydersen et al. (1992), Neshumova et al. (1983), Neshumova and Cherepanova (1984), Noren et al. (2005), Ponganis (2011), Ponganis et al. (1993, 1997), Thorson and Le Boeuf (1994)

The blood O_2 store is often the largest in phocids, although in a few species, blood and muscle O_2 stores are similar (Fig. 4.2). Both blood volume and hemoglobin concentration (Hb), the molecule in blood that binds with O_2, are elevated compared to terrestrial mammals. Mass-specific blood volumes and Hb concentrations are 2–3 and 1.3–2 times greater than in humans, respectively. As a result, phocids store 2–6 times more O_2 in blood than non-diving mammals. The greatest blood volumes and Hb concentrations occur in the longest duration divers, although there are a few exceptions, likely linked to the environment. For example, although the Baikal seal (*Pusa sibirica*) typically performs short dives (<8 min, Watanabe et al. 2004, 2006) in winter, it may be necessary to perform long dives while searching for breathing holes in the ice. This, combined with its small body size, may explain the elevated blood and muscle O_2 stores compared to other short-duration divers. A unique anatomical feature in seals is a large spleen (~1–7% of body mass compared to 0.2–0.4% in most terrestrial mammals) which can sequester 30% of blood volume or 50% of red blood cells when seals are not diving. The spleen contracts during dives, increasing hematocrit, and thereby increasing blood O_2 stores. The ability to sequester red blood cells during periods of inactivity results in lower hematocrit, decreasing blood viscosity, thereby reducing cardiac workload. There is also evidence that pinnipeds can enhance their blood O_2 stores by arterialization of venous blood before or during the beginning of dives (McDonald and Ponganis 2013; Meir et al. 2009). These elevated pre- and early-dive venous O_2 saturation values suggest the use of arterial-venous shunts that direct arterial blood directly to the venous system, bypassing active tissues (McDonald and Ponganis 2013). This shunting could occur in the well-described arteria-venous anastomoses in the skin and blubber of elephant seals (Molyneux and Bryden 1978) or potentially in undescribed shunts in other locations.

Muscle is a significant O_2 store in phocids. Myoglobin (Mb), the molecule that binds O_2 in muscle, is ~15–30 times higher than in terrestrial mammals (reviewed in Ponganis 2015). The Mb concentration is highest in the long-duration divers such as ribbon (*Histriophoca fasciata*), hooded, Weddell, and elephant seals.

4.2.2.2 Reduce O_2 Requirements

Phocids have physiological and behavioral adaptations that reduce the rate at which they use O_2 on a dive, which extends the duration an animal can dive aerobically. The dive response is a crucial mechanism allowing breath-hold divers to perform deep and long dives via cessation of breathing, bradycardia (decrease in heart rate to below resting), and peripheral vasoconstriction, conserving O_2 for the heart and brain (Blix et al. 1983; Scholander 1940). Cardiovascular regulation is critical during diving because changes in heart rate and cardiac output not only affect blood O_2 uptake from the lung but also the rate and magnitude of blood O_2 delivery to tissues. Heart rate, blood flow distribution, and muscle workload are the primary determinants of the rate and pattern of O_2 store use and, ultimately, breath-hold duration (Davis and Kanatous 1999; Ponganis 2015; Scholander 1940). Additionally, heart rate may be necessary for limiting N_2 uptake from non-collapsed lungs and enabling critical unloading of super-saturated tissues during ascent and at the surface (Fahlman et al. 2006; Hooker et al. 2009, 2012).

The dive response was first characterized in forcibly submerged animals by an immediate and intense bradycardia and peripheral vasoconstriction, which limits blood flow to non-essential organs and working muscle, retaining O_2 for critical tissues such as the brain and heart. This initially led to the concept of a diving reflex and the conclusion that diving was primarily anaerobic (Scholander 1940; Zapol et al. 1979). Later studies on trained and freely diving animals documented that while the decrease in heart rate is immediate, cardiovascular regulation is often more moderate and variable, with decline in heart rate depending on species, dive duration, activity, and expectations (Fig. 4.3; Davis and Williams 2012; Jobsis et al. 2001; Kooyman and Campbell 1972; McKnight et al. 2019; Thompson and Fedak 1993). For example, in wild gray seals (*Halichoerus grypus*) the mean dive heart rate was 25% of that at the surface during short 5-min dives. However, in dives longer than 15 min, mean dive heart rates were <10% of surface heart rates, occasionally less than 5 beats min^{-1} for several minutes (Thompson and Fedak 1993). These low-diving heart rates are similar to those during forced submersions, indicating that seals can exhibit extreme bradycardia even under natural diving conditions. A similar negative relationship between dive duration and heart rate was observed in elephant seals and Weddell seals (Fig. 4.3; Andrews et al. 1997; Hill et al. 1987; Kooyman and Campbell 1972). Heart rate is also influenced by exercise despite remaining low throughout the dive, as indicated by the positive relationship between stroke rate and heart rate in Weddell

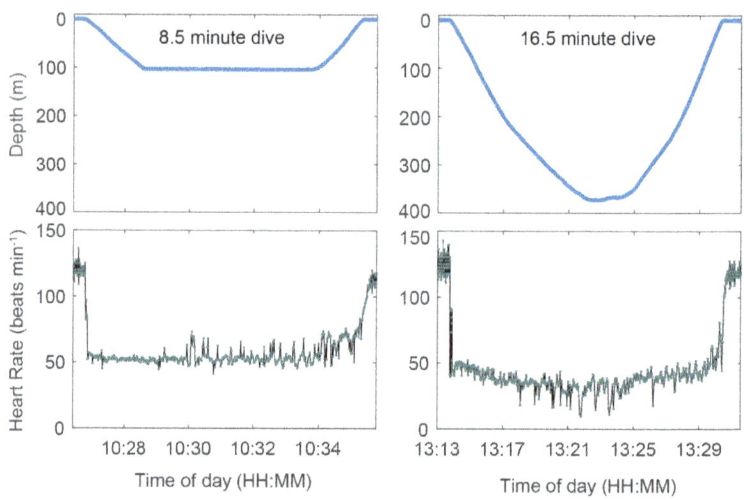

Fig. 4.3 Dive and heart rate profiles from a translocated juvenile elephant performing an 8.5-min shallow dive and a 16.5-min deeper dive. In both the shallow and deeper dive, heart rate drops rapidly at the initiation of the dive, stays relatively constant during the bottom of the dive, and slowly increases during ascent. In the longer, deeper dive, bottom heart rate is lower than in the shorter shallower dives as observed in most phocids studied (unpublished data, NSF grant #'s: 1656077 (B. McDonald), 1656822 (C. Williams), and 1921742 (A Hindle))

seals (Davis and Williams 2012). In addition to diving behavior influencing heart rate, seals may modify heart rate based on expectations. Volition control of heart rate was observed in the early forced submersion studies in seals (Grinnell et al. 1942). For example, seals increased heart rate before forced dives when the experimenter raised their hand (Grinnell et al. 1942). Since these early studies, researchers have continued to document the ability of wild and captive freely diving pinnipeds to fine-tune heart rate based on expectations (Kooyman and Campbell 1972; Jobsis et al. 2001; McDonald and Ponganis 2014). Recently, preparatory peripheral vasoconstriction accompanied by increased cerebral blood volume before submersion has been documented in freely diving harbor seals (*Phoca vitulina*), confirming that cardiovascular adjustments associated with diving are under some degree of cognitive control (McKnight et al. 2019).

While muscle is a vital oxygen store in diving animals, little is known about Mb desaturation in freely diving animals. In forcibly submerged animals, Mb desaturation was steady and rapid (Scholander et al. 1942), suggesting isolation from the central circulation. Mb saturation has only been measured in one phocid, the Weddell seal. Desaturation was generally slower in longer dives with evidence of muscle re-perfusion towards the end of some of the longer dives; however, there was no relationship between swimming effort and desaturation rate. Unfortunately, the sensor was not located on a primary swimming muscle, where this relationship would be more apparent (Guyton et al. 1995). Such a combination of sensory modulation and volitional control of the diving response has likely been selected for during the evolution of marine mammals by enabling them to fine-tune the management of their O_2 and N_2 levels to maximize net energy gains during foraging dives. Please see Ponganis (2015) and Davis (2019) for more detailed reviews of the cardiovascular response to diving.

Phocids also use behavioral strategies to minimize O_2 consumption when diving. Rather than stroking continuously during a dive, they use burst-and-glide swimming and extended glides to minimize locomotion's energetic costs (Williams et al. 2000, 2015). Two factors that facilitate the use of these efficient swimming strategies are drag reduction and buoyancy control. Seals have a streamlined fusiform that minimizes drag. This shape, combined with the buoyancy changes associated with lung compression at depth, allows them to maintain forward motion without active stroking. Weddell and northern elephant seals glide for over 70% of the descent in dives deeper than 200 m (Williams et al. 2000). In Weddell seals, gliding reduces the energetic costs of diving by 9–60%, depending on dive depth (Williams et al. 2015). In elephant seals, the round-trip strokes-per-meter reached a minimum value when seals were neutrally buoyant, allowing them to stay longer at foraging depths (Adachi et al. 2014).

4.2.2.3 Aerobic Dive Limit

The aerobic dive limit (ADL) has become an important concept used by ecologists and physiologists to interpret the foraging ecology and estimate the diving capacity of breath-hold divers (e.g., Boyd et al. 1994; Burns 1999; Costa et al. 2004). The ADL is defined as the maximum dive duration supported by aerobic metabolism, beyond which anaerobic metabolism is required, which is associated with an increase in post-dive blood lactate (Kooyman et al. 1980). This concept was first documented in Weddell seals diving at an isolated dive hole (Chap. 13). Dives greater than ~20 min in duration resulted in a post-dive increase in blood lactate, with the longest dives having the highest lactate levels. After these long dives, the seals often spent an extended time at the surface, likely to clear the lactate. Free-ranging Weddell seal dives rarely exceeded the ADL, supporting the hypothesis that most dives are aerobic, enabling repetitive dives with a short recovery between dives to replenish blood and muscle O_2 stores.

While the ADL is one of the most frequently cited concepts in diving physiology, it has only been measured in two seal species (Weddell and Baikal seals) due to the technical challenges of obtaining a blood sample at the end of dives (Kooyman et al. 1980; Ponganis et al. 1997). However, cADL has been estimated by dividing total body O_2 stores by diving metabolic rate (calculated ADL—cADL). While this equation is useful for estimating the onset of lactate accumulation, it does not indicate complete depletion of total body O_2 stores, but that muscle O_2 is depleted (Butler 2006). For many species, dives exceeding the cADL are rare and followed by an extended surface interval as first observed in the Weddell seal. However, gray and elephant seals occasionally exceed the cADL without exhibiting extended surface intervals. These long dives are often followed by a series of short shallow dives in gray seals (Thompson and Fedak 1993). However, elephant seals can repeatedly exceed their cADL without extended surface intervals (Hindell et al. 1992; Hassrick et al. 2010), which implies that lactate can be metabolized during subsequent aerobic dives. These species' differences in breath-hold capacity impact their habitat use and access to preferred prey.

4.3 Fasting Physiology

Phocid seals forage on marine prey but reproduce on land (or ice). This has led to extreme capital breeding strategies in several species where females support high energy density milk synthesis entirely from stored body reserves (Costa 1991; Chap. 8). Similarly, in species where mating occurs on land (e.g., elephant seals and gray seals), males fast for extended periods while defending groups of females. Despite fasting, successful males can approach sustained energy expenditure levels, suggested as 'metabolic ceilings' in species that feed during reproduction (Crocker et al. 2012a). Lastly, weaned pups may undergo development after provisioning that requires extended periods of fasting while developing prior to the initiation of independent foraging. These behavioral strategies lead to separation of energy acquisition from energy allocation for maternal investment, competition for mates, and during juvenile development. The intensity of this separation and the resultant importance of fasting ability varies among species. This life-history characteristic is associated with metabolic adaptations for extended fasting while supporting the energetic and nutrient requirements of necessary behaviors.

Despite the importance of fasting ability for fitness in many phocids, detailed studies of fasting adaptations are rare in adults and primarily limited to two land breeding phocids, northern elephant seals and gray seals. Fasting under high nutrient demands is sustained by large body size, high fat reserves, high lipid mobilization rates, and reduced oxidation of carbohydrates and proteins to support energy metabolism (Champagne et al. 2012a). Large body size has potential impacts on fasting endurance due to metabolic scaling. Metabolic intensity (energy requirements per unit of mass) exhibits negative metabolic scaling with body size (Glazier 2014), while fat reserves vary as the 1.19 power of mass (Lindstedt and Boyce 1985), suggesting that enhanced fasting endurance may be a driver in the evolution of large body size in phocids and influence the evolution of species lactation strategies (Schulz and Bowen 2005). The importance of lipid reserves to phocids is evident in strong impacts of adiposity on reproductive parameters, including reproductive effort, mating success, parental investment, and milk composition (Crocker et al. 2001, 2012a; Lidgard et al. 2005; Chap. 8).

Several key features are associated with sustaining fasting during high rates of nutrient demands (Champagne et al. 2012a). The most important of these are efficient protein sparing and the predominant use of mobilized lipids for energy metabolism. Direct impact of adiposity on the ability to spare protein while fasting has been demonstrated in several species (Adams and Costa 1993; Bennett et al. 2007; Carlini et al. 2001; Crocker et al. 1998; Lidgard et al. 2005). Protein sparing helps protect vital organs and skeletal muscle from catabolism during fasting. Even during the most extreme example of male elephant seals maintaining high metabolic rates for several months of competitive female defense, protein catabolism supports only ~7% of energy expenditure (Crocker et al. 2012a). This lipid-based metabolism is facilitated by maintaining diabetic-like conditions that promote lipid mobilization and reduce the use of carbohydrates, including low insulin levels, low glucose clearance, and limited tissue-specific insulin sensitivity (Houser et al. 2013).

The primary challenge to sparing body protein while fasting is related to the limited ability to store carbohydrates as glycogen. During extended fasting, providing carbohydrates to glucose-dependent tissues like the nervous system and erythrocytes and essential glucose-dependent processes like wound healing or immune responses (Banuet-Martínez et al. 2017) requires the production of new sugar or gluconeogenesis. The primary substrate used to provide carbon for glucose synthesis is amino acids mobilized from vital organs or skeletal muscle. It is this protein catabolism that ultimately leads to starvation and death. To avoid this loss of vital body protein, most domestic animals respond to fasting by reducing gluconeogenesis (Champagne et al. 2012a). In contrast, fasting phocids have high plasma glucose concentrations and high glucose production rates that greatly exceed the needs of glucose-dependent tissues and are similar to post-adsorptive values in other species (Fig. 4.4; Champagne et al. 2012b). This surprising feature is maintained through high glucose recycling, where most glucose is committed to glycolysis and converted to lactate, probably

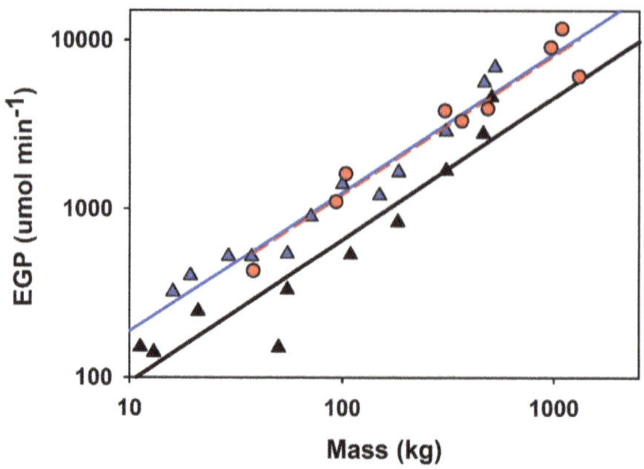

Fig. 4.4 Northern elephant seals lack the characteristic suppression of endogenous glucose production (EGP) seen in non-fasting adapted species. Blue triangles and line are post-absorptive measurements. Black triangles and line are from fasting measurements. Red circles and dotted line are group means from elephant seal studies. Other data and solid lines are from domestic animals. Data are from Champagne et al. (2012a) and references therein

by adipocytes and erythrocytes, and then reconverted to glucose by the liver (Davis 1983; Crocker et al. 2017; Tavoni et al. 2013).

In many domestic animal species, sustained fat-based metabolism can lead to the accumulation of ketoacids, a by-product of fatty acid catabolism, and eventually ketoacidosis and associated pathologies (Burge et al. 1993). Despite high rates of beta-oxidation of fatty acids to support metabolism, phocid seals exhibit comparatively low levels of ketoacids (Champagne et al. 2012a). Plasma concentrations of beta-hydroxybutyrate, the predominant ketoacid, increase significantly during the fast in most phocid species studied, but these changes are small compared to other species exposed to moderate or even short fasting durations. In elephant seals, avoidance of ketoacidosis was associated with several unique features of carbohydrate metabolism. High rates of glucose production were associated with high levels of pyruvate cycling and tricarboxylic acid cycle (TCA) cycle activity, the primary source of oxidative ATP production. Glucose production, pyruvate cycling, and TCA cycle activity were tightly correlated throughout the fast, much more so than has been observed in other species (Champagne et al. 2012b). High pyruvate cycle activity may provide a regulatory point for glucose production and replace TCA cycle intermediates without using carbon from amino acids derived from lean tissue stores, thereby helping to spare protein. This feature also may facilitate greater ATP production from fatty acids while minimizing the production and accumulation of ketoacids (Crocker et al. 2014). This system is reflected in a strong negative association between glucose production and circulating ketoacids in breeding male elephant seals.

Together, these findings suggest dramatic alterations in metabolism and endocrine regulation during fasting compared to domestic species and other wildlife systems. These features are likely critical to life-history strategies that to variable degrees separate foraging from reproduction. Large body size and exceptional fasting ability enable pure capital breeding in some species and allow foraging at larger spatial and temporal scales.

4.4 Thermoregulation

Thermoregulation in phocids is achieved through a combination of physiological and behavioral mechanisms. In adults, heat loss in cold water or air is minimized by low surface-to-volume ratios, subcutaneous and insulative blubber, and countercurrent heat exchangers (Liwanag et al. 2012). For this reason, adult phocids have broad thermal neutral zones (Hokkanen 1990), and calculated lower critical temperatures are often below the freezing temperature of water (Noren 2002). This means adults are well adapted for thermoregulation in cold water. However, most phocid pups are born with little or no blubber and rely instead on a lanugo (natal fur) that provides more insulation in the air than for a similar thickness of blubber (Ryg et al. 1993). In species where there is potential for early immersion in cold water, as in some pack ice seals or seals that enter the water early in development (e.g., spotted seals or hooded seals), lanugo is shed in utero, and pups are born with a more developed blubber layer (Oftedal et al. 1991). Habitat and early water immersion may also be associated with differences in non-shivering thermogenesis capacity as reflected in brown adipose tissue stores, expression of uncoupling proteins, and metabolic enzyme concentrations in skeletal muscle (Pearson et al. 2014).

Molting of fur requires skin perfusion and appropriate temperatures for the function of epidermal cells that may increase heat loss in cold water and force some species to haul out during molting (Boily 1995). This need to return to shore may influence foraging and haulout patterns (Paterson et al. 2019; Reder et al. 2003). In some species, this constraint has led to a 'catastrophic' molt, where the entire pelage and epidermis are rapidly shed and replaced on-shore. In Weddell seals, thermoregulatory costs were more than double during molting than pre-molt (Walcott et al. 2020). In sub-Antarctic environments, southern elephant seals may exhibit social aggregation and huddling to reduce convective heat loss and maintain skin temperature during the molting process (Chaise et al. 2019). In this species, local weather and body size may influence thermoregulatory costs during molting, resulting in trade-offs between movement and thermoregulation (Chaise et al. 2018).

In temperate environments, the adaptations that facilitate thermoregulation in cold water may impose difficulties that constrain behavior when on land. This may be particularly true of breeding seals, where the necessity of males to remain with female harems or constraints on female movements with dependent pups may limit opportunities for behavioral thermoregulation. In gray seals, proximity to water, including tide pools, is one of the most important environmental features influencing female density, distribution, and behavior (Twiss et al. 2000; Chap. 9). Females risk separation from their pups and incur locomotor energy costs and aggressive interactions with conspecifics to enable weather-dependent access to tidepools (Twiss

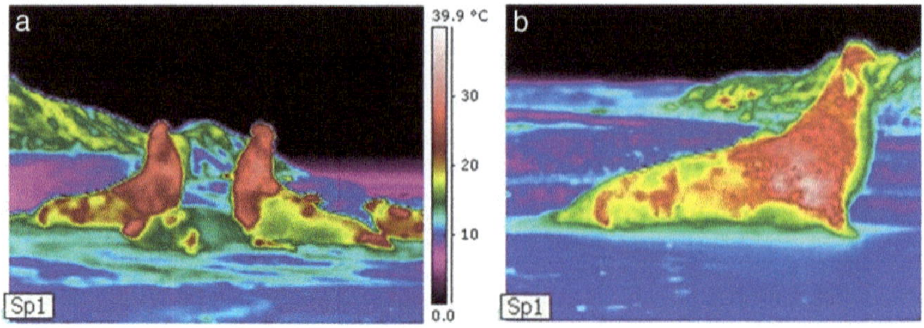

Fig. 4.5 Thermal images of northern elephant seals identify 'thermal windows', regions of the body where arterial-venous anastomoses pass large amounts of blood to superficial veins to bypass insulation. **a** Males have elevated skin temperature during competitive dominance interactions. In contrast, a resting female only has elevated temperature on her head and flippers. **b** Immediately after combat a male shows thermal windows and skin temperatures in excess of 40 °C

et al. 2002). Similarly, elephant seal females breeding on a protected cove beach with dark substrate undertook long movements into the water with their pups on warm days with high solar radiation, risking drowning of their pups and impacting their growth efficiency (Codde et al. 2016; Hooper et al. 2019).

Large body size and insulation present thermal challenges during agonistic encounters in temperate environments. The high rate of metabolic heat production in competitive male elephant seals resulted in substantial impacts of environmental features on behavior. On warm days with high solar radiation and low wind speeds, males were less likely to exhibit locomotion, dominance interactions, or copulate (Norris et al. 2010). In addition, thermoregulatory costs appeared to be an essential component of dominance interactions. Males that won fights exhibited thermal stress, actively released heat through arterial venous-anastomoses or 'thermal windows' that bypassed the blubber layer, and exhibited skin temperatures >41 °C (Fig. 4.5). In contrast, males that lost fights frequently moved directly into the water. Together, these findings suggest that thermoregulation and heat stress were important components of dominance interactions between breeding males.

4.5 Behavioral Endocrinology

Implicit in any animal behavior especially those that require discrimination of environmental or internal states is the organization of behavior through the nervous system and hormones' actions. Hormones play vital roles in organizing behaviors through organizational effects during development and activational effects on specific suites of behaviors (Fig. 4.6). Hormones help organize behaviors that can range from complex sensory processing to simple fixed action patterns that occur in response to environmental stimuli (releasers). For example, individual behavior may be markedly different during reproduction relative to other life-history stages due to the effects of

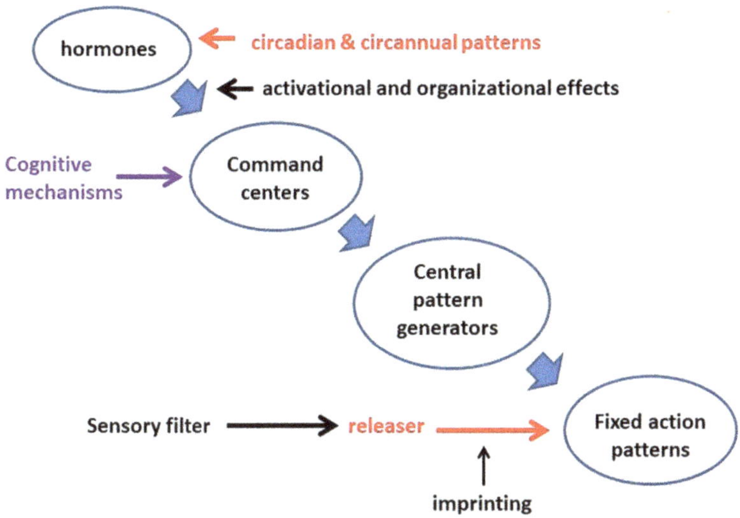

Fig. 4.6 Proximate mechanisms organizing behavior. Activational (immediate and temporary) and organizational (developmental) effects of hormones help structure behavioral phenotypes

reproductive hormones that modulate social behavior and aggressiveness. In addition to organizing behavior, hormones regulate metabolism to provide appropriate nutrients and energy substrates to support behaviors. Hormones are in turn affected by circadian and circannual patterns that help control the temporal patterning of behaviors. These patterns can vary from diel changes in foraging behavior or activity to annual breeding and estrous synchrony associated with polygyny in phocids.

Hormones are chemical messengers released into circulation by endocrine tissues. These chemical messengers coordinate individuals' physiology and behavior by altering gene expression and, in some cases, exerting non-genomic effects on regulatory proteins. Hormones exert activational and organizational effects on the nervous system that influence behavior to ensure reproductive success. For example, testosterone and estradiol are involved in sexual differentiation and gamete maturation, promote mating behavior and influence maternal behavior during lactation. Despite the importance of hormones in organizing behavior, behavioral endocrinology studies in phocid seals are mostly lacking.

4.5.1 *Metabolism*

Because of the importance of metabolic adaptation for phocid life-history strategies, considerable work has been done on endocrine regulation of metabolism, especially in land breeding species. While published studies emphasize regulation of metabolism, several associations suggest important links to behavior. For example, thyroid hormones regulate tissue energy expenditure and are associated with basal metabolic rate and in some cases field metabolic rate (FMR) of animals (Kim 2008). Furthermore, thyroid hormones are associated with activity patterns that vary with the life-history stage and may help regulate energy allocation and life-history trade-offs (e.g., self-maintenance vs. mating effort; Ketterson and Nolan 1992; Wilsterman et al. 2015). Because energy expenditure rates have important implications for diving and fasting capacity, thyroid hormones are likely to have an impact on phocid behavior.

In northern elephant seals, sex differences in FMR and activity were evident in yearlings and strongly associated with sex differences in thyroid hormones (Kelso et al. 2012). In breeding adult males, daily energy expenditure (DEE) was strongly associated with mating success (Crocker et al. 2012a). In turn, the ability to elevate triiodothyronine (T3), the active form of thyroid hormone, was positively associated with DEE, but at the expense of higher lean tissue catabolism, suggesting a trade-off between mating activity and reproductive costs that were mediated by thyroid hormone (Crocker et al. 2012b; Lee et al. 2017). In harbor seals, thyroid hormones impacted diving metabolic rate and aerobic breath-hold capacity (Weingartner et al. 2012), suggesting potential impacts of this hormone on foraging behavior.

4.5.2 Reproductive Endocrinology

Reproduction is associated with large magnitude changes in sex hormone concentrations. While these hormones' primary role is sexual development and reproduction, here we focus on associations of sex hormones with behavior. All phocid seals are seasonal breeders, and their estrous cycles are typically synchronized by parturition except for primiparous females or females that skip breeding. Behavioral estrous occurs late in lactation or after weaning and coincides with a surge in estradiol (Atkinson 1997). The estrous period is longer and more variable in phocid seals than otariids and may facilitate multiple matings and sperm competition (Boness 2009). Phocid seals are thought to exhibit obligate delayed implantation. Implantation timing and gestational length set the timing of parturition. Implantation is most likely influenced by photoperiod, although body condition may play a role (Boyd 1984). A recent study in northern elephant seals suggested that prolactin is a key regulator of implantation that allows assessment of body reserves prior to implantation in capital breeding phocids. The release of prolactin is influenced by body condition and circulating cortisol concentrations. Prolactin then modulates the daylength stimulated changes in estradiol and progesterone required for implantation (Sperou 2020).

Oxytocin and prolactin are vital hormones in mammary gland activation and lactation. However, they also play essential roles in initiating and modulating maternal behavior. Despite the importance of these behaviors to fitness, these maternal hormones have received little attention in phocids. Prolactin increased across the short lactation of hooded seals and influenced mammary gland physiology (Mellish et al. 1999), but no behavioral effects were investigated. Maternal oxytocin levels were positively associated with mother–pup proximity in gray seals (Robinson et al. 2015). Experimental manipulation of oxytocin concentrations resulted in alterations in proximity behavior and aggression between stranger conspecifics (Robinson et al. 2017). Despite the rapid physiological clearance of oxytocin, these effects persisted for 48 h. This study provided strong experimental evidence for oxytocin's effects on social behavior and showed the potential for feedback loops where social behavior influences subsequent hormone release. Future comparative studies on the impacts of oxytocin on the maternal and social behavior of phocids are warranted.

Testosterone is an important regulator of aggression, dominance, and sexual behavior in mammals. Male phocid seals exhibit significant seasonal changes in concentrations of circulating testosterone. Testosterone concentrations are increased for 1–3 months at the start of the breeding season but decrease to baseline or non-detectable concentrations before breeding behavior ends (Griffiths 1984; Noonan et al. 1991; Bartsh et al. 1992). These low or undetectable testosterone concentrations coincide with periods of peak copulatory behavior, suggesting dissociated behavioral effects for the hormone. Mean salivary testosterone concentrations have been related to dominance rank or territory holding in male Weddell and Hawaiian monk seals, *Neomonachus schauinslandi* (Bartsch et al. 1992; Atkinson 1997), but are not associated with dominance rank in northern elephant seals, being highest in subadult and peripheral males during breeding (D. Crocker, unpublished data). The increase in circulating testosterone concentrations at the start of the breeding season is followed by an increase in cortisol concentrations (see Sect. 4.5.3 below) in male Weddell seals and elephant seals (Bartsh et al. 1992; D. Crocker, unpublished data). Experimental elevation in cortisol suppressed testosterone concentrations at 48 h after perturbation (Ensminger et al. 2014). These results suggest that competition with conspecifics may suppress concentrations of circulating testosterone across breeding through the action of cortisol. In gray seals; however, testosterone increased during physical restraint, suggesting transient testosterone modifications in response to agonistic interactions that may facilitate reproductive behavior (Lidgard et al. 2008).

4.5.3 Stress Hormones

Cortisol, the primary stress hormone in mammals, has been investigated in several phocids (reviewed in Atkinson et al. 2015). These studies focused on the physiological effects of cortisol release; however, in many wildlife species cortisol is a powerful regulator of behavior, influencing aggression, dispersal, migration, reproductive behavior, foraging, and other aspects of behavioral phenotype. In gray seals, baseline cortisol was inversely associated with on-shore tenure, an important determinant of mating success (Lidgard et al. 2008). Elevation of cortisol may initiate foraging after the perinatal period in otariid seals (Guinet et al. 2004), but manipulative experiments in phocids have suggested no effects on behavior except through potential indirect effects on suppression of thyroid hormone or testosterone (Bennett et al. 2013; Ensminger et al. 2014). In seabirds, the dominant stress hormone corticosterone demonstrates complex and varied impacts on foraging effort and behavior (e.g., Cottin et al. 2014; Kroeger et al. 2019). Future studies should explore the potential impacts of cortisol on behavior and physiology in phocid seals as these may be critical components of the effects of anthropogenic stressors on seals.

4.6 Future Directions

Phocids have served as crucial research systems for the study of marine mammal physiology. These studies have focused on the more easily accessible species that breed on land or fast ice, especially elephant seals, gray seals, and Weddell seals. For example, cutting-edge biomedical metabolic tracer techniques have been used in elephant seals that have not been used in any other wildlife species (Champagne et al. 2012b). The ubiquity of the features found in these species is unknown, and the difficulty of accessing and studying polar species means that the basic physiology of some of the most numerous top predators on the planet is largely unstudied. Moving forward, comparative studies with polar species are essential to understanding the range of physiological adaptations and the role they have played in the evolution of behavioral and life-history differences among phocid species.

An essential emerging approach to understanding conservation issues is conservation physiology (Cooke et al. 2014). Conservation physiology examines physiology to predict how organisms, populations, and ecosystems respond to anthropogenic stressors and environmental change. Some modeling frameworks, such as the Population Consequences of Disturbance (PCoD) model, have been developed to model the effects of acute and chronic physiological and behavioral responses to a disturbance on individual and population health (Pirotta et al. 2018). These models have also been modified to include the cumulative effects of multiple natural and anthropogenic stressors (NAS 2017). Physiological studies on the responses to stressors that integrate diving physiology, stress physiology, endocrinology, and immunology are critical to these efforts. The unique research tractability of some phocids makes them important research species for designing, implementing, and refining these models, including elephant seals, gray seals, and harp seals (Costa et al. 2016). Recent elephant seal research has evaluated the physiological effects of acute and repeated stress and led to the development of biomarkers to detect chronic stress (McCormley et al. 2018; Pujade Busqueta et al. 2020). These techniques represent an exciting interface of physiology, behavior, and population biology that will play an essential role in marine mammal conservation.

One crucial lens into the physiology of phocids is the extraordinary power of modern functional genomic approaches. The rapid explosion of molecular tools allows investigators to make interspecific comparisons in the key macroevolutionary changes that enabled terrestrial animals to move into the marine environment and the differences between fully pelagic marine mammals and species still tied to land for reproduction. Whole-genome sequencing, array-based sequencing of single nucleotide polymorphisms, and target sequence probes (e.g., exomes), epigenomic studies, microRNA studies, and transcriptome sequencing represent a rapidly expanding tool kit for studies of phocid physiology and evolution (Cammen et al. 2016). These tools not only provide information on phylogenomics and population structure, but they also have the potential to give great insight into the physiological function and molecular adaptation to the specific physiological problems solved by phocids to exercise while holding their breath, fast for weeks or months under high nutrient demands, and transition between strikingly different thermal environments (Hindle 2020; Weitzner et al. 2020). For example, microRNA profiling has provided insight into the cellular mechanisms underlying hypoxia tolerance and vasoregulation during diving (Penso-Dolfin et al. 2020), and transcriptome analysis has provided novel information about the regulatory roles of adipose-derived hormones while fasting (Khudyakov et al. 2019). These tools may provide important insights into the physiological differences between phocids, including species for which tissue sampling is possible but manipulative experiments and behavioral studies are difficult. Though nascent in their use in phocids, these emerging techniques hold the promise of giving tremendous insight into the evolution of phocid behavior, its integration with physiological adaptation, and their expression as differences in life-history strategies among the most globally successful marine mammal taxa.

Acknowledgements DEC was supported on ONR grant # N000141812822. BIM was supported on ONR grant # N000141612852 and NSF grant # 1656077.

References

Adachi T, Maresh JL, Robinson PW, Peterson SH, Costa DP, Naito Y, Watanabe YY, Takahashi A (2014) The foraging benefits of being fat in a highly migratory marine mammal. Proc Roy Soc B Biol Sci 281:214–2020

Adams SH, Costa DP (1993) Water conservation and protein metabolism in northern elephant seal pups during the postweaning fast. J Comp Phys B 163:367–373

Andrews RD, Jones DR, Williams JD, Thorson PH, Oliver GW, Costa DP, Le Boeuf BJ (1997) Heart rates of northern elephant seals diving at sea and resting on the beach. J Exp Biol 200(15):2083–2095

Atkinson S (1997) Reproductive biology of seals. Rev of Reprod 2:175–194

Atkinson S, Crocker D, Houser D, Mashburn K (2015) Stress physiology in marine mammals: how well do they fit the terrestrial model? J Comp PhysB 185:463–486

Banuet-Martínez M, Espinosa-de Aquino W, Elorriaga-Verplancken FR, Flores-Morán A, García OP, Camacho M, Acevedo-Whitehouse K (2017) Climatic anomaly affects the immune competence of California sea lions. PLoS One 12(6):e0179359

Bartsh SS, Johnston SD, Siniff DB (1992) Territorial behavior and breeding frequency of male Weddell seals (*Leptonychotes weddelli*) in relation to age, size, and concentrations of serum testosterone and cortisol. Can J Zool 70(4):680–692

Bennett KA, Speakman JR, Moss SE, Pomeroy P, Fedak MA (2007) Effects of mass and body composition on fasting fuel utilisation in grey seal pups (*Halichoerus grypus* Fabricius): an experimental study using supplementary feeding. J Exp Biol 210(17):3043–3053

Bennett KA, Fedak MA, Moss SE, Pomeroy PP, Speakman JR, Hall AJ (2013) The role of glucocorticoids in naturally fasting grey seal (*Halichoerus grypus*) pups: dexamethasone stimulates mass loss and protein utilisation, but not departure from the colony. J Exp Biol 216(6):984–991
Blix AS, Elsner R, Kjekshus JK (1983) Cardiac output and its distribution through capillaries and A-V shunts in diving seals. Acta Phys Scand 118(2):109–116
Boily P (1995) Theoretical heat flux in water and habitat selection of phocid seals and beluga whales during the annual molt. J Theor Biol 172:235–244
Boness DJ (2009) Estrus and estrous behavior. In: Perrin WF, Wursig B, Thewissen JGM (eds) Encyclopedia of marine mammals. Academic Press, San Diego, pp 392–396
Boyd I (1984) The relationship between body condition and the timing of implantation in pregnant grey seals (*Halichoerus grypus*). J Zool 203(1):113–123
Boyd IL, Barton AJP, T, Croxall JP, (1994) Foraging behaviour of Antarctic fur seals during periods of contrasting prey abundance. J Anim Ecol 63:703–713
Burge MR, Hardy KJ, Schade DS (1993) Short-term fasting is a mechanism for the development of euglycemic ketoacidosis during periods of insulin deficiency. J Clin Endo Metab 76:1192–1198
Burns JM (1999) The development of diving behavior in juvenile Weddell seals: pushing physiological limits in order to survive. Can J Zool 77(5):737–747
Burns JM, Costa DP, Frost K, Harvey JT (2005) Development of body oxygen stores in harbor seals: effects of age, mass, and body composition. Physiol Biochem Zool 78(6):1057–1068
Butler PJ (2006) Aerobic dive limit. what is it and is it always used appropriately? Comp Biochem Phys A Mol Integ Phys 145:1–6
Cammen KM, Carroll AKR, EL et al (2016) Genomic methods take the plunge: Recent advances in high-throughput sequencing of marine mammals. J Hered 107(6):481–495. https://doi.org/10.1093/jhered/esw044
Carbone C, Houston AI (1996) The optimal allocation of time over the dive cycle: an approach based on aerobic and anaerobic respiration. Anim Behav 51(6):1247–1255
Carlini AR, Marquez MEI, Ramdohr S, Bornemann H, Panarello HO, Daneri GA (2001) Post-weaning duration and body composition changes in southern elephant seal (*Mirounga leonina*) pups at King George Island. Physiol Biochem Zoo 74(4):531–540
Chaise LL, Prinetl TC, Gallon SL, Paterson W, McCafferty DJ, Théry M, Ancel A, Gilbert C (2018) Local weather and body condition influence habitat use and movements on land of molting female southern elephant seals (*Mirounga leonina*). Ecol Evol 8(12):6081–6090
Chaise LL, McCafferty DJ, Krellenstein A et al (2019) Environmental and physiological determinants of huddling behavior of molting female southern elephant seals (*Mirounga leonina*). Phys Behav 199:182–190
Champagne CD, Crocker DE, Fowler MA, Houser DS (2012a) Fasting physiology of the pinnipeds: the challenges of fasting while maintaining high energy expenditure and nutrient delivery for lactation. In: McCue MD (ed) Comparative physiology of fasting, starvation, and food limitation. Springer, Berlin Heidelberg, pp 309–336
Champagne CD, Houser DS, Fowler MA, Costa DP, Crocker DE (2012b) Gluconeogenesis is associated with high rates of tricarboxylic acid and pyruvate cycling in fasting northern elephant seals. Am J Phys Reg Integ Comp Phys 303(3):R340–R352
Codde SA, Allen SG, Houser DS, Crocker DE (2016) Effects of environmental variables on surface temperature of breeding adult female northern elephant seals, *Mirounga angustirostris*, and pups. Therm Biol 61:98–105
Cooke SJ, Blumstein DT, Buchholz R, Caro T, Fernandez-Juricic E, Franklin CE, Metcalfe J, O'Connor CM, St. Clair CC, Sutherland WJ (2014) Physiology, behavior, and conservation. Physiol Biochem Zool 87(1):1–14
Costa DP (1991) Reproductive and foraging energetics of pinnipeds: implications for life history paterns. In: Renouf D (ed) The behavior of pinnipeds. Springer, Dordrecht, pp 300–344
Costa DP, Gales NJ, Goebel ME (2001) Aerobic dive limit: how often does it occur in nature? Comp Biochem Phys a: Mol Integ Phys 129(4):771–783

Costa DP, Kuhn CE, Weise MJ, Arnould SSA, JP, (2004) When does physiology limit the foraging behaviour of freely diving mammals? Int Cong Ser 1280:359–366

Costa DP, Huckstadt LA, Crocker DE, McDonald BI, Goebel ME, Fedak MA (2010) Approaches to studying climatic change and its role on the habitat selection of Antarctic pinnipeds. Integ Comp Biol 50(6):1018–1030

Costa DP, Schwarz L, Robinson P, Schick RS, Morris PA, Condit R, Crocker DE, Kilpatrick AM (2016) A bioenergetics approach to understanding the population consequences of disturbance: elephant seals as a model system. In: Popper AN, Hawkins A (cds) The effects of noise on aquatic life II. Springer, Dordrecht, pp 161–169

Cottin M, MacIntosh AJ, Kato A, Takahashi A, Debin M, Raclot T, Ropert-Coudert Y (2014) Corticosterone administration leads to a transient alteration of foraging behaviour and complexity in a diving seabird. Mar Ecol Prog Ser 496:249–262

Crocker DE, Webb PM, Costa DP, Le Boeuf BJ (1998) Protein catabolism and renal function in lactating northern elephant seals. Physiol Zool 71(5):485–491

Crocker DE, Williams JD, Costa DP, Le Boeuf BJ (2001) Maternal traits and reproductive effort in northern elephant seals. Ecol 82(12):3541–3555

Crocker DE, Houser DS, Webb PM (2012a) Impact of body reserves on energy expenditure, water flux, and mating success in breeding male northern elephant seals. Physiol Biochem Zool 85(1):11–20

Crocker DE, Ortiz RM, Houser DS, Webb PM, Costa DP (2012b) Hormone and metabolite changes associated with extended breeding fasts in male northern elephant seals (*Mirounga angustirostris*). Comp Biochem Phys A: Mol Integ Phys 161(4):388–394

Crocker DE, Champagne CD, Fowler MA, Houser DS (2014) Adiposity and fat metabolism in lactating and fasting northern elephant seals. Adv Nutr 5(1):57–64

Crocker DE, Wenzel BK, Champagne CD, Houser DS (2017) Adult male northern elephant seals maintain high rates of glucose production during extended breeding fasts. J Comp Phys B 187(8):1183–1192

Davis RW (1983) Lactate and glucose metabolism in the resting and diving harbor seal (Phoca vitulina). J Comp Phys 153:275–288

Davis RW (2019) Marine mammals: adaptations for the aquatic life. Springer Nature, Cham

Davis RW, Kanatous SB (1999) Convective oxygen transport and tissue oxygen consumption in Weddell seals during aerobic dives. J Exp Biol 202(9):1091–1113

Davis RW, Williams TM (2012) The marine mammal dive response is exercise modulated to maximize aerobic dive duration. J Comp Phys A 198(8):583–591

de Quirós YB, Seewald JS, Sylva SP, Greer B, Niemeyer M, Bogomolni AL, Moore MJ (2013) Compositional discrimination of decompression and decomposition gas bubbles in bycaught seals and dolphins. PloS One 8(12):e83994

Ensminger DC, Somo DA, Houser DS, Crocker DE (2014) Metabolic responses to adrenocorticotropic hormone (ACTH) vary with life-history stage in adult male northern elephant seals. Gen Comp Endo 204:150–157

Fahlman A, Olszowka A, Bostrom B, Jones DR (2006) Deep diving mammals: dive behavior and circulatory adjustments contribute to bends avoidance. Resp Phys Neurobiol 153(1):66–77. https://doi.org/10.1016/j.resp.2005.09.014

Fahlman A, Loring SH, Ferrigno, et al (2011) Static inflation and deflation pressure-volume curves from excised lungs of marine mammals. J Exp Biol 214(22):3822–3828

Fahlman A, Loring SH, Johnson SP, Haulena M, Trites AW, Fravel VA, van Bonn WG (2014) Inflation and deflation pressure-volume loops in anesthetized pinnipeds confirms compliant chest and lungs. Front Physiol 5:433. https://doi.org/10.3389/fphys.2014.00433

Fedak MA, Pullen MR, Kanwisher J (1988) Circulatory responses of seals to periodic breathing: heart rate and breathing during exercise and diving in the laboratory and open sea. Can J Zool 66(1):53–60

Fowler SL, Costa DP, Arnould JPY, Gales NJ, Burns JM (2007) Ontogeny of oxygen stores and physiological diving capability in Australian sea lions. Funct Ecol 21(5):922–935

Garland T Jr, Losos JB (1994) Ecological morphology of locomotor performance in squamate reptiles. In: Wainright PC, Reilly S (eds) Ecological morphology: integrative organismal biology. University of Chicago Press, Chicago, pp 240–302

Glazier DS (2014) Metabolic scaling in complex living systems. Systems 2:451–540

Griffiths DJ (1984) The annual cycle of the testis of the elephant seal (*Mirounga leonina*) at Macquarie Island. J Zool 203(2):193–204

Grinnel SW, Irving L, Scholander PF (1942) Experiments on the relation between blood flow and heart rate in the diving seal. J Cell Comp Physiol 19(3):341–350

Guinet C, Servera N, Mangin S, Georges JY, Lacroix A (2004) Change in plasma cortisol and metabolites during the attendance period ashore in fasting lactating subantarctic fur seals. Comp Biochem Phys a: Mol Integ Phys 137(3):523–531

Guyton GP, Stanek KS, Schneider RC et al (1995) Myoglobin saturation in free-diving Weddell seals. J Appl Phys 79(4):1148–1155

Halsey M (1982) Effects of high pressure on the central nervous system. Physiol Rev 62(4):1341–1377

Hassrick JL, Crocker DE, Teutschel NM, McDonald BI, Robinson PW, Simmons SE, Costa DP (2010) Condition and mass impact oxygen stores and dive duration in adult female northern elephant seals. J Exp Biol 213(4):585–592. https://doi.org/10.1242/jeb.037168

Hill RD, Schneider RC, Liggins GC et al (1987) Heart rate and body temperature during free diving of Weddell seals. Am J Phys Reg Integ Comp Phys 253(2):344–351

Hindell M, Slip D, Burton H (1991) The diving behavior of adult male and female southern elephant seals, *Mirounga leonina* (Pinnipedia, Phocidae). Aus J Zool 39(5):595–619

Hindell MA, Slip DJ, Burton HR, Bryden MM (1992) Physiological implications of continuous, prolonged, and deep dives of the southern elephant seal (*Mirounga leonina*). Can J Zool 70:370–379

Hindle AG (2020) Diving deep: understanding the genetic components of hypoxia tolerance in marine mammals. J Appl Phys 128(5):1439–1446

Hokkanen JEI (1990) Temperature regulation of marine mammals. J Theor Biol 145:465–485

Hooker SK, Baird RW, Fahlman A (2009) Could beaked whales get the bends? Effect of diving behaviour and physiology on modelled gas exchange for three species: *Ziphius cavirostris*, *Mesoplodon densirostris* and *Hyperoodon ampullatus*. Resp Phys Neurobiol 167(3):235–246. https://doi.org/10.1016/j.resp.2009.04.023

Hooker SK, Fahlman A, Moore MJ et al (2012) Deadly diving? physiological and behavioural management of decompression stress in diving mammals. P Roy Soc B Biol Sci 279(1731):1041–1050. https://doi.org/10.1098/rspb.2011.2088

Hooper AW, Berger RW, Rubin LS, McDonald BI, Crocker DE (2019) Maternal age influences offspring behaviour and growth efficiency during provisioning in northern elephant seals. Anim Behav 151:121–130

Houser DS, Champagne CD, Crocker DE (2013) A non-traditional model of the metabolic syndrome: the adaptive significance of insulin resistance in fasting-adapted seals. Front Endo 4:164

Jain K (1994) High-pressure neurological syndrome (HPNS). Acta Neurol Scand 90(1):45–50

Jepson P, Arbelo M, Deaville R et al (2003) Gas-bubble lesions in stranded cetaceans. Nature 425(6958):575–576

Jobsis PD, Ponganis PJ, Kooyman GL (2001) Effects of training on forced submersion responses in harbor seals. J Exp Biol 204(22):3877–3885

Kelso EJ, Champagne CD, Tift MS, Houser DS, Crocker DE (2012) Sex differences in fuel use and metabolism during development in fasting juvenile northern elephant seals. J Exp Biol 215(15):2637–2645. https://doi.org/10.1242/jeb.068833

Ketterson ED, Nolan V Jr (1992) Hormones and life histories: an integrative approach. Am Nat 140:S33–S62

Khudyakov JI, Abdollahi E, Ngo A, Sandhu G, Stephan A, Costa DP, Crocker DE (2019) Expression of obesity-related adipokine genes during fasting in a naturally obese marine mammal. Am J Phys Reg Integ Comp Phys 317(4):R521–R529
Kim B (2008) Thyroid hormone as a determinant of energy expenditure and the basal metabolic rate. Thyroid 18(2):141–144
Kooyman GL (1965) Techniques used in measuring diving capacities of Weddell seals. Polar Rec 12(79):391–394
Kooyman G (1973) Respiratory adaptations in marine mammals. Am Zool 13(2):457–468
Kooyman G, Campbell W (1972) Heart rates in freely diving Weddell seals, Leptonychotes Weddelli. Comp Biochem Phys A 43(1):31–36
Kooyman G, Schroeder J, Denison D, Hammond D, Wright J, Bergman W (1972) Blood nitrogen tensions of seals during simulated deep dives. Am J Phys 223(5):1016–1020
Kooyman G, Wahrenbrock E, Castellini M, Davis R, Sinnett E (1980) Aerobic and anaerobic metabolism during voluntary diving in Weddell seals: evidence of preferred pathways from blood chemistry and behavior. J Comp Phys 138(4):335–346
Kroeger C, Thompson CDE, DR, Torres LG, Sagar P, Shaffer SA (2019) Variation in corticosterone levels in two species of breeding albatrosses with divergent life histories: responses to body condition and drivers of foraging behavior. Physiol Biochem Zool 92(2):223–238
Kuhn CE, McDonald BI, Shaffer SA, Barnes J, Crocker DE, Burns J, Costa DP (2006) Diving physiology and winter foraging behavior of a juvenile leopard seal (*Hydrurga leptonyx*). Polar Biol 29(4):303–307. https://doi.org/10.1007/s00300-005-0053-x
Lee D, Martinez B, Crocker DE, Ortiz, RM (2017) Fasting increases the phosphorylation of AMPK and expression of sirtuin1 in muscle of adult male northern elephant seals (*Mirounga angustirostris*). Physiol Rep 5(4):e13114
Lenfant C, Johansen K, Torrance JD (1970) Gas transport and oxygen storage capacity in some pinnipeds and the sea otter. Resp Phy 9(2):277–286
Lidgard DC, Boness DJ, Bowen WD, McMillan JI (2005) State-dependent male mating tactics in the grey seal: the importance of body size. Behav Ecol 16(3):541–549
Lidgard DC, Boness DJ, Bowen WD, McMillan JI (2008) The implications of stress on male mating behavior and success in a sexually dimorphic polygynous mammal, the grey seal. Horm Behav 53(1):241–248
Lindstedt SL, Boyce MS (1985) Seasonality, fasting endurance, and body size in mammals. Am Nat 125:873–878
Liwanag HE, Berta A, Costa DP, Budge SM, Williams TM (2012) Morphological and thermal properties of mammalian insulation: the evolutionary transition to blubber in pinnipeds. Biol J Linn Soc 107(4):774–787
Lydersen C, Ryg MS, Hammill MO, O'Brien PJ (1992) Oxygen stores and aerobic dive limit of ringed seals (*Phoca hispida*). Can J Zool 70(3):458–461
McCormley MC, Champagne CD, Deyarmin JS, Stephan AP, Crocker DE, Houser DS, Khudyakov JI (2018) Repeated adrenocorticotropic hormone administration alters adrenal and thyroid hormones in free-ranging elephant seals. Conserv Phys 6(1):coy040
McDonald BI, Ponganis PJ (2013) Insights from venous oxygen profiles: oxygen utilization and management in diving California sea lions. J Exp Biol 216(17):3332–3341. https://doi.org/10.1242/jeb.085985
McDonald BI, Ponganis PJ (2014) Deep-diving sea lions exhibit extreme bradycardia in long-duration dives. J Exp Biol 217:1525–1534. https://doi.org/10.1242/jeb.098558
McKnight, JC, Bennett, KA, Bronkhorst et al (2019) Shining new light on mammalian diving physiology using wearable near-infrared spectroscopy. PLoS Biol 17(6):e3000306
Meir JU, Champagne CD, Costa DP, Williams CL, Ponganis PJ (2009) Extreme hypoxemic tolerance and blood oxygen depletion in diving elephant seals. Am J Phys Reg Int Comp Phys 297(4): R927–R939. doi:https://doi.org/10.1152/ajpregu.00247.2009

Mellish JE, Iverson SJ, Bowen WD (1999) Variation in milk production and lactation performance in grey seals and consequences for pup growth and weaning characteristics. Physiol Biochem Zool 72(6):677–690

Miller NJ, Postle AD, Orgeig S, Koster G, Daniels CB (2006) The composition of pulmonary surfactant from diving mammals. Resp Phys Neurobiol 152(2):152–168. https://doi.org/10.1016/j.resp.2005.08.001

Molyneux G, Bryden M (1978) Arteriovenous anastomoses in the skin of seals. I. The Weddell seal *Leptonychotes weddelli* and the elephant seal *Mirounga leonina* (Pinnipedia: Phocidae). Anat Rec 191(2): 239–251

Moore MJ, Hammar T, Arruda J, Cramer S, Dennison S, Montie E, Fahlman A (2011) Hyperbaric computed tomographic measurement of lung compression in seals and dolphins. J Exp Biol 214(14):2390–2397. https://doi.org/10.1242/jeb.055020

Mori Y (1999) The optimal allocation of time and respiratory metabolism over the dive cycle. Behav Ecol 10(2):155–160

National Academies of Sciences, Engineering, and Medicine (2017) Approaches to understanding the cumulative effects of stressors on marine mammals. National Academies Press

Neshumova T, Cherapanova V, Petrov E (1983) Myoglobin concentration in muscles of the seal *Pusa sibirica*. J Evol Bioch Phys Physiol 19:93–95

Neshumova T, Cherepanova V (1984) Blood supply and myoglobin stocks in muscles of the seal (*Pusa siberica*) and muskrat (*Ondatra zibethica*). J Evol Bioch Phys 20:282

Noonan LM, Ronald K, Raeside J (1991) Plasma testosterone concentrations of captive male hooded seals (*Cystophora cristata*). Can J Zool 69:2279–2282

Noren DP (2002) Thermoregulation of weaned northern elephant seal (*Mirounga angustirostris*) pups in air and water. Physiol Biochem Zool 75(5):513–523

Noren SR, Iverson SJ, Boness DJ (2005) Development of the blood and muscle oxygen stores in gray seals (*Halichoerus grypus*): implications for juvenile diving capacity and the necessity of a terrestrial postweaning fast. Physiol Biochem Zool 78(4):482–490

Norris A, Houser DS, Crocker DE (2010) Environment and activity affect skin temperature in breeding adult male elephant seals (*Mirounga angustirostris*). J Exp Biol 213(24):4205–4212

Odend'Hal S, Poulter TC (1966) Pressure regulation in the middle ear cavity of sea lions: a possible mechanism. Science 153(3737):768–769

Oftedal OT, Bowen WD, Widdowson EM, Boness DJ (1991) The prenatal molt and its ecological significance in hooded and harbor seals. Can J Zool 69(9):2489–2493

Paterson WD, Russell DJ, Wu GM, McConnell B, Currie JI, McCafferty DJ, Thompson D (2019) Post-disturbance haulout behaviour of harbour seals. Aquat Conserv 29:144–156

Pearson LE, Liwanag HE, Hammill MO, Burns JM (2014) To each its own: thermoregulatory strategy varies among neonatal polar phocids. Comp Biochem Phys a: Mol Integ Phys 178:59–67

Penso-Dolfin L, Haerty W, Hindle A, Di Palma F (2020) microRNA profiling in the Weddell Seal suggests novel regulatory mechanisms contributing to diving adaptation. BMC Genom 21:1–17

Pirotta E, Booth CG, Costa DP, Fleishman E, Kraus SD, Lusseau D, Moretti D, New LF, Schick RS, Schwarz LK (2018) Understanding the population consequences of disturbance. Ecol Evol 8:9934–9946

Ponganis PJ (2011) Diving Mammals. Compr Phys 1(1):447–465. https://doi.org/10.1002/cphy.c091003

Ponganis PJ (2015) Diving physiology of marine mammals and seabirds. Cambridge University Press, Cambridge

Ponganis PJ, Kooyman GL, Castellini MA (1993) Determinants of the aerobic dive limit of Weddell seals: analysis of diving metabolic rates, postdive end tidal PO_2's, and blood and muscle oxygen stores. Physiol Zool 66(5):732–749

Ponganis PJ, Kooyman GL, Baranov EA, Thorson PH, Stewart BS (1997) The aerobic submersion limit of Baikal seals, *Phoca Sibirica*. Can J Zool 75(8):1323–1327. https://doi.org/10.1139/z97-756

Pujade Busqueta L, Crocker DE, Champagne CD, McCormley MC, Deyarmin JS, Houser DS, Khudyakov JI (2020) A blubber gene expression index for evaluating stress in marine mammals. Conserv Phys 8(1):coaa082
Reder S, Lydersen C, Arnold W, Kovacs KM (2003) Haulout behaviour of High Arctic harbour seals (*Phoca vitulina vitulina*) in Svalbard, Norway. Polar Biol 27:6–16
Ridgway SH, Howard R (1979) Dolphin lung collapse and intramuscular circulation during free diving: evidence from nitrogen washout. Science 206(4423):1182–1183
Robinson KJ, Twiss SD, Hazon N, Pomeroy PP (2015) Maternal oxytocin is linked to close mother-infant proximity in grey seals (*Halichoerus grypus*). PLoS One 10(12): e0144577
Robinson KJ, Twiss SD, Hazon N, Moss S, Pomeroy PP (2017) Positive social behaviours are induced and retained after oxytocin manipulations mimicking endogenous concentrations in a wild mammal. P Roy Soc B Biol Sci 284:20170554
Ryg M, Lydersen C, Knutsen LO, Bjorge A, Smith TG, Oritsland NA (1993) Scaling of insulation in seals and whales. J Zool 230:193–206
Scholander PF (1940) Experimental investigations on the respiratory function in diving mammals and birds. Hvalradets Skrifter 22:1–131
Scholander P, Irving L, Grinnell S (1942) Aerobic and anaerobic changes in seal muscles during diving. J Biol Chem 142(1):431–440
Schulz TM, Bowen WD (2005) The evolution of lactation strategies in pinnipeds: a phylogenetic analysis. Ecol Monog 75:159–177
Sperou ES (2020) A comprehensive analysis of sex hormones and their role in reproductive suppression in female northern elephant seals (*Mirounga angustirostris*). Sonoma State University, Thesis
Stenfors LE, Sade J, Hellstrom S, Anniko M (2001) How can the hooded seal dive to a depth of 1000 m without rupturing its tympanic membrane? A morphological and functional study. Acta Otolaryngol 121(6):689–695
Tavoni S, Champagne CD, Houser DS, Crocker DE (2013) Lactate flux and gluconeogenesis in fasting, weaned northern elephant seals (*Mirounga angustirostris*). J Comp Phys B 183(4):537–546. https://doi.org/10.1007/s00360-012-0720-5
Thompson D, Fedak MA (1993) Cardiac responses of grey seals during diving at sea. J Exp Biol 174:139–154
Thorson PH, Le Boeuf BJ (1994) Developmental aspects of diving in northern elephant seal pups. In: Le Boeuf BJ, Laws RM (eds) Elephant seals: population ecology, behavior, and physiology. University of California Press, Berkeley, pp 271–289
Twiss SD, Caudron A, Pomeroy PP, Thomas CJ, Mills JP (2000) Finescale topographical correlates of behavioural investment in offspring by female grey seals, *Halichoerus Grypus*. Anim Behav 59(2):327–338
Twiss SD, Wright NC, Dunstrone N, Redman P, Moss S, Pomeroy PP (2002) Behavioral evidence of thermal stress from overheating in UK breeding gray seals. Mar Mamm Sci 18(2):455–468
Walcott SM, Kirkham AL, Burns JM (2020) Thermoregulatory costs in molting Antarctic Weddell seals: impacts of physiological and environmental conditions. Conserv Physiol 8(1):coaa022
Watanabe Y, Baranov EA, Sato K, Naito Y, Miyazaki N (2004) Foraging tactics of Baikal seals differ between day and night. Mar Ecol Prog Ser 279:283–289
Watanabe Y, Baranov EA, Sato K, Naito Y, Miyazaki N (2006) Body density affects stroke patterns in Baikal seals. J Exp Biol 209:3269–3280
Weingartner GM, Thornton SJ, Andrews RD, Enstipp MR, Dorota Barts A, Hochachka PW (2012) The effects of experimentally induced hyperthyroidism on the diving physiology of harbor seals (*Phoca vitulina*). Fronts Phys 3:380
Weitzner EL, Fanter CE, Hindle AG (2020) Pinniped ontogeny as a window into the comparative physiology and genomics of hypoxia tolerance. Int Comp Biol 60:1414–1424
Williams TM, Kooyman GL (1985) Swimming performance and hydrodynamic characteristics of harbor seals *Phoca vitulina*. Physiol Zool 58(5):576–589

Williams TM, Davis RW, Fuiman LA et al (2000) Sink or swim: Strategies for cost-efficient diving by marine mammals. Science 288(5463):133–136. https://doi.org/10.1126/science.288.5463.133

Williams TM, Fuiman LA, Davis RW (2015) Locomotion and the cost of hunting in large, stealthy marine carnivores. Int Comp Biol 55(4):673–682. https://doi.org/10.1093/icb/icv025

Wilsterman K, Buck CL, Barnes BM, Williams CT (2015) Energy regulation in context: free-living female arctic ground squirrels modulate the relationship between thyroid hormones and activity among life history stages. Horm Behav 75:111–119

Zapol WM, Liggins GC, Schneider RC, Qvist J, Snider MT, Creasy RK, Hochachka PW (1979) Regional blood flow during simulated diving in the conscious Weddell seal. J Appl Phys 47(5):968–973

Chapter 5
Habitat Utilization and Behavior of Phocid Seals in Relation to Oceanography

Luis A. Hückstädt and Ryan R. Reisinger

Contents

Abstract Phocids (true seals) occur in oceans across the globe, from the tropics to polar oceans, and a few species have even colonized freshwater systems. As a group, they experience a wide range of local environmental conditions and oceanographic regimes that have shaped their behavior, allowing them to better take advantage of the specific patterns of prey distribution and abundance. Despite their suite of adaptations to life in the ocean, seals are still constrained to return to a solid substrate (land or ice) and haul out to reproduce, rest, and molt. This separation between their prey (ocean) and solid substrate (land/ice) shapes their life cycles and inextricably determines their at-sea movements and habitat preferences. In this chapter, we examine the biogeographic patterns of phocid seals in the context of past and current global oceanography, explore the use of animal-borne instruments (biologging) to study movements and the environments utilized. We review the relationships between phocid seal habitat utilization and behavior and the biophysical oceanographic characteristics that directly or indirectly influence the distribution of seals and their prey. The physical properties of the environment utilized by seals (water, sea bottom, and ice) are fundamental to understanding not only the behavior and ecology of true seals but also how they may be affected by annual and long-term changes in ocean climate.

L. A. Hückstädt · R. R. Reisinger
Institute of Marine Sciences, University of California Santa Cruz, Santa Cruz, CA, USA

L. A. Hückstädt (✉)
Centre for Ecology and Conservation, University of Exeter, Penryn, Cornwall, United Kingdom

R. R. Reisinger
Centre d'Etudes Biologique de Chizé, UMR 7372 du Centre National de La Recherche Scientifique-La Rochelle Université, Villiers en Bois, France

D. P. Costa and E. McHuron (eds.), *Ethology and Behavioral Ecology of Phocids*, Ethology and Behavioral Ecology of Marine Mammals,
https://doi.org/10.1007/978-3-030-88923-4_5

Keywords Phocids · Seals · Oceanography · Distribution · Environment · Animal-borne instruments · Habitat preference · Diving · Sea ice

5.1 Introduction

Phocid (true) seals occur in seas and oceans across the globe, from the tropics (Hawaiian monk seal *Neomonachus schauinslandi*) to polar oceans, including freshwater systems (e.g., the Baikal seal, *Pusa sibirica*; Fig. 5.1). Consequently, they experience a wide range of local environmental conditions and oceanographic regimes that have shaped their behavior, allowing them to take advantage of different patterns of prey distribution and abundance. Their relatively large body size and associated energy demands necessitate the consumption of large amounts of prey. As such, seals have evolved a variety of foraging strategies, which vary from exploiting local resources within a limited range (e.g., harbor seal *Phoca vitulina*, Hawaiian monk seal, and Mediterranean monk seal *Monachus monachus*), to foraging over thousands of kilometers across ocean basins (e.g., elephant seals genus *Mirounga*). Likewise, they exploit resources across a wide diversity of vertical environments, from capturing prey at or near the surface (e.g., crabeater seals *Lobodon carcinophaga*) to diving hundreds of meters into mesopelagic (200–1,000 m) and upper bathypelagic

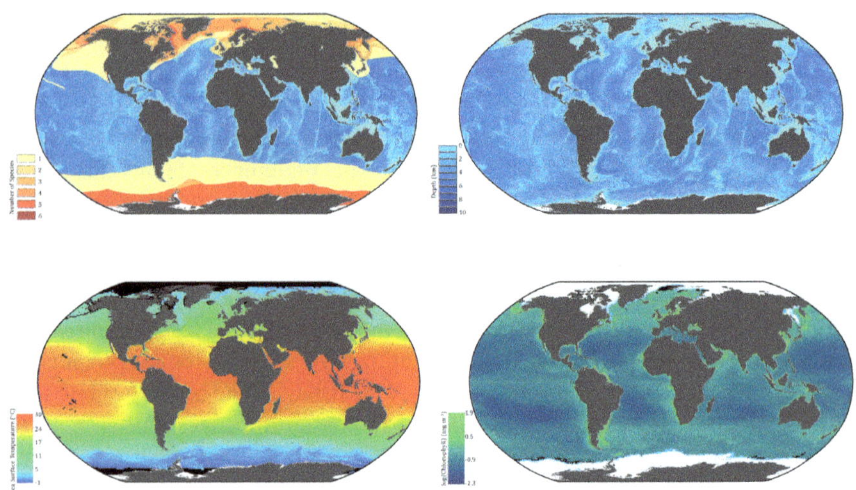

Fig. 5.1 Phocid species richness distribution and characteristics of the global oceans. The left upper panel shows the distribution and species richness for all species of phocid seals (Data from the IUCN Red List of Threatened Species). The upper right panel corresponds to the bathymetric depth of the oceans. The lower left panel shows the average Sea Surface Temperature for the 2000–2010 decade. The lower right panel is a composite image showing spring values of chlorophyll concentration for the northern (April–June) and southern (September–December) hemispheres for the same decade. The white color indicates sea ice maximum extent for the northern (February) and southern (October) hemispheres. Black indicates no data

(1,000 to >2,000 m) zones (e.g., Weddell seal *Leptonychotes weddellii* and hooded seal *Cystophora cristata*).

Despite their suite of adaptations to life in the ocean, seals are still constrained to haul out to reproduce, rest, and molt. Their at-sea movements and habitat utilization patterns are thus bound since they must return to a solid substrate (either land or sea ice) at least twice a year (e.g., elephant seals), with many species frequently returning to solid substrates weekly or daily (e.g., harbor seals). This constraint has shaped their foraging behavior because they have to face the physical separation of their reproduction/resting/molting substrate and the areas at sea where their prey occurs (Chap. 8). Consequently, phocid seals have evolved strategies to minimize the amount of time they spend on a solid substrate. For example, this varies from seals with a short lactation duration (as short as 4 days in the hooded seal, the shortest lactation in any mammal) to those with extended lactation periods during which females feed. Some give birth to precocial newborns that can venture into the water shortly after birth (Bonner 1984; Costa 1993b; Trillmich 1996; Chaps. 6 and 8). Additionally, phocid seals usually undergo an annual molt when individuals can spend weeks hauled out without access to food or water (e.g., elephant seals) or, more commonly, seals spend relatively more time hauled out, resulting in a reduction in the time they dedicate to foraging. These characteristics of their life cycles, undoubtedly, determine their at-sea movements and habitat preferences.

As predators, the main determinants of seal movements are the distribution and abundance of their prey, which are not distributed homogeneously. Prey distribution is determined and influenced by oceanographic characteristics and features that influence biological productivity and/or increase prey availability. These features create areas where foraging is more energetically efficient (Bonadonna et al. 2001; Yen et al. 2004; Bost et al. 2009, 2015; Scheffer et al. 2012; Hazen et al. 2013; Abrahms et al. 2018b; Rivière et al. 2019). Seal prey is difficult to observe directly. We, therefore, examine the link between specific foraging behaviors and the biophysical characteristics of the ocean. This approach allows us to understand the processes that dictate patterns of higher predator abundance and distribution and predict the possible responses of top predators to variations in their physical environment associated with climate change (Bost et al. 2015; Abrahms et al. 2018a).

Responses to climatic variability and physical forcing differ among species and populations due to the local conditions experienced (Croxall 1992; Boyd 1996; Simmons et al. 2007). Among the responses that seals might exhibit under the current conditions of rapid climate change are modifications in their diet, movement patterns, and at-sea distribution (Crocker et al. 2006; Abrahms et al. 2018a). Climate change will likely primarily impact predators through changes in prey distribution (Croxall 1992). However, impacts due to the loss and changes in their physical habitats are also expected (e.g., warmer waters, sea ice loss). Therefore, changes in the foraging behavior of seals can provide insight into changes in the abundance and distribution of prey (Wall et al. 2007; Costa et al. 2010; McMahon et al. 2019).

Along with diet studies, combining accurate data on the positions of seals at sea, measurements of their foraging behavior, and oceanographic data is vital in identifying how the physical and biological environment impact the foraging strategies of phocid seals. One of the challenges in studying the ecology and behavior of marine top predators has been the difficulty of directly observing them at sea. This problem is particularly acute in marine mammals, which spend most of their lives at-sea underwater (e.g., Kooyman 1966; Le Boeuf et al. 1988; Naito et al. 1989; Costa 1993a; McIntyre et al. 2010). This previously limited our knowledge of the at-sea movement patterns and behavior of these species. However, advances in animal-borne instrumentation (hereafter 'biologging') technology have allowed the remote tracking of marine top predators, including phocid seals, providing detailed information on their movements (reviews in Costa et al. 2012; McIntyre 2014; Hussey et al. 2015; Hays et al. 2016). These data can be correlated with environmental variables—measured in-situ or by remote-sensing—and this has quickly advanced our understanding of the patterns and mechanisms of habitat utilization in phocid seals (e.g., McConnell et al. 1992; Lowry et al. 2000; Burns et al. 2004; Freitas et al. 2008b; Biuw et al. 2007; Hindell et al. 2016).

In this chapter, we (1) examine biogeographic patterns of phocid seals in the context of past and current global oceanography, (2) explore the use biologging in the study of at-sea movements and the environment animals utilize, and (3) review the relationship between the patterns of habitat utilization and the behavior of phocid seals and the biophysical oceanographic characteristics that directly or indirectly influence the distribution of seals and their prey. While little is known about the at-sea movement and habitat use of some phocids (e.g., Mediterranean monk seals, Ross seals *Ommatophoca rossii*, ringed seals *Pusa hispida*), others have been well-studied (Chaps 6, 10–11, 13). Perhaps due to their large body size and accessibility onshore, most studies using satellite tags and time-depth recorders have centered on both species of elephant seals and the Weddell seal (Chaps. 11–13). Thus, many of the examples we present here are from these three species, although where possible, we make an effort to include examples from other species of seals.

5.2 Worldwide Distribution of Phocid Seals

Modern phocid seals have a wide distribution across oceans in both hemispheres, from low tropical latitudes, habitat of species such as the recently extinct Caribbean monk seal (*Neomonachus tropicalis*), to polar waters where we find the greatest diversity (Fig. 5.1). Phocids likely originated from an ancestor that crossed from the temperate waters of the North Pacific to the North Atlantic through the Central American Seaway, giving origin to a basal phocid in the North Atlantic or Paratethys region during the early to middle Miocene (15–17 Ma). They later diversified into the northern seals (Phocinae) and, in temperate waters and the southern hemisphere, Monachinae (Fyler et al. 2005; Deméré et al. 2003; Rule et al. 2020; Chaps. 1 and 8).

During the middle Miocene (around 17 Ma), oceans were undergoing a particularly warm period that was quickly followed by a rapid decline in ocean temperatures, partly due to the Antarctic continent reaching its current position and the formation of the Antarctic Circumpolar Current (Zachos et al. 2001). As a consequence, sea ice volume increased, and the sea level dropped dramatically. Despite the global cooling, ocean surface temperatures were somewhat variable across the globe, with cooler surface waters toward higher latitudes and a localized increase in surface water temperatures in low latitudes (Savin et al. 1985; Zachos et al. 2001).

Paleoclimatic changes are related to the patterns observed in the radiation of true seals and their expansion into polar environments (Deméré et al. 2003; Berta et al. 2018; Chaps. 1 and 8). As the waters cooled in the northern hemisphere, Phocinae seals rapidly expanded across the North Atlantic to reach sub-Arctic and Arctic waters. Although the diversification of the basal Monachinae was hypothesized to also have taken place in the North Atlantic (Fyler et al. 2005; Berta et al. 2018; Deméré et al. 2003), new fossil evidence suggests that the evolution of this group of seals instead occurred in the South Pacific Ocean, as the planet was cooling throughout the Middle and Late Miocene (Rule et al. 2020). Accordingly, the tribes Monachini (monk seals), Miroungini (elephant seals), and Lobodontini (Antarctic seals) all originated and diversified in the South Pacific Ocean and, in the case of elephant seals and monk seals, later reinvaded the North Pacific around 10 Ma (Rule et al. 2020). Today, the worldwide distribution of true seals reflects these biogeographic and evolutionary patterns (Fig. 5.1).

5.2.1 Low Latitude Seals

There are currently only two extant species of low-latitude/tropical seals after the extinction of the Caribbean monk seal in the mid-1900s. The Mediterranean monk seal occurs in the Mediterranean Sea, coastal waters of Equatorial western Africa, and Eastern Atlantic islands. The other is the Hawaiian monk seal, endemic to the Hawaiian archipelago in the tropical North Pacific Ocean (Fig. 5.1; Chap. 16). These species are inhabitants of warm waters of relatively low biological productivity and have, consequently, adapted to these conditions by foraging on large benthic and demersal prey in relatively shallow waters close to the coasts. Both extant monk seals are among the most endangered species of pinnipeds (Chap. 16).

5.2.2 Subtropical and Mid-Latitude Seals

Like in low latitudes, subtropical waters of the world's oceans have a limited number of true seal species. Both species of elephant seals occupy pelagic waters in the northern hemisphere (*Mirounga angustirostris*, in the eastern North Pacific) and southern hemisphere (*M. leonina*, in the western South Atlantic; Fig. 5.1, Box 5.1). In contrast, harbor seals occur in mid-latitude coastal regions of the North Pacific and North Atlantic oceans. Elephant seals spend most of their time at sea foraging in oceanic waters, in association with, or beyond, the continental shelf break (Box 5.1), taking advantage of features where prey aggregates, such as fronts, eddies, filaments, and convergence zones. Despite the wide range of habitats they occupy, harbor seals are a generalist neritic species that feed on pelagic, demersal, and benthic prey (Chap. 10). Their foraging behavior and movement patterns are associated with a variety of static (e.g., sediment type, bottom depth) and dynamic (e.g., tidal currents, riverfronts, coastal upwelling) oceanographic features (see Sect. 5.4).

5.2.3 Subpolar Seals

The diversity of true seals increases toward the polar latitudes, where this group is most successful. Few species of seals, however, breed in subpolar latitudes, although several phocids use these productive waters to forage. Northern hemisphere subpolar seals include the harbor seal (see above) and gray seals (*Halichoerus grypus*; Fig. 5.1). Gray seals occur along both coasts of the North Atlantic, where they occupy neritic waters to feed on benthic and demersal prey (Chap. 9). As in the harbor seal, habitat utilization patterns of gray seals are related to both static (e.g., sediment type, bottom topography) and dynamic (e.g., current velocity, tidal cycles) oceanographic variables (e.g., McConnell et al. 1999; Breed et al. 2009; Huon et al. 2015). The only genuinely subpolar species of true seal is the southern elephant seal. The vast majority of the southern elephant seal colonies occur on the Sub-Antarctic Islands of the Atlantic, Indian and Pacific sectors of the Southern Ocean (Fig. 5.1; Chap. 11). Individuals of this species disperse toward polar, subpolar, and, sometimes, subtropical waters to forage deep in the water column, both as mesopelagic predators in association with frontal areas, eddies, and other mesoscale features, or as deep demersal/benthic predators that feed over plateaus, seamounts or the continental shelf break, where local oceanographic features enhance the aggregation of prey (Box 5.1; e.g., Bornemann et al. 2000; McConnell et al. 1992; Campagna et al. 2006; Hückstädt et al. 2012b; Costa et al. 2010; Biuw et al. 2007; Hindell et al. 2016; Tosh et al. 2012).

5.2.4 Polar Seals

Most species of true seals inhabit waters in polar latitudes of both hemispheres. Consequently, their life histories, distribution, and foraging behavior are closely linked to sea ice's natural cycles (Fig. 5.1). Sea ice is a pivotal determinant of the biological processes in polar latitudes, directly affecting the entire ecosystem, from primary producers to the distribution of predator communities (see Sect. 5.4.7.1). Sea ice is an essential habitat for true seals, as they use it to haul out to rest, breed, and or molt. There are eight species of seals that inhabit polar (Arctic) waters in the northern hemisphere, from the ice-associated populations of harbor and gray seals to the ice-associated species of harp (*Pagophilus groenlandicus*), hooded, ribbon (*Histriophoca fasciata*), and spotted seals (*Phoca largha*), to the ice-obligate seals, bearded seal

(*Erignathus barbatus*) and the Arctic ringed seal (*Pusa hispida hispida*; Fig. 5.1). Unlike their northern hemisphere counterparts, all four species of southern hemisphere polar (Antarctic) seals are ice-obligate. They include the crabeater, leopard (*Hydrurga leptonyx*), Ross, and Weddell seal (although some small populations of Weddell seals breed on land).

5.2.5 *Non-marine Seals*

Among the least known species of the true seals, non-marine seals include the Caspian seal (*P. caspica*), Baikal seal, several subspecies of ringed seals (Baltic ringed seal *Pusa hispida botnica*, Okhotsh ringed seal *P. h. ochotensis*, Lake Ladoga ringed seal *P. h. ladogensis*, and Saima seal *P. h. saimensis*; Chap. 15) that live in inland water bodies, a subspecies of harbor seal found in Lac des Loups Marins, Canada (*P. vitulina mellonae*), and a population of North Pacific harbor seals (*P. vitulina richardii*) in Lake Iliamna, Alaska (Chap. 10). Baikal seals are the only exclusively freshwater pinniped species, whereas Caspian seals are only found in the brackish waters of the Caspian sea. Due to the different nature of the spatial and temporal scales that operate the limnological processes, and the little information available, we will not expand on these species and their habitat in this chapter.

Box 5.1 Elephant Seals

By far, the best-studied species in terms of links between at-sea behavior and environmental characteristics are the two species of elephant seal, genus *Mirounga*. Elephant seals are one of the best model species to be outfitted with sensors and electronic dataloggers and transmitters because of their large size (which means they can be equipped with relatively large tags compared to smaller species), easy accessibility when on land, and robustness to handling and sedation, among other aspects. This has allowed us to describe their at-sea distribution and behavior, giving us a better understanding of how these two species utilize their environment with respect to its biophysical oceanographic characteristics.

Both species of elephant seals have very similar life history cycles, spending 80% of the year diving regularly to depths over 500 m (up to >2,000 m) for up to 8 months to feed on mesopelagic prey, without the need to haul out on land to rest. Adult female elephant seals have been the target of most of these studies because they are large (350–800 kg), display high site fidelity, and have high survival rates once they reach adulthood (Le Boeuf et al. 2000; Chaps. 11 and 12).

Southern elephant seals (*M. leonina*) are top predators of the Southern Ocean and adjacent ecosystems, where they potentially play an important ecological role over large spatial scales (millions of km^2) as significant consumers of squid and fish (Daneri and Carlini 2002; Bradshaw et al. 2003; Cherel et al. 2008). Female southern elephant seals have been traditionally considered predators that forage over vast pelagic areas of the Southern Ocean (Bradshaw et al. 2004; Biuw et al. 2007; McConnell et al. 1992), although they can venture into the continental shelf waters of the Antarctic Peninsula (McConnell et al. 1992; McConnell et al. 2002; Costa et al. 2010; Hückstädt et al. 2012b; Muelbert et al. 2013). The foraging activities of southern elephant seals are often associated with oceanographic features, such as mesoscale eddies, fronts within the Antarctic Circumpolar Circulation (ACC) system, continental shelf margin, high sea ice concentration, thermocline depth (Hindell et al. 1991, 2016; McConnell et al. 1992; Bradshaw et al. 2004; Campagna et al. 2006; Biuw et al. 2010; Hückstädt 2012; Massie et al. 2016), and even specific water masses in the Southern Ocean (Field et al. 2001; Biuw et al. 2007; Hindell et al. 2016). Recent studies have also found a relationship between the thermal structure of the water column and the diving behavior of elephant seals (see Box 5.2; Biuw et al. 2010; McIntyre et al. 2011).

Northern elephant seals are a wide-ranging pelagic species that range across the eastern North Pacific, where they primarily forage in the open ocean in association with the Transition Zone. There is a portion of the population that forages in association with the continental shelf along the Pacific coast of North America and the eastern Aleutian Islands (Le Boeuf et al. 2000; Robinson et al. 2012). Across their range, northern elephant seals feed primarily on small mesopelagic fishes and squid (Goetsch et al. 2018; Yoshino et al. 2020). The at-sea distribution and foraging behavior of female northern elephant seals are influenced by the location of the Transition Zone Chlorophyll Front (TCZF), thermal structure of the water column, and mesoscale features (Abrahms et al. 2018b; Simmons et al. 2007; Robinson et al. 2012). Males, on the other hand, prefer to forage in association with the continental shelf break (Robinson et al. 2012; Le Boeuf et al. 2000), along the Aleutian Islands and the Gulf of Alaska, or in the vicinities of their colonies (Baja California, Mexico).

5.3 Biologging

5.3.1 *The Relationship Between Seals and Their Environment*

Advances in animal telemetry ("the process of obtaining data remotely (via a tag secured to the animal)"; Block et al. 2016) or biologging ("the study of marine life via remote-sensing loggers"; Naito 2004) have allowed us to significantly expand our understanding of the biology of several species of cryptic marine predators, gathering information on a wide diversity of aspects that include patterns of habitat usage, migratory routes, foraging, and reproductive hotspots, navigation, areas of potential interaction with human activities, and areas of special interest for conservation and management purposes (e.g., Block et al. 2011; Costa et al. 2010). Biologging has been fundamental to our understanding of the ecology and behavior of phocid seals, which spend most of their lives at sea. For instance, recording time series of seal locations at sea has meant that we can link seals' movements with data on their biophysical environment, obtained from remote-sensing, in-situ observations, and ocean models. This, in turn, has allowed an examination of how biophysical oceanographic features and conditions influence the movement and behavior of phocid seals.

New technologies and approaches have enabled us to move beyond the classical use of surface environmental variables (most often derived from satellites) or static features (such as bathymetric topography), toward an integrated approach that incorporates dynamic subsurface data obtained from both the biologging instruments themselves (e.g., Biuw et al. 2007; Dragon et al. 2010; McIntyre et al. 2011; Hindell et al. 2016) or from three-dimensional oceanographic models (Hückstädt et al. 2020; Skov et al. 2008). Such data can describe the environment where foraging takes place, particularly for deep-diving animals. Further, sensors such as 3D accelerometers allow the reconstruction of 3D, fine-scale movements underwater, in contrast to the relatively coarse, surface-only estimates from Argos and GPS tags and the depth-only information from dive recorders. This gives us a more detailed picture of the movement behavior of phocids with respect to their prey and environment (e.g., Mitani et al. 2010; Le Bras et al. 2017; Rivière et al. 2019).

5.3.2 *Statistical Approaches for Modeling Relationships Between Movement and Environment*

The amount and type of new movement and environmental data now available have stimulated the biologging community to explore and develop more advanced statistical techniques to model the relationship between animals' patterns of habitat usage and the environment they inhabit (Redfern et al. 2006; Aarts et al. 2008). Many methods have been used, but most approaches statistically describe (1) presence, (2) time-spent, or (3) behavior of individuals as a function of a set of environmental explanatory covariates (Fig. 5.2, Table 5.1). Since the interpretation of model output is different for each of the three methods, the approach dictates what information can be gleaned about the individual's behavior. As such, we describe and provide examples of these three broad approaches below.

5.3.2.1 Presence-Availability (Selection Functions)

The first broad approach compares the environmental space that is used by animals with the available environmental space to estimate habitat preference or selection (e.g., Aarts et al. 2008). The geographic positions of recorded presences over time and the environmental covariates associated with these positions and times are contrasted with the general environmental variable space potentially available to seals. This approach broadly encompasses presence-background models or use-availability models (Resource Selection Functions; Manly et al. 2007) and Step Selection Functions (Fortin et al. 2005; Thurfjell et al. 2014). Practically, these methods are primarily distinguished by how this comparison is made, particularly how the absence points are generated and how the final estimates are interpreted. In presence-background modeling or resource selection functions, the background points may be randomly sampled ('point selection functions'; Zeller et al. 2012). The popular algorithm Maxent (Phillips et al. 2006), for instance, randomly samples background points from the input data provided (e.g., Nachtsheim et al. 2017). Alternatively, some constraints can be imposed, such as accessibility from haul-out sites (e.g., Aarts et al. 2008). Instead of selecting background points, another method is to simulate possible tracks based on the movement parameters of the observed tracks ('path selection functions'; Fig. 5.2; Zeller et al. 2012). Using this method, Staniland et al. (2018) showed that leopard seals preferred shallow shelf waters and areas with sea ice. Cameron et al. (2018) and Breed et al. (2018) used this approach to link the movements of juvenile bearded seals to sea ice conditions. Lastly, in Step Selection Functions, for every step in the movement path, environmental covariates of the observed locations are compared with covariates of a set of possible locations resulting from a set of possible steps (Fig. 5.2; Fortin et al. 2005). Lone et al. (2019) used this approach to show that ringed seals prefer to be near the contour of 50% sea ice concentration.

5.3.2.2 Time-Spent

In the second approach, the response variable is the amount of time an individual or population spends in a defined area (usually a grid cell). This is then modeled as a response to the environmental covariates in that area (e.g., Jones et al. 2017; Hamilton et al. 2019b). For example, Hamilton et al. (2019b) used this approach to show how young bearded seals initially prefer shallow, coastal waters during their first months at sea. Time-spent could be considered a particular case of behavioral metrics, discussed

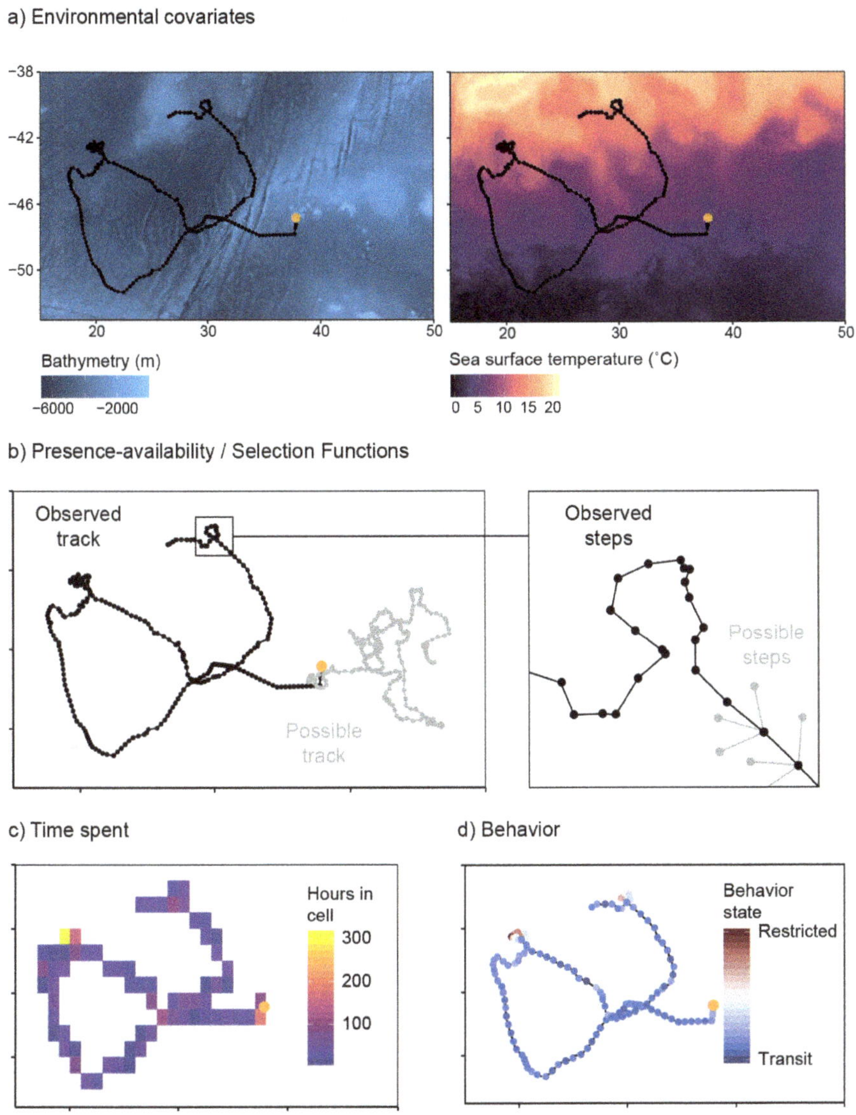

Fig. 5.2 Modeling the relationship between seal movement and environmental covariates. Shown here is the movement of a subadult male southern elephant seal, tracked on his post-molt migration from Marion Island in the Southern Indian Ocean (yellow point), with longitude and latitude (degrees) on the vertical and horizontal axes, respectively. The seal's movement path is composed of location estimates (black points) joined by 'steps' or 'displacements' (black lines). **a** First, environmental covariates are compiled at each location, for example, bathymetry and sea surface temperature. Next, a response variable is calculated; possible approaches are: **b** presence-availability or step selection functions, where the observed track or step is compared with possible tracks or steps; **c** calculating the time-spent in a predefined area; **d** calculating a behavioral state or index for each location. Lastly, the response variable is linked to the environmental covariates through a statistical model, such as regression or classification. Tracking data originally from Bester et al. (2006), compiled and processed in Reisinger et al. (2018)

Table 5.1 Examples of studies relating biologging data to environmental covariates, for each of the 18 phocid species. We chose two example studies, where available, for each species. We show whether each study modeled the relationship between movement and environmental covariates ('Quantitative model') and if this was true, we categorized the response variable broadly as 'Behavior' (B), 'Selection function' (S) or 'Time-spent' (T) (see Sect. 5.3.2). We list the explanatory covariates that were used in each case, grouped into broader categories. In one study, information on environmental covariates was given, but the relationship with phocid movement was not modeled. In four studies, no covariates were investigated; these are indicated with an asterisk in the 'Quantitative model' column. Under altimetry, SSH—sea surface height

Species Study	Quantitative model?	Response type			Bathymetry									Sea ice												Fast ice		Temperature			Salinity			Current					Biology		Hydrography				Light		Altimetry			Wind
			Distance to colony/haulout	Distance to tidal glacier front	Depth	Slope	Rugosity	Terrain ruggedness	Distance to shelf	Distance to coast	Distance to canyon	Geomorphology feature	Bottom type	Concentration	Coverage	Variability	Distance to edge	Extent	Distance to extent	Thickness	Freezing rate	Distance to polynya	Type	Time since melt	Surface heat flux	Coverage	Distance to fast ice	Surface	Bottom	Anomaly	Surface	Bottom	Difference	Surface	Bottom	Mean tidal power	Peak tidal flow	Eddy kinetic energy	Chlorophyll	Prey abundance Prey distribution	Mixed layer depth	Water mass	Front	Vertical velocity	Solar elevation	Intensity	SSH	SSH anomaly	SSH gradient	
<u>1. Mediterranean monk seal, *Monachus monachus*</u>																																																		
None																																																		
<u>2. Hawaiian monk seal, *Neomonachus schauinslandi*</u>																																																		
Wilson et al. (2017)	Y	B			Y	Y	Y	Y		Y			Y																																					

(continued)

Table 5.1 (continued)

3. Northern elephant seal, *Mirounga angustirostris*																																																			
Simmons et al. (2007)	Y	B			Y																						Y																						Y		
Block et al. (2011)		S																									Y											Y										Y		Y
4. Southern elephant seal, *Mirounga leonina*																																																			
Tosh et al. (2012)	Y	B			Y																						Y															Y								
Hindell et al. (2017)	Y	T				Y							Y																			Y					Y										Y		Y
5. Ross seal, *Ommatophoca rossi*																																																			
Arcalis-Planis et al. (2015)	N	-			Y											Y																																			
Wege et al. (2021)	Y	B			Y	Y			Y		Y			Y			Y								Y			Y		Y			Y	Y				Y				Y			Y			Y	Y	Y	Y
6. Crabeater seal, *Lobodon carcinophagus*																																																			
Burns et al. (2004)	Y	S			Y								Y																																						
Nachtsheim et al. (2017)	Y	S				Y			Y					Y						Y	Y							Y	Y		Y	Y		Y	Y																
Huckstadt et al. (2020)	Y	S				Y		Y	Y					Y		Y	Y											Y	Y		Y	Y		Y	Y							Y	Y		Y						

(continued)

Table 5.1 (continued)

7. Leopard seal, *Hydrurga leptonyx*																																																			
Meade et al. (2015)	Y	T			Y					Y							Y	Y																																	
Staniland et al. (2018)	Y	S			Y								Y														Y																								
8. Weddell seal, *Leptonychotes weddellii*																																																			
Heerah et al. (2013)	Y	B			Y		Y						Y																														Y				Y					
Nachtsheim et al. (2019)	Y	B			Y								Y								Y																						Y			Y						
9. Bearded seal, *Erignathus barbatus*																																																			
Hamilton et al. (2019)	Y	T		Y	Y					Y				Y																																						
Cameron et al. (2018)	Y	S			Y				Y	Y				Y			Y																																			
10. Hooded seal, *Cystophora cristata*																																																			
Vacquie-Garcia et al. (2017)	Y	T + B			Y								Y														Y																								
Andersen et al. (2012)	Y	T			Y	Y			Y				Y														Y											Y														

(continued)

Table 5.1 (continued)

11. Harbor seal or common seal, *Phoca vitulina*																																																			
Jones et al. (2017)	Y	T	Y		Y								Y																							Y	Y														
Bailey et al. (2014)	Y	B	Y		Y	Y							Y																																						
Grigg et al. (2012)	Y	S	Y				Y																																	Y											
12. Spotted seal, *Phoca largha*																																																			
Lowry et al. (2000)	Y	S							Y	Y							Y						Y																												
Gryba et al. (2019)	Y	B																																									Y								
13. Ringed seal, *Pusa hispida*																																																			
Freitas et al. (2008)	Y	T	Y	Y	Y				Y	Y				Y														Y																							
Lone et al. (2019)	Y	S			Y									Y																																					
14. Baikal seal, *Pusa sibirica*																																																			
Stewart et al. (1996)	N*	-																																																	
Watanabe et al. (2015)	N*	-																																																	

(continued)

Table 5.1 (continued)

15. Caspian seal, *Pusa caspica*																																																			
Dmitrieva et al. (2016)	N*	-																																																
16. Harp seal, *Pagophilus groenlandicus*																																																			
Folkow et al. (2004)	Y	B			Y																																												
Nord□y et al. (2008)	Y	T +B												Y																																					
17. Ribbon seal, *Histriophoca fasciata*																																																			
Boveng et al. (2013)	N*	-																																																
18. Grey seal, *Halichoerus grypus*																																																			
Huon et al. (2015)	Y	S	Y		Y					Y			Y																					Y																	
Van Beest et al. (2019)	Y	B	Y		Y	Y				Y			Y															Y			Y																				

below, and several behavior measurements are related to time-spent, such as transit rate (e.g., Robinson et al. 2010). Among these is the First Passage Time (Fauchald and Tveraa 2003) for each location—the time it takes for an animal to pass through a circle of a given radius (e.g., Freitas et al. 2008b; Andersen et al. 2013; Vacquie-Garcia et al. 2017). Also related to time-spent is an animal's utilization distribution, 'a probabilistic model of home range that describes the relative amount of time that an animal spends in any place' (Seaman and Powell 1996, p. 2075). The kernel density estimator is a nonparametric method for estimating the utilization distribution through data smoothing (Silverman 1986) and is frequently used to describe the distribution of animals from tracking data, although the time between locations is not explicitly included. In a preliminary study of Weddell seal movements in winter, for example, Lake et al. (2006) calculated the home range (95% kernel) and core use area (50% kernel) from tracking data.

5.3.2.3 Behavior

In the third approach, a behavioral parameter is the response variable that is explained by the environment. This behavioral parameter or state can be defined based on parameters of the movement path or might be informed by ancillary data or observation. Edelhoff et al. (2016) give an overview of the methods used for such 'path segmentation'—detecting changes in animal movement characteristics along trajectories—and the outputs of these methods can be used as inputs (the response variable) when modeling the relationship between phocid movement and environmental variables. For example, Jonsen et al. (2019) fitted a movement model that estimates a continuous (0–1) movement persistence value (autocorrelation in speed and directionality) along the tracks of southern elephant seals. They then used linear mixed-effects regression to model the relationship between this movement persistence and the explanatory variables of sea ice cover, chlorophyll-a concentration, and the salinity difference between 600 and 200 m depths. This analysis showed that seals traveling to Antarctica reduced their move persistence with increasing sea ice concentration, and that seals foraging in pelagic waters lowered their movement persistence where circumpolar deep water shoaled (Jonsen et al. 2019). Yurkowski et al. (2016) fitted a behavior-switching state-space movement model (Jonsen et al. 2005) to 130 ringed seal tracks, dealing with non-regular (in time) location estimates that are observed with error. They then modeled the estimated behavioral state in response to variables, including sea ice concentration, chlorophyll-a concentration, and ocean depth (Yurkowski et al. 2016). Jonsen et al. (2013) and Patterson et al. (2008) give an overview of the use of state-space models for biologging data.

Some movement models directly incorporate environmental data (reviewed by Patterson et al. 2017 and Hooten et al. 2017), rather than fitting a separate model that relates a movement parameter or behavioral state to the environment. In Hidden Markov Models (HMMs, a special case of state-space models), the observed movement variables (e.g., step lengths and turning angles along a track) are the outcomes of an unobserved (hidden) state process (the behavioral state), which may be driven by environmental covariates (Langrock et al. 2012). Hidden Markov Models typically require regular location estimates with negligible error, such as those obtained from GPS tags, and have thus been limited in their application to phocid studies. These models can also be used to integrate different types of data. For example, McClintock et al. (2017) developed an HMM that integrated location data, dive data, and environmental data to distinguish six movement states ("hauled out on ice", "resting at sea", "hauled out on land", "mid-water foraging", "benthic foraging", and "transit") in bearded seal tracking data.

Alternatively, rather than using at-surface locations or behaviors, dive types or dive parameters could be used as the response variable. For example, whether gray seals perform shallow, benthic, or, pelagic dive types were linked to the bottom substrate (Jessopp et al. 2013). Weddell seals spent more time at the bottom of their dives, indicating greater foraging effort, in shallow areas with a low slope dominated by Antarctic Surface Water (Heerah et al. 2013). Dive information can also be used to improve the behavioral classification of surface locations (e.g., Bestley et al. 2013; Gryba et al. 2019). Carter et al. (2016) review methods for inferring foraging behavior from location and dive data in pinnipeds (see also Chap. 6).

5.3.3 Seals as Oceanographic Samplers

Relatively new applications of biologging technologies include using different environmental sensors in the instruments that are deployed on animals (e.g., temperature, salinity, fluorometry). These sensors collect environmental data from the diverse ocean environments visited by the animals being studied, providing information from areas that might be difficult, if not impossible, to reach by more traditional survey methods, such as oceanographic vessels and ARGO floats (Boehlert et al. 2001; Biuw et al. 2007; Boehme et al. 2008a, b; Charrassin et al. 2008; Costa et al. 2008; Piñones et al. 2019).

The animal-borne, Satellite-Relay Data Logger Conductivity-Temperature-Depth (SRDL-CTD) tag (Fig. 5.3) developed in 2000/2001 (Lydersen et al. 2002; Boehme et al. 2009) made it possible to use seals to gather data on two key oceanographic parameters—salinity and temperature—as they dived through the water column (Fedak 2013; Roquet et al. 2017; Harcourt et al. 2019). Through the research program SeaOS (Southern Elephant Seals as Oceanographic Samplers; Biuw et al. 2007; Charrassin et al. 2008) and later MEOP (Marine Mammals Exploring the Oceans Pole to Pole), seals have provided more than 500,000 vertical conductivity-temperature-depth profiles since 2003, which are freely available to the research community (Treasure et al. 2017). These data have contributed many oceanographic insights; two examples include the observations of warmer and saltier conditions over much of the Nordic Seas (Isachsen et al. 2014), and the identification of a new source of Antarctic Bottom Water formation (Ohshima et al. 2013). They have also improved large-scale modeling efforts of the state of the world's oceans, such as the ECCO

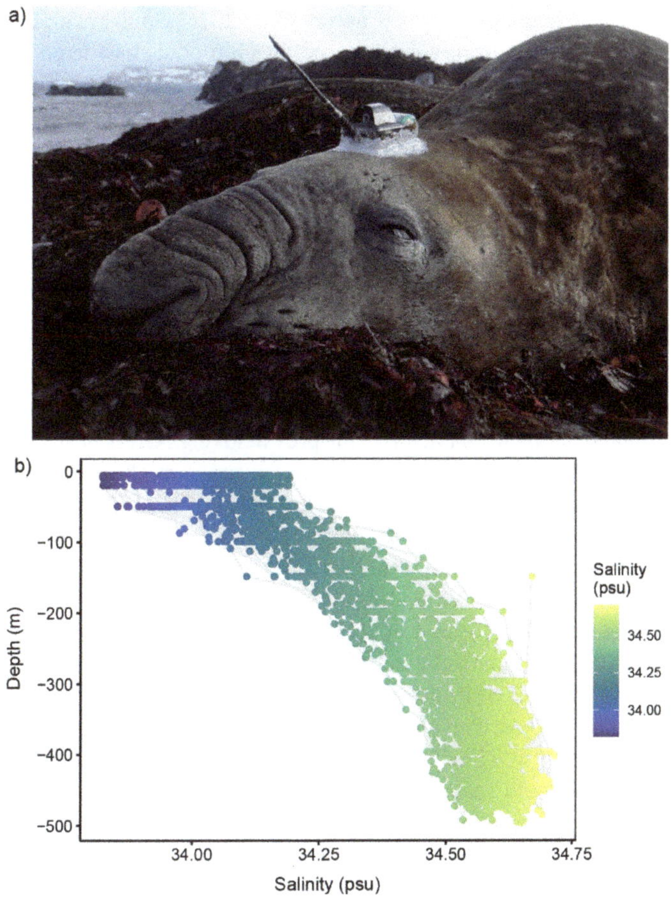

Fig. 5.3 **a** An adult male southern elephant seal equipped with a Sea Mammal Research Unit satellite-linked Conductivity-Temperature-Depth tag. **b** Shows the salinity-depth casts recorded by the instrument (and transmitted via the Argos satellite system) as the animal foraged near the South Shetland Islands, Antarctica, for ~2 months. Photograph: RR Reisinger; data from McIntyre et al. (2014)

(Estimating the Circulation and Climate of the Ocean) model, especially in marginal ice-covered seas in polar regions (e.g., Roquet et al. 2013).

Recently, a suite of new sensors has been added to animal-borne tags. For example, tags with dissolved oxygen sensors have been attached to southern elephant seals, contributing to documenting the decline in ocean oxygen concentration (Bailleul et al. 2015). In contrast, acoustic sensors have been used to estimate ocean weather, such as wind speed (Cazau et al. 2017). Fluorometers have been used to obtain in-situ measurements of chlorophyll-a concentration (e.g., Guinet et al. 2013; Keates et al. 2020), and efforts are underway to develop reliable measurements of fish biomass using animal-borne active acoustics (Lawson et al. 2015; Goulet et al. 2019). Environmental measurements from animal-borne tags are now considered an essential tool for global ocean observation (Harcourt et al. 2019) and enhance our ability to understand how animals behave as they move in their natural environment.

5.4 Effects of Environmental Characteristics on Habitat Use

Variations in prey distribution and abundance are the most important and direct factors determining the at-sea movement and distribution of seals (Chap. 6), but are challenging to measure directly. Ship-based surveys and trawls can be used concurrent with animal tracking to characterize the prey field (e.g., Grigg et al. 2012; Harvey et al. 2012; Saijo et al. 2017), but the synoptic coverage might be low (Kuhn et al. 2015). However, such surveys or sampling sites only cover a small part of the population's foraging range, necessitating interpolation and extrapolation between and beyond sampling sites (e.g., Grigg et al. 2012; Harvey et al. 2012). Onboard sensors, such as cameras (e.g., Davis et al. 1999; Watanabe et al. 2003; Naito et al. 2013, 2017; Wilson et al. 2017; Yoshino et al. 2020) and sonar tags (Lawson et al. 2015; Goulet et al. 2019), can characterize the prey field encountered by the animal, or provide a proxy thereof. Still, most studies focus on biophysical covariates from remote-sensing or models. The wide availability of biophysical data products revolutionized oceanography. They have allowed significant advances in understanding the habitat use of phocid seals. The rationale is that the features, conditions, and gradients captured by these products stimulate production and/or concentrate prey (or producers), thereby enhancing the local abundance and/or availability of prey to seals. As such, we expect that these biophysical covariates largely influence the movement and distribution of seals at different scales (Fig. 5.4).

The spatiotemporal scales of oceanographic processes and features vary greatly, from vertical mixing at centimeters to meters over time-scales of seconds to hours or days, to climate features such as the El-Niño Southern Oscillation that occurs over hundreds to thousands of kilometers at time scales of years to decades (Fig. 5.4; Crocker et al. 2006). The complex interactions of these biophysical processes at different spatiotemporal scales also produce highly heterogeneous seascapes (Scales et al. 2017) that seals must navigate to find prey. This hierarchically structured and heterogeneous arrangement of environmental features and prey drives the hierarchical and scale-dependent movements of phocids (Bailleul et al. 2008; Thums et al. 2011; Adachi et al. 2017; cf. Rodriguez et al. 2017; and in marine predators more broadly, e.g., Fauchald et al. 2000; Pinaud and Weimerskirch 2007; Benoit-Bird et al. 2013). At the same time, spatial and temporal scales also influence analyses of the relationship between seal behavior and prey or environmental covariates, something that must be considered when studying these relationships (e.g., Grigg et al.

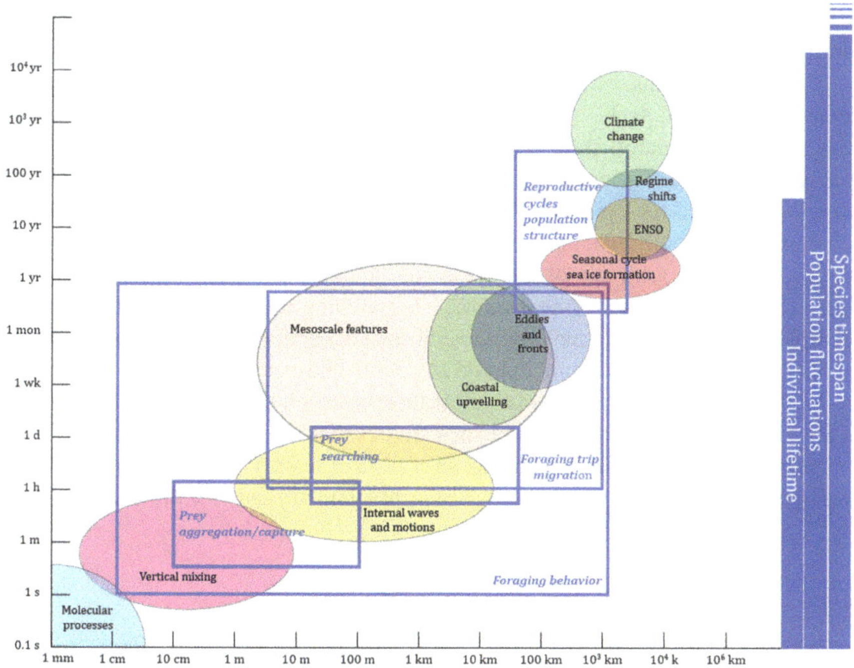

Fig. 5.4 Spatio-temporal scales of ocean features in relation to phocid life history. Spatio-temporal scales of various biophysical oceanographic features and processes (colored ellipses) in relation to the movement behavior (blue boxes) of phocid seals. Individual, population, and species temporal scales are shown on the right (blue bars). After Stommel (1963)

2009), particularly given the often-limited resolution of available movement data and environmental data.

At very large scales, processes such as climate change and climate variability influence the movements and habitats of seals through pervasive bottom-up effects (e.g., Crocker et al. 2006; Costa et al. 2010; Hamilton et al. 2015; Hindell et al. 2017; Abrahms et al. 2018a, b). Over longer time scales, climate affects population structure and species distributions. For example, southern elephant seals during the Holocene had colonies on the Antarctic coast, far south of their present Subantarctic colonies. This allowed them to take advantage of ice-free habitats during a warm period (Hall et al. 2006).

Smaller mesoscale (~100 km, weeks-months) features influence the movements of phocids throughout their foraging trips. For example, mesoscale eddies are linked with the movements of southern (e.g., Dragon et al. 2010; Cotté et al. 2015) and northern (Abrahms et al. 2018b) elephant seals at sea. At smaller scales (~1–10 km, days-weeks), features like sub-mesoscale fronts can influence the hourly-daily movements of phocids (Siegelman et al. 2019; Rivière et al. 2019).

At scales of a few meters, seal movements within a dive are likely driven solely by prey behavior. For example, Adachi et al. (2017) described a hierarchical pattern within dives, where northern elephant seals foraged mainly in small 3D patches with radius 8–10 m that were nested in slightly larger 3D patches with radius 17–19 m.

In the next sections, we discuss some examples of features driving the movement behavior of phocid seals. As examples, for each of the 18 extant phocid species (Berta 2018), we searched for papers that have related the at-sea movement and/or diving of that species, measured using telemetry, to environmental covariates (Table 5.1). We searched for examples, first, by using papers known to us and, second, by searching on Google Scholar (https://scholar.google.com) using the species name or common name with any of the keywords 'tracking', 'movement' and 'habitat'. Third, we used a 'snowballing' approach—searching the reference lists of papers identified in the previous steps and general review papers for each species—to find additional references. We could not find any published examples of telemetry studies on Mediterranean monk seals—the most threatened phocid species (Karamanlidis et al. 2016). We could not find published studies that linked telemetry data to environmental covariates for three other species (Baikal seals, Caspian seals, and ribbon seals). Still, we have listed studies that present telemetry data for these species.

5.4.1 Wind and Surface Currents

The complex interplay between atmospheric and oceanographic processes strongly influences ocean characteristics at several spatiotemporal scales (e.g., currents, coastal upwelling, and mixing). These processes ultimately provide habitat structures that can be used by phocid seals for different behaviors, such as feeding or migration. While ocean currents result from the interplay of different forces (e.g., winds, gravity, density gradients, Coriolis, tectonics), surface currents are ultimately driven by global wind patterns (Stommel 1948; Munk 1950; Price et al. 1987). Therefore, wind patterns influence the global circulation of the surface ocean and, by extension, features of the ocean ecosystems where true seals occur. For instance, the foraging behavior of northern elephant seals is associated with the location of the North Pacific Convergence Zone, NPCZ (Robinson et al. 2012; Simmons et al. 2007; Abrahms et al. 2018b), an area of the ocean where surface water masses collide as the result of global wind and surface currents (Polovina et al. 2017; Talley et al. 2011).

On a global scale, eastern boundary current ecosystems concentrate the most important areas of coastal upwelling (Carr and Kearns 2003; Chavez and Messié 2009). Winds blowing parallel to the coast set up an offshore Ekman transport that ultimately drives the coastal upwelling of nutrient-rich deep waters to the well-illuminated euphotic zone. This intensified fertilization makes the upwelling ecosystems the most biologically productive marine ecosystems of the world, supporting elevated rates of secondary production which, in turn, sustains large populations of schooling pelagic fish and a diverse community of predators that feed on them (Carr and Kearns 2003; Chavez and Messié 2009), including many marine mammals (e.g., rorquals, dolphins, sea lions, fur seals). Phocid seals, however, are mostly absent from the epipelagic trophic webs of eastern boundary currents upwelling ecosystems. The only phocid species that inhabits and forages in waters of a wind-driven coastal upwelling ecosystem is the Pacific harbor seal in the California Current System (Grigg et al. 2009). However, harbor seals take advantage of the overall higher productivity of this system as it cascades down to benthic and demersal prey, and they do not seem to be directly reacting to large-scale coastal upwelling events.

At a smaller local scale, wind stress is related to mixing, turbulence, and localized upwelling events, all of which have an impact on biological productivity and can influence the behavior of seals. Arctic harbor seals, for instance, change their diving behavior in response to wind-stress generated upwelling events, which presumably flood the shelf waters with offshore waters and bring a different assemblage of fishes that are consumed by the harbor seals (Blanchet et al. 2015).

Although the potential effect of currents on animal movement is often ignored (but see Lambardi et al. 2008; Sleeman et al. 2010; Chapman et al. 2011; Galli et al. 2012), new evidence points toward a stronger influence of currents on the trajectory and behavior of seals as measured from satellite telemetry data (Della Penna et al. 2015). This effect can be significant in areas where seals concentrate their foraging, and the influence currents can have on satellite telemetry-derived foraging metrics should be considered (Della Penna et al. 2015).

5.4.2 *Fronts*

Fronts, the boundaries between different water masses in the ocean, are associated with water circulation and flow patterns that enhance the abundance of prey through increased primary productivity and the physical concentration of prey (Acha et al. 2015). Diverse marine predators preferentially concentrate their foraging efforts in these areas (Bost et al. 2009; Scales et al. 2014).

The association between southern elephant seals and frontal and inter-frontal areas across the Southern Ocean has been well described in the literature (McConnell and Fedak 1996; Biuw et al. 2007, 2010; Hindell et al. 2016). Southern elephant seals with a mesopelagic foraging strategy predominantly use the Antarctic Circumpolar Current (ACC), an area characterized by a complex system of fronts (Subantarctic Front, SAF; Polar Front, PF; and the Southern ACC Front, SACCF) that mark the boundaries between zones of different physical, chemical, and biological properties. Despite these fronts being areas of enhanced primary productivity, southern elephant seals do not focus their Area Restricted Search behavior along with these features, suggesting that they can successfully forage on relatively sparse and patchily distributed prey (Hindell et al. 2016). On the other hand, areas south of the PF and SAACF appear to be generally associated with enhanced foraging success of some southern elephant seal populations, as indicated by positive changes in body condition (Biuw et al. 2007).

In the North Pacific, most northern elephant seals tracked on their June-January post-molt migration concentrated their foraging effort in a narrow band corresponding with the Transition Zone where cold nutrient-rich waters of the subpolar gyre meet the warm nutrient-poor waters of the subtropical gyre, driving productivity (Simmons et al. 2010; Robinson et al. 2012). At this time of year, the Transition Zone corresponds to a specific surface feature—the Transition Zone Chlorophyll Front (TZCF) that is important for various meso- and top predators (Polovina et al. 2017). During the seals' post-breeding migration in February-May, the TZCF is located further south and is no longer associated with the gyre-gyre boundary. The seals do not follow the front, instead preferring to forage at the gyre-gyre boundary that remains further north (Robinson et al. 2012).

5.4.3 Mesoscale and Sub-mesoscale Features

Oceanographic circulation is primarily dominated by mesoscale processes that take place at scales of tens to hundreds of kilometers in space and tens of days to a few months in duration. Beyond their importance in the ocean's physical and chemical structure, mesoscale processes have a high ecological relevance as they impact the distribution of nutrients, productivity, and prey, ultimately shaping the distribution and density of marine predators (Godo et al. 2012), including phocid seals.

The foraging behavior of phocid seals operates at a variety of scales, simultaneously responding to the natural variability in oceanography (Fig. 5.4). For instance, while northern elephant seals associate with the TZCF in the North Pacific at a regional scale, at a local scale, individuals select Lagrangian mesoscale features that result in enhanced energy gains (Abrahms et al. 2018b). The foraging behavior (habitat utilization and diving behavior) of northern elephant seals is influenced by semi-permanent anticyclonic eddies around the Alaska gyre area in the eastern North Pacific (Simmons et al. 2007).

The best example of the effects that mesoscale eddies have on the behavior of seals comes from the southern elephant seal. Several studies have used satellite telemetry and diving behavior data to investigate how their at-sea behavior is influenced by the presence of cyclonic and anticyclonic eddies (e.g., Campagna et al. 2006; Bailleul et al. 2010; Dragon et al. 2010). For instance, 60% of elephant seals from the Kerguelen Archipelago population intensified their foraging activity along the edges of cyclonic eddies in the Southern Ocean (Bailleul et al. 2010). This relationship appears to be complex, as the existence of a seasonal pattern in the association between foraging behavior and mesoscale features points to a more flexible behavior by the seals, evidence of the ability to respond to changes in the distribution of their prey (Cotté et al. 2015, see also Biuw et al. 2010).

At a smaller spatial scale (on the order of 1–10 km), sub-mesoscale features, such as sub-mesoscale eddies, sub-mesoscale fronts, and thin 'filaments' formed by stirring, also influence the distribution and abundance of biota (Lévy et al. 2018). Such features have also been shown to influence the at-sea behavior of southern elephant seals. Particularly, sub-mesoscale fronts and filaments modify the at-sea behavior of elephant seals, which concentrate their foraging in areas where the local dynamics of the front create favorable conditions that facilitate access to their prey (Lévy et al. 2012; Della Penna et al. 2015; Rivière et al. 2019; Siegelman et al. 2019).

5.4.4 *Productivity*

Due to their relatively large body size and associated energy demands, phocid seals generally require elevated levels of biological productivity and biomass to sustain their populations. With only two exceptions (Hawaiian and Mediterranean monk seals that inhabit oligotrophic and subtropical waters), all other populations of seals forage in highly productive neritic or oceanic ecosystems where prey are abundant.

Phytoplankton, the dominant primary producers in the ocean, depend on light and essential nutrients, such as nitrogen, for photosynthesis and growth. Therefore, the availability of light and nutrients are the main drivers of primary productivity (Falkowski and Raven 2013; Moore et al. 2013). Temperature influences the rate of the chemical reactions associated with metabolism and therefore is an essential determinant of large-scale productivity. However, the interaction of these growth-limiting factors with oceanographic processes, such as upwelling and mixed layer dynamics, results in the phytoplankton abundance patterns observed in three dimensions (Behrenfeld et al. 2005; Sathyendranath and Platt 2001; Siegel et al. 2013). Vertically, almost all primary production occurs in the surface layer of the ocean—the euphotic zone between 0 m and the photic depth (~100 m), where light levels are about 1% of those at the surface. The availability of macronutrients, such as nitrate and phosphate, is often a growth-limiting factor for phytoplankton. In some regions, micronutrients (e.g., iron) limit phytoplankton production ('high-nutrient low-chlorophyll' areas such as the Southern Ocean; Moore et al. 2013). Primary production is higher where such nutrients are brought to the surface, such as in upwelling regions (e.g., California Current System in the North Pacific, Humboldt Current System in the South Pacific, Benguela Current System in the South Atlantic), and coastal waters are generally more productive than open-ocean waters (Falkowski et al. 1998; Sathyendranath and Platt 2001).

Chlorophyll-a concentration—measured using remote-sensing—is frequently included in habitat models as a proxy for primary productivity (phytoplankton). However, phocid seals prey on animals ranging from zooplankton (such as krill) to squid and fish (Chap. 6). There is usually some delay between elevated primary production and the effects at higher trophic levels (Verity and Smetacek 1996; Croll et al. 2005). At the same time, phytoplankton is transported by ocean currents (McManus and Woodson 2012). This transport can result in a space–time mismatch between chlorophyll-a concentrations and the presence of seals if phytoplankton is depleted by the time elevated primary production manifests at higher trophic levels. Therefore, unclear or unexpected relationships are sometimes observed between chlorophyll-a concentrations and seal habitat use or foraging behavior (e.g., Bradshaw et al. 2004; Andersen et al. 2013; Yurkowski et al. 2016). This might be alleviated to some degree by using a time lag when including chlorophyll-a concentration as an environmental covariate. Using a more sophisticated Lagrangian approach, in which advection models are used to predict the transport of chlorophyll blooms, C. Cotté et al. (unpubl. data) show that southern elephant seals closely track these water parcels as they are transported (and develop) post-bloom. A second potential problem when linking surface chlorophyll-a concentration with habitat use is that the subsurface phytoplankton (and prey) distribution may differ from that at the surface (e.g., Saijo et al. 2017). Subsurface chlorophyll patches, associated with topographically-driven internal waves, were linked with foraging hotspots in several predators, including gray seals (Scott et al. 2010). Seal-borne fluorometers and light-level sensors are beginning to reveal more fine-scale, vertically resolved links between phytoplankton concentration and seal movements (see Box 5.2; Jaud et al. 2012; O'Toole et al. 2017). For example, in-situ fluorometry showed that water column fluorescence was significantly higher where spotted seals foraged (Gryba et al. 2019).

5.4.5 *Temperature*

Temperature, along with salinity and pressure, determines the density of the ocean water and ultimately defines water masses (Fig. 5.5). The horizontal and vertical variability of the ocean's temperature is defined by the rate of solar energy input (higher temperatures at lower latitudes and surface waters), and temperature is inextricably related to the ocean's biological cycles, from rates of primary productivity and oxygen dilution in water to the distribution patterns of organisms and sea ice formation and melting. While seals are endothermic homeotherms (i.e., they maintain their internal temperature above ambient), their at-sea distribution patterns and behaviors are influenced (directly and indirectly) by water temperature since the distribution of their ectothermic prey is, in turn, determined by the prey's temperature requirements (and thus by water masses). Colder ocean waters are generally more productive and, perhaps, as a consequence, seals have thrived and are more abundant at higher latitudes (subpolar and polar latitudes; Fig. 5.1).

Across the Southern Ocean, southern elephant seals that show a pelagic foraging strategy preferentially forage in association with the Circumpolar Deep Water (CDW), a relatively warm (2–4 °C) and nutrient-rich water mass (Biuw et al. 2007; Hindell et al. 2017). Adult female elephant seals that forage in association with the CDW have high foraging success, indicated by changes in their body condition inferred from changes in their drift dives (Biuw et al. 2003, 2007). Although the association is less clear than in southern elephant seals, northern elephant seals that

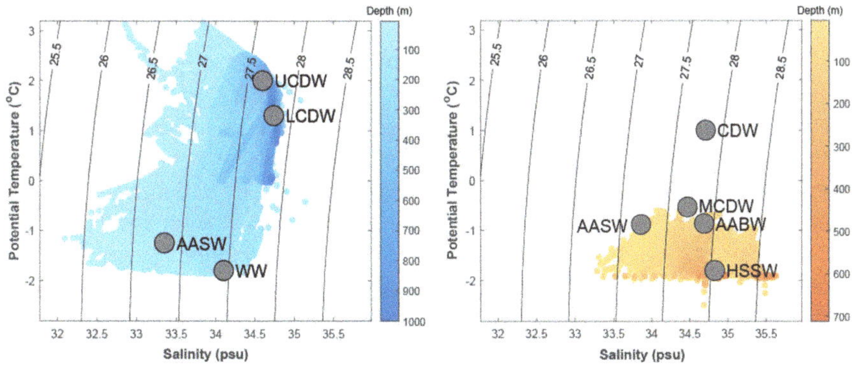

Fig. 5.5 Utilization of different water masses in the Southern Ocean by southern elephant seals (left) and Weddell seals (right). Southern elephant seals (left panel—blue) were tracked along the western Antarctic Peninsula, where seals concentrated their deep feeding (darker blue) with the two forms of Circumpolar Deep Water (Upper Circumpolar Deep Current, UCDW; Lower Circumpolar Deep Water, LCDW), while moving at the surface between Antarctic Surface Waters (AASW) and the cold Winter Water (WW). Weddell seals (right panel—orange) in the Ross Sea avoid the warm Circumpolar Deep Water (CDW) while concentrating their foraging in areas flooded with Antarctic Surface Water (AASW), Modified Circumpolar Deep Water (MCDW), Antarctic Bottom Water (AABW), and High-Salinity Shelf Water (HSSW). Southern elephant seal movement, diving, and environmental data from Costa et al. (2010) and Hückstädt et al. (2012b). Weddell seal movement, diving, and environmental data from Goetz and Costa (unpubl. data) and Piñones et al. (2019). The gray circles indicate the nucleoid potential temperature and salinity values for each particular water mass as defined for the western Antarctica Peninsula by Klinck et al. (2004) and for the Ross Sea by Rivaro et al. (2015). Environmental data for the seals were obtained from the MEOP dataset (Treasure et al. 2017)

show an oceanic foraging strategy mostly prefer to forage in an area in the North Pacific characterized by an inversion in the vertical temperature profile (Robinson et al. 2012), likely due to the convergence of different water masses.

It is unclear whether seals can use water temperature as an indicator of migration routes (e.g., McConnell et al. 2002) or to locate prey (e.g., sensing changes in water temperature indicating the presence of a front). If they do, however, we can expect modifications in their patterns of habitat utilization, movements, and behaviors as ocean temperatures rise due to global climate change (e.g., Boehme et al. 2012; Hückstädt et al. 2020). Evidence from satellite-tagged northern elephant seals suggests that during an abnormal warm event ("the blob"), seals modified their at-sea dive behavior compared to typical behavior during non-heatwave years and shifted their at-sea distribution (Holser 2020). In severe El-Niño years, female northern elephant seals had lower rates of mass gain and increased the duration of their foraging trips, although those that tracked cooler surface waters had higher rates of mass gain than those that did not (Crocker et al. 2006). Dive and movement data suggested they spent less time in prey patches and more time moving between prey patches, likely due to reduced prey availability in those patches (Crocker et al. 2006).

Box 5.2 Water Column Vertical Structure

Even though phocid seals utilize a three-dimensional environment, where foraging can be effectively separated from the surface by hundreds or even thousands of meters, attempts to understand how their behavior relates to their environment have usually relied on surface or static variables, such as sea surface temperature, altimetry or bathymetric features (previous sections). This is primarily due to our inability to effectively sample the water column at the appropriate temporal and spatial scales that match the at-sea behavior of diving seals. Nevertheless, the physical properties of the water column are fundamental to understand not only the ecology of marine top predators, but also how these organisms may be affected by annual and long-term changes in ocean climate (Boyd 1996; Hückstädt 2012; Hückstädt et al. 2020; Simmons et al. 2007; Hindell et al. 2020). During the past three decades, however, biologging technology has allowed us to move toward a more elaborate approach that incorporates data measured in depth obtained from both the instruments themselves (Biuw et al. 2007; Dragon et al. 2010; McIntyre et al. 2011; see Sect. 5.3.3).

As a direct consequence of the SeaOS and MEOP projects and the coordinated international effort to outfit southern elephant seals, among other species, with SRDL-CTD tags, several studies have addressed the effect that the subsurface oceanographic structure of the water column has on the at-sea behavior of this species (Fig. 5.5). Biuw et al. (2007) described an association between the foraging behavior of southern elephant seals, as inferred from an indirect index of foraging success, and particular water masses in the Southern Ocean, mainly the Circumpolar Deep Water (CDW). Newer evidence for the species from the Marion Island population points toward an effect of the intrusion of warmer CDW at depth, as identified from the maximum temperature deeper than 100 m or $T_{max}100$, on the foraging behavior of elephant seals (particularly females), which dived deeper and longer in areas where this temperature was higher (McIntyre et al. 2011).

As more species have been outfitted with these instruments, we are gathering a better understanding of how the water column's subsurface structure influences the at-sea behavior of seals. For example, northern elephant seals target an area of the North Pacific characterized by a shallow temperature inversion (see Sect. 5.4.2). McIntyre et al. (2013) investigated the foraging behavior of Weddell seals in relation to environmental predictors and found that seals tended to dive deeper and longer when the water column was more stratified, likely signaling a shift in targeted prey. Another example is the influence that the subsurface structure has on the foraging (diving) behavior of the highly specialized crabeater seal off the western Antarctic Peninsula. There, crabeater seals forage in areas away from deep intrusions of CDW (identified by the vertical structure of temperature) characterized by relatively shallow mixed layer depths (50–100 m; Hückstädt et al. 2020).

5.4.6 *Bathymetry and Ocean Floor Characteristics*

The ocean floor is largely out of reach for most seal species, with an average depth of about 3,700 m and 84% of the ocean floor being deeper than 2,000 m (Talley et al. 2011). The exception is continental margins (coasts and continental shelves), oceanic islands, and the top of some seamounts where depths are considerably shallower. However, the topography of the ocean floor influences the biological processes in surface waters and it can generate predictable areas with enhanced biological productivity and prey abundance. A well-known example is the topography-induced upwelling associated with seamounts and oceanic islands that force water masses flowing along the ocean floor to rise to shallower depths, bringing nutrients (e.g., nitrate and silicate) to the photic zone, resulting in enhanced primary productivity and the associated increases in prey availability (Genin 2004; Ladd et al. 2005; Morato et al. 2009).

Several species of phocids forage in association with bathymetric features (Fig. 5.6), and there is evidence that, due to their static nature, these areas are preferred by some individuals because they offer a relatively predictable source of food. In the following section, we explore how different bathymetric features affect the behavior of phocid seals.

5.4.6.1 Shelf Waters and Shelf Breaks

The continental shelf extends from the shore to the shelf break, where the continent starts to slope to the deep seafloor (Sharples and Simpson 2001; Talley et al. 2011). Continental shelf waters are the most biologically productive waters in the world's oceans, and consequently, support complex biological communities with high abundances of mid- and upper-trophic level animals. Due to their relatively shallow depths (generally <200 m, on average ~130 m; Talley et al. 2011; Simpson and Sharples 2012), nutrient enrichment from continental bedrock, river discharge, and coastal upwelling, these waters sustain most of the ocean's biological productivity. Many secondary circulation features associated with shelves and shelf breaks (e.g., shelf-break fronts [Acha et al. 2015], eddies) may aggregate prey and/or make them more readily available to diving predators. Consequently, shelves and shelf breaks are attractive to predators, including phocid seals (e.g., McConnell et al. 1992; Nordøy et al. 1995; Folkow et al. 1996; Nachtsheim et al. 2017). Cox et al. (2018) review drivers of marine mammal and seabird habitat use in shelf seas.

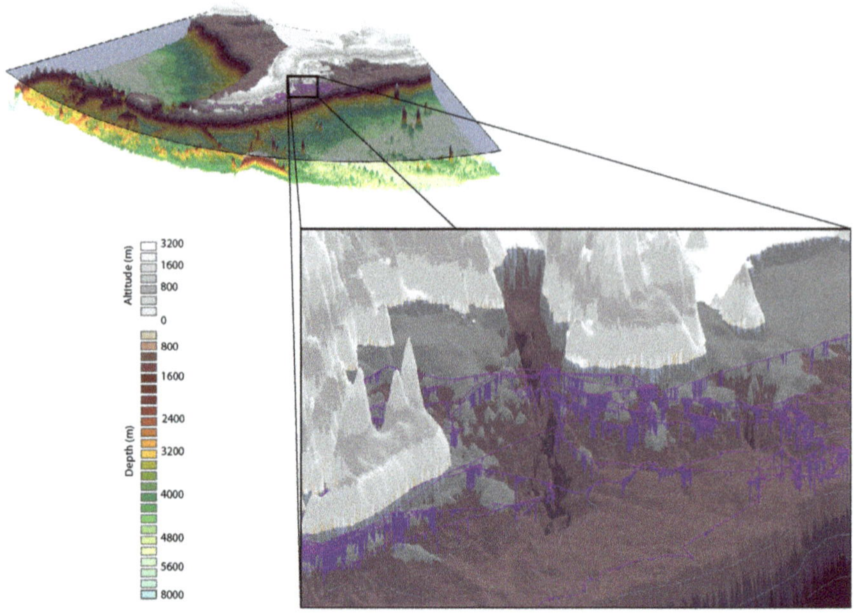

Fig. 5.6 Association between the patterns of habitat utilization of phocids and bathymetric features. The movement and diving behavior (purple) of crabeater seals (*Lobodon carcinophaga*) show how the species prefer shallow waters of the continental shelf of the western Antarctic Peninsula, foraging on their preferred prey, the Antarctic krill (*Euphausia superba*). The zoomed-in section shows a detailed view of their movements inside Marguerite Bay. It is evident that the seals avoid diving over a deep trough that cuts across the bay. Data from Hückstädt et al. (2020)

The Bering Sea shelf break is the site of a highly productive 'Green Belt', a shelf-slope front where eddies mix nutrients into the euphotic zone (Springer et al. 1996; Okkonen et al. 2004). Spotted seals have been tracked in the western Bering Sea, where they remained within 25 km of the shelf break (200 m isobath) 10–74% of the time throughout the year (Lowry et al. 2000). In the central and eastern Bering Sea, however, spotted seals did not target the 'Green Belt' associated with the shelf break, but instead, remained over the continental shelf where some species of fish concentrate (Lowry et al. 2000). Most (85%) southern elephant seals tracked in a study in the western Antarctic Peninsula moved along the outer margins of the continental shelf (Hückstädt et al. 2012b; Muelbert et al. 2013), and southern elephant seal females tracked from South Georgia showed slower movements on the shelf or shelf edge (McConnell and Fedak 1996; Biuw et al. 2007, 2010). Likewise, adult male southern elephant seals foraged predictably along the Antarctic continental shelf (Hindell et al. 2017). Male gray seals tracked from Sable Island in the western North Atlantic foraged along the shelf edge. whereas female gray seals preferred mid-shelf areas (Breed et al. 2006). Crabeater seals also associate with the shelf break during certain times of the year (Nordøy et al. 1995), while hooded seals associate with the break throughout most of their at-sea foraging periods (Folkow et al. 1996; Vacquie-Garcia et al. 2017).

5.4.6.2 Seamounts

Many pelagic predators seem to be attracted to seamounts due to higher productivity and prey biomass, driven by biophysical coupling (e.g., Santos et al. 2007; Kaschner 2007; Thompson 2007; Morato et al. 2008, 2010; Maxwell et al. 2012; Reisinger et al. 2015). Rather than locally enhanced primary productivity, seamount communities seem to be fueled mainly by the influx of food from surrounding waters, supporting zooplankton and higher trophic level organisms. The most likely mechanisms are trophic subsidy, where horizontal currents accelerate over the seamount and increase the food supply, and topographic trapping, where vertically migrating organisms are 'trapped' over the shallower waters (Genin 2004; Genin and Dower 2007; Clarke 2007; Rogers 2018). The most detailed information for phocids is from northern elephant seals in the North Pacific (Fig. 5.7). Among 179 individuals, three animals seemed to show benthic foraging behavior on seamounts or undersea ridges. Most of their foraging time was spent on seamounts, and this behavior was associated with high mass gain over the foraging trip (Maxwell et al. 2012). Some degree of association with seamounts has been observed in crabeater seals (Nordøy et al. 1995), juvenile southern elephant seals (Bornemann et al. 2000), and Hawaiian monk seals (Parrish et al. 2002).

5.4.6.3 Ocean Floor Structure and Composition

The behavior of benthic foraging seals is influenced by the type of ocean floor (e.g., rocky substrate or sediment), as these factors determine the distribution of their prey. For instance, bottom type influences habitat preference and foraging behavior of bearded, gray, harbor, and Hawaiian monk seals (Tollit et al. 1998; Parrish et al. 2002; Austin et al. 2006; Grigg et al. 2012; Jessopp et al. 2013).

Depending on their diet composition, benthic-feeding harbor and gray seals preferentially hunt in localized areas over soft sediments (sand and/or mud) or over a rocky substrate (McConnell et al. 1994; Tollit et al. 1998). The proportion of sand in bottom sediment is a strong predictor of harbor seal space-use (Bailey et al. 2014; Jones et al. 2017). For example, seals showed an increasing preference for bottom substrates with sand content up to 54%, above which usage declined (Jones et al. 2017). This association could be driven by their preference for sandeels and sand lance (*Ammodytes* spp), which live in mixed sediments of sand and gravel (Bailey et al. 2014; Jones et al. 2017).

New video technology has allowed a more detailed understanding of the influence of substrate characteristics on benthic-feeding Hawaiian monk seals. Seals have been observed searching for cryptic prey (e.g., octopus) in complex substrates, even turning

Fig. 5.7 Northern elephant seals foraging in association with seamounts. Tracking and diving data showed some individuals focused their foraging over particular flat-top seamounts (guyots) in the Northeastern Pacific Ocean, spending significant time foraging benthically over them, which positively impacted their feeding success (lipid mass gain). Data from Maxwell et al. (2012)

over rocks or digging in sandy sediments (Wilson et al. 2017). Dive data and seal-mounted cameras have revealed that some monk seals also visit coral beds to forage (Wilson et al. 2017; Parrish et al. 2002).

5.4.7 Ice

5.4.7.1 Sea Ice

Sea ice covers 16.6–27.5 million km^2 of the world's oceans, 3–6% of the earth's surface (Comiso 2010). The life history of most phocid species is inextricably linked with this ice. Sea ice is a physical obstacle that must be negotiated when foraging and surfacing to breathe, but it also provides a platform for phocids to breed, molt, and rest. Further, sea ice provides habitats for ice algae, mainly diatoms (Mock and Thomas 2005; Arrigo et al. 2010), which support zooplankton (amphipods, copepods, krill; Bluhm et al. 2010) that represent prey either directly to seals (e.g., crabeater seals and harp seals), or to their prey (e.g., planktivorous fish).

Many phocid species breed, molt, and rest on sea ice (including land fast ice), including gray seals (some populations), ribbon seals, hooded seals, ringed seals, harp seals, spotted seals, and bearded seals in the northern hemisphere (Laidre and Regehr 2018), and crabeater seals, Ross seals, Weddell seals, and leopard seals in the Antarctic (Bester et al. 2017). In Lake Baikal, which freezes over completely in winter (albeit not with sea ice, since it is a freshwater lake), the movement of Baikal seals is restricted by their need to remain near breathing holes (Stewart et al. 1996). The winter distribution of Weddell seals in the Antarctic is strongly influenced by access to breathing holes and open water (Kooyman 1975; Lake et al. 2005, 2006). Similarly, ringed seal movements are more restricted during winter, and interannual differences in winter movement extent suggest that sea ice cover has a strong direct effect on the movements of this species (Born et al. 2004; Harwood et al. 2015; Niemi et al. 2019; Chap. 15). Many phocid species favor the marginal ice zone. Freitas et al. (2008a) showed that ringed seals around Svalbard preferred sea ice concentrations around 40–80%.

Similarly, ringed seals use the marginal ice zone in the Barents Sea during summer and autumn (Lone et al. 2019). Young bearded seals are closely associated with the ice edge in the Bering Sea and the Chukchi Sea, where they track the seasonal expansion and retreat of ice (Fig. 5.8; Breed et al. 2018; Cameron et al. 2018). Greenland Sea harp seals used the marginal ice zone in the Greenland Sea between breeding and molting before migrating to the Barents Sea, where they foraged in more open water during summer and autumn (Folkow et al. 2004; Chap. 14). Harp seals from the Barents Sea population are also generally associated with the marginal ice zone but similarly use more open water in summer before retreating south as the sea ice expands in winter (Nordøy et al. 2008).

Other phocid species breed on land outside the sea ice zone but commute to the sea ice zone to forage. Southern elephant seals from Subantarctic islands, for example, travel hundreds of kilometers south into the Antarctic sea ice on foraging trips lasting months (e.g., Labrousse et al. 2018). Telemetry data often reveals new information, particularly for ice-associated species, given the challenges of direct observations in these habitats. For example, Ross seals, one of the least known phocids, had been considered ice-obligate feeders, but recent tracking studies show that they commute

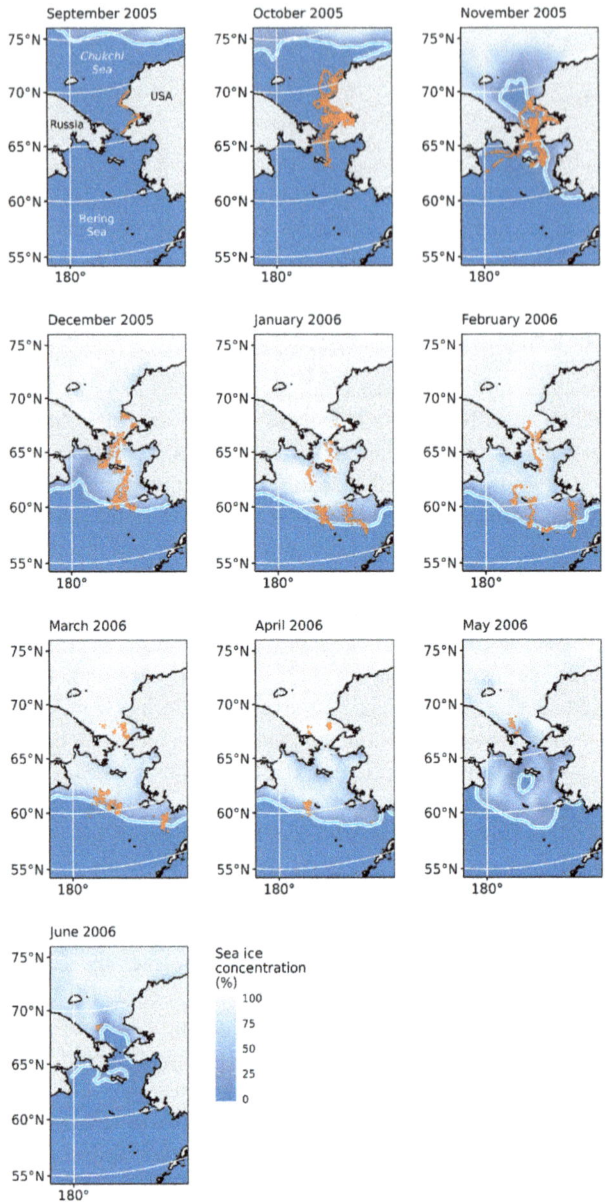

Fig. 5.8 Bearded seal movements in relation to sea ice. Argos location estimates (orange points) of 15 bearded seals in the Chukchi and Bering Seas (Cameron et al. 2018), shown in relation to monthly mean sea ice concentration. The sea ice edge (15% sea ice concentration) is represented with light blue lines. Seal tracking data are from Cameron et al. (2017), reported in Cameron et al. (2018), and sea ice concentration data are from Fetterer et al. (2017)

into ice-free pelagic waters to feed (Arcalís-Planas et al. 2015; Hückstädt 2018a; Wege et al. 2021).

In addition to overall species differences in ice association, there are also variations within species regarding their preference for sea ice. For example, among southern elephant seals, females followed the marginal ice zone as it expanded away from Antarctica in winter, switching from benthic to pelagic foraging, but juvenile males remained in the pack ice to forage benthically on the Antarctic shelf (Bailleul et al. 2007). Even among adult female southern elephant seals, individuals may be Subantarctic or high-Antarctic habitat-specialists (Hindell et al. 2017). In the Chukchi and Bering Seas, adult ringed seals remained in the ice, while subadults followed the ice edge as it contracted and expanded (Crawford et al. 2012). Ringed seals tracked from Svalbard either traveled to offshore waters with 40–80% ice coverage or remained in coastal waters near glacier fronts (Freitas et al. 2008b).

The sea ice zone is associated with high primary productivity in the Arctic and Antarctic (Engelsen et al. 2002; Qu et al. 2006; Arrigo et al. 2008; Vernet et al. 2008). When it melts in spring and summer, sea ice stimulates phytoplankton blooms and releases ice algae that use sea ice as a substrate. Freshwater input from melting sea ice stabilizes and thus stratifies the water column, allowing a shallow mixed layer to form, which effectively traps nutrients in the photic zone (Falk-Petersen et al. 2000; Arrigo et al. 2008; Vernet et al. 2008). The meltwater also seeds this mixed layer with ice algae, providing a critical food source for grazers that depend on sea ice at various life stages (Ackley and Sullivan 1994; Giesenhagen et al. 1999; Falk-Petersen et al. 2000; Schnack-Schiel 2003; Nicol 2006; Murphy 2019), such as krill (e.g., *Euphausia superba* and *E. crystallorophias* in the Southern Ocean and *Meganyctiphanes norvegica* and *Thysanoessa inermis* in the Northeast Atlantic) and copepods (e.g., *Paralabidocera antarctica* and *Calanus finmarchicus*). Phocids feed on these grazers (e.g., crabeater seals; Hückstädt 2012; Hückstädt et al. 2012a; Bengtson and Stewart 2018), or the larger predators, such as Antarctic silverfish (*Pleuragramma antarcticum*), that grazers support (e.g., Weddell seals; Hückstädt 2018b). In the Arctic, many phocids prey on Arctic cod (*Boreogadus saida*), which feed on ice-associated crustaceans that, in turn, graze on algae on the underside of ice (Ainley et al. 2003). Some species, such as harp seals and ringed seals, include krill (e.g., *Meganyctiphanes norvegica* and *Thysanoessa inermis*) and amphipods (*Themisto* sp. and *Gammarus* sp.) in their diets (e.g., Falk-Petersen et al. 2004; Lindstrøm et al. 2013; Hammill 2018; Lavigne 2018).

Future changes in sea ice extent and distribution will likely impact many phocid species; evidence for this comes from historical and current distribution changes. For example, as a result of extensive sea ice coverage of North Atlantic continental shelf waters around the Last Glacial Maximum, Boehme et al. (2012) estimate that the suitable habitat available to gray seals at that time might have been as small as 3% of their current habitat, with correspondingly low population sizes. A sea ice regime change in Svalbard, Norway shifted the marginal ice zone from over the continental shelf into deep Arctic waters. Following this change, ringed seals modified their foraging behavior, swimming further, diving longer with shorter surface intervals, resting less on sea ice, and showing Area Restricted Search behavior less frequently, suggesting increased foraging effort, which could eventually impact population dynamics (Hamilton et al. 2015). The population decrease in southern elephant seals from Macquarie Island over several decades has been related to sea ice changes in two of their core high-Antarctic foraging areas; increased sea ice extent and duration may exclude females from suitable foraging habitats (Hindell et al. 2017).

5.4.7.2 Polynyas

Polynyas are areas of persistent, recurring open water within the sea ice zone (Barber and Massom 2007). Polynyas are broadly categorized according to the way they form. Latent heat polynyas are created by prevailing winds or currents advecting sea ice away from an area, resulting in open water and heat loss to the air. Heat balance is maintained through the formation of more ice, which is also driven away by winds and currents (Morales Maqueda et al. 2004). In sensible heat polynyas, ocean heat in an area melts sea ice and prevents the formation of new sea ice (Morales Maqueda et al. 2004).

Polynyas are important to top predators in both the Antarctic and the Arctic since they are not only ice-free areas but areas of high productivity (Ainley et al. 2003; Karnovsky et al. 2007; Labrousse et al. 2018). The mechanisms supporting these food webs are similar to those discussed above for sea ice generally, but light plays an essential role in polynyas. Since they have little or no ice cover, polynyas are the first places where sunlight can stimulate production in spring, and this enhanced productivity can extend into late summer and early autumn (Arrigo and Van Dijken 2003; Deibel and Daly 2007). The high primary productivity also supports rich benthic communities via enhanced vertical carbon flux (Grebmeier and Barry 2007).

Malpress et al. (2017) showed that southern elephant seals in East Antarctica concentrated their movements in polynyas as autumn progressed to winter. Males tracked from the Kerguelen Islands spent up to 75% of their time in winter coastal polynyas in the same region; their foraging times were greater in the polynyas than outside them (Labrousse et al. 2018). Benthic-pelagic coupling plays an essential role in creating these 'winter oases' for southern elephant seals (Labrousse et al. 2018; Malpress et al. 2017). Winter tracks of ringed seals in the North Water polynya, northern Baffin Bay, were concentrated near the fast ice edge, suggesting that individuals spent most of their time at the ice edge or under fast ice (Teilmann et al. 2000). This was probably because the seals' primary prey—Arctic cod and hyperiid amphipods (*Parathemisto* spp.)—are associated with stable fast ice (Teilmann et al. 2000).

5.4.7.3 Arctic Tidewater Glaciers

Tidewater (or marine-terminating) glaciers are glaciers that terminate directly in marine waters, such as fjords. Such glaciers channel surface meltwater into the ocean through their hydrological network. This freshwater may be discharged beneath the glacier (subglacial discharge), on the glacier face (englacial discharge), or the glacier surface (supraglacial discharge; Chu 2014; Straneo and Cenedese 2015). Direct melting of the glacial ice in the ocean also releases submarine meltwater (Cape et al. 2019). Subglacial discharge and/or submarine melt leads to the formation of freshwater plumes that are less dense than the surrounding seawater and thus rise toward the surface at the glacier face, or until the plumes meet water with equal density, and then flow horizontally away from the glacier front (Straneo and Cenedese 2015). These rising freshwater plumes entrain ambient waters that contain nutrients (Hopwood et al. 2018) and biota (Lydersen et al. 2014). The volume of water transported vertically may be 10–30 times the initial freshwater input (Meire et al. 2017; Cape et al. 2019), and upwelling of deep, nutrient-rich waters into the photic zone stimulates production (Leu et al. 2015; Meire et al. 2017; Hopwood et al. 2018; Cape et al. 2019), even outside the fjords (Arrigo et al. 2017). The higher productivity and prey availability near glacier fronts resulting from these processes attract the prey of seals, such as polar cod (Arendt et al. 2013; Leu et al. 2015; Meire et al. 2017; Urbanski et al. 2017).

Ringed seals tracked around Svalbard, Norway, showed two movement strategies, either moving offshore or remaining inshore (Freitas et al. 2008b; Hamilton et al. 2019a). Seals that remained inshore frequented areas near tidal glacier fronts throughout the year (Freitas et al. 2008b; Hamilton et al. 2019a). From October to January, ringed seals dived deeper, longer, and for a greater proportion of time than the rest of the year, likely because meltwater plumes cease in mid-October, making these foraging zones less profitable (Hamilton et al. 2019a). In another example of seals being used to measure in-situ oceanographic conditions (see Sect. 5.3.3), ringed seals carrying CTD tags collected hydrographic data on freshwater plumes in Kongsfjorden, Svalbard (Everett et al. 2018). The study found that seals foraged in and around the buoyant plume, while the CTD tags provided new hydrographic data on glacial plumes.

Arctic environmental change is exerting a strong influence on the habitat use of ringed seals. Seals tracked in Svalbard from 1996 to 2003 spent nearly half of their time within 5 km of tidal glacier fronts. Following dramatic environmental changes in 2006 that led to increased Atlantic water intrusion into fjords and decreased sea ice, ringed seals intensified their use of tidal glacier fronts—seals tracked in 2010–2016 spent nearly all of their time near fronts, where they can presumably still prey on Arctic cod (Hamilton et al. 2019b).

5.5 Concluding Remarks

Throughout their evolution, phocid seals have adapted to diverse marine environments to successfully forage while being bound to a solid substrate for reproduction, resting, and molting. Their reliance on marine prey has shaped their behaviors and foraging strategies, which are inherently linked to the distribution and abundance of their prey. The latter is determined by oceanographic features that promote biological production or physically force the aggregation of prey, creating areas where foraging efficiency is maximized. The physical properties of the water column are fundamental to understand not only the behavior and ecology of true seals, but also how they may be affected by annual and long-term changes in ocean climate. Several studies forecast changes in the movement patterns and at-sea distribution of phocid seals in response to global climate change (e.g., Hindell et al. 2020; Hückstädt et al. 2020; Costa et al. 2010; Croxall 1992; Hamilton et al. 2019b; Laidre and Regehr 2018). While these impacts are primarily seen as a consequence of changes in prey distribution, phocid seals in polar latitudes additionally face the challenge of reduced physical resting, breeding, and molting habitat due to the disappearance of ice.

We are at the onset of an era where biologging developments allow us to directly link the at-sea behavior of phocids with the environment they experience at sea. Furthermore, advances in biologging technology, together with progress in sensor technologies, computing power, and analytical approaches, allow us to expand our understanding of the relationship between seals' behavior and the environment beyond surface water properties. Perspectives that consider the water column structure are fundamental to understanding how phocids behave in their three-dimensional environment and measure changes in the harder-to-sample deep ocean. Thus, our ability is rapidly increasing to empirically address issues related to the influence of oceanography on the at-sea patterns of habitat utilization and behavior of true seals, and ultimately the effects on lifetime fitness and population parameters.

Acknowledgements LAH was supported by the Office of Naval Research N00014-18-1-2822, the E&P Sound and Marine Life Joint Industry Project of the International Association of Oil and Gas Producers (JIP2207–23); and the NSF Office of Polar Programs awards 1644256 and 2054963. Bathymetric data used in the figures were obtained from GEBCO Compilation Group (2020).

References

Aarts G, MacKenzie M, McConnell B, Fedak M, Matthiopoulos J (2008) Estimating space-use and habitat preference from wildlife telemetry data. Ecography 31(1):140–160

Abrahms B, Hazen EL, Bograd SJ et al (2018a) Climate mediates the success of migration strategies in a marine predator. Ecol Lett 21(1):63–71. https://doi.org/10.1111/ele.12871

Abrahms B, Scales KL, Hazen EL, Bograd SJ, Schick RS, Robinson PW, Costa DP (2018b) Mesoscale activity facilitates energy gain in a top predator. Proc R Soc B: Biol Sci 285(1885):20181101. https://doi.org/10.1098/rspb.2018.1101

Acha EM, Piola A, Iribarne O, Mianzan H (2015) Ecological processes at marine fronts: oases in the ocean. Springer, Heidelberg

Ackley SF, Sullivan C (1994) Physical controls on the development and characteristics of Antarctic sea ice biological communities—a review and synthesis. Deep Sea Res Part I 41(10):1583–1604

Adachi T, Costa DP, Robinson PW, Peterson SH, Yamamichi M, Naito Y, Takahashi A (2017) Searching for prey in a three-dimensional environment: hierarchical movements enhance foraging success in northern elephant seals. Funct Ecol 31(2):361–369

Ainley DG, Tynan CT, Stirling I (2003) Sea ice: a critical habitat for polar marine mammals and birds. In: Thomas DN, Dieckmann GS (eds) Sea ice: an introduction to its physics, chemistry, biology and geology. Blackwell, Oxford, pp 240–266

Andersen JM, Wiersma YF, Stenson GB, Hammill MO, Rosing-Asvid A, Skern-Maurizen M (2013) Habitat selection by hooded seals (*Cystophora cristata*) in the Northwest Atlantic Ocean. ICES J Mar Sci 70(1):173–185. https://doi.org/10.1093/icesjms/fss133

Arcalís-Planas A, Sveegaard S, Karlsson O, Harding KC, Wåhlin A, Harkonen T, Teilmann J (2015) Limited use of sea ice by the Ross seal (*Ommatophoca rossii*), in Amundsen Sea, Antarctica, using telemetry and remote sensing data. Polar Biol 38(4):445–461

Arendt KE, Juul-Pedersen T, Mortensen J, Blicher ME, Rysgaard S (2013) A 5-year study of seasonal patterns in mesozooplankton community structure in a sub-Arctic fjord reveals dominance of *Microsetella norvegica* (Crustacea, Copepoda). J Plankton Res 35(1):105–120

Arrigo KR, Mock T, Lizotte MP (2010) Primary producers and sea ice. In: Thomas DN, Dieckmann GS (eds) Sea ice, 2nd edn. Blackwell Publishing Ltd., West Sussex, pp 283–325

Arrigo KR, Van Dijken GL (2003) Phytoplankton dynamics within 37 Antarctic coastal polynya systems. J Geophys Res: Oceans 108(C8)

Arrigo KR, van Dijken GL, Bushinsky S (2008) Primary production in the Southern Ocean, 1997–2006. J Geophys Res: Oceans 113(C8)

Arrigo KR, van Dijken GL, Castelao RM, Luo H, Rennermalm ÅK, Tedesco M, Mote TL, Oliver H, Yager PL (2017) Melting glaciers stimulate large summer phytoplankton blooms in southwest Greenland waters. Geophys Res Lett 44(12):6278–6285

Austin D, Bowen WD, McMillan JI, Iverson SJ (2006) Linking movement, diving, and habitat to foraging success in a large marine predator. Ecology 87(12):3095–3108. https://doi.org/10.1890/0012-9658(2006)87[3095:lmdaht]2.0.co;2

Bailey H, Hammond PS, Thompson PM (2014) Modelling harbour seal habitat by combining data from multiple tracking systems. J Exp Mar Biol Ecol 450:30–39

Bailleul F, Charrassin J-Bt, Ezraty R, Girard-Ardhuin F, McMahon CR, Field IC, Guinet C (2007) Southern elephant seals from Kerguelen Islands confronted by Antarctic Sea ice. Changes in movements and in diving behaviour. Deep Sea Res Part II: Top Stud Oceanogr 54(3–4):343–355

Bailleul F, Pinaud D, Hindell M, Charrassin JB, Guinet C (2008) Assessment of scale-dependent foraging behaviour in southern elephant seals incorporating the vertical dimension: a development of the first passage time method. J Anim Ecol 77(5):948–957. https://doi.org/10.1111/j.1365-2656.2008.01407.x

Bailleul F, Cotté C, Guinet C (2010) Mesoscale eddies as foraging area of a deep-diving predator, the southern elephant seal. Mar Ecol Prog Ser 408:251–264

Bailleul F, Vacquie-Garcia J, Guinet C (2015) Dissolved oxygen sensor in animal-borne instruments: an innovation for monitoring the health of oceans and investigating the functioning of marine ecosystems. PLoS ONE 10(7):e0132681. https://doi.org/10.1371/journal.pone.0132681

Barber DG, Massom RA (2007) The role of sea ice in Arctic and Antarctic polynyas. Elsevier Oceanogr Ser 74:1–54

Behrenfeld MJ, Boss E, Siegel DA, Shea DM (2005) Carbon-based ocean productivity and phytoplankton physiology from space. Glob Biogeochem Cycles 19(1)

Bengtson JL, Stewart BS (2018) Crabeater Seal. In: Würsig B, Thewissen JGM, Kovacs KM (eds) Encyclopedia of marine mammals. Academic Press, San Diego, pp 230–232. https://doi.org/10.1016/b978-0-12-804327-1.00098-4

Benoit-Bird KJ, Battaile BC, Nordstrom CA, Trites AW (2013) Foraging behavior of northern fur seals closely matches the hierarchical patch scales of prey. Mar Ecol Prog Ser 479:283–302

Berta A (2018) Pinnipeds. In: Würsig B, Thewissen JGM, Kovacs KM (eds) Encyclopedia of marine mammals. Academic Press, San Diego, pp 733–740

Berta A, Churchill M, Boessenecker RW (2018) The origin and evolutionary biology of pinnipeds: seals, sea lions, and walruses. Annu Rev Earth Planet Sci 46:203–228

Bester MN, Bornemann H, Plötz J, Burton HR (2006) At surface behaviour at location on spot of southern elephant seal MAR2005_sel_s_m_05 from Marion Island. PANGAEA. https://doi.org/10.1594/PANGAEA.441560

Bester MN, Bornemann H, McIntyre T (2017) Antarctic marine mammals and sea ice. In: Thomas DN (ed) Sea ice, 3rd edn. Wiley, Hoboken, pp 534–555

Bestley S, Jonsen ID, Hindell MA, Guinet C, Charrassin JB (2013) Integrative modelling of animal movement: incorporating in situ habitat and behavioural information for a migratory marine predator. Proc Biol Sci 280(1750):20122262. https://doi.org/10.1098/rspb.2012.2262

Biuw M, McConnell B, Bradshaw CJ, Burton H, Fedak M (2003) Blubber and buoyancy: monitoring the body condition of free-ranging seals using simple dive characteristics. J Exp Biol 206(19):3405–3423

Biuw M, Boehme L, Guinet C et al (2007) Variations in behavior and condition of a Southern Ocean top predator in relation to in situ oceanographic conditions. Proc Natl Acad Sci USA 104(34):13705–13710. https://doi.org/10.1073/pnas.0701121104

Biuw M, Nost OA, Stien A, Zhou Q, Lydersen C, Kovacs KM (2010) Effects of hydrographic variability on the spatial, seasonal and diel diving patterns of southern elephant seals in the eastern Weddell Sea. PLoS ONE 5(11):e13816. https://doi.org/10.1371/journal.pone.0013816

Blanchet MA, Lydersen C, Ims RA, Kovacs KM (2015) Seasonal, oceanographic and atmospheric drivers of diving behaviour in a temperate seal species living in the high Arctic. PLoS ONE 10(7):e0132686. https://doi.org/10.1371/journal.pone.0132686

Block BA, Jonsen ID, Jorgensen SJ et al (2011) Tracking apex marine predator movements in a dynamic ocean. Nature 475(7354):86–90. https://doi.org/10.1038/nature10082

Block BA, Holbrook CM, Simmons SE et al (2016) Toward a national animal telemetry network for aquatic observations in the United States. Anim Biotelem 4(1):6

Bluhm BA, Gradinger RR, Schnack-Schiel SB (2010) Sea ice meio- and macrofauna. In: Thomas DN, Dieckmann GS (eds) Sea ice, 2nd edn. Blackwell Publishing Ltd., West Sussex, pp 357–393

Boehlert GW, Costa DP, Crocker DE, Green P, O'Brien T, Levitus S, Le Boeuf BJ (2001) Autonomous pinniped environmental samplers: using instrumented animals as oceanographic data collectors. J Atmos Oceanic Tech 18(11):1882–1893. https://doi.org/10.1175/1520-0426(2001)018%3c1882:Apesui%3e2.0.Co;2

Boehme L, Lovell P, Biuw M, Roquet F, Nicholson J, Thorpe SE, Meredith MP, Fedak M (2009) Animal-borne CTD-satellite relay data loggers for real-time oceanographic data collection. Ocean Sci 5:685–695

Boehme L, Meredith MP, Thorpe SE, Biuw M, Fedak M (2008a) Antarctic circumpolar current frontal system in the South Atlantic: monitoring using merged Argo and animal-borne sensor data. J Geophys Res 113(C9). https://doi.org/10.1029/2007jc004647

Boehme L, Thorpe SE, Biuw M, Fedak M, Meredith MP (2008b) Monitoring drake passage with elephant seals: frontal structures and snapshots of transport. Limnol Oceanogr 53 (5part2):2350–2360. https://doi.org/10.4319/lo.2008.53.5_part_2.2350

Boehme L, Thompson D, Fedak M, Bowen D, Hammill MO, Stenson GB (2012) How many seals were there? The global shelf loss during the last glacial maximum and its effect on the size and distribution of grey seal populations. PloS One 7(12):e53000

Bonadonna F, Lea M-A, Dehorter O, Guinet C (2001) Foraging ground fidelity and route-choice tactics of a marine predator: the Antarctic fur seal *Arctocephalus gazella*. Mar Ecol Prog Ser 223:287–297

Bonner WN (1984) Lactation strategies in pinnipeds: problems for a marine mammalian group. In: Peaker M, Vernon RG, Knight CH (eds) Physiological strategies in lactation. Academic Press, London, pp 253–272

Born EW, Teilmann J, Acquarone M, Riget FF (2004) Habitat use of ringed seals (*Phoca hispida*) in the North Water area (North Baffin Bay). Arctic 57:129–142

Bornemann H, Kreyscher M, Ramdohr S, Martin T, Carlini A, Sellmann L, Plötz J (2000) Southern elephant seal movements and Antarctic sea ice. Antarct Sci 12(1):3–15

Bost CA, Cotté C, Bailleul F et al (2009) The importance of oceanographic fronts to marine birds and mammals of the southern oceans. J Mar Syst 78(3):363–376

Bost CA, Cotté C, Terray P et al (2015) Large-scale climatic anomalies affect marine predator foraging behaviour and demography. Nat Commun 6(1):8220. https://doi.org/10.1038/ncomms9220

Boyd IL (1996) Temporal scales of foraging in a marine predator. Ecology 77(2):426–434. https://doi.org/10.2307/2265619

Bradshaw CJ, Hindell MA, Best NJ, Phillips KL, Wilson G, Nichols PD (2003) You are what you eat: describing the foraging ecology of southern elephant seals (*Mirounga leonina*) using blubber fatty acids. Proc R Soc B: Biol Sci 270(1521):1283–1292

Bradshaw CJ, Hindell MA, Sumner MD, Michael KJ (2004) Loyalty pays: potential life history consequences of fidelity to marine foraging regions by southern elephant seals. Anim Behav 68(6):1349–1360

Breed GA, Bowen WD, McMillan JI, Leonard ML (2006) Sexual segregation of seasonal foraging habitats in a non-migratory marine mammal. Proc R Soci b: Biol Sci 273(1599):2319–2326. https://doi.org/10.1098/rspb.2006.3581

Breed GA, Jonsen ID, Myers RA, Bowen WD, Leonard ML (2009) Sex-specific, seasonal foraging tactics of adult grey seals (*Halichoerus grypus*) revealed by state-space analysis. Ecology 90(11):3209–3221. https://doi.org/10.1890/07-1483.1

Breed GA, Cameron MF, Ver Hoef JM, Boveng PL, Whiting A, Frost KJ (2018) Seasonal sea ice dynamics drive movement and migration of juvenile bearded seals *Erignathus barbatus*. Mar Ecol Prog Ser 600:223–237

Burns JM, Costa DP, Fedak MA, Hindell MA, Bradshaw CJ, Gales NJ, McDonald B, Trumble SJ, Crocker DE (2004) Winter habitat use and foraging behavior of crabeater seals along the Western Antarctic Peninsula. Deep Sea Res Part II 51(17–19):2279–2303

Cameron MF, London JM, Frost K, Whiting A, Boveng P (2017) Satellite telemetry dataset (raw): juvenile bearded and spotted seals, 2004–2006, Kotzebue, Alaska. https://doi.org/10.24431/rw1k333

Cameron MF, Frost KJ, Ver Hoef JM, Breed GA, Whiting AV, Goodwin J, Boveng PL (2018) Habitat selection and seasonal movements of young bearded seals (*Erignathus barbatus*) in the Bering Sea. PLoS ONE 13(2):e0192743. https://doi.org/10.1371/journal.pone.0192743

Campagna C, Piola AR, Marin MR, Lewis M, Fernández T (2006) Southern elephant seal trajectories, fronts and eddies in the Brazil/Malvinas Confluence. Deep Sea Res Part I 53(12):1907–1924

Cape MR, Straneo F, Beaird N, Bundy RM, Charette MA (2019) Nutrient release to oceans from buoyancy-driven upwelling at Greenland tidewater glaciers. Nat Geosci 12(1):34–39

Carr M-E, Kearns EJ (2003) Production regimes in four eastern boundary current systems. Deep Sea Res Part II 50(22–26):3199–3221

Carter MI, Bennett KA, Embling CB, Hosegood PJ, Russell DJ (2016) Navigating uncertain waters: a critical review of inferring foraging behaviour from location and dive data in pinnipeds. Mov Ecol 4(1):25. https://doi.org/10.1186/s40462-016-0090-9

Cazau D, Bonnel J, Jouma'a J, Le Bras Y, Guinet C (2017) Measuring the marine soundscape of the Indian Ocean with southern elephant seals used as acoustic gliders of opportunity. J Atmos Oceanic Tech 34(1):207–223

Chapman JW, Klaassen RH, Drake VA, Fossette S, Hays GC, Metcalfe JD, Reynolds AM, Reynolds DR, Alerstam T (2011) Animal orientation strategies for movement in flows. Curr Biol 21(20):R861-870. https://doi.org/10.1016/j.cub.2011.08.014

Charrassin JB, Hindell M, Rintoul SR et al (2008) Southern Ocean frontal structure and sea-ice formation rates revealed by elephant seals. Proc Natl Acad Sci USA 105(33):11634–11639. https://doi.org/10.1073/pnas.0800790105

Chavez FP, Messié M (2009) A comparison of eastern boundary upwelling ecosystems. Prog Oceanogr 83(1–4):80–96

Cherel Y, Ducatez S, Fontaine C, Richard P, Guinet C (2008) Stable isotopes reveal the trophic position and mesopelagic fish diet of female southern elephant seals breeding on the Kerguelen Islands. Mar Ecol Prog Ser 370:239–247

Chu VW (2014) Greenland ice sheet hydrology: a review. Prog Phys Geogr 38(1):19–54

Clarke M (2007) Seamounts and cephalopods. In: Pitcher TJ, Morato T, Hart PJB, Clark M, Haggan N, Santos RS (eds) Seamounts: ecology, fisheries & conservation. Blackwell Publishing, Oxford, pp 207–229

Comiso JC (2010) Variability and trends of the global sea ice cover. In: Thomas DN, Dieckmann GS (eds) Sea ice, 2nd edn. Blackwell Publishing Ltd., West Sussex, pp 205–246

Costa D (1993a) The secret life of marine mammals: new tools for the study of their biology and ecology. Oceanography 6(3):120–128

Costa DP (1993b) The relationship between reproductive and foraging energetics and the evolution of the Pinnipedia. Symp Zool Soc Lond 66:293–314

Costa DP, Klinck JM, Hofmann EE, Dinniman MS, Burns JM (2008) Upper ocean variability in west Antarctic Peninsula continental shelf waters as measured using instrumented seals. Deep Sea Res Part II 55(3–4):323–337. https://doi.org/10.1016/j.dsr2.2007.11.003

Costa DP, Huckstadt LA, Crocker DE, McDonald BI, Goebel ME, Fedak MA (2010) Approaches to studying climatic change and its role on the habitat selection of Antarctic pinnipeds. Integr Comp Biol 50(6):1018–1030. https://doi.org/10.1093/icb/icq054

Costa DP, Breed GA, Robinson PW (2012) New insights into pelagic migrations: implications for ecology and conservation. Annu Rev Ecol Evol Syst 43:73–96

Cotté C, d'Ovidio F, Dragon A-C, Guinet C, Lévy M (2015) Flexible preference of southern elephant seals for distinct mesoscale features within the Antarctic Circumpolar Current. Prog Oceanogr 131:46–58

Cox S, Embling C, Hosegood P, Votier S, Ingram S (2018) Oceanographic drivers of marine mammal and seabird habitat-use across shelf-seas: a guide to key features and recommendations for future research and conservation management. Estuar Coast Shelf Sci 212:294–310

Crawford JA, Frost KJ, Quakenbush LT, Whiting A (2012) Different habitat use strategies by subadult and adult ringed seals (*Phoca hispida*) in the Bering and Chukchi seas. Polar Biol 35(2):241–255

Crocker DE, Costa DP, Le Boeuf BJ, Webb PM, Houser DS (2006) Impact of El Niño on the foraging behavior of female northern elephant seals. Mar Ecol Prog Ser 309:1–10

Croll DA, Marinovic B, Benson S, Chavez FP, Black N, Ternullo R, Tershy BR (2005) From wind to whales: trophic links in a coastal upwelling system. Mar Ecol Prog Ser 289:117–130. https://doi.org/10.3354/meps289117

Croxall JP (1992) Southern Ocean environmental changes: effects on seabird, seal and whale populations. Philos Trans R Soc Lond B Biol Sci 338(1285):319–328

Daneri G, Carlini A (2002) Fish prey of southern elephant seals, *Mirounga leonina*, at King George Island. Polar Biol 25(10):739–743

Davis RW, Fuiman LA, Williams TM, Collier SO, Hagey WP, Kanatous SB, Kohin S, Horning M (1999) Hunting behavior of a marine mammal beneath the Antarctic fast Ice. Science 283(5404):993–996. https://doi.org/10.1126/science.283.5404.993

Deibel D, Daly KL (2007) Zooplankton processes in Arctic and Antarctic polynyas. Elsevier Oceanogr Ser 74:271–322

Della Penna A, De Monte S, Kestenare E, Guinet C, d'Ovidio F (2015) Quasi-planktonic behavior of foraging top marine predators. Sci Rep 5(1):18063. https://doi.org/10.1038/srep18063

Deméré TA, Berta A, Adam PJ (2003) Chapter 3: Pinnipedimorph evolutionary biogeography. Bull Am Museum Nat Hist 32–76

Dmitrieva L, Jüssi M, Jüssi I, Kasymbekov Y, Verevkin M, Baimukanov M, Wilson S, Goodman SJ (2016) Individual variation in seasonal movements and foraging strategies of a land-locked, ice-breeding pinniped. Mar Ecol Prog Ser 554:241–256

Dragon A-C, Monestiez P, Bar-Hen A, Guinet C (2010) Linking foraging behaviour to physical oceanographic structures: southern elephant seals and mesoscale eddies east of Kerguelen Islands. Prog Oceanogr 87(1–4):61–71

Edelhoff H, Signer J, Balkenhol N (2016) Path segmentation for beginners: an overview of current methods for detecting changes in animal movement patterns. Mov Ecol 4(1):21. https://doi.org/10.1186/s40462-016-0086-5

Engelsen O, Hegseth EN, Hop H, Hansen E, Falk-Petersen S (2002) Spatial variability of chlorophyll-a in the Marginal Ice Zone of the Barents Sea, with relations to sea ice and oceanographic conditions. J Mar Syst 35(1–2):79–97

Everett A, Kohler J, Sundfjord A, Kovacs KM, Torsvik T, Pramanik A, Boehme L, Lydersen C (2018) Subglacial discharge plume behaviour revealed by CTD-instrumented ringed seals. Sci Rep 8(1):13467. https://doi.org/10.1038/s41598-018-31875-8

Falk-Petersen S, Hop H, Budgell WP, Hegseth EN, Korsnes R, Løyning TB, Ørbæk JB, Kawamura T, Shirasawa K (2000) Physical and ecological processes in the marginal ice zone of the northern Barents Sea during the summer melt period. J Mar Syst 27(1–3):131–159

Falk-Petersen S, Haug T, Nilssen KT, Wold A, Dahl TM (2004) Lipids and trophic linkages in harp seal (*Phoca groenlandica*) from the eastern Barents Sea. Polar Res 23(1):43–50

Falkowski PG, Raven JA (2013) Aquatic photosynthesis. Princeton University Press, Princeton

Falkowski PG, Barber RT, Smetacek VV (1998) Biogeochemical controls and feedbacks on ocean primary production. Science 281(5374):200–207. https://doi.org/10.1126/science.281.5374.200

Fauchald P, Tveraa T (2003) Using first-passage time in the analysis of area-restricted search and habitat selection. Ecology 84(2):282–288. https://doi.org/10.1890/0012-9658(2003)084[0282:Ufptit]2.0.Co;2

Fauchald P, Erikstad KE, Skarsfjord H (2000) Scale-dependent predator–prey interactions: the hierarchical spatial distribution of seabirds and prey. Ecology 81(3):773–783

Fedak MA (2013) The impact of animal platforms on polar ocean observation. Deep Sea Res Part II 88–89:7–13. https://doi.org/10.1016/j.dsr2.2012.07.007

Fetterer F, Knowles K, Meier WN, Savoie M, Windnagel AK (2017) Sea Ice Index, Version 3. NSIDC: National Snow and Ice Data Center, Boulder, Colorado USA. https://doi.org/10.7265/N5K072F8

Field I, Hindell M, Slip D, Michael K (2001) Foraging strategies of southern elephant seals (*Mirounga leonina*) in relation to frontal zones and water masses. Antarct Sci 13(4):371

Folkow LP, Mårtensson P-E, Blix AS (1996) Annual distribution of hooded seals (*Cystophora cristata*) in the Greenland and Norwegian Seas. Polar Biol 16(3):179–189

Folkow LP, Nordøy ES, Blix AS (2004) Distribution and diving behaviour of harp seals (*Pagophilus groenlandicus*) from the Greenland Sea stock. Polar Biol 27(5):281–298

Fortin D, Beyer HL, Boyce MS, Smith DW, Duchesne T, Mao JS (2005) Wolves influence elk movements: behavior shapes a trophic cascade in Yellowstone National Park. Ecology 86(5):1320–1330

Freitas C, Kovacs KM, Ims RA, Fedak MA, Lydersen C (2008a) Ringed seal post-moulting movement tactics and habitat selection. Oecologia 155(1):193–204

Freitas C, Kovacs KM, Ims RA, Lydersen C (2008b) Predicting habitat use by ringed seals (*Phoca hispida*) in a warming Arctic. Ecol Model 217(1–2):19–32

Fyler C, Reeder T, Berta A, Antonelis G, Aguilar A, Androukaki E (2005) Historical biogeography and phylogeny of monachine seals (Pinnipedia: Phocidae) based on mitochondrial and nuclear DNA data. J Biogeogr 32(7):1267–1279

Galli S, Gaspar P, Fossette S, Calmettes B, Hays GC, Lutjeharms J, Luschi P (2012) Orientation of migrating leatherback turtles in relation to ocean currents. Anim Behav 84(6):1491–1500

Genin A (2004) Bio-physical coupling in the formation of zooplankton and fish aggregations over abrupt topographies. J Mar Syst 50(1–2):3–20

Genin A, Dower JF (2007) Seamount plankton dynamics. In: Pitcher TJ, Morato T, Hart PJB, Clark M, Haggan N, Santos RS (eds) Seamounts: ecology, fisheries & conservation. Blackwell Publishing, Oxford, pp 85–100

Giesenhagen HC, Detmer AE, de Wall J, Weber A, Gradinger RR, Jochem FJ (1999) How are Antarctic planktonic microbial food webs and algal blooms affected by melting of sea ice? Microcosm simulations. Aquat Microb Ecol 20(2):183–201

Godo OR, Samuelsen A, Macaulay GJ, Patel R, Hjollo SS, Horne J, Kaartvedt S, Johannessen JA (2012) Mesoscale eddies are oases for higher trophic marine life. PLoS ONE 7(1):e30161. https://doi.org/10.1371/journal.pone.0030161

Goetsch C, Conners MG, Budge SM, Mitani Y, Walker WA, Bromaghin JF, Simmons SE, Reichmuth C, Costa DP (2018) Energy-rich mesopelagic fishes revealed as a critical prey resource for a deep-diving predator using quantitative fatty acid signature analysis. Front Mar Sci 5:430

Goulet P, Guinet C, Swift R, Madsen PT, Johnson M (2019) A miniature biomimetic sonar and movement tag to study the biotic environment and predator-prey interactions in aquatic animals. Deep Sea Res Part I 148:1–11

Grebmeier J, Barry J (2007) Benthic processes in polynyas. Elsevier Oceanogr Ser 74:363–390

Grigg EK, Allen SG, Craven-Green DE, Klimley AP, Markowitz H, Elliott-Fisk DL (2012) Foraging distribution of Pacific harbor seals (*Phoca vitulina richardii*) in a highly impacted estuary. J Mammal 93(1):282–293

Grigg EK, Klimley AP, Allen SG, Green DE, Elliott-Fisk DL, Markowitz H (2009) Spatial and seasonal relationships between Pacific harbor seals (*Phoca vitulina richardii*) and their prey, at multiple scales. Fish Bull 107(3):359–372

Group GC (2020) GEBCO 2020 Grid. https://doi.org/10.5285/a29c5465-b138-234d-e053-6c86abc040b9

Gryba RD, Wiese FK, Kelly BP, Von Duyke AL, Pickart RS, Stockwell DA (2019) Inferring foraging locations and water masses preferred by spotted seals *Phoca largha* and bearded seals *Erignathus barbatus*. Mar Ecol Prog Ser 631:209–224

Guinet C, Xing X, Walker E et al (2013) Calibration procedures and first dataset of Southern Ocean chlorophyll a profiles collected by elephant seals equipped with a newly developed CTD-fluorescence tags. Earth Syst Sci Data 5(1):15–29

Hall BL, Hoelzel AR, Baroni C, Denton GH, Le Boeuf BJ, Overturf B, Topf AL (2006) Holocene elephant seal distribution implies warmer-than-present climate in the Ross Sea. Proc Natl Acad Sci USA 103(27):10213–10217. https://doi.org/10.1073/pnas.0604002103

Hamilton CD, Lydersen C, Ims RA, Kovacs KM (2015) Predictions replaced by facts: a keystone species' behavioural responses to declining arctic sea-ice. Biol Lett 11(11):20150803. https://doi.org/10.1098/rsbl.2015.0803

Hamilton CD, Kovacs KM, Lydersen C (2019a) Sympatric seals use different habitats in an Arctic glacial fjord. Mar Ecol Prog Ser 615:205–220

Hamilton CD, Vacquié-Garcia J, Kovacs KM, Ims RA, Kohler J, Lydersen C (2019b) Contrasting changes in space use induced by climate change in two Arctic marine mammal species. Biol Lett 15(3):20180834

Hammill MO (2018) Ringed seal. In: Würsig B, Thewissen JGM, Kovacs KM (eds) Encyclopedia of marine mammals. Academic Press, San Diego, pp 822–824. https://doi.org/10.1016/b978-0-12-804327-1.00218-1

Harcourt R, Sequeira AMM, Zhang X et al (2019) Animal-borne telemetry: an integral component of the ocean observing toolkit. Front Mar Sci 6:326

Harvey V, Hammill M, Swain D, Breed G, Lydersen C, Kovacs K (2012) Winter foraging by a top predator, the grey seal *Halichoerus grypus*, in relation to the distribution of prey. Mar Ecol Prog Ser 462:273–286

Harwood LA, Smith TG, Auld J, Melling H, Yurkowski DJ (2015) Seasonal movements and diving of ringed seals, *Pusa hispida*, in the Western Canadian Arctic, 1999–2001 and 2010–11. Arctic 68:193–209

Hays GC, Ferreira LC, Sequeira AMM et al (2016) Key questions in marine megafauna movement ecology. Trends Ecol Evol 31(6):463–475. https://doi.org/10.1016/j.tree.2016.02.015

Hazen EL, Jorgensen S, Rykaczewski RR et al (2013) Predicted habitat shifts of Pacific top predators in a changing climate. Nat Clim Chang 3(3):234–238

Heerah K, Andrews-Goff V, Williams G, Sultan E, Hindell M, Patterson T, Charrassin J-B (2013) Ecology of Weddell seals during winter: influence of environmental parameters on their foraging behaviour. Deep Sea Res Part II 88–89:23–33. https://doi.org/10.1016/j.dsr2.2012.08.025
Hindell MA, Burton HR, Slip DJ (1991) Foraging areas of southern elephant seals, *Mirounga leonina*, as inferred from water temperature data. Mar Freshw Res 42(2):115–128
Hindell MA, McMahon CR, Bester MN et al (2016) Circumpolar habitat use in the southern elephant seal: implications for foraging success and population trajectories. Ecosphere 7(5):e01213
Hindell MA, Sumner M, Bestley S, Wotherspoon S, Harcourt RG, Lea MA, Alderman R, McMahon CR (2017) Decadal changes in habitat characteristics influence population trajectories of southern elephant seals. Glob Change Biol 23(12):5136–5150. https://doi.org/10.1111/gcb.13776
Hindell MA, Reisinger RR, Ropert-Coudert Y et al (2020) Tracking of marine predators to protect Southern Ocean ecosystems. Nature 580(7801):87–92. https://doi.org/10.1038/s41586-020-2126-y
Holser RR (2020) A top predator in hot water: effects of a marine heatwave on foraging and reproduction of the northern elephant seal. Dissertation, University of California Santa Cruz
Hooten MB, Johnson DS, McClintock BT, Morales JM (2017) Animal movement: statistical models for telemetry data. CRC Press, Boca Raton
Hopwood MJ, Carroll D, Browning TJ, Meire L, Mortensen J, Krisch S, Achterberg EP (2018) Non-linear response of summertime marine productivity to increased meltwater discharge around Greenland. Nat Commun 9(1):3256. https://doi.org/10.1038/s41467-018-05488-8
Hückstädt LA (2012) Dealing with a fast changing environment: the trophic ecology of the southern elephant seal (*Mirounga leonina*) and crabeater seal (*Lobodon carcinophaga*) in the western Antarctica Peninsula. Doctoral Dissertation, UC Santa Cruz
Hückstädt LA (2018a) Ross Seal. In: Würsig B, Thewissen JGM, Kovacs KM (eds) Encyclopedia of marine mammals. Academic Press, San Diego, pp 835–837. https://doi.org/10.1016/b978-0-12-804327-1.00222-3
Hückstädt LA (2018b) Weddell Seal. In: Würsig B, Thewissen JGM, Kovacs KM (eds) Encyclopedia of marine mammals. Academic Press, San Diego, pp 1048–1051. https://doi.org/10.1016/b978-0-12-804327-1.00267-3
Hückstädt L, Burns J, Koch P, McDonald B, Crocker DE, Costa D (2012a) Diet of a specialist in a changing environment: the crabeater seal along the western Antarctic Peninsula. Mar Ecol Prog Ser 455:287–301
Hückstädt LA, Koch PL, McDonald BI, Goebel ME, Crocker DE, Costa DP (2012b) Stable isotope analyses reveal individual variability in the trophic ecology of a top marine predator, the southern elephant seal. Oecologia 169(2):395–406. https://doi.org/10.1007/s00442-011-2202-y
Hückstädt LA, Piñones A, Palacios DM, McDonald BI, Dinniman MS, Hofmann EE, Burns JM, Crocker DE, Costa DP (2020) Projected shifts in the foraging habitat of crabeater seals along the Antarctic Peninsula. Nat Clim Chang 10(5):472–477
Huon M, Jones EL, Matthiopoulos J, McConnell B, Caurant F, Vincent C (2015) Habitat selection of gray seals (*Halichoerus grypus*) in a marine protected area in France. J Wildl Manag 79(7):1091–1100. https://doi.org/10.1002/jwmg.929
Hussey NE, Kessel ST, Aarestrup K et al (2015) Aquatic animal telemetry: a panoramic window into the underwater world. Science 348(6240):1255642
Isachsen PE, Sørlie SR, Mauritzen C, Lydersen C, Dodd P, Kovacs KM (2014) Upper-ocean hydrography of the Nordic Seas during the International Polar Year (2007–2008) as observed by instrumented seals and Argo floats. Deep Sea Res Part I 93:41–59
Jaud T, Dragon A-C, Garcia JV, Guinet C (2012) Relationship between chlorophyll a concentration, light attenuation and diving depth of the southern elephant seal *Mirounga leonina*. PLoS One 7(10):e47444
Jessopp M, Cronin M, Hart T (2013) Habitat-mediated dive behavior in free-ranging grey seals. PLoS ONE 8(5):e63720. https://doi.org/10.1371/journal.pone.0063720

Jones EL, Sparling CE, McConnell BJ, Morris CD, Smout S (2017) Fine-scale harbour seal usage for informed marine spatial planning. Sci Rep 7(1):11581. https://doi.org/10.1038/s41598-017-11174-4
Jonsen ID, Flemming JM, Myers RA (2005) Robust state–space modeling of animal movement data. Ecology 86(11):2874–2880
Jonsen I, Basson M, Bestley S, Bravington M, Patterson T, Pedersen MW, Thomson R, Thygesen UH, Wotherspoon S (2013) State-space models for bio-loggers: a methodological road map. Deep Sea Res Part II 88:34–46
Jonsen I, McMahon C, Patterson T, Auger-Méthé M, Harcourt R, Hindell M, Bestley S (2019) Movement responses to environment: fast inference of variation among southern elephant seals with a mixed effects model. Ecology 100(1):e02566
Karamanlidis AA, Dendrinos P, De Larrinoa PF, Gücü AC, Johnson WM, Kiraç CO, Pires R (2016) The Mediterranean monk seal *Monachus monachus*: status, biology, threats, and conservation priorities. Mammal Rev 46(2):92–105
Karnovsky N, Ainley D, Lee P (2007) The impact and importance of production in polynyas to top-trophic predators: three case histories. Elsevier Oceanogr Ser 74:391–410
Kaschner K (2007) Air-breathing visitors to seamounts: marine mammals. In: Pitcher TJ, Morato T, Hart PJB, Clark M, Haggan N, Santos RS (eds) Seamounts: ecology, fisheries & conservation. Blackwell Publishing, Oxford, pp 230–238
Keates TR, Kudela RM, Holser RR, Hückstädt LA, Simmons SE, Costa DP (2020) Chlorophyll fluorescence as measured in situ by animal-borne instruments in the northeastern Pacific Ocean. J Mar Syst 203:103265. https://doi.org/10.1016/j.jmarsys.2019.103265
Klinck JM, Hofmann EE, Beardsley RC, Salihoglu B, Howard S (2004) Water-mass properties and circulation on the west Antarctic Peninsula Continental Shelf in Austral Fall and Winter 2001. Deep Sea Res Part II 51(17):1925–1946. https://doi.org/10.1016/j.dsr2.2004.08.001
Kooyman GL (1966) Maximum diving capacities of the Weddell seal, *Leptonychotes Weddelli*. Science 151(3717):1553–1554
Kooyman G (1975) A comparison between day and night diving in the Weddell seal. J Mammal 56(3):563–574
Kuhn CE, Sterling JT, Zeppelin TK (2015) Linking northern fur seal behavior with prey distributions: the impact of temporal mismatch between predator studies and prey surveys. Anim Biotelem 3(1):26
Labrousse S, Williams G, Tamura T et al (2018) Coastal polynyas: winter oases for subadult southern elephant seals in East Antarctica. Sci Rep 8(1):3183. https://doi.org/10.1038/s41598-018-21388-9
Ladd C, Jahncke J, Hunt GL Jr, Coyle KO, Stabeno PJ (2005) Hydrographic features and seabird foraging in Aleutian Passes. Fish Oceanogr 14:178–195
Laidre KL, Regehr EV (2018) Arctic marine mammals. In: Würsig B, Thewissen JGM, Kovacs KM (eds) Encyclopedia of marine mammals, 3rd edn. Academic Press, San Diego, pp 34–40. https://doi.org/10.1016/B978-0-12-804327-1.00049-2
Lake S, Wotherspoon S, Burton H (2005) Spatial utilisation of fast-ice by Weddell seals *Leptonychotes weddelli* during winter. Ecography 28(3):295–306
Lake S, Burton H, Wotherspoon S (2006) Movements of adult female Weddell seals during the winter months. Polar Biol 29(4):270–279
Lambardi P, Lutjeharms JR, Mencacci R, Hays GC, Luschi P (2008) Influence of ocean currents on long-distance movement of leatherback sea turtles in the Southwest Indian Ocean. Mar Ecol Prog Ser 353:289–301
Langrock R, King R, Matthiopoulos J, Thomas L, Fortin D, Morales JM (2012) Flexible and practical modeling of animal telemetry data: hidden Markov models and extensions. Ecology 93(11):2336–2342. https://doi.org/10.1890/11-2241.1
Lavigne DM (2018) Harp Seal. In: Würsig B, Thewissen JGM, Kovacs KM (eds) Encyclopedia of marine mammals. Academic Press, San Diego, pp 455–457. https://doi.org/10.1016/b978-0-12-804327-1.00146-1

Lawson GL, Hückstädt LA, Lavery AC, Jaffré FM, Wiebe PH, Fincke JR, Crocker DE, Costa DP (2015) Development of an animal-borne "sonar tag" for quantifying prey availability: test deployments on northern elephant seals. Anim Biotelem 3(1):22. https://doi.org/10.1186/s40317-015-0054-7

Le Boeuf BJ, Costa DP, Huntley AC, Feldkamp SD (1988) Continuous, deep diving in female northern elephant seals, *Mirounga Angustirostris*. Can J Zool 66(2):446–458

Le Boeuf B, Crocker DE, Costa DP, Blackwell SB, Webb PM, Houser DS (2000) Foraging ecology of northern elephant seals. Ecol Monogr 7(3):353–382

Le Bras Y, Jouma'a J, Guinet C (2017) Three-dimensional space use during the bottom phase of southern elephant seal dives. Mov Ecol 5(1):18

Leu E, Mundy C, Assmy P, Campbell K, Gabrielsen T, Gosselin M, Juul-Pedersen T, Gradinger R (2015) Arctic spring awakening–Steering principles behind the phenology of vernal ice algal blooms. Prog Oceanogr 139:151–170

Lévy M, Ferrari R, Franks PJ, Martin AP, Rivière P (2012) Bringing physics to life at the submesoscale. Geophys Res Lett 39(14)

Lévy M, Franks PJS, Smith KS (2018) The role of submesoscale currents in structuring marine ecosystems. Nat Commun 9(1):4758. https://doi.org/10.1038/s41467-018-07059-3

Lindstrøm U, Nilssen KT, Pettersen LM, Haug T (2013) Harp seal foraging behaviour during summer around Svalbard in the northern Barents Sea: diet composition and the selection of prey. Polar Biol 36(3):305–320

Lone K, Hamilton CD, Aars J, Lydersen C, Kovacs KM (2019) Summer habitat selection by ringed seals (*Pusa hispida*) in the drifting sea ice of the northern Barents Sea. Polar Res 38. https://doi.org/10.33265/polar.v38.3483

Lowry L, Burkanov V, Frost K, Simpkins M, Davis R, DeMaster D, Suydam R, Springer A (2000) Habitat use and habitat selection by spotted seals (*Phoca largha*) in the Bering Sea. Can J Zool 78(11):1959–1971

Lydersen C, Nøst OA, Lovell P, McConnell BJ, Gammelsrød T, Hunter C, Fedak MA, Kovacs KM (2002) Salinity and temperature structure of a freezing Arctic fjord-monitored by white whales (*Delphinapterus leucas*). Geophys Res Lett 29(23):34–31–34–34. https://doi.org/10.1029/2002gl015462

Lydersen C, Assmy P, Falk-Petersen S et al (2014) The importance of tidewater glaciers for marine mammals and seabirds in Svalbard, Norway. J Mar Syst 129:452–471

Malpress V, Bestley S, Corney S, Welsford D, Labrousse S, Sumner M, Hindell M (2017) Bio-physical characterisation of polynyas as a key foraging habitat for juvenile male southern elephant seals (*Mirounga leonina*) in Prydz Bay, East Antarctica. PloS One 12(9):e0184536

Manly B, McDonald L, Thomas DL, McDonald TL, Erickson WP (2007) Resource selection by animals: statistical design and analysis for field studies. Springer Science & Business Media, Berlin Heidelberg

Massie PP, McIntyre T, Ryan PG, Bester MN, Bornemann H, Ansorge IJ (2016) The role of eddies in the diving behaviour of female southern elephant seals. Polar Biol 39(2):297–307

Maxwell SM, Frank JJ, Breed GA, Robinson PW, Simmons SE, Crocker DE, Gallo-Reynoso JP, Costa DP (2012) Benthic foraging on seamounts: a specialized foraging behavior in a deep-diving pinniped. Mar Mamm Sci 28(3):E333

McClintock BT, London JM, Cameron MF, Boveng PL (2017) Bridging the gaps in animal movement: hidden behaviors and ecological relationships revealed by integrated data streams. Ecosphere 8(3):e01751. https://doi.org/10.1002/ecs2.1751

McConnell B, Fedak M (1996) Movements of southern elephant seals. Can J Zool 74(8):1485–1496

McConnell B, Chambers C, Fedak M (1992) Foraging ecology of southern elephant seals in relation to the bathymetry and productivity of the Southern Ocean. Antarct Sci 4(4):393–398

McConnell B, Fedak M, Lovell P, Hammond P (1994) The movements and foraging behaviour of grey seals. Grey seals in the North Sea and their interactions with fisheries. Sea Mammal Research Unit, Cambridge Final Report to the Ministry of Agriculture, Fisheries and Food, under Contract MF 503:88–148

McConnell B, Fedak M, Lovell P, Hammond P (1999) Movements and foraging areas of grey seals in the North Sea. J Appl Ecol 36(4):573–590
McConnell B, Fedak M, Burton HR, Engelhard G, Reijnders PJ (2002) Movements and foraging areas of naive, recently weaned southern elephant seal pups. J Anim Ecol 71(1):65–78
McIntyre T (2014) Trends in tagging of marine mammals: a review of marine mammal biologging studies. Afr J Mar Sci 36(4):409–422
McIntyre T, De Bruyn P, Ansorge I, Bester M, Bornemann H, Plötz J, Tosh C (2010) A lifetime at depth: vertical distribution of southern elephant seals in the water column. Polar Biol 33(8):1037–1048
McIntyre T, Bornemann H, Plötz J, Tosh CA, Bester MN (2011) Water column use and forage strategies of female southern elephant seals from Marion Island. Mar Biol 158(9):2125–2139
McIntyre T, Stansfield L, Bornemann H, Plötz J, Bester M (2013) Hydrographic influences on the summer dive behaviour of Weddell seals (*Leptonychotes weddellii*) in Atka Bay, Antarctica. Polar Biol 36(11):1693–1700
McIntyre T, Bornemann H, de Bruyn PN, Reisinger RR, Steinhage D, Márquez ME, Bester MN, Plötz J (2014) Environmental influences on the at-sea behaviour of a major consumer, *Mirounga leonina*, in a rapidly changing environment. Polar Res 33(1):23808
McMahon CR, Hindell MA, Charrassin JB et al (2019) Finding mesopelagic prey in a changing Southern Ocean. Sci Rep 9(1):19013. https://doi.org/10.1038/s41598-019-55152-4
McManus MA, Woodson CB (2012) Plankton distribution and ocean dispersal. J Exp Biol 215(Pt 6):1008–1016. https://doi.org/10.1242/jeb.059014
Meade J, Ciaglia M, Slip D, Negrete J, Márquez M, Mennucci J, Rogers T (2015) Spatial patterns in activity of leopard seals *Hydrurga leptonyx* in relation to sea ice. Mar Ecol Prog Ser 521:265–275
Meire L, Mortensen J, Meire P, Juul-Pedersen T, Sejr MK, Rysgaard S, Nygaard R, Huybrechts P, Meysman FJR (2017) Marine-terminating glaciers sustain high productivity in Greenland fjords. Glob Chang Biol 23(12):5344–5357. https://doi.org/10.1111/gcb.13801
Mitani Y, Andrews RD, Sato K, Kato A, Naito Y, Costa DP (2010) Three-dimensional resting behaviour of northern elephant seals: drifting like a falling leaf. Biol Let 6(2):163–166. https://doi.org/10.1098/rsbl.2009.0719
Mock T, Thomas DN (2005) Recent advances in sea-ice microbiology. Environ Microbiol 7(5):605–619. https://doi.org/10.1111/j.1462-2920.2005.00781.x
Moore C, Mills M, Arrigo K et al (2013) Processes and patterns of oceanic nutrient limitation. Nat Geosci 6(9):701–710
Morales Maqueda M, Willmott A, Biggs N (2004) Polynya dynamics: a review of observations and modeling. Rev Geophys 42(1)
Morato T, Varkey DA, Damaso C, Machete M, Santos M, Prieto R, Santos RS, Pitcher TJ (2008) Evidence of a seamount effect on aggregating visitors. Mar Ecol Prog Ser 357:23–32
Morato T, Bulman C, Pitcher TJ (2009) Modelled effects of primary and secondary production enhancement by seamounts on local fish stocks. Deep Sea Res Part II 56(25):2713–2719
Morato T, Hoyle SD, Allain V, Nicol SJ (2010) Seamounts are hotspots of pelagic biodiversity in the open ocean. Proc Natl Acad Sci USA 107(21):9707–9711. https://doi.org/10.1073/pnas.0910290107
Muelbert MM, de Souza RB, Lewis MN, Hindell MA (2013) Foraging habitats of southern elephant seals, *Mirounga leonina*, from the Northern Antarctic Peninsula. Deep Sea Res Part II 88:47–60
Munk WH (1950) On the wind-driven ocean circulation. J Meteorol 7(2):80–93
Murphy EJ (2019) Krill. In: Cochran JK, Bokuniewicz HJ, Yager PL (eds) Encyclopedia of ocean sciences, 3rd ed. Academic Press, Oxford, pp 553–560. https://doi.org/10.1016/B978-0-12-409548-9.11634-3
Nachtsheim DA, Jerosch K, Hagen W, Plötz J, Bornemann H (2017) Habitat modelling of crabeater seals (*Lobodon carcinophaga*) in the Weddell Sea using the multivariate approach Maxent. Polar Biol 40(5):961–976

Nachtsheim DA, Ryan S, Schröder M, Jensen L, Oosthuizen WC, Bester MN, Hagen W, Bornemann H (2019) Foraging behaviour of Weddell seals (*Leptonychotes weddellii*) in connection to oceanographic conditions in the southern Weddell Sea. Prog Oceanogr 173:165–179

Naito Y (2004) New steps in bio-logging science. Mem Natl Inst Polar Res Spec Iss 58:50–57

Naito Y, Le Boeuf BJ, Asaga T, Huntley AC (1989) Long-term diving records of an adult female northern elephant seal. Antarct Rec 33(1):1–9

Naito Y, Costa DP, Adachi T, Robinson PW, Fowler M, Takahashi A (2013) Unravelling the mysteries of a mesopelagic diet: a large apex predator specializes on small prey. Funct Ecol 27(3):710–717

Naito Y, Costa DP, Adachi T, Robinson PW, Peterson SH, Mitani Y, Takahashi A (2017) Oxygen minimum zone: an important oceanographic habitat for deep-diving northern elephant seals, *Mirounga Angustirostris*. Ecol Evol 7(16):6259–6270. https://doi.org/10.1002/ece3.3202

Nicol S (2006) Krill, currents, and sea ice: *Euphausia superba* and its changing environment. Bioscience 56(2):111–120

Niemi M, Liukkonen L, Koivuniemi M, Auttila M, Rautio A, Kunnasranta M (2019) Winter behavior of Saimaa ringed seals: non-overlapping core areas as indicators of avoidance in breeding females. PLoS ONE 14(1):e0210266. https://doi.org/10.1371/journal.pone.0210266

Nordøy ES, Folkow L, Blix AS (1995) Distribution and diving behaviour of crabeater seals (*Lobodon carcinophagus*) off Queen Maud Land. Polar Biol 15(4):261–268

Nordøy ES, Folkow LP, Potelov V, Prischemikhin V, Blix AS (2008) Seasonal distribution and dive behaviour of harp seals (*Pagophilus groenlandicus*) of the White Sea-Barents Sea stock. Polar Biol 31(9):1119

O'Toole M, Guinet C, Lea M-A, Hindell MA (2017) Marine predators and phytoplankton: how elephant seals use the recurrent Kerguelen plume. Mar Ecol Prog Ser 581:215–227

Ohshima KI, Fukamachi Y, Williams GD et al (2013) Antarctic Bottom Water production by intense sea-ice formation in the Cape Darnley polynya. Nat Geosci 6(3):235–240

Okkonen SR, Schmidt G, Cokelet E, Stabeno P (2004) Satellite and hydrographic observations of the Bering Sea 'Green Belt.' Deep Sea Res Part II 51(10–11):1033–1051

Parrish FA, Abernathy K, Marshall GJ, Buhleier BM (2002) Hawaiian monk seals (*Monachus schauinslandi*) foraging in deep-water coral beds. Mar Mamm Sci 18(1):244–258

Patterson TA, Thomas L, Wilcox C, Ovaskainen O, Matthiopoulos J (2008) State-space models of individual animal movement. Trends Ecol Evol 23(2):87–94. https://doi.org/10.1016/j.tree.2007.10.009

Patterson TA, Parton A, Langrock R, Blackwell PG, Thomas L, King R (2017) Statistical modelling of individual animal movement: an overview of key methods and a discussion of practical challenges. AStA Adv Stat Anal 101(4):399–438

Phillips SJ, Anderson RP, Schapire RE (2006) Maximum entropy modeling of species geographic distributions. Ecol Model 190(3–4):231–259

Pinaud D, Weimerskirch H (2007) At-sea distribution and scale-dependent foraging behaviour of petrels and albatrosses: a comparative study. J Anim Ecol 76(1):9–19. https://doi.org/10.1111/j.1365-2656.2006.01186.x

Piñones A, Hofmann EE, Costa DP, Goetz K, Burns JM, Roquet F, Dinniman MS, Klinck JM (2019) Hydrographic variability along the inner and mid-shelf region of the western Ross Sea obtained using instrumented seals. Prog Oceanogr 174:131–142. https://doi.org/10.1016/j.pocean.2019.01.003

Polovina JJ, Howell EA, Kobayashi DR, Seki MP (2017) The Transition Zone Chlorophyll Front updated: advances from a decade of research. Prog Oceanogr 150:79–85

Price JF, Weller RA, Schudlich RR (1987) Wind-driven ocean currents and Ekman transport. Science 238(4833):1534–1538. https://doi.org/10.1126/science.238.4833.1534

Qu B, Gabric AJ, Matrai PA (2006) The satellite-derived distribution of chlorophyll-a and its relation to ice cover, radiation and sea surface temperature in the Barents Sea. Polar Biol 29(3):196–210

Redfern J, Ferguson MC, Becker E et al (2006) Techniques for cetacean–habitat modeling. Mar Ecol Prog Ser 310:271–295

Reisinger RR, Keith M, Andrews RD, De Bruyn P (2015) Movement and diving of killer whales (*Orcinus orca*) at a Southern Ocean archipelago. J Exp Mar Biol Ecol 473:90–102
Reisinger RR, Raymond B, Hindell MA et al (2018) Habitat modelling of tracking data from multiple marine predators identifies important areas in the Southern Indian Ocean. Divers Distrib 24(4):535–550. https://doi.org/10.1111/ddi.12702
Rivaro P, Ianni C, Magi E, Massolo S, Budillon G, Smethie WM (2015) Distribution and ventilation of water masses in the western Ross Sea inferred from CFC measurements. Deep Sea Res Part I 97:19–28. https://doi.org/10.1016/j.dsr.2014.11.009
Rivière P, Jaud T, Siegelman L, Klein P, Cotté C, Le Sommer J, Dencausse G, Guinet C (2019) Submesoscale fronts modify elephant seals foraging behavior. Limnol Oceanogr Lett 4(6):193–204
Robinson PW, Simmons SE, Crocker DE, Costa DP (2010) Measurements of foraging success in a highly pelagic marine predator, the northern elephant seal. J Anim Ecol 79(6):1146–1156. https://doi.org/10.1111/j.1365-2656.2010.01735.x
Robinson PW, Costa DP, Crocker DE et al (2012) Foraging behavior and success of a mesopelagic predator in the northeast Pacific Ocean: insights from a data-rich species, the northern elephant seal. PLoS ONE 7(5):e36728. https://doi.org/10.1371/journal.pone.0036728
Rodriguez JP, Fernandez-Gracia J, Thums M et al (2017) Big data analyses reveal patterns and drivers of the movements of southern elephant seals. Sci Rep 7(1):112. https://doi.org/10.1038/s41598-017-00165-0
Rogers AD (2018) The biology of seamounts: 25 Years on. In: Sheppard C (ed) Advances in marine biology, vol 79. Academic Press, London, pp 137–224
Roquet F, Wunsch C, Forget G et al (2013) Estimates of the Southern Ocean general circulation improved by animal-borne instruments. Geophys Res Lett 40(23):6176–6180
Roquet F, Boehme L, Block B et al (2017) Ocean observations using tagged animals. Oceanography 30(2):139. https://doi.org/10.5670/oceanog.2017.235
Rule JP, Adams JW, Marx FG, Evans AR, Tennyson AJD, Scofield RP, Fitzgerald EMG (2020) First monk seal from the Southern Hemisphere rewrites the evolutionary history of true seals. Proc R Soc B: Biol Sci 287(1938):20202318. https://doi.org/10.1098/rspb.2020.2318
Saijo D, Mitani Y, Abe T, Sasaki H, Goetsch C, Costa DP, Miyashita K (2017) Linking mesopelagic prey abundance and distribution to the foraging behavior of a deep-diving predator, the northern elephant seal. Deep Sea Res Part II 140:163–170
Santos MA, Bolten AB, Martins HR, Riewald B, Bjorndal KA (2007) Air-breathing visitors to seamounts: sea turtles. In: Pitcher TJ, Morato T, Hart PJB, Clark M, Haggan N, Santos RS (eds) Seamounts: ecology, fisheries & conservation. Blackwell Publishing, Oxford, pp 239–244
Sathyendranath S, Platt T (2001) Primary production distribution. In: Steele JH (ed) Encyclopedia of ocean sciences. Academic Press, Oxford, pp 2272–2277. https://doi.org/10.1006/rwos.2001.0204
Savin SM, Abel L, Barrera E, Hodell D, Kennett J, Murphy M, Keller G, Killingley JS, Vincent E (1985) The evolution of Miocene surface and near-surface marine temperatures: oxygen isotopic evidence. Geol Soc Am Mem 163:49–82
Scales KL, Miller PI, Hawkes LA, Ingram SN, Sims DW, Votier SC (2014) On the front line: frontal zones as priority at-sea conservation areas for mobile marine vertebrates. J Appl Ecol 51(6):1575–1583
Scales KL, Alvarez-Berastegui D, Embling C, Ingram S (2017) Pelagic seascapes. In: Pittman SJ (ed) Seascape ecology. Wiley-Blackwell, Hoboken, pp 57–88
Scheffer A, Bost C-A, Trathan PN (2012) Frontal zones, temperature gradient and depth characterize the foraging habitat of king penguins at South Georgia. Mar Ecol Prog Ser 465:281–297
Schnack-Schiel SB (2003) The macrobiology of sea ice. In: Thomas DN, Dieckmann GS (eds) Sea ice: an introduction to its physics, chemistry, biology and geology. Blackwell, Oxford, pp 211–239
Scott B, Sharples J, Ross ON, Wang J, Pierce GJ, Camphuysen C (2010) Sub-surface hotspots in shallow seas: fine-scale limited locations of top predator foraging habitat indicated by tidal mixing and sub-surface chlorophyll. Mar Ecol Prog Ser 408:207–226

Seaman DE, Powell RA (1996) An evaluation of the accuracy of kernel density estimators for home range analysis. Ecology 77(7):2075–2085. https://doi.org/10.2307/2265701
Sharples J, Simpson J (2001) Shelf-sea and slope fronts. In: Steele JH, Thorpe SA, Turekian KK (eds) Encylopedia of ocean sciences. Academic Press, San Diego, pp 2760–2768
Siegel D, Behrenfeld MJ, Maritorena S, McClain CR, Antoine D, Bailey SW, Bontempi PS, Boss ES, Dierssen HM, Doney SC (2013) Regional to global assessments of phytoplankton dynamics from the SeaWiFS mission. Remote Sens Environ 135:77–91
Siegelman L, O'Toole M, Flexas M, Rivière P, Klein P (2019) Submesoscale ocean fronts act as biological hotspot for southern elephant seal. Sci Rep 9(1):1–13
Silverman BW (1986) Density estimation for statistics and data analysis, vol 26. CRC Press
Simpson JH, Sharples J (2012) Introduction to the physical and biological oceanography of shelf seas. Cambridge University Press, Cambridge. https://doi.org/10.1017/cbo9781139034098
Simmons SE, Crocker DE, Kudela RM, Costa DP (2007) Linking foraging behaviour of the northern elephant seal with oceanography and bathymetry at mesoscales. Mar Ecol Prog Ser 346:265–275
Simmons SE, Crocker DE, Hassrick JL, Kuhn CE, Robinson PW, Tremblay Y, Costa DP (2010) Climate-scale hydrographic features related to foraging success in a capital breeder, the northern elephant seal *Mirounga angustirostris*. Endanger Species Res 10:233–243
Skov H, Gunnlaugsson T, Budgell W, Horne J, Nøttestad L, Olsen E, Søiland H, Víkingsson G, Waring G (2008) Small-scale spatial variability of sperm and sei whales in relation to oceanographic and topographic features along the Mid-Atlantic Ridge. Deep Sea Res Part II 55(1–2):254–268
Sleeman JC, Meekan MG, Wilson SG, Polovina JJ, Stevens JD, Boggs GS, Bradshaw CJ (2010) To go or not to go with the flow: environmental influences on whale shark movement patterns. J Exp Mar Biol Ecol 390(2):84–98
Springer AM, McROY CP, Flint MV (1996) The Bering Sea Green Belt: shelf-edge processes and ecosystem production. Fish Oceanogr 5(3–4):205–223
Staniland IJ, Ratcliffe N, Trathan PN, Forcada J (2018) Long term movements and activity patterns of an Antarctic marine apex predator: the leopard seal. PLoS ONE 13(6):e0197767. https://doi.org/10.1371/journal.pone.0197767
Stewart BS, Petrov EA, Baranov E, Ivanov ATM (1996) Seasonal movements and dive patterns of juvenile Baikal seals, *Phoca Sibirica*. Mar Mamm Sci 12(4):528–542
Stommel H (1948) The westward intensification of wind-driven ocean currents. Trans Am Geophys Union 29(2):202–206
Stommel H (1963) Varieties of oceanographic experience: the ocean can be investigated as a hydrodynamical phenomenon as well as explored geographically. Science 139(3555):572–576. https://doi.org/10.1126/science.139.3555.572
Straneo F, Cenedese C (2015) The dynamics of Greenland's glacial fjords and their role in climate. Ann Rev Mar Sci 7:89–112
Talley LD, Pickard GL, Emery WJ, Swift JH (2011) Descriptive physical oceanography : an introduction. Academic Press, Boston
Teilmann J, Born EW, Acquarone M (2000) Behaviour of ringed seals tagged with satellite transmitters in the North Water polynya during fast-ice formation. Can J Zool 77(12):1934–1946
Thompson DR (2007) Air-breathing visitors to seamounts: importance of seamounts to seabirds. In: Pitcher TJ, Morato T, Hart PJB, Clark M, Haggan N, Santos RS (eds) Seamounts: ecology, fisheries & conservation. Blackwell Publishing, Oxford, pp 245–251
Thums M, Bradshaw CJ, Hindelli MA (2011) In situ measures of foraging success and prey encounter reveal marine habitat-dependent search strategies. Ecology 92(6):1258–1270. https://doi.org/10.1890/09-1299.1
Thurfjell H, Ciuti S, Boyce MS (2014) Applications of step-selection functions in ecology and conservation. Mov Ecol 2(1):4. https://doi.org/10.1186/2051-3933-2-4
Tollit D, Black A, Thompson P, Mackay A, Corpe H, Wilson B, Van Parijs S, Grellier K, Parlane S (1998) Variations in harbour seal *Phoca vitulina* diet and dive-depths in relation to foraging habitat. J Zool 244(2):209–222

Tosh CA, Steyn J, Bornemann H, Van den Hoff J, Stewart BS, Plötz J, Bester MN (2012) Marine habitats of juvenile southern elephant seals from Marion Island. Aquat Biol 17(1):71–79

Treasure AM, Roquet F, Ansorge IJ, Bester MN, Boehme L, Bornemann H, Charrassin J-B, Chevallier D, Costa DP, Fedak MA (2017) Marine mammals exploring the oceans pole to pole: a review of the MEOP consortium. Oceanography 30(2):132–138

Trillmich F (1996) Parental investment in pinnipeds. In: Rosenblatt JS, Snowdon CT (eds) Parental care: evolution, mechanisms, and adaptive significance. Advances in the Study of Behavior, vol 25. Elsevier, pp 533–577

Urbanski JA, Stempniewicz L, Weslawski JM, Draganska-Deja K, Wochna A, Goc M, Iliszko L (2017) Subglacial discharges create fluctuating foraging hotspots for sea birds in tidewater glacier bays. Sci Rep 7:43999. https://doi.org/10.1038/srep43999

Vacquie-Garcia J, Lydersen C, Biuw M, Haug T, Fedak MA, Kovacs KM (2017) Hooded seal *Cystophora cristata* foraging areas in the Northeast Atlantic Ocean-Investigated using three complementary methods. PLoS ONE 12(12):e0187889. https://doi.org/10.1371/journal.pone.0187889

Van Beest FM, Mews S, Elkenkamp S et al (2019) Classifying grey seal behaviour in relation to environmental variability and commercial fishing activity—a multivariate hidden Markov model. Sci Rep 9(1):5642. https://doi.org/10.1038/s41598-019-42109-w

Verity PG, Smetacek V (1996) Organism life cycles, predation, and the structure of marine pelagic ecosystems. Mar Ecol Prog Ser 130:277–293

Vernet M, Martinson D, Iannuzzi R, Stammerjohn S, Kozlowski W, Sines K, Smith R, Garibotti I (2008) Primary production within the sea-ice zone west of the Antarctic Peninsula: I—Sea ice, summer mixed layer, and irradiance. Deep Sea Res Part II 55(18–19):2068–2085

Wall SM, Bradshaw CJ, Southwell CJ, Gales NJ, Hindell MA (2007) Crabeater seal diving behaviour in eastern Antarctica. Mar Ecol Prog Ser 337:265–277

Watanabe Y, Mitani Y, Sato K, Cameron MF, Naito Y (2003) Dive depths of Weddell seals in relation to vertical prey distribution as estimated by image data. Mar Ecol Prog Ser 252:283–288

Watanabe YY, Baranov EA, Miyazaki N (2015) Drift dives and prolonged surfacing periods in Baikal seals: resting strategies in open waters? J Exp Biol 218(Pt 17):2793–2798. https://doi.org/10.1242/jeb.125898

Wege M, Bornemann H, Blix AS, Nordøy ES, Biddle L, Bester MN (2021) Distribution and habitat suitability of Ross seals in a warming ocean. Front Mar Sci 8:468

Wilson K, Littnan C, Halpin P, Read A (2017) Integrating multiple technologies to understand the foraging behaviour of Hawaiian monk seals. R Soc Open Sci 4(3):160703. https://doi.org/10.1098/rsos.160703

Yen PP, Sydeman WJ, Hyrenbach KD (2004) Marine bird and cetacean associations with bathymetric habitats and shallow-water topographies: implications for trophic transfer and conservation. J Mar Syst 50(1–2):79–99

Yoshino K, Takahashi A, Adachi T, Costa DP, Robinson PW, Peterson SH, Hückstädt LA, Holser RR, Naito Y (2020) Acceleration-triggered animal-borne videos show a dominance of fish in the diet of female northern elephant seals. J Exp Biol 223(5):jeb212936

Yurkowski DJ, Semeniuk CA, Harwood LA et al (2016) Influence of sea ice phenology on the movement ecology of ringed seals across their latitudinal range. Mar Ecol Prog Ser 562:237–250

Zachos J, Pagani M, Sloan L, Thomas E, Billups K (2001) Trends, rhythms, and aberrations in global climate 65 Ma to present. Science 292(5517):686–693. https://doi.org/10.1126/science.1059412

Zeller KA, McGarigal K, Whiteley AR (2012) Estimating landscape resistance to movement: a review. Landscape Ecol 27(6):777–797

Chapter 6
Foraging Ecology and Behavior

W. Don Bowen and Ian D. Jonsen

Contents

Abstract Phocid seals exhibit a range of body sizes and life-history traits. They forage at a wide range of spatial scales from long-distance oceanic movements (e.g., northern elephant seals) to short local trips of resident coastal species (e.g., harbor seals). Our current understanding of the foraging behavior of phocids comes mainly from northern and southern elephant seals, gray seals, harbor seals, ringed seals, and Hawaiian monk seals. Foraging tactics appear to develop rapidly within the first year of life, but the effect of age on the foraging of older animals appears to be weak. Sex differences in seasonal foraging distribution and diving behavior are evident in several dimorphic species but less evident or absent otherwise. Body size, condition, and reproductive status also influence foraging trip characteristics and dive effort. Recent studies reveal the consequences of foraging on mass gain and reproductive success and how those consequences vary with interannual and longer-term environmental change. More accurate location tags, improved dataloggers, and new analysis tools are being used to measure and infer in-situ patterns of foraging behavior and foraging success. Animal-borne instruments can also collect data on the ocean environment, which have proven useful in describing foraging behavior.

Keywords Phocids · Intrinsic and environmental drivers · Movement · Diving · Foraging success · Life-history consequences · New tools · Analysis methods

W. Don Bowen (✉)
Biology Department, Dalhousie University, Halifax, NS, Canada
e-mail: wbowen@dal.ca

I. D. Jonsen
Department of Biological Sciences, Macquarie University, Sydney, Australia

D. P. Costa and E. McHuron (eds.), *Ethology and Behavioral Ecology of Phocids*, Ethology and Behavioral Ecology of Marine Mammals,
https://doi.org/10.1007/978-3-030-88923-4_6

6.1 Introduction

Phocids are large-bodied and long-lived species. Their large size means they tend to be wide-ranging predators that encounter habitats differing in quality, which may, in turn, influence their foraging behavior. Their longevity suggests that foraging performance may benefit from experience gained over a long life, as prey in the marine environment are often spatially aggregated and temporally variable (Mann and Lazier 1996). This may lead to the evolution of foraging behaviors that benefit individual reproductive output and survival at interannual and longer timescales.

These and several other factors are likely, therefore, to influence the foraging behavior of phocids. These factors may be intrinsic to the individual (e.g., age, sex, physiological condition), whereas others are expressed through changes in the environment at a range of spatial and temporal scales (e.g., seasonal and longer-term changes in prey availability, competitors, and predators). Although the effects of single factors are often evident, multiple factors generally interact to shape the behavior we observe. For example, gray seals (*Halichoerus grypus*) exhibit seasonal sex segregation in at-sea foraging distribution (Breed et al. 2006) and sex-specific seasonal differences in diving characteristics at the level of individual dives and bouts of diving (Beck et al. 2003a, b). Males and females also differ in feeding frequency (Austin et al. 2006a), seasonal patterns of energy storage and expenditure (Beck et al. 2003c), and seasonal diet (Beck et al. 2007). Foraging tactics develop rapidly with age, presumably reflecting the effects of experience (Breed et al. 2011). Variables that predict gray seal foraging success differ across temporal scales indicating that environmental variation becomes important at some scales and not others (Austin et al. 2006b). The results of studies on the foraging behavior of other phocid species also provide evidence for the interaction of intrinsic and environmental factors (Hassrick et al. 2013; Dmitrieva et al. 2016; Hindell et al. 2016; Yurkowski et al. 2016a, b, c; Rodriguez et al. 2017; Crawford et al. 2019; Chap. 5).

Foraging is critical to survival. More successful foragers may increase their fitness by surviving longer and passing their genes to the next generation. Thus, foraging ecology is an important driver in the evolution of behavior, life histories, and population dynamics. In securing food, phocid seals face several decisions. The first is where and how to search for food. Second, among the many possible species that could be consumed, which ones should be eaten to maximize energy gain. Third, how to balance the risk of predation with the need to find and capture food. Fourth, how to distribute foraging effort seasonally to meet the demands of reproduction and other components of an annual life cycle, and finally, how to deal with intra- and inter-specific competition. Thus, foraging behavior emerges as individuals respond to changes in their environment, with the response directed by their physiological state and navigational capability (Nathan et al. 2008).

Optimal foraging theory, formulated by MacArthur and Pianka (1966), sought to provide a predictive model to address some of the decisions foragers make. The theory stated that natural selection favors animals whose behavioral strategies maximize their net energy intake per unit time spent foraging (i.e., searching for prey and handling it) and sought to explain why, of the wide range of foods available, animals often consume just a subset of prey species. The theory predicted that animals should balance two contrasting strategies: spending a long time searching for highly "profitable" food items (so-called energy maximizers) or minimizing the time spent searching by consuming more common but less profitable food items (so-called time minimizers). Profitability here is expressed as the energy gained by consuming a prey item relative to energy expended during searching and handling the prey item. Many factors cause animals to deviate from foraging optimally. For example, predation risk may cause an animal to forage in a relatively safe location where food is less available or of lower quality, rather than opting for a more efficient but riskier feeding strategy where predator density may be higher. Foraging decisions may also depend on the physical state of the individual and the presence of other foragers (Ydenberg et al. 2007). Optimal diving (Kramer 1988; Thompson et al. 1993; Thompson and Fedak 2001; Houston 2011), optimal search strategies (Zollner and Lima 1999), and the marginal value theorem (Charnov 1976) are related concepts that predict how individuals should behave to maximize net energy intake. Davidson and Hady (2019) attempted to develop a more realistic model of the patch-leaving problem, in which a forager must decide to leave a current resource in the search for another, by considering how the imperfect and noisy information constrained by neurobiological mechanisms influences individual foraging decisions. Although animals may not always forge optimally, these concepts serve to remind us that foraging strategies are embedded within life histories and evolve by natural selection.

Foraging decisions of individuals also play a role in determining how the energetic cost of reproduction and provisioning of young are financed (Chap. 8). Animals are capital or income breeders based on their capacity to store energy during feeding for use in future reproduction (Sibly and Calow 1986). This capacity is largely a function of body size as mass-specific metabolic expenditure decreases with body mass. In contrast, fat storage increases with body mass to the power of 1.1 (Prothero 1995). In capital breeders, the energy and nutrients needed for reproduction are stored during foraging prior to the breeding season and then used during the breeding season. Income breeders, on the other hand, continue to forage during the breeding season. These two strategies should be regarded as either end of a continuum and markedly influence foraging behaviors (Stephens et al. 2009; Chaps. 7 and 8). Among phocids, capital breeders include elephant seals (*Mirounga leonina*, *M. angustirostris*), Hawaiian monk seals (*Neomonachus schauinslandi*), gray seals, hooded seals (*Cystophora cristata*; Schulz and Bowen 2004), and most likely the crabeater seal (*Lobodon carcinophgus*; Boyd 2000; Southwell 2004). In four species, (Weddell, *Leptonychotes weddellii*; bearded, *Erignathus barbatus*; harp *Pagophilus groenlandicus*; and harbor seals, *Phoca vitulina*), females initially support lactation from body stores but supplement energy expenditure by foraging as lactation progresses. In Weddell seals, this supplemental energy is nominal, and they are fundamentally capital breeders (Wheatley et al. 2006, but see Chap. 13). The ringed seal (*Pusa hispida*) is an income breeder, and given their similar small body sizes, presumably, the Caspian (*Pusa capsica*), ribbon (*Histriophoca fasciata*), spotted (*Phoca largha*), and Baikal (*Pusa sibirica*) seals are as well, but this has not been established.

Not enough is known about the leopard (*Hydrurga leptonyx*), Ross (*Ommatophoca rossii*), and Mediterranean monk seal (*Monachus monachus*) seals to classify them along the capital-income continuum. However, their long lactation period strongly suggests that Mediterranean monk seals are income breeders.

Except for the crabeater seal, phocids are generalist predators (Bowen et al. 2002a), with foraging behavior variable in time and space. This means that to make inferences about the factors (intrinsic and extrinsic) influencing foraging behavior, a large sample of individuals should be studied to have any confidence that population variability has been reasonably identified. As the number of individuals studied increases, the probability that recorded data will reveal individual variability in the behavior also generally increases such that ecologically valuable inferences about populations become feasible (Hindell et al. 2003; Sequeira et al. 2019). The appropriate sample size is dependent on the question and the level of between-individual variation in the study population. Hindell et al. (2003) showed that the relative increase in the relationship of adult female southern elephant seals between area occupied and sample size diminished substantially after 25 seals had been tracked. Because foraging behavior may often differ in males and females by age, and in response to environmental variability, large sample sizes are generally needed to make reliable inferences. Studies on small numbers of individuals can be informative where a combination of instruments (e.g., satellite tags combined with accelerometers, oceanographic sensors, or animal-borne video) is used to gain insight into individuals' foraging behavior (see Sequeira et al. 2019 for examples). Nevertheless, inferences from such studies should be viewed as tentative as they might not be representative of the population.

6.2 Statistical Approaches for Inferring Foraging Behavior

The foraging behavior of aquatic animals usually cannot be observed directly. Therefore, most insights on the foraging ecology and behavior of phocids have come from biotelemetry devices capturing a variety of in-situ data, including horizontal location, depth, 3D acceleration, and various environmental measures such as temperature, salinity, and light level. These data are often complex, high-volume time-series with varying levels of measurement error. As a result, statistical and/or heuristic approaches are often employed to summarize the data (Heerah et al. 2014; Photopoulou et al. 2014), account for measurement error (Jonsen et al. 2005; Johnson et al. 2008), and make inferences about behavioral processes hidden within the data (e.g., Thums et al. 2011; Bestley et al. 2013; Schick et al. 2013; Heerah et al. 2019).

There are two key and inter-related areas where statistical modeling has been instrumental in advancing the understanding of phocid foraging ecology. The first is to identify where and when foraging may occur and the environmental correlates of this activity (Chap. 5). The second is estimating the success of inferred foraging activity. Because seal foraging behavior is rarely observed directly (cf. Bowen et al. 2002b; Heaslip et al. 2014), it often can be inferred only from indirect measurements such as changes in horizontal movement pattern (e.g., Jonsen et al. 2005), dive pattern (e.g., Dragon et al. 2012; Heerah et al. 2014), acceleration (e.g., Gallon et al. 2013), or some combination of these (e.g., Adachi et al. 2017; McClintock et al. 2013).

6.2.1 Foraging Inferred from Horizontal Location Data

Horizontal location data are commonly determined by Doppler effect measurements of microwave radio transmissions to polar-orbiting satellites from externally attached tags on individual animals. These transmissions are processed by Service Argos (https://www.argos-system.org/). Several continuous state-space models have been developed to extract behavioral information from Argos data. Jonsen et al. (2005) developed a Bayesian continuous, state-space model to account for measurement errors typical of Argos horizontal location data and estimate switches between movements related to putative foraging behavior and those related to transiting or migratory behavior. Breed et al. (2009) used this approach to infer dynamic networks of intensely used local foraging sites that differed markedly between male and female adult gray seals. The "behaviorally-informed" spatial use maps arising from this approach revealed fine-scale foraging activity that was not obvious from previous analyses of the same location data (Breed et al. 2006). Bestley et al. (2013) built on this approach by incorporating a binomial regression into the state-space model so that the probability of switching between behavioral states could be related to environmental and other covariates to understand factors that influence foraging patterns. McClintock et al. (2012) used a broadly similar but more generalizable Bayesian approach to estimate activity centers (presumably profitable foraging sites) and their strengths of attraction for gray seals foraging in the North Sea.

Hidden Markov models, a class of discrete state-space models, are similarly used to infer behavioral states from animal tracking data (Langrock et al. 2012). These models are generally easier to compute than continuous state-space models, facilitating greater flexibility in estimating behavioral states and resulting in potentially more realistic interpretations of seal foraging and activity budgets. For example, Michelot et al. (2017) used a 4-state hidden Markov model to infer movement behaviors of southern elephant seals on their long, post-molt foraging trips. The model was constructed so that transitions between behavioral states reflected the sequential pattern of elephant seal movements during outbound transit from the breeding colony, repeated switches between searching and foraging (or resting) states, followed by an inbound transit back to the breeding colony (Fig. 6.1).

An alternative approach is to treat movement patterns as a continuum rather than as discrete behavioral states. Jonsen et al. (2019) showed how time-varying autocorrelation (move persistence) in southern elephant seal movements could be related to environmental covariates in a mixed-effects model. The flexible random effects allow relationships between movement patterns and environmental or intrinsic factors to be

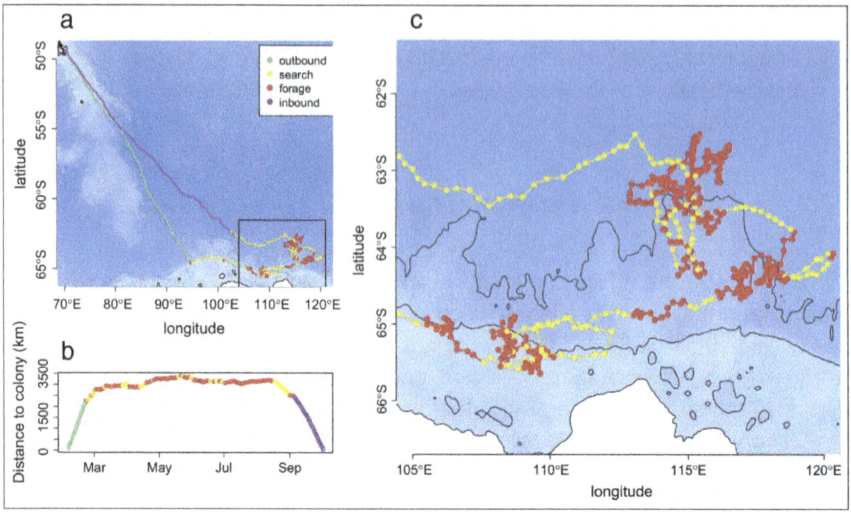

Fig. 6.1 **a** An example track of a southern elephant seal. **b** The distance from the colony (i.e., Kerguelen Island) over time. **c** Expanded view of box in panel A. The four colors correspond to the most probable states from the Viterbi algorithm. **a** and **b** clearly illutrate the phases of fast and directed movement attributed to the outbound and inbound trips. **c** Detailed movement when near the sea ice, with search (yellow) and foraging (red). From Michelot et al. (2017)

estimated so that variability across animals is explicitly quantified, thereby delving deeper into the context of movements that underlie foraging habitat usage. Such an approach may prove particularly advantageous when behavioral sequences (as per Michelot et al. 2017) are unknown.

6.2.2 *Foraging Inferred from Dive Data*

The methods highlighted in the previous section all implicitly assume a reasonably direct relationship between horizontal movement and foraging activity. Because phocids forage at depth, more direct and accurate information about foraging activity likely comes from their dive records. Early attempts to infer foraging behavior from dive data involved the classification of dive shapes on the assumption that shape encoded information on functionality (e.g., Le Boeuf et al. 1988, Chap. 12). Dives were classified visually or using classification algorithms (e.g., Scheer and Testa 1996) into four to seven types depending on angles of descent and ascent, and the fraction of time spent at the bottom of the dive (reviewed in Schreer et al. 2001). Foraging dives were thought to be mainly those where much of the time at depth was spent at the bottom of the dive (often termed U-shape or square-shape) where prey were presumably more frequently encountered. Dive shape analysis has been widely used to describe individual dives or bouts of diving derived from time-depth recorders as a means of inferring behavior (e.g., Boyd et al. 1994; Schreer et al. 2001; Fedak et al. 2001, Beck et al. 2003a).

Validation of the functional inference of dive shapes with respect to foraging has come from stomach temperature data on feeding (Lesage et al. 1999; Kuhn et al. 2009), water flux estimates of food intake (Baechler et al. 2002), animal-borne video (Bowen et al., unpublished data; Yoshino et al. 2020), and jaw accelerometers (Naito et al. 2013). Despite these validations, caution is needed when making inferences about behavior from dive shape alone, as these shapes may also be exhibited during behaviors unrelated to foraging. For example, animal-borne video of adult male harbor seals revealed that U-shape (i.e., bottom-time dives) dives also were performed during transiting and while making reproductive vocalizations (Boness et al. 2006). Similarly, Kuhn and Costa (2006) found that only 28% of feeding events based on stomach temperature data from northern elephant seals were associated with bottom-time dives, whereas some 65% were associated with V-shaped dives.

Shape is simply a two-dimensional representation of behavior performed in three dimensions, so studies have taken the next logical step by investigating the three-dimensional properties of diving behaveor (Harcourt et al. 2000; Simpkins et al. 2001; Davis et al. 1999, 2003; Adachi et al. 2017; Le Bras et al. 2017). Although three-dimensional analysis is informative, inferences about the function of dives still needed to be validated by video (Davis et al. 2003). More recently, Heerah et al. (2014, 2019) quantified southern elephant seal and Weddell seal dive behavior using an automated broken-stick algorithm to identify two behaviors, with highly sinuous segments indicating "hunting" and less sinuous segments indicating "transiting". They verified these functions using accelerometry data during prey capture attempts. A key finding was that hunting time in these species differed from bottom time, which was previously used as a proxy for time spent foraging in a dive.

Several studies have simultaneously used information from horizontal and vertical dimensions to infer foraging behavior. Bestley et al. (2015) used five common dive metrics as covariates of the rate of switching into foraging (or resident) behavioral state as estimated from a state-space model (Bestley et al. 2013). Heerah et al. (2017) used a heuristic approach based on a combination of movement data expressed as First Passage Time (Fauchald and Tveraa 2003) and a hunting time index derived from dive profiles to investigate foraging behavior. Recently, van Beest et al. (2019) combined horizontal GPS locations and summarized dive data in a multivariate hidden Markov model to infer the spatial foraging activity of gray seals. Together, these studies indicate the value of simultaneously using the information on horizontal movements and diving behavior to gain deeper insight into the foraging behavior of phocids.

6.2.3 *Inference of Foraging Success*

Statistical modeling of Argos and GPS location data from tagged phocids has provided insight into movement and foraging behavior's spatial and temporal characteristics. However, the identification of behavior without an indication of energy

Table 6.1 Methods used to infer overall foraging success or individual prey capture attempts and success. Methods are often used in combination or with other methods such as time-depth location tags. References in text

Method	Resolution	Prey capture	
		Attempts	Success
Mass gain	Integrated over entire trip	No	No
Buoyancy changes/drift diving	Integrated over entire trip	No	No
Body composition—isotope dilution	Integrated over short or partial trips	No	No
Water turnover (food intake)—isotope dilution	Integrated over short or partial trips	No	No
Accelerometry	Individual events, short duration of data collection	Yes	Sometimes
Stomach temperature telemetry	Individual feeding events/meals, short duration of data collection	No	Yes
Animal-borne video/stills	Individual feeding events, short duration of data collection	Yes	Yes
Coded acoustic tags in prey, acoustic transceiver on predator	Individual feeding events	No	Yes

gain provides an incomplete picture. Several approaches have been used to fill this gap (Table 6.1). Some integrate success over an entire or part of a foraging trip (e.g., mass gain). In contrast, others provide resolution of behavior at the level of individual events (i.e., prey capture attempt or success) at timescales of only hours or a few days (e.g., animal-borne video) due to limited battery life. The approaches listed in Table 6.1 have been used alone but more often are used in combination with other instruments, particularly time-depth and location tags.

Mass gained during foraging trips is commonly used to measure overall foraging success (e.g., Coltman et al. 1998; Le Boeuf et al. 2000; Beck et al. 2003c; Crocker et al. 2006). Mass gain ultimately must reflect foraging success, but what it really measures is the fraction of energy stored from food intake. Differences in the diet (e.g., prey size and energy content) and cost of foraging (e.g., prey behavior) influence the relationship between mass gain and foraging success. The advantage of mass gain is that it is easy to measure. However, the disadvantage is that the most accurate and informative data come from repeated measurements of the same individual, which can be difficult to acquire.

Two approaches use changes in body composition to infer foraging success. The first uses characteristics of drift dives to infer foraging success based on an individual's change in buoyancy (Biuw et al. 2003). Changes in rates of descent during elephant seal drift dives can be inferred from high- and low-resolution (i.e., broken-stick dive summaries) dive data (Arce et al. 2019). These data can be used to identify how much energy seals acquire while at sea, and where profitable foraging regions are located and how these may vary in space and time (Biuw et al. 2007). Several modeling studies have integrated drift-dive-estimates of foraging success with movement patterns/behaviors to infer rates of lipid gain along individual elephant seal foraging trips (Schick et al. 2013; New et al. 2014). Pirotta et al. (2019) adapted the Bayesian state-space model of Jonsen et al. (2005) to include feeding and body-condition components that change over time. The amount of feeding on each day was assumed to be a function of the behavioral state (transiting vs. foraging) and daily averaged dive metrics. Daily change in body condition was a function of previous body condition, informed by drift dive rates, and amount of feeding estimated for that day.

In the second method, changes in body composition (e.g., percentage of fat) of an individual are estimated by use of a hydrogen isotope dilution at departure and again on return from a foraging trip or series of trips (e.g., Coltman et al. 1998; Beck et al. 2003c; Thums et al. 2011; Hassrick et al. 2013). This provides more information on foraging success than mass gain alone because it estimates the total and individual components of energy gain during foraging. Closely related to body composition changes are estimates of food intake based on water flux measurements from isotope dilution (Costa 1987; Coltman et al. 1998; Bowen et al. 2001; Muelbert et al. 2003; Lidgard et al. 2005). An estimate of food intake, combined with data on changes in body composition, enables inferences about the net energy gain from foraging success. Disadvantages of these methods include the requirement for the recapture of the same individual, which may not be possible, and that water turnover measurements must be made at timescales (a few days to several weeks) that are often much shorter than individual foraging trips.

The methods discussed above integrate foraging success over a trip or portion of a trip. Therefore, they fall short of providing information on the spatial and temporal pattern for prey encounters and of foraging success as a function of prey species or other prey characteristics. To fill these gaps, ecologist have fitted seals with accelerometers (Naito et al. 2010, 2013; Volpov et al. 2015), stomach temperature tags (Austin et al. 2006a; Kuhn and Costa 2006), coded acoustic tags (Lidgard et al. 2014), and cameras (stills—Hooker et al. 2002; video—Parrish et al. 2000, 2002; Yoshino et al. 2020).

Accelerometers provide indirect data on foraging behavior, but if calibrated (e.g., simultaneous animal-borne video or observation) can be used to estimate prey capture attempts (e.g., Naito et al. 2010, 2013; Volpov et al. 2015; Wilson et al. 2017a; Adachi et al. 2018). However, accelerometry data alone cannot confirm if attempts were successful. Goulet et al. (2019) fitted a miniature biomimetic sonar and movement tag to a southern elephant seal to partially deal with this shortcoming. The sonar detection of prey and movement sensors provided information to determine if the predator attempted to capture the detected prey. Stomach temperature tags operate on the principle that the ingestion of prey will lower stomach temperature (Wilson et al. 1992). Although these instruments can accurately record meals, it is doubtful that individual prey captures within a meal can be reliably resolved (Bekkby and Bjørge 1998; Austin et al. 2006a; Kuhn and Costa 2006). A challenge to be overcome with respect to stomach telemetry is the difficulty in keeping the stomach temperature transmitter from being passed by the animal. Kuhn et al. (2009) obtained an average of 8 days of records (range 2–21 days) from northern elephant seals. By placing the transmitter within an ethofoam "fish", Austin et al. (2006a) obtained data for 2–40 days with a mean of 16 days from adult gray seals.

Coded acoustic tags have recently been used to study predation in gray seals (Lidgard et al. 2014). Seals were fitted with a Vemco Mobile Transceiver (VMT) and a GPS tag to determine the spatial and temporal pattern of interactions between gray seals and several fish species that had previously been surgically implanted with coded acoustic transmitters. The coded transmitters allowed the identification of individual fish. Prey encounters could be distinguished from predation by the pattern of acoustic detections recorded by the VMT. Depending on the size used, acoustic fish tags transmitted for several years, and VMT recorded transmissions for about 12 months, providing data over an ecologically relevant period.

Camera images, both still but more commonly video, provide information on prey encounters, capture attempts, and capture success. However, sampling durations are typically short (hours to days) because of limitations of battery life and instrument weight (Parrish 2000; Bowen et al. 2002b; Davis et al. 2013; Krause et al. 2015). However, this limitation was overcome by triggering the video camera to record only during periods when the animal was at depth and or triggered by jaw movements (Yoshino et al. 2020). Given the large size of cameras relative to most other externally-mounted instruments, there has been concern about the potential for adverse effects on instrumented animals' behavior or energetics. For camera deployments of about three days, Bowen et al. (2002b) found no evidence for increased rate of mass loss and dive duration in male harbor seals that were equipped with the camera compared to those in previous years that wore a small time-depth recorder. Littnan et al. (2004) also found no evidence for adverse effects on diving behavior and trip duration of short-term camera deployments on subadult monk seals. Although short-term deployments may not have measurable adverse impacts, their large size has limited the use of cameras to larger species and age classes.

6.3 Foraging Behavior

6.3.1 Spatial and Temporal Scales of Foraging

Environmental variation at different temporal and spatial scales often affects the survival and reproductive success of individuals through its impact on foraging success. Thus, to understand the ecology of phocid seals in large marine ecosystems, we need information on the scales at which foraging decisions are made (e.g., Block et al. 2011; Sequeira et al. 2018).

Conceptually, a foraging trip entails transit to a prey patch or cluster of patches, movement within and between those prey patches, and then typically a return transit to land or ice (e.g., Breed et al. 2009). Spatial and temporal scales of foraging are nested within the ranging behavior of individuals from mesoscale coastal species (harbor seal) to large-scale and oceanic ones (hooded seal, elephant seals, Table 6.2). Although data are not available for all species, enough species have been studied to provide a reasonable understanding of the spatial and temporal scales involved while foraging. Nevertheless, the values in Table 6.2 should be regarded as a rough guide, as studies use different metrics to report the spatial scales of movement, and usually report considerable individual variation as well as season and sex effects (e.g., Breed et al. 2009; Sharples et al. 2012; Hindell et al. 2016; Yurkowski et al. 2016a).

Movement associated with travel to and from foraging areas is usually at the largest scale. The magnitude and duration of those movements vary greatly among phocid species. For example, large body size and the ability to store large amounts of energy in the form of blubber enables some species (e.g., harp seals, hooded seals, northern and southern elephant seals) to exploit foraging areas 1000 s of km from their breeding colonies. Harp and hooded seals exhibit a single foraging trip consisting of return migration of >4000 km to the eastern Canadian Arctic and Greenland in Northwest Atlantic populations (Sergeant 1965; Andersen et al. 2013a, Chap. 14). Northern and southern elephant seals make just two foraging trips involving long-distance migrations each year between breeding and molting sites (McConnell et al. 1992; Stewart and DeLong 1995; Le Boeuf et al. 2000; Hindell et al. 2016; Chap.11). During these extended trips, some foraging occurs along transits in northern elephant seals (Kuhn et al. 2009) as well as at distant prey patches. Thus, it appears that the spatial and temporal scales of foraging are more complex than had been assumed in the conventional view of a foraging trip. There is no reason to imagine that similar complexity will not be found in the other species. Presumably, this complexity is primarily driven by the availability of prey and the trade-off associated with increased transit time versus time spent in more profitable prey patches, but this remains to be studied.

Rather than undertaking long migrations, most other phocids seasonally disperse throughout their range to forage. In the United Kingdom, gray seals use offshore areas (up to 100 km) connected to their haul-out sites by prominent corridors. In contrast, harbor seals primarily stay within 50 km of the coastline (Fig. 6.2, Jones et al. 2015). In the northwest Atlantic population, gray seals exhibit seasonal and sex-specific changes in foraging distribution within the continental shelf limits, and therefore, outbound transits are typically 50–80 km or less (Breed et al. 2006, 2009). There is also evidence from stomach temperature telemetry (Austin et al. 2006a) and coded acoustic tags (Lidgard et al. 2014) that gray seals forage along transits in addition to times where movements indicate area-restricted searching behavior (an indicator of increased likelihood of foraging), underscoring the complexity of foraging. Relatively short outbound transits to foraging areas are also evident in Hawaiian monk seals (Stewart et al. 2006; Wilson et al. 2017b) and Weddell and crabeater seals foraging in fjords or along the inner continental shelf of the Antarctic Peninsula (Costa et al. 2010).

Foraging trip duration also varies greatly among species, with the longest trips generally observed in long-distance migratory species and the shortest trips exhibited by coastal species. For example, trip durations of Hawaiian monk seals and harbor seals average less than a day to one week. In contrast, elephant seals may spend

Table 6.2 Representative temporal and spatial scales characteristic of foraging trips in adult phocid species. Values represent means range of seasonal means and are separated by sex (F = female, M = male) when applicable. N is the number of individuals studied

Species	N	Location	Trip duration (days)	One-way straight-line distance (km)	Dive depth (m)	Source
Northern elephant seal	37, 40	North Pacific	73, 243 F 125 M	18,000 F[a] 21,000 M	486–522 F 366 M	Stewart and Delong (1995), Le Boeuf (2000)
Southern elephant seal	287, 42	Southern Ocean	77	1,372, 2975	~430	Hindell et al. (2003, 2016)
Hooded seal	21, 51	North Atlantic		~2,600 (14,142)[a]	255	Andersen et al. (2013a, b)
Hawaiian monk seal	19	Main Hawaiian Islands	0.6	18	12–32	Wilson et al. (2017b)
Gray seal	84, 81	Scotian Shelf, NW Atlantic	7–13	50–80	50–150	Breed et al. (2009), Lidgard et al. (2020)
	11	Baltic Sea	1.5		<40	Sjoberg and Ball (2000)
Harbor seal	115	United Kingdom	<1.0–3.5	10–60	25–65	Sharples et al. (2012)
	49	Prince William Sound, Alaska		61		Lowry et al. (2001)
	37	Moray Firth, Scotland	<1.0–10.7	30 (F) 61 (M)		Thompson et al. (1998)
Ringed seal	130	Canadian Arctic		3–50	30–100	Yurkowski et al. (2016c)
	26	Baltic Sea		254	13	Oksanen et al. (2015)
Caspian seal	75	Caspian Sea			80% <15	Dmitrieva et al. (2016)
Bearded seal	7	Svalbard			24	Hamilton et al. (2018)

[a]Estimated total distance traveled

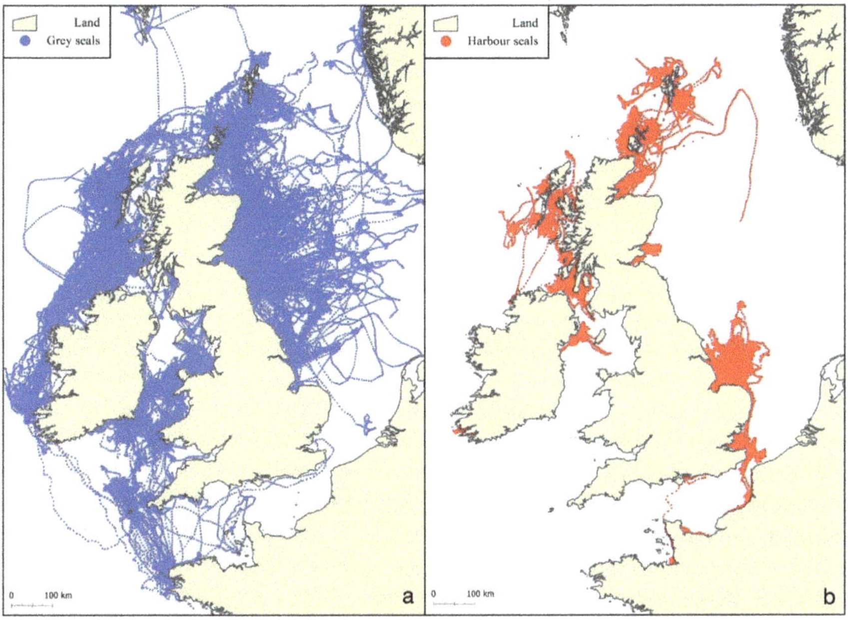

Fig. 6.2 Movement tracks of 259 Gy seals (**a**) and of 277 harbor seals (**b**) from different colonies around the United Kingdom from 1991 to 2013 and 2003 to 2013, respectively. From Jones et al. (2015)

months on a single foraging trip (Table 6.2). Both intrinsic and extrinsic factors can influence trip duration in some species. In harbor seals, males undertake longer and more distant trips than females (Thompson et al. 1998; Sharples et al. 2012). Although both sexes made trips of similar duration in gray seals, average trip duration varied seasonally from about one week in summer to two weeks in fall and winter (Breed et al. 2009). Despite shortened trips in summer, both sexes gain body mass (Beck et al. 2003a), suggesting prey may be more accessible during summer than at other times of the year.

Although there is some evidence that foraging may occur along transits, most foraging is thought to occur during periods of area-restricted search (ARS), where movements are characterized by frequent turning such that the animal stays within an area for an extended period. These areas are putative prey patches and are now routinely identified using behavioral switching state-space models (see Sect. 6.2.1). The spatial and temporal scale of ARS again varies widely among species, presumably reflecting the type and density of prey being consumed and the number of competing predators. Estimates of the spatial extent of ARS are difficult to compare across species as the methods used are often not comparable. Breed et al. (2009) used convex polygons to show that adult gray seals foraged over areas of 200–400 km^2 during ARS for periods of 3–7 days. The spatial extent of ARS differed by sex and season, being most significant in the fall and smallest in winter. Males foraged over larger areas than females in all seasons. Although regions of ARS along movement tracks are evident in all phocid species, few studies have quantified the size, number, and residence time per putative prey patch. Analysis of ARS behavior could provide insight into the characteristics of seasonal, inter-annual, and longer-term variations in prey availability.

6.3.2 *Intrinsic Factors*

6.3.2.1 Sex

The foraging behavior of males and females differs in many taxa (Clutton-Brock et al. 1982; Boinski 1988; Petit et al. 1990; Lewis et al. 2002; Neuhaus and Ruckstuhl 2002; Bearhop et al. 2006). Among the explanations for these differences, the three most prominent hypotheses are sexual body-size dimorphism, reduced intersexual competition for food, and the differential roles of each sex during reproduction and parental care. Among phocids, relatively few species have been studied in enough detail to test for sex differences robustly. Still, sex differences in foraging behavior have been found in northern and southern elephant seals, gray seals, and hooded seals (Table 6.2). These four species exhibit body-size dimorphism whereby males are larger and heavier than females. By contrast, sex differences in movement behavior, and therefore presumably foraging, are not evident in harbor (Russell et al. 2015) or ringed seals (Luque et al. 2014); species exhibiting little or moderate body-size dimorphism (males are about 25% heavier than females) or are monomorphic, respectively.

Foraging behavior of males and females may differ in spatial distribution, characteristics of diving, seasonal effort as manifested by changes in body energy storage (e.g., Biuw et al. 2007), and differences in diet (Beck et al. 2003b, 2007). The effect of sex on the spatial distribution of foraging is evident in several body-size dimorphic phocids. In northern elephant seals, males appear to primarily forage benthically along the west coast of North America from Oregon to the Gulf of Alaska. In contrast, females predominately forage pelagically in the mid-Pacific Ocean between the latitudes of about 35–50° N (Stewart and DeLong 1995; LeBoeuf et al. 2000; Robinson et al. 2010, 2012). This segregation pattern is evident in individuals during both the post-breeding and post-molt foraging trips (Stewart and DeLong 1995, Chap. 11), presumably indicating a level of predictability of profitable foraging areas. Sex segregation of foraging distribution is less evident in southern elephant seals as both sexes forage in shelf habitats. However, with the advance of sea ice in late autumn or early winter, female seals move northwards, resulting in some degree of sex segregation (Hindell et al. 2016).

In the northwest Atlantic, adult gray seals, another size-dimorphic species, exhibit strikingly different foraging distributions during the three months following the winter breeding season and again during several months leading up to the breeding season (Fig. 6.3). Males are more likely to undertake long-distance foraging trips (Austin et al. 2004). The result is that males have larger home ranges than females

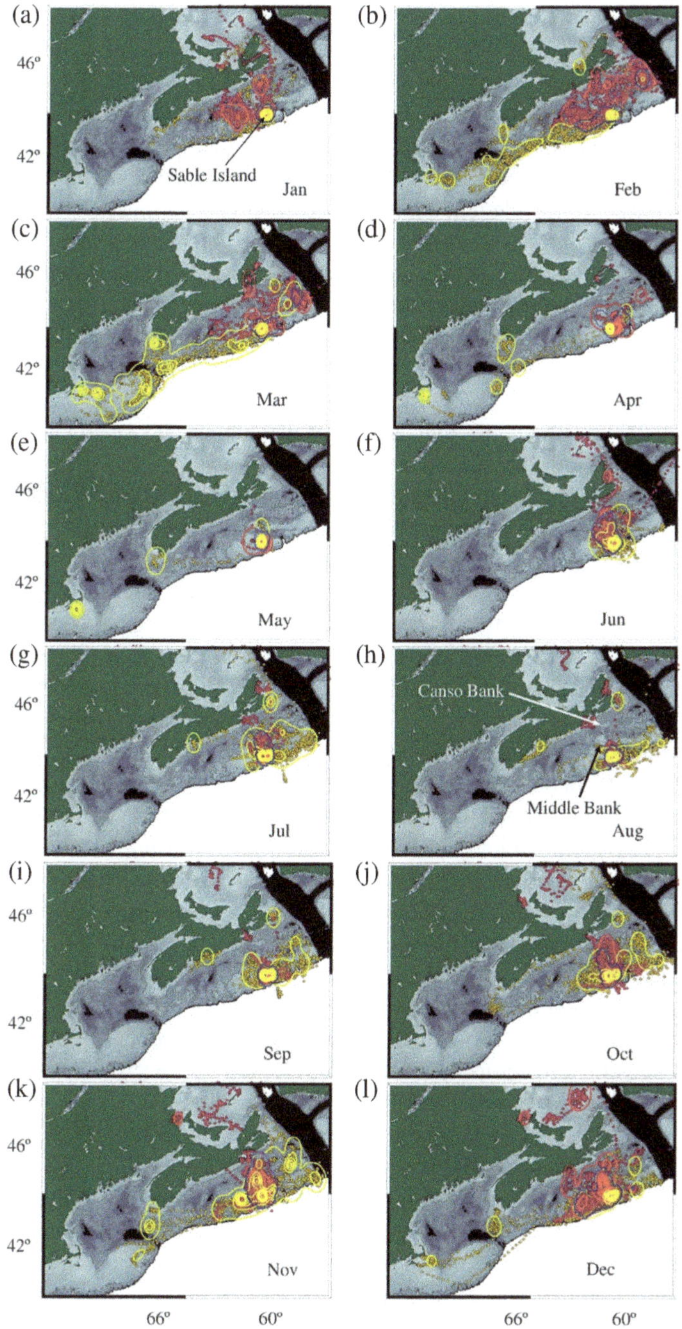

Fig. 6.3 Monthly sex segregation in adult gray seals on the Scotian Shelf. Red symbols are females, and yellow symbols are males. From Breed et al. (2006)

throughout the year (Lidgard et al. 2020). These differences in movement characteristics and foraging distribution are reflected in diet and patterns of body energy storage and expenditure (Beck et al. 2003c, 2007). Based on stomach telemetry data, adult gray seals fed on only 58% of days at sea, but when feeding occurred, males fed on average twice per day, whereas females fed only once (Austin et al. 2006a). Sex-specific, seasonal differences in the proportion of time spent foraging and traveling are also evident in juvenile and adult gray seals in the Northeast Atlantic (Russell et al. 2015). Less is known about the body-size dimorphic hooded seal, but Andersen et al. (2013b) found that males and females were spatially segregated within the same overall range based on regional differences in intensity of habitat use.

Based on current information, species exhibiting sex differences in at-sea distribution also show sex differences in diving behavior. Sex-specific diving behavior is evident in gray seals at the level of individual dives, bouts of diving, and dive effort (Beck et al. 2003a, b), and in both species of elephant seals (Le Boeuf et al. 2000; Hindell et al. 2016). Although female gray seals are smaller than males, they dived longer (5.5 vs. 4.9 min) and spent more time at depth (3.4 vs. 2.7 min), whereas males dived deeper (57 vs. 49 m; Beck et al. 2003b). Males dived consistently to near the seafloor throughout the day. In contrast, females showed strong diurnal patterns in dive depth, duration, and frequency, consistent with feeding on prey in the water column that vertically migrates. Similar findings have been reported in southern elephant seals where males had shorter dives than females, but there was no sex difference in dive depth (Hindell et al. 2016). However, in hooded seals, the effect of sex on diving characteristics was weak compared with geographic location, water depth, and season (Andersen et al. 2013a).

Sex differences in movements and foraging behavior appear to develop early in life based on studies of gray and elephant seals. Breed et al. (2011) found that although foraging in the same habitats, male and female gray seal pups showed small but significant differences in movement behavior and trip structure. However, in the UK, Carter et al. (2017) found no significant difference in trip duration between males and females in either region. However, sex differences in dive depth were apparent in pups from Celtic\Irish Seas colonies, with males diving significantly deeper than females. Similarly, sex differences in diving behavior are evident in northern elephant seals, where by their third foraging trip males exhibited more flat-bottom dives than females suggesting greater use of near-bottom foraging. Sex differences in foraging location, similar to those seen in adults, are evident before the seals reach 2 years of age (Le Boeuf et al. 1996).

Phocid species that do not exhibit sex differences in at-sea distribution also appear not to differ in diving characteristics. Sex was not a predictor of diving behavior in harbor seals (Frost et al. 2001) or ringed seals (Oksanen et al. 2015; Crawford et al. 2019). However, Luque et al. (2014) found an interaction between sex and ice concentration in ringed seals. Adult males were more likely to forage in areas with greater ice concentration than females, but the effect was relatively weak.

Based on studies to date, sex differences in seasonal patterns of at-sea distribution, dive behavior, and diet, are associated with differences in the seasonal patterns of energy storage and expenditure of males and females. These differences, in turn, reflect differences in life-history trade-offs associated with the reproductive strategies of each sex rather than the influence of body-size dimorphism alone.

6.3.2.2 Age

As long-lived species, age may influence the foraging behavior of phocids through experience. Age-related variation in foraging behavior has been reported for seabirds, which have some similar life histories as phocids (Riotte-Lambert and Weimerskirch 2013; Votier et al. 2017). However, physiology also plays an essential role in diving ability as oxygen stores in blood and muscles are needed to support aerobic diving, both of which scale with age and body size (Burns et al. 2007). The rapid development of body oxygen stores is crucial for phocid young to transition from nursing to independent foraging. Studies on phocids reveal a general pattern of increasing blood oxygen stores as a function of age until maturity (Burns 1999; Burns et al. 2005, 2007; Noren et al. 2005; Bennett et al. 2010). This increase is associated with an increase in diving ability with body mass (a general proxy for age) measured by dive depth and duration in phocids (Schreer and Kovacs 1997).

Despite its importance, the influence of age on foraging behavior in phocids has only recently begun to emerge from studies on the ontogeny of at-sea behavior. As phocid fathers provide no parental care, and mothers generally terminate parental care at weaning, newly weaned offspring receive no parental instruction on where and how to forage. The rapid development of foraging ability is critical to survival as newly weaned young have little time to learn how to capture prey before their blubber reserves are depleted (McConnell et al. 2002; Muelbert et al. 2003). Studies of phocid pups on their first and subsequent trips to sea, therefore, provide an insight into how foraging behavior develops in naïve animals during the transition to independence. Harbor seal pups, tagged during lactation, exhibited rapid increases in diving depth and duration and in the amount of time spent diving post-weaning; they exhibited an abrupt shift in movement patterns near 50 days of age, after which longer, deeper dives gradually became the norm (Blanchet et al. 2016). Gjertz et al. (2000) observed a rapid increase in diving depth of bearded seal pups, such that by 60 days of age, pups dived deeper than 448 m. Hooded seal pups began diving within days of weaning. Both dive depth and duration increased rapidly such that by 3 weeks of age, pups were diving to >100 m and for >15 min per dive, and by the fall maximum dive depths and durations had increased to 700 m and 30 min, respectively (Folkow et al. 2010).

Characteristics of foraging trips also appear to develop quickly. The initial direction of migration appears to be established during the first foraging trip in male and female northern elephant seals, but the differential migration and sexual segregation characteristic of adults do not develop until puberty when growth rates of males are substantially greater than those of females (Stewart 1997). By contrast, temporal and spatial segregation in at-sea distribution develops over the first 4 years of life in the southern elephant seal, which is thought to reduce intra-specific competition for resources (Field et al. 2005). The initial foraging trip of newly weaned southern elephant seals exhibits elements of such trips in adults. This initial trip is characterized by a rapid outbound movement (>100 km/d), slower travel rates for several months during which foraging is believed to occur, followed by a rapid inbound leg back to the birth colony (McConnell et al. 2002). Naïve southern elephant seal pups also performed shallow (100 m) and short (5.7 min) dives without a diurnal pattern during the first 60–80 days at sea. However, on the return phase of the trip, diving behavior changed abruptly with deeper and longer dives exhibiting a diurnal pattern like that of adults (Hindell et al. 1999). Together, these studies suggest that some aspects of foraging are primarily innate, but that foraging behavior develops with experience.

A comparison of the movement behavior of young phocids with that of adults provides further insight into the ontogeny of foraging tactics. Breed et al. (2011) compared the movement behavior of 5-month-old pups with subadult and adult gray seals. Tortuosity and speed of outbound travel on foraging trips were not different in pups, subadults, and adults, suggesting that pups, even at this young age, navigate as well as older animals. However, pup trips lasted 1.2–3.5 days longer. They required up to twice as much transit time to reach foraging areas that were 1.5–3 times farther from haul-out sites than those of subadults and adults. This suggests a higher foraging effort for pups than older animals. In all groups, apparent foraging effort decreased in the summer. It increased in the winter, which is consistent with seasonal changes in prey distribution and energy content. Adult foraging patterns showed complex seasonal patterns influenced by reproductive cycles and seasonal environmental variation. In contrast, annual foraging patterns of pups and subadults, neither of which invest in reproduction, were simpler and appeared more closely tied to seasonal changes in prey availability and condition.

Recently, Bowen, Nowak, Lidgard, and den Heyer (unpublished data) studied the movement patterns of naïve gray seal pups over the first 60 days at sea. Unlike older pups, the movements of naïve pups were not well defined, such that a behavioral switching model was unable to identify discrete behavioral states. Few pups made return trips characteristic of adults in this population, a striking contrast to naïve pups' behavior in the UK (Carter et al. 2017), perhaps suggesting that geography may play a role in shaping early behavior. Pups from North Sea colonies undertook a prolonged exploratory phase shortly after leaving the colony, with 69% of pups spending >20 days offshore without returning to the coast (Fig. 6.4). Some pups had trips exceeding 60 days, which is greater than typical trip durations seen in adults (McConnell et al. 1999). After this initial period of exploration, pups exhibited a pattern of return trips like that observed in adults.

Fewer studies have examined the effects of age on foraging behavior of adults. Hassrick et al. (2013) used data from post-breeding and post-molt females to investigate the extent to which intrinsic (body mass and age) and environmental factors influence foraging tactics, foraging efficiency, and foraging success in northern elephant seals. Age (4–14 years) had little impact on behavior, except for the structure of dives during the post-molt foraging trip. Age-related variation in dive structure suggested that older females were more efficient foragers as measured by rates of energy gain

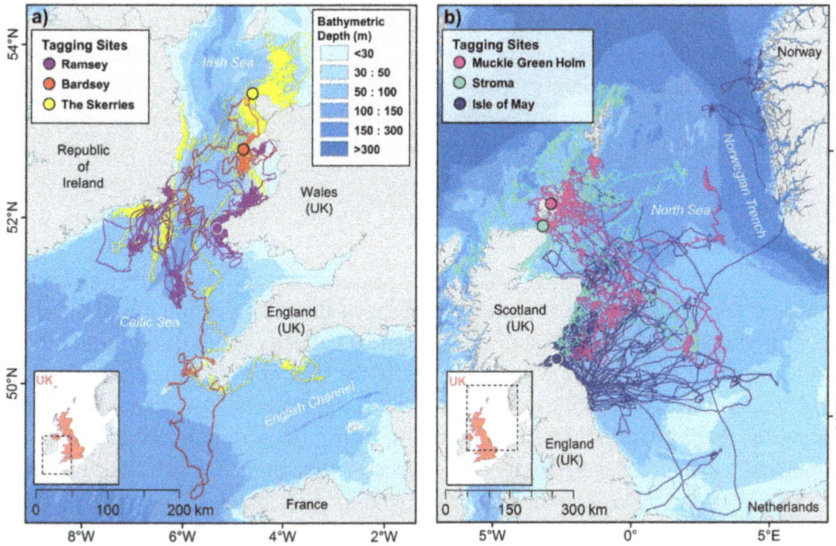

Fig. 6.4 Tagging locations. **a** Celtic and Irish Seas. **b** North Sea and trips at sea of 52 Naïve gray seal pups during the initial four months after leaving their colonies. From Carter et al. (2017)

(Hassrick et al. 2013). Recent work on a large sample of southern elephant seals from six colonies suggests that an effect of age on foraging is manifest through spatial memory (Rodriguez et al. 2017). Although the effect of age was not explicit in their analysis, the authors argue that foraging decisions are likely associated with prior experience of prey location (i.e., memory) as tracking data showed clear evidence of directed movements to areas of high use. Fine-scale analysis of gray seal GPS locations revealed differences in the frequency and duration occupancy of areas within the home range (Lidgard et al. 2020). Regions between haulout locations and apparent foraging areas were used infrequently and for short periods and thus were characterized by rapid directional movement, which may reflect memory of the location of profitable foraging areas.

Although a comparison of the movements of individuals of different ages may be the most direct means of probing the effects of age on foraging performance, age-related changes in the physiological condition of individuals can also provide insight. For example, any decline in oxygen handling or muscular performance with advancing age in a diving animal could reduce foraging ability. Hindle et al. (2011) compared oxygen stores and aerobic dive limit (ADL) of prime (9–16 years) and beyond peak-reproductive age (17–27 years) Weddell seals and found no evidence of reduced ADL late in life that would imply a decrease in foraging ability. However, Hindle et al. (2009) found that increased collagen deposition and decreased muscle myocyte suggested a probable effect of aging on sprint capacity and contractile efficiency, decreasing foraging efficiency. Nevertheless, evidence that muscular senescence reduces the foraging ability of free-ranging seals remains to be studied.

Although we have gained some insight into the ontogeny of foraging behavior in phocids, much remains to be learned. We still have little understanding of the influence of environmental variability or population size on the development of foraging behavior or how stable foraging tactics are over the life of these long-lived species (but see Sect. 6.3.2.4).

6.3.2.3 Body Size

Body size can influence the foraging behavior of phocids, presumably through its effect on energy requirements and physiological capacity. Thompson et al. (1998) found positive relationships between adult body size, trip duration, and distance in harbor seals. These findings supported the marginal value theorem's theoretical predictions that variation in the total duration of foraging trips should be positively related to the distance traveled to feed. In Prince Willian Sound, Alaska, larger harbor seals exhibited a broader isotopic niche than smaller ones, suggesting that larger seals exhibit more diverse foraging through much of the year. Smaller individuals were more likely to forage in tidewater glacial habitat throughout the year (Smith et al. 2019). In northern elephant seals, body mass was the sole predictor of residence time in foraging zones, which was the strongest behavioral predictor of foraging efficiency and success during the post-breeding foraging trip (Hassrick et al. 2013). Lidgard et al. (2020) found that heavier male and female gray seals had smaller core areas (30% home range isopleth). One possible explanation for this pattern is that heavier individuals might have competitively greater access to localized high-quality foraging habitats. However, since core area size was not a predictor of body mass gain during foraging, other explanations may be needed.

The effects of body size are expressed early in life. Dive duration and depth were positively related to body mass departure from the breeding colony in southern elephant seal pups (Irvine et al. 2000). Nevertheless, despite making shorter and shallower dives, foraging success, based on mass gain, was independent of body size. Burns et al. (1997) found that body-size differences explained about one-half of the variation in the diving behavior of yearling Weddell seals. Larger individuals made longer and shallower dives than smaller ones, suggesting that large yearlings foraged primarily on small shallow-water prey items. In contrast, small yearlings concentrated on energy-dense deep-water prey.

Body size also appears to have had strong influences on the evolution of foraging strategies of pinnipeds during the breeding season (Boness and Bowen 1996; Boyd 1998; Chap. 8). Harbor seal females, which are relatively small for a phocid, give birth to a single young that they nurse for an average of 24 days (Chap. 10). Unlike larger phocids, most females cannot support the entire cost of lactation from body reserves (Bowen et al. 1992) and begin to undertake regular foraging trips after about one week of fasting (Boness et al. 1994). However, there is a strong inverse relationship between maternal body mass and time spent foraging and food intake during mid to late lactation. By contrast, adult male harbor seals devote considerable time to foraging early in the season (Coltman et al. 1997). Most males maintain or increase body mass during this period (Walker and Bowen 1993). During the latter part of the season, when estrous females become increasingly available, males reduce foraging and devote more time to reproductive behavior while at sea. As a result, males may lose up to one kg of mass daily (Reilly and Fedak 1991; Walker and Bowen 1993). Given their relatively small body size, Coltman et al. (1998) show that most males could fast for only about 19 days or about 60% of the period when estrous females are available. Thus, males appear to balance foraging and reproductive behavior in such a way as to maximize potential encounter rates with estrous females (Coltman et al. 1997).

6.3.2.4 State-Dependent Foraging

Although body size is an important life-history trait, body condition also influences an organism's ability to survive and reproduce, where condition may include territory quality, energy reserves, foraging ability, parasite load, and the quality of its immune system (McNamara and Houston 1996). Body mass, a useful proxy for body condition, represents a state variable that accounts for much of the variability in the occurrence of pregnancy in capital breeding phocid seals (Boyd 2000). In Weddell seals, females that gave birth the following spring began the winter foraging period with significantly longer and deeper dives, as compared to non-reproductive individuals (Shero et al. 2018). By mid-to-late winter, reproductive females spent a significantly greater proportion of the day in one long daily foraging bout without rest compared to more frequent and shorter bouts by non-reproductive females. The increase in dive activity and reduction in diving metabolic rate were similar to the estimated energetic requirements of supporting a fetus, suggesting that increased foraging effort was associated with successful pregnancy (Shero et al. 2018). In northern elephant seals, Huckstadt et al. (2018) found that dives of pregnant females became shorter (accounting for dive depth) after week 17 compared to non-pregnant females, suggesting that increased fetal demand for oxygen could be the cause.

Mass gain during foraging and the resulting change in body condition can also influence diving behavior. Northern and southern elephant seals perform dives during which they spend a large proportion of time drifting passively through the water column. The rate of vertical change in depth during these "drift" dives is mostly a result of the proportion of lipid tissue in the body, with fatter seals having higher (more positive or less negative) drift rates compared with leaner seals (Crocker et al. 1997; Webb et al. 1998; Biuw et al. 2003). Several studies have used temporal changes in drift dive rate to estimate changes in body composition (a proxy for energy intake) throughout foraging trips as a measure of foraging success (Bailleul et al. 2007; Biuw et al. 2010; Robinson et al. 2010; Arce et al. 2019). In gray seals, a paired comparison of female diving behavior after molting with that just prior to the breeding season showed that the more buoyant pre-breeding females descended less rapidly, made longer dives, and spent more time at the bottom of dives than they had just after the molt when they were in poorer body condition (Beck et al. 2000). Together, these studies demonstrate that body condition plays a significant role in shaping diving behavior and that inferences about foraging success can be gleaned from such changes.

6.3.2.5 Individual Variation

There is growing evidence that individual variation in foraging ability is a critical component of natural selection, driving the evolution of life histories and individual responses to changing environmental conditions (Bolnick et al. 2003). A specialist may be defined as an individual whose niche is substantially narrower than its population's niche for reasons not attributable to its sex, age, or discrete morphological group (Bolnick et al. 2003). Individual specialization indicates that there are multiple ways of making a living. Specialization may allow animals to capture specific prey more efficiently but may also limit their ability to adapt to changes in prey resources. Individual variation can affect intra-specific competition, population persistence, species diversity, and ecological dynamics (Bolnick et al. 2003, 2011).

Although there is wide-spread taxonomic evidence for foraging specialization within populations, including marine mammals (Ford et al. 1998; Estes et al. 2003), the extent to which this occurs within phocids is not well studied. Individuals of several species (harbor seals, gray seals, and Hawaiian monk seals) use different feeding tactics to capture and consume different sizes or types of prey (Bowen et al. 2002b; Gallon et al. 2017; Kienle et al. 2019). However, the extent to which individuals specialize on a subset of prey over time to satisfy their energy requirements is still largely unknown. At a larger spatial scale, Oksanen et al. (2015) found that the number and location of apparent foraging areas of ringed seals were quite variable among individuals, suggesting different foraging tactics within the population. Some individuals foraged locally using only one area or several areas separated by short distances. However, most were long-distance foragers, with foraging areas separated by 150–825 km. Similarly, in a closely related land-locked species, the Caspian seal, Dmitrieva et al. (2016) identified three spatially exclusive types of summer dive behaviors (shallow, intermediate, and deep divers), suggesting that individual variation in behavior may play a role in niche partitioning to reduce intra-specific competition.

Estimates of the diets of individuals provide further evidence of foraging specialization in phocids. For example, Huckstadt et al. (2012) used stable isotope signatures within and between individuals to identify sources of variation in adult southern elephant seals' diet. They found that between-individual variability of δ^{13}C explained most of the total observed variance, suggesting that most individuals were specialists during the year they were studied. Similar findings are suggested by within and between individual comparison of isotopic levels in fur, skin, and blood of a small sample of gray seals (Hernandez et al. 2019).

Interannual repeatability in foraging behavior may be another expression of foraging specialization, although at a somewhat larger spatial and longer temporal scale. Long-term fidelity to foraging regions in variable environments may confer ecological benefits to individuals, such as familiarity with resources, even when energy gain is not consistently high in all years. Foraging site fidelity has been studied in several species where repeated tagging of the same individual has been possible. The most extensive data come from elephant seals, where individuals differ in the extent to which they show fidelity, while fidelity appears to be a fixed trait within individuals (Bradshaw et al. 2004; Costa et al. 2012; McIntyre et al. 2017; Abrahms et al. 2018a). Based on overlap among years in kernel density utilization distributions of 30 adult female northern elephant seals, Abrahms et al. (2018a) found that those individuals exhibiting the highest fidelity had a more consistent mass gain. In contrast, those showing lower fidelity had more variable mass gain among years. Repeated tracking of 21 adult gray seals also suggests that individuals exhibit strikingly similar movements and foraging distributions among years (W. D. Bowen and C. E. den Heyer unpublished data). Evidence of foraging site fidelity in observations of the same individual made up to 19 years apart suggests that, like elephant seals, this behavior is relatively fixed. Further studies are needed to determine at what age fidelity is developed and how fidelity to foraging areas responds to environmental variation.

Variation in foraging ability also may represent differences in the quality of individuals, where quality here is often thought to capture characteristics that are indicative of differences in fitness (e.g., Cam et al. 2002; Hamel et al. 2008). There is every reason to expect that phocids will show heterogeneity in foraging ability, given evidence of heterogeneity in reproductive performance (Weddell seal, Chambert et al. 2013; gray seals, Badger et al. 2020), but this has not been studied. Therefore, in addition to foraging specialists, there may also be heterogeneity in the foraging ability of individuals. Both specialization and heterogeneity of foraging ability may influence diet, how individuals respond to environmental variation, and, in turn, reproduction and survival. Nevertheless, for the most part, these are research questions that remain to be studied in phocids.

6.3.3 Extrinsic Factors

6.3.3.1 Environmental Variability

Physical and biological oceanographic features are dynamic over a range of spatio-temporal scales (Simpson and Sharples 2012, Chap. 5). This results in some areas having disproportionately high levels of primary productivity (Sathyendranath et al. 1995; Muller-Karger et al. 2005) that support assemblages of species at higher trophic levels (Richardson et al. 2000; Platt et al. 2003; Stevick et al. 2008; Benoit-Bird et al. 2013; Davoren 2013). The heterogeneous nature of oceanographic conditions results in a three-dimensionally patchy distribution of prey species available to upper-trophic level predators (Cox et al. 2018).

Phocids often have broad geographic ranges, providing the opportunity for within and among-population differences in foraging behavior that reflect regional environments. Regional environments may differ in geographic complexity, prey characteristics, potential competitors, and predators. There is ample evidence that the diets of phocids are influenced by environmental variation among populations, seasons, years, and over longer temporal periods (Bowen 1997; Brown and Pierce 1998; Bowen et al. 2002a; Wilson and Hammond 2019; Brown et al. 2001). Much of this variation likely reflects differences in the prey assemblages encountered spatially and temporally that may affect the way individuals forage.

Studies on several phocid species have revealed within-population differences in foraging behavior presumably driven by regional environmental conditions (e.g., Muelbert et al. 2013). In Prince William Sound, Alaska, Iverson et al. (1997) found mesoscale partitioning of harbor seal foraging habitats among different regions of the Sound (about 80 km) and finer-scale differences among sites 9–15 km apart based on differences in fatty acid profiles of harbor seals sampled at haulout locations within the Sound. These differences in fatty acid signatures undoubtedly reflected differences in the composition of diets and indicate a level of habitat partitioning that was unexpected as individual harbor seals could forage throughout the Sound. Satellite telemetry studies of harbor seals within the Sound have confirmed the scale of foraging activity indicated by the fatty acid signature analysis (Frost et al. 1998). Similar partitioning in foraging habitats also is evident among harbor seal populations in the UK. This suggests that harbor seals may benefit from increased foraging success through knowledge of local prey distribution or reduced risk of exposure to predators such as killer whales (Sharples et al. 2012; Russell et al. 2015).

The behavior of another circumpolar species provides further evidence of the influence of environmental factors on foraging behavior. At Svalbard, Norway, during years of late formation and the limited extent of land-fast ice, ringed seals made longer dives and spent less time at the bottom of dives compared to almost a decade earlier, suggesting increased foraging effort in response to environmental conditions (Hamilton et al. 2016). In the Canadian Arctic, Yurkowski et al. (2016b) found strong support for among-population differences in foraging characteristics using movement data from six tagging locations. Regional ice coverage had a significant effect on movements and foraging behavior in these populations, underscoring the need to study large numbers of individuals throughout the species range to draw reliable conclusions. As sea ice phenology may affect prey distribution, these data suggest plasticity in foraging decisions across the variable Arctic. Among-population differences in foraging behavior are supported by the results of a companion study, where the isotopic trophic position of ringed seals generally decreased with latitude and isotopic niche size increased over time, likely due to recent circumpolar increases in subarctic forage fish distribution and abundance (Yurkowski et al. 2016a).

Because many phocids are generalist predators, local changes in relative prey abundance are reflected in their foraging behavior. This is apparent for harbor seals, a generalist predator, where interannual variation in winter abundance of clupeid fishes such as Atlantic herring (*Clupea harengus*) and spratt (*Sprattus sprattus*) resulted in changes in the composition of harbor seal diets (Thompson et al. 1991, 1996). In years of high clupeid abundance, there was an increase in the proportion of these species in the diet. This change in diet was associated with a significant difference in the condition of seals caught in the spring following good and bad clupeid years, suggesting that there are energetic advantages and disadvantages to foraging on certain prey types (Thompson et al. 1996).

Large-scale variability in ocean conditions that presumably affect food supply also influences foraging behavior of pinnipeds (Field et al. 2001; Guinet et al. 2001; Dragon et al. 2010; Kuhn 2011; Hazen et al. 2013; O'Toole et al. 2015; Heerah et al. 2017; Abrahms et al. 2018b; van Beest et al. 2019; Nowak et al. 2020). For example, the period from 1975 to the late 1990s in the North Pacific Ocean was characterized by a large-scale, basin-wide warm regime that included multiple strong or long-duration El Niños, with a return to cooler conditions in the late 1990s. Le Boeuf and Crocker (2005) found that adult female northern elephant seals increased foraging effort and decreased body mass gain during this warm period compared to females foraging during the cool period. Comparing the foraging behavior of northern elephant seals during the strong 1997–98 El Niño to earlier non-El Niño years, Crocker et al. (2006) found that although the spatial behavior of females did not differ, the rate of mass gain was lowest during the El Niño year. Females attempted to compensate for decreases in foraging success by increasing trip duration. During the severe 1998 El Niño, diving data also suggested reduced residence time in prey patches and increased travel time between patches. Hassrick et al. (2013) found that age and body mass affect different components of foraging behavior in adult female northern elephant seals, both of which were influenced by environmental variation. Age influenced dive shape and dive bout characteristics, while body mass influenced the amount of time allocated to foraging zones. Interannual and seasonal impacts on distance (vertical and horizontal) that females traveled, thought to be a measure of annual variation in prey availability, influenced the rate and magnitude of foraging success.

Consequences of large-scale oceanographic variability on foraging success and demography are also evident in Hawaiian monk seals. The Transition Zone Chlorophyll Front (TZCF) is a large-scale oceanographic feature separating low surface subtropical chlorophyll waters and the high chlorophyll Transition Zone waters of the North Pacific Ocean. The TZCF annually migrates over 1000 km in latitude, and its southern extent in winter varies. Baker et al. (2007) hypothesized that movement of the front southward would bring more productive waters into monk seal foraging range, resulting in measurable positive effects at seal populations situated farthest north and nearest the TZCF. They found a nonlinear relationship between the winter position of the TZCF and survival of monk seals through 4 years of age at the most northerly atolls. This relationship did not extend to subpopulations located farther south or among adult animals at any site. They concluded that variation in ocean productivity might mediate prey availability in monk seal foraging habitat and consequently influence juvenile survival in the northern portion of their range. In a follow-up study, Baker et al. (2012) found that periods of growth and decline in monk seal abundance at four subpopulations were associated with positive and negative phases, respectively, of the Pacific Decadal Oscillation (PDO). They proposed that the PDO was a proxy for varying productivity in the northern NWHI, which propagated vertically through the food web and reflected in top predators such as the monk seal. As a result, the long-term dynamics of monk seal populations may be primarily driven by climate-ocean variability and its effects on foraging success.

At a finer scale, in-situ measurements of oceanographic properties collected at high temporal resolution by animal-borne data loggers provide insight into the conditions encountered that directly or indirectly influence behavioral decisions. Bailleul et al. (2007) used changes in the sinuosity of movement, diving frequency, and changes in body condition to show that estimated foraging success of juvenile and adult southern elephant seals was associated with different water temperatures depending on the habitat used. Dragon et al. (2010) showed that southern elephant seals preferentially exhibited foraging behavior in the presence of eddies. In northern elephant seal females, sea surface temperature was the best single predictor of estimated foraging areas (Simmons et al. 2007). In addition to water temperature, estimates of chlorophyll-a concentration have improved predictions of the spatial and temporal pattern of foraging behavior in several phocid species (Guinet et al. 2014; Vacquié-Garcia et al. 2015; Nowak et al. 2020). Vacquié-Garcia et al. (2015) studied 12 southern elephant seal females simultaneously equipped with accelerometers and with a range of physical sensors. They found that foraging environments were structured according to the primary frontal systems of the Southern Ocean and were characterized by different combinations of temperature, depth, and light level, suggesting the use of different prey communities. Nowak et al. (2020) found that chlorophyll-a concentration, estimated from animal-borne light level attenuation, varied seasonally and was positively associated with the chances of observing apparent foraging behavior in adult male and female gray seals, particularly during fall phytoplankton blooms when spatial variation in chlorophyll-a concentration was greater.

The presence of ice is another environmental factor that can influence the foraging behavior of phocids. Breed et al. (2018) found that sea ice concentration was highly predictive of juvenile bearded seal habitat use, with juveniles preferring intermediate concentrations and areas closer to the ice edge. Seasonal movements of bearded seals appeared to result from tracking the sea ice edge as it seasonally expanded and receded over the Bering and Chukchi continental shelves. Southern elephant seals making foraging trips to Antarctic sea ice consistently reduced move persistence as sea ice coverage increased but had highly variable responses to chlorophyll-a concentration. In contrast, seals foraging in the open ocean generally reduced move persistence where circumpolar deep water shoaled (Fig. 6.5, Jonsen et al. 2019).

Overall, these studies indicate that oceanographic features at various scales influence the behavior of phocids, presumably through their influence on the availability of prey. However, this presumption remains to be confirmed in most cases as simultaneous in-situ measurements of the spatial distribution of prey are generally not available.

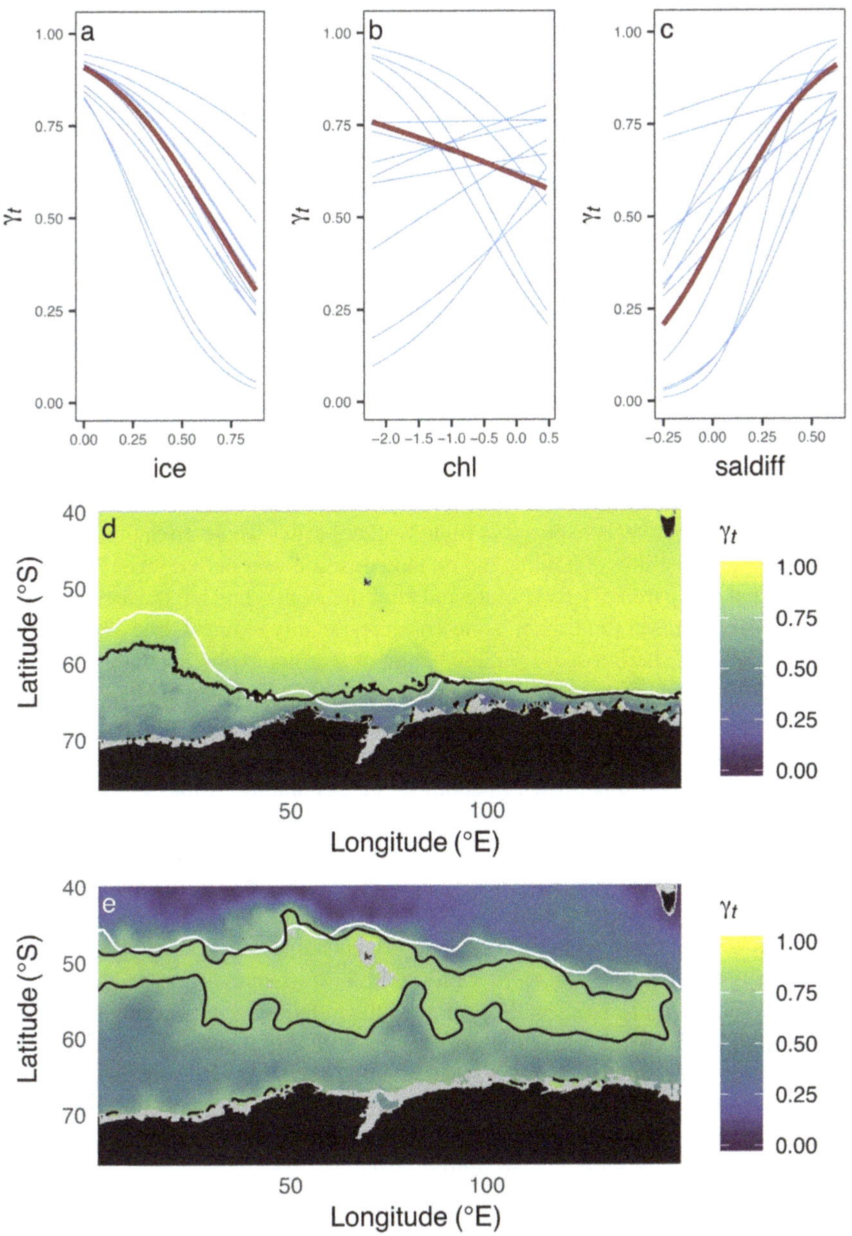

Fig. 6.5 Fixed (the mean response, red) and random (individual seals, blue) effects relationships between move persistence γ_t and (**a**) the proportion of ice cover (ice) and (**b**) chlorophyll *a* concentration (chl) for ice-foraging seals and (**c**) between γ_t and the salinity difference (saldiff) between 600 and 200 m for pelagic foraging seals. The models were used to generate spatial predictions of move persistence for ice (**d**) and pelagic (**e**) foraging seals. Areas of high move persistence (ie. $\gamma_t > 0.75$) are delineated by the black lines. The southern boundaries of the Antarctic Circumpolar Current (**d**) and the Subantarctic Front (**e**) are indicated as white lines. From Jonsen et al. (2019)

6.3.3.2 Predators and Competition

Predators can have important effects on foraging behavior and success of individuals by influencing where and when feeding occurs (Lima and Dill 1990; Peacor and Werner 2000; Preisser and Bolnick 2008; Wirsing et al. 2008). The direct effects of predation are widely recognized as a strong selective force; however, non-lethal impacts of predators may often have greater effects on their prey's behavior (Preisser et al. 2005). Relatively little research has been conducted on either the direct or indirect effect of predators on the foraging behavior of phocid species. One predator–prey interaction has provided some insight—killer whales, *Orcinus orca,* and harbor seals. Transient (Bigg's) killer whales from the US/Canada Pacific Northwest feed primarily, if not solely, on marine mammals, and harbor seals are the most frequently consumed (Baird and Dill 1995; Ford et al. 1998). The occurrence and residence time of transient killer whale pods peaked during harbor seal pupping when the energy intake by whales (mainly seal pups) also peaked. Exposure to transient killer whales can influence harbor seal behavior, such as activity budgets. In Hood Canal, Washington, harbor seals exposed to high levels of killer whale predation exhibited an increased probability of hauling out on land over a 2-year period when the whales were present compared to years before and after the whales had left the area (London et al. 2012). Increased time spent hauled out presumably reduced time available for foraging, although harbor seal foraging behavior was not explicitly studied.

A modeling study of several harbor seal predators (killer whales and Pacific sleeper sharks, *Somniosus pacificus*) provides further insight concerning non-lethal effects on harbor seal foraging behavior. Killer whales hunt near the surface, overlapping with Pacific herring (*Clupea pallasii*). In contrast, sleeper sharks primarily use depths of >100 m, overlapping with walleye pollock (*Gadus chalcogrammus*), a less energy-dense prey than herring. Both species are prey of harbor seals (Frid et al. 2006). Frid et al. (2007) predicted net energy gain and predation risk per foraging dive, parameterizing an analytical model with field data on the diving behavior of harbor seals, prey distributions, and use of depth by sleeper sharks and killer whales. Their analyses suggested that net energetic gain was highest when seals foraged for pollock in deep waters (>70 m). Nevertheless, plots of the individuals' predicted energy gain against predicted predation risk showed that seals underutilized relatively abundant prey found in deep waters where encounters with sleeper sharks were greater.

There is an increasingly accurate resolution of the movements and diving behavior of phocid seals. However, we still have relatively little understanding of how conspecifics and competitors may influence the foraging behavior of individuals. Nevertheless, we can gain some insight concerning the effects of intra-specific competition on behavior from the spatial distribution of different age classes. Patterns of habitat use and behavioral time budgets suggest that gray seal pups are likely displaced from foraging areas near the Sable Island, Canada breeding colony by adult females (Breed et al. 2013). This displacement was most pronounced in summer (Fig. 6.6). As young seals are less capable divers than adults, this may limit the habitat available to them. Juveniles born between 1998 and 2002 at Sable Island had reduced survivorship compared with cohorts born in the late 1980s, while adult survivorship

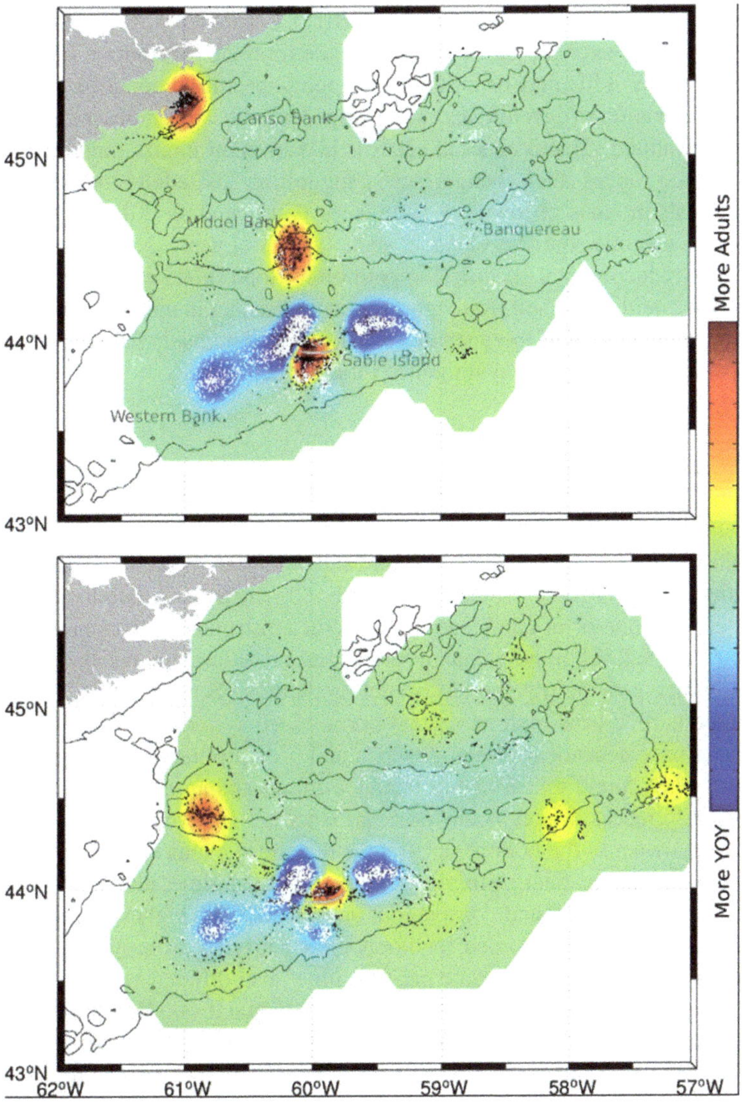

Fig. 6.6 July to September kernel density anomaly for **a** adult females versus young of the year gray seal (YOY) and **b** adult males versus YOY. White points are foraging locations of YOY, and black points are foraging locations from adults. Blue areas indicate regions used more heavily by YOY, while yellow and red areas indicate those used more heavily by adults. Green areas were used equally, while white areas were not used. From Breed et al. (2013)

has remained steady (den Heyer and Bowen 2017). Combined with behavioral observations, these survivorship data suggest that the foraging distribution of gray seals in their first year of life may be influenced by intra-specific competition from adults. Similar segregation of juvenile and adult ringed seals has been observed at glacial fronts in Svalbard (Hamilton et al. 2016). In Hawaiian monk seals, adults have smaller home ranges than younger seals, suggesting that adults may outcompete younger seals in areas of low prey abundance (Curtice et al. 2011). Age segregation among juvenile southern elephant seals (ages 1–4 years) has been interpreted as a way of reducing intra-specific competition for food (Field et al. 2005), but such segregation could also result from ontogenetic differences in diet. Although they are informative, the results of these studies provide only circumstantial support for how intra-specific competition may influence foraging behavior.

6.3.3.3 Foraging Success

Predators face decisions about which prey to include in their diet to maximize fitness. Many theoretical models attempt to explain and predict the behavior of air-breathing divers that exploit a food resource underwater. Telemetry and data-logging instruments, either alone or in combination with other instruments, have provided fine-scale (i.e., at the level of individual dives) resolution of foraging behavior of several species, which provide ways to test model predictions. Bowen et al. (2002b) fitted adult male harbor seals with an animal-borne video system and dive recorders to investigate prey-dependent foraging tactics and prey profitability in a free-ranging pinniped. They found that foraging tactics differed among and within prey types based on differences in prey behavior. Sand lance (*Ammodytes dubius*) was both a cryptic prey when in the bottom substrate and a conspicuous schooling prey when in the water column. Seal swimming speed, handling time, and capture success differed between cryptic and conspicuous sand lance. The highest capture success and handling time was recorded for flounders (family Pleuronectidae). Estimated profitability, i.e., net energy intake per unit time, also differed with prey type and prey size. Using the same data at the level of individual dives, Heaslip et al. (2014) tested nine predictions from optimal diving theory. They found support for seven of the nine predictions. As predicted, prey encounters increased with bottom duration; dive duration increased with dive depth; and travel duration, bottom duration, and percent bottom duration decreased over a wide range of travel durations. Descent duration increased with dive depth, and seals terminated dives earlier when prey were not encountered or when prey were encountered later in a dive. However, contrary to prediction, bottom duration did not increase and then decrease for short travel durations, and dives were not terminated earlier when travel durations were short, and the prey encounter rate was low.

The effect of prey density (or prey patch quality) on foraging behavior has been studied in several species. Sparling et al. (2007) conducted experiments in a large pool where gray seals had to swim horizontally to simulate diving to depth, to test the effect of prey patch density and distance from the surface on diving behavior. As predicted from theory, seals responded to prey density, leaving low-quality patches earlier. This pattern was still evident when seals had to travel far to reach the patch (simulating diving to deeper depths), contrary to the prediction that during deep dives, seals should stay at a patch regardless of prey density. Seals maximized dive durations at high prey densities, and longer distance traveled (again simulating longer dive durations), but they did not do so at short distances (simulating shallower dives). The giving-up strategy of the seals always produced higher profitability than would have been achieved if they had remained at the foraging site up to their aerobic dive limit on every dive. Thums et al. (2013) tested predictions of several optimal diving models in free-ranging, deep-diving southern elephant seals. Dive durations and bottom times varied with patch quality, but seals did not schedule their diving to maximize time spent in the foraging zone in patches of relatively higher quality. This observation is consistent with the predictions of the marginal value theorem (Charnov 1976), which states that animals foraging in a patch with better-than-average resources should spend relatively less time there. They also found higher descent and ascent rates in high-quality patches, suggesting that seals minimized travel time to the foraging patch when quality was high. Jouma'a et al. (2016) used time-depth recorder and 3D accelerometer data on southern elephant seals to test four predictions. Almost three-quarters of prey capture attempts detected by accelerometry occurred at the bottom of the dive. For dives <300 m, seals spent more time at the bottom of dives where prey were encountered. Seals had shorter bottom times and higher swimming speeds for dives where prey were encountered at deeper depths. When only dives associated with prey encounters were considered, the time spent at the bottom increased with the number of encounters. Body density, that is buoyancy, was a critical factor in controlling variations in dive duration. Finally, as predicted, post-dive surface intervals increased with dive duration and swimming effort in the previous dive. This study revealed a complex adjustment of foraging behavior relative to body density, prey encounter rate, and prey accessibility.

Fine-scale data from the combined use of animal-borne cameras and tri-axial accelerometers have permitted further tests of optimal diving theory. By integrating data from accelerometers with video and GPS, Wilson et al. (2017a) were able to identify variables associated with each foraging dive of Hawaiian monk seals. Dive depth, body motion (mean overall dynamic body acceleration during the dive), and proximity to the seafloor were the best predictors of prey searching. Search events typically occurred on long and deep dives, with more than 50% of the dive spent at the bottom.

Stomach temperature telemetry has revealed the spatial and temporal pattern of feeding in several free-ranging phocids. Austin et al. (2006a) studied feeding frequency in gray seals to test if previously reported sex differences in diving, movement, and diet were reflected in the temporal pattern of foraging success. Over 500 feeding events were recorded in 21 gray seals over periods of up to 40 days. Overall, seals fed on about 58% of days sampled with an average of 1.7 meals per day, but there were sex-specific differences and a high level of individual variability. Kuhn et al. (2009) used stomach temperature telemetry to show that northern elephant seal females fed within hours of departure and about 60 km from the rookery. Bouts of feeding began about 8 days into their foraging trip. Although associated with all dive types, they occurred most often during deep bottom-time dives. Females successfully fed only 18–24% of the time when they displayed the foraging type dive shape. This study showed that females feed extensively during early migration and that there was considerable individual variation in foraging locations and success.

Lidgard et al. (2014) used a combination of acoustic tags (Vemco Mobile Transceiver, VMT) and GPS locations to determine the spatial and temporal pattern of interactions between gray seals and potential prey that had been surgically implanted with a coded acoustic tag (Fig. 6.7). During a year study, the VMTs on 9 of 64 adult gray seals recorded detections from three species of fish, namely adult Atlantic cod *Gadus morhua*, Atlantic salmon *Salmo salar*, and American eel *Anguilla rostrata*. An examination of these seal-fish acoustic detections' temporal and spatial pattern suggested that predation on salmon and cod might have occurred. This study provided proof-of-concept that the foraging behavior of seals fitted with VMT and GPS tags can provide new insights into the nature of interspecies interactions in otherwise inaccessible environments.

Animal-borne cameras on a small sample of leopard seals have provided evidence of previously unknown foraging behavior of phocids. Video imagery showed that leopard seals mainly consumed Antarctic fur seals (*Arctocephalus gazella*, older animals and pups), demersal notothen fishes, and Pygoscelid penguins (Krause et al.

Fig. 6.7 A 30-year-old female gray seal fitted with a satellite-linked/GPS, time-depth recorder tag, a VHF radio tag, an accelerometer (head-mounted) and an acoustic transceiver (back) to study temporal and spatial encounters with acoustically tagged prey species. Photo by W. D. Bowen

2015). Seals used both chase and ambush tactics when hunting fur seals. However, successful hunting was driven by a subset of individuals using an ambush tactic, suggesting that individuals may specialize (see Sect. 6.3.2.4). Capture success was high for both Antarctic fur seal pups (76.5%), the most frequently targeted prey, and notothen fishes (64.3%). Video data also recorded kleptoparasitism, whereby other females stole 6 of 10 fur seal pups captured by one female, and food scavenging/caching of carcasses on the seafloor. These behaviors are not known to occur in other phocid species, but may be profitable in leopard seals because of the relatively large prey that are routinely captured compared to that consumed by other phocids.

6.4 Life History Consequences

Tagging of adult seals has dramatically increased our understanding of the factors that influence the inferred foraging behavior of phocid species. However, less is known about the life-history consequences of foraging behavior. We know that body mass plays a critical role in reproductive performance in male and female phocids (e.g., Iverson et al. 1993; Crocker et al. 2001; Lidgard et al. 2005; Haley et al. 1994). However, we still know relatively little about how foraging behavior influences mass gain and, thus, the amount of energy available for reproduction and survival. Studies on elephant seals and several other phocids are beginning to fill this gap. In southern elephant seals, despite relatively strong fidelity to foraging areas across years, the relative mass gain of individual seals was highly variable (Bradshaw et al. 2004). Seals returned to forage in the same regions even when they did relatively poorly in the previous year. The authors concluded that returning to the same foraging area may not necessarily provide short-term benefits because seasonal variation in ocean productivity is high. However, fidelity to regions that show higher productivity on average over an individual's lifetime may maximize net energy gain over the longer term. This finding is consistent with the results of Thums et al. (2011), whereby females did not appear to regulate broadscale movement in response to foraging success, suggesting that movement patterns represent a response to the prior expectation of the location of predictable and profitable resources. Schick et al. (2013) used a state-space model to infer daily body lipid content from drift dives of foraging northern and southern elephant seals. Estimates of body condition in individual elephant seals at fine time scales allowed inferences to be made about the areas and times in which these predators profitably exploited their environment. They found that southern elephant seals performed fewer drift dives and gained lipid at a lower rate than northern elephant seals (Fig. 6.8), but that physiological condition and foraging location were significantly associated with lipid gain in both species. In northern elephant seals, Abrahms et al. (2018a) found that despite exhibiting alternative patterns of site fidelity among females, strong and weak fidelity performed equally well in terms of body condition over a 10-year period, but the success of each strategy varied interannually and was mediated by climate conditions. While these

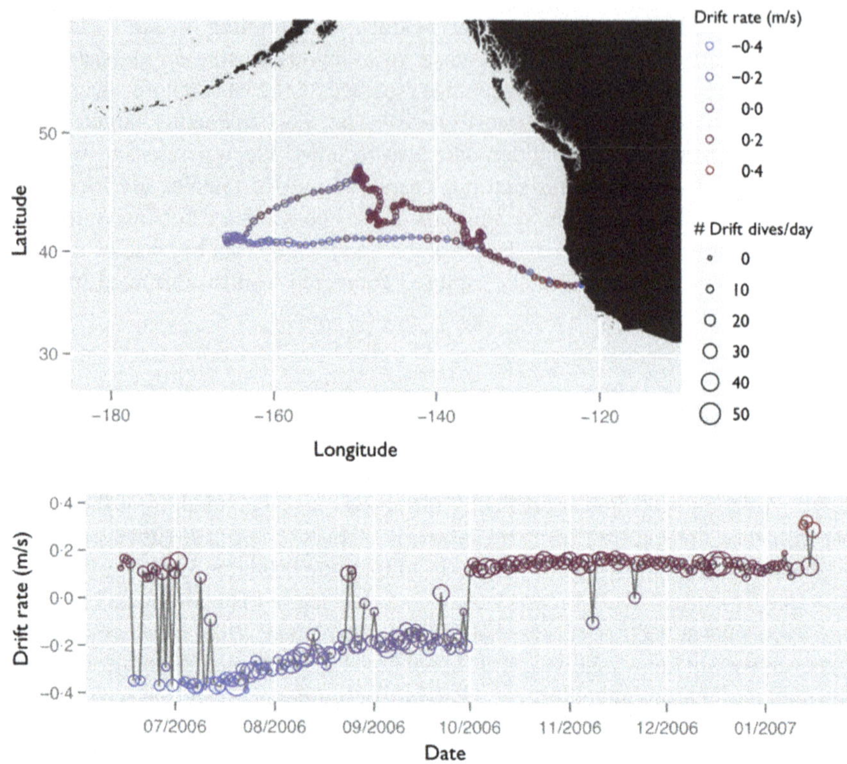

Fig. 6.8 Foraging trip map and drift rate time-series for one northern elephant seal tagged at Año Nuevo, California in 2006. The large shift to positive buoyancy occurs in early October 2006. From Schick et al. (2013)

studies provide insight on the relationship between foraging and body condition, they did not evaluate the life-history consequences of differences in foraging behavior.

A few studies have attempted to link reproductive performance and population dynamics to foraging behavior. Le Boeuf and Crocker (2005) took advantage of the fact that the weaning mass of northern elephant seal pups is strongly linked to maternal condition to examine the effects of decadal changes in prey availability and foraging success in pregnant northern elephant seal females in deep waters of the northeastern Pacific. They found evidence for a declining trend in the mean annual weaning mass of pups from 1975 to the late 1990s, a period characterized by a large-scale, basin-wide warm decadal regime that included multiple strong or long-duration El Niños. During a return to a cool regime from about 1999–2004, pup weaning mass then increased, reflecting improved foraging success of females. Robinson et al. (2012) analyzed diving, tracking, foraging success, and natality data for 297 adult female northern elephant seal migrations from 2004 to 2010. Seals focused their foraging effort along a narrow band corresponding to the boundary between the subarctic and subtropical gyres throughout the year. They found that mass gain during foraging migrations varied seasonally, but not among years. Annual mean rates of energy gain during post-molting foraging trips were good predictors of pregnancy status, but were not significant predictors of natality, perhaps due to the generally high natality rate in this population.

Several studies on southern elephant seals have provided further insight into the consequences of foraging behavior on reproductive performance. Authier et al. (2012) modeled pup weaning mass using blood stable isotope ratios to discriminate between maternal foraging behavior previously identified from movement studies. The carbon isotope ratio was a strong predictor of weaning mass, but there was an interaction between weaning mass and where females foraged. Females foraging in the inter-frontal zone weaned pups with a smaller mass compared with females foraging in Antarctic waters. McMahon et al. (2017) used the relationships between Southern Annular Mode and maximum sea-ice extent (both linked to foraging success) and maternal mass in southern elephant seals to infer consequences of changes in foraging behavior on pup weaning mass and subsequent survival. Maternal condition varied by as much as 59 kg among years. Environmental conditions increased predicted survival of pups with small mothers by 7% in 'good' versus 'bad' years compared to 1% for female pups of large mothers. New et al. (2014) used a modeling approach to simulate the effect of environmental change on the long-term fitness of female southern elephant seals. Simulations implied that single "bad" foraging years result in a relatively small (0.4%) population decline but that persistently poor foraging conditions (over 30 years) result in up to 10% population decline. Modeled reduction in maternal mass resulted in lighter pups at weaning, which in turn resulted in reduced pup survival. Using data from northern elephant seals, Pirotta et al. (2019) modeled the fitness implications of different foraging tactics. Their model simulated realistic patterns of environmental variation at different stages of a foraging trip, examined how the accumulation of lipid reserves would vary as a result of changes in foraging success, and the consequences for body condition, which in turn would affect offspring growth and survival.

6.5 Conclusions

Our understanding of foraging behavior of phocid seals has advanced considerably in the past decade or so. Most of this improved understanding results from studies on several of the more accessible species (i.e., land-breeders). More studies are needed on ice-breeding species to provide a more firm basis for generalization about phocid foraging ecology. The foraging behavior of phocids is influenced by multiple intrinsic factors (e.g., age, sex, body size, reproductive status) modified by environmental variability at various spatial and temporal scales (season, annual, inter-annual). We are just beginning to understand the ontogeny of foraging behavior and how spatial foraging patterns are influenced by early experience, individual heterogeneity, and environmental variation. Recent studies illustrate the power of statistical process-based models for integrating diverse information gleaned from animal-borne telemetry and data loggers to shed new light on phocid foraging behavior, foraging success, and, ultimately, potential fitness implications. However, more studies are needed to validate estimates of the spatial and temporal patterns of foraging success derived from statistical models of movement and diving behavior and understand the life-history consequences of changes in foraging behavior.

Acknowledgements We are grateful for helpful comments on earlier drafts of the chapter by M. A. Fedak, E. McHuron, and D. P. Costa.

References

Abrahms B, Hazen EL, Bograd SJ, Brashares JS, Robinson PW, Scales KL, Crocker DE, Costa DP (2018a) Climate mediates the success of migration strategies in a marine predator. Ecol Lett 21(1):63–71. https://doi.org/10.1111/ele.12871

Abrahms B, Scales KL, Hazen EL, Bograd SJ, Schick RS, Robinson PW, Costa DP (2018b) Mesoscale activity facilitates energy gain in a top predator. Proc Biol Sci 285 (1885). https://doi.org/10.1098/rspb.2018.1101

Adachi T, Costa DP, Robinson PW, Peterson SH, Yamamichi M, Naito Y, Takahashi A (2017) Searching for prey in a three-dimensional environment: hierarchical movements enhance foraging success in northern elephant seals. Funct Ecol 31(2):361–369. https://doi.org/10.1111/1365-2435.12686

Adachi T, Hückstädt LA, Tift MS, Costa DP, Naito Y, Takahashi A (2018) Inferring prey size variation from mandible acceleration in northern elephant seals. Mar Mamm Sci 35:893–908

Andersen JM, Skern-Mauritzen M, Boehme L, Wiersma YF, Rosing-Asvid A, Hammill MO, Stenson GB (2013a) Investigating annual diving behaviour by hooded seals (*Cystophora cristata*) within the Northwest Atlantic Ocean. Plos One 8(11). https://doi.org/10.1371/journal.pone.0080438

Andersen JM, Wiersma YF, Stenson GB, Hammill MO, Rosing-Asvid A, Skern-Maurizen M (2013b) Habitat selection by hooded seals (*Cystophora cristata*) in the Northwest Atlantic Ocean. ICES J Mar Sci 70(1):173–185. https://doi.org/10.1093/icesjms/fss133

Arce F, Bestley S, Hindell MA, McMahon CR, Wotherspoon S (2019) A quantitative, hierarchical approach for detecting drift dives and tracking buoyancy changes in southern elephant seals. Sci Rep 9. https://doi.org/10.1038/s41598-019-44970-1

Austin D, Bowen WD, McMillan JI (2004) Intraspecific variation in movement patterns: modeling individual behaviour in a large marine predator. Oikos 105(1):15–30. https://doi.org/10.1111/j.0030-1299.1999.12730.x

Austin D, Bowen WD, McMillan JI, Boness DJ (2006a) Stomach temperature telemetry reveals temporal patterns of foraging success in a free-ranging marine mammal. J Animal Ecol 75(2):408–420. https://doi.org/10.1111/j.1365-2656.2006.01057.x

Austin D, Bowen WD, McMillan JI, Iverson SJ (2006b) Linking movement, diving, and habitat to foraging success in a large marine predator. Ecology 87(12):3095–3108. https://doi.org/10.1890/0012-9658(2006)87[3095:lmdaht]2.0.co;2

Authier M, Bentaleb I, Ponchon A, Martin C, Guinet C (2012) Foraging fidelity as a recipe for a long life: foraging strategy and longevity in male southern elephant seals. Plos One 7(4). https://doi.org/10.1371/journal.pone.0032026

Badger JJ, Bowen WD, den Heyer CE, Breed GA (2020) Variation in individual reproductive performance amplified with population size in a long-lived carnivore. Ecology. https://doi.org/10.1002/ecy.3024

Baechler J, Beck CA, Bowen WE (2002) Dive shapes reveal temporal changes in the foraging behaviour of different age and sex classes of harbour seals (*Phoca vitulina*). Can J Zool 80(9):1569–1577. https://doi.org/10.1139/z02-150

Bailleul F, Charrassin JB, Monestiez P, Roquet F, Biuw M, Guinet C (2007) Successful foraging zones of southern elephant seals from the Kerguelen Islands in relation to oceanographic conditions. Philos Trans Royal Soc B-Biol Sci 362(1487):2169–2181. https://doi.org/10.1098/rstb.2007.2109

Baird RW, Dill LM (1995) Occurrence and behaviour of transient killer whales: seasonal and pod-specific variability, foraging behaviour, and prey handling. Can J Zool 73:1300–1311

Baker JD, Howell EA, Polovina JJ (2012) Relative influence of climate variability and direct anthropogenic impact on a sub-tropical Pacific top predator, the Hawaiian monk seal. Mar Ecol Prog Ser 469:175–189. https://doi.org/10.3354/meps09987

Baker JD, Polovina JJ, Howell EA (2007) Effect of variable oceanic productivity on the survival of an upper trophic predator, the Hawaiian monk seal *Monachus schauinslandi*. Mar Ecol Prog Ser 346:277–283. https://doi.org/10.3354/meps06968

Bearhop S, Phillips RA, McGill R, Cherel Y, Dawson DA, Croxall JP (2006) Stable isotopes indicate sex-specific and long-term individual foraging specialisation in diving seabirds. Mar Ecol Prog Ser 311:157–164. https://doi.org/10.3354/meps311157

Beck CA, Bowen WD, Iverson SJ (2000) Seasonal changes in buoyancy and diving behaviour of adult grey seals. J Exp Biol 203(15):2323–2330

Beck CA, Iverson SJ, Bowen WD, Blanchard W (2007) Sex differences in grey seal diet reflect seasonal variation in foraging behaviour and reproductive expenditure: evidence from quantitative fatty acid signature analysis. J Anim Ecol 76(3):490–502. https://doi.org/10.1111/j.1365-2656.2007.01215.x

Beck CA, Bowen WD, Iverson SJ (2003a) Sex differences in the seasonal patterns of energy storage and expenditure in a phocid seal. J Animal Ecol 72(2):280–291. https://doi.org/10.1046/j.1365-2656.2003.00704.x

Beck CA, Bowen WD, McMillan JI, Iverson SJ (2003b) Sex differences in diving at multiple temporal scales in a size-dimorphic capital breeder. J Animal Ecol 72(6):979–993. https://doi.org/10.1046/j.1365-2656.2003.00761.x

Beck CA, Bowen WD, McMillan JI, Iverson SJ (2003c) Sex differences in the diving behaviour of a size-dimorphic capital breeder: the grey seal. Animal Behav 66:777–789. https://doi.org/10.1006/anbe.2003.2284

Bekkby T, Bjorge A (1998) Variation in stomach temperature as indicator of meal size in harbor seals, *Phoca vitulina*. Mar Mamm Sci 14(3):627–637. https://doi.org/10.1111/j.1748-7692.1998.tb00750.x

Bennett KA, McConnell BJ, Moss SEW, Speakman JR, Pomeroy PP, Fedak MA (2010) Effects of age and body mass on development of diving capabilities of gray seal pups: costs and benefits of the postweaning fast. Physiol Biochem Zool 83(6):911–923. https://doi.org/10.1086/656925

Benoit-Bird KJ, Battaile BC, Heppell SA, Hoover B, Irons D, Jones N, Kuletz KJ, Nordstrom CA, Paredes R, Suryan RM, Waluk CM, Trites AW (2013) Prey patch patterns predict habitat use by top marine predators with diverse foraging strategies. Plos One 8(1). https://doi.org/10.1371/journal.pone.0053348

Bestley S, Jonsen ID, Hindell MA, Harcourt RG, Gales NJ (2015) Taking animal tracking to new depths: synthesizing horizontal-vertical movement relationships for four marine predators. Ecology 96(2):417–427. https://doi.org/10.1890/14-0469.1

Bestley S, Jonsen ID, Hindell MA, Guinet C, Charrassin JB (2013) Integrative modelling of animal movement: incorporating in situ habitat and behavioural information for a migratory marine predator. Proc Royal Soc B-Biol Sci 280(1750). https://doi.org/10.1098/rspb.2012.2262

Biuw M, Boehme L, Guinet C, Hindell M, Costa D, Charrassin JB, Roquet F, Bailleul F, Meredith M, Thorpe S, Tremblay Y, McDonald B, Park YH, Rintoul SR, Bindoff N, Goebel M, Crocker D, Lovell P, Nicholson J, Monks F, Fedak MA (2007) Variations in behavior and condition of a

Southern Ocean top predator in relation to in situ oceanographic conditions. Proc Natl Acad Sci USA 104(34):13705–13710. https://doi.org/10.1073/pnas.0701121104

Biuw M, McConnell B, Bradshaw CJA, Burton H, Fedak M (2003) Blubber and buoyancy: monitoring the body condition of free-ranging seals using simple dive characteristics. J Exp Biol 206(19):3405–3423. https://doi.org/10.1242/jeb.00583

Biuw M, Nost OA, Stien A, Zhou Q, Lydersen C, Kovacs KM (2010) Effects of hydrographic variability on the spatial, seasonal and diel diving patterns of southern elephant seals in the eastern Weddell Sea. PLoS One 5(11):e13816–e13816. https://doi.org/10.1371/journal.pone.0013816

Blanchet MA, Lydersen C, Ims RA, Kovacs KM (2016) Making it through the first year: Ontogeny of movement and diving behavior in harbor seals from Svalbard, Norway. Mar Mamm Sci 32(4):1340–1369. https://doi.org/10.1111/mms.12341

Block BA, Jonsen ID, Jorgensen SJ, Winship AJ, Shaffer SA, Bograd SJ, Hazen EL, Foley DG, Breed GA, Harrison AL, Ganong JE, Swithenbank A, Castleton M, Dewar H, Mate BR, Shillinger GL, Schaefer KM, Benson SR, Weise MJ, Henry RW, Costa DP (2011) Tracking apex marine predator movements in a dynamic ocean. Nature 475(7354):86–90. https://doi.org/10.1038/nature10082

Boinski S (1988) Sex-differences in the foraging behavior of squirrel-monkeys in a seasonal habitat. Behav Ecol Sociobiol 23(3):177–186. https://doi.org/10.1007/bf00300352

Bolnick DI, Amarasekare P, Araujo MS, Burger R, Levine JM, Novak M, Rudolf VHW, Schreiber SJ, Urban MC, Vasseur DA (2011) Why intraspecific trait variation matters in community ecology. Trends Ecol Evol 26(4):183–192. https://doi.org/10.1016/j.tree.2011.01.009

Bolnick DI, Svanback R, Fordyce JA, Yang LH, Davis JM, Husley CD, Forister ML (2003) The ecology of individuals: incidence and implications of individual specialization. Am Nat 161:1–28

Boness DJ, Bowen WD (1996) The evolution of maternal care in Pinnipeds. Bioscience 46(9):645–654. https://doi.org/10.2307/1312894

Boness DJ, Bowen WD, Oftedal OT (1994) Evidence of a maternal foraging cycle resembling that of Otariid seals in a small phocid, the harbor seal. Behav Ecol Sociobiol 34:95–104

Boness DJ, Bowen WD, Buhleier BM, Marshall GJ (2006) Mating tactics and mating system of an aquatic-mating pinniped: the harbor seal, *Phoca vitulina*. Behav Ecol Sociobiol 61(1):119–130. https://doi.org/10.1007/s00265-006-0242-9

Bowen WD (1997) Role of marine mammals in aquatic ecosystems. Mar Ecol Prog Ser 158:267–274

Bowen WD, Boness DJ, Iverson SJ, Oftedal OT (2001) Foraging effort, food intake, and lactation performance depend on maternal mass in a small phocid seal. Funct Ecol 15:325–334

Bowen WD, Oftedal OT, Boness DJ (1992) Mass and energy transfer during lactation in a small phocid, the harbor seal (*Phoca vitulina*). Physiol Zool 65:844–866

Bowen WD, Read A, Estes J (2002a) Feeding ecology. In: Hoelzel R (ed) Marine mammals: an evolutionary approach. Blackwell, pp 217–246

Bowen WD, Tully D, Boness DJ, Bulheier BM, Marshall GJ (2002b) Prey-dependent foraging tactics and prey profitability in a marine mammal. Mar Ecol Progr Ser 244:235–245. https://doi.org/10.3354/meps244235

Boyd IL (1998) Time and energy constraints in pinniped lactation. Am Nat 152(5):717–728. https://doi.org/10.1086/286202

Boyd IL (2000) State-dependent fertility in pinnipeds: contrasting capital and income breeders. Funct Ecol 14:623–630

Boyd IL, Arnould JPY, Barton T, Croxall JP (1994) Foraging behaviour of Antarctic fur seals during periods of contrasting prey abundance. J Anim Ecol 63:703–713

Bradshaw CJA, Hindell MA, Summer MD, Kelvin JM (2004) Loyalty pays: potential life history consequences of fidelity to marine foraging patterns by southern elephant seals. Anim Behav 68:1349–1360

Breed GA, Bowen WD, Leonard ML (2011) Development of foraging strategies with age in a long-lived marine predator. Mar Ecol Prog Ser 431:267–279. https://doi.org/10.3354/meps09134

Breed GA, Bowen WD, Leonard ML (2013) Behavioral signature of intraspecific competition and density dependence in colony-breeding marine predators. Ecol Evol 3(11):3838–3854. https://doi.org/10.1002/ece3.754
Breed GA, Bowen WD, McMillan JI, Leonard ML (2006) Sex segragation of seasonal foraging habitats in a non-migratory marine mammal. Proc Royal Soc B 273:2319–2326
Breed GA, Cameron MF, Hoef JMV, Boveng PL, Whiting A, Frost KJ (2018) Seasonal sea ice dynamics drive movement and migration of juvenile bearded seals *Erignathus barbatus*. Mar Ecol Prog Ser 600:223–237. https://doi.org/10.3354/meps12659
Breed GA, Jonsen ID, Myers RA, Bowen WD, Leonard ML (2009) Sex-specific, seasonal foraging tactics of adult grey seals (*Halichoerus grypus*) revealed by state-space analysis. Ecology 90(11):3209–3221. https://doi.org/10.1890/07-1483.1
Brown EG, Pierce GJ (1998) Monthly variation in the diet of harbour seals in inshore waters along the southeast Shetland (UK) coastline. Mar Ecol Prog Ser 167:275–289
Brown EG, Pierce GJ, Hislop JRG, Santos MB (2001) Interannual variation in the summer diets of harbour seals *Phoca vitulina* at Mousa, Shetland (UK). J Mar Biol Assoc UK 81(2):325–337. https://doi.org/10.1017/s0025315401003812
Burns JM (1999) The development of diving behavior in juvenile Weddell seals: pushing physiological limits in order to survive. Can J Zool 77(5):737–747. https://doi.org/10.1139/cjz-77-5-737
Burns JM, Costa DP, Frost K, Harvey JT (2005) Development of body oxygen stores in harbor seals: effects of age, mass, and body composition. Physiol Biochem Zool 78(6):1057–1068. https://doi.org/10.1086/432922
Burns JM, Lestyk KC, Folkow LP, Hammill MO, Blix AS (2007) Size and distribution of oxygen stores in harp and hooded seals from birth to maturity. J Comp Physiol B-Biochem Syst Environ Physiol 177(6):687–700. https://doi.org/10.1007/s00360-007-0167-2
Burns JM, Schreer JF, Castellini MA (1997) Physiological effects on dive patterns and foraging strategies in yearling Weddell seals (*Leptonychotes weddellii*). Can J Zool 75(11):1796–1810. https://doi.org/10.1139/z97-809
Cam E, Link WA, Cooch EG, Monnat JY, Danchin E (2002) Individual covariation in life-history traits: Seeing the trees despite the forest. Am Nat 159(1):96–105
Carter MID, Russell DJF, Embling CB, Blight CJ, Thompson D, Hosegood PJ, Bennett KA (2017) Intrinsic and extrinsic factors drive ontogeny of early-life at-sea behaviour in a marine top predator. Sci Rep 7. https://doi.org/10.1038/s41598-017-15859-8
Chambert T, Rotella JJ, Garrott RA (2013) An evolutionary perspective on reproductive individual heterogeneity in a marine vertebrate. J Anim Ecol 83(5):1158–1168. https://doi.org/10.1111/1365-2656.12211
Charnov EL (1976) Optimal foraging: the marginal value theorem. Theor Populat Biol 9:129–136
Clutton-Brock TH, Guinness FE, Albon SD (1982) Red deer: behaviour and ecology of two sexes. Chicago University Press, Chicago
Coltman DW, Bowen WD, Boness DJ, Iverson SJ (1997) Balancing foraging and reproduction in male harbour seals: an aquatically mating pinniped. Anim Behav 54:663–678
Coltman DW, Bowen WD, Iverson SJ, Boness DJ (1998) The energetics of male reproduction in an acquatically mating pinniped, the harbour seal. Physiol Zool 71(4):387–399. https://doi.org/10.1086/515418
Costa DP, Breed GA, Robinson PW (2012) New insights into pelagic migrations: implications for ecology and conservation. Annu Rev Ecol Evol Syst 43(43):73–96. https://doi.org/10.1146/annurev-ecolsys-102710-145045
Costa DP, Huckstadt LA, Crocker DE, McDonald BI, Goebel ME, Fedak MA (2010) Approaches to studying climatic change and its role on the habitat selection of Antarctic pinnipeds. Integr Comp Biol 50(6):1018–1030. https://doi.org/10.1093/icb/icq054
Costa DP (1987) Isotopic methods for quantifying material and energy balance of free ranging marine mammals. In: Huntley AD, Costa DP, Worthy GAJ, Castellini MA (eds) Aproaches to marine mammal energetics. Allen Press, Lawrence, Kansas

Cox SL, Embling CB, Hosegood PJ, Votier SC, Ingram SN (2018) Oceanographic drivers of marine mammal and seabird habitat-use across shelf-seas: a guide to key features and recommendations for future research and conservation management. Estuar Coast Shelf Sci 212:294–310. https://doi.org/10.1016/j.ecss.2018.06.022

Crawford JA, Frost KJ, Quakenbush LT, Whiting A (2019) Seasonal and diel differences in dive and haul-out behavior of adult and subadult ringed seals (*Pusa hispida*) in the Bering and Chukchi seas. Polar Biol 42(1):65–80. https://doi.org/10.1007/s00300-018-2399-x

Crocker DE, Costa DP, Le Boeuf BJ, Webb PM, Houser DS (2006) Impact of El Nino on the foraging behaviour of female northern elephant seals. Mar Ecol Prog Ser 309:1–10

Crocker DE, LeBoeuf BJ, Costa DP (1997) Drift diving in female northern elephant seals: implications for food processing. Can J Zool 75(1):27–39. https://doi.org/10.1139/z97-004

Crocker DE, Williams JD, Costa DP, Le Boeuf BJ (2001) Maternal traits and reproductive effort in northern elephant seals. Ecology 82(12):3541–3555. https://doi.org/10.1890/0012-9658(2001)082[3541:mtarei]2.0.co;2

Curtice C, Schick RS, Dunn DC, Halpin PN (2011) Home range analysis of Hawaiian monk seals (*Monachus schauinslandi*) based on colony, age, and sex. Aquat Mamm 37(3):360–371. https://doi.org/10.1578/am.37.3.2011.360

Davidson JD, El Hady A (2019) Foraging as an evidence accumulation process. Plos Comput Biol 15(7). https://doi.org/10.1371/journal.pcbi.1007060

Davis RW, Fuiman LA, Madden KM, Williams TM (2013) Classification and behavior of free-ranging Weddell seal dives based on three-dimensional movements and video-recorded observations. Deep-Sea Res Part II—Top Stud Oceanogr 88–89:65–77. https://doi.org/10.1016/j.dsr2.2012.07.006

Davis RW, Fuiman LA, Williams TM, Collier SO, Hagey WP, Kanatous SB, Kohin S, Horning M (1999) Hunting behavior of a marine mammal beneath the Antarctic fast ice. Science 283:993–996

Davis RW, Fuiman LA, Williams TM, Horning M, Hagey W (2003) Classification of Weddell seal dives based on 3-dimensional movements and video-recorded observations. Mar Ecol Prog Ser 264:109–122. https://doi.org/10.3354/meps264109

Davoren G (2013) Distribution of marine predator hotspots explained by persistent areas of prey. Mar Biol 160(12):3043–3058. https://doi.org/10.1007/s00227-013-2294-5

Dmitrieva L, Jussi M, Jussi I, Kasymbekov Y, Verevkin M, Baimukanov M, Wilson S, Goodman SJ (2016) Individual variation in seasonal movements and foraging strategies of a land-locked, ice-breeding pinniped. Mar Ecol Prog Ser 554:241–256. https://doi.org/10.3354/meps11804

Dragon AC, Bar-Hen A, Monestiez P, Guinet C (2012) Horizontal and vertical movements as predictors of foraging success in a marine predator. Mar Ecol Prog Ser 447:243–257. https://doi.org/10.3354/meps09498

Dragon AC, Monestiez P, Bar-Hen A, Guinet C (2010) Linking foraging behaviour to physical oceanographic structures: Southern elephant seals and mesoscale eddies east of Kerguelen Islands. Prog Oceanogr 87(1–4):61–71. https://doi.org/10.1016/j.pocean.2010.09.025

den Heyer CE, Bowen DW (2017) Estimating changes in vital rates of Sable Island grey seals using mark-recapture analysis. Canadian Science Advisory Secretartiat, Ottawa, Research Document, 2017/054:27p

Estes JA, Riedman ML, Staedler MM, Tinker MT, Lyon BE (2003) Individual variation in prey selection by sea otters: patterns, causes and implications. J Anim Ecol 72:144–155

Fauchald P, Tveraa T (2003) Using first-passage time in the analysis of area-restricted search and habitat selection. Ecology 84(2):282–288. https://doi.org/10.1890/0012-9658(2003)084[0282:ufptit]2.0.co;2

Fedak MA, Lovell P, Grant SM (2001) Two approaches to compressing and interpreting time-depth information as collected by time-depth recorders and satellite-linked data recorders. Mar Mamm Sci 17(1):94–110. https://doi.org/10.1111/j.1748-7692.2001.tb00982.x

Field IC, Bradshaw CJA, Burton HR, Sumner MD, Hindell MA (2005) Resource partitioning through oceanic segregation of foraging juvenile southern elephant seals (*Mirounga leonina*). Oecologia 142(1):127–135. https://doi.org/10.1007/s00442-004-1704-2

Field I, Hindell M, Slip D, Michael K (2001) Foraging strategies of southern elephant seals (*Mirounga leonina*) in relation to frontal zones and water masses. Antarct Sci 13(4):371–379. https://doi.org/10.1017/s0954102001000529

Folkow LP, Nordoy ES, Blix AS (2010) Remarkable development of diving performance and migrations of hooded seals (*Cystophora cristata*) during their first year of life. Polar Biol 33(4):433–441. https://doi.org/10.1007/s00300-009-0718-y

Ford JKB, Ellis GM, Barrett-Lennard LG, Morton AB, Palm RS, Balcomb KC (1998) Dietary specialization in two sympatric populations of killer whales (*Orcinus orca*) in coastal British Columbia and adjacent waters. Can J Zool 76(8):1456–1471. https://doi.org/10.1139/cjz-76-8-1456

Frid A, Baker GG, Dill LM (2006) Do resource declines increase predation rates on North Pacific harbor seals? A behavior-based plausibility model. Mar Ecol Prog Ser 312:265–275. https://doi.org/10.3354/meps312265

Frid A, Dill LM, Thorne RE, Blundell GM (2007) Inferring prey perception of relative danger in large-scale marine systems. Evol Ecol Res 9(4):635–649

Frost KJ, Simpkins MA, Lowry LF (2001) Diving behavior of subadult and adult harbor seals in Prince William Sound, Alaska. Mar Mamm Sci 17(4):813–834. https://doi.org/10.1111/j.1748-7692.2001.tb01300.x

Frost KJ, Lowry LF, Ver Hoef JM, Iverson SJ, Gotthardt T (1998) Monitoring habitat use and trophic interactions of harbor seals in Prince William Sound, Alaska. In: Exxon Valdez Oil Spill Restoration Project Annual R (ed) Alaska Department of fish and Game, Division of Wildlife Conservation, Fairbanks, Alaska

Gallon S, Bailleul F, Charrassin JB, Guinet C, Bost CA, Handrich Y, Hindell M (2013) Identifying foraging events in deep diving southern elephant seals, *Mirounga leonina*, using acceleration data loggers. Deep-Sea Res Part II—Top Stud Oceanogr 88–89:14–22. https://doi.org/10.1016/j.dsr2.2012.09.002

Gallon SL, Thompson D, Middlemas SJ (2017) What should I eat? Experimental evidence for prey selection in grey seals. Anim Behav 123:35–41. https://doi.org/10.1016/j.anbehav.2016.09.012

Gjertz I, Kovacs KM, Lydersen C, Wiig O (2000) Movements and diving of bearded seal (*Erignathus barbatus*) mothers and pups during lactation and post-weaning. Polar Biol 23(8):559–566

Goulet P, Guinet C, Swift R, Madsen PT, Johnson M (2019) A miniature biomimetic sonar and movement tag to study the biotic environment and predator-prey interactions in aquatic animals. Deep-Sea Res Part I—Oceanogr Res Pap 148:1–11. https://doi.org/10.1016/j.dsr.2019.04.007

Guinet C, Dubroca L, Lea MA, Goldsworthy S, Cherel Y, Duhamel G, Bonadonna F, Donnay J-P (2001) Spatial distribution of foraging in female Antarctic fur seals *Arctocephalus gazella* in relation to oceanographic variables: a scale-dependent approach using geographic information systems. Mar Ecol Prog Ser 219:251–264

Guinet C, Vacquie-Garcia J, Picard B, Bessigneul G, Lebras Y, Dragon AC, Viviant M, Arnould JPY, Bailleul F (2014) Southern elephant seal foraging success in relation to temperature and light conditions: insight into prey distribution. Mar Ecol Prog Ser 499:285–301. https://doi.org/10.3354/meps10660

Haley MP, Deutsch CJ, Le Boeuf BJ (1994) Size, dominance and copulatory success in male northern elephant seals, *Mirounga angustirostris*. Anim Behav 48:1249–1260

Hamel S, Cote SD (2008) Trade-offs in activity budget in an alpine ungulate: contrasting lactating and nonlactating females. Anim Behav 75:217–227. https://doi.org/10.1016/j.anbehav.2007.04.028

Hamilton CD, Lydersen C, Ims RA, Kovacs KM (2016) Coastal habitat use by ringed seals *Pusa hispida* following a regional sea-ice collapse: importance of glacial refugia in a changing Arctic. Mar Ecol Prog Ser 545:261–277. https://doi.org/10.3354/meps11598

Hamilton CD, Kovacs KM, Lydersen C (2018) Individual variability in diving, movement and activity patterns of adult bearded seals in Svalbard, Norway. Sci Rep 8. https://doi.org/10.1038/s41598-018-35306-6

Harcourt RG, Hindell MA, Bell DG, Waas JR (2000) Three-dimensional dive profiles of free-ranging Weddell seals. Polar Biol 23:479–487

Hassrick JL, Crocker DE, Costa DP (2013) Effects of maternal age and mass on foraging behaviour and foraging success in the northern elephant seal. Funct Ecol 27(4):1055–1063. https://doi.org/10.1111/1365-2435.12108

Hazen EL, Jorgensen S, Rykaczewski RR, Bograd SJ, Foley DG, Jonsen ID, Shaffer SA, Dunne JP, Costa DP, Crowder LB, Block BA (2013) Predicted habitat shifts of Pacific top predators in a changing climate. Nat Clim Chang 3(3):234–238. https://doi.org/10.1038/nclimate1686

Heaslip SG, Bowen WD, Iverson SJ (2014) Testing predictions of optimal diving theory using animal-borne video from harbour seals (*Phoca vitulina concolor*). Can J Zool 92(4):309–318. https://doi.org/10.1139/cjz-2013-0137

Heerah K, Hindell M, Andrew-Goff V, Field I, McMahon CR, Charrassin JB (2017) Contrasting behavior between two populations of an ice-obligate predator in East Antarctica. Ecol Evol 7(2):606–618. https://doi.org/10.1002/ece3.2652

Heerah K, Hindell M, Guinet C, Charrassin JB (2014) A new method to quantify within dive foraging behaviour in marine predators. Plos One 9(6). https://doi.org/10.1371/journal.pone.0099329

Heerah K, Cox SL, Blevin P, Guinet C, Charrassin JB (2019) Validation of dive foraging indices using archived and transmitted acceleration data: the case of the Weddell Seal. Front Ecol Evol 7(30). https://doi.org/10.3389/fevo.2019.00030

Hernandez KM, Bogomolni AL, Moxley JH, Waring GT, DiGiovanni RA, Hammill MO, Johnston DW, Sette L, Polito MJ (2019) Seasonal variability and individual consistency in gray seal (*Halichoerus grypus*) isotopic niches. Can J Zool 97(11):1071–1077. https://doi.org/10.1139/cjz-2019-0032

Hindell MA, Bradshaw CJA, Sumner MD, Michael KJ, Burton HR (2003) Dispersal of female southern elephant seals and their prey consumption during the austral summer: relevance to management and oceanographic zones. J Appl Ecol 40(4):703–715. https://doi.org/10.1046/j.1365-2664.2003.00832.x

Hindell MA, McConnell BJ, Fedak MA, Slip DJ, Burton HR, Reijnders PJH, McMahon CR (1999) Environmental and physiological determinants of successful foraging by naive southern elephant seal pups during their first trip to sea. Can J Zool 77(11):1807–1821. https://doi.org/10.1139/cjz-77-11-1807

Hindell MA, McMahon CR, Bester MN, Boehme L, Costa D, Fedak MA, Guinet C, Herraiz-Borreguero L, Harcourt RG, Huckstadt L, Kovacs KM, Lydersen C, McIntyre T, Muelbert M, Patterson T, Roquet F, Williams G, Charrassin JB (2016) Circumpolar habitat use in the southern elephant seal: implications for foraging success and population trajectories. Ecosphere 7(5). https://doi.org/10.1002/ecs2.1213

Hindle AG, Horning M, Mellish JAE, Lawler JM (2009) Diving into old age: muscular senescence in a large-bodied, long-lived mammal, the Weddell seal (*Leptonychotes weddellii*). J Exp Biol 212(6):790–796. https://doi.org/10.1242/jeb.025387

Hindle AG, Mellish JAE, Horning M (2011) Aerobic dive limit does not decline in an aging pinniped. Journal of Experimental Zoology Part a-Ecological Genetics and Physiology 315A(9):544–552. https://doi.org/10.1002/jez.703

Hooker SK, Boyd IL, Jessopp M, Cox O, Blackwell J, Boveng PL, Bengtson JL (2002) Monitoring the prey-field of marine predators: combining digital imaging with datalogging tags. Mar Mamm Sci 18(3):680–697. https://doi.org/10.1111/j.1748-7692.2002.tb01066.x

Houston AI (2011) Assessing models of optimal diving. Trends Ecol Evol 26:292–297

Huckstadt LA, Koch PL, McDonald BI, Goebel ME, Crocker DE, Costa DP (2012) Stable isotope analyses reveal individual variability in the trophic ecology of a top marine predator, the southern elephant seal. Oecologia 169(2):395–406. https://doi.org/10.1007/s00442-011-2202-y

Huckstadtl LA, Holserl RR, Tift MS, Costa DP (2018) The extra burden of motherhood: reduced dive duration associated with pregnancy status in a deep-diving mammal, the northern elephant seal. Biol Lett 14(2). https://doi.org/10.1098/rsbl.2017.0722

Irvine LG, Hindell MA, van den Hoff J, Burton HR (2000) The influence of body size on dive duration of underyearling southern elephant seals (*Mirounga leonina*). J Zool Lond 251:463–471
Iverson SJ, Bowen WD, Boness DJ, Oftedal OT (1993) The effect of maternal size and milk energy output on pup growth in grey seals (*Halichoerus grypus*). Physiol Zool 66:61–88
Iverson SJ, Frost KJ, Lowry LF (1997) Fatty acid signatures reveal fine scale structure of foraging distribution of harbor seals and their prey in Prince William Sound, Alaska. Marine Ecol Progr Ser 151(1–3):255–271. https://doi.org/10.3354/meps151255
Johnson DS, London JM, Lea MA, Durban JW (2008) Continuous-time correlated random walk model for animal telemetry data. Ecology 89(5):1208–1215. https://doi.org/10.1890/07-1032.1
Jones EL, McConnell BJ, Smout S, Hammond PS, Duck CD, Morris CD, Thompson D, Russell DJF, Vincent C, Cronin M, Sharples RJ, Matthiopoulos J (2015) Patterns of space use in sympatric marine colonial predators reveal scales of spatial partitioning. Mar Ecol Prog Ser 534:235–249. https://doi.org/10.3354/meps11370
Jonsen ID, Mills Flemming J, Myers RA (2005) Robust state-space modeling of animal movement data. Ecology 86:2874–2880
Jonsen ID, McMahon CR, Patterson TA, Auger-Methe M, Harcourt R, Hindell MA, Bestley S (2019) Movement responses to environment: fast inference of variation among southern elephant seals with a mixed effects model. Ecology 100(1). https://doi.org/10.1002/ecy.2566
Jouma'a J, Le Bras Y, Richard G, Vacquie-Garcia J, Picard B, El Ksabi N, Guinet C (2016) Adjustment of diving behaviour with prey encounters and body condition in a deep diving predator: the Southern Elephant Seal. Funct Ecol 30(4):636–648. https://doi.org/10.1111/1365-2435.12514
Kienle SS, Cacanindin A, Kendall T, Richter B, Ribeiro-French C, Castle L, Lentes G, Costa DP, Mehta RS (2019) Hawaiian monk seals exhibit behavioral flexibility when targeting prey of different size and shape. J Exp Biol 222(5). https://doi.org/10.1242/jeb.194985
Kramer DL (1988) The behavioral ecology of air breathing by aquatic animals. Can J Zool 66:89–94
Krause DJ, Goebel ME, Marshall GJ, Abernathy K (2015) Novel foraging strategies observed in a growing leopard seal (*Hydrurga leptonyx*) population at Livingston Island, Antarctic Peninsula. Anim Biotelem 24(3). https://doi.org/10.1186/s40317-015-0059-2
Kuhn CE (2011) The influence of subsurface thermal structure on the diving behavior of northern fur seals (*Callorhinus ursinus*) during the breeding season. Mar Biol 158(3):649–663. https://doi.org/10.1007/s00227-010-1589-z
Kuhn CE, Costa DP (2006) Identifying and quantifying prey consumption using stomach temperature change in pinnipeds. J Exp Biol 209(22):4524–4532. https://doi.org/10.1242/jeb.02530
Kuhn CE, Crocker DE, Tremblay Y, Costa DP (2009) Time to eat: measurements of feeding behaviour in a large marine predator, the northern elephant seal *Mirounga angustirostris*. J Anim Ecol 78(3):513–523. https://doi.org/10.1111/j.1365-2656.2008.01509.x
Langrock R, King R, Matthiopoulos J, Thomas L, Fortin D, Morales JM (2012) Flexible and practical modeling of animal telemetry data: hidden Markov models and extensions. Ecology 93(11):2336–2342. https://doi.org/10.1890/11-2241.1
Le Boeuf BJ, Crocker DE (2005) Ocean climate and seal condition. BMC Biol 3. https://doi.org/10.1186/1741-7007-3-9
Le Bras Y, Jouma'a J, Guinet C (2017) Three-dimensional space use during the bottom phase of southern elephant seal dives. Mov Ecol 5. https://doi.org/10.1186/s40462-017-0108-y
Le Boeuf BJ, Costa DP, Huntley AC, Feldkamp SD (1988) Continuous, deep diving in female northern elephant seals, *Mirounga angustirostris*. Can J Zool 66:446–458
Le Boeuf BJ, Crocker DE, Costa DP, Blackwell SB, Webb PM, Houser DS (2000) Foraging ecology of northern elephant seals. Ecol Monogr 70:353–382
Le Boeuf BJ, Morris PA, Blackwell SB, Crocker DE, Costa DP (1996) Diving behavior of juvenile northern elephant seals. Can J Zool 74:1632–1644
Lesage V, Hammill MO, Kovacs KM (1999) Functional classification of harbor seal (*Phoca vitulina*) dives using depth profiles, swimming velocity, and an index of foraging success. Can J Zool 77(1):74–87. https://doi.org/10.1139/z98-199

Lewis S, Benvenuti S, Dall'Antonia L, Griffiths R, Money L, Sherratt TN, Wanless S, Hamer KC (2002) Sex-specific foraging behaviour in a monomorphic seabird. Proc Royal Soc Lond Ser B—Biol Sci 269(1501):1687–1693

Lidgard DC, Boness DJ, Bowen WD, McMillan JI (2005) State-dependent male mating tactics in the grey seal: the importance of body size. Behav Ecol 16(3):541–549. https://doi.org/10.1093/beheco/ari023

Lidgard DC, Bowen WD, Jonsen ID, Iverson SJ (2014) Predator-borne acoustic transceivers and GPS tracking reveal spatiotemporal patterns of encounters with acoustically tagged fish in the open ocean. Mar Ecol Prog Ser 501:157–168. https://doi.org/10.3354/meps10670

Lidgard DC, Bowen WD, Iverson SJ (2020) Sex-differences in fine-scale home-range use in an upper-trophic level marine predator. Mov Ecol 8(1). https://doi.org/10.1186/s40462-020-0196-y

Lima SL, Dill LM (1990) Behavioural decisions made under the risk of predation: a review and prospectus. Can J Zool 68:619–640

Littnan CL, Baker JD, Parrish FA, Marshall GJ (2004) Effects of video camera attachment on the foraging behavior of immature Hawaiian monk seals. Mar Mamm Sci 20(2):345–352. https://doi.org/10.1111/j.1748-7692.2004.tb01164.x

London JM, Hoef JMV, Jeffries SJ, Lance MM, Boveng PL (2012) Haul-Out behavior of harbor seals (*Phoca vitulina*) in Hood Canal, Washington. Plos One 7(6). https://doi.org/10.1371/journal.pone.0038180

Lowry LF, Frost KJ, Ver Hoef JM, DeLong RA (2001) Movements of satellite-tagged subadult and adult harbor seals in Prince William Sound Alaska. Mar Mamm Sci 17(4):835–861. https://doi.org/10.1111/j.1748-7692.2001.tb01301.x

Luque SP, Ferguson SH, Breed GA (2014) Spatial behaviour of a keystone Arctic marine predator and implications of climate warming in Hudson Bay. J Exp Mar Biol Ecol 461:504–515. https://doi.org/10.1016/j.jembe.2014.10.002

MacArthur RH, Pianka ER (1966) On optimal use of a patchy environment. Am Nat 100:603–609

Mann KH, Lazier RRN (1996) Dynamics of marine ecosystems. Blackwell Science, Malden, MA, USA

McClintock BT, King R, Thomas L, Matthiopoulos J, McConnell BJ, Morales JM (2012) A general discrete-time modeling framework for animal movement using multistate random walks. Ecol Monogr 82(3):335–349

McClintock BT, Russell DJF, Matthiopoulos J, King R (2013) Combining individual animal movement and ancillary biotelemetry data to investigate population-level activity budgets. Ecology 94(4):838–849

McConnell BJ, Chambers C, Fedak MA (1992) Foraging ecology of southern elephant seals in relation to the bathymetry and productivity of the Southern Ocean. Antarct Sci 4:393–398

McConnell B, Fedak M, Burton HR, Engelhard GH, Reijnders PJH (2002) Movements and foraging areas of naive, recently weaned southern elephant seal pups. J Anim Ecol 71(1):65–78. https://doi.org/10.1046/j.0021-8790.2001.00576.x

McConnell BJ, Fedak MA, Lovell P, Hammond PS (1999) Movements and foraging areas of grey seals in the North Sea. J Appl Ecol 36(4):573–590. https://doi.org/10.1046/j.1365-2664.1999.00429.x

McIntyre T, Bester MN, Bornemann H, Tosh CA, de Bruyn PJN (2017) Slow to change? Individual fidelity to three-dimensional foraging habitats in southern elephant seals, *Mirounga leonina*. Anim Behav 127:91–99. https://doi.org/10.1016/j.anbehav.2017.03.006

McMahon CR, Harcourt RG, Burton HR, Daniel O, Hindell MA (2017) Seal mothers expend more on offspring under favourable conditions and less when resources are limited. J Anim Ecol 86(2):359–370. https://doi.org/10.1111/1365-2656.12611

McNamara JM, Houston AI (1996) State-dependent life histories. Nature 380(6571):215–221. https://doi.org/10.1038/380215a0

Michelot T, Langrock R, Bestley S, Jonsen ID, Photopoulou T, Patterson TA (2017) Estimation and simulation of foraging trips in land-based marine predators. Ecology 98(7):1932–1944. https://doi.org/10.1002/ecy.1880

Muelbert MMC, Bowen WD, Iverson SJ (2003) Weaning mass affects changes in body composition and food intake in harbour seal pups during the first month of independence. Physiol Biochem Zool 76(3):418–427. https://doi.org/10.1086/375427

Muelbert MMC, de Souza RB, Lewis MN, Hindell MA (2013) Foraging habitats of southern elephant seals, *Mirounga leonina*, from the Northern Antarctic Peninsula. Deep-Sea Res Part II—Top Stud Oceanogr 88–89:47–60. https://doi.org/10.1016/j.dsr2.2012.07.009

Muller-Karger FE, Varela R, Thunell R, Luerssen R, Hu C, Walsh JJ (2005) The importance of continental margins in the global carbon cycle. Geophys Res Lett 32:L01602

Naito Y, Bornemann H, Takahashi A, McIntyre T, Plotz J (2010) Fine-scale feeding behavior of Weddell seals revealed by a mandible accelerometer. Polar Sci 4(2):309–316. https://doi.org/10.1016/j.polar.2010.05.009

Naito Y, Costa DP, Adachi T, Robinson PW, Fowler M, Takahashi A (2013) Unravelling the mysteries of a mesopelagic diet: a large apex predator specializes on small prey. Funct Ecol 27(3):710–717. https://doi.org/10.1111/1365-2435.12083

Nathan R, Getz WM, Revilla E, Holyoak M, Kadmon R, Saltz D, Smouse PE (2008) A movement ecology paradigm for unifying organismal movement research. Proc Natl Acad Sci USA 105(49):19052–19059. https://doi.org/10.1073/pnas.0800375105

Neuhaus P, Ruckstuhl KE (2002) Foraging behaviour in Alpine ibex (*Capra ibex*): consequences of reproductive status, body size, age and sex. Ethol Ecol Evol 14(4):373–381

New LF, Clark JS, Costa DP, Fleishman E, Hindell MA, Klanjscek T, Lusseau D, Kraus S, McMahon CR, Robinson PW, Schick RS, Schwarz LK, Simmons SE, Thomas L, Tyack P, Harwood J (2014) Using short-term measures of behaviour to estimate long-term fitness of southern elephant seals. Mar Ecol Prog Ser 496:99-U344. https://doi.org/10.3354/meps10547

Noren SR, Iverson SJ, Boness DJ (2005) Development of the blood and muscle oxygen stores in grey seals (*Halichoerus grypus*): implications for juvenile diving capacity and the necessity of a terrestrial postweaning fast. Physiol Biochem Zool 78:482–490

Nowak BVR, Bowen WD, Whoriskey K, Lidgard DC, Mills Flemming JE, Iverson SJ (2020) Foraging behaviour of a continental shelf marine predator, the grey seal (*Halichoerus grypus*), is associated with *in situ*, subsurface oceanographic conditions. Mov Ecol 8(1). https://doi.org/10.1186/s40462-020-00225-7

Oksanen SM, Niemi M, Ahola MP, Kunnasranta M (2015) Identifying foraging habitats of Baltic ringed seals using movement data. Mov Ecol 3. https://doi.org/10.1186/s40462-015-0058-1

O'Toole MD, Lea MA, Guinet C, Schick R, Hindell MA (2015) Foraging strategy switch of a top marine predator according to seasonal resource differences. Front Mar Sci 2. https://doi.org/10.3389/fmars.2015.00021

Parrish FA, Abernathy K, Marshall GJ, Buhleier BM (2002) Hawaiian monk seals (*Monachus schauinslandi*) foraging in deep-water coral beds. Mar Mamm Sci 18(1):244–258. https://doi.org/10.1111/j.1748-7692.2002.tb01031.x

Parrish FA, Craig MP, Ragen TJ, Marshall GJ, Buhleier BM (2000) Identifiying diurnal foraging habitat of endangered Hawaiian monk seals using a seal-mounted video camera. Mar Mamm Sci 16:392–412

Peacor SD, Werner EE (2000) Predator effects on an assemblage of consumers through induced changes in consumer foraging behavior. Ecology 81(7):1998–2010. https://doi.org/10.2307/177288

Petit LJ, Petit DR, Petit KE, Fleming WJ (1990) Intersexual and temporal variation in foraging ecology of prothonotary warblers during the breeding season. Auk 107:133–145

Photopoulou T, Fedak MA, Thomas L, Matthiopoulos J (2014) Spatial variation in maximum dive depth in gray seals in relation to foraging. Mar Mamm Sci 30(3):923–938. https://doi.org/10.1111/mms.12092

Pirotta E, Schwarz LK, Costa DP, Robinson PW, New L (2019) Modeling the functional link between movement, feeding activity, and condition in a marine predator. Behav Ecol 30(2):434–445. https://doi.org/10.1093/beheco/ary183

Platt T, Fuentes-Yaco C, Frank KT (2003) Spring algal bloom and larval fish survival. Nature 423(6938):398–399

Preisser EL, Bolnick DI, Benard MF (2005) Scared to death? The effects of intimidation and consumption in predator-prey interactions. Ecology 86(2):501–509. https://doi.org/10.1890/04-0719

Preisser EL, Bolnick DI (2008) The many faces of fear: comparing the pathways and impacts of nonconsumptive predator effects on prey populations. Plos One 3(6). https://doi.org/10.1371/journal.pone.0002465

Prothero J (1995) Bone and fat as a function of body weight in adult mammals. Comp Biochem Physiol 111A:633–639

Reilly JJ, Fedak MA (1991) Rates of water turnover and energy expenditure of free-living males common seals (*Phoca vitulina*). J Zool Lond 223:461–468

Richardson K, Visser AW, Pedersen FB (2000) Subsurface phytoplankton blooms fuel pelagic production in the North Sea. 2000;22:1663–71. J Plankton Res 22:1663–1671

Riotte-Lambert L, Weimerskirch H (2013) Do naive juvenile seabirds forage differently from adults? Proc Royal Soc B-Biol Sci 280(1768). https://doi.org/10.1098/rspb.2013.1434

Robinson PW, Simmons SE, Crocker DE, Costa DP (2010) Measurements of foraging success in a highly pelagic marine predator, the northern elephant seal. J Anim Ecol 79(6):1146–1156. https://doi.org/10.1111/j.1365-2656.2010.01735.x

Robinson PW, Costa DP, Crocker DE, Gallo-Reynoso JP, Champagne CD, Fowler MA, Goetsch C, Goetz KT, Hassrick JL, Huckstadt LA, Kuhn CE, Maresh JL, Maxwell SM, McDonald BI, Peterson SH, Simmons SE, Teutschel NM, Villegas-Amtmann S, Yoda K (2012) Foraging behavior and success of a mesopelagic predator in the Northeast Pacific Ocean: insights from a data-rich species, the northern elephant seal. Plos One 7(5). https://doi.org/10.1371/journal.pone.0036728

Rodriguez JP, Fernandez-Gracia J, Thums M, Hindell MA, Sequeira AMM, Meekan MG, Costa DP, Guinet C, Harcourt RG, McMahon CR, Muelbert M, Duarte CM, Eguiluz VM (2017) Big data analyses reveal patterns and drivers of the movements of southern elephant seals. Sci Rep 7. https://doi.org/10.1038/s41598-017-00165-0

Russell DJF, McClintock BT, Matthiopoulos J, Thompson PM, Thompson D, Hammond PS, Jones EL, MacKenzie ML, Moss S, McConnell BJ (2015) Intrinsic and extrinsic drivers of activity budgets in sympatric grey and harbour seals. Oikos 124(11):1462–1472. https://doi.org/10.1111/oik.01810

Sathyendranath S, Longhurst A, Caverhill CM, Platt T (1995) Regionally and seasonally differentiated primary production in the North Atlantic. Deep-Sea Res Part I—Oceanogr Res Pap 42(10):1773–1802. https://doi.org/10.1016/0967-0637(95)00059-f

Schick RS, New LF, Thomas L, Costa DP, Hindell MA, McMahon CR, Robinson PW, Simmons SE, Thums M, Harwood J, Clark JS (2013) Estimating resource acquisition and at-sea body condition of a marine predator. J Anim Ecol 82(6):1300–1315. https://doi.org/10.1111/1365-2656.12102

Schreer JF, Kovacs KM (1997) Allometry of diving capacity in air-breathing vertebrates. Can J Zool 75(3):339–358. https://doi.org/10.1139/z97-044

Schreer JF, Kovacs KM, Hines RJO (2001) Comparative diving patterns of pinnipeds and seabirds. Ecol Monogr 71(1):137–162. https://doi.org/10.1890/0012-9615(2001)071[0137:cdpopa]2.0.co;2

Schreer JF, Testa JW (1996) Classification of Weddell seal diving behavior. Mar Mamm Sci 12:227–250

Schulz TM, Bowen WD (2004) Pinniped lactation strategies: evaluation of data on maternal and offspring life history traits. Mar Mamm Sci 20(1):86–114. https://doi.org/10.1111/j.1748-7692.2004.tb01142.x

Sequeira AMM, Rodriguez JP, Eguiluz VM, Harcourt R, Hindell M, Sims DW, Duarte CM, Costa DP, Fernandez-Gracia J, Ferreira LC, Hays GC, Heupel MR, Meekan MG, Avenn A, Bailleul F, Baylis AMM, Berumen ML, Braun CD, Burns J, Caley MJ, Campbell R, Carmichael RH, Clua E,

Einoder LD, Friedlaender A, Goebel ME, Goldsworthy SD, Guinet C, Gunn J, Hamer D, Hammerschlag N, Hammill M, Huckstadt LA, Humphries NE, Lea MA, Lowther A, Mackay A, McHuron E, McKenzie J, McLeay L, McMahond CR, Mengersenv K, Muelbert MMC, Pagano AM, Page B, Queiroz N, Robinson PW, Shaffer SA, Shivji M, Skomal GB, Thorrold SR, Villegas-Amtmann S, Weise M, Wells R, Wetherbee B, Wiebkin A, Wienecke B, Thums M (2018) Convergence of marine megafauna movement patterns in coastal and open oceans. Proc Natl Acad Sci USA 115(12):3072–3077. https://doi.org/10.1073/pnas.1716137115

Sequeira AMM, Heupel MR, Lea MA, Eguiluz VM, Duarte CM, Meekan MG, Thums M, Calich HJ, Carmichael RH, Costa DP, Ferreira LC, Fernandez-Gracia J, Harcourt R, Harrison AL, Jonsen I, McMahon CR, Sims DW, Wilson RP, Hays GC (2019) The importance of sample size in marine megafauna tagging studies. Ecol Appl 29. https://doi.org/10.1002/eap.1947

Sergeant DE (1965) Migrations of harp seals *Pagophilus groenlandicus* (Erxleben) in the Northwest Atlantic. J Fish Res Bd Can 22(2):433–463f

Sharples RJ, Moss SE, Patterson TA, Hammond PS (2012) Spatial variation in foraging behaviour of a marine top predator (*Phoca vitulina*) determined by a large-scale satellite tagging program. Plos One 7(5). https://doi.org/10.1371/journal.pone.0037216

Shero MR, Goetz KT, Costa DP, Burns JM (2018) Temporal changes in Weddell seal dive behavior over winter: are females increasing foraging effort to support gestation? Ecol Evol 8(23):11857–11874. https://doi.org/10.1002/ece3.4643

Sibly RM, Calow P (1986) Physiological ecology of animals. Blackwell Scientific Publications, Oxford

Simmons SE, Crocker DE, Kudela RM, Costa DP (2007) Linking foraging behaviour of the northern elephant seal with oceanography and bathymetry at mesoscales. Mar Ecol Prog Ser 346:265–275. https://doi.org/10.3354/meps07014

Simpkins MA, Kelly BP, Wartzok D (2001) Three-dimensional movements within individual dives by ringed seals (*Phoca hispida*). Can J Zool 79:1455–1464

Simpson JH, Sharples J (2012) Introduction to the physical and biological oceanography of shelf seas. Cambridge University Press, England, Cambridge

Sjoberg M, Ball JP (2000) Grey seal, *Halichoerus grypus*, habitat selection around haulout sites in the Baltic Sea: bathymetry or central-place foraging? Can J Zool 78(9):1661–1667

Smith J, Karpovich S, Horstmann L, McIntyre J, O'Brien DM (2019) Seasonal differences in foraging and isotopic niche width related to body size in Gulf of Alaska harbor seals. Can J Zool 97(12):1156–1163. https://doi.org/10.1139/cjz-2019-0108

Southwell C (2004) Satellite-linked dive recorders provide insights into the reproductive strategies of crabeater seals (*Lobodon carcinophagus*). J Zool 264:399–402. https://doi.org/10.1017/s0952836904005928

Sparling CE, Georges JY, Gallon SL, Fedak M, Thompson D (2007) How long does a dive last? Foraging decisions by breath-hold divers in a patchy environment: a test of a simple model. Anim Behav 74:207–218

Stephens PA, Boyd IL, McNamara JM, Houston AI (2009) Capital breeding and income breeding: their meaning, measurement, and worth. Ecology 90(8):2057–2067. https://doi.org/10.1890/08-1369.1

Stevick PT, Incze LS, Kraus SD, Rosen S, Wolff N, Baukus A (2008) Trophic relationships and oceanography on and around a small offshore bank. Mar Ecol Prog Ser 363:15–28. https://doi.org/10.3354/meps07475

Stewart BS (1997) Ontogeny of differential migration and sexual segregation in northern elephant seals. J Mamm 78(4):1101–1116. https://doi.org/10.2307/1383053

Stewart BS, DeLong RL (1995) Double migrations of the northern elephant seal, *Mirounga angustirostris*. J Mammal 76:196–205

Stewart BS, Antonelis GA, Baker JD (2006) Foraging biogeography of Hawaiian monk seals in the northwestern Hawaiian Islands. Atoll Res 131–145

Thompson D, Fedak MA (2001) How long should a dive last? A simple model of foraging decisions by breath-hold divers in a patchy environment. Anim Behav 61:287–296

Thompson D, Hiby AR, Fedak MA (1993) How fast should I swim? Behavioural implications of diving physiology. Symp Zool Soc Lond 66:349–368

Thompson PM, Mackay A, Tollit DJ, Enderby S, Hammond PS (1998) The influence of body size and sex on the characteristics of harbour seal foraging trips. Can J Zool 76(6):1044–1053. https://doi.org/10.1139/cjz-76-6-1044

Thompson PM, Pierce GJ, Hislop JRG, Miller D, Diack JSW (1991) Winter foraging by common seals (*Phoca vitulina*) in relation to food availability in the inner Moray Firth N. E. Scotland. J Anim Ecol 60:283–294

Thompson PM, Tollit DJ, Greenstreet SPR, MacKay A, Corpe HM (1996) Between-year variations in the diet and behaviour of harbour seals, *Phoca vitulina* in the Moray Firth; causes and consequences. In: Greenstreet SPR, Tasker ML (eds) Aquatic predators and their prey. Blackwell Science Publishing, Oxford, England, pp 44–52

Thums M, Bradshaw CJA, Hindell MA (2011) In situ measures of foraging success and prey encounter reveal marine habitat-dependent search strategies. Ecology 92(6):1258–1270. https://doi.org/10.1890/09-1299.1

Thums M, Bradshaw CJA, Sumner MD, Horsburgh JM, Hindell MA (2013) Depletion of deep marine food patches forces divers to give up early. J Anim Ecol 82(1):72–83. https://doi.org/10.1111/j.1365-2656.2012.02021.x

Vacquié-Garcia J, Guinet C, Laurent C, Bailleul F (2015) Delineation of the southern elephant seal's main foraging environments defined by temperature and light conditions. Deep-Sea Res Part II—Top Stud Oceanogr 113:145–153. https://doi.org/10.1016/j.dsr2.2014.10.029

van Beest FM, Mews S, Elkenkamp S, Schuhmann P, Tsolak D, Wobbe T, Bartolino V, Bastardie F, Dietz R, von Dorrien C, Galatius A, Karlsson O, McConnell B, Nabe-Nielsen J, Olsen MT, Teilmann J, Langrock R (2019) Classifying grey seal behaviour in relation to environmental variability and commercial fishing activity—a multivariate hidden Markov model. Sci Rep 9. https://doi.org/10.1038/s41598-019-42109-w

Volpov BL, Hoskins AJ, Battaile BC, Viviant M, Wheatley KE, Marshall G, Abernathy K, Arnould JPY (2015) Identification of prey captures in Australian fur seals (*Arctocephalus pusillus doriferus*) using head-mounted accelerometers: field validation with animal-borne video cameras. Plos One 10(6). https://doi.org/10.1371/journal.pone.0128789

Votier SC, Fayet AL, Bearhop S, Bodey TW, Clark BL, Grecian J, Guilford T, Hamer KC, Jeglinski JWE, Morgan G, Wakefield E, Patrick SC (2017) Effects of age and reproductive status on individual foraging site fidelity in a long-lived marine predator. Proc Royal Soc B—Biol Sci 284(1859). https://doi.org/10.1098/rspb.2017.1068

Walker BG, Bowen WD (1993) Changes in body mass and feeding behaviour in male harbour seals, *Phoca vitulina*, in relation to female reproductive status. J Zool Lond 231:423–436

Webb PM, Crocker DE, Blackwell SB, Costa DP, Le Boeuf BJ (1998) Effects of buoyancy on the diving behavior of northern elephant seals. J Exp Biol 201(16):2349–2358

Wheatley KE, Bradshaw CJA, Davis LS, Harcourt RG, Hindell MA (2006) Influence of maternal mass and condition on energy transfer in Weddell seals. J Anim Ecol 75:724–733

Wilson RP, Cooper J, Plotz J (1992) Can we determine when marine endotherms feed? A case study with seabirds. J Exp Biol 167:267–275

Wilson LJ, Hammond PS (2019) The diet of harbour and grey seals around Britain: examining the role of prey as a potential cause of harbour seal declines. Aquatic Conserv—Mar Freshw Ecosyst 29:71–85. https://doi.org/10.1002/aqc.3131

Wilson K, Littnan C, Halpin P, Read A (2017a) Integrating multiple technologies to understand the foraging behaviour of Hawaiian monk seals. Royal Soc Open Sci 4(3). https://doi.org/10.1098/rsos.160703

Wilson K, Littnan C, Read AJ (2017b) Movements and home ranges of monk seals in the main Hawaiian Islands. Mar Mamm Sci 33(4):1080–1096. https://doi.org/10.1111/mms.12429

Wirsing AJ, Heithaus MR, Frid A, Dill LM (2008) Seascapes of fear: evaluating sublethal predator effects experienced and generated by marine mammals. Mar Mamm Sci 24(1):1–15. https://doi.org/10.1111/j.1748-7692.2007.00167.x

Ydenberg RC, Butler RW, Lank DB (2007) Effects of predator landscapes on the evolutionary ecology of routing, timing and molt by long-distance migrants. J Avian Biol 38(5):523–529. https://doi.org/10.1111/j.2007.0908-8857.04202.x

Yoshino K, Takahashi A, Adachi T, Costa DP, Robinson PW, Peterson SH, Huckstadt LA, Holserl RR, Naito Y (2020) Acceleration-triggered animal-borne videos show a dominance of fish in the diet of female northern elephant seals. J Exp Biol 223(jeb212936)

Yurkowski DJ, Ferguson S, Choy ES, Loseto LL, Brown TM, Muir DCG, Semeniuk CAD, Fisk AT (2016a) Latitudinal variation in ecological opportunity and intraspecific competition indicates differences in niche variability and diet specialization of Arctic marine predators. Ecol Evol 6(6):1666–1678. https://doi.org/10.1002/ece3.1980

Yurkowski DJ, Ferguson SH, Semeniuk CAD, Brown TM, Muir DCG, Fisk AT (2016b) Spatial and temporal variation of an ice-adapted predator's feeding ecology in a changing Arctic marine ecosystem. Oecologia 180 (3):631–644. https://doi.org/10.1007/s00442-015-3384-5

Yurkowski DJ, Semeniuk CAD, Harwood LA, Rosing-Asvid A, Dietz R, Brown TM, Clackett S, Grgicak-Mannion A, Fisk AT, Ferguson SH (2016c) Influence of sea ice phenology on the movement ecology of ringed seals across their latitudinal range. Marine Ecol Progr Ser 562:237–250. https://doi.org/10.3354/meps11950

Zollner PA, Lima SL (1999) Search strategies for landscape-level interpatch movements. Ecology 80(3):1019–1030. https://doi.org/10.1890/0012-9658(1999)080[1019:ssflli]2.0.co;2

Chapter 7
Breeding Behavior

Kelly J. Robinson and Patrick P. Pomeroy

Contents

Abstract A successful breeding episode requires individuals to locate suitable mates, produce young, and rear them to independence within a suitable environment, time, and place. All this must be done while avoiding predators and competing for essential, finite resources, such as access to mates or breeding sites. Phocid seal species show great diversity in the behaviors that have evolved to overcome these challenges. Although all phocids forage in marine environments, they give birth on various substrate types, the temporal and spatial separation of which imposes significant energetic pressures on breeding individuals. Most existing detailed studies of phocid breeding behavior have been on species that breed in large, land-based colonies where most of their reproductive behavior is visible to observers. There are significant gaps in our knowledge of seals breeding in remote environments, small groups, and those that forage while breeding or display and mate underwater. Improvements in telemetry devices, ways of examining links between behavior and physiology in free-ranging seals, and remote surveying techniques will be essential for continuing to expand our understanding of breeding behavior in phocids and to predict how individuals or populations may respond to changing environments and anthropogenic activities.

K. J. Robinson (✉)
Centre for Biological Diversity, School of Biology, University of St Andrews, Sir Harold Mitchell Building, St Andrews, Fife KY16 9TF, UK
e-mail: kjr33@st-andrews.ac.uk

K. J. Robinson · P. P. Pomeroy
Sea Mammal Research Unit, Scottish Oceans Institute, School of Biology, University of St Andrews, Fife KY16 8LB, UK

D. P. Costa and E. McHuron (eds.), *Ethology and Behavioral Ecology of Phocids*, Ethology and Behavioral Ecology of Marine Mammals,
https://doi.org/10.1007/978-3-030-88923-4_7

Keywords Phocid seals · Reproductive behavior · Mating · Aggression · Maternal behavior · Philopatry · Allosuckling · Pup behavior · Post-weaning fast

7.1 Introduction

Reproductive behavior has evolved to facilitate the production of the next generation and their survival. Mating behaviors enable individuals from opposite sexes to locate appropriate partners in time and space, sometimes across vast areas, and may also enable selecting high-quality mates. Parental expenditure of resources and time are required to raise offspring to independence, but reproductive success also requires complementary offspring behavior.

Like all mammals, phocid mothers lactate to nourish their pups during early life. They must also face the challenge of nursing a largely non-aquatic pup despite living most of their lives in aquatic environments (Fig. 7.1). Phocids solve this problem by leaving the water to give birth and raise their pups on solid substrates. This separation of breeding and foraging areas for mothers means that fasting during lactation occurs in all species, ranging from weeks in capital breeding species to a matter of hours or days in income breeding species (Chap. 8). As females typically come into estrus at the end of the lactation period, males attend females on or around these breeding sites. The substantial energetic strain this places on adult males and females plays a significant role in shaping their life histories and the reproductive behaviors seen during breeding attempts in phocid seals.

A variety of reproductive behaviors have evolved in phocids to cope with the need to fast during reproduction, the energetic demands of rearing young, and the different

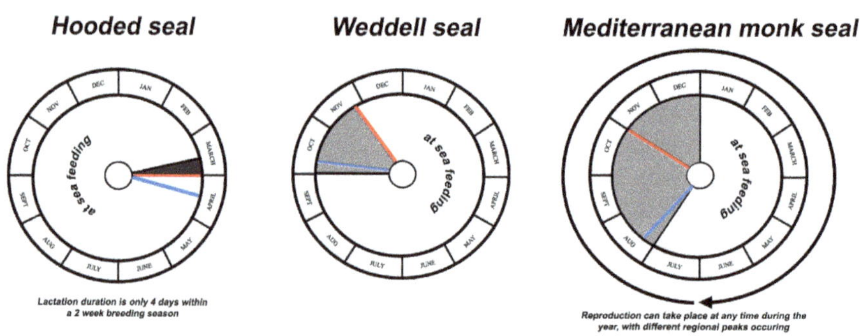

Fig. 7.1 Phocids have remarkably different life-history strategies to cope with rearing pups, fasting, and returning to the sea to feed. Adults and pups must either be able to sustain themselves away from feeding areas or must be able to move from breeding sites into the water to enable foraging behaviors. In this figure, dark gray shaded areas indicate lactation periods with fasting, light gray shaded areas indicate lactation periods with foraging trips, red lines indicate estrus and blue lines indicate when pups enter the water for the first time. Data used from the following sources: hooded seals, (Kovacs 1990, 2017; Lydersen et al. 1997; Weddell seals, Mansfield 1958; Stirling 1969; Mediterranean monk seals, Pastor and Aguilar 2003; Gucu et al. 2004; Aguilar et al. 2007)

morphology and breeding environments of each species. Across the family, there is remarkable diversity in mating systems, ranging from serial monogamy (Southwell 2004) to extreme polygyny (Le Boeuf 1974; Haley et al. 1994; Fabiani et al. 2004). Breeding aggregations range from solitary mother–pup pairs (Rogers 2017) to large colonies of thousands of individuals (Bowen et al. 2003). Phocid lactation durations are short in comparison to other marine and terrestrial mammals. All species give birth to highly precocial offspring to compensate for constrained dependent periods and abrupt weaning events (Lydersen and Kovacs 1999). The duration of lactation also varies across phocid species, ranging from the shortest lactation period of any mammal seen in hooded seals (*Cystophora cristata,* 4 days; Bowen et al. 1985) to Mediterranean monk seals (*Monachus monachus*) that nurse pups for up to five months (Aguilar et al. 2007). However, characteristics common to the majority of species are a complete lack of paternal care of offspring (Lydersen and Kovacs 1999), highly synchronized breeding seasons (Stirling 1983), and abrupt weaning events, after which pups are entirely independent of mothers (Atkinson 1997).

A primary evolutionary driver of the variety of phocid breeding behaviors is the predictability of the environment in which seals mate, give birth, and rear pups. The duration of breeding site availability, predation risk, number of individuals that use a site at one time, and how seals access water while using a site all influence a species' reproductive behavior. Phocid mothers rear pups on ice or land, with the stability of these substrates influencing the duration of lactation in each species (Oftedal et al. 1988; Costa 1991; Chap. 8). A minority of species breed on land and ice, and their breeding behaviors differ according to the substrate (Hall and Russell 2017). Ice breeding seals utilize pack ice (Bajzak et al. 2011) or fast ice (Croxall and Hiby 1983), with some species digging ice lairs to rear pups (Smith and Stirling 1975). Terrestrially breeding phocids use a variety of environments, including sandy beaches, grassy islands, caves, and intertidal zones (Boyd et al. 1962; Campagna and Lewis 1992; Thompson et al. 1994; Gucu et al. 2004). Male mating behaviors also show great diversity across the family, with mating occurring on land and in the water. Physical conflict between males competing for females occurs in all phocid species to some degree; however, many species also have breeding behaviors that involve visual or acoustic displays to attract females or deter male rivals (Stirling and Thomas 2003; Hayes et al. 2004a; Kovacs 2017). The environmental constraints that determine how many females a male can monopolize during the breeding season heavily influence breeding behavior (Chap. 8).

While evolutionary pressures generate behaviors to achieve reproductive success within a species, substantial individual differences occur in their expression. In phocids, individual factors that influence reproductive behavior include physiology, morphology, and cognitive processes. Size differences typically enable the largest individuals to be more dominant in breeding colonies and occupy the best sites for mating (Haley et al. 1994). Variation in hormone expression has been linked to differences in mating strategies (e.g., testosterone; Bartsh et al. 1992), time spent on breeding sites (e.g., cortisol; Lidgard et al. 2008), and maternal behavior (e.g., oxytocin; Robinson et al. 2015a, 2019a). Individually consistent differences in reproductive behavior, or 'personalities' (Chap. 9), have been shown in some phocid species over breeding seasons and constrain behavioral responses in certain contexts. For example, while different gray seal (*Halichoerus grypus*) mothers show a variety of responses to threats on breeding colonies, individuals tend to react in the same way across different breeding years (Twiss et al. 2012; Bubac et al. 2018). Therefore, observed reproductive behavior in phocids may depart from the 'ideal' behavioral responses to maximize mating opportunities, prevent conflicts, and ensure pup survival due to several individual-based behavioral or physiological variables. These factors can be subtle and difficult to study in natural populations, but documenting their influence on behavior is crucial as they often shed light on how suboptimal behavior can remain in a species' behavioral repertoire. Even maladaptive behaviors, such as hyper aggression resulting in the death of potential mates (Starfield et al. 1995; Le Boeuf and Mesnick 1991) or non-filial pup adoption (Boness et al. 1992), can persist in natural populations given the appropriate evolutionary and physiological context. Understanding their presence and resistance to selection pressure is as important as documenting adaptive behavioral responses during breeding episodes. This is especially crucial in species whose populations are declining or under threat due to their inability to adapt to reproducing in changing environments, which is increasingly caused by anthropogenic threats such as disturbance (Gucu et al. 2004) or climate change (Kovacs et al. 2011).

This chapter explores the variety of behaviors that different phocids exhibit while reproducing. The greater part of the chapter documents behavioral traits that adult males and females exhibit while mating and rearing offspring, focusing on topics that principally involve one of the sexes and then covering the potential interactions different breeding individuals have while reproducing. We highlight the widespread behaviors observed in breeding phocids and unusual or maladaptive ones that researchers are still attempting to understand. An additional section focuses on pup behaviors until weaning and leaving their birth colonies, as mammalian reproduction cannot be successful without coordination of behavior between mothers and offspring.

7.2 Male Breeding Behavior

Male breeding behavior and mating systems vary greatly across the 18 living species of phocids and occur in terrestrial and aquatic environments. With no role in pup-rearing, adult males focus their energies on competing for females and mating opportunities. Breeding seasons in almost all phocids are highly synchronized, with the females in a population giving birth, rearing pups, and then coming into estrus during one short period each year. Thus, a species' mating system is dictated to a large degree by the spatial and temporal aggregation patterns of receptive females and males' ability to monopolize mating opportunities while guarding against rivals. Together, these evolutionary pressures have shaped a number of male mating systems and strategies in different species, with varying degrees of sexual dimorphism in males (Fig. 7.2).

Fig. 7.2 The degree of sexual dimorphism in phocids ranges from subtle to extreme **a** Elephant seals are the most sexually dimorphic phocid species, as in this male–female pair of southern elephant seals **b** Many other phocid species show fewer differences in size, as in these harbor seals where the female (left) is only slightly smaller than the male (right). Photo credit: **a** Lars Boehme; **b** Kelly Robinson

7.2.1 Mating Systems

Phocid breeding aggregations range from single animals to colonies of thousands, giving rise to a diverse range of mating systems in this family (Fig. 7.3). Monogamy is unusual in mammalian species but is likely in species with widely dispersed females that occupy a small range when sexually receptive (Komers and Brotherton 1997). Some polar phocids fit this description, with solitary females giving birth and rearing pups on ice floes (crabeater *Lobodon carcinophaga*, hooded, ross *Ommatophoca rossii*, and spotted seals *Phoca largha*). Serial monogamy can occur in response to these circumstances, with a single male guarding one female while she rears her pup and mating when she is in estrus. When she returns to the sea, the male searches for another breeding female. Difficulty in observing polar breeding species has prevented definitive classification of their mating systems, but likely candidates for serial monogamy include crabeater (Siniff et al. 1979; Southwell 2004), Ross (Curtis et al. 2011), and spotted seals, which may be annually monogamous (Frost and Burns 2017). Hawaiian monk seals (*Neomonachus schauinslandi*) may also

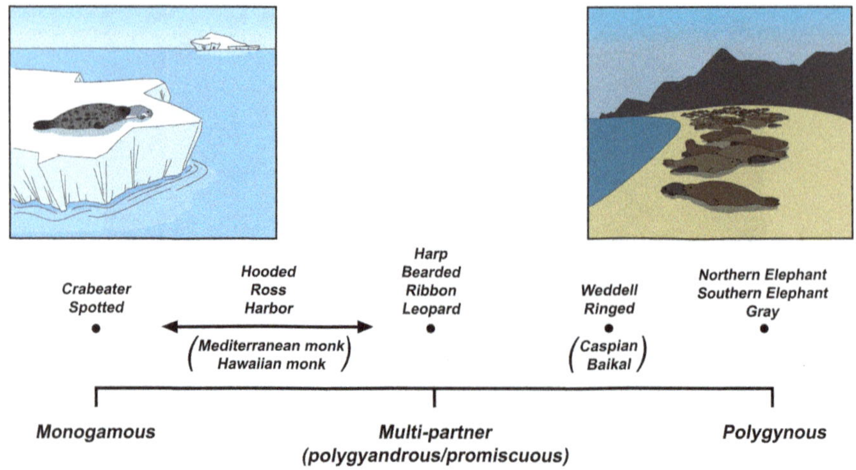

Fig. 7.3 A simplified summary of the major mating systems in phocids. Several species use different mating systems depending on local environmental conditions. Species with insufficient data to identify their major mating systems are given in parentheses

exhibit a form of serial monogamy, as males have been observed attending and guarding single females after pups have been weaned (Johanos et al. 1994). In this species, estrus is poorly synchronized, and serial monogamy may have developed due to temporal and spatial separation of mating opportunities. Mediterranean monk seals also show poor breeding synchrony, with pups in some populations born year-round (Pastor and Aguilar 2003). However, while serial monogamy may occur in some parts of this species' distribution, there is evidence for polygynous systems in place around crucial cave breeding sites (Gucu et al. 2004; Chap. 16; see below). As mating behavior is rarely observed in either of these endangered species, their mating systems' precise nature is unknown (Stirling and Thomas 2003).

In stark contrast to monogamy, polygyny is present in many phocid species, most notably in the northern and southern elephant seal species (*Mirounga leonina and M. angustirostris*) that are the most sexually dimorphic and polygynous of all mammals (Fig. 7.4; Fabiani et al. 2004). Polygynous species are characterized by multiple females mating with one male, typically with males defending some sort of territory containing females or groups of females from rivals. Elephant seals breed terrestrially and maintain beach harems containing hundreds of breeding females (Carrick et al. 1962; Van Aarde 1980). Males advertise possession of their harems with acoustic and visual displays and engage in physical conflict when these displays fail to deter challengers (Bartholomew and Collias 1962; Sandegren 1976; Casey et al. 2015, 2019; Frouin-Mouy and Hamill 2021). Weddell (*Leptonychotes weddelli*; Thomas and Kuechle 1982) and ringed seals (*Phoca hispida*; Smith et al. 1991) are believed to use acoustic displays to maintain underwater territories based around breathing holes or cracks in the ice on which they breed. Male ringed seals maintain their breathing holes and build subnivean lairs above them, which provide protection

Fig. 7.4 Different densities of breeding phocids have different mating systems associated with them: **a** Hooded seal trio, a female rears her pup on pack ice and is attended by a single male who guards the female until she comes into estrus. Once mating occurs, the male seeks another female, resulting in a serially monogamous mating system; **b** A dominant male northern elephant seal with a harem of females and their pups. Females rear pups in dense groups on terrestrial sites, and individual males attempt to monopolize mating opportunities within their territories, resulting in a polygynous mating system. Photo credit: **a** Héloïse Frouin-Mouy, taken under permit IML-2018–01; **b** Dan Costa

from harsh Arctic conditions and predators such as polar bears (*Ursus maritimus*; Furgal et al. 1996). Aquatic mating in ringed and Weddell seals reduces males' ability to physically dominate females, as some degree of cooperation from the female is required for mating to occur. Therefore, the degree of polygyny in these species is much lower than in terrestrially mating species (Harcourt et al. 2007a).

The remaining phocids either fall between the two extremes or use a combination of these systems depending on local conditions, with multi-partner or polygyandrous (often termed 'promiscuous') mating systems present in at least five species (harp *Pagophilus groenlandicus*, bearded *Erignathus barbatus*, ribbon *Phoca fasciata*, harbor *Phoca vitulina*, and leopard seals *Hydrurga leptonyx*; Stirling and Thomas 2003). While some phocids are classed as 'promiscuous' in the existing literature, in recent years, there has been some criticism of this terms use in animal behavior studies (Elgar et al. 2013) and an emphasis on using it only for species that show no discrimination in their mate choices (Elgar et al. 2013; Garcia-Gonzalez 2017; Szala and Shackelford 2019). While several species lack sufficient evidence to determine whether mate choice occurs, it is an essential component of many species' reproductive behavior (Thomas and Kuechle 1982; Kovacs 1995a; Van Parijs et al. 1997; Hayes et al. 2004b, 2006; McMahon and Bradshaw 2004; Galimberti et al. 2000; Boness et al. 2006). 'Polygyandrous' or 'polygamous,' a general term for multi-partner mating systems (McEachern et al. 2009), may therefore be the more appropriate term in many cases. However, it should be noted that such terms are not used consistently throughout the literature on animal mating systems (e.g., Wolff and Macdonald 2004; Bertram and Gorelick 2009; McEachern et al. 2009; Parreira and Chikhi 2015), including studies on phocids.

Multiple, contrasting mating systems have been described for some phocids, such as the hooded seal (Boness et al. 1988; Kovacs 1990). In reality, species with a wide distribution that use different breeding substrates for mating and rearing pups will show variations of mating systems that best fit their environment. Ice breeding populations of harbor seals have been described as serially monogamous (Burns et al. 1972), similar to breeding crabeater seals (Siniff et al. 1979), while populations breeding on terrestrial sites feature males displaying vocally underwater to multiple females (Boness et al. 2006). The display locations used by males from different harbor seal populations show substantial variation, and it has been proposed that 'lek' style mating systems are present in some regions (Hayes et al. 2006). Some species show flexible breeding systems according to the availability of females. Hooded seals are serially monogamous (Fig. 7.4; Kovacs 1990) unless females aggregate in small groups on the ice, in which case males become polygynous (Boness et al. 1988). Individual males within each species can also exhibit strikingly different mating strategies to subvert the dominant mating system, allowing males to gain copulations via alternative behavioral routes.

7.2.2 *Mating Strategies*

Males may use different mating strategies to achieve reproductive success. Resource defense, territorial, and 'resident' or 'tenured' strategies are the most obvious on breeding colonies as they typically involve dominant males using overt visual and acoustic displays or physical conflicts to repel rivals. Male breeding site fidelity is present in species with terrestrial (Twiss et al. 1994) and aquatic (Van Parijs et al. 2004; Cameron et al. 2007; Terhune et al. 2008) territories. While territorial strategies can, in some circumstances, lead to high individual reproductive success (Fabiani et al. 2004), this comes at a cost, as a male must spend years gaining the mass and experience required to be dominant enough to successfully hold a territory (Haley et al. 1994). The amount of time a dominant male can remain on a breeding colony in a season using a resident mating strategy varies across species and individuals but can be as long as three months in larger phocids (northern elephant seals; Le Boeuf 1974).

Males may never achieve the physical condition needed to succeed using a resident strategy. Although larger than average body size is a predictor of reproductive success in some systems, it also carries a greater absolute energetic overhead that must be sustained from reserves during the reproductive fast common to many phocids (Anderson and Fedak 1985; Crocker et al. 2012). Males in relatively poor condition cannot defend territories for as long as those in better condition, and males gaining

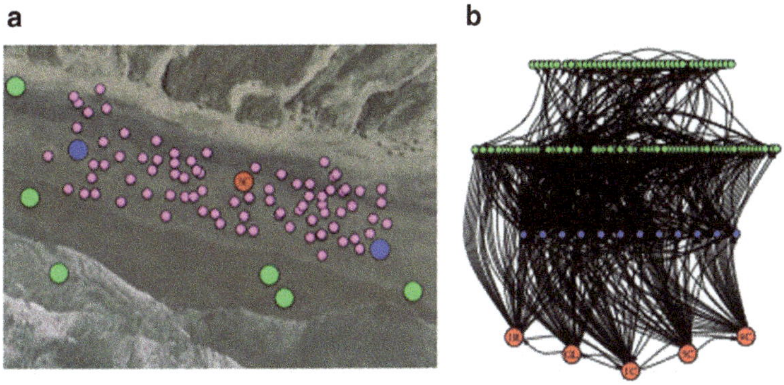

Fig. 7.5 Dominance hierarchies, male mating strategies, and interactions on a northern elephant seal breeding colony: **a** aerial schematic of a single harem, with pink dots representing females, and other colored dots representing males of different hierarchy status and mating behavior. The red dot represents an 'alpha' male who controls the majority of mating opportunities, the blue dots represent 'beta' males that consistently hold flank positions on the harem to access females, and the green dots represent 'peripheral' males that typically cannot access harem females. **b** the social network of male northern elephant seal interactions over one breeding season. When the same males interact, in >97% of cases these were won by the same individual within that pair. Figure reproduced from Casey et al. (2015)

reproductive success through intensive competition rarely maintain this advantage for more than a few seasons (Twiss et al. 1998). Alternative attendance strategies can alleviate some of these pressures, such as leaving territory to forage (Kelly and Wartzok 1996; Coltman et al. 1997, 1998a; Lidgard et al. 2005; Harcourt et al. 2007b) or altering the timing of their reproductive effort within a breeding season (Bishop et al. 2017). Subordinate males can also adopt 'peripheral,' 'transient' or 'sneaky' strategies that minimize confrontations with dominant territorial males and give them access to females on an opportunistic basis (Fig. 7.5; Le Boeuf 1974; Twiss 1991; Amos et al. 1993; Lidgard 2003; Lidgard et al. 2004; Hoelzel et al. 1999, Casey et al. 2015). Aquatic (Pastor et al. 2011) and terrestrial (Fabiani et al. 2004) mating species both show these alternative strategies, with males in some species able to switch between territorial and transient strategies within the same breeding season (e.g., Baldi et al. 1996; Van Parijs and Clark 2006). The success rates of alternate strategies vary greatly with species and the topography of the breeding colony, and their relative importance can only be quantified by extensive multi-year representative genetic comparison of pups and breeding males (Coltman et al. 1998b, 1999; Worthington Wilmer et al. 1999; Hoelzel et al. 1999) interpreted alongside detailed data on individual presence, movements, and behavior (e.g., Twiss et al. 2006).

7.2.3 *Breeding Displays*

Behavioral displays are a hallmark of territorial mating strategies and advertise the presence and in some instances the fitness (Sanvito et al. 2007) of the resident male to rivals and potential mates. Behavioral displays can convey information through a variety of sensory modalities underwater or within terrestrial territories.

7.2.3.1 Acoustic Displays

While vocal signaling is involved in all types of phocid territorial behavior (Chap. 3), aquatic breeders rely heavily on acoustic displays that can be extremely complex. Typically, underwater acoustic displays occur adjacent to breeding sites, near access points or locations where females are present; however, individuals may be widely dispersed geographically, and male vocal displays may be audible to receptive females over large distances (Rogers 2003). Lone male leopard seals display vocally underwater for hours daily over a 2-month period (Rogers 2003) and have never been documented approaching female-pup pairs on the ice (Rogers 2014). Male call sequences have individually distinct acoustic features that may aid conspecific identification of different callers (Cato and Rogers 2002). Vocalizations ascribed to breeding male Ross, hooded and crabeater seals have been documented throughout the months that reproduction takes place, however, the role of these vocalizations in gaining mating opportunities or influencing rivals is unknown (McCreery and Thomas 2009; Klinck et al. 2010; Van Opzeeland et al. 2010; Frouin-Mouy and Hamill 2021). Males of three phocid species form underwater territories around highly limited access points to breeding sites, such as breathing holes or cracks in ice (Weddell seals, ringed seals; Stirling 1969; Kelly and Wartzok 1996) or entrances to caves used for breeding (Mediterranean monk seals; Pastor et al. 2011). Males of species that display underwater adjacent to breeding colonies without such bottlenecks can do so in discrete territories or may simply call in the vicinity of receptive females (e.g., harbor seals, bearded seals, Weddell seals; Boness et al. 2006; Van Parijs et al. 2003a, 2004; Russel et al. 2016). In Weddell seals, there is also evidence that male calls produced underwater are audible to females hauled out on the ice directly above them (Terhune 2017). The complexity and diversity of acoustic displays of each species are related to the breeding system, display site, and distance over which the vocal signal must travel (Rogers 2003; Van Parijs et al. 2003b; Klinck et al. 2010; Van Opzeeland et al. 2010). Polygynous Weddell seals produce over 30 types of underwater calls during the breeding season (Fig. 7.6; Thomas and Kuechle 1982), whereas serially monogamous male crabeater seals only produce two call types when reproducing (Klinck et al. 2010). High call diversity or complexity may indicate systems in which females exercise some degree of mate choice (Rogers 2003).

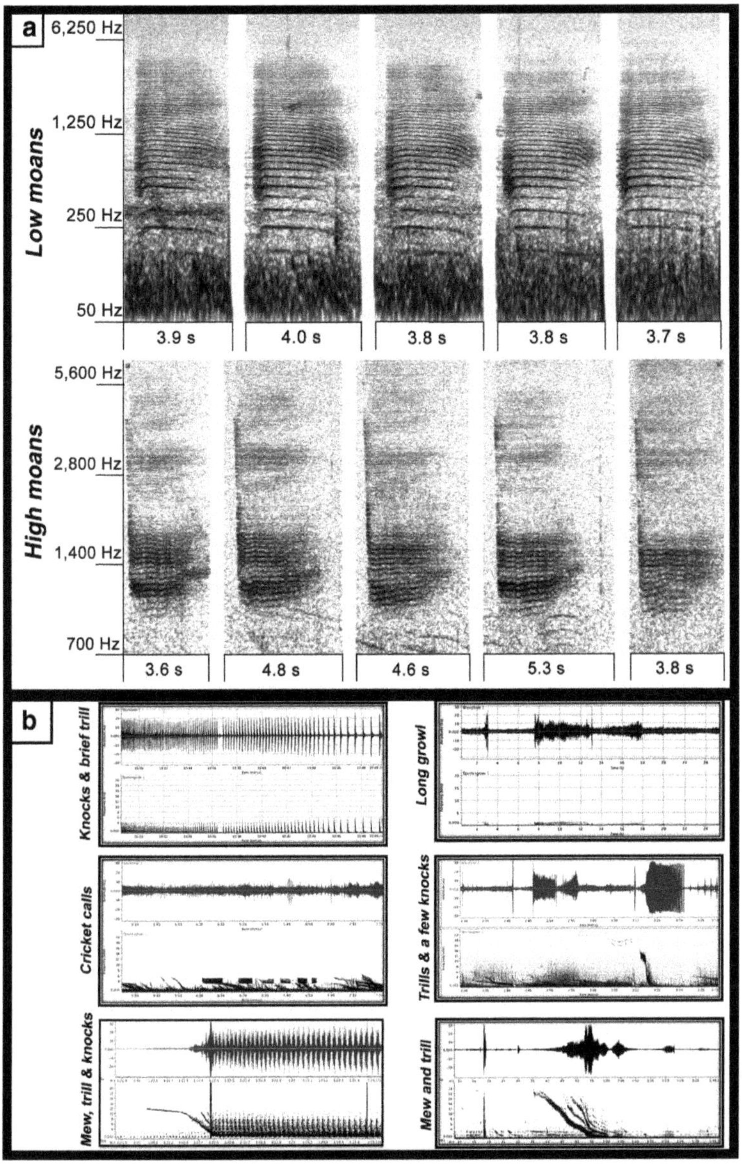

Fig. 7.6 Differences in breeding male acoustic displays from two phocid species. **a** spectrograms of two crabeater seal vocalizations (high and low moans) that are thought to be produced by breeding males **b** waveforms and spectrograms of six vocalizations made by a territorial male Weddell seal. Figures reproduced from **a** Klinck et al. (2010) and **b** Russell et al. (2016)

7.2.3.2 Combination Displays

Breeding phocids typically use displays that function across a range of sensory modalities, including visual, auditory, and tactile components (Chap. 3). Sexually dimorphic phocids can also have secondary sexual characteristics that can be used in such displays, the most dramatic being the male hooded seal's inflatable nasal septum (Kovacs 1990, 1995a; Ballard and Kovacs 1995, Frouin-Mouy and Hamill 2021). Male hooded and elephant seals have a distinct 'proboscis' structure (Fig. 7.7), and this is used in agonistic displays as a visual signal and to produce acoustic cues that can be, but are not always, honest indicators of a male's size (southern elephant seals; Sanvito et al. 2007). Male vocalizations in elephant seals are an essential mechanism by which rivals can recognize and assess each other before engaging in aggressive behaviors (Casey et al. 2015). Male elephant seal vocalizations develop into individually stable calls as males mature, containing individually distinct features that enable males to identify each other and, with experience, their place within the dominance hierarchy (Casey et al. 2019). The diversity of individual male calls and the complexity of their acoustic structure has dramatically increased over the last five decades in northern elephant seals and has eliminated regionally distinct vocal dialects present in this species previously (Casey et al. 2018).

Other visual and acoustic combination displays involve distinctive postures or gestures and specific calls that are produced in combination. The 'open mouth threat' (Fig. 7.8) and associated hissing vocalization is a common threat displayed in both sexes of many phocids including harbor, gray, and elephant seals (Sullivan 1982; Anderson et al. 1975; Boness and James. 1979; Bartholomew and Collias 1962; McCann 1981). Competing males on breeding colonies frequently use this behavior in the initial phases of interactions with rivals and other aggressive behaviors that can escalate to physical conflict (see Sect. 7.2.4). Locally distinct display behaviors can also arise. Male gray seals occupying a relatively new, expanding UK breeding colony on a large, flat sandy beach area developed a 'body slap' display that is not only

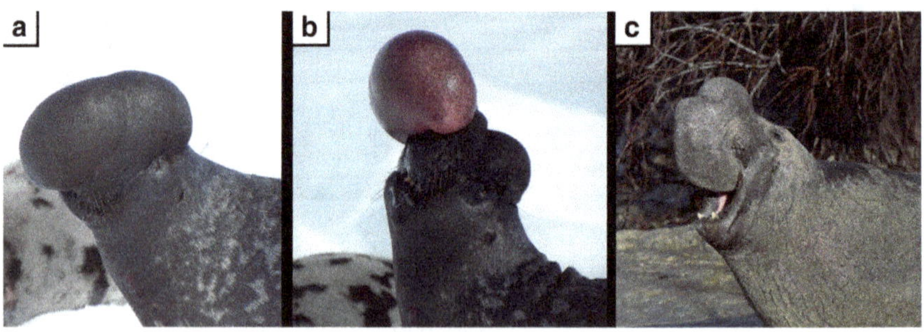

Fig. 7.7 Secondary sexual characteristics involved in visual and acoustic display behavior in breeding male phocids **a** male hooded seal with inflated proboscis **b** male hooded seal with inflated nasal septum **c** male northern elephant seal with inflated proboscis. Photo credits: **a** and **b** Héloïse Frouin-Mouy, taken under permit IML-2018–01; **c** Dan Costa

Fig. 7.8 Different male–male phocid interactions on breeding colonies: **a** Two breeding male gray seals, the male on the left displays a typical 'open mouth threat' while the male on the right attempts to retreat; **b** Two breeding male northern elephant seals vocally displaying to each other; **c** Male gray seals lunging and wrestling on the breeding colony **d** Male northern elephant seals rearing to push and shove each other; **e** Male gray seals after a fight. The victor is attempting to bite the rear flippers and hindquarters of the defeated male, who is attempting to flee to the water. Photo credit: **a, c, e** Kelly Robinson; **b** Caroline Casey; **d** Ari Friedlaender

audible but generates informative vibrations that may be detected by rivals through the colony substrate (Bishop et al. 2014). Displays using visual or tactile components on terrestrial colonies are effective over limited ranges so that males using them are frequently in close proximity and less able to escape quickly in the event of physical confrontations. Males of some species are able to distinguish acoustically between neighboring rival males they have encountered previously and novel challengers (Casey et al. 2015), so that unnecessary fighting, with its associated injury risk, can be avoided (Insley et al. 2003).

7.2.4 Interactions on the Breeding Colony

7.2.4.1 Male–Male Interactions

Breeding males should exclude other rivals from mating opportunities while avoiding unnecessary conflicts to reduce the chance of injury or displacement. As all phocid seals have formidable teeth and strong jaws, males can inflict significant wounds on each other (Deutsch et al. 1994). Physical fights between males typically occur between evenly matched rivals (Sandegren 1976) or at critical times in the breeding season, such as when many females are in estrus. In the more polygynous phocids, the risks that accompany physical conflicts may be worth taking for individuals capable of competing with dominant males, as a few individuals monopolize the majority of mating opportunities at the top of the hierarchy (Le Boeuf 1974). As their energetic resources are limited, the costs of engaging in physical combat, energy expended, and resources used in healing any resulting wounds are generally better directed toward sustaining males in holding territory or patrolling the territories of rival dominant males (Cox 1981). Therefore, display and avoidance behaviors occur alongside agonistic interactions.

When two rivals come into close proximity, they exchange display signals before escalating to a physical confrontation. On land, these involve vocalizations and physical posturing such as rearing and the 'clap-threat' in elephant seals (Southall et al. 2003). In the water, males may display by repeatedly slapping the water's surface with pectoral or hind flippers (e.g., harbor seals; Hayes et al. 2004b) or by generating violent bubble streams underwater (e.g., harp seals; Merdsoy and Renouf 1978). In some species, a male can display submissive behaviors after threat displays have been performed, enabling the 'losing' male to retreat without any physical retaliation from the victor (e.g., northern elephant seal, Sandegren 1976). If neither male retreats, then the two will fight until one shows submission and attempts to flee.

Terrestrial fights involve rearing the head away from an opponent, lunging to bite the other male, chest to chest pushing, and using the pectoral flippers to claw the opponent (Sandegren 1976; Sullivan 1982; Kovacs 1990; Twiss et al. 1998). In aquatic breeding species, fights can occur underwater and at access points to the water, such as at breathing holes in the ice (Kaufman et al. 1975; Doiron et al. 2012). The rear flippers are a target for biting during male fights, both on land and in water, and the hind flippers and neck of males show the most scarring from intraspecific fighting (Mansfield 1958; Siniff et al. 1979; Walker and Bowen 1993; Doiron et al. 2012). The pineal opening of Weddell seals is often a targeted area (D. Costa, Pers. Obs.; Chap. 13). Physical characteristics that increase the likelihood of a male winning fights differ across phocid species. Fights between male elephant seals involve pushing actions, and success is typically linked to larger body size (Le Boeuf 1974; Carlini et al. 2002). However, other phocids' success may be more dependent on males engaging in wrestling or biting specific targets, which requires a degree of agility rather than relying on body size (Fig. 7.8; Twiss 1991; Lidgard et al. 2005). Substantial external injuries can be sustained during these fights, but deaths are rarely observed directly (Deutsch et al. 1994). Defeated males attempt to retreat after a physical fight, and in some species, the victor attempts to pursue the other both on land and in the water, attempting to bite the hind flippers and rear quarters of the fleeing male (Fig. 7.8; Mansfield 1958; Le Boeuf 1971; Harcourt et al. 2007a).

In the species that typically have tenured males such as elephant and gray seals, dominant individuals defend an area or group of females. Adjacent territory holding males usually tolerate each other's presence as long as consistent separation distances are maintained (e.g., 5–7 m in dominant male northern elephant seals; Le Boeuf 1972). As neighboring tenured males can recognize each other in some species (Insley and Holt 2012), differences occur in their responses to approaching rivals. The most dominant male will challenge every male he encounters, while males lower in the dominance hierarchy modulate their responses based on previous experiences with that male (Fig. 7.5; Casey et al. 2015). Dominant males also violently interrupt rival males attempting to copulate with females within their territory (Le Boeuf 1972). Transient males that occupy the periphery of territories are usually physically smaller and younger than the tenured males (Twiss et al. 1998; Lidgard et al. 2005) and attempt to flee when approached by tenured males (Le Boeuf 1972). These interactions also occur in species that use underwater territories for breeding rather than terrestrial ones (e.g., harbor seals; Hayes et al. 2004b). Dominant male Weddell seals establish territories and display vocally underneath breathing holes in the ice at breeding colonies (Kaufman et al. 1975; Bartsh et al. 1992; Terhune and Dell'Apa 2006; Russell et al. 2016), with multiple males calling in the same area and potentially masking each other's vocalizations (Terhune 2016). Males with more experience at their particular breeding colony have greater reproductive success (Harcourt et al. 2007a). While male Weddell seals may share breathing holes and tolerate the presence of rivals accessing this crucial resource (Harcourt et al. 2007b), there are accounts of aggressive territory holding males repelling all males from breathing holes (Russell et al. 2016). Few studies have detailed interactions between conspecifics of other polar phocids. However, breeding male crabeater seals in Antarctica attempt to guard individual mother–pup pairs on ice floes and are challenged by rivals throughout lactation and after the pup is weaned (Siniff et al. 1979).

7.2.4.2 Male–Female Interactions

Most multiparous female phocids typically come into estrus just before or soon after weaning a pup. Mediterranean and Hawaiian monk seals are exceptions. Based on sparse data, estrus in Mediterranean monk seals is thought to occur approximately 3 months after parturition. (Pastor and Aguilar 2003). In contrast, female Hawaiian monk seals do not come into estrus until approximately 3 weeks after weaning their pups (Johanos et al. 1994). During lactation, the occurrence of estrus creates a conflict of interest between the sexes for most of the breeding season. As males seek mating opportunities, most females are unreceptive, protective of their pups, and resist males approaching them, sometimes very aggressively (Fig. 7.9; Merdsoy and Renouf 1978; Boness et al. 1982). Only when females enter estrus are males tolerated

Fig. 7.9 Gray seal mothers bite and harass an approaching male to drive him away. Their pups' appearance indicates they are in an early phase of lactation and are not in estrus. Photo credit: Kelly Robinson

to approach and mate. Despite this, males approach females and attempt copulations throughout lactation until the male's advances are resisted less strenuously (Anderson et al. 1975; Siniff et al. 1979; Cox 1981). Phocids do not show overt signs of entering estrus (Kovacs 1990; Atkinson 1997) and continued 'testing' may be the most reliable way for a male to gain first access to receptive females. Male harassment is present in many phocids, is not restricted to adults, can cross species boundaries (Wilson 1975), and can have severe negative consequences for targeted females (Le Boeuf and Mesnick 1991; Hiruki et al. 1993a, b; Boness et al. 1995).

Male harassment has been documented most extensively in species that breed colonially, but it also occurs in species that reproduce in isolation (e.g., crabeater seals; Siniff et al. 1979). In almost all instances of harassment prior to weaning, males that approach mother–pup pairs are either threatened by the female until they retreat or attempt to mount and copulate while the female responds aggressively until she can drive the male away (Boness et al. 1982). Female northern and southern elephant seals are too small to repel unwanted males, and instead, females vocalize and flip sand at the male's faces, alerting other males to the harassment and potentially attracting a more dominant male to disrupt the copulation attempt (Le Boeuf 1971; Le Boeuf and Mesnick 1991; Galimberti et al. 2000).

When a female enters estrus, threat and aggressive behaviors toward an approaching male decrease, and a female may cooperate, to varying extents according to species, to enable mating. Even in elephant seals, where females cannot deter unwanted mountings physically, males are four times more likely to achieve intromission if females are not protesting against the males' actions (Cox 1981). It is common for males to bite the female on the back of the neck while attempting to mate, holding the female parallel to him, so her back is flush with his ventral surface (Le Boeuf 1971; Boness and James 1979; Anderson and Harwood 1985). This positioning and behavior are consistent throughout terrestrially mating phocids and occurs in underwater mating (e.g., Cline et al. 1971). The male will then arch his pelvic region while extruding the penis to attempt intromission. Pelvic thrusts occur in some species once penetration is achieved (Cline et al. 1971), while in others, the pair remains relatively motionless until they disengage from each other (Le Boeuf 1971).

By the time colonially breeding females depart the breeding site, they have usually mated with at least one dominant male. However, it is common for females to endure significant male harassment and additional copulations with peripheral males while attempting to return to sea (Le Boeuf and Mesnick 1991; Ambs et al. 1999). This alternative male mating strategy is typically not as successful as the tenured strategy but results in some success for the males engaging in this behavior (Amos et al. 1993; Lidgard et al. 2004). In species with extreme sexual dimorphism, males are more likely to cause serious injuries or fatalities to departing females due to this behavior (Le Boeuf and Mesnick 1991). Such maladaptive behavior may be attributed to weak selection pressure discouraging aggressive mating attempts and large numbers of young, inexperienced, or subordinate males occupying the breeding colony periphery. Such males are highly motivated to achieve copulation but cannot access females in the main colony. Young inexperienced male southern elephant seals kill females during attempted copulations by biting females' heads rather than the backs of the necks and fracturing their skulls (Carrick et al. 1962).

The most extreme example of male harassment causing harm to breeding females comes from a species that shows no sexual dimorphism (Littnan et al. 2017). In the Hawaiian monk seal, extreme male aggression behavior known as 'mobbing' or 'multiple male aggression' became so prevalent and severe it likely contributed to the decline of this endangered species in the 1990s (Starfield et al. 1995; Johanos et al. 2010; Baker et al. 2011; Chap. 16). Males engaging in this behavior compete aggressively to gain access to a female. In this situation, males bite and wound the female repeatedly, tearing chunks of flesh and blubber from her back (Fig. 7.10). Females were mortally injured during these events in the 1990s (Hiruki et al. 1993a), and even if females survived, their subsequent ability to raise pups was compromised (Hiruki et al. 1993b). Although estrous females are the primary target (Atkinson et al. 1994), males also attack young or unreceptive females and young males (Starfield et al. 1995). This behavior occurred in populations with male-skewed sex ratios, and the situation was only alleviated by efforts to remove males from the affected areas, which has reduced serious mounting injuries and female fatalities to a much lower level in the subsequent years (Johanos et al. 2010; Baker et al. 2011).

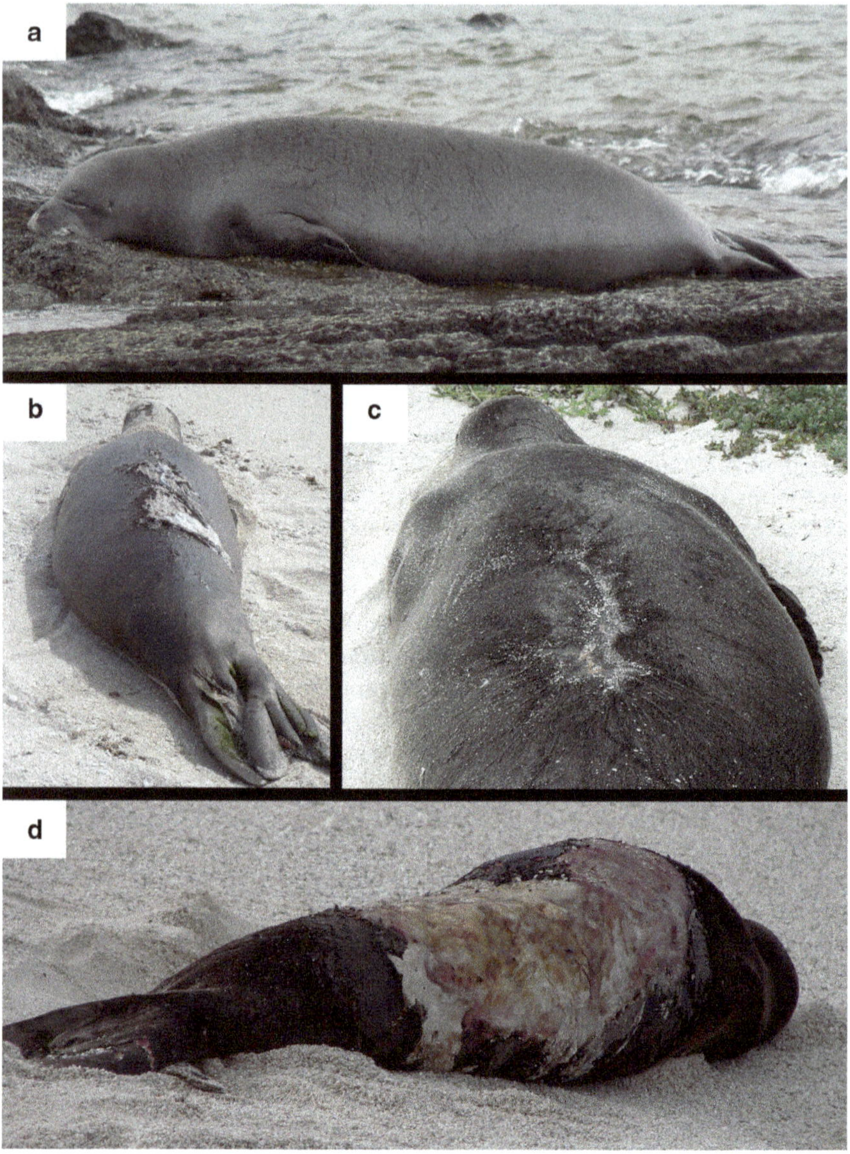

Fig. 7.10 Examples of dorsal injuries in Hawaiian monk seals typically caused by mounting during incidents of male aggression **a** a weaned male pup with multiple dorsal lacerations typical of single male aggression **b** an adult female with large, relatively fresh dorsal injuries typical of multiple male aggression **c** an adult male with a large healed dorsal injury typical of multiple male aggression **d** a subadult female with severe injuries typical of multiple male aggressive behavior, with the majority of the dorsal surface exposed. All photos courtesy of the Hawaiian Monk Seal Research Program, taken under permit MMPA/ESA permit 16,632–02. Photo credit: **a** Tracy Mercer; **b** Whitney Taylor; **c** Megan McKinzie; **d** USFWS field staff

7.2.4.3 Male–Pup Interactions

Male phocids display no form of parental behavior, and their interactions with pups are limited and often harmful. A possible exception is related to predator defense in crabeater seals, where a solitary male attending a single mother–pup pair on the ice defends the female aggressively from approaching threats, typically leopard seals (Siniff et al. 1979), which are known predators of young crabeater seals (Siniff and Bengtson 1977). However, any benefit the pup derives from this behavior is coincidental, as the male's objective is to defend the mating opportunity the solitary female represents (Siniff et al. 1979), and he is unlikely to be the father of the pup in question. As males of this species have been reported to force pups to wean early from their mothers in response to threats, pup protection is unlikely to be a strong priority for male crabeater seals (Siniff et al. 1979).

The interactions most commonly reported between breeding males and pups tend to be when males injure or kill pups. This can be accidental, with large breeding males harming young pups as they move around a colony or fight with others, or targeted when males direct aggressive and sexual behavior specifically at pups. For the pups, resulting wounds provide routes for infections that can be fatal even if the original wounding is not (Baker and Baker 1988). Accidental injuries and deaths from trauma are common in densely packed colonies and with species that exhibit sexual dimorphism. Crowding in elephant seal colonies and the considerable mass disparity between breeding males and pups can result in crush injuries accounting for 40% of pup mortalities (Le Boeuf et al. 1972). Male aggression toward pups includes biting and tossing pups out of the way (P. Pomeroy, Pers. Obs.). Misdirected sexual behavior, with males biting the necks of pups and mounting them, has been reported in northern elephant seals (Reiter et al. 1978), Hawaiian monk seals (Baker 2008), and gray seals (D. Lidgard, Pers. Obs.) with dependent and weaned pups. Young male northern elephant seals have also been seen biting and mounting harbor seal pups (Hayward 2003).

A rare form of purposeful harmful behavior is breeding males cannibalizing pups, which has been documented in two phocids. There is a single published report of a male southern elephant seal attacking and eating weaned pups (see Wilkinson et al. 2000). However, in gray seals there are detailed accounts of individual breeding males killing and consuming weaned pups (Bédard et al. 1993; Bishop et al. 2016). It has been suggested that this behavior is not isolated to a few individuals, but is widespread geographically and has been occurring over the past two decades (Brownlow et al. 2016). Such behavior's functional significance is difficult to determine due to the scarcity of observations and the discrepancies among individual gray seal males performing this behavior. Of the three detailed reports that document cannibalistic gray seal males, two males consumed large quantities of flesh from the pups they killed (Canadian male, Bédard et al. 1993; Welsh male, Boyle 2009), while an adult male on the Isle of May, Scotland ate only small amounts from each kill, then predated other pups while the fresh carcasses had substantial flesh remaining (Bishop et al. 2016 and Brownlow et al. 2016 describe the same male). Additional reports of cannibalism events from Scottish gray seals are accumulating (P. Pomeroy,

Pers. Comm.). Cannibalism in mammalian species is best documented in carnivores, rodents, and primates, but most cases involve infanticide as part of reproductive strategies, typically to bring females into estrus (Polis et al. 1984). This is unlikely to be a factor in phocid cannibalism cases given the rapid weaning process of pups. The only other marine mammal species that have been documented to cannibalize conspecific pups to date is the New Zealand sea lion (*Phocarctos hookeri*), however, there are records of interspecific pup predation across otariid and phocid species (e.g., Wilkinson et al. 2000; Penry et al. 2013).

7.2.4.4 Interspecific Interactions

Seals that advertise their presence to attract mates or compete for them are at risk of alerting and attracting predators. Despite this, predators of seals typically seem to prefer targeting pups or juvenile individuals rather than adult males that are easily located, even in the smaller species of phocids (e.g., the ringed seal; Furgal et al. 1996). Predatory mammalian (Stirling and Archibald 1977; Lowry and Fay 1984; Baird and Dill 1995) and shark (Bertilsson-Friedman 2006) species that regularly target phocids all show a preference for younger cohorts, potentially due to their lack of experience and the relative ease of capturing them compared to adults. It is possible that breeding males may be aggressive enough to deter predatory attacks (e.g., leopard seals attempting to predate crabeater seal breeding trios; Siniff et al. 1979). Polar bears and arctic foxes (*Vulpes lagopus*) avoid snow lairs constructed by breeding male ringed seals despite their distinctive scent (Krafft et al. 2007), preferring to target lairs used by pups (Furgal et al. 1996). It has been proposed that breeding male harbor seals use low-frequency calls when displaying and are silent when patrolling underwater territories in response to predation pressure from killer whales (*Orcinus orca*, Hayes et al. 2004b). This adaptation of male breeding behavior may have occurred only in harbor seals as they display and mate underwater and are the preferred prey of certain killer whale populations (Baird and Dill 1995). The other phocid commonly taken by killer whales is the southern elephant seal, a terrestrial breeder (Jefferson et al. 1991); therefore, males could not be regularly targeted while reproducing.

7.3 Female Breeding Behavior

Female phocid seals face many challenges when attempting to breed. They bear the entire cost of rearing pups while maintaining and defending themselves and their offspring from unwanted male attention, aggression from neighboring females, and harassment from predators. All phocid mothers fast for at least part of the lactation period, placing additional energetic demands on them during this crucial time (Chap. 8). Individuals that intersperse foraging trips with nursing visits on land, or those breeding in dense colonies, face challenges associated with maintaining contact with their pup throughout lactation. In some phocids, individuals may be coerced into mating with dominant males or may have to endure multiple mating attempts or physical injury to leave the breeding site. All energetic and behavioral expenditure put into a mother's current breeding attempt is a trade-off between maximizing the likelihood of the current pup surviving to independence and negatively impacting her survival or subsequent breeding attempts (Pomeroy et al. 1999).

7.3.1 Synchronicity of Pupping and Estrus

Nearly all phocids show some degree of synchronicity when giving birth and coming into estrus, which occurs in the days or weeks following parturition (Stirling 1983). This synchronization concentrates breeding into one period, and the temporal and spatial distribution of females during this time dramatically influences male mating systems and strategies. Species at high latitudes show greater synchronicity than those breeding in subtropical regions, with Hawaiian (Johanos et al. 1994) and Mediterranean (Gucu et al. 2004) monk seals showing the least synchronicity and, in some populations, pupping throughout the year (Gazo et al. 1999; Pastor and Aguilar 2003). Annual pupping at the same time of year is made possible by delayed implantation of the developing embryo at the blastocyst stage, extending total gestation time (Fisher 1954; Stewart et al. 1989). Growth of the embryo is very slow during this embryonic diapause that lasts between 1 and 4 months, depending on species (Atkinson 1997). The cue for implantation is unclear, but seals within the same species breeding at different geographical locations are observed to breed at different times. Variation in the timing of implantation has been linked to photoperiod (Boyd 1991; Temte 1994), climatic cues (Coulson 1981; Boyd 1991; Atkinson 1997), and maternal body condition (Boyd 1984). Synchronization of breeding is thought to have evolved as a response to short suitable time periods for breeding in polar species or to utilize favorable environmental conditions for reproduction (Lydersen and Kovacs 1999).

While female seals typically breed each year, individuals can skip breeding in some years, but detecting this is difficult. Females may not pup due to failing to mate, conceive or carry a pregnancy to term. Females may be mistaken as skipping if they bear a pup that is undetected, dies, or is abandoned at or soon after birth; they may leave the breeding area undetected or remain ashore without a pup. Maternal experience and environmental conditions can influence the probability of pupping through impacts on foraging success and body condition (e.g., northern elephant seals, Huber 1987; Hückstädt et al. 2018; Weddell seals, Shero et al. 2015; gray seals, Pomeroy et al. 1999; Smout et al. 2020; southern elephant seals, McMahon et al. 2017; Desprez et al. 2018). Intermittent breeding represents a compromise solution for females with low body mass, temporarily favoring survival over reproduction and giving the opportunity to benefit their lifetime reproductive success, particularly for phocids faced with replenishing low body reserves in unpredictable or challenging environments (Griffen 2018; Desprez et al. 2018; Smout et al. 2020). Females who have skipped a breeding year often return to pup at their usual breeding colony, but where they are mated is unclear. In elephant seals that normally mate ashore, some females haul out on colonies only to mate (Huber 1987), and it is suggested that others could mate opportunistically at sea, on or near foraging grounds (De Bruyn et al. 2011). Nevertheless, skipping females must come into estrus and be mated by fertile males wherever this takes place. For many phocids, it is unclear if skipping breeding occurs and whether any underlying physiological mechanisms enable the process, as obtaining detailed breeding histories over the lifetime of individual females is challenging (Pomeroy et al. 1999; De Bruyn et al. 2011; Smout et al. 2011, 2020; Desprez et al. 2018).

7.3.2 Pupping Site Selection

Before females can give birth and rear their pups successfully, they must leave the water and choose where to pup. Species breeding on temporary substrates, such as drifting pack ice, prefer the best ice conditions when ice coverage is high and ice is at its thickest and most stable (Øritsland and Øien 1995; Bajzak et al. 2011; Miksis-Olds and Parks 2011). When suitably thick ice cannot be found, females pup on any available ice, even if it is too thin to support breeding efforts (e.g., harp seals; Stenson and Hammill 2014). Several species are limited to breeding around holes or cracks in thick sea ice, and females maintain these openings with their claws or teeth or modify ice ledges to enable pups to access the water more readily (e.g., Weddell seals; Stirling 1977; see Fig. 13.7 in Chap. 13). Fidelity to particular breeding sites has been described in several species. Seals breeding on terrestrial (e.g., gray seals, Pomeroy et al. 1994, 2000; northern elephant seals, Reiter et al. 1981; southern elephant seals, McMahon and Bradshaw 2004; Fabiani et al. 2006) and fast ice (e.g., Weddell seals; Croxall and Hiby 1983) colonies return to the same site to breed over many years, sometimes with remarkable accuracy. Gray seals at Sable Island (Canada) do not show site fidelity, which has been attributed to the instability of the sand dunes that alter the landscape (Weitzman et al. 2017). While pack ice breeding species cannot show fine-scale philopatry, hooded and harp seals return to the same approximate area to breed (Davis et al. 2008). Philopatry, where seals return to their natal colony to breed, has been documented in gray seals (Pomeroy et al. 2000) and southern elephant seals (McMahon et al. 1999; McMahon and Bradshaw 2004). Evidence for philopatry in many species is lacking (Croxall and Hiby 1983), but this may be due to the difficulty of studying this behavioral phenomenon in a long-lived species that often have high first-year mortality and several years before recruitment to a breeding population (Hindell and Little 1988).

The vast majority of species give birth to pups on exposed ice or shore, but two species modify their environment to protect their offspring. To protect pups from terrestrial and aerial threats (Furgal et al. 1996; Lydersen and Smith 1989) and the cold conditions of the Arctic, ringed seal mothers use their foreclaws to scratch out birth lairs within the snow covering the sea ice, above ice holes the seals use for breathing (Smith and Stirling 1975; Lydersen and Gjertz 1986). Females of the world's only exclusively freshwater phocid, the Baikal seal (*Pusa sibirica*), also build snow/ice caves to protect their pups from predatory threats (Moore et al. 2009).

7.3.3 *Birth and the Dependent Period*

7.3.3.1 Birth and Immediate Postpartum Behavior

Pregnant seals that breed in colonies come ashore at various times before birth (Reijnders et al. 1990), with some individuals spending days resting in their chosen breeding site while others come ashore only briefly prior to pupping. Species that construct birthing lairs prior to parturition haul out to build dens before birth, although these time frames are currently undefined. The behavior of solitary breeding phocid females approaching birth is unknown.

Birth has been observed in a number of phocid species and is usually described as rapid, with the pup expelled in minutes (Fig. 7.11; Mansfield 1958; Lydersen and Kovacs 1999). Abdominal contractions are sometimes seen just before birth (Mansfield 1958; Le Boeuf et al. 1972). Females are typically either stationary during contractions or move in slow, small circles with regular pauses. When the pup is protruding from the vagina, the female parts her hind flippers and sometimes raises her hindquarters slightly off the ground while pushing. The amniotic membrane usually remains intact while the pup emerges, and bursts suddenly before the pup is fully born. When the pup is finally expelled, the umbilical cord usually breaks, or it is broken by the mother, who typically turns around immediately to sniff the newborn (Le Boeuf et al. 1972; Lydersen and Kovacs 1999).

The first hour after birth is a critical time for a phocid mother to establish a functional bond with her pup to nurture and protect her offspring until weaning (Fogden 1971; Lawson and Renouf 1987; Miller and Murray 1995). Without this bond, it is more likely that a mother abandons her pup (Burton et al. 1975) or exhibits suboptimal or maladaptive maternal behaviors (Robinson et al. 2015a) that detract from the quality of care a pup receives and can result in the pup's death (Anderson et al. 1979). Solitary breeding mothers bond with their pups without interference from neighbors, but colonially breeding females may experience or anticipate interference and become highly aggressive to nearby or encroaching seals, regardless of their sex or age (Mansfield 1958; Le Boeuf et al. 1972; Aguilar et al. 2007). A distracted or agitated mother may fail to spend time and attention on her pup during the crucial time for bond formation when the newborn is particularly vulnerable (Robinson et al. 2015a).

The placenta is passed in the hours following the pup's birth, and most phocid species sniff and then ignore it. For two arctic phocids, the blood and tissue from birth are dangerous to the mother–pup pair as it can attract predators. To prevent this, ringed seal mothers remove their placentae from their birth lairs, while bearded seal mothers move their pups to different ice floes (Lydersen and Kovacs 1999). In other species, the presence of the placenta can attract scavengers such as gulls (Fig. 7.12;

Fig. 7.11 A sequence of photographs documenting a female gray seal giving birth on the Isle of May, Scotland. **a** Female arrives at her breeding site and gives birth soon afterward **b** First contractions seen one hour later **c** First sighting of the pup three minutes later **d** Female continues to push, both on her side and her belly **e** Female raises hind flippers while pushing the widest part of the pup, the shoulders, through the birth canal **f** Once the shoulders emerge the rest of the pup follows quickly **g** Within less than a minute the entire pup is outside of the mother **h** The pup, still inside the amniotic sack, opens its eyes and moves **i** The amniotic sac is broken **j** The female begins to turn to investigate her pup **k** Turning motion tears the remaining amniotic membranes away and breaks the umbilical cord **l** Female starts sniffing her pup within 30 s of birth to begin bonding. Photo credit: Kelly Robinson

Le Boeuf et al. 1972; Burton et al. 1975), and some mothers attempt to defend both their pup and the placenta from attack. This behavior, alongside aggression directed toward neighboring seals, can distract a mother from bonding with her newborn pup and can contribute to the poor bond formation (Robinson et al. 2015a). Despite these potential disruptions, most mothers sniff and interact with their pups, encouraging the pup to suckle for the first time.

Fig. 7.12 Various gull species scavenging placentas from **a** a gray seal birth; **b** a northern elephant seal birth. While a mother remains vigilant to approaching gulls, her attention is divided between the birds and her newborn during the critical bonding period postpartum Photo credit: **a** Kelly Robinson; **b** Elizabeth McHuron

7.3.3.2 Maternal Behaviors During the Dependent Period

Phocid mothers spend most of their time during lactation resting, either on land next to their pup (Lydersen and Kovacs 1999; Aguilar et al. 2007; reviewed in Robinson et al. 2015a) or nearby in the water (Lydersen and Kovacs 1993). The exception to this is those species that undertake foraging trips to supplement the energetic costs of lactation (see Sect. 7.3.3.3). Underwater activities, other than foraging, include resting and avoiding other aggressive breeding females, and the time each mother spends in the water can be highly variable (Sato et al. 2002).

Different species vary in how often they nurse their pups, and species with the shortest dependent periods feed their pups more frequently (Lydersen and Kovacs 1999). Nursing bouts in hooded seals occur every 0.4 h, whereas ringed seals only nurse their pups once every 8 h (Lydersen and Kovacs 1999 and references therein). When feeding pups, mothers roll onto their sides to allow pups to access their nipples. A mother may encourage her pup to feed by gently rubbing its head and back with her fore flipper, which typically directs the pup to move down her body toward the nipples to initiate a suckling bout (Robinson et al. 2015a).

Aside from resting and nursing, mothers engage in vigilance behaviors (Anderson and Harwood 1985; Haller et al. 1996; Twiss et al. 2000, 2012; reviewed in Robinson et al. 2015a), although there is considerable variation in how watchful individual females are (Twiss et al. 2012). Phocid mothers typically become more aggressive toward approaching individuals directly after giving birth (Christenson and Le Boeuf 1978). Despite this, mothers in breeding colonies spend relatively little time overall in aggressive interactions (Anderson and Harwood 1985; Haller et al. 1996; Twiss et al. 2000, 2012; Aguilar et al. 2007; reviewed in Robinson et al. 2015a).

7.3.3.3 Mother–pup Separations and Reunions

Mother–pup separations can occur for a variety of reasons and are common in many species. The generalization that all phocid seals fast during the entirety of lactation has been revised in the past two decades with the advancement of telemetry and observational studies on maternal behavior (Boness and Bowen 1996). Species in which some mothers forage to some degree while rearing pups include the bearded seal (Lydersen et al. 1996), Caspian seal (*Pusa capsica;* Wilson et al. 2017b), harbor seal (Boness et al. 1994; Thompson et al. 1994), harp seal (Lydersen and Kovacs 1993), Mediterranean monk seal (Gazo and Aguilar 2005), spotted seal (Frost and Burns 2017), ringed seal (Hammill et al. 1991), and Weddell seal (Sato et al. 2002). Even when they are not foraging, mothers can spend large portions of their time either in the ocean adjacent to pupping sites on ice (Lydersen and Kovacs 1999) or, in the case of some terrestrially breeding phocids, commuting to pools of water or access points to the sea (Twiss et al. 2000). In species that do not leave their pups to access water, densely crowded colony conditions can separate mother–pup pairs (Riedman and Le Boeuf 1982), and bad weather can increase the likelihood of mother–pup separations (Boness et al. 1992).

Mother–pup pairs must stay together to ensure early pup survival, and while separations occur in most phocids (see below), coordinated movements tend to maintain contact between the two. 'Chaperoning' or 'following' behaviors have been documented in species such as Weddell seals (Rogers et al. 2005), Caspian seals (Wilson et al. 2017a, b), and harbor seals, where mothers and pups follow each other to maintain contact while on breeding sites and swimming (Renouf et al. 1983; Lawson and Renouf 1987; Wilson and Jones 2018). Pups can use the following behavior to stay with their mothers, but mothers may take the lead in chaperoning their pups, especially when perceived threats are nearby (Lawson and Renouf 1987).

The ability of mothers and pups to reunite depends greatly on the context and duration of the separation. There is no evidence that mothers can recognize their pups in species such as hooded seals, where mothers and pups generally do not separate (Kovacs and Lavigne 1992). Mothers in species that breed in groups, forage during lactation, or regularly separate to access water have varying degrees of recognition abilities and include terrestrially breeding species (e.g., northern elephant seals, Riedman and Le Boeuf 1982; Linossier et al. 2021; gray seals, McCulloch and Boness 2000) and species with pups that venture into the water (e.g., harbor seals, Perry and Renouf 1988; Weddell seals, Van Opzeeland et al. 2012). Mothers use combinations of auditory, olfactory, visual, and geographical information to locate and identify offspring (Kovacs 1995b; reviewed in Insley et al. 2003; Chap. 3). Pups can play an active role in this process. In Weddell seals pups are the main instigators of reunions, producing contact calls and locomoting while mothers vocalize to help their pups find them (Collins et al. 2005, 2011). Recognition abilities in at least one seal species vary across different populations. Gray seal mothers on the Sable Island breeding colony (Canada) were capable of recognizing their pups acoustically (McCulloch and Boness 2000), while those on the Isle of May colony (Scotland) were not despite their pups producing individually distinct vocalizations (McCulloch et al. 1999). It is unclear why such population-level differences in recognition abilities should occur (McCulloch and Boness 2000). Selection pressure to reduce fostering or allo-suckling of non-filial pups is considered a driver of recognition abilities in phocids, given the high costs of lactation (McCulloch et al. 1999; Perry et al. 1998). However, poor or non-existent (Hawaiian monk seals; Job et al. 1995), mother–pup recognition abilities persist in several group breeding phocids. Even if offspring recognition abilities are lacking, the benefits of colony breeding may simply outweigh the energetic costs of allo-suckling behaviors for some phocid mothers (Perry et al. 1998).

7.3.3.4 Maladaptive Maternal Behavior

Maladaptive traits are defined as any aspect of an organism's phenotype that does not result in higher relative fitness than the other possible traits within the species (Crespi 2000). Maladaptive maternal behaviors in phocids can either reduce the female's fitness, reduce the likelihood her pup survives, and thus her overall reproductive success, or both. Despite strong selective pressure on phocid mothers to exhibit optimal maternal behavior, many behaviors that negatively impact a mother and her pup's welfare persist in seal species.

Mothers can harm their pups by neglect or by inflicting physical injury. Mother–pup separation leading to starvation is the largest cause of pup death in some colonially breeding phocids (Anderson et al. 1979; Le Boeuf and Briggs 1977), affecting up to a quarter of breeding female northern elephant seals (Le Boeuf et al. 1972). Many separations are due to events beyond the female's control; however, some are due to the mothers' apparent indifference, where females do not stay with or ignore their pups (Robinson et al. 2015a). Without their mothers, pups usually starve and incur injuries from other seals. Wounds quickly become infected due to the pup's weakened condition, and the pup dies before leaving the colony (Anderson et al. 1979; Riedman and Le Boeuf 1982; Spraker et al. 2014; Salogni et al. 2016). To survive, an abandoned or lost pup has to approach unrelated females to either steal milk from them or be adopted. It has been suggested that, due to the high energetic cost of producing fat-rich milk, mothers abandon pups if they do not have the body reserves to rear them successfully (Jemison and Kelly 2001); however, there is little evidence to support this theory. Young, inexperienced mothers are more prone to abandon their pups than older females (Reiter et al. 1981; Riedman and Le Boeuf 1982; Pistorius et al. 2001), and younger females are often smaller. Irrespective of her prior breeding experience, the success of hormone-mediated bonding within the first hours postpartum has been shown to be a good predictor of a mother's likelihood of abandoning her pup (Robinson et al. 2015a). Mothers may also abandon pups after disturbance from predators or humans (Osinga et al. 2012) or harm them directly through physical contact (Fig. 7.13). Such harmful interactions are common when females are disturbed by some threat, but occasionally, mothers bite, shake, or crush their pups in apparently undisturbed circumstances for unknown reasons (Mansfield 1958; P Pomeroy and K Robinson, Pers. Obs.).

While abandonment leads to a failed reproductive episode for a mother, feeding unrelated pups represents an additional, energetic cost that diminishes their own pup's resources and depletes the mother's reserves. This can occur in some contexts and is a common form of maladaptive maternal behavior in many colonially breeding seals (Perry et al. 1998). Mothers may feed unrelated pups sporadically while trying to feed their pup, or can 'adopt' a non-filial pup and rear it to weaning (Riedman and Le Boeuf 1982). These 'failure to discriminate errors' can result from misidentification of pups (Boness et al. 1992) or the persistence of starving or weaned pups that approach and solicit suckling opportunities (Reiter et al. 1978). Mothers typically reject the unrelated pup violently once the error is detected, and resume feeding behaviors toward their own pup (Reiter et al. 1978). Adoption or fostering of a non-filial pup occurs in some circumstances. If mothers are separated permanently from their pup, then some may adopt and rear a non-filial pup (Riedman and Le Boeuf 1982); there are also records of females adopting multiple pups (Fig. 7.13) or adopting a pup alongside her own (Riedman and Le Boeuf 1982; Gelatt et al. 2001). Allo-suckling is widespread in Hawaiian monk seals, with mothers appearing unable to distinguish among pups on a beach (Boness 1990; Job et al. 1995). As several species of phocids show philopatry or site fidelity to their pupping sites, average relatedness might be expected to be high, and kin selection may explain this costly behavior in phocid mothers (Pomeroy et al. 2000). However, genetic studies of harbor (Schaeff et al. 1999) and gray (Perry et al. 1998; Poland et al. 2008) seals found no evidence that fostered non-filial pups were closely related to their foster-mothers. Fostering behavior may persist in colonially breeding phocids as there is insufficient selection pressure to remove it (e.g., Hawaiian monk seals; Boness 1990).

Fig. 7.13 Examples of maladaptive maternal behavior **a** Multiple adoptions of pups (feeding and remaining with pups over a number of days) by a female gray seal on the Isle of May, Scotland. It was unknown whether any of these pups were her own **b** 'Crushing' behavior in a gray seal mother from Donna Nook, England. While the majority of pups survive, in at least one case, a mother was observed to drown her pup in a pool as a result of this behavior. Photo credit: **a** Izzy Langley; **b** Kelly Robinson

7.3.4 Interactions on the Breeding Colony

7.3.4.1 Female–Female Interactions

Colonial breeders typically experience some competition for resources such as space and access to water, which often leads to interactions between mothers (Cassini 2000). The choice of the pupping site may produce the first potential interactions between females on a colony in a breeding season. Pregnant females must come ashore and move to their chosen site, passing other pregnant seals and mother–pup pairs on the way (Pomeroy et al. 1994, 2005). Where access is limited, interactions with those already ashore can influence the destination of newly arrived females and may establish structure within the colony (Pomeroy et al. 2005). While the density and size of these female groups vary across species and colonies, mothers with pups generally react aggressively to any approaching female coming within one adult body length of the pair (Baldi et al. 1996; van Polanen Petel et al. 2007;

Robinson et al. 2015b). The pup's position relative to the intruder also influences the level of aggression from a mother (Boness et al. 1982; Stewart 1987). High-density colonies have higher female-female aggression rates than low-density ones (Baldi et al. 1996). Seals show consistent individual differences in behavioral traits across different contexts, including aggressive responses to encroaching intruders and perceived threats (Twiss et al. 2012; Bubac et al. 2018). Environmental conditions can also affect the number of interactions a mother may experience. In dry seasons, gray seal mothers commuting to pools of water or the ocean during lactation may come into conflict with many females they would not normally encounter (Twiss et al. 2002).

Agonistic interactions between females use similar aggressive behaviors as those seen in male-male and male–female encounters. Mothers vocalize and display the 'open mouth threat' posture to each other, and if this does not succeed in making an opponent retreat, then the females lunge, bite, and claw at each other (Fig. 7.14; Christenson and Le Boeuf 1978; Robinson et al. 2015a). Such aggression prevents encroaching females from injuring pups or driving them away, which could result in separation and abandonment (Christenson and Le Boeuf 1978). However, agonistic interactions are costly and risk injury to all parties. As discussed above, some seal species show fidelity to their breeding sites and even philopatry across many years. As neighboring mothers encounter each other over multiple breeding seasons, recognition and reduced aggressive interactions might occur between familiars (Pomeroy et al. 2005). Indeed, Robinson et al. (2015b) found that gray seals can recognize and modulate their aggression toward individuals with whom they have previously co-habited.

Fig. 7.14 Breeding female gray seals in a three-way dispute for space around themselves and their pups. In crowded colonies, fights between females can quickly escalate to involve more individuals if they encroach on others' space (Photo credit: Kelly Robinson)

Other species of phocid mothers can recognize neighboring females and adjust their aggressive behaviors toward them based on their social ranking. It is not known whether mothers in these instances truly recognize individuals or whether they simply recognize the dominance level a neighbor occupies, but in either case, the overall benefit of avoiding agonistic encounters still accrues. In southern and northern elephant seals, social hierarchies are not strictly linear, but multiple dominant/subordinate dyads exist between different females with age rather than body size determining rank (Christenson and Le Boeuf 1978; Reiter et al. 1981). Dominant mothers display more aggression, while subordinate mothers are first to retreat from agonistic interactions (Christenson and Le Boeuf 1978). High-ranking mothers can have greater pup survival rates (Reiter et al. 1981), and lower rates of pup injuries and pup separations (Reiter et al. 1981; McCann 1982; Christenson and Le Boeuf 1978).

7.3.4.2 Female–Male Interactions

Females of many phocid species are capable of exerting at least some degree of mate choice when breeding. In some species, females' ability to choose their mates and the opportunities to do so are obvious (e.g., harbor seals; Hayes et al. 2004b). However, even females in the most sexually dimorphic, polygynous species can exert some control over who they are likely to mate with via their choices during the course of reproducing (e.g., southern elephant seals; Galimberti et al. 2000; McMahon and Bradshaw 2004).

Many phocids mate underwater, which facilitates females escaping or avoiding unwanted male attention, leading to greater degrees of female mate choice. Male aquatic vocal displays can attract females who may choose among candidate males, especially in species where individuals are widely dispersed across large geographic areas (e.g., leopard, Ross, bearded, and ribbon seals; Rogers 2003). Long-distance displays tend to be simple in structure and highly repetitive to enable them to travel undistorted to distant receptive females. In contrast, males of some species display vocally in underwater territories adjacent to pupping sites with multiple breeding females nearby, who can move freely through or around these areas (Hayes et al. 2006). This breeding system enables females to choose an individual male from several potential mates, all of whom are advertising simultaneously. In these circumstances, males are competing for female attention, and in these species, such displays can be acoustically complex and accompanied by underwater displays of 'acrobatics' (see sect. 7.2.2).

In species that mate on land, males are more able to physically dominate and harass females to secure mating opportunities, potentially reducing females' ability to select mates. Most seals in terrestrial environments are not agile enough to evade males quickly to prevent mating as in aquatic mating species. Therefore the evolutionary pressure to attract females that influences male vocal display behavior underwater is mostly absent in land breeding species (Rogers 2003). While less likely, some degree of female mate choice is possible even for species subjected to male harassment and forced copulations upon departure. Some female gray seals leave the male home ranges they have pupped within specifically to mate with other males before returning to care for their pup (Twiss et al. 2006). Even female southern elephant seals, whose comparatively small size makes overt mate choice difficult, can make it more likely they will mate with a dominant, or high-quality, male during estrus by choosing to join large harems rather than small ones, as it is more likely that a high-quality male becomes the dominant male in that location (McMahon and Bradshaw 2004). They also object vocally to copulation attempts by subordinate males, which directly attracts the dominant male's attention and enables him to protect the female, and the mating opportunity she represents, from the intruder (Galimberti et al. 2000). In this manner, the choice of harem size is a proxy for the male's fitness level, who is likely to mate with her when she is in season, even if choosing specific individual males is impossible in that species.

7.3.4.3 Female–non-Filial Pup Interactions

Interactions between females and non-filial pups are governed by the female's perception of the pup's identity. As outlined above, in many species of phocid, some mothers nurse or adopt non-filial pups, and this process is facilitated by female uncertainty or inability to recognize her own offspring (Boness et al. 1992; Job et al. 1995). Pups that are physically well developed or have successfully weaned can also use deceptive tactics to steal milk from inattentive mothers (Reiter et al. 1978). However, if a phocid mother realizes that an approaching pup is not her own, her reaction is almost always aggressive (Fig. 7.15). Females reject non-filial pups by biting them and in some instances pursuing them over short distances to drive them away, stopping when the pup is sufficiently far away or when the pup's own mother intervenes to defend it (Christenson and Le Boeuf 1978).

7.3.4.4 Interspecific Interactions

Predators and scavengers are attracted to phocid breeding events. A mother's response to these threats depends on whether she is physically capable of repelling an attack. Mothers react aggressively to avian species that attack their pups or try to scavenge their placentae (Le Boeuf et al. 1972; Burton et al. 1975; Robinson et al. 2015a), and can seriously injure or kill birds. Seal mothers facing large predators such as polar bears cannot defend their pups against such threats, and would likely be predated themselves in the attempt (Lydersen and Kovacs 1999). Phocids that regularly encounter terrestrial predators flee rapidly to water (Burns and Frost 1979), and it has been suggested that some species may aggregate to enable defense against such threats (e.g., Caspian seals; Wilson et al. 2017b). Mothers of species that breed under severe predation risk also flee in response to small predators, such as ringed seal mothers responding to arctic foxes attacking their birth lairs (Kelly et al. 1988). Bearded and ringed seal mothers rear their pups alongside ice holes or on isolated

Fig. 7.15 A gray seal mother attacks and bites a non-filial pup that has approached her. Photo credit: Kelly Robinson

floes to provide them with constant access to water (Lydersen and Kovacs 1999). As the pups of these two species are competent swimmers soon after birth, mothers and pups can attempt to escape into the ocean simultaneously. Mothers can also spend time in the water rather than hauled out next to their pups to reduce their visibility (Lydersen and Kovacs 1999). While pups are usually the primary target of predators, reproducing adults can also be targeted by predators drawn to breeding colonies, such as killer whales who predate harbor seal colonies (Baird and Dill 1995). In one population of harbor seals off Sable Island, Canada, reproducing females were predated more frequently than males by sharks, although the reason for this is unknown (Lucas and Stobo 2000). Aquatic predator attacks are best avoided by not entering the water, which is not possible for those species that forage during lactation.

Hybridization in phocid seals is extremely rare, and only one report exists of a female hooded seal giving birth to a hybrid pup, fathered by a harp seal (Kovacs et al. 1997). The size difference between the two species and the contrasting methods of attracting mates makes it challenging to understand how the hybrid was conceived (Kovacs et al. 1997). It is plausible that such births have happened in other species, but documenting their frequency is tricky. Genetic analysis is needed to confirm a true hybrid. In turn, it will only be accomplished if researchers can identify a potential hybrid, usually by sighting individuals that display physical characteristics of two species (Kovacs et al. 1997) and obtaining a suitable DNA sample. Hybridization is an unusual occurrence in phocids, but changing environmental conditions that alter optimal times to reproduce or make locating suitable mates or breeding areas difficult may provide more opportunities for mating between different species (Attard et al. 2012).

7.3.5 Lactation Duration and Weaning

The duration of lactation and the weaning process in phocids varies across species and within populations. The evolution of these characteristics is influenced heavily by the breeding substrate each species uses to rear pups, the proximity of feeding areas, and maternal body size. Ice breeding seals that use pack ice floes tend to have short lactation periods, limiting the time that mothers are reliant on unstable substrates to keep pups out of the water (e.g., the four-day dependent period in hooded seals; Oftedal et al. 1993). Seals that breed in lower latitudes typically have longer lactation periods, with lactation duration of 5 weeks in the Hawaiian monk seal and up to 5 months in the Mediterranean monk seal (Johanos et al. 1994; Aguilar et al. 2007). Other terrestrially breeding phocids have lactation durations between these two extremes, with durations of between 2 and 6 weeks for northern elephant seals, harbor seals, and gray seals (Riedman and Le Boeuf 1982; Fedak and Anderson 1982; Muelbert and Bowen 1993). Of the phocids breeding on fast ice, only Weddell seals have been well studied, with longer lactation durations compared to terrestrially breeding species (6–7 weeks; Wheatley et al. 2006). Outliers from this general pattern have also evolved. Bearded seals breed on unstable ice floes, but their pups enter the water soon after birth, rapidly becoming adept swimmers, and weaning occurs at just over three weeks (Gjertz et al. 2000).

When weaning their offspring, mothers of some species abandon their pups abruptly after a certain number of days postpartum (Riedman and Le Boeuf 1982; Kovacs 1987a) while in other species, pups are weaned gradually, start to forage independently, and take prey items while still suckling (bearded and ringed seals; Lydersen and Kovacs 1999). There is no evidence that post-weaning associations between mothers and pups occur, with the possible exception of bearded seal mother–pup pairs (Gjertz et al. 2000).

7.4 Pup Behavior

There are fewer studies of pup behavior when they are dependent on their mothers (Lydersen and Kovacs 1999) compared with adults, despite the critical role pups have in their survival. While maternal behavior is essential to ensure pups are fed and protected, mothers cannot successfully rear pups to independence without some degree of coordination between the two (Robinson et al. 2019a). Pups develop their physical (Mansfield 1958) and cognitive abilities over the lactation period, and this can limit their ability to respond optimally to encounters with their mothers, other conspecifics, and predators (Christenson and Le Boeuf 1978). Some pups also exhibit behaviors that increase the risk of being separated from their mothers (Riedman and Le Boeuf 1982), often leading to the pup's death (Le Boeuf et al. 1972; Anderson et al. 1979). Without the pup acting to facilitate nurturing behaviors, even an experienced female displaying high-quality maternal behavior is likely to fail at rearing her offspring.

7.4.1 Behaviors During the Dependent Period

Phocid pups are usually described as precocial (Lydersen and Kovacs 1999); however, as newborns they can struggle to coordinate their limbs, and it can take several days for pups to suckle successfully from their mothers (Christenson and Le Boeuf 1978). Newborn pups have a thin blubber layer to insulate them from the surrounding environment, and pups of some species in cold, polar regions shiver in the hours after birth to keep warm (Blix et al. 1979; Pearson et al. 2014). Providing that pups have an attentive mother, they nurse, gain weight, and develop physically so that they can move about the breeding site and, in some species, swim (Lydersen and Kovacs 1999). While still dependent on their mothers, pups in many species spend large portions of time resting beside her on terrestrial breeding sites (Kovacs 1987a, b). This facilitates effective mass and energy transfer from the mother to the pup and potentially aids mothers to relocate pups if they become separated (Kovacs 1987b, 1995b). While resting is the optimal tactic for maximizing mass gain and maintaining the mother–pup pair, some pups wander around the breeding colony without any apparent motivation, frequently encountering neighboring seals or becoming separated from their mothers in the process (Riedman and Le Boeuf 1982). In some species, pups produce distinct calls that stimulate their mothers to find them (Riedman and Le Boeuf 1982; Collins et al. 2005). Pups usually only interact with their mothers as other females will attack them (Christenson and Le Boeuf 1978; Kovacs 1995b). While hauled out, some pups engage in the object or self-directed play (Kovacs 1987a, b), and while they are capable of biting and aggressive behavior, they cannot defend themselves against adult seals or predators.

Pups of many phocid species typically spend little time in the water, but if they are forced into the sea are physically able to survive and haul out onto the breeding site again (Lydersen et al. 1997). In several species, pups enter the ocean in the days after birth, with some pups swimming only hours postpartum (Fig. 7.16) and becoming proficient swimmers and divers prior to weaning (e.g., bearded seals, Lydersen et al. 1994, Gjertz et al. 2000; harbor seals, Bowen et al. 1999, Jørgensen et al. 2001; ringed seals, Lydersen and Hammill 1993; Weddell seals, Burns et al. 1999). Young pups who cannot follow their diving mothers remain at the ocean's surface until she returns, but as pups mature, they can join their mothers at depth (Bowen et al. 1999; Gjertz et al. 2000). These pups subsequently spend large amounts of the lactation period in the water, hauling out to nurse and rest (Lydersen and Kovacs 1999).

Ideally, phocid pups only interact with their mother prior to weaning, but they can be forced into contact with other breeding seals or hunted by predators. In

Fig. 7.16 **a** A harbor seal mother with her pup, less than 24 h old, entering the sea off its birth site **b** and at one-week-old swimming with its mother. Photo credit: Andy Law

both cases, these interactions are dangerous for pups, and injury or death can come from reacting inappropriately in these situations. Ringed seal pups face a variety of predatory threats, including polar bears, arctic foxes, human hunters (Furgal et al. 1996), ravens (*Corvus corax*), and glaucous gulls (*Larus hyperboreus*; Lydersen and Smith 1989), and dwell in subnivean birth lairs to reduce the risk of attack. Avian predators cannot dig pups out of these lairs (Lydersen and Smith 1989), and pups within lairs attacked by bears, foxes, or humans attempt to escape by exiting the lair via breathing hole within it, swimming to safety (Furgal et al. 1996). Harp seal breeding colonies are also subject to polar bear predation on occasion. Pups of this species exhibit a distinctive 'freeze' behavior in response to threats, where they draw in their heads, stiffen their bodies, stop breathing, and lower their heart rate (Lydersen and Kovacs 1999; Chap. 14). Harbor seal pups at various geographic locations face different predation risks that require opposing behavioral strategies to counter. Coyotes (*Canis latrans*) target pups in certain colonies within Puget Sound, and pups must flee to the water to avoid capture (Steiger et al. 1989), while nearby off the coast of Vancouver Island, pups must go to shore to avoid killer whale attacks (Baird and Dill 1995). Other predators of seal pups include wolves (*Canis lupus*) and large eagles (e.g., white-tailed eagles, *Haliaeetus albicilla*), that are thought to take large numbers of Caspian seal pups (Krylov 1990; Harkonen et al. 2008), and sharks, that target Hawaiian monk seal pups (Bertilsson-Friedman 2006) and harbor seal pups (Lucas and Stobo 2000) swimming off breeding colonies. Walrus (*Odobenus rosmarus*) have also been observed ambushing pups on ice floes, and stomach content analysis indicates that this feeding tactic is becoming more common in the species (Seymour et al. 2014).

Pups interacting with other conspecifics are usually met with aggression, and there is evidence that pups raised on crowded parts of a breeding colony are more aggressive once they are weaned than pups reared on less densely populated areas (Robinson et al. 2017). Interacting with other breeding females is risky for pups but can result in obtaining extra milk. Pups that have been abandoned or separated from their mothers can steal or solicit feeds from surrounding mothers or attempt to repeatedly approach one mother so that adoption occurs if the pup is accepted. Approaching unrelated females for milk is a dangerous tactic that frequently ends in the pup being bitten (Riedman and Le Boeuf 1982); however, starving pups have few survival options. These individuals attempt to approach females to gain feeding opportunities, often unsuccessfully, until they become moribund.

7.4.2 Weaning and Post-Weaning Fasts

All phocid seals feed away from their breeding sites, and this division of foraging and reproductive behavior impacts pups as much as it affects breeding adults. In species where mothers wean pups abruptly, the ocean may be a complete novelty to youngsters on their first foraging trip. In these species, pups undergo a post-weaning fast on the breeding colony that can last weeks (Worthy and Lavigne 1983; Reilly 1991; Noren et al. 2008) or, in some cases, months (Houser and Costa 2001). During this time, weaned pups are expelled to the periphery of colonies (Reiter et al. 1978), either alone or in groups (Fig. 7.17), and undergo physiological changes to develop their physical coordination and diving abilities prior to going to sea (Noren et al. 2005; Bennett et al. 2010). Some northern elephant seal pups in the post-wean fast attempt to steal additional feeds from females still raising their pups on the breeding colony. These 'super weaners' are typically male and become much larger than pups reared by a single mother (Fig. 7.17; Reiter et al. 1978). Pups of species with gradual weaning transitions accompany mothers to develop their swimming and diving capabilities. In some species, this leads to pups catching their prey successfully before weaning (e.g., bearded and ringed seals; Lydersen and Kovacs 1999), while in others, there is no evidence of pups feeding on anything but milk prior to weaning (e.g., harbor seal; Muelbert and Bowen 1993).

Fig. 7.17 Weaned phocid will pups gather together on the peripheries of breeding colonies during post-wean fasts; **a** Northern elephant seal pups **b** Gray seal pups **c** Two weaned northern elephant seal pups, a 'super weaner' that has solicited additional milk from non-filial mothers to gain additional weight post-weaning on the left compared to a more typical weaner on the right. Photo credit: **a, c** Dan Costa; **b** Kelly Robinson

7.5 Future Directions

The availability of terrestrial breeding species has aided the current understanding of phocid breeding behavior. There is still much to understand, especially in species with underwater components to their reproductive behavior. The nature of the mating systems for several phocids, particularly ice breeding species, is largely unknown (Stirling and Thomas 2003), and species breeding in remote regions of the world, or in low densities, are typically poorly represented in the literature. The environments seals use while breeding, or feeding in preparation to reproduce, are changing at unprecedented rates due to a variety of anthropogenic causes. Without a thorough knowledge of their breeding behaviors, it is difficult to know whether seals can cope with such changes. Intraspecific variation in breeding behaviors indicates some degree of plasticity is already present for some species, but it unknown whether this will help facilitate adaptation to new environments and constraints at short timescales.

Additionally, our ability to provide interventions to mitigate species declines will depend on our knowledge of breeding behaviors and what population imbalances can do to the remaining individuals trying to reproduce (e.g., Hawaiian monk seals; Starfield et al. 1995; Johanos et al. 2010). Environmental change can take many forms, and climate change (Sundqvist et al. 2012; Stenson and Hammill 2014), disturbance (Güçlüsoy et al. 2004; Costa et al. 2016), exposure to novel or historic pollutants (Robinson et al. 2019b) changes to prey abundance (DeMaster et al. 2001; Lowry et al. 2011), increased hunting effort by humans (Harding et al. 2007), and entanglement in human-made debris (Lowry et al. 2011) impact phocid breeding. Improvements in telemetry devices, new methods to collect, model, or analyze data from wild populations, study links between behavior and physiology in free-ranging seals, and remote surveying techniques are essential to expand our knowledge of phocid breeding behavior and to implement effective management and protection measures.

References

Aguilar A, Cappozzo LH, Gazo M, Pastor T, Forcada J, Grau E (2007) Lactation and mother–pup behaviour in the Mediterranean monk seal *Monachus monachus*: an unusual pattern for a phocid. J Mar Biol Assoc UK 87:93–99

Ambs SM, Boness DJ, Bowen WD, Perry EA, Fleischer RC (1999) Proximate factors associated with high levels of extraconsort fertilization in polygynous gray seals. Anim Behav 58:527–535

Amos WB, Twiss S, Pomeroy PP, Anderson SS (1993) Male mating success and paternity in the gray seal, *Halichoerus grypus*: a study using DNA fingerprinting. P Roy Soc Lond B Bio 252:199–207

Anderson SS, Fedak MA (1985) Grey seal males: energetic and behavioural links between size and sexual success. Anim Behav 33:829–838

Anderson SS, Harwood J (1985) Time budgets and topography: how energy reserves and terrain determine the breeding behaviour of gray seals. Anim Behav 33:1343–1348

Anderson SS, Burton RW, Summers CF (1975) Behaviour of gray seals (*Halichoerus grypus*) during a breeding season at North Rona. J Zool 177:179–195

Anderson SS, Baker JR, Prime JH, Baird A (1979) Mortality in gray seal pups: incidence and causes. J Zool 189:407–417

Atkinson S (1997) Reproductive biology of seals. Rev Reprod 2:175–194

Atkinson S, Becker BL, Johanos TC, Pietraszek JR, Kuhn BCS (1994) Reproductive morphology and status of female Hawaiian monk seals (*Monachus schauinslandi*) fatally injured by adult male seals. Reproduction 100:225–230

Attard CR, Beheregaray LB, Jenner KCS, Gill PC, Jenner MN, Morrice MG, Robertson KM, Möller LM (2012) Hybridization of Southern Hemisphere blue whale subspecies and a sympatric area off Antarctica: impacts of whaling or climate change? Mol Ecol 21:5715–5727

Baird RW, Dill LM (1995) Occurrence and behaviour of transient killer whales: seasonal and pod-specific variability, foraging behaviour, and prey handling. Can J Zool 73:1300–1311

Bajzak CE, Hammill MO, Stenson GB, Prinsenberg S (2011) Drifting away: implications of changes in ice conditions for a pack-ice-breeding phocid, the harp seal (*Pagophilus groenlandicus*). Can J Zool 89:1050–1062
Baker JD (2008) Variation in the relationship between offspring size and survival provides insight into causes of mortality in Hawaiian monk seals. Endanger Species Res 5:55–64
Baker JR, Baker R (1988) Effects of environment on gray seal (*Halichoerus grypus*) pup mortality. Studies on the Isle of May. J Zool 216:529–537
Baker JD, Becker BL, Wurth TA, Johanos TC, Littnan CL, Henderson JR (2011) Translocation as a tool for conservation of the Hawaiian monk seal. Biol Conserv 144:2692–2701
Baldi R, Campagna C, Pedraza S, Le Boeuf BJ (1996) Social effects of space availability on the breeding behaviour of elephant seals in Patagonia. Anim Behav 51:717–724
Ballard KA, Kovacs KM (1995) The acoustic repertoire of hooded seals (*Cystophora cristata*). Can J Zool 73:1362–1374
Bartholomew GA, Collias NE (1962) The role of vocalization in the social behaviour of the northern elephant seal. Anim Behav 10:7–14
Bartsh SS, Johnston SD, Siniff DB (1992) Territorial behavior and breeding frequency of male Weddell seals (*Leptonychotes weddelli*) in relation to age, size, and concentrations of serum testosterone and cortisol. Can J Zool 70:680–692
Bédard C, Kovacs K, Hammill M (1993) Cannabalism by gray seals, *Halichoerus grypus*, on Amet Island, Nova Scotia. Mar Mammal Sci 9:421–424
Bennett KA, McConnell BJ, Moss SE, Speakman JR, Pomeroy PP, Fedak MA (2010) Effects of age and body mass on development of diving capabilities of gray seal pups: costs and benefits of the postweaning fast. Physiol Biochem Zool 83:911–923
Bertilsson-Friedman P (2006) Distribution and frequencies of shark-inflicted injuries to the endangered Hawaiian monk seal (*Monachus schauinslandi*). J Zool 268:361–368
Bertram SM, Gorelick R (2009) Quantifying and comparing mating systems using normalized mutual entropy. Anim Behav 77:201–206
Bishop AM, Lidstone-Scott R, Pomeroy P, Twiss SD (2014) Body slap: an innovative aggressive display by breeding male gray seals (*Halichoerus grypus*). Mar Mammal Sci 30:579–593
Bishop AM, Onoufriou J, Moss S, Pomeroy PP, Twiss SD (2016) Cannibalism by a male gray seal (*Halichoerus grypus*) in the North Sea. Aqaut Mamm 42:137
Bishop AM, Stewart JE, Pomeroy P, Twiss SD (2017) Intraseasonal temporal variation of reproductive effort for male gray seals. Anim Behav 134:167–175
Blix AS, Grav HJ, Ronald K (1979) Some aspects of temperature regulation in newborn harp seal pups. Am J Physiol Reg I 236:188–197
Boness DJ (1990) Fostering behavior in Hawaiian monk seals: is there a reproductive cost? Behav Ecol Sociobiol 27:113–122
Boness DJ, Bowen WD (1996) The evolution of maternal care in pinnipeds: new findings raise questions about the evolution of maternal feeding strategies. Bioscience 46:645–654
Boness DJ, James H (1979) Reproductive behaviour of the gray seal (*Halichoerus grypus*) on Sable Island, Nova Scotia. J Zool 188:477–500
Boness DJ, Anderson SS, Cox CR (1982) Functions of female aggression during the pupping and mating season of gray seals, *Halichoerus grypus*. Can J Zool 60:2270–2278
Boness DJ, Bowen WD, Oftedal OT (1988) Evidence of polygyny from spatial patterns of hooded seals (*Cystophora cristata*). Can J Zool 66:703–706
Boness DJ, Bowen D, Iverson SJ, Oftedal OT (1992) Influence of storms and maternal size on mother–pup separations and fostering in the harbor seal, *Phoca vitulina*. Can J Zool 70:1640–1644
Boness DJ, Bowen WD, Oftedall OT (1994) Evidence of a maternal foraging cycle resembling that of otariid seals in a small phocid, the harbor seal. Behav Ecol Sociobiol 34:95–104
Boness DJ, Bowen WD, Iverson SJ (1995) Does male harassment of females contribute to reproductive synchrony in the gray seal by affecting maternal performance? Behav Ecol Sociobiol 36:1–10

Boness DJ, Bowen WD, Buhleier BM, Marshall GJ (2006) Mating tactics and mating system of an aquatic-mating pinniped: the harbor seal, *Phoca vitulina*. Behav Ecol Sociobiol 61:119–130

Bowen WD, Oftedal OT, Boness DJ (1985) Birth to weaning in 4 days: remarkable growth in the hooded seal, *Cystophora cristata*. Can J Zool 63:2841–2846

Bowen WD, Boness DJ, Iverson SJ (1999) Diving behaviour of lactating harbor seals and their pups during maternal foraging trips. Can J Zool 77:978–988

Bowen WD, McMillan J, Mohn R (2003) Sustained exponential population growth of gray seals at Sable Island, Nova Scotia. ICES J Mar Sci 60:1265–1274

Boyd IL (1984) The relationship between body condition and the timing of implantation in pregnant gray seals (*Halichoerus grypus*). J Zool 203:113–123

Boyd IL (1991) Environmental and physiological factors controlling the reproductive cycles of pinnipeds. Can J Zool 69:1135–1148

Boyd JM, Lockie JD, Hewer HR (1962) The breeding colony of gray seals on North Rona, 1959. P Zool Soc Lond 138:257–277

Boyle D (2009) Gray seal breeding census: skomer Island 2009. In: Wildlife trust of South and West Wales, CCW Regional Report CCW/WW/10/4

Brownlow A, Onoufriou J, Bishop A, Davison N, Thompson D (2016) Corkscrew seals: gray seal (*Halichoerus grypus*) infanticide and cannibalism may indicate the cause of spiral lacerations in seals. PLoS One 11:e0156464

Bubac CM, Coltman DW, Bowen WD, Lidgard DC, Lang SL, den Heyer CE (2018) Repeatability and reproductive consequences of boldness in female gray seals. Behav Ecol Sociobiol 72:100

Burns JJ, Frost KJ (1979) The natural history and ecology of the bearded seal, *Erignathus barbatus*. Environ Assess Alaskan Contin Shelf Final Rep 19:311–392

Burns JJ, Ray GC, Fay FH, Shaughnessy PD (1972) Adoption of a strange pup by the ice-inhabiting harbor seal *Phoca Vitulina Largha*. J Mammal 53(3):594–598

Burns JM, Castellini MA, Testa JW (1999) Movements and diving behavior of weaned Weddell seal (*Leptonychotes weddellii*) pups. Polar Biol 21:23–36

Burton RW, Anderson SS, Summers CF (1975) Perinatal activities in the gray seal (*Halichoerus grypus*). J Zool 177:197–201

Cameron MF, Siniff DB, Proffitt KM, Garrott RA (2007) Site fidelity of Weddell seals: the effects of sex and age. Antarct Sci 19:149–155

Campagna C, Lewis M (1992) Growth and distribution of a southern elephant seal colony. Mar Mammal Sci 8:387–396

Carlini AR, Poljak S, Daneri GA, Márquez ME, Plötz J (2002) Dynamics of male dominance of southern elephant seals (*Mirounga leonina*) during the breeding season at King George Island. Pol Polar Res 23:153–159

Carrick R, Csordas SE, Ingham SE (1962) Studies on the southern elephant seal, *Mirounga leonina* (L.). IV. Breed Dev CSIRO Wildlife Res 7:161–197

Casey C, Reichmuth C, Costa DP, Le Boeuf B (2018) The rise and fall of dialects in northern elephant seals. Proc R Soc B 285:20182176

Casey CB, Charrier I, Mathevon N, Nasr C, Forman P, Reichmuth C (2019) The genesis of giants: behavioral, social, and vocal development of northern elephant seals. J Acous Soc Am 146:2768

Casey C, Charrier I, Mathevon N, Reichmuth C (2015) Rival assessment among northern elephant seals: evidence of associative learning during male–male contests. R Soc Open Sci 2:150228

Cassini MH (2000) A model on female breeding dispersion and the reproductive systems of pinnipeds. Behav Process 51:93–99

Cato D, Rogers T (2002) Individual variation in the acoustic behaviour of the adult male leopard seal, *Hydrurga leptonyx*. Behaviour 139(10):1267–1286

Christenson TE, Le Boeuf BJ (1978) Aggression in the female northern elephant seal, *Mirounga angustirostris*. Behaviour 64:158–171

Cline DR, Siniff DB, Erickson AW (1971) Underwater copulation of the Weddell seal. J Mammal 52:216–218

Collins KT, Rogers TL, Terhune JM, McGreevy PD, Wheatley KE, Harcourt RG (2005) Individual variation of in-air female "pup contact" calls in Weddell seals, *Leptonychotes weddellii*. Behaviour 142:167–190

Collins KT, McGreevy PD, Wheatley KE, Harcourt RG (2011) The influence of behavioural context on Weddell seal (*Leptonychotes weddellii*) airborne mother-pup vocalisation. Behav Process 87:286–290

Coltman DW, Bowen WD, Boness DJ, Iverson SJ (1997) Balancing foraging and reproduction in the male harbour seal, an aquatically mating pinniped. Anim Behav 54:663–678

Coltman DW, Bowen WD, Iverson SJ, Boness DJ (1998) The energetics of male reproduction in an aquatically mating pinniped, the harbour seal. Physiol Zool 71:387–399

Coltman DW, Bowen WD, Wright JM (1998) Male mating success in an aquatically mating pinniped, the harbour seal (*Phoca vitulina*), assessed by microsatellite DNA markers. Mol Ecol 7:627–638

Coltman DW, Bowen WD, Wright JM (1999) A multivariate analysis of phenotype and paternity in male harbor seals, *Phoca vitulina*, at Sable Island, Nova Scotia. Behav Ecol 10:69–177

Costa DP (1991) Reproductive and foraging energetics of pinnipeds: implications for life history patterns. In: Renouf D (ed) The behaviour of pinnipeds. Springer, Dordrecht, pp 300–344

Costa DP, Schwarz L, Robinson P, Schick RS, Morris PA, Condit R, Crocker DE, Kilpatrick AM (2016) A bioenergetics approach to understanding the population consequences of disturbance: Elephant seals as a model system. In: Popper AN, Hawkins (eds) The effects of noise on aquatic life II. Springer, New York, pp 161–169

Coulson JC (1981) A study of the factors influencing the timing of breeding in the gray seal *Halichoerus grypus*. J Zool 194:553–571

Cox CR (1981) Agonistic encounters among male elephant seals: frequency, context, and the role of female preference. Am Zool 21:197–209

Crespi BJ (2000) The evolution of maladaptation. Heredity 84:623–629

Crocker DE, Houser DS, Webb PM (2012) Impact of body reserves on energy expenditure, water flux, and mating success in breeding male northern elephant seals. Physiol Biochem Zool 85:11–20

Croxall JP, Hiby L (1983) Fecundity, survival and site fidelity in Weddell seals, *Leptonychotes weddelli*. J Appl Ecol 20:9–32

Curtis C, Stewart BS, Karl SA (2011) Genetically effective population sizes of Antarctic seals estimated from nuclear genes. Conserv Genet 12:1435–1446

Davis CS, Stirling IAN, Strobeck C, Coltman DW (2008) Population structure of ice-breeding seals. Mol Ecol 17:3078–3094

De Bruyn PJN, Tosh CA, Bester MN, Cameron EZ, McIntyre T, Wilkinson IS (2011) Sex at sea: alternative mating system in an extremely polygynous mammal. Anim Behav 82:445–451

DeMaster DP, Fowler CW, Perry SL, Richlen MF (2001) Predation and competition: the impact of fisheries on marine-mammal populations over the next one hundred years. J Mammal 82:641–651

Desprez M, Gimenez O, McMahon CR, Hindell MA, Harcourt RG (2018) Optimizing lifetime reproductive output: intermittent breeding as a tactic for females in a long-lived, multiparous mammal. J Anim Ecol 87:199–211

Deutsch CJ, Crocker DE, Costa DP, Le Boeuf BJ (1994) Sex-and age-related variation in reproductive effort of northern elephant seals. In: Le Boeuf BJ, Laws RM (eds) Elephant seals: population ecology, behavior, and physiology, pp 169–210

Doiron EE, Rouget PA, Terhune JM (2012) Proportional underwater call type usage by Weddell seals (*Leptonychotes weddellii*) in breeding and nonbreeding situations. Can J Zool 90:237–247

Elgar MA, Jones TM, McNamara KB (2013) Promiscuous words. Front Zool 10:66

Fabiani A, Galimberti F, Sanvito S, Hoelzel AR (2004) Extreme polygyny among southern elephant seals on Sea Lion Island, Falkland Islands. Behav Ecol 15:961–969

Fabiani A, Galimberti F, Sanvito S, Hoelzel AR (2006) Relatedness and site fidelity at the southern elephant seal, *Mirounga leonina*, breeding colony in the Falkland Islands. Anim Behav 72:617–626

Fedak MA, Anderson SS (1982) The energetics of lactation: accurate measurements from a large wild mammal, the gray seal (*Halichoerus grypus*). J Zool 198:473–479
Fisher HD (1954) Delayed implantation in the harbor seal *Phoca Vitulina L.*. Nature 173:879
Fogden SCL (1971) Mother-young behaviour at gray seal breeding beaches. J Zool 164:61–92
Frost KJ, Burns JJ (2017) Spotted seal: *Phoca largha*. In: Würsig B, Thewissen JGM, Kovacs KM (eds) Encyclopedia of marine mammals, 3rd edn. Academic Press, London, pp 928–931
Frouin-Mouy HC, Hammill MO (2021) In-air and underwater sounds of hooded seals during the breeding season in the Gulf of St. Lawrence. J Acoust Soc Am 150(1):281–293. https://doi.org/10.1121/10.0005478
Furgal C, Kovacs KM, Innes S (1996) Characteristics of ringed seal, *Phoca hispida*, subnivean structures and breeding habitat and their effects on predation. Can J Zool 74:858–874
Galimberti F, Boitani L, Marzetti I (2000) Female strategies of harassment reduction in southern elephant seals. Ethol Ecol Evol 12:367–388
Garcia-Gonzalez F (2017) Promiscuity. In: Vonk J, Shackelford T (eds) Encyclopedia of animal cognition and behavior. Springer, Cham
Gazo M, Aguilar A (2005) Maternal attendance and diving behavior of a lactating Mediterranean monk seal. Mar Mammal Sci 21:340–345
Gazo M, Layna JF, Aparicio F, Cedenilla MA, González LM, Aguilar A (1999) Pupping season, perinatal sex ratio and natality rates of the Mediterranean monk seal (*Monachus monachus*) from the Cabo Blanco colony. J Zool 249:393–401
Gelatt TS, Davis CS, Siniff DB, Strobeck C (2001) Molecular evidence for twinning in Weddell seals (*Leptonychotes weddellii*). J Mammal 82:491–499
Gjertz I, Kovacs KM, Lydersen C, Wiig Ø (2000) Movements and diving of bearded seal (*Erignathus barbatus*) mothers and pups during lactation and post-weaning. Polar Biol 23:559–566
Griffen BD (2018) Reproductive skipping as an optimal life history strategy in the southern elephant seal, *Mirounga leonina*. Ecol Evol 8:9158–9170
Güçlüsoy H, Kiraç CO, Veryeri NO, Savas Y (2004) Status of the Mediterranean monk seal, *Monachus monachus* in the coastal waters of Turkey. EU J Fish Aquat Sci 21:201–210
Gucu AC, Gucu G, Orek H (2004) Habitat use and preliminary demographic evaluation of the critically endangered Mediterranean monk seal (*Monachus monachus*) in the Cilician Basin (Eastern Mediterranean). Biol Conserv 116:417–431
Haley MP, Deutsch CJ, Le Boeuf BJ (1994) Size, dominance and copulatory success in male northern elephant seals, *Mirounga angustirostris*. Anim Behav 48:1249–1260
Hall AJ, Russell DJ (2017) Gray seal: *Halichoerus grypus*. In: Würsig B, Thewissen JGM, Kovacs KM (eds) Encyclopedia of marine mammals, 3rd edn. Academic Press, London, pp 420–422
Haller MA, Kovacs KM, Hammill MO (1996) Maternal behaviour and energy investment by grey seals (*Halichoerus grypus*) breeding on land-fast ice. Can J Zool 74:1531–1541
Hammill MO, Lydersen C, Ryg M, Smith TG (1991) Lactation in the ringed seal (*Phoca hispida*). Can J Fish Aquat Sci 48(12):2471–2476
Harcourt RG, Kingston JJ, Cameron MF, Waas JR, Hindell MA (2007a) Paternity analysis shows experience, not age, enhances mating success in an aquatically mating pinniped, the Weddell seal (*Leptonychotes weddellii*). Behav Ecol Sociobiol 61:643–652
Harcourt RG, Kingston JJ, Waas JR, Hindell MA (2007b) Foraging while breeding: alternative mating strategies by male Weddell seals? Aquat Conserv 17:68–78
Harding KC, Härkönen T, Helander B, Karlsson O (2007) Status of Baltic gray seals: population assessment and extinction risk. NAMMCO Sci Publ 6:33–56
Harkonen T, Jüssi M, Baimukanov M, Bignert A, Dmitrieva L, Kasimbekov Y, Verevkin M, Wilson S, Goodman SJ (2008) Pup production and breeding distribution of the Caspian seal (*Phoca caspica*) in relation to human impacts. Ambio 37:356–361
Hayes SA, Kumar A, Costa DP, Mellinger DK, Harvey JT, Southall BL, Le Boeuf BJ (2004a) Evaluating the function of the male harbour seal, *Phoca vitulina*, roar through playback experiments. Anim Behav 67:1133–1139

Hayes SA, Costa DP, Harvey JT, Le Boeuf BJ (2004b) Aquatic mating strategies of the male Pacific harbor seal (*Phoca vitulina richardii*): are males defending the hotspot? Mar Mammal Sci 20:639–656
Hayes SA, Pearse DE, Costa DP, Harvey JT, Le Boeuf BJ, Garza JC (2006) Mating system and reproductive success in eastern Pacific harbor seals. Mol Ecol 15:3023–3034
Hayward JL (2003) Sexual aggression by a male northern elephant seal on harbor seal pups in Washington. Northwest Nat 84:148–150
Hindell MA, Little GJ (1988) Longevity, fertility and philopatry of two female southern elephant seals (*Mirounga leonina*) at Macquarie Island. Mar Mammal Sci 4:168–171
Hiruki LM, Gilmartin WG, Becker BL, Stirling I (1993a) Wounding in Hawaiian monk seals (*Monachus schauinslandi*). Can J Zool 71:458–468
Hiruki LM, Stirling I, Gilmartin WG, Johanos TC, Becker BL (1993b) Significance of wounding to female reproductive success in Hawaiian monk seals (*Monachus schauinslandi*) at Laysan Island. Can J Zool 71:469–474
Hoelzel AR, Le Boeuf BJ, Reiter J, Campagna C (1999) Alpha-male paternity in elephant seals. Behav Ecol Sociobiol 46:298–306
Houser D, Costa D (2001) Protein catabolism in suckling and fasting northern elephant seal pups (*Mirounga angustirostris*). J Comp Physiol B 171:635–642
Huber HR (1987) Natality and weaning success in relation to age of first reproduction in northern elephant seals. Can J Zool 65:1311–1316
Hückstädt LA, Holser RR, Tift MS, Costa DP (2018) The extra burden of motherhood: reduced dive duration associated with pregnancy status in a deep-diving mammal, the northern elephant seal. Biol Let 14:20170722
Insley SJ, Holt MM (2012) Do male northern elephant seals recognize individuals or merely relative dominance rank? J Acoust Soc Am 131:35–41
Insley S, Phillips AV, Charrier I (2003) A review of social recognition in pinnipeds. Aquat Mamm 29:181–201
Jefferson TA, Stacey PJ, Baird RW (1991) A review of killer whale interactions with other marine mammals: predation to co-existence. Mammal Rev 21:151–180
Jemison LA, Kelly BP (2001) Pupping phenology and demography of harbor seals (*Phoca vitulina richardsi*) on Tugidak Island, Alaska. Mar Mammal Sci 17:585–600
Job DA, Boness DJ, Francis JM (1995) Individual variation in nursing vocalizations of Hawaiian monk seal pups, *Monachus schauinslandi* (*Phocidae, Pinnipedia*), and lack of maternal recognition. Can J Zool 73:975–983
Johanos TC, Becker BL, Ragen TJ (1994) Annual reproductive cycle of the female Hawaiian monk seal (*Monachus schauinslandi*). Mar Mammal Sci 10:13–30
Johanos TC, Becker BL, Baker JD, Ragen TJ, Gilmartin WG, Gerrodette T (2010) Impacts of sex ratio reduction on male aggression in the critically endangered Hawaiian monk seal *Monachus schauinslandi*. Endanger Species Res 11:123–132
Jørgensen C, Lydersen C, Brix O, Kovacs KM (2001) Diving development in nursing harbor seal pups. J Exp Biol 204:3993–4004
Kaufman GW, Siniff DB, Reichle R (1975) Colony behavior of Weddell seals, *Leptonychotes weddelli*, at Hutton Cliffs, Antarctica. Rapp. P.-V. Reun Cons Int Explor Mer 169:228–246
Kelly BP, Wartzok D (1996) Ringed seal diving behavior in the breeding season. Can J Zool 74:1547–1555
Kelly BP, Burns JJ, Quakenbush LT (1988) Responses of ringed seals (*Phoca hispida*) to noise disturbance. Port Ocean Eng Under Arctic Cond 2:27–38
Klinck H, Mellinger DK, Klinck K, Hager J, Kindermann L, Boebel O (2010) Long-range underwater vocalizations of the crabeater seal (*Lobodon carcinophaga*). J Acoust Soc Am 128:474–479
Komers PE, Brotherton PN (1997) Female space use is the best predictor of monogamy in mammals. P R Soc Lond B Bio 264:1261–1270

Kovacs KM (1987) Maternal behaviour and early behavioural ontogeny of gray seals (*Halichoerus grypus*) on the Isle of May, UK. J Zool 213:697–715
Kovacs KM (1987) Maternal behaviour and early behavioural ontogeny of harp seals, *Phoca groenlandica*. Anim Behav 35:844–855
Kovacs KM (1990) Mating strategies in male hooded seals (*Cystophora cristata*)? Can J Zool 68:2499–2502
Kovacs KM (1995) Mother–pup reunions in harp seals, *Phoca groenlandica*: cues for the relocation of pups. Can J Zool 73:843–849
Kovacs KM (1995) Harp and hooded seals–a case study in the determinants of mating systems in pinnipeds. Dev Mar Biol 4:329–335
Kovacs KM (2017) Hooded seal: *Cystophora cristata*. In: Würsig B, Thewissen JGM, Kovacs KM (eds) Encyclopedia of marine mammals, 3rd edn. Academic Press, London, pp 477–480
Kovacs KM, Lavigne DM (1992) Mass-transfer efficiency between hooded seal (*Cystophora cristata*) mothers and their pups in the Gulf of St Lawrence. Can J Zool 70:1315–1320
Kovacs KM, Lydersen C, Hammill MO, White BN, Wilson PJ, Malik S (1997) A harp seal × hooded seal hybrid. Mar Mammal Sci 13:460–468
Kovacs KM, Lydersen C, Overland JE, Moore SE (2011) Impacts of changing sea-ice conditions on Arctic marine mammals. Mar Biodivers 41:181–194
Krafft BA, Kovacs KM, Lydersen C (2007) Distribution of sex and age groups of ringed seals *Pusa hispida* in the fast-ice breeding habitat of Kongsfjorden, Svalbard. Mar Ecol Prog Ser 335:199–206
Krylov VI (1990) Ecology of the Caspian seal. Finnish Game Res 47:32–36
Lawson JW, Renouf D (1987) Bonding and weaning in harbor seals, *Phoca vitulina*. J Mammal 68:445–449
Le Boeuf BJ (1971) The aggression of the breeding bulls. Nat Hist 80:83–94
Le Boeuf BJ (1972) Sexual behavior in the northern elephant seal *Mirounga angustirostris*. Behaviour 41:1–26
Le Boeuf BJ (1974) Male-male competition and reproductive success in elephant seals. Am Zool 14:163–176
Le Boeuf BJ, Briggs KT (1977) The cost of living in a seal harem. Mammalia 41:128–195
Le Boeuf BJ, Mesnick S (1991) Sexual behavior of male northern elephant seals: I Lethal injuries to adult females. Behaviour 116:143–162
Le Boeuf BJ, Whiting RJ, Gantt RF (1972) Perinatal behavior of northern elephant seal females and their young. Behaviour 43:121–156
Lidgard DC, Boness DJ, Bowen WD, McMillan JI, Fleischer RC (2004) The rate of fertilization in male mating tactics of the polygynous grey seal. Mol Ecol 13:3543–3548
Lidgard DC, Boness DJ, Bowen WD, McMillan JI (2005) State-dependent male mating tactics in the gray seal: the importance of body size. Behav Ecol 16:541–549
Lidgard DC, Boness DJ, Bowen WD, McMillan JI (2008) The implications of stress on male mating behavior and success in a sexually dimorphic polygynous mammal, the gray seal. Horm Behav 53:241–248
Lidgard DC (2003) Intraspecific variation in male mating behaviour in the grey seal: linking phenotype, behaviour and success. PhD dissertation, Universite de La Rochelle, La Rochelle, France
Linossier J, Casey C, Charrier I, Mathevon N, Reichmuth C (2021) Maternal responses to pup calls in a high-cost lactation species. Biol Lett 17(12):20210469. https://doi.org/10.1098/rsbl.2021.0469
Littnan C, Karamanlidis A, Dendrinos P (2017) Monk Seals: *Monachus monachus*, *Neomonachus schauinslandi* and *N. tropicalis*. In: Würsig B, Thewissen JGM, Kovacs KM (eds) Encyclopedia of marine mammals, 3rd edn. Academic Press, London, pp 420–422
Lowry LF, Fay FH (1984) Seal eating by walruses in the Bering and Chukchi Seas. Polar Biol 3:11–18

Lowry LF, Laist DW, Gilmartin WG, Antonelis GA (2011) Recovery of the Hawaiian monk seal (*Monachus schauinslandi*): a review of conservation efforts, 1972 to 2010, and thoughts for the future. Aquat Mamm 37:397–419

Lucas Z, Stobo WT (2000) Shark-inflicted mortality on a population of harbor seals (*Phoca vitulina*) at Sable Island, Nova Scotia. J Zool 252:405–414

Lydersen C, Gjertz I (1986) Studies of the ringed seal (*Phoca hispida*) in its breeding habitat in Kongsfjorden, Svalbard. Polar Res 4:57–63

Lydersen C, Hammill MO (1993) Diving in ringed seal (*Phoca hispida*) pups during the nursing period. Can J Zool 71:991–996

Lydersen C, Kovacs KM (1993) Diving behaviour of lactating harp seal, *Phoca groenlandica*, females from the Gulf of St Lawrence, Canada. Anim Behav 46:1213–1221

Lydersen C, Kovacs KM (1999) Behaviour and energetics of ice-breeding, North Atlantic phocid seals during the lactation period. Mar Ecol Prog Ser 187:265–281

Lydersen C, Smith TG (1989) Avian predation on ringed seal *Phoca hispida* pups. Polar Biol 9:489–490

Lydersen C, Hammill MO, Kovacs KM (1994) Diving activity in nursing bearded seal (*Erignathus barbatus*) pups. Can J Zool 72:96–103

Lydersen C, Kovacs KM, Hammill MO, Gjertz I (1996) Energy intake and utilisation by nursing bearded seal (*Erignathus barbatus*) pups from Svalbard, Norway. J Comp Physiol B 166:405–411

Lydersen C, Kovacs KM, Hammill MO (1997) Energetics during nursing and early postweaning fasting in hooded seal (*Cystophora cristata*) pups from the Gulf of St Lawrence, Canada. J Comp Physiol B 167:81–88

Mansfield AW (1958) The breeding behaviour and reproductive cycle of the Weddell Seal (*Leptonychotes weddelli*). Falkland Islands Depend Sur Sci Rep 18:1–41

McCann TS (1981) Aggression and sexual activity of male southern elephant seals, *Mirounga leonina*. J Zool 195:295–310

McCann TS (1982) Aggressive and maternal activities of female southern elephant seals (*Mirounga leonina*). Anim Behav 30:268–276

McCreery L, Thomas JA (2009) Acoustic analysis of underwater vocalizations from crabeater seals (*Lobodon carcinophagus*): not so monotonous. Aquat Mamm 35:490–501

McCulloch S, Boness DJ (2000) Mother–pup vocal recognition in the gray seal (*Halichoerus grypus*) of Sable Island, Nova Scotia, Canada. J Zool 251:449–455

McCulloch S, Pomeroy PP, Slater PJ (1999) Individually distinctive pup vocalizations fail to prevent allo-suckling in gray seals. Can J Zool 77:716–723

McEachern MB, McElreath RL, Van Vuren DH, Eadie JM (2009) Another genetically promiscuous 'polygynous' mammal: mating system variation in *Neotoma fuscipes*. Anim Behav 77:449–455

McMahon CR, Bradshaw CJ (2004) Harem choice and breeding experience of female southern elephant seals influence offspring survival. Behav Ecol Sociobiol 55:349–362

McMahon CR, Burton HR, Bester MN (1999) First-year survival of southern elephant seals, *Mirounga leonina*, at sub-Antarctic Macquarie Island. Polar Biol 21:279–284

McMahon CR, Harcourt RG, Burton HR, Daniel O, Hindell MA (2017) Seal mothers expend more on offspring under favourable conditions and less when resources are limited. J Anim Ecol 86:359–370

Merdsoy BR, Renouf D (1978) Preliminary underwater observations of the breeding behavior of the harp seal (*Pagophilus groenlandicus*). J Mammal 59:181–185

Miksis-Olds JL, Parks SE (2011) Seasonal trends in acoustic detection of ribbon seal (*Histriophoca fasciata*) vocalizations in the Bering Sea. Aquat Mamm 37:464–471

Miller EH, Murray AV (1995) Structure, complexity, and organization of vocalizations in harp seal (*Phoca groenlandica*) pups. Sens Syst Aqu Mamm 237–264

Moore MV, Hampton SE, Izmest'Eva LR, Silow EA, Peshkova EV, Pavlov BK (2009) Climate change and the world's "sacred sea" Lake Baikal, Siberia. Bioscience 59:405–417

Muelbert MMC, Bowen WD (1993) Duration of lactation and postweaning changes in mass and body composition of harbor seal, *Phoca vitulina*, pups. Can J Zool 71:1405–1414

Noren SR, Iverson SJ, Boness DJ (2005) Development of the blood and muscle oxygen stores in gray seals (*Halichoerus grypus*): implications for juvenile diving capacity and the necessity of a terrestrial postweaning fast. Physiol Biochem Zool 78:482–490

Noren SR, Boness DJ, Iverson SJ, McMillan J, Bowen WD (2008) Body condition at weaning affects the duration of the postweaning fast in gray seal pups (*Halichoerus grypus*). Physiol Biochem Zool 81:269–277

Oftedal OT, Boness DJ, Bowen WD (1988) The composition of hooded seal (*Cystophora cristata*) milk: an adaptation for postnatal fattening. Can J Zool 66:318–322

Oftedal OT, Bowen WD, Boness DJ (1993) Energy transfer by lactating hooded seals and nutrient deposition in their pups during the four days from birth to weaning. Physiol Zool 66:412–436

Øritsland T, Øien N (1995) Aerial surveys of harp and hooded seal pups in the Greenland Sea pack-ice. In Dev Mar Bio 4:77–87

Osinga N, Nussbaum SB, Brakefield PM, de Haes HAU (2012) Response of common seals (*Phoca vitulina*) to human disturbances in the Dollard estuary of the Wadden Sea. Mamm Biol 77:281–287

Parreira BR, Chikhi L (2015) On some genetic consequences of social structure, mating systems, dispersal, and sampling. P Natl Acad Sci USA 112:E3318–E3326

Pastor T, Aguilar A (2003) Reproductive cycle of the female Mediterranean monk seal in the western Sahara. Mar Mammal Sci 19:318–330

Pastor T, Cappozzo HL, Grau E, Amos W, Aguilar A (2011) The mating system of the Mediterranean monk seal in the Western Sahara. Mar Mammal Sci 27:302–320

Pearson LE, Liwanag HE, Hammill MO, Burns JM (2014) Shifts in thermoregulatory strategy during ontogeny in harp seals (*Pagophilus groenlandicus*). J Therm Biol 44:93–102

Penry GS, Baartman AC, Bester MN (2013) Vagrant elephant seal predation on Cape fur seal pups, Plettenberg Bay, South Africa. Polar Biol 36:1381–1383

Perry EA, Renouf D (1988) Further studies of the role of harbor seal (*Phoca vitulina*) pup vocalizations in preventing separation of mother–pup pairs. Can J Zool 66:934–938

Perry EA, Boness DJ, Fleischer RC (1998) DNA fingerprinting evidence of nonfilial nursing in gray seals. Mol Ecol 7:81–85

Pistorius PA, Bester MN, Kirkman SP, Taylor FE (2001) Pup mortality in southern elephant seals at Marion Island. Polar Biol 24:828–831

Poland VF, Pomeroy PP, Twiss SD, Graves JA (2008) Fine-scale study finds limited evidence of kin clustering in a gray seal colony. Mar Mammal Sci 24:371–387

Polis GA, Myers CA, Hess WR (1984) A survey of intraspecific predation within the class Mammalia. Mammal Rev 14:187–198

Pomeroy PP, Anderson SS, Twiss SD, McConnell BJ (1994) Dispersion and site fidelity of breeding female gray seals (*Halichoerus grypus*) on North Rona, Scotland. J Zool 233:429–447

Pomeroy PP, Fedak MA, Rothery P, Anderson S (1999) Consequences of maternal size for reproductive expenditure and pupping success of gray seals at North Rona, Scotland. J Anim Ecol 68:235–253

Pomeroy PP, Twiss SD, Redman P (2000) Philopatry, site fidelity and local kin associations within gray seal breeding colonies. Ethology 106:899–919

Pomeroy PP, Redman PR, Ruddell SJS, Duck CD, Twiss SD (2005) Breeding site choice fails to explain interannual associations of female gray seals. Behav Ecol Sociobiol 57:546–556

Reijnders PJ, Plötz J, Zegers J, Gräfe M (1990) Breeding biology of Weddell seals (*Leptonychotes weddellii*) at Drescher Inlet, riiser larsen ice shelf, Antarctica. Polar Biol 10:301–306

Reilly JJ (1991) Adaptations to prolonged fasting in free-living weaned gray seal pups. Am J Physiol Reg I 260:267–272

Reiter J, Stinson NL, Le Boeuf BJ (1978) Northern elephant seal development: the transition from weaning to nutritional independence. Behav Ecol Sociobiol 3:337–367

Reiter J, Panken KJ, Le Boeuf BJ (1981) Female competition and reproductive success in northern elephant seals. Anim Behav 29:670–687

Renouf D, Lawson J, Gaborko L (1983) Attachment between harbour seal (*Phoca vitulina*) mothers and pups. J Zool 199:179–187

Riedman ML, Le Boeuf BJ (1982) Mother-pup separation and adoption in northern elephant seals. Behav Ecol Sociobiol 11:203–215

Robinson KJ, Twiss SD, Hazon N, Moss S, Lonergan M, Pomeroy PP (2015) Conspecific recognition and aggression reduction to familiars in newly weaned, socially plastic mammals. Behav Ecol Sociobiol 69:1383–1394

Robinson KJ, Pomeroy PP, Hazon N, Moss S, Twiss SD (2017) Individual size, sex and rearing environment impact on aggression in newly weaned seals. Mar Mammal Sci 33:621–629

Robinson KJ, Hall AJ, Scholl G, Debier C, Thomé JP, Eppe G, Adam C, Bennett KA (2019) Investigating decadal changes in persistent organic pollutants in Scottish grey seal pups. Aquat Conserv 29:86–100

Robinson KJ, Twiss SD, Hazon N, Pomeroy PP (2015a) Maternal oxytocin is linked to close mother-infant proximity in gray seals (*Halichoerus grypus*). PLoS One 10:e0144577

Robinson KJ, Hazon N, Twiss SD, Pomeroy PP (2019a) High oxytocin infants gain more mass with no additional maternal energetic costs in wild gray seals (*Halichoerus grypus*). Psychoneuroendocrinolo 110:104423

Rogers TL (2003) Factors influencing the acoustic behaviour of male phocids seals. Aquat Mamm 29:247–260

Rogers TL (2014) Source levels of the underwater calls of a male leopard seal. J Acoust Soc Am 136:1495–1498

Rogers TL (2017) Leopard seal, *Hydrurga leptonyx*. In: Würsig B, Thewissen JGM, Kovacs KM (eds) Encyclopedia of marine mammals, 3rd edn. Academic Press, London, pp 550–552

Rogers TL, Collins KT, Harcourt RG, Wheatley KE, McGreevy PD, Terhune JM (2005) Individual variation of in-air female "pup contact" calls in Weddell seals, *Leptonychotes weddellii*. Behaviour 142:167–189

Russell LR, Purdy JE, Davis RW (2016) Social context predicts vocalization use in the courtship behaviors of Weddell seals (*Leptonychotes weddellii*): a case study. Anim Behav Cognition 3:95–119

Salogni E, Sanvito S, Galimberti F (2016) Postmortem examination and causes of death of northern elephant seal (*Mirounga angustirostris*) pups of the San Benito Islands, Baja California, Mexico. Mar Mammal Sci 32:743–752

Sandegren FE (1976) Agonistic behavior in the male northern elephant seal. Behaviour 57:136–157

Sanvito S, Galimberti F, Miller EH (2007) Vocal signalling of male southern elephant seals is honest but imprecise. Anim Behav 73:287–299

Sato K, Mitani Y, Cameron MF, Siniff DB, Watanabe Y, Naito Y (2002) Deep foraging dives in relation to the energy depletion of Weddell seal (*Leptonychotes weddellii*) mothers during lactation. Polar Biol 25:696–702

Schaeff CM, Boness DJ, Bowen WD (1999) Female distribution, genetic relatedness, and fostering behaviour in harbor seals, *Phoca vitulina*. Anim Behav 57:427–434

Seymour J, Horstmann-Dehn L, Wooller MJ (2014) Inter-annual variability in the proportional contribution of higher trophic levels to the diet of Pacific walruses. Polar Biol 37:597–609

Shero MR, Krotz RT, Costa DP, Avery JP, Burns JM (2015) How do overwinter changes in body condition and hormone profiles influence Weddell seal reproductive success? Funct Ecol 29:1278–1291

Siniff DB, Bengtson JL (1977) Observations and hypotheses concerning the interactions among crabeater seals, leopard seals, and killer whales. J Mammal 58:414–416

Siniff DB, Stirling I, Bengston JL, Reichle RA (1979) Social and reproductive behavior of crabeater seals (*Lobodon carcinophagus*) during the austral spring. Can J Zool 57:2243–2255

Smith TG, Stirling I (1975) The breeding habitat of the ringed seal (*Phoca hispida*). The birth lair and associated structures. Can J Zool 53:1297–1305

Smith TG, Hammill MO, Taugbøl G (1991) A review of the developmental, behavioural and physiological adaptations of the ringed seal, *Phoca hispida*, to life in the Arctic winter. Arctic 44:124–131

Smout S, King R, Pomeroy P (2011) Integrating heterogeneity of detection and mark loss to estimate survival and transience in UK gray seal colonies. J Appl Ecol 48:364–372
Smout S, King R, Pomeroy P (2020) Environment-sensitive mass changes influence breeding frequency in a capital breeding marine top predator. J Anim Ecol 89:384–396
Southall BL, Schusterman RJ, Kastak D (2003) Acoustic communication ranges for northern elephant seals (*Mirounga angustirostris*). Aquat Mamm 29:202–213
Southwell C (2004) Satellite-linked dive recorders provide insights into the reproductive strategies of crabeater seals (*Lobodon carcinophagus*). J Zool 264:399–402
Spraker TR, Lyons ET, Kuzmina TA, Tift MS, Raverty S, Jaggi N, Crocker DE (2014) Causes of death in preweaned northern elephant seal pups (*Mirounga angustirostris*, Gill, 1866), Año Nuevo State Reserve, California, 2012. J Vet Diagn Invest 26:320–326
Starfield AM, Roth JD, Ralls K (1995) "Mobbing" in Hawaiian monk seals (*Monachus schauinslani*): the value of simulation modeling in the absence of apparently crucial data. Conserv Biol 9:166–174
Steiger GH, Calambokidis J, Cubbage JC, Skilling DE, Smith AW, Gribble DH (1989) Mortality of harbor seal pups at different sites in the inland waters of Washington. J Wildlife Dis 25:319–328
Stenson GB, Hammill MO (2014) Can ice breeding seals adapt to habitat loss in a time of climate change? ICES J Mar Sci 71:1977–1986
Stewart RE (1987) Behavioral reproductive effort of nursing harp seals, *Phoca groenlandica*. J Mammal 68:348–358
Stewart REA, Stewart BE, Lavigne DM, Miller GW (1989) Fetal growth of Northwest Atlantic harp seals *Phoca groenlandica*. Can J Zool 67:2147–2157
Stirling I (1969) Ecology of the weddell seal in McMurdo Sound, Antarctica. Ecology 50:573–586
Stirling I (1983) The evolution of mating systems in pinnipeds. Adv Study Mamm Behav 7:489–527
Stirling I, Archibald WR (1977) Aspects of predation of seals by polar bears. J Fish Res Board Can 34:1126–1129
Stirling I, Thomas JA (2003) Relationships between underwater vocalizations and mating systems in phocid seals. Aqaut Mamm 29:227–246
Stirling I (1977) Adaptations of Weddell and ringed seals to exploit the polar fast ice habitat in the absence or presence of surface predators. In: Adaptations within Antarctic ecosystems: proceedings of the third SCAR (Scientific Committee on Antarctic Research) symposium on Antarctic biology. Gulf Publishing, Houston, p 26–30
Sullivan RM (1982) Agonistic behavior and dominance relationships in the harbor seal, *Phoca vitulina*. J Mammal 63:554–569
Sundqvist L, Harkonen T, Svensson CJ, Harding KC (2012) Linking climate trends to population dynamics in the Baltic ringed seal: impacts of historical and future winter temperatures. Ambio 41:865–872
Szala A, Shackelford TK (2019) Polygynandry. In: Vonk J., Shackelford T. (eds) Encyclopedia of animal cognition and behavior. Springer, Cham
Temte JL (1994) Photoperiod control of birth timing in the harbor seal (*Phoca vitulina*). J Zool 233:369–384
Terhune JM (2016) Weddell seals do not lengthen calls in response to conspecific masking. Bioacoustics 25:75–88
Terhune JM (2017) Through-ice communication by Weddell seals (*Leptonychotes weddellii*) is possible. Polar Biol 40:2133–2136
Terhune JM, Quin D, Dell'Apa A, Mirhaj M, Plötz J, Kindermann L, Bornemann H (2008) Geographic variations in underwater male Weddell seal trills suggest breeding area fidelity. Polar Biol 31:671–680
Terhune JM, Dell'Apa A (2006) Stereotyped calling patterns of a male Weddell seal (*Leptonychotes weddellii*). Aquat Mamm 32:175
Thomas JA, Kuechle VB (1982) Quantitative analysis of Weddell seal (*Leptonychotes weddelli*) underwater vocalizations at McMurdo Sound, Antarctica. J Acoust Soc Am 72:1730–1738

Thompson PM, Miller D, Cooper R, Hammond PS (1994) Changes in the distribution and activity of female harbor seals during the breeding season: implications for their lactation strategy and mating patterns. J Anim Ecol 63:24–30

Twiss SD, Pomeroy PP, Anderson SS (1994) Dispersion and site fidelity of breeding male gray seals (*Halichoevus grypus*) on North Rona, Scotland. J Zool 233:683–693

Twiss SD, Anderson SS, Monaghan P (1998) Limited intra-specific variation in male gray seal (*Halichoerus grypus*) dominance relationships in relation to variation in male mating success and female availability. J Zool 246:259–267

Twiss SD, Caudron A, Pomeroy PP, Thomas CJ, Mills JP (2000) Finescale topographical correlates of behavioural investment in offspring by female gray seals, *Halichoerus grypus*. Anim Behav 59:327–338

Twiss SD, Wright NC, Dunstone N, Redman P, Moss S, Pomeroy PP (2002) Behavioral evidence of thermal stress from overheating in UK breeding gray seals. Mar Mamm Sci 18:455–468

Twiss SD, Poland VF, Graves JA, Pomeroy PP (2006) Finding fathers: spatio-temporal analysis of paternity assignment in gray seals (*Halichoerus grypus*). Mol Ecol 15:1939–1953

Twiss SD, Cairns C, Culloch RM, Richards SA, Pomeroy PP (2012) Variation in female gray seal (*Halichoerus grypus*) reproductive performance correlates to proactive-reactive behavioural types. PLos One 7:e49598

Twiss SD (1991) Behavioural and energetic determinants of individual mating success in male grey seals (*Halichoerus grypus*, Fabricus 1791. PhD dissertation, University of Glasgow

Van Aarde RJ (1980) Harem structure of the southern elephant seal *Mirounga leonina* at Kerguelen Island. Rev D'écol 34:31–44

Van Parijs SM, Clark CW (2006) Long-term mating tactics in an aquatic-mating pinniped, the bearded seal, *Erignathus barbatus*. Anim Behav 72:1269–1277

Van Opzeeland I, Van Parijs S, Bornemann H, Frickenhaus S, Kindermann L, Klinck H, Plötz J, Boebel O (2010) Acoustic ecology of Antarctic pinnipeds. Mar Ecol Prog Ser 414:267–291

Van Opzeeland ICV, Parijs SMV, Frickenhaus S, Kreiss CM, Boebel O (2012) Individual variation in pup vocalizations and absence of behavioral signs of maternal vocal recognition in Weddell seals (*Leptonychotes weddellii*). Mar Mammal Sci 28:158–172

Van Parijs SM, Thompson PM, Tollit DJ, Mackay ANN (1997) Distribution and activity of male harbor seals during the mating season. Anim Behav 54:35–43

Van Parijs SM, Lydersen C, Kovacs KM (2003a) Vocalizations and movements suggest alternative mating tactics in male bearded seals. Anim Behav 65:273–283

Van Parijs SM, Corkeron PJ, Harvey J, Hayes SA, Mellinger DK, Rouget PA, Thompson PM, Kovacs WKM (2003b) Patterns in the vocalizations of male harbor seals. J Acoust Soc Am 113:3403–3410

Van Parijs SM, Lydersen C, Kovacs KM (2004) Effects of ice cover on the behavioural patterns of aquatic-mating male bearded seals. Anim Behav 68:89–96

van Polanen Petel TD, Giese MA, Wotherspoon S, Hindell MA (2007) The behavioural response of lactating Weddell seals (*Leptonychotes weddellii*) to over-snow vehicles: a case study. Can J Zool 85:488–496

Walker BG, Bowen WD (1993) Behavioural differences among adult male harbor seals during the breeding season may provide evidence of reproductive strategies. Can J Zool 71:1585–1591

Weitzman J, Den Heyer C, Bowen DW (2017) Factors influencing and consequences of breeding dispersal and habitat choice in female grey seals (*Halichoerus grypus*) on Sable Island, Nova Scotia. Oecologia 183:367–378

Wheatley KE, Bradshaw CJ, Davis LS, Harcourt RG, Hindell MA (2006) Influence of maternal mass and condition on energy transfer in Weddell seals. J Anim Ecol 75:724–733

Wilkinson IS, Childerhouse SJ, Duignan PJ, Gulland FMD (2000) Infanticide and cannibalism in the New Zealand sea lion, *Phocarctos hookeri*. Mar Mammal Sci 16:494–500

Wilson SC (1975) Attempted mating between a male gray seal and female harbor seals. J Mamm 56:531–534

Wilson SC, Trukhanova I, Dmitrieva L, Dolgova E, Crawford I, Baimukanov M, Baimukanov T, Ismagambetov B, Pazylbekov M, Jüssi M, Goodman SJ (2017) Assessment of impacts and potential mitigation for icebreaking vessels transiting pupping areas of an ice-breeding seal. Biol Conserv 214:213–222

Wilson SC, Dolgova E, Trukhanova I, Dmitrieva L, Crawford I, Baimukanov M, Goodman SJ (2017) Breeding behavior and pup development of the Caspian seal, *Pusa caspica*. J Mammal 98:143–153

Wilson SC, Jones KA (2018) Behaviour of harbour seal (*Phoca vitulina vitulina*) mother-pup pairs in Irish Sea intertidal habitats. Biol Environ P R Irish Acad B 118:13–27

Wolff JO, Macdonald DW (2004) Promiscuous females protect their offspring. Trends Ecol Evol 19:127–134

Worthington Wilmer J, Allen PJ, Pomeroy PP, Twiss SD, Amos W (1999) Where have all the fathers gone? An extensive microsatellite analysis of paternity in the gray seal (*Halichoerus grypus*). Mol Ecol 8:1417–1429

Worthy GA, Lavigne DM (1983) Energetics of fasting and subsequent growth in weaned harp seal pups, *Phoca groenlandica*. Can J Zool 61:447–456

Chapter 8
Reproductive Energetics of Phocids

Daniel P. Costa and Jennifer L. Maresh

Contents

Abstract Phocid seals have a highly derived reproductive strategy characterized by the separation of feeding and reproduction, oftentimes across considerable spatial and temporal scales. Seals spend an extended period of time foraging far from the breeding colony, building up their energy reserves for a relatively brief pupping season onshore where access to food is limited. During lactation, enormous amounts of energy are quickly transferred from mother to pup before an abrupt weaning. This unusual reproductive strategy enables phocids to utilize habitats unavailable to most other mammals that must feed during lactation, from the unproductive waters of the tropics to the unstable pack ice of the world's polar seas. In this chapter, we examine: the physiological, ecological, and behavioral adaptations behind the phocid breeding system; the ways in which resources are acquired and allocated to understand the importance of body size, lactation interval, milk energy content, and lipid stores in support of phocid reproduction; the connections between phocid mating systems and pupping substrate; and, finally, the origins and evolutionary implications of this extraordinary life history pattern.

D. P. Costa (✉)
Institute of Marine Science, Department of Ecology & Evolutionary Biology, University of California Santa Cruz, Santa Cruz, CA 95060, USA
e-mail: costa@ucsc.edu

J. L. Maresh
Department of Biology, West Chester University, 730 S Church St, West Chester, PA 19383, USA

D. P. Costa and E. McHuron (eds.), *Ethology and Behavioral Ecology of Phocids*, Ethology and Behavioral Ecology of Marine Mammals,
https://doi.org/10.1007/978-3-030-88923-4_8

Keyword Metabolic overhead · Evolution · Body size · Parental investment · Capital · Breeding · Lactation

8.1 Introduction

The initial utilization of the marine environment by ancestral pinnipeds occurred at a time when coastal upwelling was at a cyclic high, resulting in an abundant, diverse, and essentially untapped food resource (Lipps and Mitchell 1976; Vermeij and Dudley 2000). While they were able to take advantage of the abundant marine resources, their terrestrial ancestry dictated they remain tied to shore for pupping. Phocid pups are born onshore or on ice and, for most species, largely remain there until they are weaned. In some species, the pups make short forays into the waters adjacent to the breeding colony, prior to weaning. As in most mammals, lactation limits females to be the sole caregiver, thus, lactating females must balance the time spent on shore with their pup with the need to acquire prey energy at sea. As male reproductive behavior is associated with optimizing the number of females inseminated, spatial and temporal separation of feeding from reproduction is necessary for both males and females (Bartholomew 1970). The constraints associated with balancing these conflicting energetic demands play a major role in determining the reproductive behavior of phocids (Costa 1993; Boyd 1998; Trillmich 2010; Stephens et al. 2014). Phocids are an ideal study system because reproduction occurs over a very defined interval and, for many phocids, all or most of the resources needed for maternal investment are contained within the female when she gives birth to her pup. As a result, the energy and material transferred between mother and pup are readily measured, making phocid reproduction one of the best-studied systems of all large vertebrates.

In the sections that follow, we consider the connection between habitat diversity and the diversity of phocid lactation patterns; the fitness advantages and energetic challenges of the capital breeding system; the origins and evolutionary implications of this reproductive strategy; and future directions for research into phocid reproductive energetics.

8.2 The Capital–Income Continuum

Phocids have evolved the highly derived "capital breeding" system, a reproductive strategy that is characterized by the complete separation of feeding and lactation (Costa 1991a, b; Schulz and Bowen 2005). For those species that are "pure" capital breeders, females conceive young during the previous reproductive season, and then spend an extended period of time foraging at sea for their own metabolic needs and those of the developing fetus, as well as laying down reserves (capital) for the short, but energetically intense lactation period of 4–35 days that follow parturition

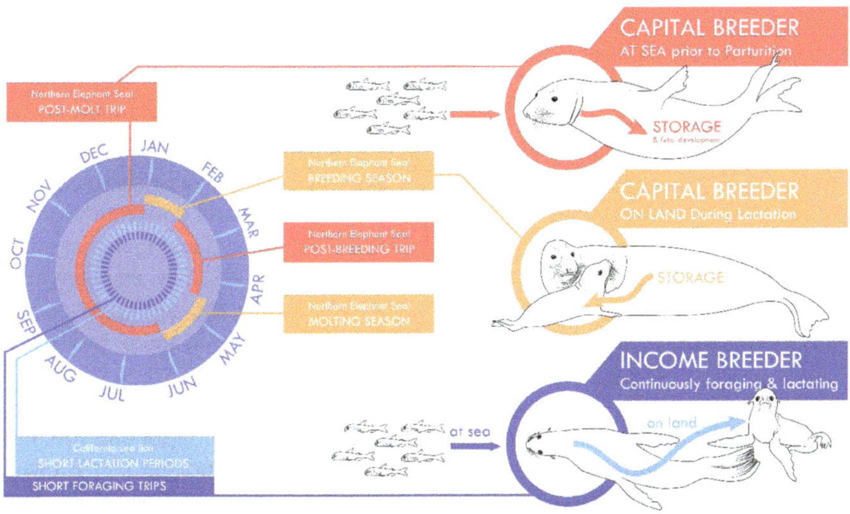

Fig. 8.1 The annual cycle of a capital breeder is compared to that of an income breeder. An income breeder feeds throughout the year and supports the growth of the pup from prey resources captured during intermittent foraging trips. Capital breeders spend extended periods at sea away from the breeding colony acquiring resources and then return to the colony to pup. During this period, the mother fasts and produces milk from stored resources acquired over the long foraging trip. Figure by Jessica Kendall-Barr

(Fig. 8.1; Schulz and Bowen 2004). For example, female northern elephant seals (*Mirounga angustirostris*) undertake one long foraging trip to sea, traveling over 3000 km from their breeding colonies to their feeding grounds. After 8 months of foraging at sea, they return to shore to give birth, where they are able to support a ~1-month-long lactation period without the need to return to sea to forage (Robinson et al. 2012). In this regard, female phocids are truly remarkable; lactation is so energetically demanding that, with the exception of bears and some marine mammals, no other mammal is capable of producing milk without feeding (Oftedal 2000).

The nutritional provisioning of young under the capital breeding system uncouples the need to travel to and from the foraging grounds during lactation, which allows individuals to utilize prey that is more dispersed, patchy, unpredictable, or distant from the breeding grounds (Costa 1993; Stephens et al. 2014). This uncoupling buffers populations against short-term fluctuations in habitat quality and resource availability. Not all phocids are "true" capital breeders and many species fall along a continuum between pure capital and income breeding. While elephant (*Mirounga* spp.), gray (*Halichoerus grypus*), Hawaiian monk (*Neomonachus schauinslandi*), and hooded (*Cystophora cristata*) seals are true capital breeders, species like Weddell (*Leptonychotes weddelli*), Mediterranean monk (*Monachus monachus*), harp (*Pagophilus groenlandicus*), ringed (*Pusa hispida*), and harbor (*Phoca vitulina*) seals include some degree of supplemental feeding during lactation, therefore relying on a certain amount of "income" during lactation (Hammill et al. 1991; Boness et al.

1994; Gazo and Aguilar 2005; Wheatley et al. 2006; Chaps. 9–10, , 12–14 and 16). As these species are constrained to remain nearby the nursing grounds, they cannot access all the advantages of capital breeding.

While some phocids exhibit supplemental feeding, the energy expended during lactation is primarily supplied by stored body reserves for all phocids except Mediterranean monk seals and possibly ringed seals. Body reserves are necessary as phocid mothers could not acquire sufficient energy foraging during the short lactation period to provide all of the energy necessary for both milk production and maternal maintenance metabolism (Schulz and Bowen 2005; Wheatley et al. 2008; Stephens et al. 2014). Supplemental foraging might be a way of meeting the additional cost of maternal metabolism associated with an extended lactation period and/or small body size.

The capital and/or supplemented capital breeding system made it possible for phocid seals to inhabit the less productive waters of the Caribbean and Mediterranean Seas, as well as the Hawaiian Islands, and to reproduce on ice, allowing them to colonize both polar oceans. The ability to remain with the pup throughout lactation influences a variety of behaviors discussed elsewhere, such as mother–pup interactions. For example, there is less selective pressure for females to be able to recognize their pups (Chap. 3), which may lead to the alloparental care reported in Hawaiian monk seals (Boness et al. 1998), and higher pup mortality when pups become separated from their mothers, as reported for elephant seals (Briggs and Le Boeuf 1975).

In contrast to the comparatively rare capital breeding strategy, the necessity of feeding during lactation and acquiring most if not all the energy required for lactation during this time characterizes most mammals and all other pinnipeds as "income breeders." Under this strategy, sustained milk production for young is supported by many short foraging bouts across a relatively long lactation period, constraining mothers to forage on resources close to their breeding colonies (Costa and Valenzuela-Toro 2021). The income breeding strategy thus couples reproductive success with local prey abundance (Costa 1993; Boyd 1998), creating a stronger link between local environmental fluctuations and population status (e.g., El Niño events; Trillmich et al. 1991; Costa 2008). As a result, income breeders tend to be limited to breeding in highly productive regions (Costa 1993; Costa and Valenzuela-Toro 2021). Otariids exemplify this income-breeding strategy, with lactation durations that can last upward of 2 + years (Trillmich 1990). For example, female California sea lions (*Zalophus californianus*) can make 50 trips to sea during an 11-month lactation period, typically spending 3–6 days foraging within 50–100 km of their breeding colonies (Fig. 8.2; Kuhn and Costa 2014; Costa and Valenzuela-Toro 2021).

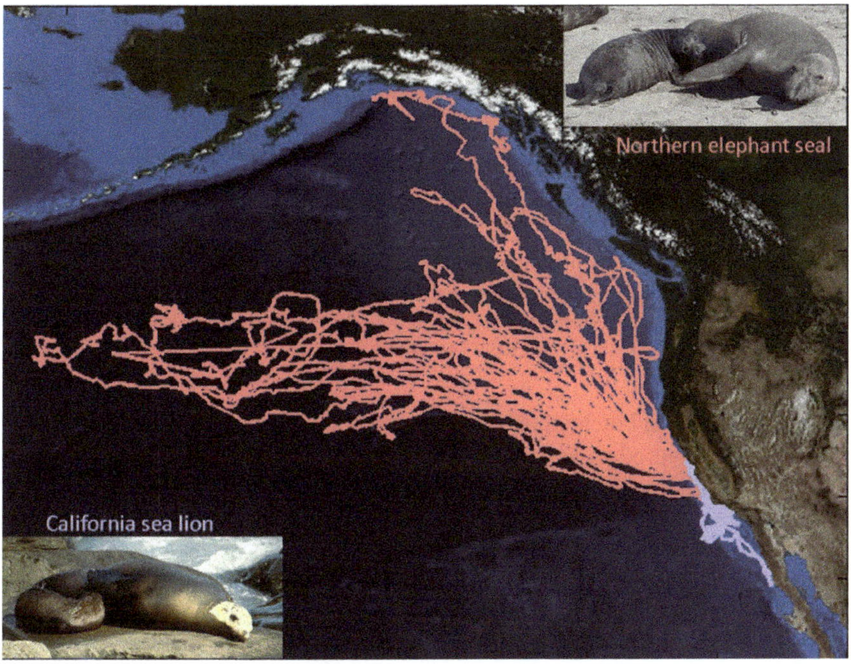

Fig. 8.2 Tracks of female northern elephant seals on their post-molt foraging trips traveling over 3000 km from the Año Nuevo colony (pink) compared to foraging trips of female California sea lions (purple) that are typically 85 km away from the San Nicolas Island colony

8.3 Habitat and Breeding Systems: Environmental Potential for Polygyny

Phocids can access prey resources in most habitat types found in the marine environment, including open ocean (e.g., elephant seals), coastal inshore habitats (e.g., monk and harbor seals), freshwater (harbor, Caspian *Pusa capsica*, and Baikal seals *Pusa sibirica*) and pack and fast ice habitats (e.g., hooded, harp, ringed, Weddell, and crabeater seals *Lobodon carcinophagus*), (Costa 1993; Schulz and Bowen 2004, 2005). In addition, the ability to rely on body stores to offset or entirely support the cost of lactation allows phocids to utilize a diversity of substrates for reproduction, from relatively fixed landmasses to unstable pack ice, giving rise to a variety of lactation strategies (see also Chap. 7).

Phocid seals exhibit the complete range of breeding systems observed within pinnipeds, ranging from the highly polygynous system of elephant seals, which exhibit some of the greatest sexual dimorphism observed in mammals (Fig. 8.3; Chap. 11), to serial monogamy with little or no dimorphism (many ice seals), to the polygynous reverse sexual dimorphism observed in leopard (*Hydrurga leptonyx*) and Weddell seals (Chap. 13). These different breeding systems follow the predictions

Fig. 8.3 The top image shows the extreme sexual dimorphism of a male seal copulating with a female northern elephant seal at the Año Nuevo Colony. Males weigh 1360–2270 kg compared to 400–770 kg for females. The bottom image shows two crabeater seals on an ice-floe. It is hard to tell the difference between a male and a female crabeater seal. Photo credits: D. Costa

of the Environmental Potential for Polygyny (EPP), where different habitat types are coupled with the reproductive behaviors best suited for the challenges and advantages of that environment (Emlen and Oring 1977; Trillmich 2010). Specifically, EPP predicts that breeding systems reflect the ability of males to control access to, and mate with, estrous females, which, in turn, is determined by the degree to which estrous females aggregate in space and time around critical resources. Phocids use land or ice for pupping, and the different characteristics of these breeding substrates—including abiotic factors such as substrate stability and access to the water, as well as biotic factors such as prey availability and exposure to predators—influences the degree to which females aggregate (Fig. 8.4; Le Boeuf 1991; Chap. 7).

8.3.1 Land-Breeding Seals

Islands and mainland beaches generally offer stable and relatively safe breeding grounds for phocids compared with some other habitat types; however, the availability of sites favorable for pupping can be limited (Fig. 8.4). Also, land-breeding females often have species-specific or even individualistic preferences for habitat characteristics, resulting in a tendency to aggregate in the most prime locations (Reiter et al. 1981; Cameron et al. 2007; Hadley et al. 2008). Together with estrous synchrony, these predictable aggregations allow males to mate with many females as well as defend them from rival males (Bartholomew 1970). Thus, the EPP is highest for species that breed on land, and we observe a breeding system that reflects this.

Land-breeding is relatively uncommon in phocids but includes both species of elephant seals and the gray seal—all excellent examples of high EPP for land-breeding phocids (Deutsch et al. 1990; Chaps. 9 and 11). The remaining species of land-breeding seals, which includes harbor seals and both species of monk seals, instead exhibit a slightly polygynous or serially monogamous breeding system. For these species, pupping occurs on land, but copulation occurs in the water (Johanos et al. 1994), so males have a more difficult time controlling access to estrous females.

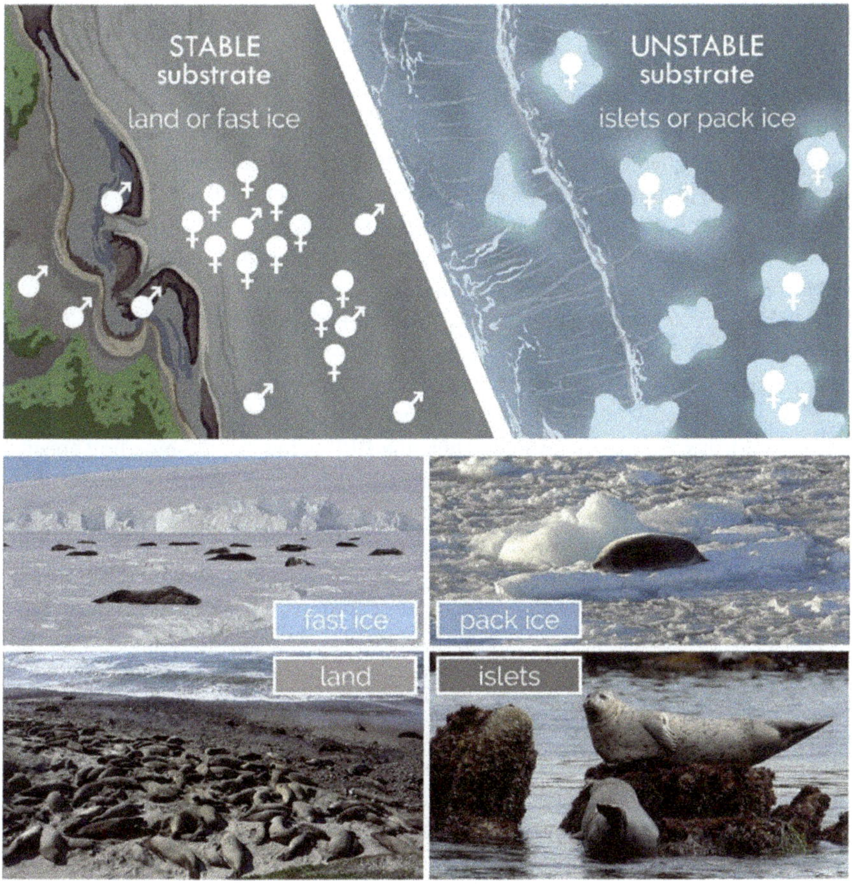

Fig. 8.4 Top left, a hypothetical example of a land-based breeding site where females form aggregations or "harems" in the center of the colony. A single male typically guards each harem. Subadult and subordinate males aggregate around the periphery of the harem. Top right, a hypothetical example of a pack ice breeding situation where the disbursed breeding substrate causes females to be disbursed. In this situation, a male can only guard one female. The bottom four figures are examples of fast ice breeding Weddell seals (upper left), a pack ice breeding crabeater seal (upper right), land breeding northern elephant seals (lower left), and an islet breeding harbor seal (lower right). Photo credits: D. Costa. Figure by Jessica Kendall-Bar

This results in a low EPP (Hayes et al. 2004), with reduced or even reversed sexual size dimorphism (females are the same size or larger).

8.3.2 Ice-Breeding Seals

Most phocids pup on ice, which is less stable and predictable than land, although the degree to which that is the case varies substantially between different types of ice (Fig. 8.4). At one extreme is pack ice, which consists of many small pieces of drifting ice that are widely dispersed and somewhat ephemeral. Most phocids breed on pack ice, including crabeater, leopard, hooded, harp, spotted (*Phoca largha*), Ross (*Ommatophoca rossii*), ribbon (*Histriophoca fasciata*), and bearded (*Erignathus barbatus*) seals. The other form is fast ice, expansive sheets of ice relatively fixed to land. Fewer species breed on fast ice but include Weddell, Baikal, Caspian, and ringed seals. Both types of ice—pack and fast—reduce the tendency of females to aggregate, although for different reasons. In addition, copulation between ice-breeding seals usually takes place in the water, where the ability of males to control access to estrous females is limited. Thus, the EPP is low for ice-breeding species, and sexual dimorphism is, in most cases, reduced or absent; however, the details of each breeding system are dictated by the specific characteristics of the substrate.

In species that breed on pack ice, females are widely dispersed and are generally observed alone with their pup or in a triad (a male and female with her pup). However, some species form large, loose breeding aggregations (Lydersen and Kovacs 1999). As pack ice is unstable and may disappear, females give birth during the short time when the pack ice is most stable and have a very short lactation interval to reduce the threat to newborn offspring. For example, hooded seals give birth and wean their pups in 4 days, while harp seals wean their pups in 12 days (Lydersen and Kovacs 1999). Upon weaning, females come into estrus synchronously, but briefly, which, together with the degree of spatial separation, limits the ability of males to mate with multiple females. However, serial monogamy may be possible where a male guards a female until she comes into estrus, mates with her, and then deserts her as he attempts to find unescorted, pre-estrous females (Bowen et al. 1985; Boness et al. 1988; Kovacs 1990). The difficulty of observing pack ice seals has led to far fewer observations of their breeding and copulatory behaviors when compared with phocids that breed on land. As a result, the degree of polygyny is challenging to assess. However, the consistent trend is for males to be slightly larger than females, suggesting slight polygyny or serial monogamy (Le Boeuf 1991).

Seals that breed on more stable fast ice are generally found in well-spaced colonies where females aggregate in small-to-moderate-sized groups with their pups (Fig. 8.4; Siniff et al. 1977; Bartsh et al. 1992). The limited availability of reliable cracks and holes in the fast ice that provides females access to the water facilitates the formation of these reproductive colonies. Individual males can establish an underwater territory and "resource guard" a crack or hole against rival males, thereby increasing the EPP compared with pack ice-breeding species. However, reproductive synchrony, in combination with these loose, well-spaced aggregations, makes it more difficult for males to monopolize either females or reproductive habitat when compared to land-breeding seals. This is generally the situation in the Antarctic; in the Arctic, predation pressure from polar bears further limits the aggregation of females on the fast ice, bringing the EPP closer to the pack ice end of the scale. Regardless of hemisphere, the predominant mating strategy for fast ice-breeding seals is serial monogamy or moderate polygyny.

8.4 Energy Acquisition and Expenditure Relative to Reproductive Patterns

An examination of how energy flows through an organism provides a mechanism to quantitatively assess the effort an animal spends acquiring resources and how those resources are allocated. Energy flow can be described as what goes into the animal as food and what comes out in the form of growth, reproduction, repair, mechanical work, and waste (Fig. 8.5). In the steady state, animals maintain constant mass, and the energy consumed is equal to the energy expended. In this situation, ingested energy remaining after fecal and urinary losses—metabolic energy (ME)—is expended in support of homeostatic maintenance of body functions (basal metabolism, thermoregulation) and in the further acquisition and processing of prey (mechanical work for

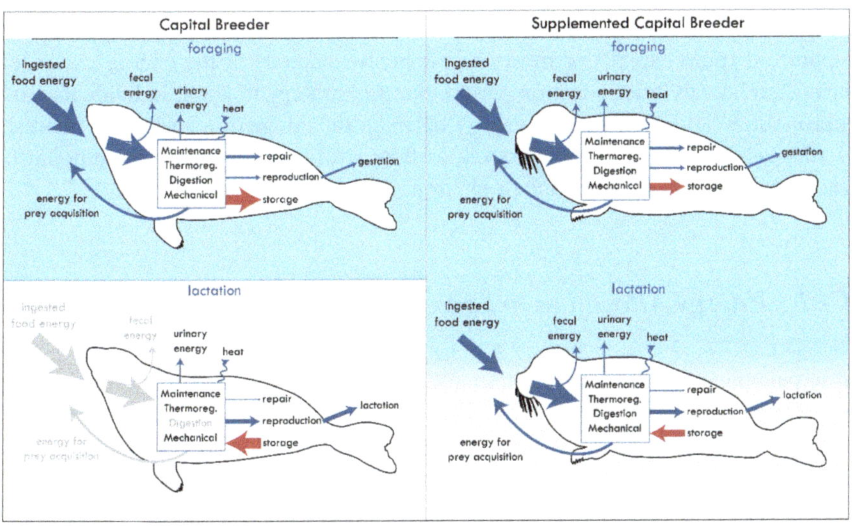

Fig. 8.5 A comparison of energy flow in reproductive female phocids, showing the differences between capital breeding and supplemented capital strategies. On the left, a capital-breeding phocid seal alternates between (top panel) positive energy balance while foraging at sea, and (bottom panel) negative energy balance while fasting onshore during the breeding season. While foraging, surplus energy is used in support of fetal gestation and in the buildup of maternal fat stores (capital; red arrows) that are later used in support of milk synthesis and metabolic overhead during a brief but energetically intense lactation period. On the right is a phocid mother who supplements maternal fat stores (capital; red arrows) with some degree of feeding (income) during the breeding season to sustain milk production across a relatively long lactation period

foraging, digestion of tissue). When more energy is consumed than is expended as ME, the surplus is used for growth, storage, or reproduction; this surplus is stored as an increase in body tissue, and the animal gains mass (positive energy balance). When the energy consumed does not meet ME, an animal will utilize the energy stored as body tissue and lose mass (negative energy balance). There is thus an ongoing tradeoff between energy in and energy out, and whether there is excess energy that can be used for storage, growth, or reproduction. Therefore, growth and reproduction require an animal to acquire more energy and nutrients than are necessary to support maintenance.

The rate of prey energy acquisition is directly related to the availability and quality of prey, and how it is distributed in time and space. As prey availability and quality decrease, the cost of finding prey increases, and a greater proportion of time (and therefore energy) is expended searching for prey (Costa 2008; Benoit-Bird et al. 2013a, b). Capital and supplemental capital breeders undergo periodic fluctuations in this dynamic equilibrium. They gain significant amounts of energy while feeding in highly productive environments over short periods, or while feeding on patchy or more dispersed prey over more extended periods, followed by prolonged negative energy balance while fasting during migration or reproduction (Costa 1993; Trillmich and Weissing 2006; Stephens et al. 2014).

As capital or supplemented capital breeders, phocids draw from stored energy reserves (i.e., metabolize fat) accumulated during an extensive foraging season to support all (pure capital) or most (supplemented capital) of the energy expended during a relatively brief lactation period. For this strategy to work, individuals must appropriately allocate acquired energy between the competing biological interests of survival, growth, and reproduction, as they are limited by the finite amount of resources available to them in their environment.

8.4.1 Energy Allocation to Growth and Reproduction

In female phocids, as in other mammals, reproductive costs are associated with an investment in offspring growth through gestation and lactation. Gestation costs come in the form of the energy contained in fetal tissue, as well as the energy to fuel the metabolic processes associated with gestation—a.k.a., the "heat of gestation" (Brody 1945). Lactation costs come in the form of the energy and nutrients contained in milk, as well as the energy to fuel milk synthesis. Maternal energy reserves must also be used in support of the mother's maintenance metabolism during lactation—a.k.a., maternal "metabolic overhead" (Fedak and Anderson 1982). While there are no measurements on the heat of gestation in marine mammals, maternal metabolic overhead is relatively well studied, and we devote an entire section to discussing it below (see sect. 8.3.3). As metabolic overhead is considered a cost separate from energy spent on lactation per se, here we focus our discussion of lactation costs associated with the energy and nutrients contained in milk.

8.4.2 Milk Production and Lactation Duration

The relative investment of phocid mothers in their young via gestation and lactation can be appreciated by comparing pup mass at birth and at weaning to that of other carnivores. After a relatively long gestation period, phocid pups are born larger as a proportion of maternal mass (3.5–16.5%) than other carnivores (0.5–3%), indicating that phocid mothers invest more time and energy into fetal development (Figs. 8.6 and 8.7). Despite similar gestation durations, phocid mothers devote much less time toward offspring growth post-parturition than other carnivores (Fig. 8.7). For their body size, phocids have the shortest lactation period of any mammal, ranging from 4 days (hooded seals) to about 6–7 weeks (Weddell seals) before the pup is abruptly weaned (Bowen et al. 1985; Gazo and Aguilar 2005). Interestingly, Mediterranean monk seals are exceptional in that they wean their pups gradually over 4–5 months, more like an otariid than other phocids (Aguilar et al. 2007). Over the short lactation period, mothers transfer considerable amounts of energy to their pups, losing up to one-third of their postpartum mass to support accelerated pup growth (Crocker et al. 2001; Lydersen and Kovacs 1996; Mellish et al. 2000; Wheatley et al. 2006). For example, during their brief 4-day lactation period, hooded seal mothers transfer 10 kg of milk per day to their pups, which in turn, gain 7 kg per day in mass (Iverson et al. 1995). As a result of rapid energy transfer and an accelerated pup growth rate, phocid mothers can wean their young at a size comparable to that of other carnivores despite the short lactation interval (Fig. 8.7).

Lipid-rich milk facilitates the rapid delivery of energy to the pup, allowing species with the shortest lactation intervals to have the highest rates of milk energy delivery (Fig. 8.8). The fat content of phocid milk ranges from 24.5% in harbor seals to 65% in hooded seals (Oftedal et al. 1988), whereas most terrestrial mammals produce milk relatively low in fat. For example, cows and humans produce milk containing 4 and 8% fat, respectively. There are disadvantages to the high-fat content of phocid milk, as it is achieved by a reciprocal decrease in water content, with no corresponding change in protein content. Therefore, the protein-to-energy ratio of pinniped milk is lowest in the most energy-dense milk, and, while pups may receive enough energy to support their metabolism over the postweaning period, relatively little protein is provided for the growth of the lean body. As a result, by the time of weaning, the protein content of hooded, harbor, and gray seal pups has increased by 3, 10, and 11%, respectively, compared to an increase in fat stores of 82, 52, and 58% (Oftedal et al. 1993). Thus, species with extremely short lactation intervals have the least amount of lean tissue growth. For many species, such as elephant seals and Weddell seals, once pups are weaned, they undergo a postweaning fast that allows them to further develop their physiological capabilities, including their osmoregulatory and diving abilities (Adams and Costa 1993; Thorson and Le Boeuf 1994; Burns and Castellini 1996; Burns et al. 1997; Burns 1999). In addition to needing to develop their oxygen stores, the high proportion of adipose tissue would make pups relatively buoyant, further decreasing their diving ability. Finally, it is unclear how suckling pups maintain water balance given the low water content of their milk (Fig. 8.9).

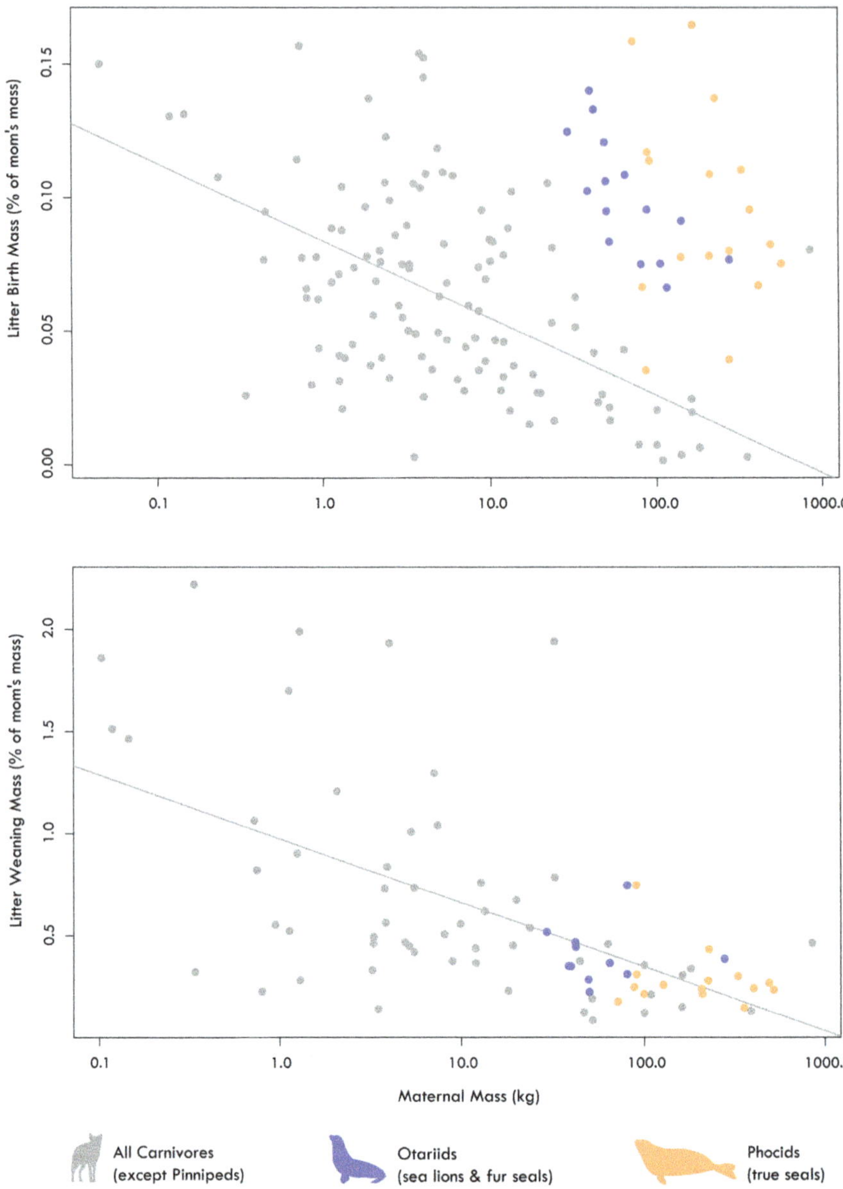

Fig. 8.6 The top panel shows the litter mass of Carnivora pups at birth compared to maternal mass. The bottom panel shows the litter mass at weaning for carnivores relative to maternal mass. Litter mass is the combined mass of all pups born in one parturition. Phocid (and otariid) females invest more in their pups at birth, but weaning occurs at similar weights to other carnivores

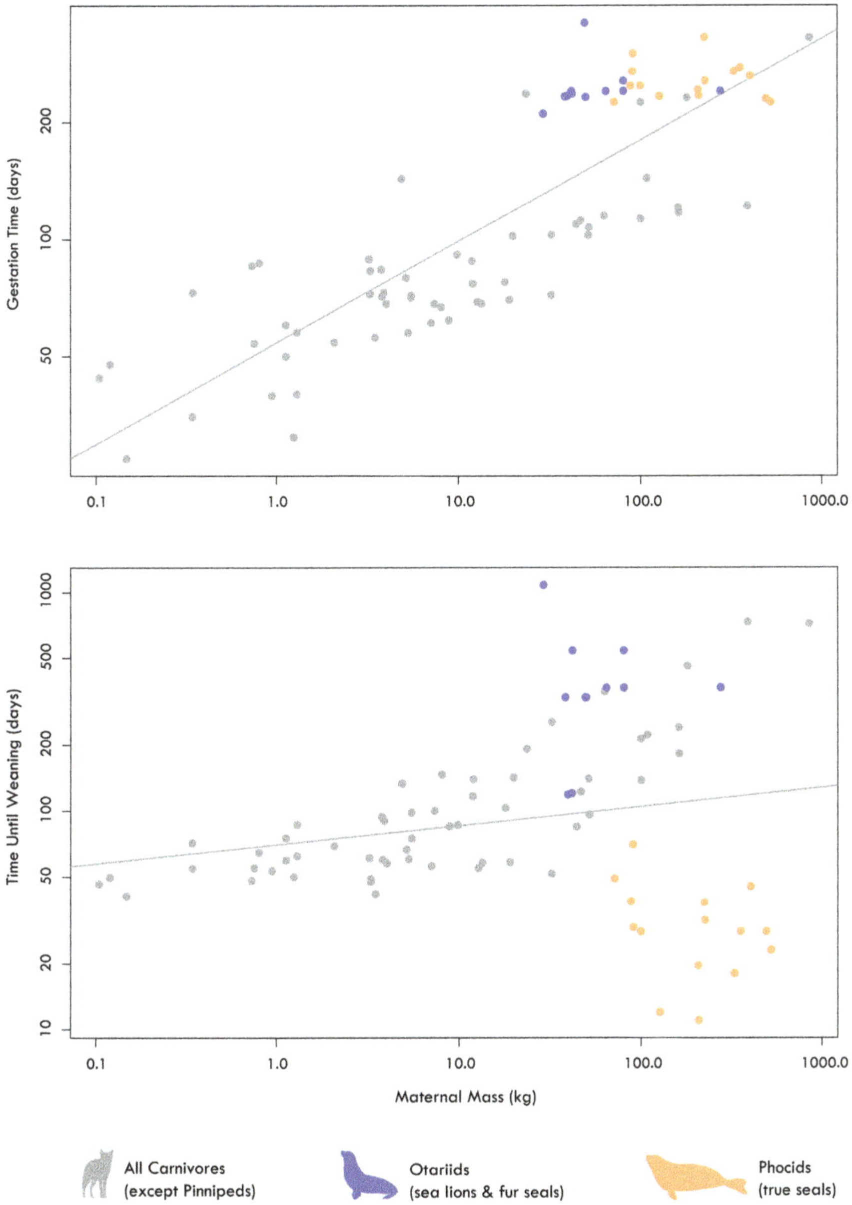

Fig. 8.7 The top panel shows the duration of gestation, and the bottom panel shows the lactation duration relative to maternal mass for carnivores. Compared to other carnivore species, phocid mothers have a longer gestation duration (top), but a much shorter lactation duration (bottom)

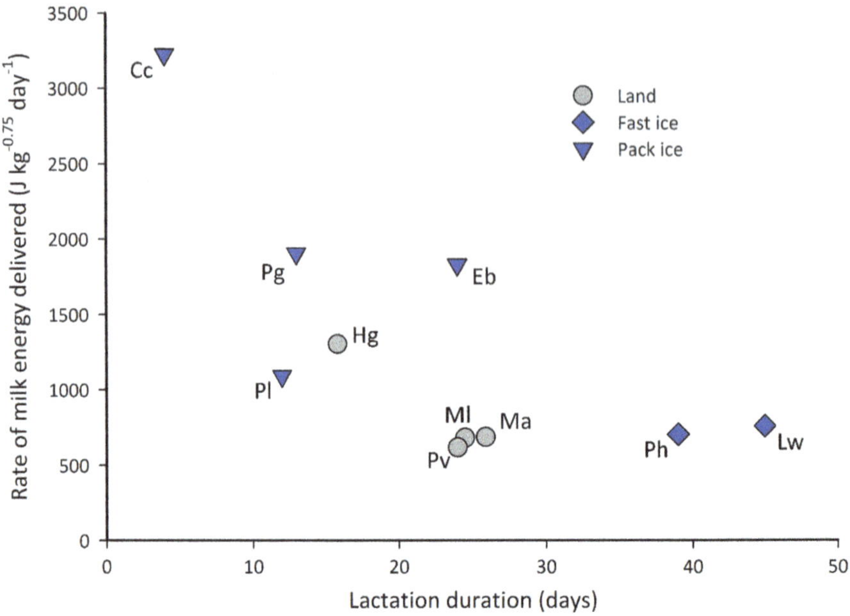

Fig. 8.8 The rate of milk energy provided to the pup compared to the duration of lactation. The rate of energy delivery normalized for differences in female body mass by dividing the absolute amount of energy in J delivered per day by female body mass raised to 0.75. The type of breeding substrate is also delineated for each species and are Ma = *Mirounga angustirostris*, Ml = *Mirounga leonna*, Pv = *Phoca vitulina*, Ph = *Pusa hispida*, Pg = *Pagophilus groenlandicus*, Hg = *Halichoerus grypus*, Pl = *Phoca largha*, Eb = *Erignathus babatus*, Cc = *Cystophora cristata*, Lw = *Leptonychotes weddelli*

Given the different lactation patterns exhibited in phocids, we might expect the total energy gained by pups to be quite variable; in fact, the energy gained by phocid pups is quite similar. What is different is how that energy is provided. For example, hooded seal pups are born with almost twice as much fat as other seal pups (14%) compared to harbor (10%), elephant (9%), gray (6%), and harp (3%) seal pups. They receive slightly less energy during lactation (13 MJ/kg$^{0.75}$ compared to 15 MJ/kg$^{0.75}$ for gray and harbor seal pups), but are still weaned with a fat content of 44%, which is similar to that of elephant (48%), harp (47%), gray (40%), and harbor (37%) seal pups (Kretzmann et al. 1993; Oftedal et al. 1993, 1996).

8.4.3 Body Size and Maternal Resources: The Role of Maternal Overhead

The ability of a female to fast while providing milk to her offspring is related to the size of her energy and nutrient reserves, and the rate at which she utilizes them.

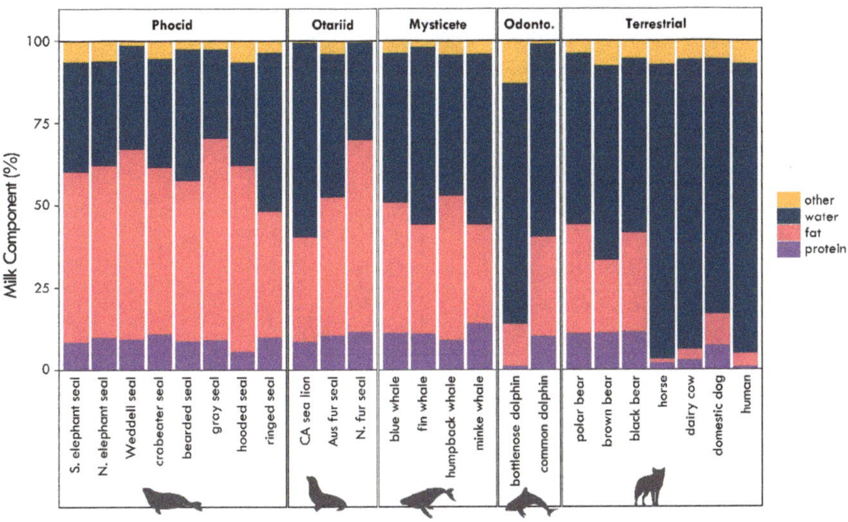

Fig. 8.9 A comparison of milk content across mammalian groups. Fat content is highest in capitally-breeding phocids, mysticetes, and bears. The high-fat content of phocid milk allows for the transfer of considerable amounts of energy to pups during a brief lactation interval. Note: polar bears (*Ursus maritimus*) are considered marine mammals but are shown in comparison to other bears

Reserves come in the form of fat stores, and female seals average 34% fat at or near parturition, with a range from 24.3% in harbor seals to 40.2% in Weddell seals (Bowen et al. 1992; Wheatley et al. 2006). This compares to a fat content ranging from a low of 9% to a high of 28% in terrestrial carnivores (Pond and Ramsay 1992). Fedak and Anderson (1982) coined the term "metabolic overhead" to define the amount of energy a female seal expends on herself while on land nursing her pup (Fig. 8.10). Metabolic overhead is relatively lower in larger animals because maintenance metabolism (energy demand) scales as mass$^{0.75}$, while fat stores (energy supply) scale as mass$^{1.0}$ (Kleiber 1975). Thus, large females can provision their young from stored body reserves for longer than small females. For example, the larger body size of older, more experienced northern and southern elephant seal mothers is associated with lower maternal metabolic overhead (see below), which allows them to provide more milk to their pups (Fedak et al. 1996; Crocker et al. 2001). Similarly, younger, smaller Weddell seals that have a higher proportional metabolic overhead spend more time foraging away from their pups than older, larger females (Wheatley et al. 2008).

The relationship between metabolic overhead, milk production, and lactation duration in a capital breeder was examined using a data-driven empirical model based on elephant seals (Costa 1993). This work was followed by mathematical optimization models that compared a fur seal to an elephant seal (Boyd 1998), an extension to a range of body sizes (Trillmich and Weissing 2006), and an examination of the role of food supply (Stephens et al. 2014). Together, these models show that a phocid mother can reduce her metabolic overhead by reducing the duration

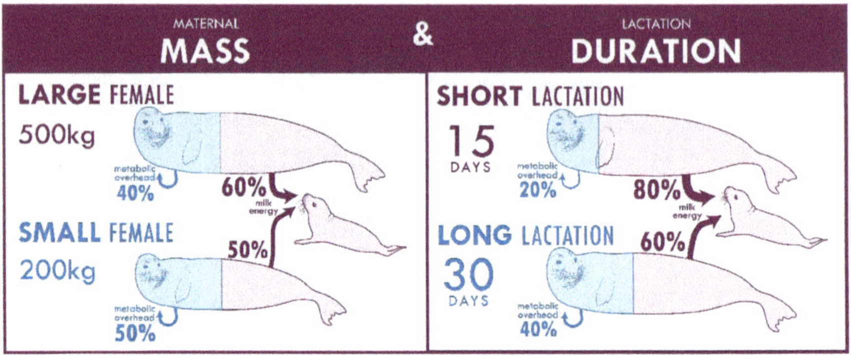

Fig. 8.10 Illustration of the importance of body size and lactation duration on the ability of the mother to invest her stored resources as milk energy to the pup. Larger females, and females with a shorter lactation period, have lower metabolic overhead. They are, therefore, able to provide a higher proportion of their stored energy to the pup as milk. Figure by Jessica Kendall-Bar

of lactation and/or by increasing her body size, thereby increasing the proportion of stored energy she provides her pup (Fig. 8.10). Thus, shorter lactation intervals allow a phocid mother to devote more energy to milk production rather than maintenance and, given the same lactation interval, larger females should have a lower metabolic overhead than smaller females.

The interaction between body size and lactation duration and its effect on metabolic overhead can be seen by comparing species along a continuum of lactation intervals. For example, the extremely short 4-day lactation interval of hooded seals results in a very low metabolic overhead of 27% (Mellish et al. 1999b). In contrast, elephant seals, the largest seal, have the next lowest metabolic overhead of 40% (lactation duration ~26 days (Costa et al. 1986; Hindell and Slip 1997). Gray seals are intermediate in size and have a metabolic overhead of 45% across the 16-day lactation period (Mellish et al. 1999a). Weddell seals have the most prolonged lactation interval studied so far (with respect to metabolism). They have a metabolic overhead of 54.9%, at least during the first 3 weeks of lactation when they are fasting (Wheatley et al. 2008).

As large body size confers an advantage in fasting ability, it follows that females that are ecologically constrained to being small must have short lactation intervals or must feed during lactation. For example, the need to avoid predators likely constrains ringed seals to be small so that they can hide from polar bears (Chap. 15). Similarly, harbor seals must be small enough to occupy small coastal islets and intertidal rocks to avoid predation by bears and wolves (Schulz and Bowen 2004). Both ringed and harbor seals feed during lactation, but as resources are nearby where they pup, they can do this. However, when ecological conditions favor large body size (or at least do not disfavor it), a longer lactation interval is possible while still fasting, if not adaptive. For example, as elephant seals breed on islands devoid of land predators, there is no pressure to be small and hide. This coupled with the ability to forage far from the colony at great depths, makes their large body size quite adaptive.

While there are no land predators in the Antarctic, killer whales (*Orcinus orca*) and leopard seals are effective marine predators (Siniff and Bengtson 1977; Guinet 1992; Pitman and Durban 2012). The potential impact of aquatic predators on the life history patterns of phocids requires further investigation. Nevertheless, the fast ice breeding colonies of Weddell seals are free from any form of predation, at least during the reproductive period until the fast ice breaks up in the late summer (Chap. 13). Large body size also improves diving ability and is discussed below (see Sect. 8.5).

8.5 Constraints on Body Size

Large body size is generally an advantage under the capital breeding system; however, there are factors outside of reproductive considerations that limit phocid body size, such as thermoregulation, locomotion, and patterns of prey availability. Thermoregulatory constraints put a lower and upper limit on body size due to the relationship between an animal's surface area to volume ratio and rates of heat transfer (Schulz and Bowen 2005). Smaller organisms lose heat faster than large organisms due to a higher surface area to volume ratio; therefore, given the high heat capacity of water, there is a limit to how small an endothermic, aquatic mammal can be (Clauset 2013). Conversely, as pinnipeds must periodically return to land, they face the opposite problem while onshore of needing to be able to offload heat, which is constrained by a large body with excellent insulation optimized for reducing heat loss while at sea (Bartholomew and Wilke 1956; Boness and Francis 1991; Norris et al. 2010; Codde et al. 2016). Thus, given the low heat capacity of air, there is a limit to how large a well-insulated mammal on land can be.

Locomotory considerations also constrain phocid body size. On land, the reproductive advantages of being large come at the expense of locomotor performance due to the morphological compromises associated with their amphibious lifestyle (Fish 2000). Biomechanical models show that the mechanical energy expended while traveling over land is disproportionately greater ("exorbitantly high") in the largest male elephant seals compared to that of smaller phocids such as gray and harbor seals (Garrett and Fish 2015; Tennett et al. 2018). Conversely, in the water, extreme size is generally associated with improved swimming economy and greater diving ability; however, the deposition of the fat stores necessary to sustain reproduction on land (i.e., fattening) can cause animals to deviate from the optimal, streamlined shape that minimizes drag in swimming animals (Fish 1994; Stephens et al. 2014). Fat stores also impact swimming costs via their effect on buoyancy (Stephens et al. 2014). For example, female elephant seals exert greater effort during diving when they deviate from neutral buoyancy as they increase or decrease their fat stores (Adachi et al. 2014).

Finally, patterns of prey availability also place limits on body size. While larger animals are more energetically efficient on a mass-specific basis (i.e., they require less prey for a given mass), they nevertheless require more prey as they are larger. This relationship has been examined in optimality models (Boyd 1998; Trillmich and Weissing 2006), which have shown that, for a given travel time, small animals could successfully subsist on a lower prey availability than large animals. In contrast, a more recent modeling effort found that increases in food availability, seasonality, and to a lesser extent, the unpredictability of prey, all favor the capital breeding system and, thus, large body size (Stephens et al. 2014). While these modeling efforts are quite insightful and have provided support for some of the factors related to food availability that lead to the evolution of a capital breeding system, they are limited in the processes they can examine (Trillmich and Weissing 2006). For example, the physiological processes that determine the rate at which prey and milk can be processed will constrain time at sea and lactation duration, respectively. Mothers can only deliver milk as fast as the pup can consume and process it, and the maximum rate at which a female can feed is limited by her ability to digest and process prey. Interestingly, compared to terrestrial carnivores, pinnipeds have longer intestines that allow them to process prey faster, making it possible for them to forage more frequently (Williams et al. 2001; Williams and Yeates 2004).

The effect of prey availability on body size has been well studied in ringed seals (Ferguson et al. 2018). This species generally exhibits a reproductive strategy with more income breeding traits than most phocids, having a smaller body size and, therefore, less capital available to transfer to their young. They also inhabit a large latitudinal range with highly varying climate and resource conditions; individuals at the northern end of the range experience greater variability in climatic conditions, which should favor larger (more efficient) body size to buffer against poor food availability and unpredictable timing of ice break-up (Ferguson et al. 2018). As predicted, a latitudinal gradient in body size has been reported for ringed seals, with an overall larger body size and reverse sexual dimorphism seen in more northern populations (Ferguson et al. 2018, 2019). In essence, ringed seals exhibit a within-species latitudinal gradient along with the capital–income spectrum whereby northern individuals exhibit stronger capital breeding traits than their more southerly counterparts (i.e., more frequent skipping of a reproductive cycle, slower growth rates, and larger size at maturity; Ferguson et al. 2019).

8.6 Foraging and Reproduction: An Economical Lifestyle

Optimization models have shown that capital breeding will emerge when the correct set of environmental parameters are in place (Boyd 1998; Trillmich and Weissing 2006; Stephens et al. 2014). These models assumed that the cost of existence for capital and income breeders is equivalent, but empirical data suggest otherwise. Income breeding otariids have higher maintenance and field metabolic rates than capital breeding phocids (Maresh et al. 2014; Costa and Maresh 2018). Costa (1991a) hypothesized that capital breeders would have lower metabolic costs (mostly associated with activity costs) because, with spatial and temporal separation of breeding from feeding, the mother is released from having to return to the pup in a timely manner. This system promotes energetically conservative foraging behaviors in capital breeders. For example, the ability to utilize highly dispersed or distant prey resources is improved by traveling to and from those resources economically (e.g., slow and deliberate flipper stroking, (Adachi et al. 2014; Maresh et al. 2014). The conservative lifestyle of phocids extends to the way that they quietly surface and descend while swimming (Costa 1991a; Williams and Maresh 2015), a pattern quite different than that observed in otariids. Economical travel reduces at-sea metabolic rate and overall energy expenditure, thereby increasing the amount of energy available for reproduction (Maresh et al. 2015).

Moreover, making a single long foraging trip increases the time available to forage and reduces the time spent in transit. Thus, by spreading the cost of reproduction (milk production and maternal metabolism) over a long trip to sea, a northern elephant seal female, for example, only has to increase her daily prey consumption by 12% (Costa et al. 1986) to gather sufficient energy to support herself and her pup on shore. This compares to an increase of 1.6 times (160%) the prey intake of a lactating compared to non-lactating female northern fur seal (*Callorhinus ursinus*; Perez and Mooney 1986).

Together with large body size, economical foraging, and a reduced at-sea metabolic rate increase diving ability by increasing a seal's aerobic dive limit (Ponganis 2016), allowing seals access to prey at greater depths (Chap. 4). As air breathers, foraging marine mammals are physiologically constrained during breath-hold by the amount of oxygen they have stored in their tissues compared to the rate at which this oxygen is utilized. Oxygen storage capacity increases with body size; therefore, just as metabolic overhead is constrained by body size, so is diving ability. Like the scaling of energy supply and demand described above, metabolic rate (oxygen demand) scales to body mass$^{0.75}$ whereas oxygen stores (supply) scale with body mass$^{1.0}$. More massive seals, therefore, have a lower mass-specific metabolism for a relatively constant proportion of oxygen storage capacity (Ponganis 2016). All things being equal, large mammals should be able to dive longer and deeper compared to smaller mammals based on body size alone.

Pure capital breeding phocids, such as elephant seals, have a reproductive pattern that is ideally suited for dealing with dispersed or unpredictable prey or prey that is located at great distances from the rookery or at great depth. Access to these resources is optimized by their large body size and elevated oxygen stores, coupled with the adoption of energetically conservative foraging and traveling behaviors that keep at-sea metabolic rates low. As a result, elephant seals are exceptional divers, comparing favorably with beaked and sperm whales as the deepest and longest diving mammals (Ponganis 2016).

8.7 Male Reproductive Energetics

In most mammals, males do not invest energy in parental care of offspring; instead, the cost of reproduction is associated with finding and maintaining access to estrous females and with inseminating as many females as possible (Deutsch et al. 1990;

Tinker et al. 1995; Kovacs et al. 1996; Coltman et al. 1997, 1998; Beck et al. 2003). For some phocids, this necessitates a substantial investment in growth, as large body size confers an advantage in antagonistic interactions (Anderson and Fedak 1985; Haley 1994). In addition, large body size is advantageous for territory maintenance as larger animals can fast longer (Bartholomew 1970). For these reasons, sexual size dimorphism is most pronounced in land-breeding seals where the EPP is highest (e.g., elephant and gray seals). While defending estrous females from rival males is not as costly as lactation, males can use their energy stores to sustain a longer fast. This is advantageous for both longer attendance of estrous females as well as for territory defense (Crocker et al. 2012). For example, adult male elephant seals haul out on breeding beaches weeks before the arrival of the first females to establish a dominance hierarchy with rival males and claim the best-pupping sites as territory (Haley et al. 1994). While the lactation interval for each elephant seal mother is only ~26 days, the arrival and departure dates of individual females are staggered across an almost 3-month-long breeding season. The dominant harem master males are larger, which allows them to expend more energy defending the harem as well as prolonging their harem tenure (Deutsch et al. 1990). During this period, they rely solely on their stored energy reserves to sustain them through the entire breeding season losing, 7.8 kg day^{-1} over 89 days, or 39% of their initial mass (Crocker et al. 2012). Interestingly, with relation to male body size, it has been hypothesized that as females are under the greatest energetic constraints, their body size patterns are most optimal and that sexual selection drives male body size relative to that of females (Kienle et al. 2022).

Large body size is less critical and the selective pressures less straightforward for males in species that copulate in the water, as there are advantages and disadvantages to large body size in the maintenance of underwater territories. For these phocids, larger males may have an advantage over smaller males as they would not have to feed as often and would have more time available to compete for and mate with females (e.g., harbor seals; Coltman et al. 1998). However, in species where males maintain territories and copulate in the water, greater fasting ability may not be as important, because males may have access to prey as they patrol their underwater territories, or they might be able to leave on short foraging trips. Moreover, large size may diminish agility when competing for mates underwater. Some of these species even exhibit reverse sexual size dimorphism, where males are smaller than females (e.g., Weddell seals). While little is known about the breeding system of leopard seals, they also exhibit reverse sexual dimorphism, suggesting a breeding system like other aquatically-breeding seals.

8.8 Origins and Evolutionary Implications

It is interesting to consider the origins and selective pressures that shaped the highly derived capital breeding system of phocid seals, as well as when and how it diverged from the ancestral condition. The appearance of key aspects of the life history patterns of phocids, such as increased body size, sexual dimorphism, and diving ability, can be identified in the fossil record (Debey and Pyenson 2013; Cullen et al. 2014; Chap. 1).

The ancestral pinniped, *Enaliarctos* spp., evolved in the North Pacific during the late Oligocene (Berta et al. 1989; Chap. 1). This early pinniped was sexually dimorphic, suggesting that the ancestral state included a polygynous breeding system (Bartholomew 1970; Wyss 1994; Cullen et al. 2014). This small (150–170 cm), fur seal-like primitive carnivore (Churchill et al. 2015) likely made short feeding forays in the coastal ocean when upwelling of cold, nutrient-rich water was greater (Costa and Valenzuela-Toro 2021). Heightened upwelling resulted in greater biological productivity (Lipps and Mitchell 1976), making prey more available. Increased availability of cold ectothermic prey would give endothermic tetrapods an advantage over other predators (Vermeij and Dudley 2000). As a mammal, *Enaliarctos* spp. would have had a high, constant core body temperature which, given a good coat of fur, could be maintained even in cold water. As an air breather, this ancestral seal could maintain an elevated body temperature with a high aerobic metabolism that, in turn, allowed its muscles to perform optimally even when operating in cold water (Costa and Valenzuela-Toro 2021). As upwelled water is cold, the muscles of ectothermic prey are not as efficient, and these organisms would, therefore, be susceptible to predation from faster, more efficient endothermic predators (Cairns et al. 2008; Grady et al. 2019). Thus, the congruence of abundant prey, and the predatory advantage of air-breathing endothermy, likely made it possible for this ancestral seal to invade the coastal marine environment (Costa and Valenzuela-Toro 2021).

Over time, these early pinnipeds diverged into two major lineages: one that refined this coastal intermittent foraging pattern, leading to the otariid and odobenid lifestyles; and a second that increased their diving ability, allowing them to forage farther and deeper offshore, thereby reducing competition with other coastal inshore marine mammals (Costa and Valenzuela-Toro 2021). This second lineage led to the phocid lifestyle. Foraging further away from the colony required an increase in the duration of foraging trips, coupled with a corresponding decrease in the number and frequency of trips, which required a greater energy return per trip. This could be achieved by an increased reliance on stored maternal reserves, allowing a transition from an income to a capital breeding system and, therefore, a separation of feeding from lactation.

As detailed in this chapter, body size is critical to enabling the capital breeding system, and an increase in body size would be consistent with the ability to employ the capital breeding system, as observed in some modern phocid seals. A trend seen in the early evolution of pinnipeds is an initial increase in body size, which first appeared in the early Miocene (23 Ma) as *Allodesmus kelloggi*, a 258-cm Desmatophocid (Mitchell 1975; Barnes 1987) within the 250–350 cm range of northern elephant seal females. Further, seals with large eye orbits tend to be deep divers, and the eye orbits of *Allodesmus* spp. approximate those of modern elephant seals, suggesting that they were also deep divers (Debey and Pyenson 2013). Thus, their large body would not only allow them to dive deep but also enable the separation of feeding from lactation that would have been necessary for utilization of distant offshore foraging grounds, highly dispersed prey, or environments of lower productivity. This life-history pattern would have allowed these early seals to cross the equator and invade the South Pacific Ocean, as well as pass through the region associated with the Central American Seaway and establish breeding populations throughout the Atlantic Ocean (Muizon and Bond 1982; Wyss 1994; Cozzuol 2001; Churchill et al. 2015). Once in the Atlantic, phocids could move north and south as they would have no competition from otariids (see below).

Similarly, phocids would not have been precluded from invading the South Pacific as they could cross the equator (Valenzuela-Toro et al. 2013; Churchill et al. 2014). Large body size and attenuated lactation allowed some species to fully inhabit the predator-free ice and island habitats of the Southern Ocean. As the phocines invaded the Arctic, they became smaller (Chap. 1), most likely in response to the threat of predation (Schulz and Bowen 2004).

It is interesting that, except for elephant seals, there are no modern-day phocids in the temperate to sub-polar regions of the southern hemisphere (Costa et al. 2006) despite the presence of several fossil phocids in these areas (Valenzuela-Toro et al. 2013). It is tempting to speculate that, as income breeders, early otariids would have faced an oceanographic barrier associated with the Central American Seaway (Briscoe et al. 2017). This barrier would cease to exist once the hydrographic conditions changed with the closure of the Central American Seaway around 2.8 Ma (O'Dea et al. 2016). This would have created a migration corridor that allowed otariids to cross and reach the highly productive upwelling regions of the Southern Hemisphere, where they may have outcompeted phocid seals. As otariids have a more energy-intensive lifestyle in highly productive regions, they might have an ability to obtain and invest more resources into reproduction than phocids, who have a more economical lifestyle. This would be consistent with the hypothesis that the reproductive output is greater for species with higher metabolic rates and/or that have a faster pace of life (Hennemann 1983; Schmitz and Lavigne 1984; Dmitriew 2011; Wright et al. 2018). The presence of elephant seals is not surprising as their foraging behavior allows them to utilize resources that are far offshore and very deep, well out of the reach of any otariid.

8.9 Future Directions for Research

Most of the research on phocid reproductive patterns and energetics has been carried out on only a few species of seals, predominately Weddell, elephant, harbor, and gray seals. We have some exciting but limited information on ringed, harp, and hooded seals (Bowen et al. 1985; Lydersen and Hammill 1993; Oftedal et al. 1996; Schulz and Bowen 2004; Ferguson et al. 2018), and we know almost nothing about monk seals and the remaining ice-breeding species such as crabeater, leopard, Ross, ribbon, bearded, Caspian, and Baikal seals. The prolonged lactation interval of Mediterranean monk seals is particularly interesting as it lies significantly outside the range observed for other phocids. The reasons for the lack of data are obvious: as both species of monk seals are endangered, any form of manipulative research is limited, and seals that are found deep in the pack ice or other remote regions require considerable logistical support and are extremely difficult to study. However, research carried out on harp and hooded seals has shown that, given the right resources, even species in challenging environments can be studied (Bowen et al. 1985, 1992; Oftedal et al. 1988, 1996).

We lack fundamental bioenergetics measurements for many species, such as prey assimilation efficiencies and metabolic rates. We also lack data on the birth mass, weaning mass, growth rate, and lactation duration for most phocids. Milk intake measurements and milk composition data have been completed on only a few species, and we have only scratched the surface of the physiological adaptations that allow a phocid pup to process such a vast quantity of lipid-rich milk. A basic understanding of bioenergetics, lactation patterns, and breeding strategies is critical for understanding the mechanistic constraints on behavior and their evolutionary origins. These are also crucial inputs in modeling efforts to predict how environmental disturbance will affect behavior and, ultimately, population dynamics (Costa 2012; McHuron et al. 2017; Pirotta et al. 2018).

While the capital breeding life history pattern typified by some phocids is evolutionarily more derived, the broader distribution of income breeders, as typified by otariids, especially in the southern hemisphere, suggests that when food is more available, the income strategy prevails. Is the phocid life history pattern only advantageous in regions where otariids cannot go? For example, northern and Antarctic fur seals (*Arctocephalus gazella*) overlap with northern and southern elephant seals in the North Pacific and the Southern Ocean, but they forage differently. Fur seals make short, shallow dives in the epipelagic zone, while elephant seals make long, deep dives to the mesopelagic zone (Costa et al. 2010). Further comparisons between different species in different habitats would help elucidate the eco-evolutionary dynamics that drove the emergence of a reproductive strategy as remarkable as the phocid capital breeding system. A better understanding of the energetic constraints faced by species with different reproductive patterns would allow us to predict how they might respond to changing ocean conditions as well as the resilience of these different reproductive patterns to anthropogenic disturbance.

Acknowledgements We thank the many collaborators, students, and colleagues who, over the years, have shaped our thinking on reproductive patterns, energetics, and life history patterns. These individuals have freely shared their ideas, thoughts, and data. DPC thanks Mary I Zavanelli, whose support over his career made his work possible; he only wishes she was here to read this. We thank A. Valenzuela-Toro, R. Holser, S. Kienle, and P. Pomeroy for comments on drafts of this paper, and Jessica Kendall-Barr for assistance in the production of the figures. The writing of this chapter was supported by the Office of Naval Research N00014-18-1-2822, the E&P Sound and Marine Life Joint Industry Project of the International Association of Oil and Gas Producers (JIP2207–23); and the NSF Polar Programs and Environments award 1644256.

References

Adachi T, Maresh JL, Robinson PW, Peterson SH, Costa DP, Naito Y, Watanabe YY, Takahashi A (2014) The foraging benefits of being fat in a highly migratory marine mammal. Proc Biol Sci 281:20142120

Adams SH, Costa DP (1993) Water conservation and protein metabolism in northern elephant seal pups during the postweaning fast. J Comp Physiol B 163:367–373

Aguilar A, Cappozzo LH, Gazo M, Pastor T, Forcada J, Grau E (2007) Lactation and mother–pup behaviour in the Mediterranean monk seal *Monachus monachus*: an unusual pattern for a phocid. J Mar Biol Assoc UK 87:93

Anderson SA, Fedak MA (1985) Grey seals males: energetics and behavioural links between size and sexual success. Anim Behav 33:829–838

Barnes LG (1987) An early Miocene pinniped of the genus *Desmatophoca* (Mammalia: Otariidae) from Washington. Natural History of Los Angeles County, Los Angeles

Bartholomew GA (1970) A model for the evolution of pinniped polygyny. Evolution 24:546–559

Bartholomew GA, Wilke F (1956) Body temperature in the northern fur seal, *Callorhinus ursinus*. J Mammal 37:327–337

Bartsh SS, Johnston SD, Siniff DB (1992) Territorial behavior and breeding frequency of male Weddell seals (*Leptonychotes weddelli*) in relation to age, size, and concentrations of serum testosterone and cortisol. Can J Zool 70:680–692

Beck CA, Bowen WD, Iverson SJ (2003) Sex differences in the seasonal patterns of energy storage and expenditure in a phocid seal. J Anim Ecol 72:280–291

Benoit-Bird KJ, Battaile BC, Nordstrom CA, Trites AW (2013b) Foraging behavior of northern fur seals closely matches the hierarchical patch scales of prey. Mar Ecol Prog Ser 479:283–302

Benoit-Bird KJ, Battaile BC, Heppell SA, Hoover B, Irons D, Jones N, Kuletz KJ, Nordstrom CA, Paredes R, Suryan RM, Waluk CM, Trites AW (2013a) Prey patch patterns predict habitat use by top marine predators with diverse foraging strategies. PLoS One 8:e53348

Berta A, Ray CE, Wyss AR (1989) Skeleton of the oldest known pinniped, *Enaliarctos mealsi*. Science 244:60–62

Boness DJ, Francis JM (1991) The effect of thermoregulatory behaviour on the mating system of the Juan Fernández fur seal, *Arctocephalus philippii*. Behaviour 119:104–126

Boness DJ, Bowen WD, Oftedal OT (1988) Evidence of polygyny from spatial patterns of hooded seals (*Cystophora cristata*). Can J Zool 66:703–706

Boness DJ, Bowen WD, Oftedal OT (1994) Evidence of a maternal foraging cycle resembling that of otariid seals in a small phocid, the harbor seal. Behav Ecol Sociobiol 34:95–104

Boness DJ, Craig MP, Honigman L, Austin S (1998) Fostering behavior and the effect of female density in Hawaiian monk seals, *Monachus schauinslandi*. J Mammal 79:1060–1069

Bowen WD, Oftedal OT, Boness DJ (1985) Birth to weaning in 4 days–remarkable growth in the hooded seal, *Cystophora cristata*. Can J Zool 63:2841–2846

Bowen WD, Oftedal OT, Boness DJ (1992) Mass and energy-transfer during lactation in a small phocid, the harbor seal (*Phoca vitulina*). Physiol Zool 65:844–866

Boyd IL (1998) Time and energy constraints in pinniped lactation. Am Nat 152:717–728

Briggs KT, Le Boeuf BJ (1975) Pup mortality in a growing colony of elephant seals. Am Zool 15:821–821

Briscoe DK, Hobday AJ, Carlisle A, Scales K, Eveson JP, Arrizabalaga H, Druon JN, Fromentin JM (2017) Ecological bridges and barriers in pelagic ecosystems. Deep Sea Res Pt II Top St Oce 140:182–192

Brody S (1945) Bio-Energetics and growth With special Reference to the efficiency complex in domestic animals. Reinhold Publishing Corp, New York

Burns JM (1999) The development of diving behavior in juvenile Weddell seals: pushing physiological limits in order to survive. Can J Zool 77:737–747

Burns JM, Castellini MA (1996) Physiological and behavioral determinants of the aerobic dive limit in Weddell seal (*Leptonychotes weddellii*) pups. J Comp Physiol B 166:473–483

Burns JM, Castellini MA, Schreer JF (1997) Physiological effects on dive patterns and foraging strategies in yearling Weddell seals (*Leptonychotes weddellii*). Can J Zool 75:1796–1810

Cairns DK, Gaston AJ, Huettmann F (2008) Endothermy, ectothermy and the global structure of marine vertebrate communities. Mar Ecol Prog Ser 356:239–250

Cameron MF, Siniff DB, Proffitt KM, Garrott RA (2007) Site fidelity of Weddell seals: the effects of sex and age. Antarct Sci 19:149–155

Churchill M, Boessenecker RW, Clementz MT (2014) Colonization of the Southern Hemisphere by fur seals and sea lions (Carnivora: Otariidae) revealed by combined evidence phylogenetic and Bayesian biogeographical analysis. Zool J Linn Soc 172:200–225

Churchill M, Clementz MT, Kohno N (2015) Cope's rule and the evolution of body size in Pinnipedimorpha (Mammalia: Carnivora). Evolution 69:201–215

Clauset A (2013) How large should whales be? PLoS One 8:e53967

Codde SA, Allen SG, Houser DS, Crocker DE (2016) Effects of environmental variables on surface temperature of breeding adult female northern elephant seals, *Mirounga angustirostris*, and pups. J Therm Biol 61:98–105

Coltman DW, Bowen WD, Boness DJ, Iverson SJ (1997) Balancing foraging and reproduction in the male harbour seal, an aquatically mating pinniped. Anim Behav 54:663–678

Coltman DW, Bowen WD, Iverson SJ, Boness DJ (1998) The energetics of male reproduction in an aquatically mating pinniped, the harbour seal. Physiol Zool 71:387–399

Costa DP (1991) Reproductive and foraging energetics of high-latitude penguins, albatrosses and pinnipeds–implications for life-history patterns. Am Zool 31:111–130

Costa DP (1991) Reproductive and foraging energetics of pinnipeds: implications for life history patterns. In: Renouf D (ed) Behaviour of pinnipeds. Chapman and Hall, Cambridge, pp 300–344

Costa DP (1993) The relationship between reproductive and foraging energetics and the evolution of the Pinnipedia. In: Boyd IL (ed) Marine mammals: advances in behavioural and population biology, Book 66. Oxford University Press, Symposium Zoological Society of London, pp 293–314

Costa DP (2008) A conceptual model of the variation in parental attendance in response to environmental fluctuation: foraging energetics of lactating sea lions and fur seals. Aquat Conserv Mar Freshwat Ecosyst 17:S44–S52

Costa DP, Maresh JL (2018) Energetics. In: Würsig B, Thewissen JGM, Kovacs K (eds) Encyclopedia of marine mammals. Academic Press, pp 329–335

Costa DP, Valenzuela-Toro AM (2021) When physiology and ecology meet: the interdependency between foraging ecology and reproduction in Otariids. In: Campagna C, Harcourt R (eds) Ethology and behavioral ecology of Otariids and the Odobenid. Springer International Publishing, Switzerland, pp 21–50

Costa DP, Boeuf BJL, Huntley AC, Ortiz CL (1986) The energetics of lactation in the northern elephant seal, *Mirounga angustirostris*. J Zool 209:21–33

Costa DP, Weise MJ, Arnould JPY (2006) Worldwide pinniped population status and tTrends in whales, whaling and ocean ecosystems. In: Estes JA, Williams TM, Doak D, DeMaster D (eds) Whales, whaling and ocean ecosystems. University of California Press, Berkeley, pp 342–357

Costa DP, Huckstadt LA, Crocker DE, McDonald BI, Goebel ME, Fedak MA (2010) Approaches to studying climatic change and its role on the habitat selection of antarctic pinnipeds. Integr Comp Biol 50:1018–1030

Costa DP (2012) A bioenergetics approach to developing the PCAD model. In: Popper AN, Hawkins A (eds) The effects of noise on aquatic life advances in experimental medicine and biology. Springer Science+Business Media, pp 423–426

Cozzuol MA (2001) A "northern" seal from the miocene of Argentina: implications for phocid phylogeny and biogeography. J Vertebr Paleontol 21:415–421

Crocker DE, Williams JD, Costa DP, Le Boeuf BJ (2001) Maternal traits and reproductive effort in northern elephant seals. Ecology 82:3541–3555

Crocker DE, Houser DS, Webb PM (2012) Impact of body reserves on energy expenditure, water flux, and mating success in breeding male northern elephant seals. Physiol Biochem Zool 85:11–20
Cullen TM, Fraser D, Rybczynski N, Schroder-Adams C (2014) Early evolution of sexual dimorphism and polygyny in Pinnipedia. Evolution 68:1469–1484
Debey LB, Pyenson ND (2013) Osteological correlates and phylogenetic analysis of deep diving in living and extinct pinnipeds: what good are big eyes? Mar Mammal Sci 29:48–83
Deutsch CJ, Haley MP, Le Boeuf BJ (1990) Reproductive effort of male northern elephant seals–estimates from mass-loss. Can J Zool 68:2580–2593
Dmitriew CM (2011) The evolution of growth trajectories: what limits growth rate? Biol Rev 86:97–116
Emlen ST, Oring LW (1977) Ecology, sexual selection, and the evolution of mating systems. Science 197:215–223
Fedak MA, Anderson SS (1982) The energetics of lactation–accurate measurements from a large wild mammal, the grey seal (*Halichoerus grypus*). J Zool 198:473–479
Fedak MA, Arnbom T, Boyd IL (1996) The relation between the size of southern elephant seal mothers, the growth of their pups, and the use of maternal energy, fat and protein during lactation. Physiol Zool 69:887–911
Ferguson SH, Zhu X, Young BG, Yurkowski DJ, Thiemann GW, Fisk AT, Muir DCG (2018) Geographic variation in ringed seal (*Pusa hispida*) growth rate and body size. Can J Zool 96:649–659
Ferguson SH, Yurkowski DJ, Young BG, Willing C, Zhu X, Muir DCG, Fisk AT, Thiemann GW (2019) Do intraspecific life history patterns follow interspecific predictions? A test using latitudinal variation in ringed seals. Popul Ecol 61:371–382
Fish FE (1994) Influence of hydrodynamic-design and propulsive mode on mammalian swimming energetics. Aust J Zool 42:79–101
Fish FE (2000) Biomechanics and energetics in aquatic and semiaquatic mammals: Platypus to whale. Physiol Biochem Zool 73:683–698
Garrett JN, Fish FE (2015) Kinematics of terrestrial locomotion in harbor seals and gray seals: Importance of spinal flexion by amphibious phocids. Mar Mammal Sci 31:459–478
Gazo M, Aguilar A (2005) Maternal attendance and diving behavior of a lactating Mediterranean monk seal. Mar Mammal Sci 21:340–345
Grady JM, Maitner BS, Winter AS, Kaschner K, Tittensor DP, Record S, Smith FA, Wilson AM, Dell AI, Zarnetske PL, Wearing HJ, Alfaro B, Brown JH (2019) Metabolic asymmetry and the global diversity of marine predators. Science 363:366
Guinet C (1992) Predation behavior of killer whales (*Orcinus orca*) around Crozet Islands. Can J Zool 70:1656–1667
Hadley GL, Rotella JJ, Garrott RA (2008) Spatial variation in age-specific probabilities of first reproduction for Weddell seals. Oikos 117:1165–1174
Haley MP (1994) Resource-holding power asymmetries, the prior residence effect, and reproductive payoffs in male northern elephant seal fights. Behav Ecol Sociobiol 34:427–434
Haley MP, Deutsch CJ, Le Boeuf BJ (1994) Size, dominance and copulatory success in male northern elephant seals, *Mirounga angustirostris*. Anim Behav 48:1249–1260
Hammill MO, Lydersen C, Ryg M, Smith TG (1991) Lactation in the ringed seal (*Phoca hispida*). Can J Fisheries Aquat Sci 48:2471–2476
Hayes SA, Costa DP, Harvey JT, Le Boeuf BJ (2004) Aquatic mating strategies of the male Pacific harbor seal (*Phoca vitulina richardii*): are males defending the hotspot? Mar Mammal Sci 20:639–656
Hennemann WW (1983) Relationship among body mass, metabolic rate and the intrinsic rate of natural increase in mammals. Oecologia 56:104–108
Hindell MA, Slip DJ (1997) The importance of being fat: Maternal expenditure in the southern elephant seal *Mirounga leonina*. In: Hindell M, Kemper C (eds) Marine mammal research in

the southern hemisphere, vol 1. Status, ecology and medicine. Surrey Beatty and Sons Pty Ltd., Chipping Norton, New South Wales, Australia, pp 72–77
Iverson SJ, Oftedal OT, Bowen WD, Boness DJ, Sampugna J (1995) Prenatal and postnatal transfer of fatty acids from mother to pup in the hooded seal. J Comp Physiol B 165:1–12
Johanos TC, Becker BL, Ragen TJ (1994) Annual reproductive-cycle of the female Hawaiian monk seal (*Monachus schauinslandi*). Mar Mammal Sci 10:13–30
Kienle SS, Friedlaender, AS, Crocker, Daniel, Mehta SR, Costa, DP (2022) Trade-offs between foraging reward and mortality risk drive sex-specific foraging strategies in sexually dimorphic northern elephant seals. Royal Society Open Science, in press
Kleiber M (1975) The fire of life: an introduction to animal energetics. R.E. Krieger Publishing Co., Huntington
Kovacs KM (1990) Mating strategies in male hooded seals (*Cystophora cristata*). Can J Zool 68:2499–2502
Kovacs KM, Lydersen C, Hammill M, Lavigne DM (1996) Reproductive effort of male hooded seals (*Cystophora cristata*): estimates from mass loss. Can J Zool 74:1521–1530
Kretzmann MB, Costa DP, Boeuf BJ (1993) Maternal energy investment in elephant seal pups: evidence for sexual equality? Am Nat 141:466–480
Kuhn CE, Costa DP (2014) Interannual variation in the at-sea behavior of California sea lions (*Zalophus californianus*). Mar Mammal Sci 30:1297–1319
Le Boeuf BJ (1991) Pinniped mating systems on land, ice and in the water: emphasis on the Phocidae. In: Renouf D (ed) Behaviour of pinnipeds. Chapman and Hall, Cambridge, pp 45–65
Lipps JH, Mitchell E (1976) Trophic model for the adaptive radiations and extinctions of pelagic marine mammals. Paleobiology 2:147–155
Lydersen C, Hammill MO (1993) Activity, milk intake and energy consumption in free-living ringed seal (*Phoca hispida*) pups. J Comp Physiol B 163:433–438
Lydersen C, Kovacs KM (1996) Energetics of lactation in harp seals (*Phoca groenlandica*) from the Gulf of St. Lawrence Canada. J Comp Physiol B 166:295–304
Lydersen C, Kovacs KM (1999) Behaviour and energetics of ice-breeding, North Atlantic phocid seals during the lactation period. Mar Ecol Prog Ser 187:265–281
Maresh JL, Simmons SE, Crocker DE, McDonald BI, Williams TM, Costa DP (2014) Free-swimming northern elephant seals have low field metabolic rates that are sensitive to an increased cost of transport. J Exp Biol 217:1485–1495
Maresh J, Adachi T, Takahashi A, Naito Y, Crocker D, Horning M, Williams T, Costa D (2015) Summing the strokes: energy economy in northern elephant seals during large-scale foraging migrations. Mov Ecol 3:22
McHuron EA, Costa DP, Schwarz L, Mangel M (2017) State-dependent behavioural theory for assessing the fitness consequences of anthropogenic disturbance on capital and income breeders. Methods Ecol Evol 8:552–560
Mellish JA, Iverson SJ, Bowen WD (1999a) Variation in milk production and lactation performance in grey seals and consequences for pup growth and weaning characteristics. Physiol Biochem Zool 72:677–690
Mellish JE, Iverson SJ, Bowen WD, Hammill MO (1999) Fat transfer and energetics during lactation in the hooded seal: the roles of tissue lipoprotein lipase in milk fat secretion and pup blubber deposition. J Comp Physiol B 169:377–390
Mellish JA, Iverson SJ, Bowen WD (2000) Metabolic compensation during high energy output in fasting, lactating grey seals (*Halichoerus grypus*): metabolic ceilings revisited. Proc Biol Sci 267:1245–1251
Mitchell ED (1975) Parallelism and convergence in the evolution of Otariidae and Phocidae. Rapp P-v Reun Cons Int Explor Mer 169:12–26
Muizon Cd, Bond ME (1982) Le Phocidae (Mammalia) Miocene de la formation Parana (Entre Rios, Argentine). Bulletin du Muséum National d'Histoire Naturelle, section C, 4eme série 4:165–207
Norris AL, Houser DS, Crocker DE (2010) Environment and activity affect skin temperature in breeding adult male elephant seals (*Mirounga angustirostris*). J Exp Biol 213:4205–4212

O'Dea A, Lessios HA, Coates AG, et al (2016) Formation of the Isthmus of Panama. Sci Adv 2:e1600883

Oftedal OT (2000) Use of maternal reserves as a lactation strategy in large mammals. Proc Nutr Soc 59:99–106

Oftedal OT, Boness DJ, Bowen WD (1988) The composition of hooded seal (*Cystophora cristata*) milk–an adaptation for postnatal fattening. Can J Zool 66:318–322

Oftedal OT, Bowen WD, Boness DJ (1993) Energy transfer by lactating hooded seals and nutrient deposition in their pups during the 4 days from birth to weaning. Physiol Zool 66:412–436

Oftedal OT, Bowen WD, Boness DJ (1996) Lactation performance and nutrient deposition in pups of the harp seal, *Phoca groenlandica*, on ice floes off southeast labrador. Physiol Zool 69:635–657

Perez MA, Mooney EE (1986) Increased food and energy consumption of lactating northern fur seals *Callorhinus ursinus*. Fish Bull 84:371–382

Pirotta E, Booth CG, Costa DP et al (2018) Understanding the population consequences of disturbance. Ecol Evol 8:9934–9946

Pitman RL, Durban JW (2012) Cooperative hunting behavior, prey selectivity and prey handling by pack ice killer whales (*Orcinus orca*), type B, in Antarctic Peninsula waters. Mar Mammal Sci 28:16–36

Pond CM, Ramsay MA (1992) Allometry of the distribution of adipose tissue in Carnivora. Can J Zool 70:342–347

Ponganis PJ (2016) Diving physiology of marine mammals and seabirds. Cambridge University Press, Cambridge, UK

Reiter J, Panken KJ, Le Boeuf BJ (1981) Female competition and reproductive success in northern elephant seals. Anim Behav 29:670–687

Robinson PW, Costa DP, Crocker DE, et al (2012) Foraging behavior and success of a mesopelagic predator in the northeast Pacific Ocean: insights from a data-rich species, the northern elephant seal. PLoS One 7:e36728

Schmitz OJ, Lavigne DM (1984) Intrinsic rate of increase, body size, and specific metabolic rate in marine mammals. Oecologia 62:305–309

Schulz TM, Bowen WD (2004) Pinniped lactation strategies: evaluation of data on maternal and offspring life history traits. Mar Mammal Sci 20:86–114

Schulz TM, Bowen WD (2005) The evolution of lactation strategies in pinnipeds: a phylogenetic analysis. Ecol Monogr 75:159–177

Siniff DB, Bengtson JL (1977) Observations and hypotheses concerning the interactions among crabeater seals, leopard seals, and killer whales. J Mammal 58:414–416

Siniff DB, Demaster DP, Hofman RJ, Eberhardt LL (1977) Analysis of dynamics of a Weddell seal population. Ecol Monogr 47:319–335

Stephens PA, Houston AI, Harding KC, Boyd IL, McNamara JM (2014) Capital and income breeding: the role of food supply. Ecology 95:882–896

Tennett KA, Costa DP, Nicastro AJ, Fish FE (2018) Terrestrial locomotion of the northern elephant seal (*Mirounga angustirostris*): limitation of large aquatically adapted seals on land? J Exp Biol 221

Thorson PH, Le Boeuf BJ (1994) Developmental aspects of diving in northern elephant seal pups. In: Le Boeuf BJ, Laws RM (eds) Elephant seals: population ecology, behavior, and physiology. University of California Press, Berkeley, pp 271–289

Tinker MT, Kovacs KM, Hammill MO (1995) The reproductive-behavior and energetics of male gray seals (*Halichoerus grypus*) breeding on a land-fast ice substrate. Behav Ecol Sociobiol 36:159–170

Trillmich F (1990) The behavioral ecology of maternal effort in fur seals and sea lions. Behaviour 114:3–20

Trillmich F (2010) Parental care: adjustments to conflict and cooperation. In: Kappeler P (ed) Animal behaviour: evolution and mechanisms. Springer, Berlin, Heidelberg, pp 267–298

Trillmich F, Weissing FJ (2006) Lactation patterns of pinnipeds are not explained by optimization of maternal energy delivery rates. Behav Ecol Sociobiol 60:137–149

Trillmich F, Ono KA, Costa DP et al (1991) The effects of El Niño on pinniped populations in the Eastern Pacific. In: Trillmich F, Ono K (eds) Pinnipeds and El Niño. Springer, Berlin, Heidelberg, pp 247–270

Valenzuela-Toro AM, Gutstein CS, Varas-Malca RM, Suarez ME, Pyenson ND (2013) Pinniped turnover in the South Pacific Ocean: new evidence from the Plio-Pleistocene of the Atacama Desert, Chile. J Vertebr Paleontol 33:216–223

Vermeij GJ, Dudley R (2000) Why are there so few evolutionary transitions between aquatic and terrestrial ecosystems? Biol J Linn Soc 70:541–554

Wheatley KE, Bradshaw CJ, Davis LS, Harcourt RG, Hindell MA (2006) Influence of maternal mass and condition on energy transfer in Weddell seals. J Anim Ecol 75:724–733

Wheatley KE, Bradshaw CJ, Harcourt RG, Hindell MA (2008) Feast or famine: evidence for mixed capital-income breeding strategies in Weddell seals. Oecologia 155:11–20

Williams TM, Maresh JL (2016) Exercise energetics. In: Castellini MA, Mellish JA (eds) Marine mammal physiology: requisites for ocean living. CRC Press, Bacon Raton, FL, pp 47–68

Williams TM, Yeates L (2004) The energetics of foraging in large mammals: a comparison of marine and terrestrial predators. Anim Environ 1275:351–358

Williams TM, Haun J, Davis RW, Fuiman LA, Kohin S (2001) A killer appetite: metabolic consequences of carnivory in marine mammals. Comp Biochem Physiol A Mol Integ Physiol 129:785–796

Wright J, Bolstad GH, Araya-Ajoy YG, Dingemanse NJ (2018) Life-history evolution under fluctuating density-dependent selection and the adaptive alignment of pace-of-life syndromes. Biol Rev Camb Philos Soc 94:230–247

Wyss AR (1994) The evolution of body size in phocids: some ontogenetic and phylogenetic observations. Proc San Diego Soc Nat Hist 29:69–75

Part II
Examples of Phocid Ethology and Behavioral Ecology: Insights from Data-Rich Species

Chapter 9
The Gray Seal: 80 Years of Insight into Intrinsic and Extrinsic Drivers of Phocid Behavior

Sean Twiss, Amy Bishop, and Ross Culloch

Contents

Abstract The gray seal is a data-rich species with behavioral studies dating back to the 1940s. The reasons for the wealth of knowledge are partly fortuitous; pioneering naturalists ventured forth to remote island colonies around the UK and Canada to observe the 'hook-nosed sea-pig' during their annual breeding seasons. These early qualitative treatises on gray seal behavior ignited further, more quantitative, research interest, which has continued to expand to this day. Several gray seal traits enhance its suitability as a study system for understanding drivers of behavior, such as their ease of observation and site fidelity during breeding, and unique pelage patterns enabling the collection of long-term data on known individuals. Gray seals also inhabit a remarkable variety of habitats, both on land and at sea, facilitating comparative studies of environmental drivers of behavior. These traits have enabled pioneering

S. Twiss (✉)
Animal Behaviour and Behavioural Ecology, Department of Biosciences, Durham University, Durham D1 3LE, UK
e-mail: s.d.twiss@durham.ac.uk

A. Bishop
Water and Environmental Research Center, UAF Centennial Postdoc, University of Alaska Fairbanks, Fairbanks, AK 99775, USA
e-mail: ambishop2@alaska.edu

R. Culloch
Scottish Government, Marine Laboratory, Marine Scotland Science, 375 Victoria Road, Aberdeen AB11 9DB, UK
e-mail: ross.culloch@gov.scot

D. P. Costa and E. McHuron (eds.), *Ethology and Behavioral Ecology of Phocids*, Ethology and Behavioral Ecology of Marine Mammals,
https://doi.org/10.1007/978-3-030-88923-4_9

behavioral research across a wide range of life-history stages. This chapter aims to capture the diversity of behavioral knowledge derived from gray seals, revealing how they interact with one another and their environment. We also highlight new research areas likely to present research opportunities and challenges in the near future, especially in the context of rapidly changing terrestrial and marine environments.

Keywords *Halichoerus grypus* · Social interactions · Maternal care · Polygyny · Mating patterns · Foraging behavior · Novel behaviors · Individual differences · Personality

Some early behavioral observations of gray seals from 'Martin Martin: A Description Of The Western Islands Of Scotland Circa 1695*' (Concerning a debate on the eating of seals);

> a debate between a Protestant gentleman and a Papist of my acquaintance: the former alleged that the other had transgressed the rules of his church, by eating flesh in Lent: the latter answered that he did not ; for, says he, I have eat a sea-creature, which only lives and feeds upon fish. The Protestant replied, that this creature is amphibious, lies, creeps, eats sleeps, and so spends much of its time on land, which no fish can do and live. It hath also another faculty that no fish has, that is, it breaks wind backward so loudly, that one may hear it at a great distance.

*Edited by D.J. Macleod, Birlinn Ltd, 1994

9.1 Introduction

There is a long and productive history of research on gray seals (*Halichoerus grypus*) in both the Eastern and Western Atlantic populations, encompassing demography, life history, reproduction, and foraging ecology. Behavioral studies are integral to a thorough understanding of gray seal biology and feed into all aspects of gray seal research. In this chapter, we focus on research where ethological aspects are foremost, or where behavioral observations have resolved outstanding questions, provided new insights, or indeed, highlighted new paths for scientists to explore.

Behavior is essentially about interactions between individuals and their environment, whether abiotic or biotic, including conspecifics and heterospecifics. Therefore, one should not view behavior in isolation. One needs to quantify the context of behavior at an appropriate biologically relevant spatial, temporal, or social scale (or scales), to get at the meaning of, and scope for, behavioral choices. Ideally, this should link with information on energetic and physiological constraints that set the intrinsic scope for behavior within and across individuals (Chaps. 4 and 8). As behavior is basically about interactions, it seems artificial to impose a chapter structure that separates males, females, juveniles, and pups. For example, male-male interactions are inevitably modified by the availability of sexually receptive females. However, there are some key areas of gray seal behavior that have been studied in detail and that illustrate these interactions well: male-male reproductive competition, female-pup interactions, and female–habitat interactions. We deal with these interactions, then look at some new avenues emerging in the literature and consider potentially valuable future directions to promote a deeper understanding of the behavior of gray seals, if certain practical considerations can be overcome. First, however, we reflect on the long history of gray seal ethological studies, with the aims of illustrating the development of gray seal behavioral research, and how the qualities of this system have allowed researchers to contribute to the wealth of knowledge about gray seals, making this a 'data-rich' species.

9.2 History

Early ethological studies of gray seals centered on terrestrial life-history phases, and in particular, breeding behavior. Gray seal colonial breeding behavior provides excellent observational access to breeding adults, pups of the year, and in some cases, juveniles of pre-breeding age. Consequently, most gray seal ethology is conducted on wild populations rather than captive individuals. This has the advantage of allowing examination of behavior in the context of natural selective pressures under which seals operate, though limiting the scope for experimental manipulation of behavior. Frank Fraser-Darling pioneered ethological observations of breeding gray seals based on his prolonged stay in the North Rona (Scotland, UK) colony (Fraser-Darling 1939) that described much of the basic breeding behavior of gray seals. In the 1950s, other intrepid researchers followed Fraser-Darling's example. They ventured to various UK colonies during the autumn breeding season, providing detailed descriptive, but often rather anecdotal, accounts of behavior (e.g., Davies 1949; Matthews 1950; Hewer 1957, 1960; Fogden 1971) that are engagingly collected in a book by Hewer (1974) along with his extensive observations. In the Western Atlantic, Cameron (1967, 1969, 1970) braved winter conditions to provide comparative behavioral descriptions of gray seals that breed in mid-winter. The 1960s also saw a growth in more quantitative studies among Eastern Atlantic gray seal populations (e.g., Hewer and Backhouse 1960; Boyd and Laws 1962; Boyd et al. 1962; Coulson and Hickling 1961, 1964), which focused primarily on establishing basic population parameters, such as the timing of breeding, numbers of seals, sex ratios, pup production, and mortality with the aim of understanding the drivers of population dynamics. The ethological aspects of these studies remained largely anecdotal and descriptive. It was not until the mid-1970s that extensive, formally structured quantitative behavioral observations came to the fore as a key component of gray seal biology. Studies such as those by Anderson et al. (1975), Burton et al. (1975), and Anderson and Fedak (1985) followed known individuals on North Rona (Scotland, UK). Parallel studies of the Western Atlantic gray seal population at Sable Island (Canada) provided valuable comparisons across populations (e.g., Boness and James 1979; Miller and Boness 1979; Boness et al. 1982; Boness 1984). These studies integrated behavioral information about specific individuals into the broader breeding biology of gray seals. They framed their research in the context of emerging theories within behavioral and evolutionary ecology, such as drivers of variation in female and male lifetime reproductive success. Much of this early behavioral focus was on mother–pup interactions, with a particular interest in maternal investment and pup development, again primarily with a focus on the impacts at the level of the population. Studies elsewhere complemented these efforts, with Kovacs (1987) producing detailed ethological work on mothers and pups in the Isle of May (Scotland, UK) colony.

These foundational quantitative ethological studies of breeding gray seals established a route that many have followed, exploring in detail the processes that drive patterns of behavior. The breadth and depth of understanding of gray seal behavior have been extended by the integration of new research techniques from emerging disciplines, such a molecular ecology (Ambs et al. 1999; Amos et al. 1993, 1995, 2001; Allen et al. 1995; Boskovic et al. 1996; Perry et al. 1998; Worthington-Wilmer et al. 2000; Pomeroy et al. 2001; Bean et al. 2004; Twiss et al. 2006; Tollit et al. 2009) and spatial ecology (e.g., Twiss et al. 2000; Pomeroy et al. 2001; Stewart et al. 2014); and by asking questions from emerging perspectives such as the concept of animal personalities (Twiss and Franklin 2010; Twiss et al. 2012a, 2020). Throughout much of this research, a notable strength of the gray seal study system has been the ability to identify individuals consistently across successive encounters (e.g., breeding seasons), enabling detailed longitudinal studies of known individuals (e.g., Pomeroy et al., 1999; Bubac et al. 2018; Weitzman et al., 2017). In earlier studies, identification was primarily by artificial marks (e.g., flippers tags, brands). However, with the advent of more dedicated ethological studies, pelage patterns were used by skilled observers as a reliable and non-invasive means to recognize known individuals (Redman et al. 2001). This process came into much wider use with the application of digital photography and the development of computer-aided matching (e.g., Hiby and Lovell 1990; Vincent et al. 2001; Karlsson et al. 2005; Gerondeau et al. 2007; Hiby et al. 2013). The ability to reliably identify known individuals has yielded new insights into individual differences in behavior (e.g., Twiss and Franklin 2010; Twiss et al. 2012a; Bubac et al. 2018). Consequently, ethological studies of gray seals are asking ever more detailed behavioral questions, particularly at the level of within-population and individual variation; delving into the mechanisms underlying mother–pup interactions (e.g., Robinson et al. 2015a, 2019), male competitive and mating behavior (e.g., Bishop et al. 2014, 2015a, b, Lidgard et al. 2012), female mate choice (Amos et al. 1995, 2001; Ambs et al. 1999; Twiss et al. 2006), social associations (e.g., Pomeroy et al. 2005), fine-scale environmental drivers of habitat use (e.g., Twiss et al. 2000; Pomeroy et al. 2001; Matthiopoulos et al. 2005; Stewart et al. 2014) and how behavior responds to environmental change (e.g., Twiss et al. 2007; Weitzman et al. 2017).

The depth and breadth of ethological research into gray seals has shown that long-term behavioral observations are integral to understanding the biology and life history of breeding seals. Among the Eastern and Western Atlantic gray seal populations, the breeding colonies of North Rona and the Isle of May off the Scottish coast, and Sable Island off the Canadian East coast, have been three of the most important long-term study sites yielding much of the behavioral research on breeding gray seals. In more recent years, these long-term study sites have been complemented by a growing number of behavioral studies among populations and colonies in different contexts and habitats. For example, cave breeding seals around the UK coast (e.g., Leeney et al. 2010), land-fast and pack ice breeding colonies (e.g., Lydersen et al. 1994; Haller et al. 1996), seals in the Baltic that alternate between land and ice breeding habitats dependent upon prevailing weather conditions (Karlsson et al. 2005; Jussi et al. 2008), and newly emerging study sites as populations of gray seals shift over time, such as parts of the North Sea (e.g., Reijnders et al. 1995; Bishop et al. 2014; Abt and Engler 2009; Brasseur et al. 2015), Greenland (Rosing-Asvid et al. 2010) and the Northeast US coast (Lerner et al. 2018).

Breeding is only part of the life history of a gray seal. While the annual breeding attempt is arguably the culmination of the previous year's foraging effort, understanding the behavior of gray seals at sea is also critical. Gathering such data has been challenging due to limited opportunities to directly observe seals at sea, and especially to observe interactions between individuals, an essential aspect of behavioral studies. Observational studies at non-breeding haul-out sites can provide some insights into at-sea behavior patterns, in particular spatial and temporal movement patterns between haul-outs (e.g., Karlsson et al. 2005; Leeney et al. 2010; Sjöberg and Ball 2000; Survilienė et al. 2016). However, it is the application of biotelemetry devices that has revolutionized our understanding of gray seal behavior at sea (Chap. 6), though such 'behavior' is typically inferred from indirect measures such as space use (e.g., Thompson et al. 1991; McConnell et al. 1992, 1999; Beck et al. 2000; Lidgard et al. 2003; Austin et al. 2004; Harvey et al. 2008; Breed et al. 2009; Cronin et al. 2012; Carter et al. 2016) and motion (e.g., Shuert et al. 2018). Being dependent upon expensive technology, studies at sea can suffer from more restricted sample sizes than observational studies of seals on land (Hazekamp et al. 2010). However, the wealth of biotelemetry studies of gray seals provides some of the most comprehensive and informative insights into what seals are capable of once they disappear under the waves (Chap. 6).

Our overview of gray seal ethology starts by examining behavior during the main terrestrial phase of the gray seal lifecycle: the breeding season. We then head underwater and explore the insights provided by a growing number of studies that track seals at sea. Finally, we focus on some intriguing and challenging emerging areas of gray seal ethological research, aiming to highlight potentially fruitful, but certainly exciting, avenues for future research.

9.3 Interactions on Land: The Breeding Season

9.3.1 Interactions Among Males

Male gray seals reach sexual maturity at 2.6 years of age, but social maturity, in terms of achieving reproductive success, typically is not achieved until 8–10 years

(Hammill and Gosselin 1995). There are two predominant behavioral phenotypes, or reproductive strategies, exhibited by male gray seals during the breeding season (Anderson et al. 1975; Boness 1984; Anderson and Fedak 1985; Twiss et al. 1994; Lidgard et al. 2005, Bishop et al. 2015b). Males that maintain access to females in a 'come early and stay long' strategy are typically classified as 'tenured'. In contrast, males that remain on the periphery of the colony, roaming and attempting to mate with females opportunistically, are considered 'transient' (Boness and James 1979; Lidgard et al. 2001, 2005). Although the length of tenure varies among individuals, tenured males can remain in the breeding colony for up to 60 days and fast throughout this time (Boness and James 1979; Anderson and Fedak 1985; Twiss 1991; Lidgard et al. 2003; Bishop et al. 2017). While there is some evidence of aquatic mating, there is debate whether this represents a third discrete mating strategy or is an extension of the opportunistic transient strategy (Gemmell et al. 2001; Lidgard et al. 2005; Twiss et al. 2006). The relative success of transient and tenured strategies has been assessed at several colonies (Amos et al. 1993; Twiss et al. 1998, 2006, 2007; Ambs et al. 1999; Worthington-Wilmer et al. 1999; Lidgard et al. 2001, 2004; Bishop et al. 2017). Most studies rely on visual observations of consortship or copulations as measures of male success, and there are likely to be discrepancies between observed mating success and realized reproductive success, i.e., fertilizations (Amos et al. 1993; Ambs et al. 1999; Worthington-Wilmer et al. 1999; Lidgard et al. 2004; Twiss et al. 2006). There is, however, evidence of significant polygynous reproductive skew at most gray seal colonies, with tenured males having a high probability of fertilization relative to transient males (Lidgard et al. 2004; Twiss et al. 2006).

During the breeding season, tenured males engage in agonistic interactions to defend non-exclusive, often loosely defined territories in which they attempt to maintain position among shifting groups of females (Anderson et al. 1975; Boness and James 1979; Twiss et al. 1994). Males engage in two forms of agonistic contest interactions. Non-contact interactions consist of open-mouth threats (Miller and Boness 1979; Boness 1984), rolls (Cameron 1967), and other posturing behaviors (Boness and James 1979; Bishop 2015; Bishop et al. 2014; Fig. 9.1). In cases where two males are similarly matched or neither retreats after such exchanges of threats, then aggression can escalate, taking the form of contact interactions or fights. Contact interactions consist of males exchanging bites and lunges to each other's neck and wrestling (Fig. 9.1). Fights typically conclude when a winner maneuvers and is able to bite, or threaten to bite, the opponent's hind flippers or tail or when one combatant begins to back away in an attempt to turn and flee, often pursued by the other and chased out of the area (Boness and James 1979; Twiss 1991; Bishop 2015). Very rarely does a contest end in a mortal injury (Miller 1991). This pattern of escalating interactions is generally well-conserved across other polygynous, sexually size-dimorphic pinnipeds (Miller 1991). However, for gray seals, 23–60% of non-contact and contact interactions result in a draw outcome, where neither male cedes position, which results in non-linear dominance hierarchies (Anderson and Fedak 1985, Tinker et al. 1995, Twiss et al. 1998, Bishop et al. 2015b, Fig. 9.2). This behavioral mechanism of 'not losing' for gray seals is linked with mating success (Anderson and Fedak 1985), but how individuals make decisions during contests

Fig. 9.1 The most ubiquitous threat display for male gray seals is the non-vocal open-mouth threat wherein a male opens his mouth to a wide gape oriented toward or perpendicular to opponent (**a**–**b**). When contests escalate into an interaction with physical contact, males exchange lunges and bites during bouts of wrestling (**c**–**d**). Both males typically attempt to maneuver toward his opponent's hind flippers or tail. The male that successfully bites his opponents' hind flippers is usually, but not exclusively, the winner of the interaction Image credits: AM Bishop

Fig. 9.2 An example of the non-linear male dominance hierarchy from Donna Nook breeding colony, England (Bishop 2015). Individual males' dominance scores were calculated using David's Score where high positive values represent high dominance (Gammell et al. 2003; de Vries et al. 2006) and included the outcome of draws. Apart from a few clearly dominant and subordinate individuals, most males have similar dominance scores. Similar hierarchy structures have been identified at North Rona, Scotland (Twiss et al. 1998) Image credits: AM Bishop

regarding whether to de-escalate, settle for a draw, or push for a win–lose outcome is poorly understood. Considering that many studies of animal contests do not even mention draws as a possible outcome (e.g., Haley 1994; O'Donnell 1998; Colléter and Brown 2011), the gray seal system represents a prime opportunity to explore a relatively understudied facet of animal decision making, game theory, and information gathering (Riechert and Hedrick 1993; Whiting 1999; Carlin et al. 2005; Whiting et al. 2006; Bishop 2015).

In addition to observing patterns of behavioral interactions, considerable work has been carried out to link intrinsic and extrinsic factors with male success. For example, like other sexually size-dimorphic pinnipeds (Bartholomew 1970, Chaps. 7 and 11), the 'tenured' male gray seal strategy is typically adopted by older and larger individuals (Anderson and Fedak 1985; Godsell 1991; Carlini et al. 2006; Lidgard et al. 2012). Within this category, there is some evidence that larger males are more likely to win fights and maintain their position (Anderson and Fedak 1985). However, linked to the prevalence of draws was the finding that 'not losing' is more important than winning (Anderson and Fedak 1985), and males that can stay longer in a breeding colony have more matings and, therefore, overall reproductive success (Anderson et al. 1975; Anderson and Fedak 1985; Lidgard et al. 2005; Twiss et al. 2006; Bishop et al. 2017). To maximize length of stay, stabilizing selection has likely favored individual tenured males that achieve a fine balance between energy conservation (by minimizing activity) and the importance of not losing position among females. This trade-off between endurance for long tenure and agility during fights is evidenced by the findings that 60–90% of male time in breeding colonies (whether land or ice) is spent resting or idle (Boness 1984; Twiss 1991; Tinker et al. 1995, Bishop et al. 2015c), and by the inverted 'U' shape relationship between size and mating success observed on Sable Island, in which intermediate-sized males were most successful (Lidgard et al. 2005).

Extrinsic factors also influence the underlying relationship between the length of tenure and reproductive success. Colony topography and environmental factors (see Sect. 9.3.6) can play a role in the behavioral mechanisms required to achieve extended stay, such as the necessary levels of aggression, which in turn can shape the degree of polygyny (Anderson and Harwood 1985; Twiss et al. 1998, Bishop et al. 2015b). Gray seals breed on a variety of substrates, including rock, sand, and ice (Stirling 1975; Boness and James 1979; Anderson and Harwood 1985; Tinker et al. 1995; Twiss et al. 1998; Bishop et al. 2015b). In colonies with open access and wide, uniform, tidal beach topography, sex ratios are typically less skewed, and individual tenured males engage in more aggressive interactions per day compared to males at colonies with restricted access and heterogeneous topographic features (Twiss et al. 1998, Fig. 9.3). Local neighborhood social stability has also been linked to a reduced number of aggressive interactions at the individual level. However, the localized stability experienced by individuals can be disrupted by environmental perturbations such as tidal surges or storms (Bishop et al. 2015b). Thus, while historical studies at long-term monitoring sites provide a valuable baseline, it will be essential to track how male-male behaviors and subsequent colony dynamics respond under

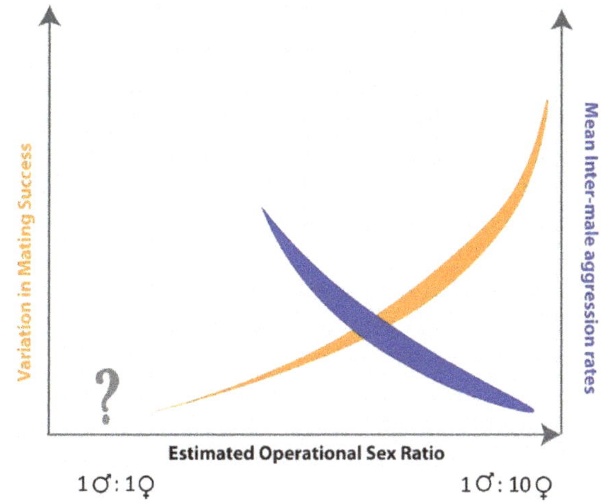

Fig. 9.3 Schematic illustrating the general relationships between operational sex ratio in breeding colonies (the average number of reproductive females per male), the degree of variation in male mating success and the average levels of male-male aggression. Operational sex ratio is strongly influenced by the physical structure of the colony in terms of access to/from the sea and the degree of female aggregation driven by the availability of suitable pupping habitat within the colony. Generally, colonies with more open access (e.g., ice breeding colonies or colonies with extensive sandy shores) and more uniform terrain (ice, sand) tend to have more even sex ratios, as more males are able to gain positions ashore. Rocky colonies with more restricted access and more varied terrain tend to have more female biased sex ratios. At such sites, subordinate males are less able to gain or maintain positions in the colony, hence the generally lower level of male-male aggression observed in the colony. Examples of different colony habitats can be seen in Fig. 9.6 (adapted from Twiss et al. 1998; Bishop 2015)

current environmental conditions, particularly relative to predictions of increased stochasticity in weather related to global climate change.

9.3.2 Interactions Between Males and Females

The primary interactions between males and females relate to mating. Males (tenured or transient) do not actively herd females but often ‘test’ females during lactation (Boness and James 1979; Boness et al. 1982; Anderson and Fedak 1985; Lidgard et al. 2001) by approaching them, appearing to sniff at females, and in some instances attempting to mount the female. Females in early lactation typically respond with aggressive rebuffs (Boness et al. 1982), but the intensity of this response tends to decline as females approach estrus at the end of lactation (Twiss et al. 2006). In some cases, females actively solicit males when in estrus, approaching males and rubbing their body against the male. Typically, only secure tenured males can repeatedly approach females(s) in their ‘territory’. Transient males are more opportunistic and are often very aggressively rebuffed (Boness et al. 1982, 1995). Transients, however, may attempt to force copulations (Boness et al. 1982, 1995), but often with limited success. In fact, females can express choice of mating and reproductive partner (Amos et al. 1995, 2001; Twiss et al. 2006, though also see Ambs et al. 1999), actively seeking specific males, though such behavior is often hidden as dominant, and presumably 'desirable' males tend to occupy territories encompassing most females' locations anyway. Whether the initial selection of pupping sites by females is influenced by the identity or status of males already present in the colony is unknown. Females may also copulate multiple times (Anderson et al. 1975; Boness and James 1979; Twiss et al. 2006), either with the same or different males, raising the possibility, yet unproven, of post-copulatory mate choice. The duration of successful copulations is typically in excess of 20 min (and can extend to over 60 min); therefore, the probability of subordinate, transient males achieving successful copulations in the vicinity of more dominant males' territories is small. Such mating attempts are likely to be met with both aggression from the female, and/or the local dominant, tenured male(s). Transient males are more likely to gain success in isolated or peripheral parts of colonies, where they may 'guard' or at least consort with individual females in an attempt to secure a mating opportunity once the female enters estrus (Ambs et al. 1999; Lidgard et al. 2001).

Compared to some other phocid species, such as elephant seals (*Mirounga* spp), gray seals exhibit moderate sexual size dimorphism and polygyny (González-Suárez and Cassini 2014). The extent to which specific males can monopolize mating opportunities is a key driver of the degree of polygyny observed at any colony. The ability of different male mating strategies to monopolize access to females is essentially driven by the spatial and temporal distribution of females in estrus (Twiss et al. 2007). The spatial distribution of mating opportunities is dictated primarily by the colony's physical structure and local environment (see Sect. 9.3.6). However, because males typically fast during their stay in the colony, the temporal distribution of mating opportunities is equally important to consider. The number and density of females in estrus changes during the breeding season. Therefore, males must also time their reproductive effort carefully, adaptively balancing the rewards of access to multiple females against the costs of long tenure and male-male competition (Bishop et al. 2017). However, females are not passive recipients of male attention and play an essential role in determining variation in male reproductive success. Female gray seals can react aggressively to male sexual approaches. They exhibit active mate choice, preferentially selecting the same mating partner in successive years (Amos et al. 1995), or selecting different fathers for their pups that are significantly more genetically diverse than expected from random mating (possibly via post-copulatory mate choice mechanisms, Amos et al. 2001).

While male–female interactions have been observed during the breeding season, less is known about their interactions outside the breeding season. Telemetry studies reveal some evidence of sexual differences in foraging habits and habitats (see Sect. 9.4, Chap. 6), but very little is known about inter-sexual social behavioral interactions outside of the breeding season. Information on at-sea associations may help illuminate the importance or strength of mate choice and partner fidelity during breeding.

9.3.3 Interactions Between Females and Their Pups

Interactions between mother and pup have been a primary focus of studies into gray seal behavior since early observations (e.g., Fogden 1971; Anderson et al. 1975). The mother is the sole provider for and protector of the pup during the 2–3 weeks of dependence (lactation periods tend to be shorter in the Western Atlantic populations compared to Eastern Atlantic; Lydersen and Kovacs 1999). Therefore, researchers have endeavored to understand the behavioral mechanisms involved in maternal care and the drivers of variation in the quality and success of pup rearing within and across colonies.

The immediate post-parturition period, within approximately 1 h of birth, is a critical time for mothers and pups to establish a social bond (Davies 1949; Fogden 1971; Burton et al. 1975; Fig. 9.4), and failure to achieve this likely contributes to poor maternal care. Postpartum females bond with their pup through early behavioral interaction (Davies 1949; Fogden 1971; Burton et al. 1975), which triggers an increase in basal plasma oxytocin concentrations in the mother (Robinson et al. 2015a, 2019). These elevated plasma oxytocin concentrations likely initiate essential maternal behavior and maintenance of mother–pup proximity. Failure to achieve this early interaction interrupts the feedback between behavior and physiology, and without the elevation of oxytocin levels in the mother, appropriate maternal behavior may not be triggered. Consequently, the probability of pup mortality is much greater in the first few days postpartum (Coulson and Hickling, 1964; Burton et al. 1975; Anderson et al. 1979; Baker 1984; Baker and Baker 1988; Twiss et al. 2003). Stochastic events around the time of parturition can dramatically affect this process. For example, Robinson et al. (2015a) showed that, even in highly experienced mothers that have raised pups successfully in previous breeding seasons, disturbances during this critical time window that distract the mother from making early social interactions can disrupt the process, leading to a failure of maternal care, and likely abandonment of the pup. Distractions can include aggression from neighboring conspecifics, especially prevalent where mothers are more densely aggregated (see Sect. 9.3.6), disturbance from gulls that seek to scavenge the placenta, separation of mother and pup such as during tidal inundation in beach breeding colonies (see Sect. 9.3.6), and potential anthropogenic disturbance where humans come into close contact with seals during pupping (Burton et al. 1975). While some mothers may be more resilient to such distractions (see Sect. 9.5), the chance element means that unsuccessful pupping events can occur even among typically successful mothers. This complicates predictions of lifetime reproductive success, especially in times of rapid environmental change.

In addition to protecting her pup from harm, the mothers' other essential task is to provision her pup with milk (Chap. 8). Provisioning is a balance between maximizing

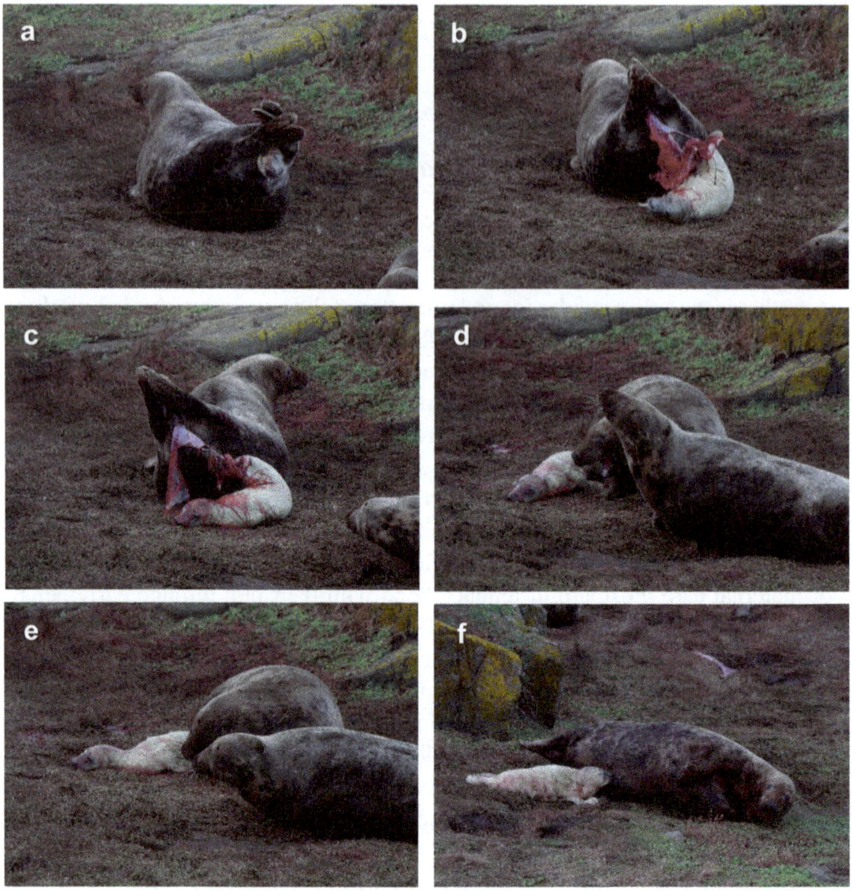

Fig. 9.4 Depiction of a birth sequence on the Isle of May, Scotland, illustrating some of the typical challenges a mother faces immediately postpartum. Mother giving birth (**a**–**c**; time 17:14:04–17:14:32 GMT) attracts the attention of a neighboring mother (**c**; 17:14:32) and requires the new mother to engage in female-female aggressive interactions (**d**; 17:14:54). However, the new mother also needs to establish contact with the neonate, and she succeeds in making her first visual and olfactory contact despite the proximity of the neighbor (**e**; 17:14:56). The new mother then moves approximately 2 m further away from the neighbor, with the pup following, and positions herself to nurse the new pup (**f**; 17:15:22) Image credits: SD Twiss

the pup's chances of survival post-weaning (Hall et al. 2001, 2002; Bennett et al. 2007, 2010) while avoiding over-investing in the pup, which risks incurring longer-term adverse effects, such as having to skip the subsequent breeding season (Pomeroy et al. 1999). A standard component of an ethogram for mother–pup observational studies is nursing and/or presenting, where the female lies on her flank, exposing her nipples to the pup (presenting), at which point the pup typically brings its nose to the mother's nipples. The mother is often considered to be nursing when the pup makes oral contact with a nipple. It is difficult to observe when a pup is consuming milk or how much; therefore, measures of time spent in nursing and/or presenting are not well correlated with energy transfer. However, they are a measure of the behavioral effort a mother is investing in her pup and how this may change over the course of lactation. There appear to be differences across colonies in the temporal patterns of presenting and nursing. For example, Kovacs (1987) found no change in the daily proportion of time mothers at the Isle of May spent presenting and nursing as their pup grew. Yet, on North Rona, Culloch (2012) identified an increase in time devoted to presenting and nursing across lactation. At Sable Island, Lang et al. (2011) also reported that the proportion of time spent nursing increased significantly between early and late lactation. This concurs with Mellish et al.'s (1999) study, where rates of milk intake and milk fat content, which are related to pup growth rates, increased significantly during lactation, although with considerable inter-individual variation among mothers. These behavioral studies were confined to daylight observations and are not likely representative of night-time behavior (Culloch et al. 2016; Fraser et al. 2019). Using animal-mounted triaxial accelerometers, Shuert et al. (2018, 2020a.) overcame the challenges of night-time observations to identify the presenting/nursing posture during both day and night. From these studies, Shuert et al. (2020a) found that time spent presenting/nursing did not vary across lactation for 38 mothers during two consecutive breeding seasons at the Isle of May, although mothers often engaged in more presenting/nursing behavior at night than in daylight.

A theoretical expectation is that mothers, particularly those that are in better condition, should invest more in male offspring. Some studies have provided behavioral or energetic (e.g., maternal expenditure measured as the rate of mass loss) evidence of differential investment between the sexes (e.g., Kovacs and Lavigne 1986; Anderson and Fedak 1987; Kovacs 1987; Baker et al. 1995), while others have failed to find such a difference (e.g., Bowen et al. 1992; Smiseth and Lorensten 1995a; Pomeroy et al. 1999; Shuert et al. 2020a). It remains unclear whether the differences are due to local random sampling biases, or if some populations exhibit differential investment while others do not.

Despite the remaining uncertainty about whether mothers routinely differ in their behavioral and/or energetic investment in pups based on the pup's sex, individual mothers vary in attentiveness toward their pups. Some mothers regularly check on their pup, with the mother giving a definite, distinct and directed look to her pup (Twiss et al. 2012a, b; Culloch 2012; James 2013), irrespective of the pup's behavior or the behavior of conspecifics; while other mothers do so more infrequently (see Sect. 9.5). Mothers adjust attentiveness as their pup ages. In the Isle of May and North Rona colonies, Kovacs (1987) and Culloch (2012) reported that pup-checking rates decreased as the pup aged. Furthermore, on the Isle of May, young pups were accompanied by their mothers significantly more than older pups (Kovacs, 1987), suggesting that females tend to be more vigilant while the pups are most vulnerable (Anderson et al. 1979; Coulson and Hickling 1964). Mothers generally react to their pup's behavior, commonly as a result of movement or begging behavior (e.g., Fogden 1971; Kovacs 1987; Smiseth and Lorentsen 2001), although with considerable variation across mothers (James 2013).

Whether mothers or pups initiate nursing bouts has been an intriguing question that touches on parent–offspring conflict; mothers seek to provision their current pup adequately but not over-invest in any one reproductive event, while each pup seeks to maximize its gain from its mother. On the Isle of May, Kovacs (1987) reported that nursing was often preceded by the mother nosing her pup. The mother then typically presented to the pup and moved several meters, with the pup following. This was interpreted as the mother leading her pup away from conspecifics to nurse without disturbance. However, nosing behavior is performed by the pup as well, with the pup often repeatedly nosing the mother's abdominal region in an attempt to persuade the mother to adopt the presenting/nursing posture. Similar behavioral interactions between mother and pup occur in other colonies, and, in most cases, it is the pup that initiates the majority of suckling bouts (e.g., Fogden 1971; Smiseth and Lorentsen 2001; Culloch, 2012; James 2013).

Pups also produce vocalizations, which may be an essential behavioral cue for attracting maternal attention (Davies 1949; Fogden 1971). However, some observational studies have placed less weight on the importance of pup vocalizations in establishing contact when a mother comes ashore to reunite with her pup (Burton et al. 1975). A more quantitative study on these begging calls was undertaken by Smiseth and Lorentsen (2001) in the Froan archipelago, Norway, where mother–pup pairs are widely dispersed, and the mother spends considerable time in the water. There, pups begged at significantly higher rates when hungry than when satiated. Pups vocalized more often than expected by chance before their mothers came ashore and, in turn, the mothers were more likely to approach their pup shortly after it gave a begging call, indicating that mothers respond to pup vocalizations. However, it was unclear whether mothers were able to recognize and respond to the call of their pup or were simply responding to the distinctive sound of pup vocalizations in general. Caudron et al. (1998) found high levels of pup call individuality in the harmonics, but they argued that such calls were used too infrequently to play a significant role in mother–pup reunions. On the Isle of May, McCulloch et al. (1999) also found pup calls to be individually distinctive, but playback experiments revealed that mothers in this colony did not respond more to calls from their pups than to calls from non-filial pups. When conducting the same tests at Sable Island (Canada), McCulloch and Boness (2000) found that mothers responded more strongly to vocalizations from their pup compared to those of other pups, suggesting an ability to discriminate and respond preferentially to calls of their dependant pup. These studies also highlighted much higher levels of allo-suckling at the Isle of May colony compared to Sable Island. The importance of pup vocal cues with respect to ensuring a successful reunion with the mother may vary across colonies, and it remains unclear what selective pressures might enhance individual recognition capabilities in one population, or colony, compared to others.

Irrespective of stimuli that mothers respond to, mothers need to provide adequate protection for their pup throughout lactation. The majority of pup mortality during dependence results from permanent separation of mother and pup (Coulson and Hickling 1964; Burton et al. 1975; Anderson et al. 1979; Baker 1984; Baker and Baker 1988; Twiss et al. 2003). Therefore, the most critical aspect of maternal protection is to prevent the pup from straying or sustaining injuries from neighboring adults (Burton et al. 1975). Essential to such protection is the maintenance of mother–pup proximity, and recent studies have shown that proximity seeking behavior in gray seals is facilitated by elevated maternal plasma oxytocin (Robinson et al. 2015a, 2017a, 2019).

Energetic investment and protection from harm are maternal activities that all mothers engage in, but social investment in the pup is not universally essential. Maternal social investment involves affiliative physical interactions with her pup, which includes (but is not limited to) nosing (touching her pup with her nose), flippering (using her flipper to 'stroke' her pup), and play behaviors. Kovacs (1987) and Culloch (2012) both found that such social interactions increased throughout lactation, which coincides with pups becoming more active and exploratory as they become older. It is not clear whether social investment has a longer-term impact upon the behavior of pups post-weaning (see Sect. 9.6). Mothers vary markedly in their tendency to perform such behaviors, with some engaging in social interactions far more than others (James 2013), which suggests distinct mothering styles. A significant challenge in furthering knowledge of the impact of such social interactions on pup development is the current inability to conduct longitudinal studies on weaned pups through to breeding age.

9.3.4 *Interactions Between Females*

Pregnant females often lie in close proximity (even within 1–2 m) on the periphery of the colony (Pomeroy et al. 1994) and seem very tolerant of each other. However, postpartum females tend to become intolerant of conspecifics that come within 3–4 m (Boness et al. 1982; Caudron 1998; Twiss et al. 2000). Therefore, any activity that places females close to others in the colony is likely to result in an aggressive interaction (e.g., Fogden 1971; Boness et al., 1982). Reasons why mothers may come into conflict typically involve pups straying, for example, while a mother is inattentive, which often leads to aggression toward the pup from neighboring adults. Mothers must then take remedial action to divert their pup from the potential threat or deter the neighboring adult (e.g., Fogden 1971, Boness et al. 1982, Fig. 9.5a). Conflict between females can occur as individuals commute from their pupping site to available pools of water in the colony (see Sect. 9.3.6), especially where availability of pools is limited (Fig. 9.5b), or where other environmental conditions initiate more movement, such as breeding beaches that are subject to tidal inundation (Coulson and Hickling 1964). Mothers have been reported to reduce their levels of aggression toward females in the latter stages of lactation (Boness et al. 1982), which supports the supposition that female aggression, particularly during the earlier stages of lactation when the pup is more vulnerable, is to protect the pup from conspecifics. Nonetheless, aggression between females is seldom physically damaging. Most interactions involve low-level threat behaviors such as wailing, flippering, open-mouth threats, and lunging, often with erect vibrissae, ending the lunge just as the vibrissae approach

Fig. 9.5 **a** Female-female aggression in the North Rona colony, Scotland, in the context of protection of the pup from neighbors. The mother on the left was previously at rest, meanwhile her pup strayed too close to the neighboring mother on the right, who then lunged at the pup. This action alerted the pup's mother who approached rapidly and can be seen exchanging open-mouth threats with the neighbor, as the pup flees back to safety. We can clearly see the erect vibrissae on both mothers, and no physical contact was made between the two. Image credits: SD Twiss. **b** Female 'sparring' aggression in the context of competition over limited access to pools in the Isle of May breeding colony, Scotland. All three females are mothers, who have left their pups (up to c. 50 m away in this case) to bathe and drink in this muddy wallow. All three are performing open-mouth threats, with vibrissae erect, and the right-hand female is also aggressively flippering (without contact). Such interactions tend to be less prevalent where pools are more abundant or larger, whereas they tend to increase in colonies with few, smaller, or more ephemeral pools Image credits: SD Twiss

or touch the opponent's body (e.g., Fogden 1971; Boness et al. 1982; Kovacs 1987; Culloch 2012). In most cases, female-female aggressive encounters avoid biting. Even in more intense aggressive encounters, most physical contact tends to consist of aggressive flippering.

One aspect of female-female social interactions that has not been adequately quantified is that of dominance relations. While quantitative measures of relative dominance among males are readily determined from observations of aggressive encounters (e.g., Twiss et al. 1998, Bishop et al. 2015b, 2017), no one has yet achieved the same for females. The problem is that female-female aggression is far more context-dependent than among males. For tenured males, the objective is to avoid being defeated by a rival and thereby losing access to potential mating opportunities (see Sect. 9.3.1). For a female, the objective is to protect her pup, and so the outcome of a female-female aggressive interaction can be determined by the actions of a third party, the pup. If the pup actively or inadvertently moves out of harm's way, then its mother often retires from a conflict with her neighbor, even though she may not necessarily be subordinate to the opponent. Therefore, simple win–loss metrics cannot be used to assess female dominance relations. In addition, an individual's willingness to engage in aggression, and the intensity of aggression, once initiated, is potentially related to behavioral type (Twiss et al. 2012a, see Sect. 9.5).

Interactions among postpartum females are not solely aggressive. Often, neighboring females inspect each other non-aggressively, typically by placing their noses close together or even touching. This 'nosing' behavior (e.g., Fogden 1971; Culloch 2012) has not been examined quantitatively. Such female-female interactions may represent the behavioral process for individuals becoming familiar and/or reacquainting themselves with neighbors from previous breeding seasons. In UK breeding colonies, females can use the same breeding colony for up to 25 years (Pomeroy et al. 1999), and depending on the colony, individuals often return to within a few meters of their previous years' pupping site (median measures for levels of interannual site fidelity range from 25 to 55 m, Pomeroy et al. 1994, 2000a, 2005). Despite this, Poland et al. (2008) found limited evidence of fine-scale kin clustering in the North Rona colony. Furthermore, neighboring mothers who were likely to interact socially because they were in the colony at the same time and in the same location, were no more related to one another than by random. Therefore, it is unlikely that the social interactions of mothers on North Rona are influenced by kin selection. However, there is evidence of non-kin-based social associations occurring at North Rona. With the high degree of site fidelity at this site, individual mothers tend to have the same neighbors over multiple years, raising the possibility of pro-social behaviors such as reduced aggression among familiar neighbors (Ruddell et al. 2007).

A likely mechanism for reduced aggression lies in the elevated oxytocin levels of mothers (see Sect. 9.3.3), a possibility highlighted by Robinson et al.'s (2015a, 2017a; b) work showing reduced levels of costly interactions among familiar weaned pups (see Sect. 9.3.5). Repeated proximity does not necessarily demonstrate active choice to be near familiar neighbors across breeding seasons and could simply be a by-product of the observed site fidelity. However, at North Rona, among females that showed less site fidelity by pupping more than the median distance from the previous year's pupping sites, the co-occurrence of neighbors across years was much greater than expected based on random relocations (Pomeroy et al. 2005). This indicates that at least some females formed long-term active associations and were 'choosing' to be near one another in successive years. At other colonies, however, such high degrees of site fidelity do not occur. Only 2.9% of females returned to their previous pupping site at Sable Island, with a median dispersal distance of 5.1 km between years (Weitzman et al. 2017). Therefore, site fidelity on Sable Island is rare; however, females spend time on the island prior to giving birth, often traveling more than 10 km over several days before deciding on a pupping site. Based on these observations, Weitzman et al. (2017) suggested that females are making decisions on their breeding site based on external cues. Of the potential mechanisms not tested in their study, they considered local density and social interactions with other gray seals as possible explanations. Since gray seals are gregarious, females may be attracted to other females as cues for appropriate breeding locations (e.g., Pomeroy et al. 2000b, 2001). Therefore, patterns of female-female interactions are likely to differ across sites dependent upon the distribution and familiarity of neighboring females.

9.3.5 Interactions Between Pups

Most ethological research on gray seal pups has been of mother–pup interactions, with less on pup–pup interactions, either during the phase of maternal dependency or post-weaning. Generally, dependent pups show little motivation to interact with each other, apart from occasions when abandoned pups attempt to suckle from another pup's mother, or follow dependent or weaned pups, again seeking (in this case, mistakenly) nursing opportunities (Smith 1968; Fogden 1971; Perry et al. 1998; McCulloch et al. 1999; McCulloch and Boness 2000). However, few quantitative studies have detailed the occurrence or consequences of these behaviors for the pup or mother.

There have been studies of interactions between weaned gray seal pups during their prolonged (up to c.40 days, Reilly 1991; Lydersen and Kovacs 1999; Bennett et al. 2007; Noren et al. 2008) post-weaning fast in the breeding colony. Building on preliminary evidence that adult females might recognize specific conspecifics (Pomeroy et al. 2005), Robinson et al. (2015b) conducted behavioral tests on wild, newly weaned gray seal pups. The authors examined investigative (olfactory or visual) and aggressive behaviors in pairs of weaners that were either strangers or had encountered each other previously. Even at the age of 3–4 weeks old, gray seal pups can recognize individuals they have previously encountered and reduce the levels of costly interactions with these familiars. Robinson et al. (2017b) went on to investigate some of the drivers of aggressive behavior between unfamiliar weaned pups, showing that larger male pups tended to exhibit more aggression, while relationships between size and aggressiveness were less clear for female weaners. The social or physical environment in which a pup was raised during its dependent period strongly influenced aggressiveness, with pups that nursed in more crowded parts of the colony engaging in more aggression as weaners. These results provide evidence of possible early social environmental influences on individual behavioral profiles (see Sects. 9.5 and 9.6). Robinson et al. (2017a) also explored the role of oxytocin in behavioral expression among weaned pups by experimentally elevating plasma oxytocin concentrations. Weaned pups that received additional oxytocin tended to seek others out and remain close to one another, with reduced frequencies of aggressive and investigative behaviors. Together, these studies provide intriguing suggestions of early pro-social development in gray seals.

Although gray seals are gregarious on breeding grounds and haul-out sites, it is clear they do not exhibit high levels of sociality, with most of the interactions between conspecifics, except for mother–pup behavior, being agonistic rather than affiliative. However, they do show relatively simple pro-social behaviors, such as reduced aggression among familiars. Such behavioral mechanisms are likely to be key stepping-stones for the evolution of more complex forms of sociality (Pomeroy et al. 2005; Ruddell et al. 2007). Therefore, more in-depth studies of social interactions among individual gray seals have the potential to shed light on the early stages of social evolution.

Dependent and weaned pups engage in play and exploratory behavior (Wilson 1974; Kovacs 1987; Hunter et al. 2002), which tend to increase as dependent pups age (Kovacs 1987). However, this is typically self-directed or object play, and rarely takes the form of social play with conspecific peers (Kovacs 1987). The need to conserve energy by minimizing activity during the nursing period seems to be the main driver that constrains play behavior, at least during the earlier stages of development (Kovacs 1987). Observations of juvenile gray seals show that social play behavior occurs in the early years post-weaning (Survilienė et al. 2016). However, there has been little effort to measure this behavior, likely because juveniles are irregular attendants around the periphery of breeding colonies, making it challenging to observe. Published information on interactions between young gray seals is therefore limited and has probably been regarded as a topic of little relevance to understanding gray seal biology. However, given increasing knowledge of individual variation in behavioral strategies in adults (see Sect. 9.5), and the remarkable foraging feats of adults (see Sect. 9.4), understanding how behavior patterns develop within individuals is likely to be informative.

9.3.6 Interactions with the Physical Environment

As Stirling (1975) emphasized, gray seals breed on a remarkable range of habitats, from pack ice and land-fast ice, through to sandy beaches and rugged and broken rocky terrain (Fig. 9.6). Indeed, gray seals breed over almost the entire range of habitat types utilized by all species of phocids, which has enabled detailed comparative investigations that have provided essential insights into the factors that regulate breeding patterns in phocids. Within island colonies, fine-scale spatial and temporal variations in the physical environment are fundamental drivers of individual behavior,

Fig. 9.6 Examples of the differing terrain across gray seal breeding colonies. Images of gray seals breeding on ice show the overall dispersion of seals (**a**), but also illustrate the heterogeneity of the substrate at a fine spatial scale (**b**). Sable Island, Canada, is a sandbar approximately 43 km long but only 1.2 km across at its widest point, providing extensive breeding grounds along its shores (**c**), and also in the interior of the island (**d**). Island colonies, such as North Rona, Scotland (**e**) have limited access to the sea via narrow gullies, but relatively open terrain inland for pupping, with many widely distributed pools. By contrast the Isle of May, Scotland, (**f**) is predominated by more irregular terrain and fewer pools. Image credits: KM Kovacs (**a**, **b**), Damian C. Lidgard (**c**, **d**), SD Twiss (**e**, **f**)

both by the direct effects of an individual's immediate environment, but also by the indirect effects of the environment's impact on distribution, density, and movement of seals within colonies (Fig. 9.7). Consequently, gross differences in gray seal behavior patterns and demographics between colonies or populations can, to a large extent, be explained by colony physical habitat (Boyd et al. 1962; Twiss et al. 2000, 2003; Matthiopoulos et al. 2005; Stephenson et al. 2007).

Fig. 9.7 Examples of the contrasting topographies and seal distribution even within one breeding colony; the Isle of May, Scotland. **a** Dense breeding aggregation adjacent to inlet access to/from sea. Here, mothers regularly commute between their pup on land and the large tidal inlet. **b** Tidal boulder beach, where mothers will often enter the sea, particularly during high tide. Pups tend to be pushed to the high tide line by wave action, though some pups will inadvertently be swept into the water, while others, particularly older pups may actively enter the water. **c** Dense breeding aggregation around limited pools in an inland site. The physical terrain dictates the availability and distribution of pools, around which pupping sites tend to cluster, and mothers will commute from nearby pupping sites to and from the pools. **d** Low density breeding area with relatively uniform terrain and no pools, and the resulting more uniform distribution of mothers and pups. There is some clustering around the damper ground (darker brown substrate), but far less than in (**c**). Mothers at this site rarely commute to pools (nearest is approximately 100 m away) but will do so in unusually warm conditions. These patterns of female distribution and movement, driven by the nature of the terrain, strongly influence the behavior patterns of males as they seek to monopolize mating opportunities Image credits: SD Twiss

The structure and stability of breeding habitat influences patterns of interannual site fidelity. High fidelity is often exhibited in rocky island colonies around the UK that have temporally stable intra-colony variation in pupping-site characteristics and quality (Pomeroy et al. 1994; Twiss et al. 1994; Pomeroy et al. 2000a). Conversely, site fidelity is limited on more uniform or unstable or unpredictable substrates such as land-fast or pack ice (e.g., Lydersen et al. 1994; Haller et al. 1996), or the shifting sand of Sable Island (Weitzman et al. 2017).

Given the restricted movements of mothers during lactation, fine-scale habitat quality of pupping sites influences time-activity budgets of lactating females (Anderson and Harwood 1985; Twiss et al. 2000, Shuert et al. 2020a) and the level and quality of pup attendance (Twiss et al. 2000; Redman et al. 2001; Pomeroy et al.

2001), with probable consequences for female energetics (Shuert et al. 2020b) and pup survival (Boyd et al. 1962; Summers et al. 1975; Twiss et al. 2003). Breeding substrate also strongly influences behavioral patterns during the breeding season. Although most studies show a high proportion of time-activity budgets spent in rest (typically over 60%) by adult males, females, and pups (e.g., Anderson and Harwood 1985; Kovacs 1987; Haller et al. 1996; Twiss et al. 2000; Bishop et al. 2015c; Fraser et al. 2019; Shuert et al. 2020a), the levels of activity are driven in part by accessibility to the sea. Lactating mothers breeding on the periphery of land-fast or pack ice or tidal beaches or inlets tend to spend more time in the sea (Fig. 9.7), though still retaining proximity to their pups, who generally remain on the ice or shoreline (Cameron 1967, 1969; Anderson and Harwood 1985; Kovacs 1987; Lydersen et al. 1994; Haller et al. 1996; Twiss et al. 2000, 2001). Seals that are subject to tidal inundation spend more time in the sea (Anderson and Harwood 1985, Smiseth and Lorentsen 1995b, 2001, Twiss et al. 2000) and experience greater disruption, periodically separating mothers and pups, rearranging neighbors, and disrupting social organization (Bishop et al. 2015b). When adult seals enter the sea during the breeding period, there is the potential for both males and females to supplement their energy reserves by feeding (Lydersen et al. 1994; Lidgard et al. 2003). By contrast, when seals remain on land throughout the breeding period, they must sustain their activity, and for mothers, the provisioning of their offspring, exclusively on energy reserves acquired prior to the breeding period (Chap. 8). Unlike many breeding pinnipeds (especially among otariids but also elephant seals), gray seal mothers tend to maintain a 'personal' space between themselves and their neighbors of at least 1–2 adult body lengths, even among the densest aggregations (Boness et al. 1982; Twiss et al. 2000), allowing them to protect their pups from neighbors. However, distributions of mothers in colonies are still patchy, with varying densities and degrees of aggregation dependent upon topography (e.g., Anderson and Harwood 1985; Twiss et al. 2000, 2001; Fig. 9.7). In more uniform terrain, such as sandy beaches, mothers are more evenly distributed, whereas in more irregular terrain densities and aggregation sizes vary.

Gray seal mothers prefer to pup close to water for thermoregulation (Redman et al. 2001; Twiss et al. 2002) and drinking (Stewart et al. 2014), and the distribution of pools of fresh, brackish, or seawater consequently drives much of female spatial distribution. The location and temporal permanence of pools is a product of local topography, substrate permeability, and prevailing weather conditions. Not all females are able to gain pupping sites close to pools and may have to periodically commute to pools, especially in drier, warmer conditions (Redman et al. 2001, Shuert et al. 2020a, b). Such commuting comes with associated risks of permanent separation from offspring, as pups do not commute but may wander from their pupping site while the mother is away (Redman et al. 2001), and mother–pup separation is a significant cause of pup mortality (Summers et al. 1975; Anderson et al. 1979; Baker 1984; Baker and Baker 1988; Twiss et al. 2003). Therefore, the availability and distribution of pools relative to pupping sites strongly influence maternal behavior and, consequently, pup behavior. Where pools are few, small, or aggregated in distribution, access to pools may also be dependent upon maternal dominance, with potentially more dominant or experienced females gaining pupping sites close to pools (Twiss et al. 2000; Pomeroy et al. 2001). However, occupying such sites during pup rearing also incurs costs, due to higher densities of seals and regular disruption from incoming commuters (Redman et al. 2001, Stephenson et al. 2007, Fig. 9.5b). At colonies with limited or no pools, such as Sable Island, females tend to remain with their pups throughout lactation (Perry et al. 1998; McCulloch and Boness 2000), presumably deriving all water needs from fat metabolism (although seals have been observed eating snow at this site, Twiss pers. obs.). What impact no access to drinking water has on the reproductive physiology of gray seal mothers is unknown. However, with the breeding season spanning mid-winter, with much lower ambient temperatures, seals at Sable Island may be less thermally stressed than those occupying inland pupping sites at UK colonies in the autumn.

The spatial distribution and behavior of males in the breeding colony are driven indirectly by topography because males compete for access to aggregations of females (Boness and James 1979, Anderson and Fedak 1985, Twiss et al. 1994, 2007, Tinker et al. 1995, Haller et al. 1996, Lidgard et al. 2003, Bishop et al. 2017, see Sects. 9.3.1 and 9.3.2), and female aggregations are primarily determined by topography. More dominant males can maintain 'prime' positions among the larger or denser aggregations of females, typically located close to larger, more permanent pools. As the distribution and availability of water and prevailing ambient temperatures influence the degree of commuting of mothers, this inevitably impacts male behavior. Increased female mobility leads to reduced potential for dominant males to monopolize access to females, consequently reducing the degree of polygyny and intensity of sexual selection (Twiss et al. 2007). Notably, this link between topography (pools), ambient temperatures, female distribution, and behavior is a critical consideration in the context of environmental change. More unpredictable weather conditions during the annual gray seal breeding season, with exposure to drier, warmer conditions are likely to lead to more females commuting, potentially more mother–pup separations, more pup abandonment, and changes in the mating pattern (Twiss et al. 2007, Shuert et al. 2020b). This link between topography and weather patterns is a crucial driver of both female and male reproductive success across all gray seal breeding habitats, and understanding the effects of climate change on these processes is essential for predicting future impacts on population dynamics (Jussi et al. 2008; Kovacs and Lydersen 2008, Klimova et al. 2014).

9.4 Interactions at Sea

It is challenging to reduce knowledge gaps of at-sea behavior, particularly for long-lived, far-ranging species such as gray seals. At the surface, visual observations of behavior inevitably lack context, as most of an animal's time is spent below the surface. The development of biologging technology in the 1970s and 1980s provided new tools for investigating the at-sea behavior of marine mammals (Chaps. 5 and 6). Along with Weddell (*Leptonychotes weddellii*, Chap. 13) and elephant seals

(Chaps. 11 and 12), gray seals were the subjects of some of the first deployments of VHF and satellite telemetry devices. The first insights into movements and foraging strategies of gray seals at sea were published in the early 1990s (Thompson et al. 1991; McConnell et al. 1992). Building on this initial work, researchers have generated some of the most extensive, detailed, and informative datasets of phocid behavior at sea using the gray seal study system. There have been numerous publications using telemetry data to investigate gray seal movement patterns (Thompson et al. 1996; McConnell et al. 1999), foraging areas (Jessopp et al. 2013; Gosch et al. 2019), seasonal variation in foraging (Breed et al. 2009, Chap. 6), sexual segregation (Beck et al. 2003; Carter et al. 2017), and the potential impacts of anthropogenic disturbance on behavior (Mikkelsen et al. 2019).

Extrinsic and intrinsic factors inevitably drive at-sea behavior. For example, Carter et al. (2017) found that water depth contributed significantly toward sex differences in diving behaviors. Females foraged in shallower waters than males but spent more time closer to the seabed. Sex-specific reproductive requirements could drive seasonal variation, both broadly across the species and between sexes (e.g., Austin et al. 2004; Breed et al. 2013; Carter et al. 2017), a supposition supported by studies on gray seal diet composition (e.g., Beck et al. 2003, 2007). In adult gray seals, differences in fatty acid signatures were greatest between adult males and adult females during the post-breeding period. During this time, females foraged more selectively on higher quality prey than males, likely due to females needing to recover energy supplies to support pregnancy (Beck et al. 2003, 2007). While the interpretation of telemetry data can be challenging, the above examples highlight how complementary studies such as fatty acid analyses can provide additional information to enhance behavioral inferences from telemetry deployments.

Telemetry devices can help ascertain time-activity budgets for gray seals by using metrics that are more spatial than behavioral, such as the percentage of time spent at given distances to shore (e.g., Breed et al. 2013). Time-activity budgets can also be derived from both location and movement data (e.g., McClintock et al. 2013; Russell et al. 2015; Mikkelsen et al. 2019). Russell et al. (2015) investigated population-level time-activity budgets using telemetry data from 63 gray seals, defining four behavioral states comprising resting at sea, resting on land, and two components of diving, traveling, and foraging. Both juveniles and adults exhibited sex-specific seasonal variation in the proportion of time spent foraging and traveling. However, there was no evidence of sex, age, or seasonal trends in time spent in the broader behavioral categorizations of resting (on land and at sea) or diving (combining both traveling and foraging). The authors hypothesized that gray seals may exhibit little flexibility in the proportion of time spent resting, suggesting possible constraints on the levels of activity at sea. Alternatively, the seals may be able to vary the allocation of resting or diving into their respective parts (i.e., resting on land, resting at sea, traveling, and foraging) in response to varying intrinsic and extrinsic drivers to obtain their energetic requirements. Studies such as these illustrate how telemetry data can provide time-activity budgets for seals at sea and how that time partitioning varies across individuals.

The focus of recent telemetry studies has often been driven by management or conservation concerns, particularly in the context of interactions with, or consequences of, anthropogenic activities in the marine environment. However, such studies still yield important behavioral insights. One area to receive substantial attention is interactions with commercial fisheries (e.g., Cronin et al. 2016; Jessopp et al. 2013; Königson et al. 2013; Stavenow et al. 2016; Vincent et al. 2016). In many areas where gray seals and fisheries co-exist, there is often a management need to assess overlap and interaction (e.g., Jessopp et al. 2013; Stavenow et al. 2016). In the Baltic sea, evidence from underwater cameras showed that some adult male gray seals have exploited a niche to become specialized in raiding salmon traps (Königson et al. 2013). Observations of 600 seal visits to two salmon traps revealed that 426 were made by just 11 individuals, four of whom regularly returned throughout the two-year study. The authors concluded that these seals had developed a characteristic behavior pattern that has persisted over a long time.

In Irish waters, Cronin et al. (2016) provided evidence that gray seal space use overlapped with passive fishery locations (i.e., fisheries using static gear such as tangle, trammel, and gillnetters) much more than with active gear fisheries, such as trawls. The authors suggest that physical interactions at the net may be by a small number of individuals that have learned net-feeding behavior, reinforced with the reward of an 'easy meal' that requires minimal energy to acquire. In the Baltic Sea, van Beest et al. (2019) found that adult male and female gray seals favored areas with similar physical and environmental features as active fisheries. However, they stressed that overlap with fisheries was minimal (where the seal was <5 km from an active fishing net occurred c. 3% of the time), indicating that the temporal aspect is critical for inference when investigating seal interactions with fisheries when using telemetry data. Such studies provide valuable information that can improve our understanding of habitat use and potentially contentious issues, such as resource overlap and competition between top predators and fisheries, essential for the effective conservation and management of charismatic, top marine predators. These studies also give insight into how behaviorally adaptable gray seals might be (see Sects. 9.5 and 9.6), and their potential to cope with rapid changes to their environment (e.g., Russell et al. 2014).

Telemetry studies of gray seals at sea often identify a considerable degree of individual variation in behavioral patterns and, in many cases, individuals show clear preferences for foraging sites and routes to and from haul-out sites (e.g., McConnell et al. 1992, 1999; Austin et al. 2004; Breed et al. 2013; Jessopp et al. 2013; Russell et al. 2015; Lidgard et al. 2020). Whether these patterns are indicative of specific behavioral types or not (see Sect. 9.5) is unclear, but it indicates some degree of individual behavioral specialization. Often, our understanding of at-sea behavior is limited to small sample sizes, particularly in cases where expensive telemetry devices are used. Deliberate repeat tagging of individuals is rarely undertaken to avoid pseudoreplication. However, there is value in understanding how individuals use their habitat across years, as this can provide insights into the development of foraging strategies and preferences and the extent of behavioral plasticity that gray seals might exhibit, all of which are key considerations in a rapidly changing

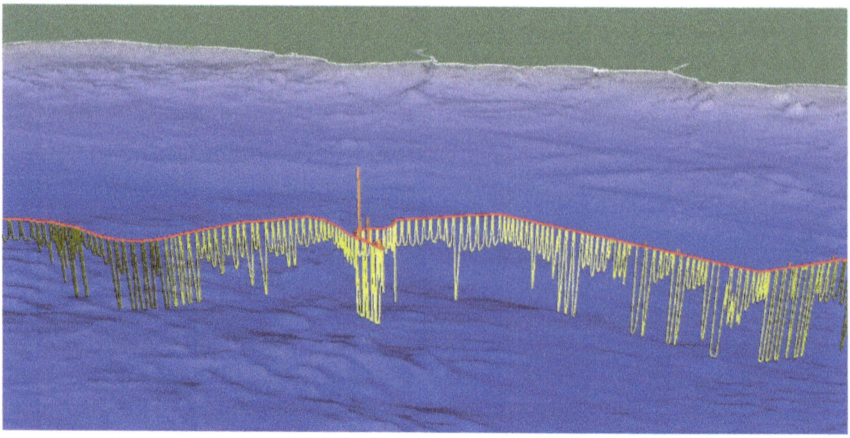

Fig. 9.8 A 3D track of a grey seal at sea (in North Sea, UK), showing not only location and depth but also behavioral events. On-board processing of accelerometer data from a GPS phone tag detects possible prey capture attempts. The vertical orange bars indicate the number of prey capture attempts in each successive phase of diving (descent, bottom, and ascent). Telemetry devices were deployed by Cecile Vincent (CNRS/ Uni. De la Rochelle), image created by Sea Mammal Research Unit using MamVisAD Bathymetry source: EMODnet 2018 DTM

environment. The telemetry studies cited in this section, among many others, have provided valuable insights into at-sea behavior (Fig. 9.8) and have ultimately helped us to better understand how seals use their marine and terrestrial habitats. There remains a paucity of information on how individuals interact with one another while at sea, primarily due to a current lack of technological solutions (Baker et al. 2014).

9.5 Personality

Despite the general patterns in behavior in relation to the breeding environment, there remains considerable individual variation in behavior at all colony habitat types (Lydersen and Kovacs 1999; Twiss et al. 2000). This individual variation is driven in part by the local habitat and environment (e.g., Lydersen et al. 1994; Tinker et al. 1995; Twiss et al. 2000, 2001, 2007; Redman et al. 2001; Stewart et al. 2014), but other factors play important roles, including individual age and experience (Godsell 1991, Haller et al. 1996, Lidgard et al. 2012), status (Boness and James 1979; Twiss et al. 1998, Bishop et al. 2015b), and condition (Anderson and Fedak 1985; Tinker et al. 1995; Mellish et al. 1999; Pomeroy et al. 1999; Lang et al. 2009). However, even among individuals of the same age, sex, and status, we still observe apparent individual differences in behavior, a level of individual variation akin to personalities.

The concept of personality in non-human animals in ethological research has seen a remarkable growth of interest over the past two decades, with much debate over what constitutes personality in non-human animals. A review of that debate is beyond the scope of this chapter, but a useful introduction to the topic can be found in Briffa and Weiss (2010), with more detailed discussions in Sih et al. (2004), Réale et al. (2007), Dingemanse et al. (2010), and Carere and Maestripieri (2013). At its heart is the concept that individuals, even within the same age and sex class, differ in their behavior and do so consistently over time, functional contexts (such as foraging or courtship), and/or situations (differing local environmental conditions). Although this does not preclude the possibility of behavioral flexibility (plasticity) in individuals, it does mean that individuals tend to maintain consistent rank order differences in behavioral expression, even if that behavioral expression is modified in response to internal or external stimuli. Evidence of such consistent individual differences (CIDs) in behavior has been revealed in a remarkably wide range of taxa, from Cnidaria (Briffa and Greenaway 2011) to Mammalia (Bell et al. 2009; Smith and Blumstein 2008), suggesting that such within-population variation is a fundamental evolutionary condition under strong or persistent selective pressure, or is a product of constraints on developmental plasticity that are widespread (Duckworth 2010). Either way, understanding the causes and consequences of inter-individual variation is critical for our comprehension of how organisms respond to their environment (Sih et al. 2012; Wolf and Weissing 2012; Dochtermann and Dingemanse 2013). There are still few studies of personality differences in marine mammals (Highfill and Kuczaj 2007, Twiss et al. 2012a, b, 2020, de Vere et al. 2017, Frick et al. 2017, DeRango et al. 2019), with the only definitive studies of CIDs in phocids being on gray seals during the breeding season.

As far back as 1949, Davies observed that gray seal pups had different 'personalities', while in 1994, Lydersen et al. commented on the distinct differences in 'manner' and 'character' of mothers. Twiss and Franklin (2010) were the first to quantitatively show CIDs in behavior by examining vigilance patterns in adult males across successive breeding seasons (Figs. 9.9 and 9.10). Subsequent studies confirmed that vigilance behaviors provide useful metrics of individual differences in behavioral types for breeding females as well (Twiss et al. 2012a, b). For example, postpartum gray seals exhibit individually different rates of pup-checking behavior that remain consistent both within and across breeding seasons. This seems to be a general pattern as it has been observed in multiple UK breeding colonies (Twiss et al. 2012a, b; Culloch 2012, James 2013). These measures of CIDs in vigilance behavior were all derived from observational studies of wild gray seals. Twiss et al. (2012a, b), however, used a remotely controlled vehicle to experimentally manipulate the exposure of wild breeding seals to a novel auditory stimulus. Again, CIDs in pup-checking behavior were observed, but the degree to which individual mothers altered their pup-checking behavior from undisturbed (natural) to disturbed (remote-controlled vehicle) situations varied. Some mothers maintained consistent pup-checking rates across the two situations, while others raised pup-check rates in response to the novel stimulus. Twiss et al. (2012a) argued that these individual differences in behavioral plasticity likely reflect pro- and reactive stress-coping styles that have a known physiological basis that drives the observed behavioral differences (Koolhaas et al. 1999; Biro and Stamps 2010; Carere et al. 2010; Coppens et al. 2010; Twiss et al. 2020; Fig. 9.11).

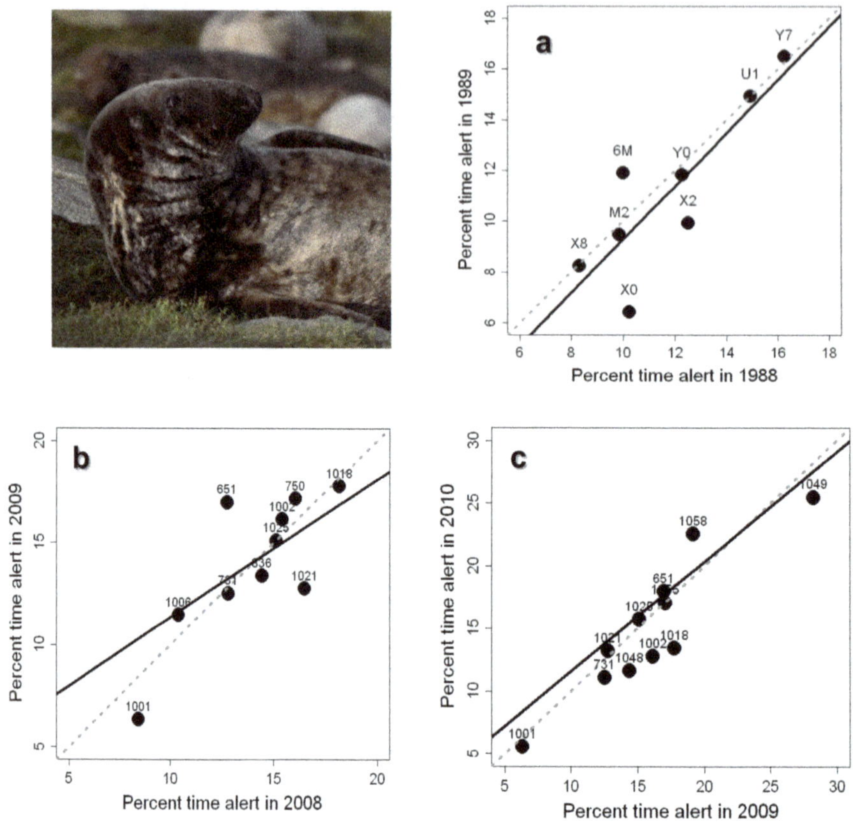

Fig. 9.9 Consistent individual differences in male gray seal behavior across multiple breeding seasons at the North Rona colony, Scotland. Tenured males exhibited interannual consistency in time spent alert across consecutive seasons (**a**–**c**), in 1988 and 1989 (Twiss and Franklin 2010), but also among males present in 2008, 2009, 2010 (Twiss; unpublished). Repeatability was highly significant in all plots (**1988/89**: ICC = 0.83, $F_{7,7} = 10.9$, p = 0.002, **2008/09**: ICC = 0.79, $F_{9,9} = 8.3$, p = 0.002, **2009/10**: ICC = 0.91, $F_{10,10} = 21.0$, p < 0.0001). These individually differing tendencies to perform vigilance behaviors were not related to levels of external stimuli but appear to be inherent differences among individuals that are consistent over time. None of the males present in 1988/1989 were present in 2008–2010, and yet the same pattern of consistent individual differences exists across these different generations of males. In all plots the solid line represents linear regression, and dashed line indicates the 1:1. Alphanumeric codes represent male identities Image credits: SD Twiss

Twiss et al. (2020) and Shuert et al. (2020a) used an integrative physiological indicator of stress-coping style (resting heart rate variability) in free-ranging gray seals to show that individual coping styles can influence both behavior and success during the breeding season. Shuert et al. (2020a) show that stress-coping styles influence time-activity budgets, especially the trade-off between time spent in rest (conserving energy) and time spent being vigilant (monitoring their local environment and/or

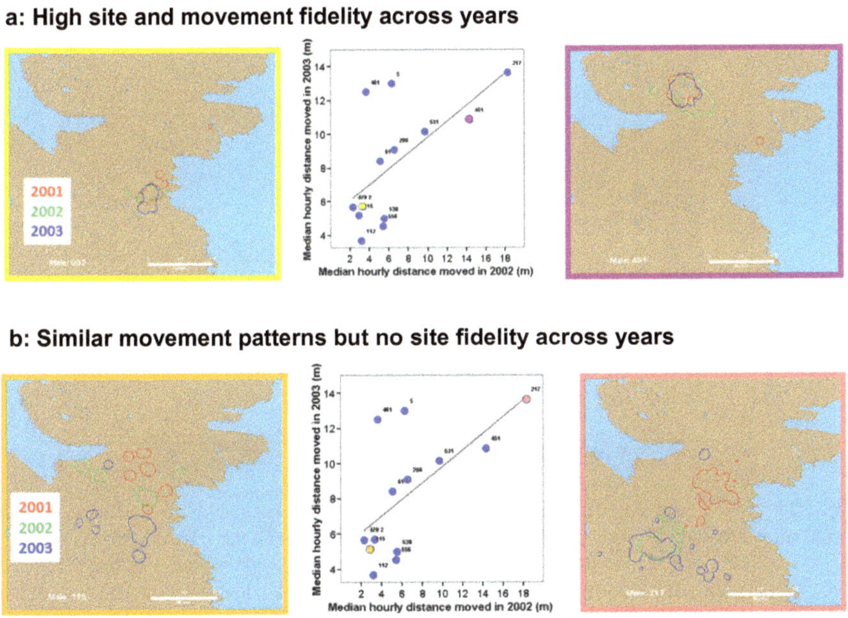

Fig. 9.10 Examples of interannual behavioral consistency in terms of space use in the breeding colony by tenured males on North Rona, Scotland, showing consistent movement patterns and high (**a**) or low site fidelity (**b**) (Twiss; unpublished). Space use metrics were derived from a sub-meter accurate GIS of male locations during the 2001–2003 breeding seasons. The scatterplots depict the median hourly distances moved by tenured males within their territories for males present in 2002 and 2003 (similar patterns exist when comparing 2001 and 2002). Individual movement patterns remained consistent across successive breeding seasons (**2001–2002**: $r = 0.81$, $p < 0.001$, **2002–2003**: $r = 0.79$, $p = 0.001$). Such consistency might be explained by male site fidelity (Twiss et al. 1994), such that individuals that occupy similar locations in each year are therefore exposed to similar levels of stimuli, but that explanation does not apply to all males. The maps depict the territories (95% kernel density estimates) of two site faithful (**a**) and two unfaithful (**b**) males across 2001–2003 and highlights their movement metrics on the scatterplots with points to match the colored borders around their respective maps

reacting to stimuli), with more proactive mothers tending to spend relatively more time vigilant. Twiss et al. (2020) provided evidence that stress-coping style influences short-term (within season) reproductive performance, with the behaviorally flexible reactive mothers exhibiting more variation in reproductive expenditure and consequent pup growth rates than proactive mothers.

Even within the wider personality literature, few studies examine the longer-term persistence of CIDs in long-lived species, especially in wild populations (Trillmich et al. 2018). Bubac et al. (2018) recorded behavioral responses of over 400 gray seal mothers to human approach and the handling of their pup, providing repeated scores of individual boldness (response to a potentially risky situation, Réale et al. 2007) over nine successive breeding seasons on Sable Island. This extensive study demonstrated individual differences in boldness that were highly repeatable between

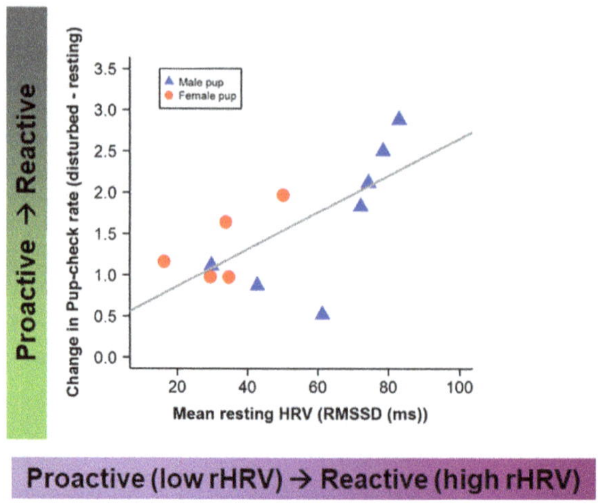

Fig. 9.11 The relationship between physiological indicators of stress-coping style and observational measures of behavioral plasticity among lactating females on the Isle of May, Scotland, in 2014 ($r_s = 0.59$, $n = 12$, $p = 0.021$; Twiss, unpublished). Coping-style is measured by the integrative physiological parameter of resting heart rate variability (rHRV); low rHRV indicates proactive individuals, high rHRV indicates reactive individuals (Koolhaas et al. 1999; Twiss et al. 2020). Behavioral flexibility in this case is based on how much individuals change their pup-checking rates from undisturbed situations to a disturbed situation (tested used a standardized stimulus delivered by remote-controlled vehicle, Twiss et al. 2012b). These data show that behavioral flexibility is linked to underlying physiological traits that dictate how individuals react to stimuli

and within years. Younger females tended to be less bold than older, more experienced mothers, which the authors argue may represent a life-history trade-off where younger mothers, with higher future reproductive potential, are more risk-averse.

Among the published literature, there is tantalizing evidence of individuality in many aspects of gray seal behavior. During the breeding season, females and males show high levels of site fidelity (Pomeroy et al. 1994, Twiss et al. 1994, though see Weitzman et al. 2017), males exhibit consistency of duration of tenure (Lidgard et al. 2012), and mothers show repeatable individual differences in their physiological capacity to provision their pups with milk (Lang et al. 2009). Outside of the breeding season, gray seals exhibit considerable variation in foraging behavior and at-sea movements (McConnell et al. 1999; Austin et al. 2004; Lidgard et al. 2020), including evidence of foraging specializations (Tucker et al. 2007). One difficulty with at-sea behavior is that few tracking studies followed the same individuals over multiple deployments of devices, and so longer-term persistence of individual differences in at-sea behavior has not been verified.

What is clear from this work is that a seal is not just a seal; there is a spectrum of behavioral types and life-history strategies within populations. This realization is essential when considering intra- and interspecific interactions and for refining models of potential population responses to environmental change. The existence of CIDs implies constrained behavioral plasticity within individuals and/or differential plasticity across individuals. Therefore, environmental changes are likely to impact specific sub-sections of the population more than others (Sih et al. 2012; Wolf and Weissing 2012).

9.6 Ontogeny of Behavior

There are few published studies of behavioral development in young gray seals beyond the interaction between the pup and mother (see Sect. 9.3.3). During the nursing period, gray seal pups spend most time resting (Kovacs 1987). Reduced activity is likely strongly selected to conserve energy, thereby maximizing growth and the deposition of energy-rich blubber (Lydersen et al. 1995; Lydersen and Kovacs 1999), as mass at weaning is a significant determinant of first-year survival (Hall et al. 2001, 2002). Once weaned, gray seals often remain in the breeding colony for several weeks, and interactions among weaned pups can influence behavior patterns, along with the earlier social environment during pup dependence (Robinson et al. 2015a, 2017a; b, b). However, it remains unknown whether these are persistent changes that influence life-long behavior patterns of individuals. What determines the length of the post-weaning fast is unclear; there are physiological limits to the post-weaning fast (Bennett et al. 2010, Chaps. 4 and 8), but local topography and particularly the availability of larger pools of water likely influence the length of stay in a colony and whether weaners gain much experience of the aquatic environment before venturing out to sea (Kovacs 1987). Ease of access to the sea from the colony also plays a role in at least providing an opportunity for early departure, and potentially earlier development of foraging skills (Smiseth and Lorentsen 1995b; Jenssen et al. 2010).

Less is known about how behavior patterns develop once young seals leave the colony. This is a critical phase for young, naïve gray seals that must learn to dive and hunt effectively, if not entirely efficiently, before their blubber and protein reserves become depleted (Bennett et al. 2010). This learning seems to be mostly independent, as there is no evidence of a maternal role post-weaning. Analysis of location and dive data from animal-borne telemetry devices deployed on weaned gray seal pups as they depart their natal colonies for the first time provide some insights into the development of dive characteristics (Bennett et al. 2010; Carter et al. 2017). During their first 40 days at sea, gray seals show rapid increases in dive duration, depth, bottom time, and benthic diving. However, Carter et al. (2017) also found sex and regional differences in diving behavior development, suggesting that both intrinsic and extrinsic factors, such as water depth, contribute to early sex differences in foraging behavior (see Sect. 9.4). Sex differences in foraging are also apparent in the latter stages of the first year of life, which is particularly interesting as size dimorphism is not fully developed at this stage of life (Breed et al. 2011; Russell et al. 2015).

Breed et al. (2011) compared the diving behavior of young-of-the-year (5 to c.15 months of age) to the behavior of sub-adult and adult gray seals in the Northwest Atlantic. The telemetry tracks of these seals suggested that young-of-the-year navigated with similar capabilities to older individuals but tended to undertake (or be forced into) longer foraging trips that were further from their haul-out sites. Unlike adults, which also have the demands of the annual breeding season to modify foraging patterns, young-of-the-year responded primarily to seasonal patterns of prey availability and condition. The development of foraging capacity is strongly tied to the development of physiology during these early phases of life (Chap. 4). Although the post-weaning fast is critical in developing the physiological capacity to commence diving behaviors, weaned pups are still developing physiologically as they depart their natal colonies (Noren et al. 2008; Bennett et al. 2010).

While biotelemetry studies can provide insights into changing dive capabilities of individuals, they, unfortunately, reveal little about the actual learning processes involved. However, experiential learning is probably fundamental to young seals continuing to refine their foraging capabilities during their first few years at sea. Whether seals learn from observing conspecifics is unknown. Observations of play in juveniles (Wilson 1974; Survilienė et al. 2016) provide some indication of mechanisms of behavioral or at least social development, but such studies are few and limited in scope. The primary logistical challenge is following specific individuals from weaning through to reproductive age and measuring not just the physical environment and food availability they encounter, but also their social environment during juvenile and sub-adult phases. These early years of independence are likely especially formative and potentially vital in determining individual behavioral profiles (Trillmich et al. 2018), and consequently, individual life-history trajectories (such as pro- vs. reactive behavioral types, see Sect. 9.5), but they remain a major gap in our knowledge.

Behavioral traits are often labile and have the potential for modification throughout an individual's lifespan. As by-products of studies with other research foci, there have been some intriguing indications of learning capacity and cognitive capabilities in gray seals, but no direct investigations of these faculties. Oliver (1977) investigated the ability of a captive juvenile male gray seal to navigate a maze. Although the focus of this study was on the sensory modalities used in detecting the presence of objects underwater in light and dark conditions, the study indicated some spatial learning and memory capacity evidenced through reduced error rates over repeat trials. Götz and Janik (2010, 2011) demonstrated both sensitization and habituation of gray seals to anthropogenic sounds with studies aimed at investigating the behavioral response to aversive sound signals (such as 'seal scarers') in water (Chap. 2).

Interestingly, these studies demonstrated long-term behavioral changes in (avoidance) behavior in both captive and wild settings. Stansbury et al. (2015a) show that juvenile gray seals can learn to use sounds from acoustic fish tags to indicate the location of a food reward, demonstrating that gray seals have the capacity for associative learning of novel cues. Further evidence of cognitive capabilities, at least in response to auditory stimuli, comes from studies investigating gray seal ability to identify and classify call types (Shapiro et al. 2004), and to generalize acoustically similar calls into classes (Stansbury et al. 2015b). Although it remains unknown how seals might discriminate different call types in terms of the auditory information the seal is accessing, such studies raise the possibility that gray seals may be able to distinguish between calls of different conspecifics. This is particularly intriguing given the evidence of social associations in this species (Pomeroy et al. 2005; Ruddell et al. 2007). Evidence of continued ability to learn in adulthood comes from Königson et al.'s (2013) study showing that adult males can alter their foraging behavior and adopt new foraging skills (targeting salmon traps, see Sect. 9.4). Russell et al. (2014) showed intriguing evidence of gray seals' abilities to adapt their foraging behavior to exploit novel opportunities provided by anthropogenic underwater structures (e.g., pipelines, wind turbine bases). Other 'new' behaviors observed among gray seal populations may further support this ability of individuals to modify their behavior patterns throughout life (see Sect. 9.7), though the degree of behavioral plasticity may vary across behavioral types (Twiss et al. 2012a, 2020, see Sect. 9.5). Studies of learning and cognition in gray seals remain scarce and are understandably based on a few individuals. However, given the evidence of considerable individual variation in a wide range of behaviors and behavioral types in this species, much remains to be discovered about the learning capacity of gray seals of all ages and behavioral types.

9.7 Novel Observations and Emergent Behaviors

Despite the long history of gray seal behavioral studies, researchers still find new behaviors, either unobserved previously or newly evolved. While the general gray seal behavioral repertoire is conserved across colonies (Hewer 1957; Boness and James 1979; Boness 1984; Anderson and Harwood 1985, Bishop et al. 2015c), novel behaviors are identified regularly but infrequently. Such behaviors are often peculiar to particular populations or colonies. On Sable Island, male gray seals were observed performing 'yodel calls' as part of their breeding aggression displays (Boness and James 1979). This vocalization was reported during the breeding season at other colonies in Nova Scotia (Schneider 1974) but appeared to be absent from male repertoires in the Eastern Atlantic (Boness and James 1979; Anderson and Harwood 1985). Similarly, in 2014 it was reported that male gray seals breeding at colonies along the eastern coast of England, UK (e.g., Donna Nook, Blakeney Point, and Horsey) were performing a 'body slap' behavior during aggressive contests that may be conveying information via acoustic signals or substrate vibrations (Bishop et al. 2014, Bishop et al. 2015a). Anecdotal observations suggest that this behavior was likely present since 1993 but had not been documented for over 20 years (Bishop et al. 2014).

In addition to breeding behaviors, isolated and novel foraging behaviors have been observed, with some suggestion of increased occurrence in recent years. In the Northeast Atlantic, there has been an increase since 2012 in reports of gray seals predating on marine mammals, including harbor porpoises, *Phocoena phocoena* (Haelters et al. 2012; Bouveroux et al. 2014; Jauniaux et al. 2014; van Bleijswijk et al. 2014; Haelters et al. 2015; Leopold et al. 2015a, b; Stringell et al. 2015), harbor seals, *Phoca vitulina* (van Neer et al. 2015), and even cannibalism of gray seal pups and juveniles (Bishop et al. 2016; Brownlow et al. 2016; van Neer et al. 2019). Observations of male gray seals exhibiting cannibalism have been reported in Canada (Bédard et al. 1993; Kovacs et al. 1996), but there were no further updates until the observations in Scotland (UK) and Germany (Bishop et al. 2016; van Neer et al. 2019).

It is tempting to suggest that the present reported surge in foraging on marine mammals reflects an increase in frequency or that the patterns reflect broader responses to ecosystem changes. Due to the opportunistic nature of the reporting of such behaviors, it is difficult to ascertain the role of sampling bias that might result from increased awareness and/or technological advances (e.g., DNA testing of wounds, Haelters et al. 2012). Nevertheless, these observations highlight a need for studies to explore the ethology of novel foraging strategies. As a data-rich species, gray seals provide considerable opportunity to explore the emergence, adaptive significance, and spread of behaviors through social or learned mechanisms. Capitalizing on the historical foundation of gray seal ethology and developing contemporary baselines for novel behaviors will be particularly important when there are direct conservation and management concerns. Recent examples are of gray seals raiding salmon traps (Königson et al. 2013), and mortalities associated with 'corkscrew lesions' that have contributed to declines in localized harbor seal populations (Brownlow et al. 2016).

9.8 Future Directions

Ethological research on the gray seal has played a vital role in understanding a wide range of aspects of phocid ethology and behavioral ecology, as evidenced by the literature cited in the various chapters in Sect. 1, such as reproductive behavior. In this chapter, we have endeavored to illustrate how research has grown from its early qualitative foundations through to quantitative analyzes addressing fundamental behavioral ecological principles that help to inform conservation and management practices, and that place gray seal behavioral research in a broader context.

Although gray seals can be regarded as a 'data-rich' species, compiling a review of the extensive literature allows one to identify limits of current knowledge and pinpoint areas of potentially productive future research (Fig. 9.12). For example, little is known about the mechanisms of learning in gray seals, the role of maternal effects on behavior beyond the direct provisioning of nutrients, or how physiological state might dictate an individual's behavioral options. Perhaps the most challenging gap in our knowledge of gray seal ethology is how individual behavior patterns develop during early independence and prior to first breeding. Ultimately, we are seeking answers to the question of how much of gray seal behavior is controlled by nature and how much by nurture.

The increased emphasis on intra- and inter-individual variation in behavior is providing a greater depth of understanding of gray seal biology, but we still have much to learn about what behavior is 'chosen', when and why, and from what suite

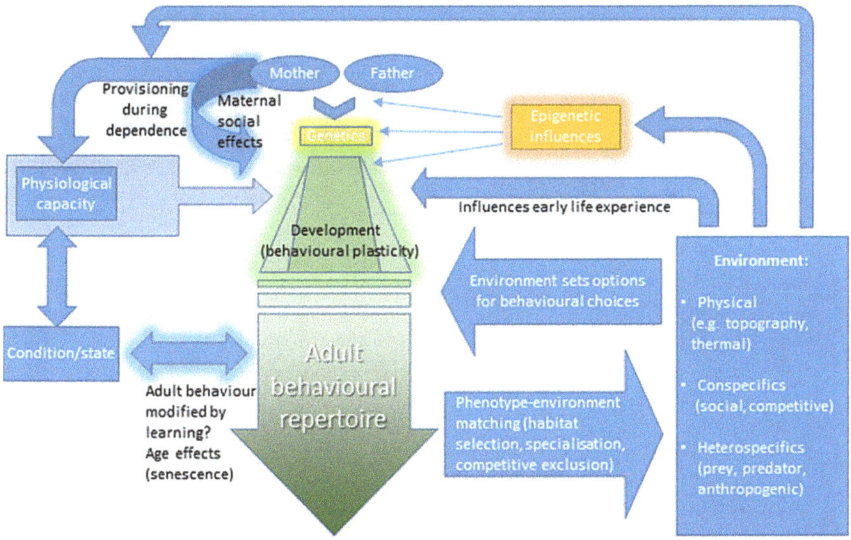

Fig. 9.12 Schematic summarizing and synthesizing the likely main drivers of an individual's behavioral repertoire. The elements with 'glowing' edges represent the areas/linkages that are currently unknown; the linkages between specific genotypes and behaviors, how behavior is shaped and modified during (post-weaning) development and the role of potential non-genetic maternal effects (e.g., social interaction). How development translates into the degree of behavioral plasticity is also unknown (represented by the width of the trapezoid; wider = more behavioral flexibility). Even once canalized, behavior is likely to be modified during adulthood, either through active learning processes, or constraint imposed by senescence. Although much is known about the environmental drivers of behavior (both on land and at sea), nothing is yet known about whether environment also plays a role through epigenetic effects

of available options. Future research will require ever more synthetic and complex studies of known individuals over longer time periods that simultaneously test more parameters; such as the role of maternal social care or early-life stress, while also framing such analyses in the context of opportunity for behavior, given physiological constraints and environmental possibilities. Studies will need to integrate more sophisticated and capable biotelemetry devices (e.g., McKnight et al. 2019) that monitor physiological change in real time with more traditional behavioral observations (Fig. 9.13), and find ways to conduct ethical experimental studies that seek to manipulate behavior to test specific hypotheses.

Behavioral adaptation is often an organism's first line of defense against environmental perturbations and can be a key indicator of fundamental shifts in ecological processes. In a time of rapid environmental change and increased anthropogenic activities in both terrestrial and marine environments, it is vital to understand the ability of seals to adapt to these changes. Gray seal populations are undergoing range expansions and/or shifts in Eastern and Western Atlantic populations. This brings new challenges to the seals themselves, such as increased human interactions (e.g., gray seals in the Thames, UK, interactions with fish farms and fishing gear, interactions

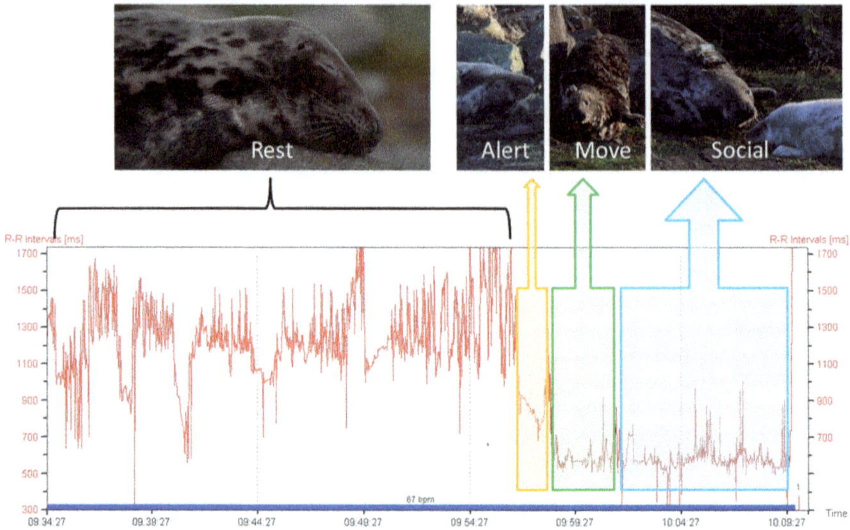

Fig. 9.13 An example of how integrating physiological and behavioral data at fine temporal resolution for individual seals can reveal more about the causes and consequences of behavior. Here, heart rate monitors deployed on lactating females on the Isle of May, Scotland, show how inter-beat interval (R-R interval in milliseconds) changes over a period of 35 min as a mother changes behavioral state. Initially, the mother is at rest near a pool approximately 15 m away from her pup, with long inter-beat intervals (equating to a low heart rate) with regular fluctuations. The female then becomes alert (yellow box), with a rapid reduction in inter-beat interval (heart rate increase), and then locomotes toward her pup (green box). Upon reunion with her pup, the mother engages in social interactions (mainly nosing; blue box) before proceeding to nurse the pup. Note the low inter-beat intervals (i.e., high heart rate) throughout the locomotion and social interaction phases (Twiss unpublished) Image credits: SD Twiss

with sub-sea marine renewable energy devices) and changing ecosystem dynamics (e.g., white sharks and gray seals on the USA east coast, gray seals consuming porpoise and harbor seals). Understanding the processes of behavioral development and how individuals differ in their abilities to modify behavior patterns will be critical for effective planning of conservation and/or management strategies and predicting potential impacts of the gray seal as a top predator within a changing ecosystem.

The gray seal is a data-rich species but also presents an excellent study system for understanding phocid behavior and ecology, and broader ecosystem dynamics. As we have seen in this chapter, there has been a wealth of information gathered on gray seal behavior over the past 80 years, providing a solid understanding of many aspects of gray seal life history, some in great detail such as their breeding behavior and ecology. The traits of this species that have enabled researchers to observe gray seals closely and examine the drivers of behavior are still relevant today: the variety of habitats they occupy, both on land and at sea, the ability to recognize and repeatedly observe and sample known individuals, and the ability to equip individuals with ever more sophisticated telemetry (e.g., McKnight et al. 2019) to provide windows into aspects of their lifecycle that are hidden from traditional behavioral observation. These traits will ensure that the gray seal remains a key study system, and the solid foundation of existing knowledge provides an ideal platform from which to investigate how individuals and populations cope with the pressing ecological and conservation issues of the twenty-first century.

References

Abt K, Engler J (2009) Rapid increase of the grey seal (*Halichoerus grypus*) breeding stock at Helgoland. Helgol Mar Res 63:177–180. https://doi.org/10.1007/s10152-008-0143-6

Allen PJ, Amos W, Pomeroy PP, Twiss SD (1995) Microsatellite variation in grey seals (*Halichoerus grypus*) shows evidence of genetic differentiation between two British breeding colonies. Mol Ecol 4:653–662. https://doi.org/10.1111/j.1365-294X.1995.tb00266.x

Ambs SM, Boness DJ, Bowen WD, Perry EA, Fleischer RC (1999) Proximate factors associated with high levels of extraconsort fertilization in polygynous grey seals. Anim Behav 58(3):527–535. https://doi.org/10.1006/anbe.1999.1201

Amos B, Twiss S, Pomeroy P, Anderson S (1993) Male mating success and paternity in the grey seal, *Halichoerus grypus*: a study using DNA fingerprinting. Proc R Soc Lond B 252:199–207. https://doi.org/10.1006/anbe.2001.1739

Amos B, Twiss S, Pomeroy P, Anderson S (1995) Evidence for mate fidelity in the gray seal. Science 268(5219):1897–1899. https://doi.org/10.1126/science.268.5219.1897

Amos W, Worthington-Wilmer J, Kokko H (2001) Do female grey seals select genetically diverse mates? Anim Behav 62:157–164. https://doi.org/10.1006/anbe.2001.1739

Anderson SS, Baker JR, Prime JH, Baird A (1979) Mortality in grey seal pups: incidence and causes. J Zool 189:407–417. https://doi.org/10.1111/j.1469-7998.1979.tb03972.x

Anderson SS, Burton RW, Summers CF (1975) Behaviour of grey seals (*Halichoerus grypus*) during a breeding season at North Rona. J Zool 177:179–195

Anderson SS, Fedak MA (1985) Grey seal males: energetic and behavioural links between size and sexual success. Anim Behav 33:829–838. https://doi.org/10.1016/S0003-3472(85)80017-8

Anderson SS, Fedak MA (1987) Grey seal, *Halichoerus grypus*, energetics: females invest more in male offspring. J Zool 211:667–679. https://doi.org/10.1111/j.1469-7998.1987.tb04478.x

Anderson SS, Harwood J (1985) Time budgets and topography–how energy reserves and terrain determine the breeding behaviour of grey seals. Anim Behav 33:1343–1348. https://doi.org/10.1016/S0003-3472(85)80196-2

Austin D, Bowen W, McMillan JI (2004) Intraspecific variation in movement patterns: modeling individual behaviour in a large marine predator. Oikos 105:15–30. https://doi.org/10.1111/j.0030-1299.1999.12730.x

Baker JR (1984) Mortality and morbidity in grey seal pups (*Halichoerus grypus*). Studies on its causes, effects of environment, the nature and sources of infectious agents and the immunological status of pups. J Zool 203:23–48. https://doi.org/10.1111/j.1469-7998.1984.tb06042.x

Baker JR, Baker R (1988) Effects of environment on grey seal (*Halichoerus grypus*) pup mortality. Studies on the Isle of May. J Zool 216:529–537. https://doi.org/10.1111/j.1469-7998.1988.tb02449.x

Baker LL, Jonsen ID, Mills Flemming JE, Lidgard DC, Bowen WD, Iverson SJ, Webber DM (2014) Probability of detecting marine predator-prey and species interactions using novel hybrid acoustic transmitter-receiver tags. PLoS ONE 9(6):e98117. https://doi.org/10.1371/journal.pone.0098117

Baker SR, Barrette C, Hammill MO (1995) Mass transfer during lactation of an ice-breeding pinniped, the grey seal (*Halichoerus grypus*), in Nova Scotia, Canada. J Zool 236:531–542. https://doi.org/10.1111/j.1469-7998.1995.tb02730.x

Bartholomew GA (1970) A model for the evolution of pinniped polygyny. Evolution 24(3):546–559. https://doi.org/10.1111/j.1558-5646.1970.tb01790.x
Bean K, Amos W, Pomeroy PP, Twiss SD, Coulson TN, Boyd IL (2004) Patterns of parental relatedness and pup survival in the grey seal (*Halichoerus grypus*). Mol Ecol 13:2365–2370. https://doi.org/10.1111/j.1365-294X.2004.02199.x
Beck CA, Bowen W, Iverson SJ (2000) Seasonal changes in buoyancy and diving behaviour of adult grey seals. J Exp Biol 203:2323–2330
Beck CA, Bowen WD, McMillan JI, Iverson SJ (2003) Sex differences in the diving behaviour of a size-dimorphic capital breeder: the grey seal. Anim Behav 66:777–789. https://doi.org/10.1006/anbe.2003.2284
Beck CA, Iverson SJ, Bowen WD, Blanchard W (2007) Sex differences in grey seal diet reflect seasonal variation in foraging behaviour and reproductive expenditure: evidence from quantitative fatty acid signature analysis. J Anim Ecol 76:490–502. https://doi.org/10.1111/j.1365-2656.2007.01215.x
Bédard C, Kovacs K, Hammill M (1993) Cannabalism by grey seals, *Halichoerus Grypus*, on Amet Island, Nova Scotia. Mar Mamm Sci 9(4):421–424. https://doi.org/10.1111/j.1748-7692.1993.tb00474.x
Bell AM, Hankison SJ, Laskowski KL (2009) The repeatability of behaviour: a meta-analysis. Anim Behav 77:771–783. https://doi.org/10.1016/j.anbehav.2008.12.022
Bennett KA, Speakman JR, Moss SEW, Pomeroy PP, Fedak MA (2007) Effects of mass and body composition on fasting fuel utilisation in grey seal pups (*Halichoerus grypus* Fabricius): an experimental study using supplementary feeding. J Exp Biol 210:3043–3053. https://doi.org/10.1242/jeb.009381
Bennett KA, McConnell BJ, Moss SEW, Speakman JR, Pomeroy PP, Fedak MA (2010) Effects of age and body mass on development of diving capabilities of gray seal pups: costs and benefits of the postweaning fast. Physiol Biochem Zool 83:911–923. https://doi.org/10.1086/656925
Biro PA, Stamps JA (2010) Do consistent individual differences in metabolic rate promote consistent individual differences in behaviour? Trends Ecol Evol 25:653–659. https://doi.org/10.1016/j.tree.2010.08.003
Bishop AM (2015) Behavioural mechanisms of conflict and conflict reduction in a wild breeding polygynous pinniped. Doctoral thesis, Durham University. oai:etheses.dur.ac.uk:11146
Bishop AM, Lidstone-Scott R, Pomeroy PP, Twiss SD (2014) Body slap: an innovative aggressive display by breeding male gray seals (*Halichoerus grypus*). Mar Mamm Sci 30(2):579–593. https://doi.org/10.1111/mms.12059
Bishop A, Denton P, Pomeroy P, Twiss S (2015a) Good vibrations by the beach boys: magnitude of substrate vibrations is a reliable indicator of male grey seal size. Anim Behav 100:74–82. https://doi.org/10.1016/j.anbehav.2014.11.008
Bishop AM, Pomeroy P, Twiss SD (2015b) Variability in individual rates of aggression in wild gray seals: fine-scale analysis reveals importance of social and spatial stability. Behav Ecol Sociobiol 69(10):1663–1675. https://doi.org/10.1007/s00265-015-1978-x
Bishop A, Pomeroy P, Twiss SD (2015c) Breeding male grey seals exhibit similar activity budgets across varying exposures to human activity. Mar Ecol Prog Ser 527:247–259. https://doi.org/10.3354/meps11254
Bishop AM, Onoufriou J, Moss S, Pomeroy PP, Twiss SD (2016) Cannibalism by a male grey seal (*Halichoerus grypus*) in the North Sea. Aquatic Mammals 42(2):137–143. http://dx.doi.org/https://doi.org/10.1578/AM.42.2.2016.137
Bishop AM, Stewart JE, Pomeroy P, Twiss SD (2017) Intraseasonal temporal variation of reproductive effort for male grey seals. Anim Behav 134:167–175. https://doi.org/10.1016/j.anbehav.2017.10.021
Boness DJ (1984) Activity budget of male gray seals, *Halichoerus grypus*. J Mammal 65(2):291–297. https://doi.org/10.2307/1381168
Boness DJ, James H (1979) Reproductive behaviour of the grey seal (*Halichoerus grypus*) on Sable Island, Nova Scotia. J Zool 188:477–500. https://doi.org/10.1111/j.1469-7998.1979.tb03430.x

Boness DJ, Anderson SS, Cox CR (1982) Functions of female aggression during the pupping and mating season of grey seals, *Halichoerus grypus* (Fabricius). Can J Zool 60:2270–2278. https://doi.org/10.1139/z82-293

Boness D, Bowen W, Iverson S (1995) Does male harassment of females contribute to reproductive synchrony in the grey seal by affecting maternal performance? Behav Ecol Sociobiol 36(1):1–10. https://doi.org/10.1007/BF00175722

Boskovic R, Kovacs KM, Hammill MO, White BN (1996) Geographic distribution of mitochondrial DNA haplotypes in grey seals (*Halichoerus grypus*). Can J Zool 74:1787–1796. https://doi.org/10.1139/z96-199

Bouveroux T, Kiszka J, Heithaus R, Jauniaux T, Pezeryl S (2014) Direct evidence for gray seal (*Halichoerus grypus*) predation and scavenging on harbor porpoises (*Phocoena phocoena*). Mar Mamm Sci 30:1542–1548. https://doi.org/10.1111/mms.12111

Bowen WD, Stobo WT, Smith SJ (1992) Mass changes of grey seal *Halichoerus grypus* pups on Sable Island-differential maternal investment reconsidered. J Zool 227:607–622. https://doi.org/10.1111/j.1469-7998.1992.tb04418.x

Boyd JM, Laws RM (1962) Observations on the grey seal (*Halichoerus grypus*) at North Rona in 1960. Proc Zool Soc Lond 139:249–260. https://doi.org/10.1111/j.1469-7998.1962.tb01829.x

Boyd JM, Lockie JD, Hewer HR (1962) The breeding colony of seals on North Rona, 1959. Proc Zool Soc Lond 138:257–277. https://doi.org/10.1111/j.1469-7998.1962.tb05697.x

Brasseur SM, Polanen Petel TD, Gerrodette T, Meesters EH, Reijnders PJ, Aarts G (2015) Rapid recovery of Dutch gray seal colonies fueled by immigration. Mar Mamm Sci 31:405–426. https://doi.org/10.1111/mms.12160

Breed GA, Bowen WD, Leonard ML (2011) Development of foraging strategies with age in a long-lived marine predator. Mar Ecol Prog Ser 431:267–279. https://doi.org/10.3354/meps09134

Breed GA, Bowen WD, Leonard ML (2013) Behavioral signature of intraspecific competition and density dependence in colony-breeding marine predators. Ecol Evol 3(11):3838–3854. https://doi.org/10.1002/ece3.754

Breed GA, Jonsen ID, Myers RA, Bowen WD, Leonard ML (2009) Sex-specific, seasonal foraging tactics of adult grey seals (*Halichoerus grypus*) revealed by state-space analysis. Ecology 90:3209–3221. https://doi.org/10.1890/07-1483.1

Briffa M, Weiss A (2010) Animal personality. Curr Biol 20:R912–R914. https://doi.org/10.1016/j.cub.2010.09.019

Briffa M, Greenaway J (2011) High *in situ* repeatability of behaviour indicates animal personality in the beadlet anemone *Actinia equina* (Cnidaria). PLoS ONE 6(7):e21963. https://doi.org/10.1371/journal.pone.0021963

Brownlow A, Onoufriou J, Bishop A, Davison N, Thompson D (2016) Corkscrew seals: grey seal (*Halichoerus grypus*) infanticide and cannibalism may indicate the cause of spiral lacerations in seals. PLoS ONE 11(6):e0156464. https://doi.org/10.1371/journal.pone.0156464

Bubac CM, Coltman DW, Bowen WD, Lidgard DC, Lang SL, den Heyer CE (2018) Repeatability and reproductive consequences of boldness in female gray seals. Behav Ecol Sociobiol 72(6):100. https://doi.org/10.1007/s00265-018-2515-5

Burton RW, Anderson SS, Summers CF (1975) Perinatal activities in the grey seal (*Halichoerus grypus*). J Zool 177:197–201. https://doi.org/10.1111/j.1469-7998.1975.tb05978.x

Cameron AW (1967) Breeding behaviour in a colony of western Atlantic gray seals. Can J Zool 45:161–173. https://doi.org/10.1139/z67-023

Cameron AW (1969) The behaviour of adult gray seals (*Halichoerus grypus*) in the early stages of the breeding season. Can J Zool 47:229–233. https://doi.org/10.1139/z69-049

Cameron AW (1970) Seasonal movements and diurnal activity rhythms of grey seal (*Halichoerus grypus*). J Zool 161(1):15–23. https://doi.org/10.1111/j.1469-7998.1970.tb02166.x

Carere C, Maestripieri D (eds) (2013) Animal personalities. Behavior, physiology, and evolution. The University of Chicago Press, Chicago (IL)

Carere C, Caramaschi D, Fawcett TW (2010) Covariation between personalities and individual differences in coping with stress: converging evidence and hypotheses. Curr Zool 56:728–740

Carlin CM, Hayden TJ, Jennings DJ, Gammell MP (2005) Win, lose or draw: a comparison of fight structure based on fight conclusion in the fallow deer. Behaviour 142(4):423–439. https://doi.org/10.1163/1568539054012001

Carlini AR, Poljak S, Daneri GA, Márquez ME, Negrete J (2006) The dynamics of male harem dominance in southern elephant seals (*Mirounga leonina*) at the South Shetland Islands. Polar Biol 29(9):796–805. https://doi.org/10.1007/s00300-006-0117-6

Carter MID, Bennett KA, Embling CB, Hosegood PJ, Russell DJF (2016) Navigating uncertain waters: a critical review of inferring foraging behaviour from location and dive data in pinnipeds. Mov Ecol 4:25. https://doi.org/10.1186/s40462-016-0090-9

Carter MID, Russell DJF, Embling CB, Blight CJ, Thompson D, Hosegood PJ, Bennett KA (2017) Intrinsic and extrinsic factors drive ontogeny of early-life at-sea behaviour in a marine top predator. Sci Rep 7(1):15505. https://doi.org/10.1038/s41598-017-15859-8

Caudron A (1998) Behavioural plasticity in function of the breeding environment in a marine mammal, the grey seal *Halichoerus grypus*. Cahiers D'ethologie 18:299–550

Caudron AK, Kondakov AA, Siryanov SV (1998) Acoustic structure and individual variation of grey seal (*Halichoerus grypus*) pup calls. J Mar Biol Assoc UK 78:651–658. https://doi.org/10.1017/S0025315400041680

Colléter M, Brown C (2011) Personality traits predict hierarchy rank in male rainbowfish social groups. Anim Behav 81(6):1231–1237. https://doi.org/10.1016/j.anbehav.2011.03.011

Coppens CM, de Boer SF, Koolhaas JM (2010) Coping styles and behavioural flexibility: towards underlying mechanisms. Philos T R Soc B 365:4021–4028. https://doi.org/10.1098/rstb.2010.0217

Coulson JC, Hickling G (1961) Variation in the secondary sex ratio of the grey seal, *Halichoerus grypus* (Fab.), during the breeding season. Nature 190:281

Coulson JC, Hickling G (1964) The breeding biology of the grey seal, *Halichoerus grypus* (Fab.), on the Farne Islands, Northumberland. J Anim Ecol 33:485–512. https://doi.org/10.2307/2568

Cronin M, Pomeroy P, Jessopp M (2012) Size and seasonal influences on the foraging range of female grey seals in the northeast Atlantic. Mar Biol 160:531–539. https://doi.org/10.1007/s00227-012-2109-0

Cronin M, Gerritsen H, Reid D, Jessopp M (2016) Spatial overlap of grey seals and fisheries in Irish waters, some new insights using telemetry technology and VMS. PLoS ONE 11(9):e0160564. https://doi.org/10.1371/journal.pone.0160564

Culloch R (2012) The application of modern statistical approaches to identify consistent individual differences in the behaviour of wild postpartum female grey seals (*Halichoerus grypus*). Doctoral thesis, Durham University. oai:etheses.dur.ac.uk:3620

Culloch RM, Pomeroy PP, Twiss SD (2016) The difference between night and day: the nocturnal and diurnal activity budget of gray seals (*Halichoerus grypus*) during the breeding season. Mar Mamm Sci 32:400–408. https://doi.org/10.1111/mms.12259

Davies JL (1949) Observations on the grey seal *Halichooerus grypus* at Ramsey Island, Pembs. Proc Zool Soc Lond 119(3):673–692. https://doi.org/10.1111/j.1096-3642.1949.tb00896.x

DeRango EJ, Schwarz JFL, Kalberer S, Piedrahita P, Páez-Rosas D, Krüger O (2019) Intrinsic and maternal traits influence personality during early life in Galápagos sea lion, *Zalophus wollebaeki*, pups. Anim Behav 154:111–120. https://doi.org/10.1016/j.anbehav.2019.06.011

de Vere AJ, Lilley MK, Highfill L (2017) Do pinnipeds have personality? Broad dimensions and contextual consistency of behavior in harbor seals (*Phoca vitulina*) and California sea lions (*Zalophus californianus*). Int J Comp Psychol 30. https://escholarship.org/uc/item/4f37d0m3

de Vries H, Stevens JMG, Vervaecke H (2006) Measuring and testing the steepness of dominance hierarchies. Anim Behav 71:585–592. https://doi.org/10.1016/j.anbehav.2005.05.015

Dingemanse NJ, Kazem AJN, Réale D, Wright J (2010) Behavioural reaction norms: animal personality meets individual plasticity. Trends Ecol Evol 25:81–89. https://doi.org/10.1016/j.tree.2009.07.013

Dochtermann NA, Dingemanse NJ (2013) Behavioral syndromes as evolutionary constraints. Behav Ecol 24:806–811. https://doi.org/10.1093/beheco/art002

Duckworth RA (2010) Evolution of personality: developmental constraints on behavioural flexibility. Auk 127(4):752–758. https://doi.org/10.1525/auk.2010.127.4.752

Fogden SCL (1971) Mother-young behaviour at grey seal breeding beaches. J Zool 164:61–92. https://doi.org/10.1111/j.1469-7998.1971.tb01298.x

Fraser ZL, Culloch RM, Twiss SD (2019) As clear as day: nocturnal activity differs from diurnal activity in a temporally constrained capital breeder. Behaviour 156(10):997–1016. https://doi.org/10.1163/1568539X-00003553

Fraser-Darling F (1939) A naturalist on Rona: essays of a biologist in isolation. Oxford University Press Clarendon Press, Oxford

Frick EE, de Vere AJ, Kuczaj SA (2017) What do we want to know about personality in marine mammals? In: Vonk J, Weiss A, Kuczaj SA (eds) Personality in nonhuman animals. Springer International Publishing, pp 237–253. https://doi.org/10.1007/978-3-319-59300-5_12

Gammell MP, de Vries H, Jennings DJ, Carlin CM, Hayden TJ (2003) David's score: a more appropriate dominance ranking method than Clutton-Brock *et al'*.s index. Anim Behav 66:601–605. https://doi.org/10.1006/anbe.2003.2226

Gemmell NJ, Burg TM, Boyd IL, Amos W (2001) Low reproductive success in territorial male Antarctic fur seals (*Arctocephalus gazella*) suggests the existence of alternative mating strategies. Mol Ecol 10:451–460. https://doi.org/10.1046/j.1365-294x.2001.01186.x

Gerondeau M, Barbraud C, Ridoux V, Vincent C (2007) Abundance estimate and seasonal patterns of grey seal (*Halichoerus grypus*) occurrence in Brittany, France, as assessed by photo-identification and capture–mark–recapture. J Mar Biol Assoc UK 87:365–372. https://doi.org/10.1017/S0025315407054586

Godsell J (1991) The relative influence of age and weight on the reproductive behaviour of male grey seals *Halichoerus grypus*. J Zool 224:537–551. https://doi.org/10.1111/j.1469-7998.1991.tb03784.x

González-Suárez M, Cassini MH (2014) Variance in male reproductive success and sexual size dimorphism in pinnipeds: testing an assumption of sexual selection theory. Mammal Rev 44(2):88–93. https://doi.org/10.1111/mam.12012

Gosch M, Cronin M, Rogan E, Hunt W, Luck C, Jessopp M (2019) Spatial variation in a top marine predator's diet at two regionally distinct sites. PLoS ONE 14(1):e0209032. https://doi.org/10.1371/journal.pone.0209032

Götz T, Janik VM (2010) Aversiveness of sounds in phocid seals: psycho-physiological factors, learning processes and motivation. J Exp Biol 213:1536–1548. https://doi.org/10.1242/jeb.035535

Götz T, Janik VM (2011) Repeated elicitation of the acoustic startle reflex leads to sensitisation in subsequent avoidance behaviour and induces fear conditioning. BMC Neurosci 12:30. https://doi.org/10.1186/1471-2202-12-30

Haelters J, Kerckhof F, Jauniaux T, Degraer S (2012) The grey seal (*Halichoerus grypus*) as a predator of harbour porpoises (*Phocoena phocoena*)? Aquat Mamm 38:343–353. https://doi.org/10.1578/AM.38.4.2012.343

Haelters J, Kerckhof F, van Neer A, Leopold M (2015) Exposing grey seals as horses and scientists as human. Aquat Mamm 41(3):351–354. https://doi.org/10.1578/AM.41.3.2015.351

Haley MP (1994) Resource-holding power asymmetries, the prior residence effect, and reproductive payoffs in male northern elephant seal fights. Behav Ecol Sociobiol 34(6):427–434. https://doi.org/10.1007/BF00167334

Hall AJ, McConnell BJ, Barker RJ (2001) Factors affecting first-year survival in grey seals and their implications for life history strategy. J Anim Ecol 70:138–149. https://doi.org/10.1111/j.1365-2656.2001.00468.x

Hall AJ, McConnell BJ, Barker RJ (2002) The effect of total immunoglobulin levels, mass and condition on the first-year survival of grey seal pups. Funct Ecol 16:462–474. https://doi.org/10.1046/j.1365-2435.2002.00649.x

Haller MA, Kovacs KM, Hammill MO (1996) Maternal behaviour and energy investment by grey seals (*Halichoerus grypus*) breeding on land-fast ice. Can J Zool 74:1531–1541. https://doi.org/10.1139/z96-167

Hammill MO, Gosselin JF (1995) Grey seal (*Halichoerus grypus*) from the Northwest Atlantic: female reproductive rates, age at first birth, and age of maturity in males. Can J Fish Aquat Sci 52(12):2757–2761. https://doi.org/10.1139/f95-864

Hazekamp AAH, Mayer R, Osinga N (2010) Flow simulation along a seal: the impact of an external device. Eur J Wildl Res 56:131–140. https://doi.org/10.1007/s10344-009-0293-0

Harvey V, Côté SD, Hammill MO (2008) The ecology of 3-D space use in a sexually dimorphic mammal. Ecography 31(3):371–380. https://www.jstor.org/stable/30244587

Hewer HR (1957) A Hebridean breeding colony of grey seals, *Halichoerus grypus* (Fab.), with comparative notes on grey seals of Ramsey Island, Pembrokeshire. Proc Zool Soc Lond 128:23–66. https://doi.org/10.1111/j.1096-3642.1957.tb00255.x

Hewer HR (1960) Behaviour of the grey seal (*Halichoerus grypus*, Fab.) in the breeding season. Mammalia 24:400–421

Hewer HR (1974) British seals, collins new naturalist library, Book 57.

Hewer HR, Backhouse KM (1960) A preliminary account of a colony of grey seals, *Halichoerus grypus* (Fab.), in the southern Inner Hebrides. Proc Zool Soc Lond 134:157–195. https://doi.org/10.1111/j.1469-7998.1960.tb05586.x

Hiby L, Lovell P (1990) Computer aided matching of natural markings: a prototype system for grey seals. Reports of the International Whaling Commission, Special Issue 12:57–61

Hiby L, Paterson WD, Redman P, Watkins J, Twiss SD, Pomeroy P (2013) Analysis of photo-id data allowing for missed matches and individuals identified from opposite sides. Methods Ecol Evol 4:252–259. https://doi.org/10.1111/2041-210x.12008

Highfill LE, Kuczaj SA (2007) Do bottlenose dolphins (*Tursiops truncatus*) have distinct and stable personalities? Aquat Mamm 33:380–389. https://doi.org/10.1578/AM.33.3.2007.380

Hunter SA, Bay MS, Martin ML, Hatfield JS (2002) Behavioral effects of environmental enrichment on harbor seals (*Phoca vitulina concolor*) and gray seals (*Halichoerus grypus*). Zoo Biol 21:375–387. https://doi.org/10.1002/zoo.10042

James HMC (2013) Individual differences in maternal behaviour in the grey seal (*Halichoerus grypus*) and the impact of disturbance at Donna Nook. Masters thesis, Durham University. oai:etheses.dur.ac.uk:7305

Jauniaux T, Garigliany MM, Loos P, Bourgain JL, Bouveroux T, Coignoul F, Haelters J, Karpouzopoulos J, Pezeril S, Desmecht D (2014) Bite injuries of grey seals (*Halichoerus grypus*) on harbour porpoises (*Phocoena phocoena*). PLoS ONE 9(12):e108993. https://doi.org/10.1371/journal.pone.0108993

Jenssen BM, Åsmul JI, Ekker M, Vongraven D (2010) To go for a swim or not? Consequences of neonatal aquatic dispersal behaviour for growth in grey seal pups. Anim Behav 80:667–673. https://doi.org/10.1016/j.anbehav.2010.06.028

Jessopp M, Cronin M, Hart T (2013) Habitat-mediated dive behavior in free-ranging grey seals. PLoS ONE 8(5):e63720. https://doi.org/10.1371/journal.pone.0063720 PMID: 23667663

Jussi M, Harkonen T, Helle E, Jussi I (2008) Decreasing ice coverage will reduce the breeding success of Baltic grey seal (*Halichoerus grypus*) females. Ambio 37:80–85. https://doi.org/10.1579/0044-7447(2008)37[80:dicwrt]2.0.co;2

Karlsson O, Hiby L, Lundberg T, Jussi M, Jussi I, Helander B (2005) Photo-identification, site fidelity, and movement of female gray seals (*Halichoerus grypus*) between haul-outs in the Baltic sea. Ambio 34:628–634. https://doi.org/10.1639/0044-7447(2005)034[0628:psfamo]2.0.co;2

Klimova A, Phillips CD, Fietz K, Olsen MT, Harwood J, Amos W, Hoffman JI (2014) Global population structure and demographic history of the grey seal. Mol Ecol 23:3999–4017. https://doi.org/10.1111/mec.12850

Königson S, Fjälling A, Berglind M, Lunneryd SG (2013) Male gray seals specialize in raiding salmon traps. Fish Res 148:117–123. https://doi.org/10.1016/j.fishres.2013.07.014

Koolhaas JM, Korte SM, De Boer SF, Van der Vegt BJ, Van Reenen CG, Hopster H, De Jong IC, Ruis MA, Blokhuis HJ (1999) Coping styles in animals: current status in behavior and stress physiology. Neurosci Biobehav Rev 23:925–935. https://doi.org/10.1016/S0149-7634(99)00026-3

Kovacs KM (1987) Maternal behaviour and early behavioural ontogeny of grey seals (*Halichoerus grypus*) on the Isle of May, UK. J Zool 213:697–715. https://doi.org/10.1111/j.1469-7998.1987.tb03735.x

Kovacs KM, Lavigne D (1986) Growth of grey seal (*Halichoerus grypus*) neonates: differential maternal investment in the sexes. Can J Zool 64:1937–1943. https://doi.org/10.1139/z86-291

Kovacs KM, Lydersen C (2008) Climate change impacts on seals and whales in the North Atlantic Arctic and adjacent shelf seas. Sci Prog 91:117–150. https://doi.org/10.3184/003685008X324010

Kovacs K, Lydersen C, Hammill M (1996) Grey seal cannibalism. Mar Mamm Sci 12(1):161

Lang SLC, Iverson SJ, Bowen WD (2009) Repeatability in lactation performance and the consequences for maternal reproductive success in grey seals. Ecology 90:2513–2523. https://doi.org/10.1086/316708

Lang SL, Boness DJ, Bowen WD, Iverson SJ (2011) Primiparous females do not exhibit reduced maternal care in gray seals (*Halichoerus grypus*). Mar Mamm Sci 27:E153–E164. https://doi.org/10.1111/j.1748-7692.2010.00443.x

Leeney R, Broderick A, Mills C, Sayer S, Witt M, Godley B (2010) Abundance, distribution and haul-out behaviour of grey seals (*Halichoerus grypus*) in Cornwall and the Isles of Scilly, UK. J Mar Biol Assoc UK 90(5):1033–1040. https://doi.org/10.1017/S0025315409991512

Leopold MF, Begeman L, van Bleijswijk JD, IJsseldijk LL, Witte HJ, Gröne A (2015a) Exposing the grey seal as a major predator of harbour porpoises. Proc Royal Soc B. 282:20142429. https://doi.org/10.1098/rspb.2014.2429

Leopold MF, Begeman L, Heße E, van der Hiele J, Hiemstra S, Keijl G, Meesters EH, Mielke L, Verheyen D, Gröne A (2015b) Porpoises: from predators to prey. J Sea Res 97:14–23. https://doi.org/10.1016/j.seares.2014.12.005

Lerner JE, Ono K, Hernandez KM, Runstadler JA, Puryear WB, Polito MJ (2018) Evaluating the use of stable isotope analysis to infer the feeding ecology of a growing US gray seal (*Halichoerus grypus*) population. PLoS ONE 13(2):e0192241. https://doi.org/10.1371/journal.pone.0192241

Lidgard DC, Boness DJ, Bowen WD (2001) A novel mobile approach to investigating mating tactics in male grey seals (*Halichoerus grypus*). J Zool 255(3):313–320. https://doi.org/10.1017/S0952836901001418

Lidgard DC, Boness DJ, Bowen WD, McMillan JI (2003) Diving behaviour during the breeding season in the terrestrially breeding male grey seal: implications for alternative mating tactics. Can J Zool 81(6):1025–1033. https://doi.org/10.1139/z03-085

Lidgard DC, Boness DJ, Bowen WD, McMillan JI, Fleischer RC (2004) The rate of fertilization in male mating tactics of the polygynous grey seal. Mol Ecol 13(11):3543–3548. https://doi.org/10.1111/j.1365-294X.2004.02334.x

Lidgard DC, Boness DJ, Bowen WD, McMillan JI (2005) State-dependent male mating tactics in the grey seal: the importance of body size. Behav Ecol 16(3):541–549. https://doi.org/10.1093/beheco/ari023

Lidgard DC, Bowen WD, Boness DJ (2012) Longitudinal changes and consistency in male physical and behavioural traits have implications for mating success in the grey seal (*Halichoerus grypus*). Can J Zool 90:849–860. https://doi.org/10.1139/z2012-053

Lidgard DC, Bowen WD, Iverson SJ (2020) Sex-differences in fine-scale home-range use in an upper-trophic level marine predator. Mov Ecol 8:11. https://doi.org/10.1186/s40462-020-0196-y

Lydersen C, Hammill MO, Kovacs KM (1994) Activity of lactating ice-breeding grey seals. *Halichoerus grypus*, from the Gulf of St. Lawrence, Canada. Anim Behav 48:1417–1425. https://doi.org/10.1006/anbe.1994.1377

Lydersen C, Hammill MO, Kovacs KM (1995) Milk intake, growth and energy consumption in pups of ice-breeding grey seals (*Halichoerus grypus*) from the Gulf of St. Lawrene, Canada. J Comp Physiol B 164:585–592. https://doi.org/10.1007/BF00389798

Lydersen C, Kovacs KM (1999) Behavior and energetics of ice-breeding, North Atlantic phocid seals during the lactation period. Mar Ecol Prog Ser 187:265–281. https://doi.org/10.3354/meps187265

Matthews LH (1950) The natural history of the grey seal, including lactation. Proc Zool Soc Lond 120:763

Matthiopoulos J, Harwood J, Thomas L (2005) Metapopulation consequences of site fidelity for colonially breeding mammals and birds. J Anim Ecol 74(4):716–727. https://doi.org/10.1111/j.1365-2656.2005.00970.x

McClintock BT, Russell DJF, Matthiopoulos J, King R (2013) Combining individual animal movement and ancillary biotelemetry data to investigate population-level activity budgets. Ecology 94:838–849. https://doi.org/10.1890/12-0954.1

McConnell BJ, Chambers C, Nicholas KS, Fedak MA (1992) Satellite tracking of grey seals *Halichoerus grypus*. J Zool 226:271–282. https://doi.org/10.1111/j.1469-7998.1992.tb03839.x

McConnell BJ, Fedak MA, Lovell P, Hammond PS (1999) Movements and foraging areas of grey seals in the North Sea. J Appl Ecol 36:573–590. https://doi.org/10.1046/j.1365-2664.1999.00429.x

McCulloch S, Pomeroy PP, Slater PJB (1999) Individually distinctive pup vocalizations fail to prevent allo-suckling in grey seals. Can J Zool 77:716–723. https://doi.org/10.1139/z99-023

McCulloch S, Boness DJ (2000) Mother-pup vocal recognition in the grey seal (*Halichoerus grypus*) of Sable Island, Nova Scotia, Canada. J Zool 251:449–455. https://doi.org/10.1111/j.1469-7998.2000.tb00800.x

McKnight JC, Bennett KA, Bronkhorst M, Russell DJF, Balfour S, Milne R, Bivins M, Moss SEW, Colier W, Hall AJ, Thompson D (2019) Shining new light on mammalian diving physiology using wearable near-infrared spectroscopy. PLoS Biol 17(6):e3000306. https://doi.org/10.1371/journal.pbio.3000306

Mellish JAE, Iverson SJ, Bowen WD (1999) Variation in milk production and lactation performance in grey seals and consequences for pup growth and weaning characteristics. Physiol Biochem Zool 72:677–690. https://doi.org/10.1086/316708

Mikkelsen L, Johnson M, Wisniewska DM, van Neer A, Siebert U, Madsen PT, Teilmann J (2019) Long-term sound and movement recording tags to study natural behavior and reaction to ship noise of seals. Ecol Evol 9:2588–2601. https://doi.org/10.1002/ece3.4923

Miller EH (1991) Communication in pinnipeds, with special reference to non-acoustic communication. In: Renouf D (ed) The behaviour of pinnipeds. Chapman & Hall, London, pp 128–235

Miller EH, Boness DJ (1979) Remarks on display functions of the snout of the grey seal, *Halichoerus grypus* (Fab.), with comparative notes. Can J Zool 57:140–148. https://doi.org/10.1139/z79-011

Noren SR, Boness DJ, Iverson SJ, McMillan J, Bowen WD (2008) Body condition at weaning affects the duration of the postweaning fast in gray seal pups (*Halichoerus grypus*). Physiol Biochem Zool 81:269–277. https://doi.org/10.1086/528777

O'Donnell S (1998) Dominance and polyethism in the eusocial wasp *Mischocyttarus mastigophorus* (Hymenoptera: Vespidae). Behav Ecol Sociobiol 43:327–331. https://doi.org/10.1007/s002650050498

Oliver GW (1977) Navigation in mazes by a grey seal, *Halichoerus grypus* (Fabricius). Behaviour 67(1/2):97–114. https://doi.org/10.1163/156853978X00279

Perry EA, Boness DJ, Fleischer RC (1998) DNA fingerprinting evidence of nonfilial nursing in grey seals. Mol Ecol 7:81–85. https://doi.org/10.1046/j.1365-294x.1998.00313.x

Poland VF, Pomeroy PP, Twiss SD, Graves JA (2008) Fine-scale study finds limited evidence of kin clustering in a gray seal colony. Mar Mamm Sci 24(2):371–387. https://doi.org/10.1111/j.1748-7692.2008.00191.x

Pomeroy PP, Anderson SS, Twiss SD, McConnell BJ (1994) Dispersion and site fidelity of breeding female grey seals (*Halichoerus grypus*) on North Rona, Scotland. J Zool 233(3):429–447. https://doi.org/10.1111/j.1469-7998.1994.tb05275.x

Pomeroy PP, Fedak MA, Rothery P, Anderson S (1999) Consequences of maternal size for reproductive expenditure and pupping success of grey seals at North Rona, Scotland. J Anim Ecol 68:235–253. https://doi.org/10.1046/j.1365-2656.1999.00281.x

Pomeroy PP, Twiss SD, Redman P (2000a) Philopatry, site fidelity and local kin associations within grey seal breeding colonies. Ethology 106(10):899–919. https://doi.org/10.1046/j.1439-0310.2000.00610.x

Pomeroy PP, Twiss SD, Duck CD (2000b) Expansion of a grey seal (*Halichoerus grypus*) breeding colony: changes in pupping site use at the Isle of May, Scotland. J Zool 250:1–12. https://doi.org/10.1111/j.1469-7998.2000.tb00573.x

Pomeroy PP, Worthington Wilmer J, Amos W, Twiss SD (2001) Reproductive performance links to fine scale spatial patterns of female grey seal relatedness. Proc R Soc Lond B 268(1468):711–717. https://doi.org/10.1098/rspb.2000.1422

Pomeroy PP, Redman PR, Ruddell SJS, Duck CD, Twiss SD (2005) Breeding site choice fails to explain interannual associations of female grey seals. Behav Ecol Sociobiol 57:546–556. https://doi.org/10.1007/s00265-004-0882-6

Réale D, Reader SM, Sol D, McDougall PT, Dingemanse NJ (2007) Integrating animal temperament within ecology and evolution. Biol Rev 82:291–318. https://doi.org/10.1111/j.1469-185X.2007.00010.x

Redman P, Pomeroy PP, Twiss SD (2001) Grey seal maternal attendance patterns are affected by water availability on North Rona, Scotland. Can J Zool 79(6):1073–1079. https://doi.org/10.1139/z01-081

Reijnders PJH, Van Dijk J, Kuiper D (1995) Recolonization of the Dutch Wadden Sea by the grey seal *Halichoerus grypus*. Biol Cons 71:231–235. https://doi.org/10.1016/0006-3207(94)00032-L

Reilly JJ (1991) Adaptions to prolonged fasting in free-living weaned gray seal pups. Am J Physiol 260(2):R267–R272. https://doi.org/10.1152/ajpregu.1991.260.2.R267

Riechert SE, Hedrick AV (1993) A test for correlations among fitness-linked behavioural traits in the spider *Agelenopsis aperta* (Araneae, Agelenidae). Anim Behav 46(4):669–675. https://doi.org/10.1006/anbe.1993.1243

Robinson KJ, Twiss SD, Hazon N, Pomeroy PP (2015a) Maternal oxytocin is linked to close mother-infant proximity in grey seals (*Halichoerus grypus*). PLoS ONE 10(12):e0144577. https://doi.org/10.1371/journal.pone.0144577

Robinson KJ, Twiss SD, Hazon N, Moss S, Lonergan M, Pomeroy PP (2015b) Conspecific recognition and aggression reduction to familiars in newly weaned, socially plastic mammals. Behav Ecol Sociobiol 69(8):1383–1394. https://doi.org/10.1007/s00265-015-1952-7

Robinson KJ, Twiss SD, Hazon N, Moss S, Pomeroy PP (2017a) Positive social behaviours are induced and retained after oxytocin manipulations mimicking endogenous concentrations in a wild mammal. Proc Royal Soc B 284:20170554. https://doi.org/10.1098/rspb.2017.0554

Robinson KJ, Pomeroy PP, Hazon N, Moss S, Twiss SD (2017b) Individual size, sex and rearing environment impact on aggression in newly weaned seals. Mar Mamm Sci 33:621–629. https://doi.org/10.1111/mms.12367

Robinson KJ, Hazon N, Twiss SD, Pomeroy PP (2019) High oxytocin infants gain more mass with no additional maternal energetic costs in wild grey seals (*Halichoerus grypus*). Psychoneuroendocrinology 110:104423. https://doi.org/10.1016/j.psyneuen.2019.104423

Rosing-Asvid A, Teilmann J, Dietz R, Olsen MT (2010) First record of grey seals in Greenland. Arctic 63:471–473. https://doi.org/10.14430/arctic3336

Ruddell SJS, Twiss SD, Pomeroy PP (2007) Measuring opportunity for sociality: quantifying social stability in a colonially breeding phocid. Anim Behav 74:1357–1368. https://doi.org/10.1016/j.anbehav.2007.01.024

Russell DJF, Brasseur SMJM, Thompson D, Hastie GD, Janik VM, Aarts G, McClintock BT, Matthiopoulos J, Moss SEW, McConnell B (2014) Marine mammals trace anthropogenic structures at sea. Curr Biol 24:R638–R639. https://doi.org/10.1016/j.cub.2014.06.033

Russell DJ, McClintock BT, Matthiopoulos J, Thompson PM, Thompson D, Hammond PS, Jones EL, MacKenzie ML, Moss S, McConnell BJ (2015) Intrinsic and extrinsic drivers of activity budgets in sympatric grey and harbour seals. Oikos 124:1462–1472. https://doi.org/10.1111/oik.01810

Schneider J (1974) Description and probable behavioral significance of grey seal (*Halichoerus grypus*) vocalizations. Doctoral dissertation, University of Rhode Island.

Shapiro AD, Slater PJB, Janik VM (2004) Call usage learning in gray seals (*Halichoerus grypus*). J Comp Psychol 118:447–454. https://doi.org/10.1037/0735-7036.118.4.447

Sih A, Bell A, Johnson JC (2004) Behavioural syndromes: an ecological and evolutionary overview. Trends Ecol Evol 19:372–377. https://doi.org/10.1016/j.tree.2004.04.009

Sih A, Cote J, Evans M, Fogarty S, Pruitt J (2012) Ecological implications of behavioural syndromes. Ecol Lett 15(3):278–289. https://doi.org/10.1111/j.1461-0248.2011.01731.x

Shuert CR, Pomeroy PP, Twiss SD (2018) Assessing the utility and limitations of accelerometers and machine learning approaches in classifying behaviour during lactation in a phocid seal. Anim Biotelemetry 6(1):14. https://doi.org/10.1186/s40317-018-0158-y

Shuert CR, Pomeroy PP, Twiss SD (2020a) Coping styles in capital breeders modulate behavioural trade-offs in time allocation: assessing fine-scale activity budgets in lactating grey seals (*Halichoerus grypus*) using accelerometry and heart rate variability. Behav Ecol Sociobiol 74:8. https://doi.org/10.1007/s00265-019-2783-8

Shuert CR, Halsey LG, Pomeroy PP, Twiss SD (2020b) Energetic limits: defining the bounds and trade-offs of successful energy management in a capital breeder. J Anim Ecol 00:1–12. https://doi.org/10.1111/1365-2656.13312

Sjöberg M, Ball JP (2000) Grey seal, *Halichoerus grypus*, habitat selection around haul-out sites in the Baltic Sea: bathymetry or central-place foraging? Can J Zool 78:1661–1667. https://doi.org/10.1139/cjz-78-9-1661

Smiseth PT, Lorentsen SH (1995a) Evidence of equal maternal investment in the sexes in the polygynous and sexually dimorphic grey seal (*Halichoerus grypus*). Behav Ecol Sociobiol 36:145–150. https://doi.org/10.1007/BF00177790

Smiseth PT, Lorentsen SH (1995b) Behaviour of female and pup grey seals *Halichoerus grypus* during the breeding period at Froan, Norway. J Zool 236:11–16. https://doi.org/10.1111/j.1469-7998.1995.tb01780.x

Smiseth PT, Lorentsen SH (2001) Begging and parent-offspring conflict in grey seals. Anim Behav 62:273–279. https://doi.org/10.1016/anbe.2001.1763

Smith BR, Blumstein DT (2008) Fitness consequences of personality: a meta-analysis. Behav Ecol 19:448–455. https://doi.org/10.1093/beheco/arm144

Smith EA (1968) Adoptive suckling in the grey seal. Nature, Lond 217:762–763

Stansbury AL, Götz T, Deecke VB, Janik VM (2015a) Grey seals use anthropogenic signals from acoustic tags to locate fish: evidence from a simulated foraging task. Proc Royal Soc B 282:20141595. https://doi.org/10.1098/rspb.2014.1595

Stansbury AL, de Freitas M, Wu GM, Janik VM (2015b) Can a gray seal (*Halichoerus grypus*) generalize call classes? J Comp Psychol 129(4):412–420. https://doi.org/10.1037/a0039756

Stavenow J, Ljungberg P, Kindt-Larsen L, Lunneryd S, Königason S (2016) What attracts Baltic sea grey seals to seal-safe cod pots and when do they attempt to attack fish in the pots? J Ocean Technol 11(4):91–107

Stephenson CM, Matthiopoulos J, Harwood J (2007) Influence of the physical environment and conspecific aggression on the spatial arrangement of breeding grey seals. Ecol Inform 2(4):308–317. https://doi.org/10.1016/j.ecoinf.2007.09.001

Stewart JE, Pomeroy PP, Duck CD, Twiss SD (2014) Finescale ecological niche modeling provides evidence that lactating gray seals (*Halichoerus grypus*) prefer access to fresh water in order to drink. Mar Mamm Sci 30(4):1456–1472. https://doi.org/10.1111/mms.12126

Stirling I (1975) Factors affecting the evolution of social behaviour in the Pinnipedia. Rapports et Procés-verbaux des Réunions–Conseil International pour l'Exploration de la Mer 169:205–212.

Stringell T, Hill D, Rees D, Rees F, Rees P, Morgan G, Morgan L, Morris C (2015) Predation of harbour porpoises (*Phocoena phocoena*) by grey seals (*Halichoerus grypus*) in Wales. Aquat Mamm 41(2):188–191. https://doi.org/10.1578/AM.41.2.2015.188

Summers CF, Burton RW, Anderson SS (1975) Grey seal (*Halichoerus grypus*) pup production at North Rona: a study of birth and survival statistics collected in 1972. J Zool 175:439–451. https://doi.org/10.1111/j.1469-7998.1975.tb01410.x

Survilienė V, Rukšėnas O, Pomeroy P (2016) Play behaviour of wild grey seals (*Halichoerus grypus*): effects of haulout group size and composition. Aquat Mamm 42(2):144–161. https://doi.org/10.1578/AM.42.2.2016.144

Thompson D, Hammond PS, Nicholas KS, Fedak MA (1991) Movements, diving and foraging behaviour of grey seals (*Halichoerus grypus*). J Zool 224:223–232. https://doi.org/10.1111/j.1469-7998.1991.tb04801.x

Thompson PM, McConnell BJ, Tollit DJ, Mackay A, Hunter C, Racey PA (1996) Comparative distribution, movements and diet of harbour and grey seals from the Moray Firth, NE Scotland. J Appl Ecol 33:1572–1584. https://doi.org/10.2307/2404795

Tinker MT, Kovacs KM, Hammill MO (1995) The reproductive behavior and energetics of male gray seals (*Halichoerus grypus*) breeding on a landfast ice substrate. Behav Ecol Sociobiol 36:159–170. https://doi.org/10.1007/BF00177792

Tollit DJ, Schulze AD, Trites AW, Olesiuk PF, Crockford SJ, Gelatt TS, Ream RR, Miller KM (2009) Development and application of DNA techniques for validating and improving pinniped diet estimates. Ecol Appl 19:889–905. https://doi.org/10.1890/07-1701.1

Trillmich F, Müller T, Müller C (2018) Understanding the evolution of personality requires the study of mechanisms behind the development and life history of personality traits. Biol Lett 14(2):20170740

Tucker S, Bowen WD, Iverson SJ (2007) Dimensions of diet segregation in grey seals *Halichoerus grypus* revealed through stable isotopes of carbon (δ^{13}C) and nitrogen (δ^{15}N). Mar Ecol Prog Ser 339:271–282. https://doi.org/10.3354/meps339271

Twiss S.D (1991) Behavioural and energetic determinants of individual mating success in male grey seals (*Halichoerus grypus*, Fabricius 1791). Zoology Department. University of Glasgow. Doctoral Dissertation.

Twiss SD, Pomeroy PP, Anderson SS (1994) Dispersion and site fidelity of breeding male grey seals (*Halichoerus grypus*) on North Rona, Scotland. J Zool 233(4):683–693. https://doi.org/10.1111/j.1469-7998.1994.tb05374.x

Twiss SD, Anderson SS, Monaghan P (1998) Limited intra-specific variation in male grey seal (*Halichoerus grypus*) dominance relationships in relation to variation in male mating success and female availability. J Zool 246(3):259–267. https://doi.org/10.1111/j.1469-7998.1998.tb00156.x

Twiss SD, Caudron A, Pomeroy PP, Thomas CJ, Mills JP (2000) Finescale topographical correlates of behavioural investment in offspring by female grey seals, *Halichoerus grypus*. Anim Behav 59(2):327–338. https://doi.org/10.1006/anbe.1999.1320

Twiss SD, Thomas CJ, Pomeroy PP (2001) Topographic spatial characterisation of grey seal *Halichoerus grypus* breeding habitat at a sub-seal size spatial grain. Ecography 24(3):257–266. https://doi.org/10.1111/j.1600-0587.2001.tb00198.x

Twiss SD, Wright NC, Dunstone N, Redman P, Moss S, Pomeroy PP (2002) Behavioral evidence of thermal stress from over-heating in UK breeding gray seals. Mar Mamm Sci 18(2):455–468. https://doi.org/10.1111/j.1748-7692.2002.tb01048.x

Twiss SD, Duck C, Pomeroy PP (2003) Grey seal (*Halichoerus grypus*) pup mortality not explained by local breeding density on North Rona, Scotland. J Zool 259(1):83–91. https://doi.org/10.1017/S0952836902003035

Twiss SD, Poland VF, Graves JA, Pomeroy PP (2006) Finding fathers: spatio-temporal analysis of paternity assignment in grey seals (*Halichoerus grypus*). Mol Ecol 15(7):1939–1953. https://doi.org/10.1111/j.1365-294X.2006.02927.x

Twiss SD, Thomas CJ, Poland VF, Graves JA, Pomeroy PP (2007) The impact of climatic variation on the opportunity for sexual selection. Biol Lett 3(1):12–15. https://doi.org/10.1098/rsbl.2006.0559

Twiss SD, Franklin J (2010) Individually consistent behavioural patterns in wild, breeding male grey seals (*Halichoerus grypus*). Aquat Mamm 36(3):234–238. https://doi.org/10.1578/AM.36.3.2010.234

Twiss SD, Cairns C, Culloch RM, Richards SA, Pomeroy PP (2012a) Variation in female grey seal (*Halichoerus grypus*) reproductive performance correlates to proactive-reactive behavioural types. PLoS ONE 7(11):e49598. https://doi.org/10.1371/journal.pone.0049598

Twiss SD, Culloch RM, Pomeroy PP (2012b) An in-field experimental test of pinniped behavioral types. Mar Mamm Sci 28(3):E280–E294. https://doi.org/10.1111/j.1748-7692.2011.00523.x

Twiss SD, Shuert CR, Brannan N, Bishop AM, Pomeroy PP (2020) Reactive stress-coping styles show more variable reproductive expenditure and fitness outcomes. Sci Rep 10(1):9550. https://doi.org/10.1038/s41598-020-66597-3

van Beest FM, Mews S, Elkenkamp S, Schuhmann P, Tsolak D, Wobbe T, Bartolino V, Bastardie F, Dietz R, von Dorrien C, Galatius A, Karlsson O, McConnell B, Nabe-Nielsen J, Olsen MT, Teilmann J, Langrock R (2019) Classifying grey seal behaviour in relation to environmental variability and commercial fishing activity - a multivariate hidden Markov model. Sci Rep 9:5642. https://doi.org/10.1038/s41598-019-42109-w

van Bleijswijk JD, Begeman L, Witte HJ, IJsseldijk LL, Brasseur SM, Gröne A, Leopold MF (2014) Detection of grey seal *Halichoerus grypus* DNA in attack wounds on stranded harbour porpoises *Phocoena phocoena*. Mar Ecol Prog Ser 513:277–281. https://doi.org/10.3354/meps11004

van Neer A, Jensen LF, Siebert U (2015) Grey seal (*Halichoerus grypus*) predation on harbour seals (*Phoca vitulina*) on the island of Helgoland, Germany. J Sea Res 97:1–4. https://doi.org/10.1016/j.seares.2014.11.006

van Neer A, Gross S, Kesselring T, Wohlsein P, Leitzen E, Siebert U (2019) Behavioural and pathological insights into a case of active cannibalism by a grey seal (*Halichoerus grypus*) on Helgoland, Germany. J Sea Res 148:12–16. https://doi.org/10.1016/j.seares.2019.03.004

Vincent C, Meynier L, Ridoux V (2001) Photo-identification in grey seals: legibility and stability of natural markings. Mammalia 65:363–372. https://doi.org/10.1515/mamm.2001.65.3.363

Vincent C, Ridoux V, Fedak MA, McConnell BJ, Sparling EE, Leaute J-P, Jouma'a J, Spitz J (2016) Foraging behaviour and prey consumption by grey seals (*Halichoerus grypus*)-spatial and trophic overlaps with fisheries in a marine protected area. ICES J Mar Sci 73:2653–2665. https://doi.org/10.1093/icesjms/fsw102

Weitzman J, den Heyer C, Bowen DW (2017) Factors influencing and consequences of breeding dispersal and habitat choice in female grey seals (*Halichoerus grypus*) on Sable Island, Nova Scotia. Oecologia 183:367–378. https://doi.org/10.1007/s00442-016-3764-5

Whiting MJ (1999) When to be neighbourly: differential agonistic responses in the lizard *Platysaurus broadleyi*. Behav Ecol Sociobiol 46(3):210–214. https://doi.org/10.1007/s002650050611

Whiting MJ, Stuart-Fox DM, O'Connor D, Firth D, Bennett NC, Blomberg SP (2006) Ultraviolet signals ultra-aggression in a lizard. Anim Behav 72(2):353–363. https://doi.org/10.1016/j.anbehav.2005.10.018

Wilson S (1974) Juvenile play of the common seal (*Phoca vitulina vitulina*) with comparative notes of the grey seal (*Halichoerus grypus*). Behavior 48:37–60. https://doi.org/10.1163/156853974X00246

Wolf M, Weissing FJ (2012) Animal personalities: consequences for ecology and evolution. Trends Ecol Evol 27(8):452–461. https://doi.org/10.1016/j.tree.2012.05.001

Worthington-Wilmer J, Allen PJ, Pomeroy PP, Twiss SD, Amos W (1999) Where have all the fathers gone? An extensive microsatellite analysis of paternity in the grey seal (*Halichoerus grypus*). Mol Ecol 8(9):1417–1429. https://doi.org/10.1046/j.1365-294x.1999.00705.x

Worthington-Wilmer J, Overall AJ, Pomeroy PP, Twiss SD, Amos W (2000) Patterns of paternal relatedness in British grey seal colonies. Mol Ecol 9:283–292. https://doi.org/10.1046/j.1365-294x.2000.00872.x

Chapter 10
The Harbor Seal: The Most Ubiquitous Phocid in the Northern Hemisphere

James T. Harvey, Magda Chudzinska, Bernie McConnell, and Gordon Hastie

Contents

Abstract True to its name, the harbor (or common) seal (*Phoca vitulina*) is found in many coastal environments in the northern hemisphere and is perhaps the most wide-ranging pinniped in the world. Harbor seals haul out on many types of coastal or ice environments where they engage principally in resting but also come ashore for pupping and nursing, temperature maintenance, predator avoidance, digestion, and molting. Males may defend underwater territories near haulout sites or foraging areas and use acoustic displays to attract mates or ward off competing males. Mating occurs underwater, and females give birth to a single pup in the boreal spring/summer that is capable of swimming and diving within minutes of birth. Pups are weaned in three to six weeks of birth and often disperse more widely than adults. Although they begin their lives eating slower swimming prey, such as some invertebrates and smaller fish, they eventually become adept predators of fishes and cephalopods in many types of coastal environments. Foraging trips might be a day's swim from the haulout site or may be two to three weeks duration. Due to their coastal distribution, harbor seals are more susceptible to anthropogenic impacts, such as contaminants, disturbance, human-made structures, and noise pollution. Harbor seals are perhaps the most adaptable phocid to deal with potential climate change issues, given their evolutionary history in the dynamic coastal environments near human populations.

J. T. Harvey (✉)
Moss Landing Marine Laboratories, San José State University, San Jose, CA, USA
e-mail: james.harvey@sjsu.edu

M. Chudzinska · B. McConnell · G. Hastie
Sea Mammal Research Unit, University of St Andrews, St Andrews, Scotland

D. P. Costa and E. McHuron (eds.), *Ethology and Behavioral Ecology of Phocids*, Ethology and Behavioral Ecology of Marine Mammals,
https://doi.org/10.1007/978-3-030-88923-4_10

Keywords Harbor seal · Haulout sites · Underwater territories · Benthic foraging · Anthropogenic stressors

10.1 Introduction

The harbor or common seal (*Phoca vitulina*) is a relatively small coastal species, and the most ubiquitous phocid in the northern hemisphere (Fig. 10.1). Its use of nearshore habitats, including bays and harbors, is what inspired its common name. Harbor seals occur as far north as 78° N in Svalbard, Norway (Prestrud and Gjertz 1990) to as far south as 26° S off Baja, Mexico (Orr et al. 2018).

Rice (1998) recognized five subspecies of harbor seals, *P.v. concolor* in the western Atlantic, *P.v. vitulina* in the eastern Atlantic, *P.v. mellonae* in a few freshwater lakes and rivers in eastern Canada, *P.v. stejnegeri* in the western North Pacific, and *P.v. richardii* in the eastern Pacific. Berta and Churchill (2012) have more recently recommended that *P.v. stejnegeri* and *P.v. concolor* not be recognized as valid subspecies. If this is the case, then there are only three subspecies, one in the Atlantic, one in the Pacific, and one in freshwater habitats in eastern Canada. Because most populations of harbor seals are contiguous and the species is widely dispersed, there are greater genetic differences with geographical distance (Westlake and O'Corry-Crowe 2002) and some fine-scale population structuring (Olsen et al. 2017).

Harbor seals have evolved numerous behavioral adaptations to cope with the extensive range of habitats they occupy, which includes semi-tropical, freshwater,

Fig. 10.1 A harbor seal atop a rocky coastline. Photo by S. Hansen

and polar environments. Their use of haulout sites depends on many environmental factors such as tides, weather, season, time of year, and reproductive status, and given their proximity to human developments along continental margins, our knowledge of this species' behavior is relatively robust. Harbor seals typically come ashore along coastal shorelines and in bays and harbors, making them easily observed. Interactions among individuals on land have been recorded for decades. Their use of nearshore haulout sites has enabled many studies of individual behavioral interactions, pupping and nursing, female-pup interactions, and reactions to disturbances.

The bulk of behavioral observations of harbor seals have occurred along the west and east coasts of North America, Alaska, and northwest Europe. Some well-known reviews of harbor seal biology and their behaviors have been produced regarding seals in Canada (Boulva and McLaren 1979; Bigg 1969), off the west coast of the USA (Scheffer and Slipp 1944), and in northern Europe (Bjørge et al. 2010). Additional information on reproductive behaviors, physiology, and foraging behaviors has been recorded for harbor seals held in captivity in various locations worldwide (e.g., Ydesen et al. 2014; Kastelein et al. 2010).

Advances in bio-logging technology have shed light on the at-sea and underwater life of harbor seals. We have learned that harbor seals feed in a variety of habitats, typically on the continental shelf, usually not venturing far or for long away from their commonly used haulout sites. Animal-borne tags also allow us to assess the human impacts of ocean-based energy and the increasing industrialization of coastal waters.

The combination of observations at haulout sites, electronic tagging, and captive research has provided a wealth of information about this phocid living along most of the northern hemisphere, often amid human developments. In this chapter, we review some of the knowledge of this expansive and adaptable species. We describe the wide variety of terrestrial and ice habitats used by harbor seals and discuss a variety of factors driving haulout behavior. After reviewing the extensive literature of male and female reproductive behavior, we discuss the drivers of foraging behavior and diet and habitat choice, and then of movement itself at a variety of scales. We emphasize the importance of an analytical modeling framework to predict behavior in a variety of scenarios. Finally, we review the potential anthropogenic threats, and conclude that harbor seals behavioral plasticity is particularly advantageous at a time of global oceanic change.

10.2 Haulout Behavior

Harbor seals typically use haulout sites for resting, social interactions, predator avoidance, digestion, thermoregulation, skin maintenance, molting, pupping, and nursing (Brown and Mate 1983; London et al. 2012; Montgomery et al. 2007; Simpkins et al. 2003; Watts 1996). Based on haulout deprivation experiments, it seems harbor seals need to haul out periodically (Brasseur et al. 1996), and they spend about 10–50% of their day resting on the shore, depending on the season (Cunningham et al. 2009;

Mikkelsen et al. 2019; Pitcher and McAllister 1981; Sullivan 1982). For instance, approximately 41% of radio-tagged seals on San Miguel Island, California hauled out each day (Yochem et al. 1987). The probability of being hauled out was 50–55% in an Alaskan fjord in August (Womble et al. 2020). Because of their limited mobility on land, harbor seals typically haul out on sites that provide easy access to water. Often these can be on offshore rocks, isolated beaches, sand or mudflats in estuaries and embayments, ice flows, or on human-made structures, such as docks and log booms (Fig. 10.2). The proximity to a channel or deep water allows the animal to seek refuge and escape potential threats (Montgomery et al. 2007). Haulout sites are often in protected areas (Brost et al. 2020), free from repeated disturbance from predators or humans, and are usually located relatively near foraging areas (Montgomery et al. 2007; see Sect. 10.4).

Energetic and environmental models of haulout behavior indicate that seals should come ashore whenever they are not foraging or mating unless they are far from haulout sites while foraging. There is an energetic cost to remaining in the water, and hauling out provides freedom from aquatic predators and energy savings compared with sleeping in the water (Watts 1996). Grouping behavior, however, causes agonistic reactions that are more frequent at intermediate densities and during the post molt, which comes with an energetic cost (Honeywell and Maher 2017). Because the molt progresses through the population, based on sex and age—yearlings molt first followed by subadults, adult females, and then adult males—age and sex class of animals using haulout sites changes throughout the molting season in late summer

Fig. 10.2 Harbor seals on various haulout site habitats, including sand bar (**a**), rock outcrop in San Francisco Bay (**b**), mudflat (**c**), and ice flows (**d**). Photos by J. Harvey and C. Young, using a telephoto lens at some distance so as not to disturb

and early autumn (Daniel et al. 2003). Although it seems clear why seals come ashore during the molt and breeding periods, it is less clear why they haul out at other times of the year, other than for predator avoidance, skin maintenance, and possibly digestion. Because digestion is energetically costly, resting ashore may be necessary to assist with efficient breakdown and assimilation of ingested food from the previous foraging bout (Sparling et al. 2007).

Floating ice (e.g., glacial ice or sea ice) provides a unique, mobile haulout surface, and greater numbers of harbor seals use this habitat when there is greater ice cover during the middle of the day, clear skies, and no precipitation (Blundell and Pendleton 2015; Blundell et al. 2011; Calambokidis et al. 1987). Ice is one of the most important haulout surfaces for harbor seals in Alaska, where 10% of seals use glacial ice. More than 5000 harbor seals (some of the world's largest groups) have been observed in these glacial ice areas (Jansen et al. 2015; Womble et al. 2010). Because ice flows provide haulout habitat throughout the day, ice flows are often used at any tide (Boveng et al. 2003), resulting in greater pup production and provide protection from land-based predators (Calambokidis et al. 1987; Womble et al. 2010). Greater numbers of seals typically are found on moderate densities of ice (Jansen et al. 2015). For harbor seals off Svalbard, the most northerly population globally, seals used terrestrial locations mostly during the austral summer and used shore-fast ice, offshore ice, or glacier ice more commonly during autumn and winter (Hamilton et al. 2014). These seals continued to haul out even during the coldest periods of the polar night (Hamilton et al. 2014).

Harbor seals also inhabit freshwater areas, in particular, one lake in Alaska and a series of lakes and connecting streams in Canada. In northern Quebec, Canada, harbor seals have lived in lake systems for at least 260 years (perhaps isolating from the marine environment 3000–8000 years ago; DFO 2017) and travel among the lakes using the connecting rivers (Smith et al. 2006). These freshwater harbor seals are considered a subspecies, *Phoca vitulina mellonae*, and haul out less frequently during the winter as the lakes begin to freeze over (Smith et al. 2006). They can be distinguished from *P.v. richardii* by stable isotopes and fatty acid profiles (Smith et al. 1996). Harbor seals also inhabit Iliamna Lake in Alaska and are considered to be related to *P.v. richardii*, but with some distinctive genetic differences (Boveng et al. 2016). Harbor seals in Iliamna Lake also haulout less frequently during the winter as the lake freezes over, and occasionally use the ice as a haulout surface (Boveng et al. 2016).

The seasonal effects of the number of seals ashore are based on several factors, such as breeding, molt, weather conditions, and prey availability. Typically, the highest number of seals resting ashore occurs near the time of pupping and molt (Fig. 10.3; Brown and Mate 1983; Daniel et al. 2003; Jansen et al. 2015). Because pups are born, nursed, and rest on shore during the spring/summer, there is a peak in their occurrence on shore during this time. A peak in numbers ashore in summer/fall corresponds to the timing of molt. Spending more time ashore during molt likely increases skin temperature, allowing the molt to proceed more rapidly and efficiently (Feltz and Fay 1966; Paterson et al. 2012). During other times of the year, harbor seals spend less than 50% of their time hauled out, spending more time at sea and

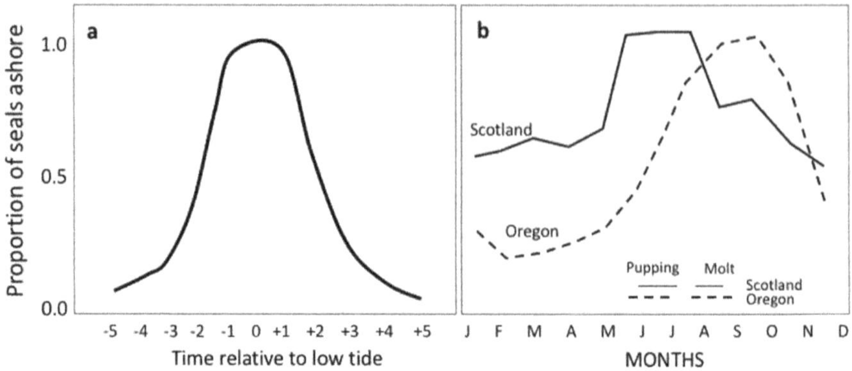

Fig. 10.3 The proportion of seal ashore relative to the time of low tide (**a**); Schneider and Payne (1983), and the proportion of the maximum number of seals ashore throughout the year in North Atlantic (Scotland; Thompson et al. 1997) and eastern Pacific (Oregon; Brown and Mate 1983) haulout sites, also showing the timing of pupping and molt for each location (**b**)

foraging farther from land and diving deeper (Cunningham et al. 2009). Harbor seals can be affected by overheating; thus, specific locations or surfaces may affect the number and timing on specific haulout sites (Watts 1992), as can winter temperatures and wind that might reduce the number of seals ashore.

In most tidally influenced haulout sites, the highest number of harbor seals occurs during low water (Fig. 10.3) in the middle of the day (Jemison et al. 2006; Roen and Bjørge 1995; Schneider and Payne 1983; Simpkins et al. 2003). However, some locations are primarily used at high tides (London et al. 2012), and some individuals haul out more at night than during the day (Yochem et al. 1987). Many haulout sites affected by tides are either surrounded by water (e.g., mudflats in estuaries, rocky outcrops offshore) or become inaccessible to predators during certain tides. For harbor seals using haulout sites that are available throughout the tidal cycle (e.g., many shoreline beaches and ice flows), time of day may be more important than the tidal state. For instance, Stewart (1984) reported the highest number of harbor seals on San Miguel Island in the early afternoon. Still, the peak counts were greater for afternoon low tides compared with afternoon high tides. The availability of a haulout site may also be affected by nearby environmental conditions. For instance, Hastie et al. (2016) found fewer seals on nearby haulout sites during flooding tides because the seals were foraging in a few local high-current locations. Although several factors can affect the number of seals hauled out, such as wind/chill, wave height (Pauli and Terhune 1987), and air temperature, the most significant factors were tide and disturbance at a location in Massachusetts, USA (Schneider and Payne 1983).

Harbor seals may haul out in isolation, although most congregate on haulout sites likely because it increases vigilance for predators or disturbances. Typically, larger group sizes allow fewer seals within the group to remain vigilant at any one time (Norris 2009). Harbor seals in the Bay of Fundy decreased scanning durations and increased interscan intervals as group size increased (Terhune 1985). By grouping during haulout periods, individual seals could reduce their surveillance time without affecting the probability of detecting a potential predator because predator detection was shared among the group (Terhune 1985). Da Silva and Terhune (1988) concluded that harbor seals in the western Atlantic that grouped together increased their vigilance and detected potential predators at a greater distance than solitary individuals. Seals on the periphery of groups and individuals that had recently hauled out scanned significantly longer than those in the center or that had been ashore for some time (Terhune and Brillant 1996). London et al. (2012) reported a 40–50% increase in the probability of being hauled out for seals that had previously been exposed to killer whales in the Hood Canal region of Washington State, USA, another indication that predators can influence haulout behavior.

Individual harbor seals will move around and use multiple haulout sites, but some have fidelity to specific locations. This is somewhat dependent on age, as newly weaned seals and juveniles are more likely to disperse and use many haulout sites than adults who use a particular region (Lowry et al. 2001; Small et al. 2005). Although younger animals may disperse widely, they can return to their natal area by year two. Females tend to show more site fidelity as they age (Harkonen and Harding 2001). Males disperse more widely as they age and are less likely to occur in their natal area (Harkonen and Harding 2001), which would decrease the chances of inbreeding. These movements by younger animals are likely based on exploration for new foraging and breeding locations. Yochem et al. (1987) reported that most (72%) radio-tagged seals on San Miguel Island only used the site where they were tagged. On Kodiak Island, Alaska, radio-tagged seals commonly used one or two principal haulout sites (Pitcher and McAllister 1981). In Alaska, seals radio-tagged on terrestrial haulout sites primarily used land-based haulout sites, whereas seals caught on glacial ice primarily used ice, indicating some fidelity to haulout locality (Blundell et al. 2011). Thus, it appears that once the newly weaned individuals find a suitable location, they become more sedentary in the area that is now familiar to them.

Although harbor seals are not particularly aggressive on land, a seal will act antagonistically to another seal entering its space. Antagonistic behaviors include extending, waving, or scratching with a foreflipper, growling, or thrusting the head towards another (Sullivan 1982). Adult males dominate in these interchanges. Typically, larger-sized animals were dominant as were males when interacting with like-sized or slightly smaller females (Sullivan 1982). Density and season can affect agonistic behaviors. Harbor seals in Humboldt County, California, were more aggressive when suitable haulout space was limited (Neumann 1999). The rate of interactions among harbor seals in Maine (USA) was greater during molt, but the intensity of agonistic behaviors was greater during post molt (Honeywell and Maher 2017). Typically, individuals that occupy a space on a haulout site are more successful in keeping their spot than an intruder (Honeywell and Maher 2017). Except for mothers and pups' close behavioral bond, there are few stable social relationships among harbor seals that come ashore for rest (Godsell 1988).

10.3 Reproductive Behavior

Harbor seals generally mate underwater, give birth and nurse their pups on land, and use a hybrid form of capital and income breeding strategies (Chaps. 7 and 8). The chronology of reproduction is such: pups are born in spring/summer (see regional differences on Fig. 10.4) and nursed for three to six weeks until weaning; shortly after weaning, females come into estrus (the estrous period may last from one to nine weeks) and mate; there is a two to three-month delay in implantation of the blastocyst; after that an eight to nine-month gestation period, until the pup is born about one year after mating (Bigg 1969, 1973). Although this pattern is repeated throughout their range, the timing of these events is specific to the locale. Harbor seal reproductive timing is mostly dependent on latitude (e.g., Mexico to Washington, along eastern North America; Fig. 10.4). There is a cline in pupping and mating (Bigg 1969; Temte et al. 1991) determined by photoperiod and energetic demands (Temte et al. 1991). In other areas, such as the eastern North Atlantic and from Washington to Alaska, there does not appear to be a clinal effect (Temte et al. 1991).

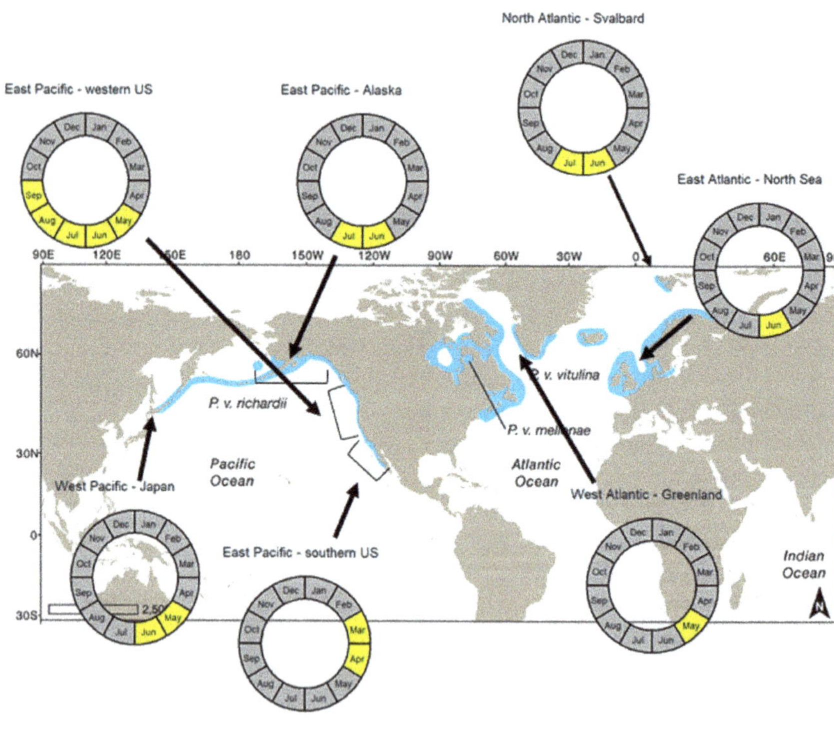

Fig. 10.4 Regional differences in the timing of the breeding season (represented in yellow). Blue area shows the distribution of harbor seals in the northern hemisphere (adapted by Nina Lisowski from Jefferson et al. 2015)

Although it is assumed that most mating occurs underwater, occasionally mating occurs on land or in the intertidal zone (Allen 1985). Those matings observed on land involved the male initiating contact and mounting the female dorsally, the male biting the nape of her neck, and the female eventually terminating copulation (Allen 1985; Sullivan 1982). Early studies determined that harbor seals formed hierarchies based on sex and size (Sullivan 1982), but because most activities occur underwater, the type of mating system has been challenging to determine. During the mating season, females come and go from a variety of haulout sites. This behavior makes it difficult for males to restrict other males from access to females. Given the aquatic nature of harbor seal mating systems, researchers have used acoustics, electronic tags, and to a lesser extent, visual observation to study breeding behavior.

10.3.1 Male Reproductive Behavior

The behavioral mating system used by males at sea is diverse. Like some other phocids, some harbor seal males may hold underwater territories, and haul out less, and spend more time nearshore as the mating season approaches (Boness et al. 2006; Hayes et al. 2004a). Aquatic territoriality makes some sense for this species because mating occurs underwater. The females use specific sites to haul out and tend to pups while frequently entering the water to travel to particular foraging locations. As a result, males hold these underwater territories either near haulout sites used by females, along aquatic corridors used by females to travel between haulout sites and foraging sites, or near foraging areas used by females (Hayes et al. 2004a; Perry 1993; Van Parijs et al. 2000a). The location of male territories depends on the topography of the area and the females' predictable movements. For instance, in a small estuary in central California, four males established adjacent underwater territories along a corridor from the ocean to the main haulout site. They maintained these same territories for 2–4 years (Hayes et al. 2006). Off Scotland, males repeatedly used the same territories for several years that were between 40 and 135 m^2 in area (Van Parijs et al. 2000b). In most of these situations, males use a type of lekking display to attract females, which is often accompanied by vocal displays to attract females and in competitive interactions with other males (Hanggi and Schusterman 1992). In defending their underwater territories, males may inflict injuries to the tail and rear flippers of other intruding males.

Males may begin displaying acoustic signaling at least one month before peak estrus and form underwater territories in areas where females are likely to be present. Maximum calling by males typically occurs slightly before or during peak estrus, depending on the location, and most calls are made at night when the females are in the water foraging (Hayes et al. 2004b; Nikolich et al. 2018; Van Parijs et al. 1999). The acoustic characteristics of male vocalizations may provide females with cues to male fitness (Hanggi and Schusterman 1992). Larger, more dominant males make lower frequency, longer duration vocalizations, which likely communicate their status to females and other males (Nicholson 2000). Although experimental playbacks suggested that females responded to dominant males' vocalizations more than control sounds, it was not a statistically significant effect (Matthews et al. 2018). The vocalizations of large, adult males often attract other males, who approach submissively and place their muzzle up to the muzzle of the vocalizing male that is making a" roar" call (Fig. 10.5). Because males have distinctive vocalizations (Van Parijs et al. 2000a, 2003), approaching a vocalizing dominant male may be a method for other males to determine the identity of the caller and assess their placement in the hierarchy of males in the area (Hayes et al. 2004b; Nicholson, 2000). Besides vocalizations, males also slap their hind flippers (lob-tailing) or fore-flippers on the surface in an apparent attempt to attract females or signal their presence to competing males (Fig. 10.5).

The evidence that males holding territories have a higher probability of siring offspring is limited (Hayes et al. 2006) and requires further investigation with a greater sample size. Coltman et al. (1999) suggested that males who sired the most pups were of moderate size and did not haul out frequently, thus having more time to search for mates. Paternity testing revealed a breeding system with little or no polygyny. Males typically sire only one pup regardless of their size, their territoriality, or their other attributes (Coltman et al. 1998; Hayes et al. 2006). Males use a variety of strategies; if they are large and have abundant stored energy, they can hold nearshore territories; if they are of moderate size, mobile, and not high in the dominance hierarchy, they can move around and display to maximize their encounters with females. During the mating season, males move to shallower areas, closer to haulout sites, and reduce feeding as they engage in slow patrolling, agonistic behavior with other males, and visual and acoustic display towards females (Coltman et al. 1997).

Fig. 10.5 Male harbor seals attending an adult male that is producing a "roar" call that attracts other males (**a**: photo by T. Nicholson), and a male flipper slapping in an aquatic territory (**b**: photo by S. Hayes)

10.3.2 Female Reproductive Behavior

Perhaps one of the more remarkable features of harbor seals is their ability to reproduce in areas that are so variable, with the pup able to dive minutes after birth, hence freeing themselves from the need to only use islands or ice flows to avoid predators while reproducing. Their flexibility also is based on using a modified maternal foraging strategy (Boness and Bowen 1996). Pregnant females often separate themselves from other seals shortly before birth, and may maintain a distance from the colony's main body for the first two weeks of lactation (Newby 1973). Females deliver their pups within about four minutes of the onset of contractions, sometimes thrashing their hindquarters from side to side once the pup's head has emerged (Lawson and Renouf 1985). The lanugo coat is shed in utero; hence the pups have a pelage similar to the adult when born. The lanugo coat is expelled after birth. Because harbor seals often haul out on rocky and sandy substrates, the spotted pelage helps with camouflage, another trait evolved for coastal living. Mothers often will immediately initiate nursing (Fig. 10.6), with the pup nursing within one hour of birth and able to follow their mothers into the water within minutes of birth (Lawson and Renouf 1985). The ability of pups to swim shortly after birth is an evolved trait that allows them to use tidal haulout sites where the substrate is periodically submerged, and to be successful in coastal environments.

Because the lactation period is relatively short, females need to be extremely attentive to the pup and protective. Females vocalize during interactions with their

Fig. 10.6 Harbor seal pup nursing from mother on a sandy beach haulout. Photo by J. Harvey

Fig. 10.7 Female and pup nuzzling. Photo by J. Harvey

pups, likely to help facilitate mother–pup recognition (Renouf et al. 1983). The pups produce a distinctive "mother attraction call," often a sound that is similar to sheep ("ma-a-a") that possibly can be recognized by the mother over long distances, even underwater (Newby 1973; Renouf et al. 1983; Sauve et al. 2015). Often mothers will nuzzle the pup making face-to-face contact, likely to create a bond with the pup and initiate recognition (Fig. 10.7; Newby 1973). In Washington State (USA), females nursed their pups for 25–160 s periods spaced three to four hours apart and associated with the tides (Newby 1973). Recent use of stomach temperature transmitters indicated that pups may suckle more underwater than while hauled out (Schreer et al. 2010). As for most mammals, female harbor seals are protective of their pups by displaying aggression if other females, males, or other species approach their pup.

Because the pups can swim at birth, mothers take their pups to sea almost immediately, thus freeing them from the birth site. This capability allows flexibility for the location of pupping and allows the mother access to foraging areas while simultaneously caring for their pup. Although the pup may not be able to dive as well as the adult mother, the female can still forage, move about, nurse, and avoid dangerous circumstances. We know female harbor seals take their pups with them on foraging bouts, but this may vary with location. At Gertrude Island off Washington State, pups accompanied the mother on 98.6% of foraging trips (D'Agnese 2015). On Sable Island (Nova Scotia, Canada), females often leave their pups on the beach during foraging bouts, but this may be dependent on the number of female-pup pairs on the beach, such that females take their pups to sea when there are fewer females to provide some sort of protection (Bowen et al. 1999). During early lactation, females can dive for a longer duration and spend more time on the bottom than their pups, but pups can generally dive to the same depth as their mothers (Bowen et al. 1999). During paired diving bouts, the pup descends with the mother but has to return for a breath if the dive exceeds about 1.5 min, with the mother often accompanying the pup to the surface. Pups increase their diving capacity as lactation progresses, spending up to 70% of their time in the water (Jorgensen et al. 2001), and dive nearly as deeply, as often, and a similar duration as the mother (Bowen et al. 1999).

Harbor seals fall somewhat in the middle of the capital breeders (e.g., many large phocids) strategy on one side and income breeders (e.g., otariids) on the other (D'Agnese 2015; Chap. 8). That is, female harbor seals have some stored energy that allows them to nurse their pup on land, initially using fat reserves as capital, but those stored reserves are used up quickly. The female soon has to go to sea to gain income by foraging (Boness and Bowen 1996; Bowen et al. 1992). Harbor seals were thought to use a maternal foraging strategy (Boness and Bowen 1996) where the mother went to sea to forage but nursed the pup back on land. We now know that they can also nurse the pup at sea (Schreer et al. 2010).

Female harbor seals mobilize nearly 80% of their stored fat and 20% of their stored protein during the first 19 days of lactation (33% loss in total mass), of which nearly 50% of that energy is deposited in the pup (Bowen et al. 2001). Later in lactation, females increase their diving activities to acquire additional energy necessary for themselves and pup growth, increasing food intake by up to six times that at the beginning of lactation (Bowen et al. 2001). Suckling bout duration and milk intake also increase as the pup ages (Hedd et al. 1995).

Harbor seal pups have a strong propensity to stay with their mothers. Although their mothers are attentive, the mother weans the pup after 3–6 weeks of nursing (Lawson and Renoud 1987). How the female weans the pup is unknown but likely involves the mother abruptly separating herself from the pup. The pup's age at weaning is likely affected by maternal age, the female's nutritional state, the sex of the pup, and the indirect effects of local seal density (Muelbert and Bowen 1993). Pups fast for an average of 15–17 days post-weaning while remaining in the natal area, losing 21% of their mass before they transition to feeding on their own (Muelbert and Bowen 1993). First-year survival of pups post-weaning increases with pup size, with a pup of 17 kg having a 63% chance of survival and a pup of 32 kg having a 96% chance of survival (Harding et al. 2005).

10.4 Foraging Behavior

10.4.1 Diving Behavior and Physiology—Seasonal and Diurnal Changes

Diving behavior of harbor seals, as in most pinnipeds, includes functions such as foraging, reproduction, communication, traveling, resting, and predator avoidance (Madden et al. 2008). Some of these behaviors are performed simultaneously. Harbor seals mainly rest on land but can also rest at sea or sleep close to the bottom. Ramasco

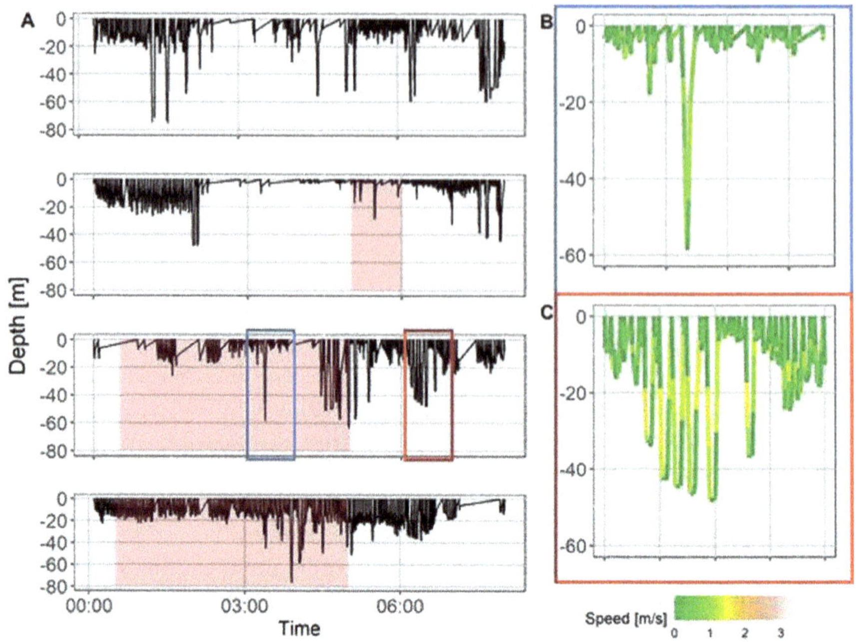

Fig. 10.8 Example of 8 h dive profiles for four consecutive days for an adult harbor seal tagged off the coast of Norway in 2009 (**a**: Ramasco et al. 2014). Dive profiles can provide insight into the behavior and activity budget of seals. On graph (**a**), marked in pink, are periods of low activity dives: dives during which seals are most likely sleeping. Such interpretation is based on coupling dive profiles with ascent and descent speed of individuals (**b–c**). For harbor seals, low activity dives are characterized by lower descent speed (**b**) than for foraging dives (**c**). Even more detailed interpretation of seal underwater behavior can be obtained by coupling dive profiles and accelerometer data as in Mikkelsen et al. (2019). *Data source* Ramasco et al. (2014)

et al. (2014) and Mikkelsen et al. (2019) reported that harbor seals rested during 8–12% of their dives (Fig. 10.8).

Harbor seals generally forage nearshore in relatively shallow waters (<200 m) using a central-place foraging mode (Eguchi and Harvey 2005; Gjertz et al. 2001; Hastings et al. 2004; Ries et al. 1997; Schreer and Kovacs 1997; Tollit et al. 1998). Dives exceeding 450–500 m and lasting more than 30 min have also been recorded (e.g., Gjertz et al. 2001; Hastings et al. 2004). The deepest recorded dive for a harbor seal is to 631 m off the coast of Greenland (Rosing-Asvid et al. 2020). The aerobic dive limit (ADL) is the maximum breath hold time without an increase of lactic acid concentration in the blood during or after a dive (Kooyman 1985; Chap. 4). The aerobic dive limit depends on the animals' metabolic rate, which determines the time before the oxygen stored in the lungs, muscle, and blood is depleted (Ponganis 2015). Juvenile seals have a lower predicted ADL due to higher metabolic rate, less efficient heart rate regulation, respiration rate, vasoconstriction, body temperature, and higher relative blubber content than adults (Hassrick et al. 2007). Estimated

ADL for a harbor seal pup is around 3 min (Jorgensen et al. 2001), whereas it is almost 9 min for an 80 kg adult female (Bowen et al. 1999; Thompson and Fedak 2001). Nevertheless, pups diving for nearly 10 min have been reported (Blanchet et al. 2016).

In general, harbor seal foraging and diving behavior differs with region, season, energetic demands, diet preferences, and occurrence. Hence, the depth of each dive is related to prey composition of seals, as well as regional differences in-depth distribution of the sea bottom (Frost et al. 2001; Hastings et al. 2004; Ramasco et al. 2017; Womble et al. 2014). The depth and duration of dives also vary with age, as defined by body size, differences in metabolic rates, and probably experience (Carter et al. 2017). For example, Hastings et al. (2004) reported that compared with adults, juveniles used a greater variety of depths, dove more frequently and more deeply despite having shorter dive durations, and displayed a stronger diurnal pattern in dive depth. Finally, the foraging intensity varies seasonally depending on the energetic demands of seals. For example, after reduced foraging during molting, seals tend to dive more frequently to replenish energy stores (e.g., Krafft et al. 2002); post-breeding foraging trips of harbor seals are more extensive in comparison to the breeding season (e.g., Womble and Gende 2013).

Preparation for reproduction may cause adult seals to increase diving frequency and energy storage (Austin et al. 2006). During the breeding season, the at-sea distribution of foraging females is influenced by the proximity of the sites where they suckle their pups. The distribution of adult males and possibly subadult males is determined by access to females, both on land and in the water, because mating usually takes place in the water (Van Parijs et al. 1997). Males tend to make shorter feeding trips during the breeding season, reduce their home range, and make shorter dives during that time, with some dives likely associated with underwater vocalization and displays during the mating period (Thompson et al. 1989, 1998; Van Parijs et al. 1997; see 10.3). Therefore, males face a trade-off between staying close to land where the probability of encountering females is higher, but the opportunity to forage is lower due to competition for prey. In contrast, the likelihood of encountering a female will be less where the possibility of finding prey is higher due to reduced competition (Van Parijs et al. 1997). As there should be local prey depletion due to the concentration of animals around the breeding sites, the non-breeding animals should be compensated for their travel costs if they encounter less depleted prey resources at greater distances (Robson et al. 2004). Males tend to forage in relatively deep waters offshore prior to mating to maintain their body mass (Coltman et al. 1997).

Before electronic-tagging studies, researchers assumed harbor seals fed at night because few seals were hauled out at nighttime, which was thought to be related to diel fish movements (Roen and Bjørge 1995). In central California, radio-tagged seals foraged mostly at night, with seals leaving the haulout site at dusk and returning at sunrise (Eguchi 1998). Additional investigations have shown that harbor seals feed during daytime and nighttime, depending on the site and situation. For instance, foraging harbor seals in an Alaska fjord system dove deeper during daylight independent of whether they were tagged at a terrestrial or glacial ice haulout site (Womble et al. 2014). Seals appear to rely on experience along with some searching to determine where to forage. They repeatedly returned to the same foraging locations, presumably because they were previously successful at those sites (Thompson and Miller 1990).

10.4.2 Diet and Habitat Preferences

Harbor seals are generalist and opportunistic predators, consuming a diversity of small to medium-sized (range 5–80 cm, average 8–40 cm) fishes and invertebrates, and even birds (Frost et al. 1999; Tallman and Sullivan 2004). Although as a species, harbor seals consume a wide range of species, individuals may specialize on different prey types (Sharples et al. 2009; Wilson and Hammond 2016). For example, harbor seals off the Oregon and Washington coast consume 120 species or genera from 37 taxonomic orders and 62 families altogether (Steingass 2017). Preferred habitat characteristics are linked to prey distribution, which is reflected in seasonal differences in movement in relation to prey distribution and availability (Grecian et al. 2020; Jansen et al. 2015; Lowry et al. 2001; Ramasco et al. 2017; Simpkins et al. 2003). In Monterey Bay, California, harbor seals foraged at locations on the continental shelf near the edge of the Monterey Canyon (Eguchi 1998; Trumble 1995), or occasionally in estuaries and embayments (Gibble and Harvey 2015; Harvey et al. 1995). Prey type and size also varies with age; newly weaned harbor seals eat smaller fish and slow-moving invertebrates, likely because they have not developed the foraging skills and dive capacities of larger individuals (Muelbert et al. 2003). During the four to six weeks post-weaning, pups lose body mass and mobilize more than 50% of their fat before they successfully transition to more energetically valuable prey that allows them to start gaining mass (Muelbert et al. 2003). As seals grow, they transition to larger fishes and cephalopods and eat fewer invertebrates.

The diet of harbor seals varies with location and season, and is likely associated with changes in the abundance and availability of prey (Lance et al. 2012; Sharples et al. 2009). For instance, harbor seals eat more gadoids in the northern North Sea in the winter and more cephalopods and sandeels (Ammodytidae) in the summer/autumn (Brown and Pierce 1998; Tollit et al. 1998). Two hundred km south, sand lance (Ammodytidae) dominates in winter/spring and salmonids in the rest of the year (Sharples et al. 2009). In subarctic fjords in northern Norway, seals' diets reflect the seasonal changes in the dominance of fish species, with small codfish (*Gadus* sp) preferred during winter when ice-covered fjords were not accessible to pelagic fish. This pattern reversed in the spring when the ice melted (Ramasco et al. 2017). Off Oregon, smaller fish are eaten in spring and summer, and larger fish are eaten during autumn and winter (Brown and Mate 1983; Harvey 1987). Another example comes from the Shetlands Island in the northern UK; seasonal movements of prey affect foraging behavior, such that when herring (*Clupea harengus)* and mackerel (*Scomber scombrus*) move inshore during summer/autumn, their presence in the harbor seal diet increases (Brown and Pierce 1998).

Because there are clinal changes in distribution and abundance of prey, we would expect the same pattern in harbor seal foraging behavior. The west coast of North America serves as an excellent example because of the broad latitudinal distribution of seals and the many studies along this cline. In Alaska, harbor seals consume mostly fishes (74.5% occurrence; Pitcher 1980), primarily walleye pollock (*Gadus chalcogrammus*), capelin (*Mallotus villosus*), euchalon (*Thaleichthys pacificus*), and Pacific herring (*Clupea pallasii*); octopus (*Octopus* sp.) is the main cephalopod consumed. In the Bering Sea, pollock is a major prey (Frost and Lowry 1986). There is evidence from fatty acid signatures that harbor seals in Alaska forage relatively close to their haulout site; thus there are fine-scale differences in prey composition among haulout sites (Iverson et al. 1997). In SE Alaska, seals of all sizes use tidewater glacial habitats for foraging, but larger animals range more widely for food the rest of the year (Smith et al. 2019). Off British Columbia, harbor seals eat cephalopods, rockfish (*Sebastes* spp.), salmon (*Oncorhynchus* spp.), and herring (Spalding 1963). In the Salish Sea, harbor seals eat mostly Pacific sand lance (*Ammodytes hexapterus*), Shiner perch (*Cymatogaster aggregata*), herring, cottids (Cottidae), rockfish, walleye pollock, northern anchovies (*Engraulis mordax*), and salmon (Bromaghin et al. 2013; Lance et al. 2012; Luxa and Acevedo-Gutiérrez 2013). In Oregon, they eat benthic prey, such as flatfishes (Pleuronectiformes), sculpins (Cottidae), sandlance, croakers (*Genyonemus* sp.) cuskeels (Chilara), and octopus, and pelagic prey, such as herring, hake (*Merluccius productus*), surfperch, and squid (Brown and Mate 1983; Harvey 1987; Orr et al. 2004). Off California, they consume octopus, surfperch, cuskeel (Ophidiidae), rockfish, flatfish, cottids (Gibble and Harvey 2015; Harvey et al. 1995; Oxman 1995; Trumble 1995). Interestingly, in San Francisco Bay, an embayment with a high degree of invasive species, the most numerous fishes in the diet of harbor seals are a few goby species introduced from China (Gibble and Harvey 2015). The general clinal trend is the consumption of gadid, clupeid, sculpinid, and hexagrammid fishes in the north, with seasonal consumption of specific runs of fishes like eulachon, to a diet dominated more by flatfish, cottids, rockfishes, octopus, and cuskeel in the southern portion of their range. The above examples of changes in diet composition along the western coast of North America demonstrate the flexible and generalist nature of harbor seals (see review by Steingass 2017 for more examples). This species can also adapt to rapid local prey changes, for example, where harbor seals explore newly created habitat and forage on species that live around the foundations of offshore wind farms off the coast of the Netherlands (Fig. 10.10; Russell et al. 2014).

One of the controversial foraging behaviors involves the consumption of commercial, threatened, or endangered salmonid species. This includes direct depredation of fishes from lines or nets or competition for the same resources as commercial and recreational fisheries (Harwood and Greenwood 1985). Foraging can occur in estuaries and river systems where the salmonids return to natal spawning habitats (Brown and Mate 1983; Harris and Northridge 2017). Harbor seals remove fish from gillnets during commercial or tribal fisheries. In Washington and Oregon, the number of seals in specific locations correlates with salmon runs, which may indicate they move into specific river systems in response to the presence of salmonids returning to their natal streams (Brown and Mate 1983; Farrer et al. 2010). Harbor seals in the freshwater Iliamna Lake also take advantage of the seasonal presence of adult sockeye salmon (*Oncorhynchus nerka*). When salmon are not present, seals prey on smaller, resident fish (Hauser et al. 2008).

10.4.3 Prey Detection and Capture

Harbor seals detect their prey using vision, tactile, hydrodynamics (vibrissae), and olfaction (Chap. 2). Active touching or detecting hydrodynamics stimuli using their vibrissae facilitates foraging at night or at depths with reduced light levels, in turbid estuarine and nearshore environments, and on cryptic benthic fishes. Besides using their vibrissae to locate prey using their tactile senses, harbor seals can determine the size of objects using just their vibrissae (Dehnhardt and Kaminski 1995). Captive-trained harbor seals can follow hydrodynamic trails produced by their swimming prey (Dehnhardt et al. 2001) and can use their vibrissae to detect prey in crevices and buried in the sediments. The hydrodynamic trails of fishes can remain detectable above background noise for up to three minutes (Niesterok and Hanke 2013), allowing harbor seals to follow the path of a swimming fish using only their vibrissae. Currents generated by the movement of water moving across flatfish gills cause a hydrodynamic stimulus detectable by harbor seal vibrissae (Niesterok et al. 2017).

To capture prey, harbor seals use biting, suction, and occasionally hydraulic jetting (Marshall et al. 2014; Ydesen et al. 2014). Pelagic prey is probably visually detected and pursued/captured with rapid bursts of speed and an extension of the neck (Bowen et al. 2002). Because harbor seals consume benthic prey (e.g., flatfishes, sculpins, octopus, cuskeels, gobies) that live on or in the sediments (Tollit and Thompson 1996), it is likely they use suction or hydraulic jetting to capture prey below the sediment's surface. Harbor seals also use their flippers to disturb prey buried in the sediments, for instance, when preying on sand lance (Bowen et al. 2002). In rocky habitats, harbor seals can remove fishes from crevices, where they have taken up refuge at night (Haaker et al. 1984). Handling time depends on the size and shape of the prey. Larger species, such as adult salmon, typically are brought to the surface and ripped apart because they are too large to consume whole underwater. Relatively short but wide species, such as flatfishes, take more handling time as the seal orients the body headfirst into their mouth (Bowen et al. 2002).

10.5 Movements

10.5.1 Large- and Small-Scale Adult Movement

Harbor seals are highly residential, with individuals remaining in a localized region. On a large regional scale, at-sea distribution of harbor seals shows small but dense aggregations of animals. Movements of tracked individuals and genetic studies indicate little migration and exchange of individuals among these aggregations, indicative of discrete regional populations (Cunningham et al. 2009; Jones et al. 2015; Olsen et al. 2017; Sharples et al. 2012). On a small local scale, harbor seals move between haulout sites and at-sea foraging areas, showing high site fidelity towards both sites (Figs. 10.9 and 10.10). While harbor seals usually haul out in large groups, they appear to forage alone, using previous experience and memory, rather than following other individuals, to localize foraging patches (Austin et al. 2004, 2006; Carter et al. 2017; Sharples et al. 2012). This solitary foraging behavior results in high inter- but low intra-individual variability in movement characteristics, site fidelity, and foraging behavior (Fig. 10.8; Cunningham et al. 2009; Sharples et al. 2012).

After hauling out for some hours, individuals travel relatively short distances (<50 km) to forage at specific areas, often returning to the haulout site after a half-day or longer foraging trip (Thompson 1993; Thompson et al. 1998). Longer foraging trips are usually associated with greater distances from the haulout sites. There are

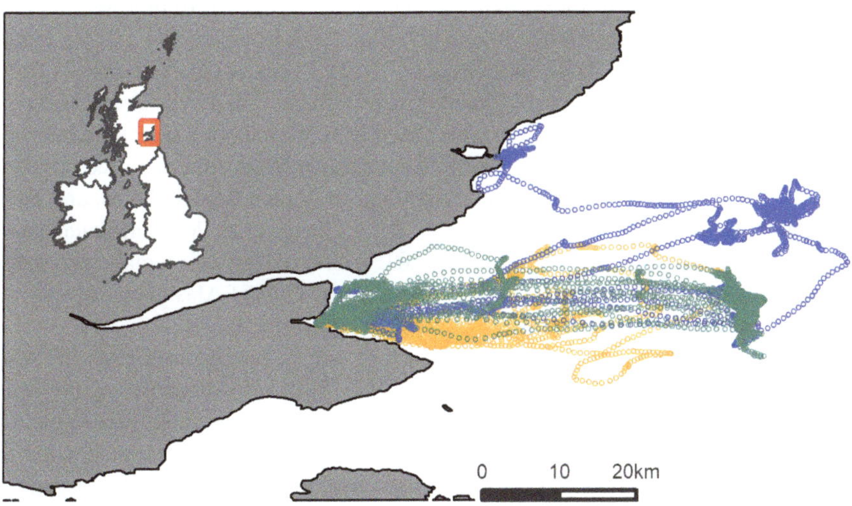

Fig. 10.9 Example of movement of three adult harbor seals tracked over three months off the east coast of the UK. Although all three individuals regularly haul out at the same place, they forage at different at-sea locations. Note that each individual shows high site fidelity to its at-sea foraging location, which then differs from other individuals' location. *Data source* Sea Mammal Research Unit, University of St Andrews, UK

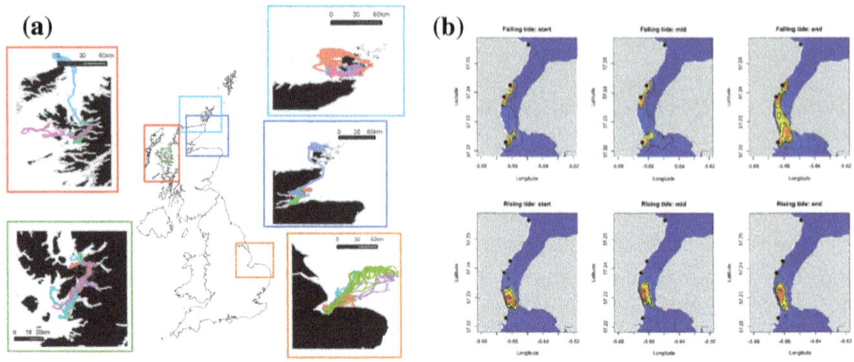

Fig. 10.10 Depiction of different movements of adult harbor seals around the UK (**a**). All individuals were tracked between 1 and 3 months outside their breeding season. It is common that harbor seals from the same location show different movement patterns in terms of trip extent and repeatability. An interesting movement is observed for seals living in strong, dynamic tidal environments like in Kyle Rhea (green square) (**b**). These seals spent a large proportion of their time around the narrowest point of a channel, with greatest number of seals during the flood tide and minimum during ebb tide. This pattern is likely related to prey availability driven by the underlying tidal pattern in the water movements through the channel (Hastie et al. 2016). *Data source* **a** dark blue: Paul Thompson, Lighthouse Field Station, University of Aberdeen, UK; orange: Hastie et al. (2015), Russell et al. (2016), green and red: (Hastie et al. 2016), light blue: Sea Mammal Research Unit, University of St Andrews, UK, unpublished data; **b** Reproduced from Hastie et al. (2016)

substantial differences in movement characteristics between different regions and even various haulout sites of the same region (Fig. 10.9; Blundell et al. 2011; Lesage et al. 2004; Lowry et al. 2001; Sharples et al. 2012). For example, harbor seals from Thames Estuary in the southeast UK and Prince William Sound, Alaska rarely travel farther than 20 km from the haulout sites. Their trips are seldom longer than one day (Lowry et al. 2001; Sharples et al. 2009). In contrast, in North America's ice-covered regions, harbor seals frequently forage >200 km from their haulout areas, and their trips often exceed four days (Blundell et al. 2011; Lesage et al. 2004; in Canada and Alaska respectively). Womble and Gende (2013) observed an adult female who spent three weeks near the continental shelf margin off the Gulf of Alaska coast foraging without returning to land.

Most harbor seals show high site fidelity to only a few haulout sites in close proximity to each other (e.g., Sharples et al. 2012). Still, individuals having multiple activity centers spread over larger areas have also been reported (e.g., Peterson et al. 2012). Therefore, the home range sizes of this species are strongly correlated with trip extents and the number and spread of the used haulout sites. Average home ranges (95% fixed kernel) of adult seals were relatively low from France and estimated to be 124 km^2 (range: 58–358 km^2) from Baie du Mont-Saint-Michel and 80 km^2 (range: 55–152 km^2) from Baie des Veys (Vincent et al. 2010). Adult harbor seals from Prince Williams Sound showed much larger home ranges that averaged 1749 km^2 (range: 311–16 970 km^2; percentage kernel not provided; Lowry et al. 2001).

However, home range sizes are difficult measures to compare between studies, as they are frequently calculated with a range of methodologies and strongly depend on the authors' definition of home ranges.

There are no consistent differences in movement characteristics between sexes, although some researchers report that males or larger individuals travel slightly further than females (e.g., Peterson et al. 2012). Thompson et al. (1998) reported that males traveled on average twice as far as females (61 and 30 km, respectively). In contrast, a study by Dietz et al. (2013) reported the opposite trend. A study by Sharples et al. (2012) indicated no statistical differences in trip distance between sexes. Thompson et al. (1998) hypothesized that movement patterns were likely to be different prior to the breeding season as this is when physiological and behavioral preparation for the breeding season occurs. Lowry et al. (2001) also suggested that sex-related differences in movements were likely associated with reproduction and molting. Differences in trip characteristics are more pronounced in sexually dimorphic species such as elephant (*Mirounga* spp.) or gray seals (*Halichoerus grypus*) than for species like harbor seals where both sexes are of relatively similar size (Breed et al. 2006; Le Boeuf et al. 2000).

Analysis of 2D surface movement data has defined harbor seals' at-sea behavior into two discrete modes: directional, and more stationary, tortuous movement (McClintock et al. 2012; Russell et al. 2015). However, when location and diving data are combined, at-sea behavior cannot be easily divided into discrete 'functional' modes such as traveling (moving in between foraging patches without consuming prey) and feeding (searching and obtaining food), as observed in other pinnipeds (e.g., gray seals; Austin et al. 2006). For example, time-depth records or accelerometers indicate that harbor seals perform benthic, U-shaped, foraging-like dives most of the time (Mikkelsen et al. 2019; Ramasco et al. 2014; Russell et al. 2015). However, harbor seals also spend considerable time resting at the sea surface or on the sea bottom sleeping or digesting their food (Fig. 10.7; McClintock et al. 2013; Mikkelsen et al. 2019; Ramasco et al. 2014; Russell et al. 2015). As resting and foraging dives exhibit minimal horizontal movement, they cannot be reliably differentiated if only tracking data are available. The factors determining whether a seal rests at sea or goes back to the haulout site are not understood. Digestion is a costly physiological process that is incompatible with the physiological adjustments to diving, therefore, can be delayed until after periods of active foraging (Sparling et al. 2007; Williams et al. 2004). Digestive constraints are related to stomach capacity and the rate at which food passes through the digestive tract (Williams et al. 2001). At-sea resting may be used by seals to pass food from their stomach into the intestines where food is further processed during longer resting breaks, or to metabolize anaerobic by-products progressively accumulated during diving (Boyd 1997). Resting at sea may also enable seals to conserve energy by not traveling to a distant haulout site. An individual-based model based on optimal foraging strategy, cognitive and physiological processes revealed that, indeed, when introducing digestive constraints, modeled harbor seals show the same pattern in resting behavior (like spatial and temporal distribution of resting) as observed (Chudzinskam et al. 2021).

10.5.2 Movement of Weaned Pups and Juveniles

Compared with adults, post-weaning and juvenile movement is usually characterized by more exploratory, dispersal movement. The extent of this dispersal differs regionally. Out of 225 pups flipper-tagged from the Orkney (UK) subpopulation of harbor seals, most of them dispersed within 50 km from their natal site (Thompson et al. 1994). Similar short-distance dispersal from the natal site was observed for pups from the Gulf of Alaska (Small et al. 2005). On the opposite end of the spectrum, newly weaned pups tagged in central California dispersed up to 802 km from the tagging location (Greig et al. 2019). Off Svalbard, the mean distance of dispersal was 178 km, but ranged from 13 to 980 km depending on the year (Blanchet et al. 2016). In this study, pups continued to swim greater distances each day until about 250–300 days of age.

The strength of site fidelity increases with age for harbor seals, as juveniles show less site fidelity towards hauling out and foraging areas than adults (Dietz et al. 2013). Although the literature seldom provides dispersal information in a uniform and comparable manner, examples from France, Scotland, and Alaska show that movement patterns vary in different regions and appear to correlate with local and regional distributions of food resources (Dietz et al. 2013). Not all juveniles disperse, but most individuals show high site fidelity to their natal sites (Small et al. 2005; Thompson et al. 1994). Adult seals may return to haulout sites on which they were born even after several years of absence in the area (Cordes and Thompson 2015; Mackey et al. 2008).

10.5.3 Modeling Movement and Distribution of Seals

Mathematical models can be used for various purposes, but in terms of modeling movement and distribution, bottom-up and top-down are two approaches of particular interest.

Spatially explicit process- and individual-based models (IBMs, a bottom-up approach) can be used to understand how various 'what-if' scenarios may affect movement, distribution, and behavior in altered conditions/environments. This approach is of particular importance in informing conservation and spatial planning. Steingass (2017), for example, built an IBM to understand how varying intensity of oxygen depletion in the water column (hypoxia), which affects prey distribution of harbor seals, affects energy balance of seals. Results suggest that during increasing hypoxia, smaller individuals could benefit by shifting from foraging on larger schooling fishes to foraging closer inshore on less energetically dense forage fish. Larger adult seals may benefit by shifting from foraging on groundfish to smaller, schooling fishes as hypoxia increases. The model suggests a mechanism by which hypoxia may increase foraging efficiency of Pacific harbor seals, resulting in increased rates of predation on coastal fishes on the continental shelf during hypoxic events. A more complex spatially explicit IBM that incorporated physiological, behavioral, and cognitive processes indicated that the main drivers behind the central-place movement of harbor seals were physiological (such as the need to rest due to digestive constraints), memory-based movement, and distribution and abundance of prey and haulout sites (Chudzinskam et al. 2021).

An example of a top-down approach is state-space modeling based on diving activity data and/or movement. State-space modeling allows for the identification of a range of behavioral states such as resting, foraging, and transiting using various aspects of movement (speed and turning angle), diving properties (depth, shape of dives), and a range of intrinsic (age, sex) and extrinsic covariates (time of day, spatial). These models can then serve as a base for establishing individual or population-level time-activity budgets. Russell et al. (2015), following the approach by McClintock et al. (2013), found that harbor seals from the UK may spend over 5% of their time resting at sea if far away (>5 km) from a haulout site. They also found that in regions where harbor seals overlap with gray seals, there was evidence of temporal separation in time hauled out, but not in time foraging. Using the same methodological approach, Grecian et al. (2020) reported that the strongest driver of harbor seal habitat preference was the distance to the haulout site. Ward et al. (2012) used a state-based approach to incorporate data on distribution, movement, and diet of harbor seals to identify and map "hot spots" of harbor seal predation risk for an endangered prey species of rockfish (*Sebastes* spp.) by seals off the coast of Washington State. The identification of high-risk areas is particularly important when designing reserves or protected areas.

Another example of a top-down approach is using location or behavioral data of harbor seals to establish habitat preferences and model distribution for areas and/or seasons for which we do not have empirical data. Jones et al. (2017b) used this approach to confirm that distance from haulout site, the proportion of sand in seabed sediment, and annual mean tidal power were important predictors of space use by harbor seals. Fine-scale usage maps were used by Jones et al. (2017a) to produce risk maps of seal co-occurrence with shipping traffic, and by Thompson et al. (2013) to estimate the number of harbor seals impacted by the noise from pile driving from offshore wind farm construction.

10.6 Anthropogenic Impacts

In the past century, subsistence and commercial hunting was a significant anthropogenic threat to harbor seal populations until hunting was banned throughout most of Europe (de Vooys et al. 2012; Reijnders 1992), North America (Baird, 2001; Gilbert et al. 2005), and Greenland in the second half of the twentieth century (Greenland Institute of Natural Resources). Subsistence harvest of harbor seals in Alaska is still authorized under the U.S. Marine Mammal Protection Act. Harbor seals are vital to traditional/subsistence use for Alaska Natives. The meat, organs, and oil from their blubber are important parts of many Alaska Natives' diet. The annual subsistence harvest of harbor seals in Alaska was about 2,900 animals annually in the early 1990s, 1700 animals annually in the early 2000s, and about 600 individuals annually in recent years (Wolfe et al. 2012). Declining seal harvests are associated with fewer seal hunters and a declining harvest rate of successful hunters. Subsistence hunting in Alaska is considered sustainable (Wolfe et al. 2006, 2012). On the other hand, ongoing hunting in Iceland has been reported to be one of the two main reasons (beside by-catch) for a substantial decrease in harbor seal populations in this region (NAMMCO 2018; Punt et al. 2020). The current population of harbor seals, estimated from haulout counts during the molting period, is close to 70% smaller than when surveys began around Iceland in 1980 (Granquist and Hauksson 2019).

In the past two decades, coastal marine environments have experienced rapid industrialization with increases in marine transportation, oil and gas exploration and extraction, and fisheries (Halpern et al. 2008; Smith 2000). Because harbor seals are abundant around many coastlines of the North Atlantic and North Pacific Oceans, the spatial overlap between marine industries and harbor seals is particularly high compared with most marine mammals. This has led to concerns about the potential impacts of anthropogenic effects on seal populations (Pirotta et al. 2018). Although harbor seals might serve as a model system for understanding potential anthropogenic impacts to other phocid species, it might also be that their adaptability would create a false signal of no-impact, when in reality, there are real stressors for other species.

Many of the activities associated with marine industries pose potential acute risks to harbor seals. For example, pinnipeds can be injured or killed as a result of vessel collisions (Stroud and Roffe 1979), fisheries gear entanglement and bycatch (Konigson et al. 2015; Luck et al. 2020), and conflicts with aquaculture (Thompson et al. 2007). In many cases, the nature and extent of these interactions can have important consequences for the demographics of affected populations (Read et al. 2006; Thompson et al. 2007). For example, Read et al. (2006) reported that significant numbers of harbor seals were by-caught in coastal gillnet fisheries off the west coast of the USA during the 1990s, prior to implemented mitigation measures. Similarly, Bjørge et al. (2002) reported that almost one-half of tagged harbor seal pups died in bottom-set nets off the coast of Norway. Further, Thompson et al. (2007) suggested that shooting in response to perceived conflicts with fisheries and aquaculture may be responsible for annual population declines (2–5% per year) of harbor seals in northeast Scotland that occurred between 1993 and 2004.

Ambitious renewable energy targets developed to mitigate impacts of climate change (Jay 2010; Toke 2011) have led to deployment of novel technologies in coastal environments that have the potential to cause injury or mortality to harbor seals. Tidal stream energy extraction is being rapidly developed in a number of countries; this is typically carried out using subsurface turbines that extract energy from tidally induced moving water. These have large moving blades that operate similarly to wind turbines. Concerns derive primarily from the potential for physical injury to marine mammals through direct contact with moving structures or parts of the devices (Wilson et al. 2007). However, there are too few data on seal movements around tidal turbines to quantify the true nature of the risks associated with potential interactions (Hastie et al. 2018; Sparling et al. 2018).

Many human activities can have less direct, chronic impacts on harbor seals. For example, disturbance at haulout sites from various recreational and industrial activities, including boats, pedestrians, and aircraft, can startle seals, causing them to prematurely enter the water (Suryan and Harvey 1998; Young et al. 2014). Although such responses are often relatively short term in nature (Andersen et al. 2012; Paterson et al. 2019), they have the potential to result in increased energetic expenditure, mother/pup separation, disruption of suckling, physiological stress, and interruption of molting (Costa 2012; Costa et al. 2016).

Exposure to chemical pollutants also has potential chronic impacts on harbor seals. For example, oil spills may affect harbor seals through inhalation, ingestion, and dermal pathways; each pathway may cause a suite of physiological responses that can compromise health, reproduction, and long-term survival (Helm et al. 2015), which may affect population dynamics (Frost et al. 1999). Levels of persistent organic pollutants (Hall and Thomas 2007; Neale et al. 2005, 2009) are often high in the blubber of harbor seals. Exposure has been linked to reproductive failure (Reijnders 1980, 1986) and immune dysfunction (de Swart et al. 1996). Therefore, these pollutants may exacerbate the impacts of pathogen exposure, particularly viral infection, by suppressing intracellular pathogen-specific immune defenses (Hall 2002; Ross et al. 1996). Elevated mercury concentrations have been detected in seals from some regions (Brookens et al. 2009; McHuron et al. 2014), although correspondingly increased selenium levels may alleviate adverse health effects by sequestering mercury, affording some degree of protection.

The impacts of anthropogenic noise on seals are of increasing conservation concern, with vessel noise being one of the major contributors to acoustic pollution. In coastal areas, seals are exposed to vessel noise up to ~20% of their time in the water (Mikkelsen et al. 2019). High vessel noise levels can cause an interruption of important functional behaviors, such as resting (Mikkelsen et al. 2019). More recently, the deployment of several thousand wind turbines has been proposed throughout coastal areas of Europe. This has led to concerns about potential impacts from noise during their construction. Pile driving during construction produces loud, impulsive sounds that may be loud enough to damage hearing (Hastie et al. 2015) and cause animals to avoid important foraging habitats, as evidenced by declines in seal abundance out to a range of 25 km during pile driving (Russell et al. 2016). In contrast, the operation of windfarms appears to pose minimal risk to seals, and may present foraging opportunities (Fig. 10.11; Russell et al. 2014).

Although many populations of harbor seals appear to be thriving in often highly urbanized marine habitats, other populations are currently exhibiting marked declines (Thompson et al. 2019). To estimate the population-level impacts of human activities, the long-term impacts of exposure to stressors on individual fitness, fecundity, and survival need to be quantified (Thompson et al. 2013). Such information is required so that offshore industries and recreational activities are developed in an environmentally sound manner, and to ensure that they can successfully coexist with harbor seals and other coastal marine mammal species (Pirotta et al. 2018).

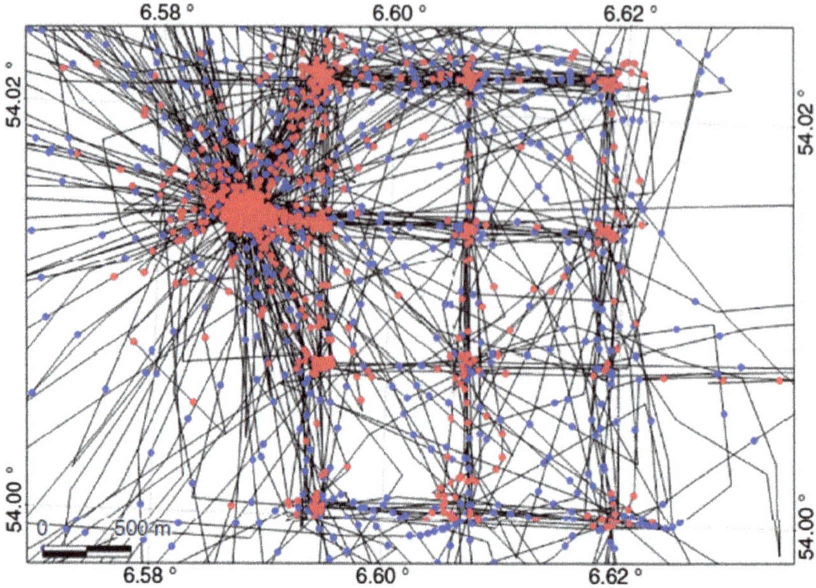

Fig. 10.11 The tracks of harbor seals around Alpha Ventus windfarm (Germany). Points show locations at 30 min intervals: red indicates higher chances of foraging as predicted by a state-space model and blue higher chances of traveling. This individual appears to forage at 12 turbines and the meteorological mast to the west of the windfarm. Reproduced from Russell et al. (2014)

10.7 Final Thoughts

In this final section, we consider the consequences of a changing marine environment and consider both the data and analytical framework required to advance our understanding of harbor seal ethology.

Like many pinniped species, harbor seals face numerous threats around the world, including climate change, fisheries interactions, habitat loss and deterioration, and pollutants (Kovacs et al. 2012). Although most harbor seal subspecies and populations are not in critical danger, they are exposed to many of the problems impacting pinnipeds worldwide. Harbor seals have evolved in dynamic and wide-ranging environments that overlap with human developments and the resultant environmental impacts. Crucially, harbor seals can adapt in response to rapid local changes. Their behavioral plasticity allows them to use many different types of haulout substrates, and they have a flexible diet and foraging strategy. Given their close proximity to human developments, they are experienced with some levels of disturbance and human activities. Of all the phocids, harbor seals may be best able to adapt to, and perhaps even take advantage of, the rapid changes occurring in the world's oceans.

Our future understanding of harbor seal biology depends on the continued development of bio-logging systems. Telemetry tags relay locational and behavioral information (over many months) from various sensors using a variety of radio channels (cell phone, Argos satellite, or UHF). However, these radio channels have limited bandwidth and may require extensive energy causing reduced battery life. Thus, on-board data processing, fusion, and compression are required (Fedak et al. 2002). Such processing of, say, accelerometry data, may reveal individual changes in body condition through time and also indicate individual foraging attempts (Aoki et al. 2011; Ydesen et al. 2014). An alternative approach is to deploy high-resolution archival tags (e.g., D-tags) that store huge amounts of data into memory, but for shorter periods (Mikkelsen et al. 2019; Heaslip et al. 2014; Yoshino et al. 2020). These may also include passive sound (the local soundscape) or active sound (that can reveal local prey density). The downside is that these tags must be recovered to retrieve the data. A hybrid approach (or double tagging) may yield exciting future insights. In particular, the relay of video sequences over periods spanning months will provide direct evidence of feeding attempts and prey field density (Heaslip et al. 2014; Yoshino et al. 2020).

Effective conservation of harbor seals in an altered habitat must be built on an understanding of the link between environment and movement and other behavior. Process-based modeling is currently the best way to synthesize behavioral data and provide robust predictions of the effect of a range of 'what-if?' scenarios on behavior, distribution, and population dynamics of harbor seals. Examples of such models are population-level consequences of disturbance models (PCoD; Pirotta et al. 2018), and agent-based models (ABM). One example of such ABM is AgentSeal—a behavior—and physiology-based ABM, which has been used to simulate harbor seal movements in relation to anthropogenic offshore activities (Chudzinska et al. 2021). Further development of such models and the joining of various modeling tools are currently the best ways to provide robust advice to organizations responsible for managing and conserving seals.

Acknowledgments We would like to take this opportunity to acknowledge the incredible contributions and persona of Bernie McConnell, our co-author on this chapter. Bernie was a friend, colleague, and amazing personality. Always self-effacing while being a witty critique, he made this endeavor so much more enriching. He has left this world of marine mammals and humans in a better and more joyful place.

References

Allen SG (1985) Mating-behavior in the harbor seal. Mar Mamm Sci 1:84–87

Andersen SM, Teilmann J, Dietz R, Schmidt NM, Miller LA (2012) Behavioural responses of harbour seals to human-induced disturbances. Aquatic Conserv-Marine Freshwater Ecosyst 22:113–121

Aoki K, Watanabe YY, Crocker DE, Robinson PW, Biuw M, Costa DP, Miyazaki N, Fedak MA, Miller PJO (2011) Northern elephant seals adjust gliding and stroking patterns with changes in buoyancy: validation of at-sea metrics of body density. J Exp Biol 214:2973–2987

Austin D, Bowen WD, McMillan JI (2004) Intraspecific variation in movement patterns: modeling individual behaviour in a large marine predator. Oikos 105:15–30

Austin D, Bowen WD, McMillan JI, Iverson SJ (2006) Linking movement, diving, and habitat to foraging success in a large marine predator. Ecology 87:3095–3108

Baird RW (2001) Status of harbour seals, *Phoca vitulina*, in Canada. Canadian Field-Naturalist 115:663–675

Berta A, Churchill M (2012) Pinniped taxonomy: review of currently recognized species and subspecies, and evidence used for their description. Mammal Rev 42:207–234

Bigg MA (1969) The harbor seal in British Columbia. Fisheries Res Board Canada Bull 172:33

Bigg MA (1973) Adaptations in the breeding of the harbour seal, *Phoca vitulina*. J Reprod Fertil Suppl 19:131–142

Bjørge A, Desportes G, Watring G, Rosing-Asvid A (2010) The harbour seal (*Phoca vitulina*)—a global perspective. NAMMCO Sci Publ 8:7–11

Bjørge A, Ølen N, Harvedt S, Bøthun G, Bekkby T (2002) Dispersal and bycatch mortality in gray, *Halichoerus grypus*, and harbor, *Phoca vitulina*, seals tagged at the Norwegian coast. Mar Mamm Sci 18:963–976

Blanchet MA, Lydersen C, Ims RA, Kovacs KM (2016) Making it through the first year: Ontogeny of movement and diving behavior in harbor seals from Svalbard, Norway. Mar Mamm Sci 32:1340–1369

Blundell GM, Pendleton GW (2015) Factors affecting haul-out behavior of harbor seals (*Phoca vitulina*) in tidewater glacier inlets in Alaska: can tourism vessels and seals coexist? PLoS One 10:e0125486

Blundell GM, Womble JN, Pendleton GW, Karpovich SA, Gende SM, Herreman JK (2011) Use of glacial and terrestrial habitats by harbor seals in Glacier Bay, Alaska: costs and benefits. Mar Ecol Prog Ser 429:277–290

Boness DJ, Bowen WD (1996) The evolution of maternal care in pinnipeds. Bioscience 46:645–654

Boness DJ, Bowen WD, Buhleier BM, Marshall GJ (2006) Mating tactics and mating system of an aquatic-mating pinniped: the harbor seal, *Phoca vitulina*. Behav Ecol Sociobiol 61:119–130

Boulva J, McLaren IA (1979) Biology of the harbor seal, *Phoca vitulina*, in Eastern Canada. Bull Fisheries Res Board Canada 200:1–24

Boveng PL, Bengtson JL, Withrow DE, Cesarone JC, Simpkins MA, Frost KJ, Burns JJ (2003) The abundance of harbor seals in the Gulf of Alaska. Mar Mamm Sci 19:111–127

Boveng PL, Dahle SP, Jansen JK, London JM, Taylor BL, Withrow DE (2016) Scientific evaluation of the distinctness of harbor seals (*Phoca vitulina*) in Iliamna Lake. Report from the NOAA Alaska Fisheries Science Center to the NOAA Fisheries Alaska Region Office. Marine Mammal Laboratory, Alaska Fisheries Science Center, 7600 Sand Point Way NE, Seattle, WA 98115

Bowen WD, Boness DJ, Iverson SJ (1999) Diving behaviour of lactating harbour seals and their pups during maternal foraging trips. Can J Zool 77:978–988

Bowen WD, Iverson SJ, Boness DJ, Oftedal OT (2001) Foraging effort, food intake and lactation performance depend on maternal mass in a small phocid seal. Funct Ecol 15:325–334

Bowen WD, Oftedal OT, Boness DJ (1992) Mass and energy-transfer during lactation in a small phocid, the harbor seal (*Phoca vitulina*). Physiol Zool 65:844–866

Bowen WD, Tully D, Boness DJ, Bulheier BM, Marshall GJ (2002) Prey-dependent foraging tactics and prey profitability in a marine mammal. Mar Ecol Prog Ser 244:235–245

Boyd IL (1997) The behavioural and physiological ecology of diving. Trends Ecol Evol 12:213–217

Brasseur S, Creuwels J, Vanderwerf B, Reijnders P (1996) Deprivation indicates necessity for haul-out in harbor seals. Mar Mamm Sci 12:619–624

Breed GA, Bowen WD, McMillan JI, Leonard ML (2006) Sexual segregation of seasonal foraging habitats in a non-migratory marine mammal. Proceedings of the Royal Society B-Biological Sciences 273:2319–2326

Bromaghin JF, Lance MM, Elliot EW, Jeffries SJ, Acevedo-Gutiérrez A, Kennish JM (2013) New insights into the diets of harbor seals (*Phoca vitulina*) in the Salish Sea revealed by analysis of fatty acid signatures. Fish Bull 111:13–26

Brookens TJ, O'Hara TM, Taylor RJ, Bratton GR, Harvey JT (2009) Total mercury body burden in Pacific harbor seal, *Phoca vitulina richardii*, pups from central California. Mar Pollut Bull 58:1953–1953

Brost BM, Hooten MB, Small RJ (2020) Model-based clustering reveals patterns in central place use of a marine top predator. Ecosphere 11:e03123

Brown EG, Pierce GJ (1998) Monthly variation in the diet of harbour seals in inshore waters along the southeast Shetland (UK) coastline. Mar Ecol Prog Ser 167:275–289

Brown RF, Mate BR (1983) Abundance, movements, and feeding-habits of harbor seals, *Phoca vitulina*, at Netarts and Tillamook bays, oregon. Fish Bull 81:291–301

Calambokidis J, Taylor BL, Carter SD, Steiger GH, Dawson PK, Antrim LD (1987) Distribution and haul-out behavior of harbor seals in Glacier Bay, Alaska. Can J Zool 65:1391–1396

Carter MID, Russell DJF, Embling CB, Blight CJ, Thompson D, Hosegood PJ, Bennett KA (2017) Intrinsic and extrinsic factors drive ontogeny of early-life at-sea behaviour in a marine top predator. Sci Rep 7:15505

Chudzinskam M, Smout S, Nabe-Nielsen J, Aarts G, Brasseur S, Thompson P, Graham I, McConnell B (2021) AgentSeal: agent-based model describing movement of marine central-place foragers. Ecol Model 440:109397

Coltman DW, Bowen WD, Boness DJ, Iverson SJ (1997) Balancing foraging and reproduction in the male harbour seal, an aquatically mating pinniped. Anim Behav 54:663–678

Coltman DW, Bowen WD, Wright JM (1998) Male mating success in an aquatically mating pinniped, the harbour seal (*Phoca vitulina*), assessed by microsatellite DNA markers. Mol Ecol 7:627–638

Coltman DW, Bowen WD, Wright JM (1999) A multivariate analysis of phenotype and paternity in male harbor seals, *Phoca vitulina*, at Sable Island, Nova Scotia. Behav Ecol 10:169–177

Cordes LS, Thompson PM (2015) Mark-resight estimates of seasonal variation in harbor seal abundance and site fidelity. Popul Ecol 57:467–472

Costa DP (2012) A bioenergetics approach to developing the PCAD model. In: Popper AN, Hawkins T (eds) The effects of noise on aquatic life. Springer, New York, pp 423–426

Costa DP, Schwarz L, Robinson P, Schick RS, Morris PA, Condit R, Crocker DE, Kilpatrick AM (2016) A bioenergetics approach to understanding the population consequences of disturbance: elephant seals as a model system. Adv Exp Med Biol 875:161–169

Cunningham L, Baxter JM, Boyd IL, Duck CD, Lonergan M, Moss SE, McConnell B (2009) Harbour seal movements and haul-out patterns: implications for monitoring and management. Aquatic Conserv-Marine Freshwater Ecosyst 19:398–407

D'Agnase ER (2015) The effect of maternal traits on rearing success in Pacific harbor seals (*Phoca vitulina richardsii*). Western Washington University, MS

Daniel RG, Jemison LA, Pendleton GW, Crowley SM (2003) Molting phenology of harbor seals on Tugidak Island, Alaska. Mar Mamm Sci 19:128–140

Da Silva J, Terhune JM (1988) Harbor seal grouping as an anti-predator strategy. Anim Behav 36:1309–1316

de Vooys KGN, Brasseur SMJM, Van Der Meer J, Reijnders PJH (2012) Analyses of four centuries of bounty hunting on seals in Zeeland, SW-Netherlands. Lutra 55:55–65

Dehnhardt G, Kaminski A (1995) Sensitivity of the mystacial vibrissae of harbor seals (*Phoca vitulina*) for size differences of actively touched objects. J Exp Biol 198:2317–2323

Dehnardt G, Mauck B, Hanke W, Bleckmann H (2001) Hydrodynamic trail-following in harbor seals (*Phoca vitulina*). Science 293:102–104

de Swart RL, Ross PS, Vos JG, Osterhaus A (1996) Impaired immunity in harbour seals (*Phoca vitulina*) exposed to bioaccumulated environmental contaminants: Review of a long-term feeding study. Environ Health Perspect 104:823–828

DFO (2017) Recovery strategy for the harbour seal, Lacs des Loups Marins subspecies (*Phoca vitulina mellonae*). Species at Risk Recovery Strategy Series. Fisheries and Oceans Canada. Ottowa

Dietz R, Teilmann J, Andersen SM, Riget F, Olsen MT (2013) Movements and site fidelity of harbour seals (*Phoca vitulina*) in Kattegat, Denmark, with implications for the epidemiology of the phocine distemper virus. ICES J Mar Sci 70:186–195

Eguchi T (1998) Morphology of the Pacific harbor seal (*Phoca vitulina richardsi*) using Elkhorn Slough, California, and their movements and diving behavior in the Monterey Bay area. California State University Fresno, MS

Eguchi T, Harvey JT (2005) Diving behavior of the Pacific harbor seal (*Phoca vitulina richardii*) in Monterey Bay, California. Mar Mamm Sci 21:283–295

Farrer J, Acevedo-Gutiérrez A (2010) Use of haul-out sites by harbor seals (*Phoca vitulina*) in Bellingham: implications for future development. Northwest Nat 91:74–79

Fedak M, Lovell P, McConnell B, Hunder C (2002) Overcoming the constraints of long range radio telemetry from animals: getting more useful data from smaller packages. Integr Comp Biol 42:3–10

Feltz ET, Fay FH (1966) Thermal requirements in vitro of epidermal cells from seals. Cryobiology 3:261–264

Frost KJ, Lowry LF (1986) Sizes of walleye pollock, *Theragra chalcogramma*, consumed by marine mammals in the Bering Sea. Fish Bull 84:192–197

Frost KJ, Lowry LF, Verhoef JM (1999) Monitoring the trend of harbor seals in Prince William Sound, Alaska, after the Exxon Valdez oil spill. Mar Mamm Sci 15:494–506

Frost KJ, Simplins MA, Lowrly LF (2001) Diving behavior of subadult and adult harbor seals in Prince William Sound, Alaska. Mar Mamm Sci 17:813–834

Gibble CM, Harvey JT (2015) Food habits of harbor seals (*Phoca vitulina richardii*) as an indicator of invasive species in San Francisco Bay, California. Mar Mamm Sci 31:1014–1034

Gilbert JR, Waring GT, Wynne KM, Guldager N (2005) Changes in abundance of harbor seals in Maine, 1981–2001. Mar Mamm Sci 21:519–535

Gjertz I, Lydersen C, Wiig Ø (2001) Distribution and diving of harbour seals (*Phoca vitulina*) in Svalbard. Polar Biol 24:209–214

Godsell J (1988) Herd formation and haul-out behavior in harbor seals (*Phoca vitulina*). J Zool 215:83–98

Granquist SM, Hauksson E (2019) Population estimate, trends and current status of the Icelandic harbour seal (*Phoca vitulina*) population in 2018. Marine and Freshwater Research in Iceland. HV 2019-36 [online]. https://www.hafogvatn.is/static/files/hv2019-36.pdf

Grecian J, Masden E, Hammond PHS, Owen E, Daunt F, Wanless S, Russell DJF (2020). Man-made structures and apex predators: spatial interactions and overlap (MAPS). Final report to INSITE. https://s3-eu-west-1.amazonaws.com/static.insitenorthsea.org/files/MAPS_INSITE_revised_report_submitted.pdf

Greig DJ, Gulland FMD, Harvey JT, Lonergan M, Hall AJ (2019) Harbor seal pup dispersal and individual morphology, hematology, and contaminant factors affecting survival. Mar Mamm Sci 35:187–209

Haaker P, Parker D, Henderson K (1984) Observations of harbor seal, *Phoca vitulina richardsi* feeding in southern California waters. Bull South Calif Acad Sci 83:259–268

Hall AJ (2002) Organohalogenated contaminants in marine mammals. In: Evans PGH, Raga JA (eds) Marine Mammals. Springer, Boston, pp 523–563

Hall AJ, Thomas GO (2007) Polychlorinated biphenyls, DDT, polybrominated diphenyl ethers, and organic pesticides in United Kingdom harbor seals (*Phoca vitulina*)—mixed exposures and thyroid homeostasis. Environ Toxicol Chem 26:851–861

Halpern BS, Walbridge S, Selkoe et al (2008) A global map of human impact on marine ecosystems. Science 319:948–952

Hamilton CD, Lydersen C, IMS RA, Kovacs KM (2014) Haul-out behaviour of the world's northernmost population of harbour seals (*Phoca vitulina*) throughout the year. PloS One 9:e86055

Hanggi EB, Schusterman RJ (1992) Underwater acoustic displays by male harbor seals (*Phoca vitulina*): initial results. In: Thomas JA, Kastelein RA, Supin AY (eds) Marine mammal sensory systems. Springer, Boston, pp 449–457

Harding KC, Fujiwara M, Axberg Y, Harkonen T (2005) Mass-dependent energetics and survival in harbour seal pups. Funct Ecol 19:129–135

Harkonen T, Harding KC (2001) Spatial structure of harbour seal populations and the implications thereof. Can J Zool 79:2115–2127

Harris R, Northridge SP (2017) Seals and wild salmon fisheries. Interim report to Scottish Government—SSI

Harvey JT (1987) Population dynamics, annual food consumption, movements, and dive behaviors of harbor seals, Phoca vitulina richardsi, in Oregon. Dissertation, Oregon State University

Harvey JT, Helm RC, Morejohn GV (1995) Food habits of harbor seals inhabiting Elkhorn Slough, California. California Fish and Game 81:1–9

Harwood J, Greenwood JJD (1985) Competition between British grey seals and fisheries. In: Beddington JR, Beverton RJH, Lavigne DM (eds) Marine mammals and fisheries. Allen and Unwin, London, pp 153–169

Hassrick JL, Crocker DE, Zeno RL, Blackwell SB, Costa DP, Le Boeuf BJ (2007) Swimming speed and foraging strategies of northern elephant seals. Deep-Sea Res Part II Top Stud Oceanogr 54:369–383

Hastie GD, Russell DJF, Benjamins S, Moss S, Wilson B, Thompson D (2016) Dynamic habitat corridors for marine predators; intensive use of a coastal channel by harbour seals is modulated by tidal currents. Behav Ecol Sociobiol 70:2161–2174

Hastie GD, Russell DJF, Lepper P, Elliott J, Wilson B, Benjamins S, Thompson D (2018) Harbour seals avoid tidal turbine noise: implications for collision risk. J Appl Ecol 55:684–693

Hastie GD, Russell DJF, McConnell B, Moss S, Thompson D, Janik VM (2015) Sound exposure in harbour seals during the installation of an offshore wind farm: predictions of auditory damage. J Appl Ecol 52:631–640

Hastings KK, Frost KJ, Simpkins MA, Pendleton GW, Swain UG, Small RJ (2004) Regional differences in diving behavior of harbor seals in the Gulf of Alaska. Can J Zool 82:1755–1773

Hauser D, Allen C, Rich H, Quinn T (2008) Resident harbor seals (*Phoca vitulina*) in Iliamna Lake, Alaska: Summer diet and partial consumption of adult sockeye salmon (*Oncorhynchus nerka*). Aquat Mamm 34:303–309

Hayes SA, Costa DP, Harvey JT, Le Boeuf BJ (2004) Aquatic mating strategies of the male Pacific harbor seal (*Phoca vitulina richardii*): are males defending the hotspot? Mar Mamm Sci 20:639–656

Hayes SA, Kumar A, Costa DP, Mellinger DK, Harvey JT, Southall BL, Le Bouef BJ (2004) Evaluating the function of the male harbour seal, *Phoca vitulina*, roar through playback experiments. Anim Behav 67:1133–1139

Hayes SA, Pearse DE, Costa DP, Harvey JT, Le Boeuf BJ, Garza JC (2006) Mating system and reproductive success in eastern Pacific harbour seals. Mol Ecol 15:3023–3034

Heaslip SG, Bowen WD, Iverson SJ (2014) Testing predictions of optimal diving theory using animal-borne video from harbour seals (*Phoca vitulina concolor*). Can J Zool 92:309–318

Hedd A, Gales R, Renouf D (1995) Use of temperature telemetry to monitor ingestion by a harbor seal mother and her pup throughout lactation. Polar Biol 15:155–160

Helm RC, Costa DP, DeBruyn TD, O'Shea TJ, Wells RS, Williams TM (2015) Overview of effects of oil spills on marine mammals. In: Finas M (ed) Handbook of oil spill science and technology. Wiley, Hoboken, pp 455–476

Honeywell AF, Maher CR (2017) Intensity, rate, and outcome of agonistic interactions in harbor seals (*Phoca vitulina concolor*) vary with density on haul-out ledges. J Mammal 98:135–142

Iverson SJ, Frost KJ, Lowry LF (1997) Fatty acid signatures reveal fine scale structure of foraging distribution of harbor seals and their prey in Prince William Sound, Alaska. Mar Ecol Prog Ser 151:255–271

Jansen JK, Boeeng PL, Hoef JMV, Dahle SP, Bengtson JL (2015) Natural and human effects on harbor seal abundance and spatial distribution in an Alaskan glacial fjord. Mar Mamm Sci 31:66–89
Jay S (2010) Planners to the rescue: Spatial planning facilitating the development of offshore wind energy. Mar Pollut Bull 60:493–499
Jefferson TA, Webber MA, Pitman RL (2015) Marine Mammals of the World: a comprehensive guide to their identification 2 nd ed. Elsevier, San Diego
Jemison LA, Pendleton GW, Wilson CA, Small RJ (2006) Long-term trends in harbor seal numbers at Tugidak Island and Nanvak Bay, Alaska. Mar Mamm Sci 22:339–360
Jones EL, Hastie GD, Smout S, Onoufriou J, Merchant ND, Brookes KL, Thompson D (2017) Seals and shipping: quantifying population risk and individual exposure to vessel noise. J Appl Ecol 54:1930–1940
Jones EL, McConnell BJ, Smout S et al (2015) Patterns of space use in sympatric marine colonial predators reveal scales of spatial partitioning. Mar Ecol Prog Ser 534:235–249
Jones EL, Sparling CE, McConnell BJ, Morris CD, Smout S (2017) Fine-scale harbour seal usage for informed marine spatial planning. Sci Rep 7:11581
Jorgensen C, Lydersen C, Brix O, Kovacs KM (2001) Diving development in nursing harbour seal pups. J Exp Biol 204:3993–4004
Kastelein RA, Staa C, Wiepkema PR (2010) Food consumption and body mass of captive harbor seals (*Phoca vitulina*). Aquat Mamm 31:34–42
Konigson S, Lovgren J, Hjelm J, Ovegard M, Ljunghager F, Lunneryd SG (2015) Seal exclusion devices in cod pots prevent seal bycatch and affect their catchability of cod. Fish Res 167:114–122
Kooyman GL (1985) Physiology without restraint in diving mammals. Mar Mamm Sci 1:166–178
Kovacs KM, Aguilar A, Aurioles D et al (2012) Global threats to pinnipeds. Mar Mamm Sci 28:414–436
Krafft BA, Lydersen C, Gjertz I, Kovacs KM (2002) Diving behaviour of sub-adult harbour seals (*Phoca vitulina*) at Prins Karls Forland, Svalbard. Polar Biol 25:230–234
Lance MM, Chang WY, Jeffries SJ, Pearson SF, Acevedo-Gutiérrez A (2012) Harbor seal diet in northern Puget Sound: implications for the recovery of depressed fish stocks. Mar Ecol Prog Ser 464:257–271
Lawson JW, Renoud D (1985) Parturition in the Atlantic harbor seal, *Phoca vitulina concolor*. J Mammal 66:395–398
Lawson JW, Renoud D (1987) Bonding and weaning in harbor seals, *Phoca vitulina*. J Mammal 68:445–449
Le Boeuf BJ, Crocker DE, Costa DP, Blackwell SB, Webb PM, Houser D (2000) Foraging ecology of northern elephant seals. Ecol Monogr 70:353–382
Lesage W, Hammill MO, Kovacs KM (2004) Long-distance movements of harbour seals (Phoca vitulina) from a seasonally ice-covered area, the St. Lawrence River estuary, Canada. Canadian J Zool 82:1070–1081
London JM, Hoef JMV, Jeffries SJ, Lance MM, Boveng PL (2012) Haul-out behavior of harbor seals (*Phoca vitulina*) in Hood Canal, Washington. PLoS One 7:e38180
Lowry LF, Frost KJ, Ver Hoef JM, DeLong RA (2001) Movements of satellite-tagged subadult and adult harbor seals in Prince William Sound, Alaska. Mar Mamm Sci 17:835–861
Luck C, Cronin M, Gosch M, Healy K, Cosgrove R, Tully O, Rogan E, Jessop M (2020) Drivers of spatiotemporal variability in bycatch of a top marine predator: first evidence for the role of water turbidity in protected species bycatch. J Appl Ecol 57:219–228
Luxa K, Acevedo-Gutiérrez A (2013) Food habits of harbor seals (*Phoca vitulina*) in two estuaries in the central Salish Sea. Aquat Mamm 39:10–22
Mackey BL, Durban JW, Middlemas SJ, Thompson PM (2008) A Bayesian estimate of harbour seal survival using sparse photo-identification data. J Zool 274:18–27
Madden KM, Fuiman LA, Williams TM, Davis RW (2008) Identification of foraging dives in free-ranging Weddell seals *Leptonychotes weddellii*: confirmation using video records. Mar Ecol Prog Ser 365:263–275

Marshall CD, Wieskotten S, Hanke W, Hanke FD, Marsh A, Kot B, Dehnhardt G (2014) Feeding kinematics, suction, and hydraulic jetting performance of harbor seals (*Phoca vitulina*). PLoS One 9:e86710

Matthews LP, Blades B, Parks S (2018) Female harbor seal (*Phoca vitulina*) behavioral response to playbacks of underwater male acoustic advertisement displays. PeerJ 6:e4547

McClintock BT, King R, Thomas L, Matthiopoulos J, McConnell BJ, Morales JM (2012) A general discrete-time modeling framework for animal movement using multistate random walks. Ecol Monogr 82:335–349

McClintock BT, Russell DJF, Matthiopoulos J, King R (2013) Combining individual animal movement and ancillary biotelemetry data to investigate population-level activity budgets. Ecology 94:838–849

McHuron EA, Harvey JT, Castellini JM, Stricker CA, O'Hara TM (2014) Selenium and mercury concentrations in harbor seals (*Phoca vitulina*) from central California: health implications in an urbanized estuary. Mar Pollut Bull 83:48–57

Mikkelsen L, Johnson M, Wisniewska DM, Van Neer A, Siebert U, Madsen PT, Teilmann J (2019) Long-term sound and movement recording tags to study natural behavior and reaction to ship noise of seals. Ecol Evol 9:2588–2601

Montgomery RA, Hoef JMV, Boveng PL (2007) Spatial modeling of haul-out site use by harbor seals in Cook Inlet, Alaska. Mar Ecol Prog Ser 341:257–264

Muelbert MMC, Bowen WD (1993) Duration of lactation and postweaning changes in mass and body composition of harbor seal, *Phoca vitulina*, pups. Can J Zool 71:1405–1414

Muelbert MMC, Bowen WD, Iverson SJ (2003) Weaning mass affects changes in body composition and food intake in harbour seal pups during the first month of independence. Physiol Biochem Zool 76:418–427

NAMMCO (2018). Report of the NAMMCO Scientific Committee 25th Meeting, 13–16 November 2018, Norway

Neale JCC, Schmelzer K, Harvey J, Berg E, Small R, Grigg E, Allen S, Tjeerdema R (2009) PCB and DDE Contamination in harbor seals (*Phoca vitulina*) from North-Central California and Bristol Bay, Alaska. Aquat Mamm 35:1–11

Neale JCC, Gulland FMD, Schmelzer KR, Harvey JT, Berg EA, Allen SG, Greig DJ, Grigg EK, Tjeerdema RS (2005) Contaminant loads and hematological correlates in the harbor seal (*Phoca vitulina*) of San Francisco Bay, California. J Toxicol Environ Health A 68:617–633

Neumann DR (1999) Agonistic behavior in harbor seals (*Phoca vitulin*a) in relation to the availability of haul-out space. Mar Mamm Sci 15:507–525

Newby TC (1973) Observations on the breeding behavior of the harbor seal in the state of Washington. J Mammal 54:540–543

Nicholson TE (2000) Social structure and underwater behavior of harbor seals in southern Monterey Bay, California. MS, San Francisco State University

Niesterok B, Hanke W (2013) Hydrodynamic patterns from fast-starts in teleost fish and their possible relevance to predator-prey interactions. J Comp Physiol Neuroethol Sensory Neural Behav Physiol 199:139–149

Niesterok B, Kruger Y, Wieskotten S, Dehnhardt G, Hanke W (2017) Hydrodynamic detection and localization of artificial flatfish breathing currents by harbour seals (*Phoca vitulina*). J Exp Biol 220:174–185

Nikolich K, Frouin-Mouy H, Acevedo-Gutiérrez A (2018) Clear diel patterns in breeding calls of harbor seals (*Phoca vitulina*) at Hornby Island, British Columbia, Canada. Can J Zool 96:1236–1243

Norris A (2009) Nocturnal behavior for the harbour seal (*Phoca vitulina*) from Prudence Island, Rhode Island. Bios 78:81–86

Olsen MT, Islas V, Graves JA, Onofriou A, Vincent C, Brasseur S, Frie AK, Hall AJ (2017) Genetic population structure of harbour seals in the United Kingdom and neighbouring waters. Aquatic Conserv-Marine Freshwater Ecosyst 27:839–845

Orr AJ, Banks AS, Mellman S, Huber HR, Delong RL, Brown RF (2004) Examination of the foraging habits of Pacific harbor seal (*Phoca vitulina richardsi*) to describe their use of the Umpqua River, Oregon, and their predation on salmonids. Fish Bull 102:108–117

Orr AJ, Norris TA, Galvez C, Deming AC, Elorriaga-Verplancken FR (2018) First extralimital record of an eastern Pacific harbor seal (*Phoca vitulina richardii*) pup at Guadalupe Island, Mexico. Aquat Mamm 44:279–284

Oxman D (1995) Seasonal abundance, movements, and food habits of harbor seals (*Phoca vitulina richardsi*) in Elkhorn Slough, California. MS, California State University Stanislaus

Paterson W, Sparling CE, Thompson D, Pomeroy PP, Currie JI, McCafferty DJ (2012) Seals like it hot: changes in surface temperature of harbour seals (*Phoca vitulina*) from late pregnancy to moult. J Therm Biol 37:454–461

Paterson WD, Russell DJF, Wu GM, McConnell B, Currie JI, McCafferty DJ, Thompson D (2019) Post-disturbance haulout behaviour of harbour seals. Aquatic Conserv-Marine Freshwater Ecosyst 29:144–156

Pauli BD, Terhune JM (1987) Meteorological influences on harbour seal haul-out. Aquat Mamm 13:114–118

Perry EA (1993) Aquatic territory defense by male harbour seals (*Phoca vitulina*) at Miquelon: relationship between defense and male reproductive success. Dissertation, Memorial University

Peterson SH, Lance MM, Jeffries SJ, Acevedo-Gutiérrez A (2012) Long distance movements and disjunct spatial use of harbor seals (*Phoca vitulina*) in the inland waters of the Pacific Northwest. PLoS One 7:e39046

Pirotta E, Booth CG, Costa DP et al (2018) Understanding the population consequences of disturbance. Ecol Evol 8:9934–9946

Pitcher KW (1980) Food of the harbor seal, *Phoca vitulina richardsi*, in the Gulf of Alaska. Fish Bull 78:544–549

Pitcher KW, McAllister DC (1981) Movements and haulout behavior of radio-tagged harbor seals, *Phoca vitulina*. Canadian Field-Naturalist 95:292–297

Ponganis PJ (2015) Diving physiology of marine mammals and seabirds. Cambridge University Press

Prestrud P, Gjertz I (1990) The most northerly harbor seal, *Phoca vitulina*, at Prins Karls Forland, Svalbard. Mar Mamm Sci 6:215–220

Punt AE, Siple M, Sigurosson GM et al (2020) Evaluating management strategies for marine mammal populations: an example for multiple species and multiple fishing sectors in Iceland. Can J Fish Aquat Sci 77:1316–1331

Ramasco V, Biuw M, Nilssen KT (2014) Improving time budget estimates through the behavioural interpretation of dive bouts in harbour seals. Anim Behav 94:117–134

Ramasco V, Lindstrom U, Nilssen KT (2017) Selection and foraging response of harbour seals in an area of changing prey resources. Mar Ecol Prog Ser 581:199–214

Read AJ, Drinker P, Northridge S (2006) Bycatch of marine mammals in US and global fisheries. Conserv Biol 20:163–169

Reijnders PJH (1980) Organochlorine and heavy-metal residues in harbor seals from the Wadden Sea and their possible effects on reproduction. Neth J Sea Res 14:30–65

Reijnders PJH (1986) Reproductive failure in common seals feeding on fish from polluted coastal waters. Nature 324:456–457

Reijnders PJH (1992) Retrospective population analysis and related future management perspectives for the harbor seal *Phoca vitulina* in the Wadden Sea. Netherlands Inst Sea Res Spec Publ 20:193–197

Renouf D, Lawson J, Gaborko L (1983) Attachment between harbor seal (*Phoca vitulina*) mothers and pups. J Zool 199:179–187

Rice DW (1998) Marine mammals of the world. The Society of Marine Mammalogy, Lawrence Kansas

Ries EH, Paffen P, Traut IM, Goehart PW (1997) Diving patterns of harbour seals (*Phoca vitulina*) in the Wadden Sea, the Netherlands and Germany, as indicated by VHF telemetry. Can J Zool 75:2063–2068

Robson BW, Goebel ME, Baker JD, Ream RR, Loughlin TR, Francis RC, Antonelis GA, Costa DP (2004) Separation of foraging habitat among breeding sites of a colonial marine predator, the northern fur seal (*Callorhinus ursinus*). Can J Zool 82:20–29

Roen R, Bjørge A (1995) Haul-out behaviour of the Norwegian harbour seal during summer. In: Blix AS, Walløe L, Ulltang Ø (eds) Whales, seals, fish and man, Proceedings of the international symposium on the biology of marine mammals in the North East Atlantic, 29 November–1 December 1994. Elsevier, pp 61–67

Rosing-Asvid A, Teilmann J, Olsen MT, Dietz R (2020) Deep diving harbor seals (*Phoca vitulina*) in South Greenland: movements, diving, haul-out and breeding activities described by telemetry. Polar Biol 43:359–368

Ross P, DeSwart R, Addison R, Van Loveren H, Vos J, Osterhaus A (1996) Contaminant-induced immunotoxicity in harbour seals: wildlife at risk? Toxicology 112:157–169

Russell DJF, Brasseur S, Thompson D et al (2014) Marine mammals trace anthropogenic structures at sea. Curr Biol 24:R638–R639

Russell DJF, Hastie GD, Thompson D, Janik VM, Hammond PS, Scott-Hayward LAS, Matthiopoulos J, Jones EL, McConnell BJ (2016) Avoidance of wind farms by harbour seals is limited to pile driving activities. J Appl Ecol 53:1642–1652

Russell DJF, McClintock BT, Matthiopoulos J, Thompson PM, Thompson D, Hammond PS, Jones EL, Mackenzie ML, Moss S, McConnell BJ (2015) Intrinsic and extrinsic drivers of activity budgets in sympatric grey and harbour seals. Oikos 124:1462–1472

Sauve CC, Beauplet G, Hammill MO, Charrier I (2015) Acoustic analysis of airborne, underwater, and amphibious mother attraction calls by wild harbor seal pups (*Phoca vitulina*). J Mammal 96:591–602

Scheffer VB, Slipp JW (1944) The harbor seal in Washington State. Am Midl Nat 32:373–416

Schneider DC, Payne PM (1983) Factors affecting haul-out of harbor seals at a site in Southeastern Massachusetts. J Mammal 64:518–520

Schreer JF, Kovacs KM (1997) Allometry of diving capacity in air-breathing vertebrates. Can J Zool 75:339–358

Schreer JF, Lapierre JL, Hammill MO (2010) Stomach temperature telemetry reveals that harbor seal (*Phoca vitulina*) pups primarily nurse in the water. Aquat Mamm 36:270–277

Sharples RJ, Arrizabalaga B, Hammond PS (2009) Seals, sandeels and salmon: diet of harbour seals in St. Andrews Bay and the Tay Estuary, southeast Scotland. Mar Ecol Prog Ser 390:265–276

Sharples RJ, Moss SE, Patterson TA, Hammond PS (2012) Spatial variation in foraging behaviour of a marine top predator (*Phoca vitulina*) determined by a large-scale satellite tagging program. PLoS One 7:e37216

Simpkins MA, Withrow DE, Cesarone JC, Boveng PL (2003) Stability in the proportion of harbor seals hauled out under locally ideal conditions. Mar Mamm Sci 19:791–805

Small RJ, Lowry LF, Hoef JMV, Frost KJ, DeLong RA, Rehberg MJ (2005) Differential movements by harbor seal pups in contrasting Alaska environments. Mar Mamm Sci 21:671–694

Smith HD (2000) The industrialisation of the world ocean. Ocean Coast Manag 43:11–28

Smith J, Karpovich S, Horstmann L, McIntyre J, O'Brien DM (2019) Seasonal differences in foraging and isotopic niche width related to body size in Gulf of Alaska harbor seals. Can J Zool 97:1156–1163

Smith RJ, Cox TM, Westage AJ (2006) Movements of harbor seals (*Phoca vitulina mellonae*) in Lacs des Loups Marins, Quebec. Mar Mamm Sci 22:480–485

Smith RJ, Hobson KA, Koopman HN, Lavigne D (1996) Distinguishing between populations of fresh- and salt-water harbour seals (*Phoca vitulina*) using stable-isotope ratios and fatty acid profiles. Can J Fish Aquat Sci 53:272–279

Spalding DJ (1963) Comparative feeding habits of the fur seal (*Callorhinus ursinus*), sea lion (*Eumetopias jubatus*) and harbour seal (*Phoca vitulina*) on the British Columbia coast. Fish Res Bull Canada 146:1–52

Sparling C, Lonergan M, McConnell B (2018) Harbour seals (*Phoca vitulina*) around an operational tidal turbine in strangford narrows: no barrier effect but small changes in transit behaviour. Aquatic Conserv-Marine Freshwater Ecosyst 28:194–204

Sparling CE, Fedak MA, Thompson D (2007) Eat now, pay later? Evidence of deferred food-processing costs in diving seals. Biol Let 3:94–98

Steingass S (2017) Dietary composition of four stocks of Pacific harbor seal (*Phoca vitulina richardii*) in the Northern California Current Large Marine Ecosystem as synthesized from historical data, 1931–2013. Northwest Nat 98:8–23

Stewart BS (1984) Diurnal hauling patterns of harbor seals at San Miguel Island, California. J Wildl Manag 48:1459–1461

Stroud RK, Roffe TJ (1979) Causes of death in marine mammals stranded along the Oregon coast. J Wildl Dis 15:91–97

Sullivan RM (1982) Agonistic behavior and dominance relationships in the harbor seal, *Phoca vitulina*. J Mammal 63:554–569

Suryan RM, Harvey JT (1998) Tracking harbor seals (*Phoca vitulina richardsi*) to determine dive behavior, foraging activity, and haul-out site use. Mar Mamm Sci 14:361–372

Tallman J, Sullivan C (2004) Harbor seal (*Phoca vitulina*) predation on a male harlequin duck (*Histrionicus histrionicus*). Northwest Nat 85:31–32

Temte JL, Bigg MA, Wiig O (1991) Clines revisited—the timing of pupping in the harbor seal (*Phoca vitulina*). J Zool 224:617–632

Terhune JM (1985) Scanning behavior of harbor seals on haul-out sites. J Mammal 66:392–395

Terhune JM, Brillant SW (1996) Harbour seal vigilance decreases over time since haul out. Anim Behav 51:757–763

Thompson D, Duck CD, Morris CD, Russell DJF (2019) The status of harbour seals (*Phoca vitulina*) in the UK. Aquatic Conserv-Marine Freshwater Ecosyst 29:40–60

Thompson D, Fedak MA (2001) How long should a dive last? A simple model of foraging decisions by breath-hold divers in a patchy environment. Anim Behav 61:287–296

Thompson PM (1993) Harbour seal movement patterns. Symposia Zool Soc London 66:225–239

Thompson PM, Fedak MA, McConnell BJ, Nicholas KS (1989) Seasonal and sex-related variation in the activity patterns of common seals (*Phoca vitulina*). J Appl Ecol 26:521–535

Thompson PM, Hastie GD, Nedwell J, Barham R, Brookes KL, Cordes LS, Bailey H, McLean N (2013) Framework for assessing impacts of pile-driving noise from offshore wind farm construction on a harbour seal population. Environ Impact Assess Rev 43:73–85

Thompson PM, Kovacs KM, McConnell BJ (1994) Natal dispersal of harbor seals (*Phoca vitulina*) from breeding sites in Orkney, Scotland. J Zool 234:668–673

Thompson PM, Mackay A, Tollit DJ, Enerby S, Hammond PS (1998) The influence of body size and sex on the characteristics of harbour seal foraging trips. Can J Zool 76:1044–1053

Thompson PM, Mackay B, Barton TR, Duck C, Butler JRA (2007) Assessing the potential impact of salmon fisheries management on the conservation status of harbour seals (*Phoca vitulina*) in north-east Scotland. Anim Conserv 10:48–56

Thompson PM, Miller D (1990) Summer foraging activity and movements of radio-tagged common seals (*Phoca vitulina*) in the Moray Firth, Scotland. J Appl Ecol 27:492–501

Thompson PM, Tollit DJ, Wood D, Corpe HM, Hammond PS, Mackay A (1997) Estimating harbour seal abundance and status in an estuarine habitat in north-east Scotland. J Appl Ecol 34:43–52

Toke D (2011) The UK offshore wind power programme: a sea-change in UK energy policy? Energy Policy 39:526–534

Tollit DJ, Black AD, Thompson PM, Mackay A, Corpe HM, Wilson B, Van Parijs SM, Grellier K, Parlane S (1998) Variations in harbour seal *Phoca vitulina* diet and dive-depths in relation to foraging habitat. J Zool 244:209–222

Tollit DJ, Thompson PM (1996) Seasonal and between-year variations in the diet of harbour seals in the Moray Firth, Scotland. Can J Zool 74:1110–1121
Trumble SJ (1995) Abundance, movements, dive behaviour, food habits, and mother-pup interactions of harbor seals (*Phoca vitulina richardsi*) near Monterey Bay, California. MS, California State University, Fresno
Van Parijs SM, Corkeron PJ, Harvey J, Hayes SA, Mellinger DK, Rouget PA, Thompson PM, Wahlberg M, Kovacs K (2003) Patterns in the vocalizations of male harbor seals. J Acoust Soc Am 113:3403–3410
Van Parijs SM, Hastie GD, Thompson PM (1999) Geographical variation in temporal and spatial vocalization patterns of male harbour seals in the mating season. Anim Behav 58:1231–1239
Van Parijs SM, Hastie GD, Thompson PM (2000a) Individual and geographical variation in display behaviour of male harbour seals in Scotland. Animal Behav 59:559–568
Van Parijs SM, Janik VM, Thompson PM (2000b) Display-area size, tenure length, and site fidelity in the aquatically mating male harbour seal, *Phoca vitulina*. Canadian J Zool 78:2209–2217
Van Parijs SM, Thompson PM, Tollit DJ, Mackay A (1997) Distribution and activity of male harbour seals during the mating season. Anim Behav 54:35–43
Vincent C, McConnell B, Delayat S, Elder J-F, Gautier G, Ridoux V (2010) Winter habitat use of harbour seals (*Phoca vitulina*) fitted with Fastloc™ GPS/GSM tags in two tidal bays in France. NAAMCO Sci Publ 8:285–302
Ward EJ, Levin PS, Lance MM, Jeffries SJ, Acevedo-Gutiérrez A (2012) Integrating diet and movement data to identify hot spots of predation risk and areas of conservation concern for endangered species. Conserv Lett 5:37–47
Watts P (1992) Thermal constraints on hauling out by harbor seals (*Phoca vitulina*). Can J Zool 70:553–560
Watts P (1996) The diel hauling-out cycle of harbour seals in an open marine environment: correlates and constraints. J Zool 240:175–200
Westlake RL, O'Corry-Crowe GM (2002) Macrogeographic structure and patterns of genetic diversity in harbor seals (*Phoca vitulina*) from Alaska to Japan. J Mammal 83:1111–1126
Williams TM, Fuiman LA, Horning M, Davis RW (2004) The cost of foraging by a marine predator, the Weddell seal Leptonychotes weddellii: pricing by the stroke. J Exp Biol 207:973–982
Williams TM, Haun J, Davis RW, Fuiman LA, Kohin S (2001) A killer appetite: metabolic consequences of carnivory in marine mammals. Comp Biochem Physiol A—Molec Integr Physiol 129:785–796
Wilson B, Batty RS, Daunt F, Carter C (2007) Collision risks between marine renewable energy devices and mammals, fish and diving birds. Report by Centre for Ecology & Hydrology. Reprot for Scottish Government
Wilson LJ, Hammond PS (2016) Harbour seal diet composition and diversity. Scottish Marine Freshwater Sci 7:84
Wolfe R, Bryant J, Hutchinson-Scarbrough L, Kookesh M, Sill LA (2012) The subsistence harvest of harbor seals and sea lions by Alaska Natives in 2006. Technical Paper No. 383 by Alaska Department of Fish and Game, Division of Subsistence and Alaska Native Harbor Seal Commission. http://www.adfg.alaska.gov/sf/Pu
Wolfe R, Fall JA, Riedel M (2006) Subsistence harvest of harbor seals and sea lions by Alaska Natives in 2006. Technical Paper No. 339 by Alaska Department of Fish and Game, Division of Subsistence and Alaska Native Harbor Seal Commission. http://www.adfg.alaska.gov/sf/Pu
Womble JN, Blundell GM, Gende SM, Horning M, Sigler MF, Csepp DJ (2014) Linking marine predator diving behavior to local prey fields in contrasting habitats in a subarctic glacial fjord. Mar Biol 161:1361–1374
Womble JN, Gende SM (2013) Post-breeding season migrations of a top predator, the harbor seal (*Phoca vitulina richardii*), from a marine protected area in Alaska. PLoS One 8:e55386
Womble JN, Pendleton GW, Mathews EA, Blundell GM, Bool NM, Gende SM (2010) Harbor seal (*Phoca vitulina richardii*) decline continues in the rapidly changing landscape of Glacier Bay National Park, Alaska 1992–2008. Mar Mamm Sci 26:686–697

Womble JN, Ver Hoef JM, Gende S, Mathews E (2020) Calibrating and adjusting counts of harbor seals in a tidewater glacier fjord to estimate abundance and trends 1992 to 2017. Ecosphere 11:e03111

Ydesen KS, Wisniewska DM, Hansen JD, Beedholm K, Johnson M, Madsen PT (2014) What a jerk: prey engulfment revealed by high-rate, super-cranial accelerometry on a harbour seal (*Phoca vitulina*). J Exp Biol 217:2239–2243

Yochem PK, Stewart BS, Delong RL, DeMaster DP (1987) Diel haul-out patterns and site fidelity of harbor seals (*Phoca vitulina richardsi*) on San Miguel island, california, in autumn. Mar Mamm Sci 3:323–332

Yoshino K, Takahashi A, Adachi T, Costa DP, Robinson PW, Peterson SH, Hückstädt LA, Holser RR, Naito Y (2020) Acceleration-triggered animal-borne videos show a dominance of fish in the diet of female northern elephant seals. J Exp Biol 22:jeb212936

Young C, Gende S, Harvey J (2014) Effects of vessels on harbor seals in Glacier Bay National Park. Tour Mar Environ 10(1–2):5–20

Chapter 11
The Elephant Seal: Linking Phenotypic Variation with Behavior and Fitness in a Sexually Dimorphic Phocid

Roxanne S. Beltran, Mark A. Hindell, and Clive R. McMahon

Contents

Abstract Southern (*Mirounga leonina*) and northern elephant seals (*M. angustirostris*) are extreme in nearly all aspects of their behavior, from fasting to fighting to diving to navigation. Behavioral research over the past six decades has provided unparalleled insight into key elephant seal traits that drive survival and reproductive success. Specifically, the extensive study of northern and southern elephant seals using innovative methodologies and technologies has shown how behavior during reproduction, molt, and foraging interact to drive the evolution of body mass. Their extreme sexual dimorphism has led to pronounced niche segregation across three-dimensional space, allowing a synthetic assessment of physiological constraints in an ecological context. Additionally, their behavioral feats have made elephant seals

R. S. Beltran (✉)
Department of Ecology and Evolutionary Biology, University of California Santa Cruz, Santa Cruz, CA, USA
e-mail: roxanne@ucsc.edu

M. A. Hindell · C. R. McMahon
Institute for Marine and Antarctic Studies, University of Tasmania, Hobart, TAS, Australia

C. R. McMahon
Department of Biological Sciences, Macquarie University, Sydney, NSW, Australia

Sydney Institute of Marine Science, Mosman, NSW, Australia

D. P. Costa and E. McHuron (eds.), *Ethology and Behavioral Ecology of Phocids*, Ethology and Behavioral Ecology of Marine Mammals,
https://doi.org/10.1007/978-3-030-88923-4_11

successful remote sensing platforms for measuring the physical and biological characteristics of the open ocean. Promising directions for future research include interactions between intrinsic and extrinsic factors and their relationship to individual quality, especially for juveniles and adult males.

Keywords Mirounga · Sexual dimorphism · Polygyny · Foraging · Diving · Fasting · Breeding · Niche segregation · Oceanography

11.1 Introduction

Southern elephant seals (*Mirounga leonina*) and northern elephant seals (*M. angustirostris*) are among the best-studied pinnipeds. Broad interest in elephant seals is partly a consequence of their critical ecological role as apex predators in both hemispheres, occupying a specialized deep-diving niche that requires well-developed physiological and anatomical adaptations. Elephant seals also exhibit extraordinary physiological abilities, including tolerance for environmental extremes experienced by few other mammals. Research is facilitated by the relative ease of observations, sample collection, and instrument attachment due to colony accessibility and large body size. Elephant seals also have incredible homing abilities (Oliver et al. 1998). They consistently return to their natal colony to give birth at around the same time each year (Le Boeuf et al. 1972), allowing repeated sampling of the same individuals (Hofmeyr et al. 2012). As such, elephant seals have become a model system for assessing the impacts of climate change and anthropogenic disturbance on marine predators' behavior (Block et al. 2011; Maxwell et al. 2013; Hindell et al. 2020).

11.1.1 History and Progress of Behavioral Ecology Research

Behavioral ecology research on northern and southern elephant seals has evolved from simple land-based studies of abundance to complex analyses of at-sea spatial utilization relative to oceanographic features. This progress is primarily due to the development of new technologies. Research in the 1960s focused predominantly on describing the geographic distribution, abundance (Orr et al. 1965), and on-land behavior of elephant seals (Le Boeuf and Peterson 1969; Fogden 1968; Bryden 1968). In the 1970s, research efforts delved deeper into social behavior (Le Boeuf et al. 1972; Le Boeuf 1974), genetic variation (Bonnell and Selander 1974), morphology (Briggs and Morejohn 1975), anatomy (Ling and Thomas 1967), ontogeny (Reiter et al. 1978), and descriptions of basic physiology, including milk composition (Le Boeuf and Ortiz 1977) and energy influx (Ortiz et al. 1978). The 1980s were characterized by advancements in field physiology, such as measurements of blood chemistry (Costa and Ortiz 1982) and at-sea energetics (Huntley and LeBoeuf 1989), along with rudimentary investigations of at-sea behavior and fidelity (Condit and Le Boeuf

1984). Research in the 1990s focused on population demography (Reiter and Le Boeuf 1991), method development such as photogrammetry (Haley et al. 1991) and body condition measurements (Slip et al. 1992a; Hindell and Slip 1997), the experimental translocation paradigm (Oliver et al. 1998), and measurements of field metabolic rates (Boyd et al. 1993).

During this same time, innovations in technology provided unprecedented insight into the cryptic lives of elephant seals (Costa 1993), including their fine-scale diving behavior (Le Boeuf et al. 1993; Campagna et al. 1998; Hindell and Lea 1998) and association with oceanographic features and bathymetry (Simmons et al. 2007). Research findings in the 2000s led to breakthroughs in at-sea physiology (Williams et al. 2000; Meir and Ponganis 2010; Meir et al. 2013; Williams 2001; Le Boeuf et al. 2000b), investigations into elephant seal predators (Le Boeuf 2004) and prey (Bradshaw et al. 2003; Lewis et al. 2006; Kuhn et al. 2009), fasting physiology (Houser and Costa 2001; Noren et al. 2003), combinations of factors such as maternal traits and reproductive effort (Crocker et al. 2001), diving physiology (Thornton et al. 2001), and a better understanding of human effects from sound sources (Costa et al. 2003) and climate change (Le Boeuf and Crocker 2005; Crocker et al. 2006). In the past decade, we have gained valuable insights into the utilization of physical habitats (Maxwell et al. 2012) and oceanographic conditions (Cotté et al. 2015; Dragon et al. 2010; Guinet et al. 2013), the mesopelagic ecosystem (McMahon et al. 2019), intraspecific variation in behavior (Robinson et al. 2012), stress physiology (Vázquez-Medina et al. 2010; Champagne et al. 2012), sensory biology and cognition (Casey et al. 2015, 2013), ontogeny (Casey et al. 2020), and conservation applications relevant to marine megafauna on a global scale (Maxwell et al. 2013; Hazen et al. 2013, 2019; Block et al. 2011; Hindell et al. 2020).

Monitoring studies initiated on elephant seals in the early years continue to provide critical behavioral insights. Long-term demographic programs on northern elephant seals initiated at Año Nuevo, California, United States of America (37.1°N, 122.3°W) in 1963 (Condit et al. 2014; Le Boeuf and Peterson 1969; Le Boeuf et al. 2019) and on southern elephant seals at Marion Island, Prince Edwards Islands, South Africa (46.9°S, 37.7°E) in 1983 (Bester and Wilkinson 1994; McMahon et al. 2000, 2003, 2005b; Oosthuizen et al. 2015, 2019; Pistorius et al. 2001a, 2001b, 2011) have been key in characterizing mechanisms driving population dynamics and intraspecific variation. In these studies, cohorts of weaned pups have been tagged annually since the programs began. This large number of known-aged individuals and annual resighting efforts have yielded a vast body of scientific literature on vital rates (survival and fecundity) and population dynamics in northern and southern elephant seals. Over the past two decades, the at-sea behavior of elephant seals has been monitored in 695 northern elephant seals tracked from Año Nuevo and 535 southern elephant seals from Îles Kerguelen (Subantarctic Desolation Islands, France).

Innovations in research methodologies along with the long-term monitoring programs in elephant seals have played a critical role in facilitating scientific investigations of other phocid species. For example, elephant seals were used to develop chemical immobilization (Bornemann et al. 2013; Carlini et al. 2009; Gales and Burton 1987; Briggs et al. 1977) and diet determination (Beltran et al. 2015; Lübcker et al. 2016; Spurlin et al. 2019) techniques that are now widely used. Elephant seals have also played an essential role in understanding the effects of biologger attachment on energetics and survival (McMahon et al. 2008; Maresh et al. 2014) and providing sample size recommendations for large marine vertebrate studies (Sequeira et al. 2019; Harrison 2012). A suite of breakthrough technologies has been developed using elephant seals, from time-depth recorders (LeBoeuf et al. 1986), acoustic recorders (Burgess et al. 1998) and playback instruments (Fregosi et al. 2016; Costa et al. 2003), to stomach temperature pills (Kuhn et al. 2009), heart rate loggers (Hindell and Lea 1998; Andrews et al. 1997), blood temperature thermistors (Meir and Ponganis 2010), respiratory frequency tags (Andrews et al. 2000), and oceanographic instruments (Boehlert et al. 2001; Campagna et al. 2000). Validating new analytical techniques such as mass estimates from photogrammetry (De Bruyn et al. 2009; Postma et al. 2013), and errors associated with ARGOS satellite (Costa et al. 2010) and light-level geolocation (Delong et al. 1992) location methods have allowed the application of new technologies to other species. Many of the technical and procedural innovations developed in elephant seals are widely used in other species and taxa (Huntley et al. 1987). These innovations were driven primarily by the need to determine the reasons for widescale decreases in the southern elephant seal populations and the recovery of the northern elephant seals from near extinction.

11.1.2 Exploitation History and Current Population Trends

Northern and southern elephant seals are similar in appearance and behavior, although southern elephant seals are slightly larger and have a less pronounced proboscis that gives the species its name (Fig. 11.1). The two species differ in their geographical ranges and population trends, except for a rogue southern elephant seal found in the Gulf of Mexico (Elorriaga-Verplancken et al. 2020). Southern elephant seals are exposed to more extreme conditions, including snow and ice, due to their use of beaches at higher latitudes (42°S–60°S) compared to northern elephant seals (27°N–41°N). Both species were heavily exploited in the nineteenth and twentieth centuries for oil, which severely depleted their populations (Hindell and Burton 1987; Bartholomew and Boolootian 1960). Northern elephant seals were deemed extinct on three occasions (Townsend 1912) and there were likely as few as 20 seals left alive (Hoelzel et al. 1993; Bartholomew and Boolootian 1960). After protection by the Mexican government, northern elephant seals expanded exponentially in population size and range, recolonizing their historic breeding range (Rick et al. 2011). They also expanded to new areas (Lowry et al. 2014; Hodder et al. 1998) via prospecting and spill-over (Le Boeuf and Laws 1994) and now inhabit the island and mainland beaches. Phenotypic variation among colonies in feeding strategies and foraging locations (Kienle 2019) has likely contributed to their current range-wide success, despite the apparent lack of genetic variation. In contrast, although southern elephant seals were hunted extensively in the past century, the population never fell below several hundred thousand individuals (Laws 1994). Therefore, genetic diversity in

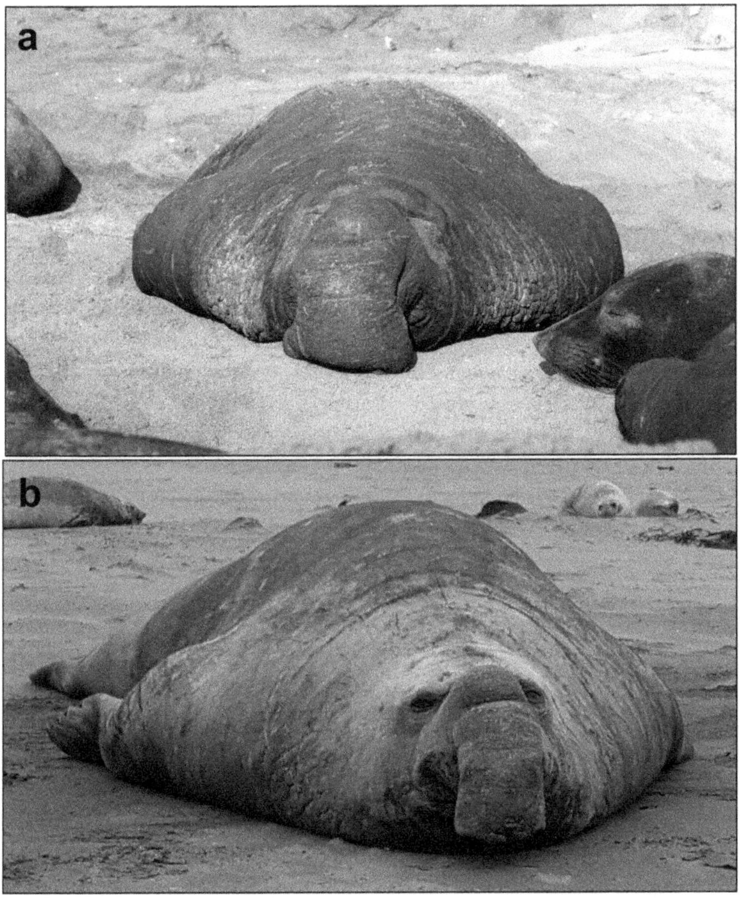

Fig. 11.1 Adult male northern **a** and southern **b** elephant seals rest on the beach. Northern elephant seal males are notably smaller than southern elephant seal males but have a larger proboscis. *Photo credit* Daniel Costa (**a**) and Mark Hindell (**b**)

northern elephant seals is extremely low while genetic diversity in southern elephant seals is comparable to other large mammals (Hoelzel et al. 2002, 1993).

Despite these population bottlenecks, both species have now recovered (Lowry et al. 2014; Hindell et al. 2017) and are considered of Least Concern by the International Union for Conservation of Nature (IUCN). Northern elephant seals are currently in a period of rapid expansion in terms of numbers and geographic range (Lowry et al. 2014). Southern elephant seals experienced a period of recovery; however, many of the major populations have experienced another period of decline since the 1960s, likely due to a complex mix of environmental change (Van den Hoff et al. 2014) and predation (McMahon et al. 2005a). Of the four genetically distinct stocks (Corrigan et al. 2016), the South Georgia, Península Valdés, and Kerguelen populations are either stable or increasing slightly while the Macquarie

Island population is decreasing at 1.2% per year (Bonner 1958; Hindell et al. 2016; Ling and Bryden 1992). Despite the declining Macquarie stock, the entire southern elephant seal species is currently stable at around 750,000 seals (Hindell et al. 2016). In contrast, the northern elephant seal population continues to increase at approximately 3.8% per year, with a current population estimate of 239,000 seals (Lowry et al. 2014).

Population trajectories are mediated by individual behaviors and their resulting fitness consequences (Hindell et al. 2017). Like all marine mammals (Barlow and Boveng 1991), elephant seals have relatively low survivorship at young ages (Condit et al. 2014). Many die during their first trip to sea (Cox et al. 2020); as few as 17% (average 25%) of female northern elephant seals survive long enough to recruit at age three or four (Huber et al. 1991b; Condit et al. 2014) despite a maximum lifespan of 23 years (Condit et al. 2014; Hindell and Little 1988; Schwarz et al. 2012). The same is true for male northern elephant seals, of which only 4% survive to prime breeding age around nine years old (Pistorius et al. 1999; Lloyd et al. 2019). Surprisingly, given their stable status, survival appears to be higher in southern elephant seals, with 30–40% of females surviving long enough to recruit (Schwarz et al. 2012; Oosthuizen et al. 2018). Both species experience intense selective pressures on phenotypic traits to optimize the size of their offspring and thereby maximize their offspring's survival and recruitment into the breeding population. Despite aggressive breeding behavior and extreme fasting on land, almost all mortality in adult elephant seals occurs at sea (Pistorius et al. 1999, 2011). Most deaths can likely be attributed to predation or starvation, but the relative importance of these factors and their relation to individual variability is currently unknown. On-land mortality occurs almost exclusively in pups (mortality rates 2–10% in southern elephant seals (Galimberti and Boitani 1999; Pistorius et al. 2001b) and 10–40% in northern elephant seals (Huber et al. 1991a; Le Boeuf and Reiter 1991), due to starvation after mother–pup separation or being crushed by large males (Le Boeuf and Briggs 1977; Spraker et al. 2014). Terrestrial mortality data for northern elephant seals were collected when the colony attendance was smaller and located on nearby Año Nuevo Island, and a follow-up study on the present mainland colony would help determine whether causes of pup mortality have changed.

11.1.3 Overview of Chapter

In the following sections, we discuss how behaviors during reproduction, molt, and foraging are driven by the extreme sexual dimorphism in elephant seals. Throughout the chapter, several common threads emerge.

A first common thread is the linkages between aquatic and terrestrial components of the annual cycle (Fig. 11.2). On the one hand, factors that facilitate success in one environment may be detrimental in the other, setting up trade-offs between life-history events. For example, large lipid stores in females can facilitate reproduction and ensure future offspring success (McMahon and Burton 2005; McMahon

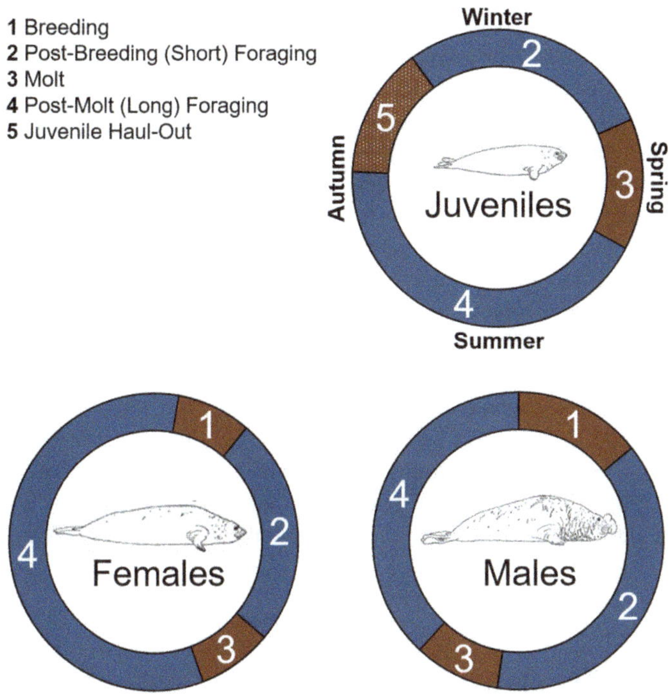

Fig. 11.2 The annual cycle of female and male elephant seals in conceptual form. Northern and southern elephant seals have similar life-history phenology after controlling for boreal versus austral seasons in the northern and southern hemispheres. For females, the breeding season starts with parturition and is followed by lactation, weaning, copulation, and fertilization and then a period of embryonic diapause that lasts through the short foraging trip and the molt. Then, female seals undergo implantation of the blastocyst and embryonic development throughout the long foraging trip. For males, the breeding season and molt are interspersed with two foraging trips of similar duration. Sub-adult males (age < 7 years) have slightly earlier timing, and haul out during the summer or fall in place of other haul-out periods. Elephant seal silhouettes were adapted from http://www.eleseal.org/survey/survey_info.html

et al. 2003), but can increase the cost of swimming as buoyancy deviates from neutral (Adachi et al. 2014). Alternatively, success in one environment can facilitate success in the other. For instance, foraging behavior affects body condition, which can influence survival, fecundity, and population dynamics (Chaps. 6 and 8).

A second common thread is differences in anatomy, physiology, and ecology between males and females that result in stark behavioral differences, both on land and at sea (Figs. 11.2 and 11.3). Intense sexual selection pressures on males have resulted in a nearly five-fold difference in body mass between adult male and female elephant seals (Figs. 11.4 and 11.7; Deutsch et al. 1990; Haley et al. 1994), making them the most sexually dimorphic of any mammalian species. The high environmental potential for polygyny in this terrestrial breeding system may explain why elephant seals exhibit such pronounced sexual dimorphism (Fig. 11.7; Krüger et al.

2014). These sex-specific size differences translate to differences in the at-sea behavioral strategies of adult elephant seals during their foraging trips (Le Boeuf et al. 2000a), as well as different rates of survival (Condit et al. 2014) and reproductive success (Le Boeuf 1974; Reiter et al. 1981).

The third common thread is variability among individual seals, which makes them ideal candidates for understanding phenotypic variation and linking behavior with fitness and population dynamics. For example, Simmons et al. (2007) found that oceanography and bathymetry influence at-sea behavior of northern elephant seals to a lesser degree than that of individual behavioral traits. Similarly, Hückstädt et al. (2012) found a large degree of individual variability in the diet of female southern elephant seals, suggesting different individual foraging strategies. Individual quality is now emerging as one of the principal drivers of survival and recruitment into the breeding population and therefore population growth rates of elephant seals (Oosthuizen et al. 2019), similar to observations from other species (Paterson et al. 2016, 2018). This clearly illustrates that not all individuals contribute equally to population growth rates and persistence, but that a small number of high-quality individuals are disproportionately responsible for the stability and viability of the population (Le Boeuf et al. 2019).

11.2 Annual Cycle and Life History

Elephant seals have a biphasic life history, with distinct phases that occur either on land (reproductive, molting, and wintering behaviors) or at sea (foraging behaviors; Fig. 11.2; Kirkman et al. 2001, 2003, 2004). The adult annual cycle consists of a terrestrial breeding season (parturition, lactation, copulation), a post-breeding oceanic foraging trip, a terrestrial molting season, and a post-molting oceanic foraging trip (Fig. 11.2). Prior to sexual maturity, juveniles haul out for the molting and breeding seasons earlier than adults. As seals reach sexual maturity (3–4 years in females and 5–6 years in males, although social maturity in males occurs much later Pistorius et al. 2008; Desprez et al. 2014; Oosthuizen et al. 2019; Le Boeuf 1974), these haul-out periods occur progressively later (and last longer) such that the male breeding haul out coincides with that of females in estrus (Le Boeuf and Laws 1994). As adults, the phenological differences between male and female elephant seals are significant, with the male molt occurring during summer and the female molt occurring during spring (Fig. 11.2). The annual cycles of northern and southern elephant seals are slightly offset from each other: whereas northern elephant seals breed in winter and molt in spring, southern elephant seals breed in austral spring and molt in austral summer. This shift may be due in part to the more extreme seasonality (and weather conditions) experienced by southern elephant seals, which attend breeding colonies at notably higher latitudes (below 42°S) than northern elephant seals (above 27°N). Of course, the life-history events occurring each month are reversed as a result of differences in seasons between the northern and southern hemispheres.

Elephant seals are capital breeders. They rely on endogenous energy stores rather than intermittent feeding to sustain them throughout their terrestrial lactation (and their subsequent molt phase; Hindell et al. 1994a). The significant energetic challenges of fasting while breeding and molting result in extensive body mass losses on land that must be recuperated during oceanic foraging trips (Table 11.1, Fig. 11.3). Broadly speaking, females from the two species have similar patterns of mass gain and loss across their annual cycle, with impressive mass gains during the two foraging trips and dramatic mass losses during the molting and breeding seasons (Postma et al. 2013). However, extreme sexual dimorphism combined with the demands of pup rearing place vastly different energetic requirements and result in vastly different mass gain patterns between males and females. While previous studies have quantified the duration, distance, and mass gain of foraging trips along with mass loss during breeding and molt, these values have yet to be directly compared across northern and southern elephant seals.

We have summarized the typical mass changes during an adult elephant seal's annual cycle (for males and females of both species; Table 11.1, Fig. 11.3). Elephant seals are heaviest at the start of the reproductive period. Northern elephant seals have a slightly longer lactation period than their southern counterparts (26 vs 24 days; Costa et al. 1986; Postma et al. 2013; McDonald and Crocker 2006), which results in greater net energy expenditure for the female and larger weaning masses than southern elephant seals (Hindell and Slip 1997; Hindell et al. 1994a). Male northern elephant seals lose 7.8 kg/day on average during the breeding season (Crocker et al. 2012b) and mass loss strongly varies by level of aggression (Galimberti et al. 2007).

Elephant seals undergo a post-breeding trip to sea to recuperate the mass lost during the breeding season (Deutsch et al. 1990). Northern elephant seal females spend a longer duration at sea (Robinson et al. 2012) and travel slightly farther than southern elephant seals (Campagna et al. 1998). Based on existing mass estimates, it appears that male southern elephant seals may lose mass during this post-breeding trip (Galimberti et al. 2007; Slip et al. 1992a; Carlini et al. 2009). Males and females of both species lose mass during the following molting period (Fig. 11.3). Again, northern elephant females undergo a longer post-molt foraging trip than southern elephant seals and travel considerably farther (Robinson et al. 2012; Campagna et al. 1998). During this long foraging trip, both species replenish energy stores lost during the molt while also gestating a pup (females) or preparing for the next breeding season (males). During the long post-molt trip, mass gain is comparable between northern and southern elephant seals, although males gain ~4.5 × more mass than females (Robinson et al. 2012; Kienle 2019; Gales and Burton 1987). Male mass gain on the focal foraging areas can exceed 12 kg per day (Le Boeuf et al. 2000a). Based on average durations of life-history events, it appears that northern elephant seals spend many more days at sea than southern elephant seals. Given that lactation length is only different by several days, the molt and/or the breeding or molting haul-out periods must be significantly longer in southern elephant seals, although this has not been formally evaluated.

Table 11.1 Body mass, mass gain, foraging trip distance (cumulative path) and foraging trip duration for male and female northern and southern elephant seals. If data do not exist to our knowledge, they are given as: "- " or, when possible, [a]calculated as the Pre-Molt mass minus Post-Breeding mass; or [b]Calculated as Pre-Breeding mass minus Post-Molt mass. Mass was measured by weighing, directly measuring, and/or photogrammetry. For foraging trips, PB = post-breeding and PM = post-molt

Parameter	Females				Males			
	Northern elephant seal		Southern elephant seal		Northern elephant seal		Southern elephant seal	
	Value	Citation	Value	Citation	Value	Citation	Value	Citation
Pre-breeding mass	504 kg	Costa et al. (1986)	494–643 kg	Arnbom et al. (1994), Carlini et al. (1997), Hindell and Slip (1997)	1573 kg	Le Boeuf et al. (2000a)	2092 kg	Galimberti et al. (2007)
Post-breeding mass	333–347 kg	Costa et al. (1986), Kienle (2019)	333–357 kg	Hindell and Slip (1997), Arnbom et al. 1993)	1133 kg	Le Boeuf et al. (2000a)	1641 kg	Galimberti et al. (2007)
Pre-molt mass	385 kg	Worthy et al. (1992)	447 kg	Hindell et al. (1994b)	1464 kg	Le Boeuf et al. (2000a)	1487 kg	Slip et al. (1992a)
Post-molt mass	270–295 kg	Worthy et al. (19920, Kienle (2019)	342 kg	Hindell et al. (1994b)	1260 kg	Le Boeuf et al. (2000a)	1331 kg	Slip et al. (1992b)
PB trip mass gain	73 kg	Kienle (2019)	132 kg	Carlini et al. (1997)	332–415 kg	Le Boeuf et al. (2000a), Kienle (2019)	−194 kg	[a]
PM trip mass gain	224 kg	Kienle (2019)	178 kg	[b]	313–483 kg	Le Boeuf et al. (2000a), Kienle)(2019)	761 kg	[b]
PB trip duration	75 days	Robinson et al. (2012)	62–72 days	Campagna et al. (1998), Boyd et al. (1993)	114–124 days	Le Boeuf et al. (2000a), Kienle (2019)	101 days	Campagna et al. (1999)

(continued)

Table 11.1 (continued)

Parameter	Females				Males			
	Northern elephant seal		Southern elephant seal		Northern elephant seal		Southern elephant seal	
	Value	Citation	Value	Citation	Value	Citation	Value	Citation
PB trip distance	4,913 km	Robinson et al. (2012)	4,438 km	Campagna et al. (1998)	–	–	2,446 km	Campagna et al. (1999)
PM trip duration	219 days	Robinson et al. (2012)	180 days	Campagna et al. (1998)	117–127 days	Le Boeuf et al. (2000a), Kienle (2019)	–	–
PM trip distance	9,850 km	Robinson et al. (2012)	5,889 km	Campagna et al. (1998)	–	–	–	–

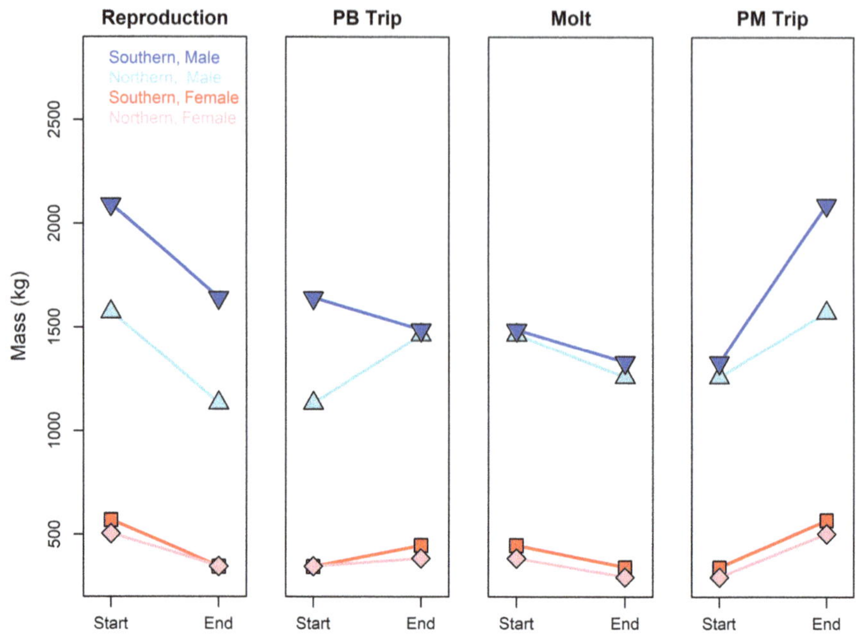

Fig. 11.3 Annual mass changes in adult male and female northern and southern elephant seals throughout the annual cycle, including the breeding and molting seasons onshore and the post-breeding (PB) and post-molting (PM) foraging trips to sea. Citations and body mass values used in this figure are provided in Table 1. If a range of body mass is given in Table 1, the points in this figure represent the midpoint. Most data are from longitudinal studies, but sample sizes vary widely

11.3 Reproductive Behavior

Despite stark differences in population trajectories, genetic variation, and global distribution, both elephant seal species exhibit remarkably similar behavior and energy dynamics associated with reproduction. Perhaps the most renowned characteristic of both species is the highly polygynous breeding system, in which one large "beachmaster" (alpha male) defends a harem of females (Fig. 11.4; Le Boeuf and Peterson 1969). In polygynous breeding systems, reproductive success is vastly more variable in males than females, and therefore competition among males for access to females is strong (Trivers 1972; Emlen and Oring 1977). The beachmaster benefits by securing mating access with many receptive females; however, this incurs several costs. For instance, males sacrifice two months of foraging to arrive at the breeding colonies early and assert dominance before females arrive (Campagna et al. 1993; Hindell and Burton 1988). The costs of maintaining the harem also include energy expenditure due to moving and fighting (Deutsch et al. 1990), physical injury from fighting (Galimberti et al. 2007), and a risk of overheating (Norris et al. 2010).

For females, the date of return to land from the long foraging trip varies by six weeks. Approximately 6–7 days after returning, females give birth and begin to nurse

Fig. 11.4 During the breeding season, an alpha male northern elephant seal vocalizes to maintain a harem as females suckle their pups. Elephant seals exhibit extremely social behavior onshore in contrast to their seemingly solitary at-sea foraging. Photo credit: Bruce Lyon

their pups (Fig. 11.5; Le Boeuf et al. 1972). Aggression between adult females and between females and non-filial pups is common in dense colonies (McCann 1982; Christenson and Le Boeuf 1978) but not in spacious colonies (Baldi et al. 1996). Generally speaking, mother and pup remain close together and do not venture into the water during this time (but see Codde et al. 2016). It appears that mothers can recognize the calls of their pups (Petrinovich 1974; Insley 1992), but mothers can fail to successfully reunite with their pups if separated (Riedman and Le Boeuf 1982). Females that lose a pup sometimes attempt to steal suckling pups, adopt a weaned pup, or indiscriminately nurse orphaned pups (Le Boeuf and Laws 1994). Suckling frequency increases from 18 to 40 min per day throughout lactation in northern elephant seals (Le Boeuf et al. 1972) and up to 60 min per day in southern elephant seals (Bryden 1968). Milk fat increases from 15 to 55% during this time (Riedman and Ortiz 1979; Hindell et al. 1994a). Female investment in offspring is extreme but short-lived; in just under one month, pups gain ~100 kg from milk energy and wean at a mean mass of 122 kg (northern; Le Boeuf et al. 1972; Le Boeuf and Crocker 2005) and 118 kg (southern; Clausius et al. 2017) while mothers lose ~157 kg (Costa et al. 1986).

Female breeding strategies are associated with drastically different mass requirements as compared to males. In females, mean size differs only slightly across northern and southern elephant seals (Table 11.1; Le Boeuf and Laws 1994),

Fig. 11.5 A female northern elephant seal nurses her pup for 26 days, which results in the pup growing from ~ 20 kg (**a**) to ~ 120 kg (**b**). The weaned pup goes on to spend about a month in various weaner pods (**c**) before departing on its first oceanic migration. Photo credit: Daniel Costa

suggesting that perhaps thermal requirements cannot fully explain the size differences between male northern and southern elephant seals. The maximum mass of adult female southern elephant seals can reach 977 kg (Arnbom et al. 1994), which is substantially larger than that of northern elephant seals at 608 kg (D. Costa, *unpublished data*). Female northern elephant seals go into estrus approximately 24 days after parturition, which lasts around 4 days (Le Boeuf et al. 1972). Females have no shortage of mating opportunities and do not court males; instead, females protest ~93% of mounts (Cox and Le Boeuf 1977), although they are less likely to protest at the end of estrus (Cox 1981), especially for dominant males with presumably higher-quality traits that may be inherited by offspring. While male-male competition occurs throughout the breeding season, most copulations occur toward the end of the breeding season. The females depart for their foraging trips long before males leave the beach (see Chap. 7 for further discussion of female choice).

Throughout the lactation period, adult males are actively engaged in the establishment and defense of harems, often using vocalizations to minimize physical altercations (Fig. 11.4; Pistorius et al. 1999, 2011; Sandegren 1976). Recent work on the social behavior of male northern elephant seals has shown that vocalizations are unique signature calls, containing information about identity rather than size or dominance of the caller (Mathevon et al. 2017; Casey et al. 2015). After engaging in a physical confrontation (Cox 1981), males associate their opponent's vocalization with the fight outcome (i.e., victory or loss) so that future fights can be avoided (Casey et al. 2015). When vocalizations from familiar rivals were recorded and experimentally played back to male elephant seals, they consistently moved toward subordinate males and retreated away from dominant males (Casey et al. 2015). Similar playbacks of unfamiliar males from a distant colony elicited no predictable responses, further suggesting that vocalizations must be paired with the outcomes of physical confrontations (Casey et al. 2015; Casey 2020).

During male-male interactions, large body size confers an advantage for establishing and maintaining harems by facilitating longer fasting durations (Haley et al. 1994). Only the largest males in the best condition can defend harems and procure the highest number of copulations (Crocker et al. 2012a; Lloyd et al. 2019). Alpha males (~1% of males) are responsible for 85–90% of successful inseminations (Le Boeuf and Peterson 1969; Fabiani et al. 2004) and therefore contribute most of the raw genetic material to the next generation (Crocker et al. 2012a).To copulate, males of both species use the bulk and weight of their bodies, a fore-flipper clasp, and a neck bite to restrain females (Fig. 11.6). Southern elephant seal males are on average heavier than northern elephant seals, reaching a maximum of 3,700 kg (Ling and Bryden 1981) compared with 2,300 kg (Deutsch et al. 1990), respectively. The greater discrepancy in sexual size dimorphism in southern elephant seals (Fig. 11.7) matches their greater reproductive success compared to northern elephant seals at similar-sized harems (Hoelzel et al. 1999). Alternatively, mass differences between northern and southern elephant seals may be driven by the greater thermal requirements of the colder habitat of southern elephant seals (Slip et al. 1992a); however, further comparative studies of body size and thermoregulation are needed to fully characterize these dynamics.

Fig. 11.6 An adult male northern elephant seal copulates with a female on the beach at Año Nuevo (**a**) and in the waters surrounding San Nicolas Island, both California (**b**). While mating on the beach is typical, aquatic mating also occurs just offshore and in the open ocean. Photo credit: Daniel Costa (**a**) and Elizabeth McHuron (**b**)

Although most individuals breed on land, there is some evidence for a non-polygynous alternate mating strategy in which elephant seals skip the breeding season and mate at sea, similar to other phocid species (De Bruyn et al. 2011). Near-shore aquatic mating has also been observed at least once in northern elephant seals (Fig. 11.6, E. McHuron *pers. comm.*). It has been postulated that most sexually immature females mate for the first time at sea (Laws 1956; Le Boeuf and Laws 1994) because the number of beach copulations observed is not sufficient to

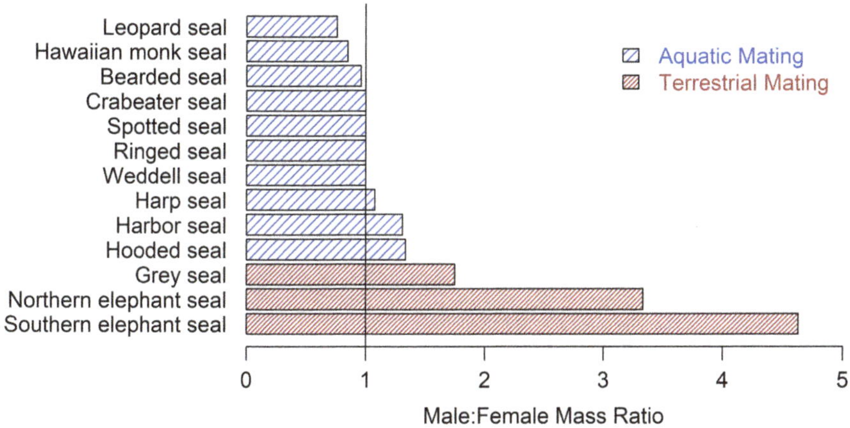

Fig. 11.7 Elephant seals are the most sexually dimorphic among phocid seals as measured by the male to female body mass ratio. This is thought to be due to their terrestrial mating system with the high environmental potential for polygyny and associated sexual selection (Le Boeuf 1991). Values for species-specific male and female mass were obtained from Folkens and Reeves (2002). A subset of species is shown

account for the number of successful pregnancies during the next breeding season. On the other hand, 5–13% of adult female northern elephant seals (Le Boeuf et al. 2011; Hückstädt et al. 2018) and 20% of southern elephant seals that attend the colonies each year do not produce a pup (Desprez et al. 2018; Robinson et al. 2012). Pregnancy status, identified from shorter dive durations during foraging trips (Le Boeuf and Laws 1994; Hückstädt et al. 2018), haul-out phenology (Robinson et al. 2012), and whisker biochemistry (Lübcker et al. 2020), has shown that most females (>85%) get pregnant, but not all bring the fetus to term (i.e., the blastocyst does not implant or the fetus is aborted) (Hückstädt et al. 2018). There is strong evidence linking female survival with past reproductive success, where skip breeding can increase survival and reproductive success (Griffen 2018), as well as age (Desprez et al. 2014, 2018). Survival also relates to individual quality such that higher-quality individuals recruit earlier and breed for longer than their poor quality conspecifics (Oosthuizen et al. 2019). While intermittent breeding increases lifetime reproductive output, high-quality individuals appear to gain more from skipping breeding than their lower quality counterparts (Desprez et al. 2018; Oosthuizen et al. 2019). Here, quality refers to both the underlying characteristics and the environmental conditions that contribute to an individual's impact on the evolutionary trajectory of the population (Desprez et al. 2018; Oosthuizen et al. 2019).

11.4 Molting Behavior

Unlike reproduction, the annual replacement of hair (hereafter, molt) is an understudied life-history event in elephant seals. Seals typically return to their natal beaches to molt on suitable habitat, including wallows and vegetated areas (Setsaas et al. 2008). Both northern and southern elephant seals have a catastrophic molt, where fur and skin are shed over one or two weeks while seals fast (Fig. 11.8; Beltran et al. 2018). The molt is energetically expensive (Worthy et al. 1992) and results in extreme physiological stress. The molt haul-out period is notably longer than the molt itself, lasting approximately 30 days in female southern elephant seals (L. DeKock, *pers. comm.*), 36 days in male southern elephant seals (Slip et al. 1992a), and 32 days in female northern elephant seals (Worthy et al. 1992). The molt likely plays an essential role in the annual cycle since its timing coincides with implantation (Le Boeuf and Laws 1994). Ethograms have yet to be completed during the molt, but individuals enter and exit the water often, which is thought to allow cooling (White and Odell 1971) or movement to adjacent beaches (Setsaas et al. 2008). Animals may spend as much as one-quarter of their time in the water while molting (R. Holser, *pers. comm.*). Molting animals also regularly flip sand onto their backs to facilitate thermoregulation (White and Odell 1971) and/or scratch an itch (Huey 1924; Heath and Schusterman 1975). A recent study suggested that stereotypical behaviors, including itching, sneezing, and yawning, may be contagious in elephant seals (Wojczulanis-Jakubas et al. 2019). Northern elephant seal weanlings molt their black natal pelage

Fig. 11.8 During their catastrophic molt, southern elephant seals cease feeding and shed their fur and skin over the course of one to two weeks. Photo credit: Daniel Costa

(lanugo) to a shiny silver coat at approximately 4–6 weeks old (Fig. 11.5c; Le Boeuf et al. 1972), whereas most southern elephant seals pups lose their lanugo before they wean. The timing of the annual molt occurs progressively later with age in southern elephant seals and, similar to other phocids (Beltran et al. 2019), is also delayed in females by reproductive success (Kirkman et al. 2003). The same pattern appears to occur in northern elephant seals (R. Beltran, *pers. obs.*) but this has yet to be formally tested. Molting may serve an additional purpose because it allows seals to offload contaminants, such as mercury that is accumulated from prey during foraging trips, that can have adverse effects on health (Cossaboon et al. 2015; Peterson et al. 2015).

11.5 At-Sea Behavior

The relatively short reproduction and molting periods allow elephant seals to spend more than 80% of their lives at sea and utilize distant habitats that are thousands of kilometers from land (Fig. 11.9; Hindell 2018; Robinson et al. 2012; McIntyre et al. 2010). Elephant seals spend 90% of their foraging trips below the surface (McIntyre et al. 2010), often at considerable depth, prompting Houston and Carbone (1992) to classify elephant seals as "surfacers rather than divers." Elephant seals dive continuously, spending only a few minutes at the surface between breath-hold dives. Both species regularly dive to 1500 m during > 60-min dives (Robinson et al. 2012; McMahon et al. 2019). Below, we summarize horizontal and vertical behavior during both the short (after breeding and before molt) and long (after molt and before breeding) foraging trips.

11.5.1 Horizontal Movement

Elephant seals are well-known for their navigational skills (Robinson 2009; Horton et al. 2017). Experimental translocations of juvenile elephant seals from the breeding colony to another land or sea site have revealed their truly remarkable homing abilities (Oliver et al. 1998). Females of both species have demonstrated exceptionally high individual foraging site fidelity (Robinson et al. 2012; Bradshaw et al. 2004; Authier et al. 2012b; McIntyre et al. 2017) despite annual variation in the quality of foraging sites (Hindell et al. 2017). Site fidelity in northern elephant seals shows some inter-annual variability mediated by climate conditions (Abrahms et al. 2018). Abrahms et al. (2018) suggest that strong site fidelity represents a safe approach (low risk and stable energetic rewards) while weak fidelity is a high risk/high reward approach. In this case, under increasing environmental variability, the optimal individual strategy would be to consistently have either weak or strong fidelity. In contrast, mixed strategies would be more beneficial to the population as a whole. There is some evidence that fidelity also varies with reproductive success (McMahon and Bradshaw 2004).

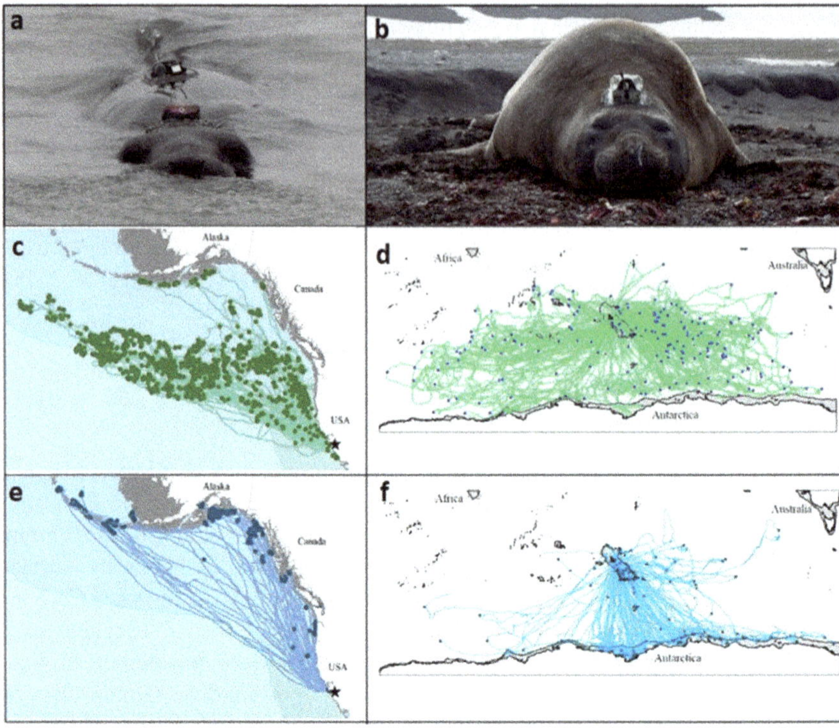

Fig. 11.9 Satellite tracking data are used to compare male (blue) and female (green) foraging habitats for northern elephant seals tagged at Año Nuevo (left, replicated with permission from Kienle 2019) and southern elephant seals tagged at Iles Kerguelen (right). Dots represent the focal foraging area of each animal. *Photo credit* Daniel Costa

Sex is arguably the most critical factor in determining the foraging routes of elephant seals. Sub-adult and adult males tend to use continental shelf habitats, while adult females use pelagic habitats (Fig. 11.9). Northern elephant seal males use the Alaskan Aleutian archipelago's coastal waters or the Alaskan and Canadian Pacific coast (Le Boeuf et al. 2000a; Kienle 2019), while females are broadly associated with the North Pacific Transition Zone chlorophyll front offshore (Robinson et al. 2012). In contrast, southern elephant seal males forage either on the Antarctic continental shelf or on shelf waters associated with sub-Antarctic islands, such as the Campbell and Kerguelen Plateaus (Hindell et al. 2016; Biuw et al. 2007). Southern elephant seal females use the span of ocean between the sub-Antarctic Front and the marginal ice zone, sometimes in polynyas (Labrousse et al. 2015, 2018). Female southern elephant seals tend to avoid sea ice by moving northwards in late autumn or early winter, which displaces them from the Antarctic shelf (Hindell et al. 2016). The foraging habitat difference between male and female southern elephant seals appears to occur at a finer spatial scale than that of northern elephant seals (Fig. 11.9).

It is unknown whether sexual segregation in foraging locations results from different energetic needs, physiological diving capabilities, intraspecific competition, or a combination of these. The consequences of sexually selected body size on aquatic behavior have yet to be described; it may be that large body size forces males to seek out high reward resources located in higher-risk habitats, as suggested for other sexually dimorphic mammals (Isaac 2005). For instance, foraging on the Kerguelen Plateau is thought to have higher mortality probability and higher trophic level prey than pelagic waters (Chaigne et al. 2013). Additionally, the core habitats of species thought to predate elephant seals, white sharks (*Carcharodon carcharias*; Carlisle et al. 2012), salmon sharks (*Lamna ditropsis*; Queiroz et al. 2019), sleeper sharks (*Somniosus antarcticus*; Van den Hoff and Morrice 2008), and killer whales (*Orcinus orca*; Ford et al. 1998), overlap with elephant seal habitat (Jorgensen et al. 2019). Fresh wounds and scars from white shark attacks are frequently observed on male and female elephant seals (LeBoeuf and Crocker 1996), suggesting that predation may play a critical role in driving population dynamics.

11.5.2 Diving Behavior and Diet

Elephant seal diving behavior, diet, and foraging success have been extensively studied for several decades (Saijo et al. 2017; Schwarz et al. 2015; Biuw et al. 2007; Robinson et al. 2010; Ferraro et al. 2017; Pirotta et al. 2019). Seals eat mostly myctophid fish, estimated at 4.8 – 16.1 kg of fish per day (Guinet et al. 2014), and some squid. Diet is highly variable across individuals (Hückstädt et al. 2012; Goetsch et al. 2018) and age classes (Lübcker et al. 2017; Walters et al. 2014). Much elephant seal prey are located in the mesopelagic zone (200 – 1000 m), which requires extensive breath-hold capacities and long dive durations (Naito et al. 2013; Adachi et al. 2014, 2017). However, in particular, males also feed near or at the bottom when on shelf habitats, presumably taking benthic prey, such as toothfish (*Dissostichus* spp.; Banks et al. 2014). Diving abilities are thought to be important for minimizing predation risk as well. Although the vertical risk landscape is not well understood, it appears that killer whales are limited in depth by their diving physiology (Miller et al. 2010). White sharks are limited by light and temperature requirements (Boustany et al. 2002). Indeed, seals often dive deeper as they approach coastal areas where shark presence is high (LeBoeuf and Crocker 1996). Exposure experiments have shown that marine mammals respond to predator vocalizations by extending breath-holds and diving deeper (Fregosi et al. 2016; Williams et al. 2017). Together these findings indicate that depth and darkness confer safety, which has critical implications for the evolution of daily activity budgets, sensory biology, and diving physiology (Minias 2020).

Bio-loggers have provided valuable insight into elephant seal diving abilities and their relationship to prey and predators. The simple depth and time data provided by biologging instruments contain a wealth of information about fundamental behavioral characteristics (Chap. 12). Studies of elephant seals have demonstrated their extraordinary diving abilities and raised stimulating questions regarding the physiological mechanisms that allow breath-holds (Le Boeuf et al. 2000a; Robinson et al. 2012; Hassrick et al. 2010; Hindell et al. 1992, 2000; Hindell and Lea 1998). Time-series depth data have provided valuable ecological insights about how males forage on the ocean floor, and how females feeding in the open ocean hunt mesopelagic prey with marked diurnal vertical migration (Naito et al. 2013; Goetsch et al. 2018; Yoshino et al. 2020; McMahon et al. 2019). Biologging data has also been used to quantify searching behavior (Adachi et al. 2017; Bailleul et al. 2008; Le Bras et al. 2017) and determine pregnancy status (Le Boeuf and Laws 1994; Hückstädt et al. 2018). Similarly, physiological inference, such as oxygen consumption, can be made because prolonged post-dive surface intervals allow for the clearance of metabolic byproducts from anaerobic metabolism (Jouma'a et al. 2016).

Accelerometers, acoustic recorders, and video cameras have provided valuable fine-scale data on flipper stroke magnitude and frequency and overall body movement to estimate energy expenditure, and head and jaw movements to quantify prey capture attempts (Naito et al. 2010, 2013; Heerah et al. 2014; Génin et al. 2015). These data have been used to link prey encounters with seal behavior and body condition gains as well as oceanographic features (Saijo et al. 2017; Biuw et al. 2007). For instance, when prey are more abundant, seals stay at depth to maximize their energy intake before returning to the surface to replenish their oxygen stores (Jouma'a et al. 2016). In contrast, Thums et al. (2013) showed that dives were the longest in areas where prey patches were of low quality. This illustrates the complex interplay of prey type, prey distribution, and predator behavior and physiology. When coupled with remotely-sensed data, location data can be used to extract information such as chlorophyll, fronts, polynyas (Malpress et al. 2017), sea ice (Labrousse et al. 2017), eddies, and their effects on the abundance of prey (Abrahms et al. 2018; Malpress et al. 2017; Cotté et al. 2015; Bailleul et al. 2010; Guinet et al. 2014). Together, these findings demonstrate the considerable power of fine-scale behavioral parameters for explaining how seals respond to their biological and physical environments (Maxwell et al. 2012).

In addition to matching seal behavior with satellite-derived oceanography products, the continuous, deep-diving behavior of elephant seals has made them ideal platforms for measuring the hydrography of their environment, including temperature and salinity to track water masses in space and time (Boehme et al. 2008; Lowther et al. 2016; Roquet et al. 2014), and chlorophyll to quantify productivity (Blain et al. 2013; Guinet et al. 2013). Custom tags have been designed to measure the ocean's properties where seals travel, including areas where ships rarely venture, such as ice-covered waters. The worldwide Southern Elephant Seals as Oceanographic Samplers (SEaOS) project found that elephant seals are likely to forage where ocean conditions are prime for squid, and that populations with declining seals are often farther from these prime areas (Hindell et al. 2016). More recently, novel tags have been developed to better characterize mesopelagic prey fields, including their precise abundance (Goulet et al. 2019) and bioluminescent behavior (Goulet et al. 2020) relative to oceanographic characteristics, such as the oxygen minimum zone (Naito et al. 2017). These data streams have been combined to explain the effects of physical, chemical, and biological oceanography on the spatial, seasonal, and diel diving patterns of elephant seals (Biuw et al. 2010; Dragon et al. 2010; Bailleul et al. 2007; Simmons et al. 2010).

11.5.3 Environmental Drivers of Reproductive Success

In capital breeders, variability in pupping success relates strongly to age, mass, and body condition (Boyd 2000). Because the latter two state variables depend upon foraging success, at-sea behavior influences the probability of successful reproduction on land. This is especially true during years of extreme environmental variability, such as El Niño (Crocker et al. 2001; Huber et al. 1991b; McMahon and Burton 2005) and positive Southern Annular Mode phases when females have more difficulty finding food and produce fewer pups in the following breeding season (McMahon et al. 2017; Van den Hoff et al. 2014). In these years, elephant seals exhibit behavioral flexibility (Crocker et al. 2006), suggesting that behavioral plasticity as a mode of adaptation to changing climates is possible.

Environmental variability does not appear to affect breeding female survival (Desprez et al. 2018, 2014) as may be expected for long-lived species living in relatively stable environments (Eberhardt 2002). However, environmental factors affect female reproductive success, as more females breed when conditions are favorable (Desprez et al. 2014; Van den Hoff et al. 2014). Additionally, the environment affects maternal investment such that southern elephant seal mothers allocate more energy to offspring when conditions are right (McMahon et al. 2017). In this way, seals can maximize their lifetime reproductive success by adjusting reproductive expenditure relative to prevailing environmental conditions and their own capabilities (McMahon et al. 2017).

11.6 Ontogeny of Behavior

The breeding, molting, and foraging behaviors of adults take many years to develop. These factors, along with inexperience and small body size, result in relatively low survivorship of juvenile elephant seals (Condit et al. 2014). Sex differences in social behavior become apparent early in development (Reiter et al. 1978). For example, differential selection pressures on body size across males and females are facilitated by a slightly heavier birth mass and significantly heavier weaning mass of male pups relative to females (Guinet 1991; Salogni et al. 2018). Male pups suckle for an extra day and are more adept at stealing milk because their teeth erupt five weeks later than females (Reiter et al. 1978; Hindell and Kemper 1998). There is evidence that mothers invest disproportionately more in male pups when they have superior energy stores (McMahon et al. 2017). Added maternal investment is known to benefit offspring fitness, including faster pup growth rates, earlier recruitment into the breeding population, higher survival (McMahon et al. 2003), and higher lifetime reproductive output (Oosthuizen et al. 2018, 2019; Desprez et al. 2018). Analysis of sex-specific size at age and the fitness consequences of deviation from species-wide averages have yet to be undertaken throughout ontogeny but would provide valuable insights.

After weaning and before departing for their first trip to sea, weanlings alternate between short bouts of shallow diving and long bouts of onshore breath-holds (Le Boeuf and Laws 1994; Modig et al. 1997). During this time, they typically join a pod of weanlings (Rasa 1971) or occasionally attempt to suckle from a second adult female. Weanlings that are successful at double-mother-suckling are referred to as "superweaners" and can weigh as much as 238 kg (compared to ~ 120 kg in typical weanlings;Reiter et al. 1978; Engelhard et al. 2001). The purpose of spending time (often sunset to sunrise) in the water is not clear, given that weanlings do not feed during the post-weaning haul out (Reiter et al. 1978). One possible explanation is increasing diving duration, diving depth, and blood and muscle oxygen stores to prepare for their first trip to sea (Thorson and Le Boeuf 1994). After losing ~ 25% of their body mass (Reiter et al. 1978), weaned elephant seal pups begin their first migration and learn to navigate and forage without any parental instruction (Reiter et al. 1978; McConnell et al. 2002). Studies of behavioral and navigational development in young seals are limited due to the high cost of satellite tags as well as low weanling survival rates (Condit et al. 2014; Reiter et al. 1978; Thorson and Le Boeuf 1994; McConnell et al. 2002). Young seals make shorter dives to shallower depths with lower foraging efficiency than adults because they need to spend more time recovering at the surface after dives (Carter et al. 2017). Heavier weaning mass appears to provide an advantage for diving duration and depth (Hindell et al. 1999; Orgeret et al. 2019) and allows individuals to survive longer without starving (McConnell et al. 2002; Cox et al. 2020). Foraging route fidelity develops around age four in southern elephant seals (Martin et al. 2011; Authier et al. 2012a). A similar pattern is likely in northern elephant seals (Stewart 1997).

Dispersal to other beaches is common as young seals prospect other colonies (Hofmeyr 2000; Van den Hoff 2001). Upon their return to land, everyday interactions between juveniles include disputes over space (all sexes), mock fights (mostly male-male), and vocal learning (Sanvito et al. 2007) in preparation for competitive interactions as adults (Rasa 1971; Reiter et al. 1978; Modig et al. 1997). Throughout reproductive development, males increase residency time and restrict fine-scale movement patterns ashore while increasing the reliability of their signature calls (Casey et al. 2020). Females begin to breed between ages 3 and 5 years old. However, younger females have lower weaning success because they are pushed to the harem periphery where they are exposed to aggressive males, dangerous surf conditions, and high tides (Reiter et al. 1981; Le Boeuf and Reiter 1988; Christenson and Le Boeuf 1978). Young females also exhibit lower lactation efficiency (Crocker et al. 2001), are more distant from their pups during lactation (Hooper et al. 2019), produce milk with lower lipid content (McDonald and Crocker 2006), and produce pups that weigh less at birth and weaning (Le Boeuf et al. 2011; Galimberti and Boitani 1999).

11.7 Future Directions

Much remains to be learned about elephant seal behavior at sea and its links to demography and population change. Most research efforts have focused on adult females due to their high survival rates, high site fidelity, and substantial impact on population dynamics (Le Boeuf et al. 2019; Oosthuizen et al. 2019). One critical question is how these adult female elephant seals maximize lifetime reproductive success (Le Boeuf et al. 2019). This question is impossible to answer without disentangling causes of mortality (e.g., predation vs. starvation) and gaining a more comprehensive understanding of the distribution of elephant seal predators and prey (Green et al. 2020). The recent development of novel biologging technologies such as the Life History Transmitter tag (Horning and Hill 2005; Horning and Mellish 2009) will help elucidate the risks and rewards associated with various behavioral strategies. Specifically, the transmission of post-mortem information will provide information on demographics that are under-represented in the research literature (Hazen et al. 2012), such as adult males and juveniles that contribute significantly to population growth rates (McMahon and Burton 2005). Similarly, satellite-relay tags with onboard acceleration and depth processing will allow researchers to study fine-scale behavior and foraging success of demographic groups with lower survival, such as juveniles and adult males (Cox et al. 2018; Heerah et al. 2019; Orgeret et al. 2019).

Novel biologging data will also help disentangle the implications of vast differences in body size resulting from sexual dimorphism on diving behavior and diet at sea (Fig. 11.9). A combination of these factors may help explain why males and females use different foraging habitats. For instance, alpha males with relatively short foraging periods and larger post-breeding energy deficits than subordinate males may need to consume energy-rich benthic prey, such as Patagonian toothfish *Dissostichus eleginoides* (Van den Hoff et al. 2017). A stomach lavage study found that male southern elephant seals consumed a large amount of squid, at least within a ~ 500 km radius of the haul-out site (Van den Hoff et al. 2003). Skull morphology suggests that these prey species may be too large for adult females to consume (Tarnawski et al. 2013); instead, females profitably hunt shoaling mesopelagic fish and squid species in the open ocean. Diet studies have yet to be undertaken on northern elephant seal males but would provide valuable insights.

The relationships between extrinsic and intrinsic factors are complex (McMahon et al. 2021), and there remains much to be learned in elephant seals. For example, it is unclear whether the individual quality is innate or behaviorally mediated. The genetic, morphological, behavioral, and physiological factors that differ between low- and high-quality individuals is a topic that warrants future study. For instance, attempts to compare consistent individual variability in foraging strategies (e.g., coastal versus pelagic routes, focused vs nomadic foraging) to behavioral responses on land are in their infancy (R. Holser, *pers. comm.*). Additionally, the degree to which behavior and quality are transferred between generations is unknown. These factors are critical for addressing the magnitude and speed of population responses to environmental and anthropogenic change.

11.8 Conclusions

Some elephant seal populations have now been studied for over 60 years, with mark-recapture studies providing critical data on population dynamics and at-sea behavior. These studies are crucial for establishing the links between behavior, intrinsic and extrinsic factors, and population trends. For the most part, research on northern and southern elephant seals has occurred in parallel; here, we show that a synthetic cross-species approach can highlight the rules of life that act similarly upon species in different hemispheres. Specifically, the polygynous breeding system in which a small number of males contribute to the gene pool provides strong selection for large body size and the resulting sexual segregation in foraging strategies and habitat use. This, in turn, exposes animals to different predators and prey that consequently impact future survival and reproductive attempts. While both sexes demonstrate individual variability and behavioral flexibility, sexual selection pressures likely constrain male at-sea behavior to a much larger degree than females. Innovations in research methodologies and technologies have played and will continue to play a critical role in facilitating scientific investigations of phocid behavior. The next several decades of research will help elucidate the ecological context and physiological constraints that link climate to behavior to vital rates, especially for lesser understood demographic groups such as males and juveniles.

Acknowledgements The authors wish to thank the pioneers of elephant seal behavior. The comments and insights of Dr. Dan Costa, Dr. Elizabeth McHuron, Dr. Nico deBruyn, Rachel Holser, Dr. Patrick Robinson, Dr. Colleen Reichmuth, and Dr. Caroline Casey significantly improved the chapter. This work was supported by National Science Foundation and President's Postdoctoral Fellowship Program postdoctoral fellowships to RSB. Photographs were taken under National Marine Fisheries Services marine mammal permit #19108 and 87–1851. The ARGOS seal tracking data were sourced from the Integrated Marine Observing System (IMOS). IMOS is a national collaborative research infrastructure supported by the Australian Government. It is operated by a consortium of institutions as an unincorporated joint venture, with the University of Tasmania as the lead agent. The International Argo Program provided the Argo floats hydrographic data (https://portal.aodn.org.au). This work was further supported by the French Polar Institute (program 109: PI. H. Weimerskirch and 1201: PI. C. Gilbert and C. Guinet), the SNO-MEMO, and CNES-TOSCA.

References

Abrahms B, Scales KL, Hazen EL, Bograd SJ, Schick RS, Robinson PW, Costa DP (2018) Mesoscale activity facilitates energy gain in a top predator. Proceed Royal Soc B Biol Sci 285(1885):20181101

Adachi T, Costa DP, Robinson PW, Peterson SH, Yamamichi M, Naito Y, Takahashi A (2017) Searching for prey in a three-dimensional environment: hierarchical movements enhance foraging success in northern elephant seals. Funct Ecol 31(2):361–369

Adachi T, Maresh JL, Robinson PW, Peterson SH, Costa DP, Naito Y, Watanabe YY, Takahashi A (2014) The foraging benefits of being fat in a highly migratory marine mammal. Proceed Royal Soc B Biol Sci 281(1797):2014–2120

Andrews RD, Costa DP, Le Boeuf BJ, Jones DR (2000) Breathing frequencies of northern elephant seals at sea and on land revealed by heart rate spectral analysis. Respir Physiol 123(1–2):71–85
Andrews RD, Jones D, Williams J, Thorson P, Oliver G, Costa D, Le Boeuf B (1997) Heart rates of northern elephant seals diving at sea and resting on the beach. J Exp Biol 200(15):2083–2095
Arnbom T, Fedak M, Boyd I, McConnell B (1993) Variation in weaning mass of pups in relation to maternal mass, postweaning fast duration, and weaned pup behaviour in southern elephant seals (*Mirounga leonina*) at South Georgia. Can J Zool 71(9):1772–1781
Arnbom T, Fedak M, Rothery P (1994) Offspring sex ratio in relation to female size in southern elephant seals, *Mirounga Leonina*. Behav Ecol Sociobiol 35(6):373–378
Authier M, Bentaleb I, Ponchon A, Martin C, Guinet C (2012a) Foraging fidelity as a recipe for a long life: foraging strategy and longevity in male southern elephant seals. PLoS One 7(4):e32026
Authier M, Dragon A-C, Richard P, Cherel Y, Guinet C (2012) O'mother where wert thou? Maternal strategies in the southern elephant seal: a stable isotope investigation. Proceed Royal Soc B Biol Sci 279(1738):2681–2690
Bailleul F, Charrassin J-B, Monestiez P, Roquet F, Biuw M, Guinet C (2007) Successful foraging zones of southern elephant seals from the Kerguelen Islands in relation to oceanographic conditions. Philosophical Trans Royal Soc B Biol Sci 362(1487):2169–2181
Bailleul F, Cotté C, Guinet C (2010) Mesoscale eddies as foraging area of a deep-diving predator, the southern elephant seal. Mar Ecol Prog Ser 408:251–264
Bailleul F, Pinaud D, Hindell M, Charrassin JB, Guinet C (2008) Assessment of scale-dependent foraging behaviour in southern elephant seals incorporating the vertical dimension: a development of the first passage time method. J Anim Ecol 77(5):948–957
Baldi R, Campagna C, Pedraza S, LeBoeuf B (1996) Social effects of space availability on the breeding behaviour of elephant seals in Patagonia. Anim Behav 51(4):717–724
Banks J, Lea M-A, Wall S, McMahon CR, Hindell MA (2014) Combining bio-logging and fatty acid signature analysis indicates spatio-temporal variation in the diet of the southern elephant seal, Mirounga leonina. J Exp Mar Biol Ecol 450:79–90. https://doi.org/10.1016/j.jembe.2013.10.024
Barlow J, Boveng P (1991) Modeling age-specific mortality for marine mammal populations. Mar Mamm Sci 7(1):50–65
Bartholomew GA, Boolootian RA (1960) Numbers and population structure of the pinnipeds on the California Channel Islands. J Mammal 41(3):366–375
Beltran RS, Burns JM, Breed GA (2018) Convergence of biannual moulting strategies across birds and mammals. Proceed Royal Soc B Biol Sci 285(1878):20180318
Beltran RS, Kirkham AL, Breed GA, Testa JW, Burns JM (2019) Reproductive success delays moult phenology in a polar mammal. Sci Rep 9(1):5221
Beltran RS, Sadou M, Condit R, Peterson S, Reichmuth C, Costa D (2015) Fine-scale whisker growth measurements can reveal temporal foraging patterns from stable isotope signatures. Mar Ecol Prog Ser 523:243–253. https://doi.org/10.3354/meps11176
Bester MN, Wilkinson IS (1994) Population ecology of southern elephant seals at Marion Island. In: LeBoeuf BJ, Laws RM (eds) Elephant seals: population ecology, behavior, and physiology. University of California Press, Berkeley, pp 85–97
Biuw M, Boehme L, Guinet C, Hindell M, Costa D, Charrassin J-B, Roquet F, Bailleul F, Meredith M, Thorpe S (2007) Variations in behavior and condition of a Southern Ocean top predator in relation to *in situ* oceanographic conditions. Proc Natl Acad Sci 104(34):13705–13710
Biuw M, Nøst OA, Stien A, Zhou Q, Lydersen C, Kovacs KM (2010) Effects of hydrographic variability on the spatial, seasonal and diel diving patterns of southern elephant seals in the eastern Weddell Sea. PLoS One 5(11)
Blain S, Renaut S, Xing X, Claustre H, Guinet C (2013) Instrumented elephant seals reveal the seasonality in chlorophyll and light-mixing regime in the iron-fertilized Southern Ocean. Geophys Res Lett 40(24):6368–6372

Block BA, Jonsen ID, Jorgensen SJ, Winship AJ, Shaffer SA, Bograd SJ, Hazen EL, Foley DG, Breed G, Harrison A (2011) Tracking apex marine predator movements in a dynamic ocean. Nature 475(7354):86

Boehlert GW, Costa DP, Crocker DE, Green P, O'Brien T, Levitus S, Le Boeuf BJ (2001) Autonomous pinniped environmental samplers: using instrumented animals as oceanographic data collectors. J Atmos Oceanic Tech 18(11):1882–1893

Boehme L, Thorpe S, Biuw M, Fedak M, Meredith M (2008) Monitoring Drake Passage with elephant seals: frontal structures and snapshots of transport. Limnol Oceanogr 53(5part2):2350–2360

Bonnell ML, Selander RK (1974) Elephant seals: genetic variation and near extinction. Science 184(4139):908–909

Bonner WN (1958) Exploitation and conservation of seals in South Georgia. Oryx 4(6):373–380

Bornemann H, de Bruyn PN, Reisinger RR, Kästner S, Márquez ME, McIntyre T, Bester MN, Plötz J (2013) Tiletamine/zolazepam immobilization of adult post-moult southern elephant seal males. Polar Biol 36(11):1687–1692

Boustany AM, Davis SF, Pyle P, Anderson SD, Le Boeuf BJ, Block BA (2002) Satellite tagging: expanded niche for white sharks. Nature 415(6867):35

Boyd I (2000) State-dependent fertility in pinnipeds: contrasting capital and income breeders. Funct Ecol 14(5):623–630

Boyd I, Arnbom T, Fedak M (1993) Water flux, body composition, and metabolic rate during molt in female southern elephant seals (*Mirounga leonina*). Physiol Zool 43–60

Bradshaw CJ, Hindell MA, Best NJ, Phillips KL, Wilson G, Nichols PD (2003) You are what you eat: describing the foraging ecology of southern elephant seals (*Mirounga leonina*) using blubber fatty acids. Proc Biol Sci 270(1521):1283–1292. https://doi.org/10.1098/rspb.2003.2371

Bradshaw CJA, Hindell MA, Sumner MD, Michael KJ (2004) Loyalty pays: potential life history consequences of fidelity to marine foraging regions by southern elephant seals. Anim Behav 68(6):1349–1360. https://doi.org/10.1016/j.anbehav.2003.12.013

Briggs GD, Henrickson RV, Le Boeuf BJ (1977) Ketamine immobilization of northern elephant seals. JAVMA 167(7):546–548

Briggs K, Morejohn GV (1975) Sexual dimorphism in the mandibles and canine teeth of the northern elephant seal. J Mammal 56(1):224–231

Bryden M (1968) Lactation and suckling in relation to early growth of the southern elephant seal, *Mirounga leonina* (L.). Aust J Zool 16(5):739–748

Burgess W, Tyack P, Le Boeuf B, Costa D (1998) A programmable acoustic recording tag and first results from free-ranging northern elephant seals. Deep Sea Res Part II 45(7):1327–1351

Campagna C, Fedak MA, McConnell BJ (1999) Post-breeding distribution and diving behavior of adult male southern elephant seals from Patagonia. J Mammal 80(4):1341–1352

Campagna C, Lewis M, Baldi R (1993) Breeding biology of southern elephant seals in Patagonia. Mar Mamm Sci 9(1):34–47

Campagna C, Quintana F, Le Boeuf BJ, Blackwell S, Crocker DE (1998) Diving behaviour and foraging ecology of female southern elephant seals from Patagonia. Aquat Mamm 24(1):1–11

Campagna C, Rivas AL, Marin MR (2000) Temperature and depth profiles recorded during dives of elephant seals reflect distinct ocean environments. J Mar Syst 24(3–4):299–312

Carlini A, Daneri G, Marquez M, Soave G, Poljak S (1997) Mass transfer from mothers to pups and mass recovery by mothers during the post-breeding foraging period in southern elephant seals (*Mirounga leonina*) at King George Island. Polar Biol 18(5):305–310

Carlini A, Negrete J, Daneri G, Rogers T, Márquez M, Ciaglia M, Mennucci J (2009) Immobilization of adult male southern elephant seals (*Mirounga leonina*) during the breeding and molting periods using a tiletamine/zolazepam mixture and ketamine. Polar Biol 32(6):915–921

Carlisle AB, Kim SL, Semmens BX, Madigan DJ, Jorgensen SJ, Perle CR, Anderson SD, Chapple TK, Kanive PE, Block BA (2012) Using stable isotope analysis to understand the migration and trophic ecology of northeastern Pacific white sharks (*Carcharodon carcharias*). PLoS One 7(2):e30492

Carter MI, Russell DJ, Embling CB, Blight CJ, Thompson D, Hosegood PJ, Bennett KA (2017) Intrinsic and extrinsic factors drive ontogeny of early-life at-sea behaviour in a marine top predator. Sci Rep 7(1):15505
Casey C (2020) Acoustic coding of information in a complex social network: Identity signaling in northern elephant seals. In: Aubin T, Mathevon N (eds) Coding strategies in vertebrate acoustic communication. Springer, pp 247–269
Casey C, Charrier I, Mathevon N, Nasr C, Forman P, Reichmuth C (2020) The genesis of giants: behavioural ontogeny of male northern elephant seals. Anim Behav 166:247–259
Casey C, Charrier I, Mathevon N, Reichmuth C (2015) Rival assessment among northern elephant seals: evidence of associative learning during male–male contests. Royal Soc Open Sci 2(8):150228
Casey C, Reichmuth C, Fregosi S, Charrier I, Mathevon N (2013) The acoustic signature of the male northern elephant seal: individual variation supports recognition during competitive interactions. J Acous Soc Am 134(5):3988–3988
Chaigne A, Authier M, Richard P, Cherel Y, Guinet C (2013) Shift in foraging grounds and diet broadening during ontogeny in southern elephant seals from Kerguelen Islands. Mar Biol 160(4):977–986
Champagne CD, Houser DS, Costa DP, Crocker DE (2012) The effects of handling and anesthetic agents on the stress response and carbohydrate metabolism in northern elephant seals. PloS One 7(5):e38442
Christenson T, Le Boeuf B (1978) Aggression in the female northern elephant seal, *Mirounga Angustirostris*. Behaviour 64(1–2):158–171
Clausius E, McMahon C, Harcourt R, Hindell M (2017) Effect of climate variability on weaning mass in a declining population of southern elephant seals *Mirounga leonina*. Mar Ecol Prog Ser 568:249–260
Codde SA, Allen SG, Houser DS, Crocker DE (2016) Effects of environmental variables on surface temperature of breeding adult female northern elephant seals, Mirounga angustirostris, and pups. J Therm Biol 61:98–105
Condit R, Le Boeuf BJ (1984) Feeding habits and feeding grounds of the northern elephant seal. J Mammal 65(2):281–290
Condit R, Reiter J, Morris PA, Berger R, Allen SG, Le Boeuf BJ (2014) Lifetime survival rates and senescence in northern elephant seals. Mar Mamm Sci 30(1):122–138. https://doi.org/10.1111/mms.12025
Corrigan L, Fabiani A, Chauke L, McMahon C, de Bruyn M, Bester M, Bastos A, Campagna C, Muelbert M, Hoelzel A (2016) Population differentiation in the context of Holocene climate change for a migratory marine species, the southern elephant seal. J Evol Biol 29(9):1667–1679
Cossaboon JM, Ganguli PM, Flegal AR (2015) Mercury offloaded in Northern elephant seal hair affects coastal seawater surrounding rookery. Proc Natl Acad Sci 112(39):12058–12062
Costa D, LeBoeuf B, Huntley A, Ortiz C (1986) The energetics of lactation in the northern elephant seal, Mirounga angustirostris. J Zool 209(1):21–33
Costa DP (1993) The secret life of marine mammals novel tools for studying their behavior and biology at sea. Oceanography 6(3):120–128
Costa DP, Crocker DE, Gedamke J, Webb PM, Houser DS, Blackwell SB, Waples D, Hayes SA, Le Boeuf BJ (2003) The effect of a low-frequency sound source (acoustic thermometry of the ocean climate) on the diving behavior of juvenile northern elephant seals, *Mirounga angustirostris*. J Acoust Soc Am 113(2):1155–1165
Costa DP, Ortiz C (1982) Blood chemistry homeostasis during prolonged fasting in the northern elephant seal. Am J Physiol-Regul Integr Comp Physiol 242(5):R591–R595
Costa DP, Robinson PW, Arnould JP, Harrison A-L, Simmons SE, Hassrick JL, Hoskins AJ, Kirkman SP, Oosthuizen H, Villegas-Amtmann S (2010) Accuracy of ARGOS locations of pinnipeds at-sea estimated using Fastloc GPS. PloS One 5 (1):e8677

Cotté C, d'Ovidio F, Dragon A-C, Guinet C, Lévy M (2015) Flexible preference of southern elephant seals for distinct mesoscale features within the Antarctic Circumpolar Current. Prog Oceanogr 131:46–58
Cox CR (1981) Agonistic encounters among male elephant seals: frequency, context, and the role of female preference. Am Zool 21(1):197–209
Cox CR, Le Boeuf BJ (1977) Female incitation of male competition: a mechanism in sexual selection. Am Nat 111(978):317–335
Cox SL, Authier M, Orgeret F, Weimerskirch H, Guinet C (2020) High mortality rates in a juvenile free-ranging marine predator and links to dive and forage ability. Ecol Evol 10(1):410–430
Cox SL, Orgeret F, Gesta M, Rodde C, Heizer I, Weimerskirch H, Guinet C (2018) Processing of acceleration and dive data on-board satellite relay tags to investigate diving and foraging behaviour in free-ranging marine predators. Methods Ecol Evol 9(1):64–77
Crocker DE, Costa DP, Le Boeuf BJ, Webb PM, Houser DS (2006) Impact of El Niño on the foraging behavior of female northern elephant seals. Mar Ecol Prog Ser 309:1–10
Crocker DE, Houser DS, Webb PM (2012) Impact of body reserves on energy expenditure, water flux, and mating success in breeding male northern elephant seals. Physiol Biochem Zool 85(1):11–20. https://doi.org/10.1086/663634
Crocker DE, Ortiz RM, Houser DS, Webb PM, Costa DP (2012) Hormone and metabolite changes associated with extended breeding fasts in male northern elephant seals (*Mirounga angustirostris*). Comp Biochem Physiol A: Mol Integr Physiol 161(4):388–394
Crocker DE, Williams JD, Costa DP, Le Boeuf BJ (2001) Maternal traits and reproductive effort in northern elephant seals. Ecology 82(12):3541–3555
De Bruyn P, Tosh CA, Bester MN, Cameron EZ, McIntyre T, Wilkinson IS (2011) Sex at sea: alternative mating system in an extremely polygynous mammal. Anim Behav 82(3):445–451
De Bruyn PJN, Bester MN, Carlini AR, Oosthuizen WC (2009) How to weigh an elephant seal with one finger: a simple three-dimensional photogrammetric application. Aquat Biol 5:31–39. https://doi.org/10.3354/ab00135
Delong RL, Stewart BS, Hill RD (1992) Documenting migrations of northern elephant seals using day length. Mar Mamm Sci 8(2):155–159
Desprez M, Gimenez O, McMahon CR, Hindell MA, Harcourt RG (2018) Optimizing lifetime reproductive output: Intermittent breeding as a tactic for females in a long-lived, multiparous mammal. J Anim Ecol 87(1):199–211
Desprez M, Harcourt R, Hindell MA, Cubaynes S, Gimenez O, McMahon CR (2014) Age-specific cost of first reproduction in female southern elephant seals. Biol Let 10(5):20140264
Deutsch CJ, Haley MP, Le Boeuf BJ (1990) Reproductive effort of male northern elephant seals: estimates from mass loss. Can J Zool 68(12):2580–2593
Dragon A-C, Monestiez P, Bar-Hen A, Guinet C (2010) Linking foraging behaviour to physical oceanographic structures: Southern elephant seals and mesoscale eddies east of Kerguelen Islands. Prog Oceanogr 87(1–4):61–71
Eberhardt L (2002) A paradigm for population analysis of long-lived vertebrates. Ecology 83(10):2841–2854
Elorriaga-Verplancken FR, Blanco-Jarvio A, Silva-Segundo CA, Paniagua-Mendoza A, Rosales-Nanduca H, Robles-Hernández R, Mote-Herrera S, Amador-Capitanachi MJ, Sandoval-Sierra J (2020) A southern elephant seal (*Mirounga leonina*) in the Gulf of California: genetic confirmation of the northernmost record to date. Aquat Mamm 46(2):137–145
Emlen ST, Oring LW (1977) Ecology, sexual selection, and the evolution of mating systems. Science 197(4300):215–223
Engelhard GH, van den Hoff J, Broekman M, Baarspul AN, Field I, Burton HR, Reijnders PJ (2001) Mass of weaned elephant seal pups in areas of low and high human presence. Polar Biol 24(4):244–251
Fabiani A, Galimberti F, Sanvito S, Hoelzel AR (2004) Extreme polygyny among southern elephant seals on Sea Lion Island, Falkland Islands. Behav Ecol 15(6):961–969

Ferraro MS, Decker RR, Costa DP, Robinson PW, Houser DS, Crocker DE (2017) Evaluating gain functions in foraging bouts using vertical excursions in northern elephant seals. Anim Behav 129:15–24

Fogden S (1968) Suckling behaviour in the grey seal (*Halichoerus grypus*) and the northern elephant seal (*Mirounga angustirostris*). J Zool 154(4):415–420

Folkens PA, Reeves RR (2002) Guide to marine mammals of the world. National Audubon Society

Ford JK, Ellis GM, Barrett-Lennard LG, Morton AB, Palm RS, Balcomb KC III (1998) Dietary specialization in two sympatric populations of killer whales (*Orcinus orca*) in coastal British Columbia and adjacent waters. Can J Zool 76(8):1456–1471

Fregosi S, Klinck H, Horning M, Costa DP, Mann D, Sexton K, Hückstädt LA, Mellinger DK, Southall BL (2016) An animal-borne active acoustic tag for minimally invasive behavioral response studies on marine mammals. Anim Biotelemetry 4(1):1

Gales NJ, Burton HR (1987) Prolonged and multiple immobilizations of the southern elephant seal using ketamine hydrochloride-xylazine hydrochloride or ketamine hydrochloride-diazepam combinations. J Wildl Dis 23(4):614–618

Galimberti F, Boitani L (1999) Demography and breeding biology of a small, localized population of southern elephant seals (*Mirounga leonina*). Mar Mamm Sci 15(1):159–178

Galimberti F, Sanvito S, Braschi C, Boitani L (2007) The cost of success: reproductive effort in male southern elephant seals (*Mirounga leonina*). Behav Ecol Sociobiol 62(2):159–171

Génin A, Richard G, Jouma'a J, Picard B, El Ksabi N, Vacquié Garcia J, Guinet C (2015) Characterization of postdive recovery using sound recordings and its relationship to dive duration, exertion, and foraging effort of southern elephant seals (*Mirounga leonina*). Mar Mamm Sci 31(4):1452–1470

Goetsch C, Conners MG, Budge SM, Mitani Y, Walker WA, Bromaghin JF, Simmons SE, Reichmuth CJ, Costa DP (2018) Energy-rich mesopelagic fishes revealed as a critical prey resource for a deep-diving predator using quantitative fatty acid signature analysis. Front Mar Sci 5:430

Goulet P, Guinet C, Campagna C, Campagna J, Tyack PL, Johnson M (2020) Flash and grab: deep-diving southern elephant seals trigger anti-predator flashes in bioluminescent prey. J Exp Biol 223(10)

Goulet P, Guinet C, Swift R, Madsen PT, Johnson M (2019) A miniature biomimetic sonar and movement tag to study the biotic environment and predator-prey interactions in aquatic animals. Deep Sea Res Part I 148:1–11

Green D, Bestley S, Trebilco R, Corney S, Lehodey P, McMahon C, Guinet C, Hindell MA (2020) Modelled mid-trophic pelagic prey fields improve understanding of marine predator foraging behaviour. Ecography 43(7):1014–1026

Griffen BD (2018) Reproductive skipping as an optimal life history strategy in the southern elephant seal, Mirounga Leonina. Ecol Evol 8(18):9158–9170

Guinet C (1991) Growth from birth to weaning in the southern elephant seal (*Mirounga leonina*). J Mammal 72(3):617–620

Guinet C, Vacquié-Garcia J, Picard B, Bessigneul G, Lebras Y, Dragon AC, Viviant M, Arnould JP, Bailleul F (2014) Southern elephant seal foraging success in relation to temperature and light conditions: insight into prey distribution. Mar Ecol Prog Ser 499:285–301

Guinet C, Xing X, Walker E, Monestiez P, Marchand S, Picard B, Jaud T, Authier M, Cotté C, Dragon A (2013) Calibration procedures and first data set of Southern Ocean chlorophyll a profiles collected by elephant seals equipped with a newly developed CTD-fluorescence tags. Earth Syst Sci Data 5:15–29

Haley MP, Deutsch CJ, Boeuf BJL (1991) A method for estimating mass of large pinnipeds. Mar Mamm Sci 7(2):157–164

Haley MP, Deutsch CJ, Le Boeuf BJ (1994) Size, dominance and copulatory success in male northern elephant seals, Mirounga Angustirostris. Anim Behav 48(6):1249–1260

Harrison A-L (2012) A synthesis of marine predator migrations, distribution, species overlap, and use of Pacific Ocean Exclusive Economic Zones. Dissertation, University of California Santa Cruz

Hassrick J, Crocker DE, Teutschel N, McDonald B, Robinson P, Simmons S, Costa D (2010) Condition and mass impact oxygen stores and dive duration in adult female northern elephant seals. J Exp Biol 213(4):585–592

Hazen EL, Abrahms B, Brodie S, Carroll G, Jacox MG, Savoca MS, Scales KL, Sydeman WJ, Bograd SJ (2019) Marine top predators as climate and ecosystem sentinels. Front Ecol Environ. https://doi.org/10.1002/fee.2125

Hazen EL, Jorgensen S, Rykaczewski RR, Bograd SJ, Foley DG, Jonsen ID, Shaffer SA, Dunne JP, Costa DP, Crowder LB (2013) Predicted habitat shifts of Pacific top predators in a changing climate. Nat Clim Chang 3(3):234

Hazen EL, Maxwell SM, Bailey H, Bograd SJ, Hamann M, Gaspar P, Godley BJ, Shillinger GL (2012) Ontogeny in marine tagging and tracking science: technologies and data gaps. Mar Ecol Prog Ser 457:221–240

Heath ME, Schusterman RJ (1975) "Displacement" sand flipping in the northern elephant seal (Mirounga angustirostris). Behav Biol 14(3):379–385

Heerah K, Cox SL, Blevin P, Guinet C, Charrassin J-B (2019) Validation of dive foraging indices using archived and transmitted acceleration data: the case of the Weddell seal. Front Ecol Evol 7:30

Heerah K, Hindell M, Guinet C, Charrassin JB (2014) A new method to quantify within dive foraging behaviour in marine predators. PLoS One 9(6):e99329. https://doi.org/10.1371/journal.pone.0099329

Hindell M, Kemper C (1998) Marine mammal research in the Southern Hemisphere. Volume 1: status, ecology and medicine. Oceanograp Literature Rev 8(45):1419

Hindell M, Slip D (1997) The importance of being fat: maternal expenditure in the southern elephant seal *Mirounga leonina*. In: Hindell M, Kemper C (eds) Marine mammal research in the southern hemisphere, vol 1. Status, ecology, and medicine. Surrey Beatty and Sons, Chipping Norton, pp 72–77

Hindell MA (2018) Elephant seals: *Mirounga angustirostris* and *M. leonina*. In: Würsig B, Thewissen JGM, Kovacs K (eds) Encyclopedia of marine mammals. Elsevier, pp 303–307

Hindell MA, Bryden M, Burton HR (1994) Early growth and milk composition in southern elephant seals (*Mirounga leonina*). Aust J Zool 42(6):723–732

Hindell MA, Burton H (1987) Past and present status of the southern elephant seal (*Mirounga leonina*) at Macquarie Island. J Zool 213(2):365–380

Hindell MA, Burton HR (1988) Seasonal haul-out patterns of the southern elephant seal (*Mirounga leonina L.*), at Macquarie Island. J Mammal 69 (1):81–88

Hindell MA, Lea M-A (1998) Heart rate, swimming speed, and estimated oxygen consumption of a free-ranging southern elephant seal. Physiol Biochem Zool 71(1):74–84

Hindell MA, Lea M-A, Morrice MG, MacMahon CR (2000) Metabolic limits on dive duration and swimming speed in the southern elephant seal *Mirounga leonina*. Physiol Biochem Zool 73(6):790–798

Hindell MA, Little GJ (1988) Longevity, fertility and philopatry of two female southern elephant seals (*Mirounga leonina*) at Macquarie Island. Mar Mamm Sci 4(2):168–171

Hindell MA, McConnell BJ, Fedak M, Slip DJ, Burton HR, Reijnders PJ, McMahon CR (1999) Environmental and physiological determinants of successful foraging by naive southern elephant seal pups during their first trip to sea. Can J Zool 77(11):1807–1821

Hindell MA, McMahon CR, Bester MN, Boehme L, Costa D, Fedak MA, Guinet C, Herraiz-Borreguero L, Harcourt RG, Huckstadt L (2016) Circumpolar habitat use in the southern elephant seal: implications for foraging success and population trajectories. Ecosphere 7(5):e01213

Hindell MA, Reisinger RR, Ropert-Coudert Y, Hückstädt LA, Trathan PN, Bornemann H, Charrassin J-B, Chown SL, Costa DP, Danis B (2020) Tracking of marine predators to protect Southern Ocean ecosystems. Nature 580(7801):87–92

Hindell MA, Slip DJ, Burton HR (1994) Body mass loss of moulting female southern elephant seals, Mirounga leonina, at Macquarie Island. Polar Biol 14(4):275–278

Hindell MA, Slip DJ, Burton HR, Bryden MM (1992) Physiological implications of continuous, prolonged, and deep dives of the southern elephant seal (*Mirounga leonina*). Can J Zool 70(2):370–379

Hindell MA, Sumner M, Bestley S, Wotherspoon S, Harcourt RG, Lea MA, Alderman R, McMahon CR (2017) Decadal changes in habitat characteristics influence population trajectories of southern elephant seals. Glob Change Biol 23(12):5136–5150

Hodder J, Brown RF, Cziesla C (1998) The northern elephant seal in Oregon: a pupping range extension and onshore occurrence. Mar Mamm Sci 14(4):873–881

Hoelzel A, Halley J, O'Brien SJ, Campagna C, Arnborm T, Le Boeuf B, Ralls K, Dover G (1993) Elephant seal genetic variation and the use of simulation models to investigate historical population bottlenecks. J Hered 84(6):443–449

Hoelzel AR, Fleischer R, Campagna C, Le Boeuf B, Alvord G (2002) Impact of a population bottleneck on symmetry and genetic diversity in the northern elephant seal. J Evol Biol 15(4):567–575

Hoelzel AR, Le Boeuf BJ, Reiter J, Campagna C (1999) Alpha-male paternity in elephant seals. Behav Ecol Sociobiol 46(5):298–306

Hofmeyr G (2000) Dispersal and dispersion of southern elephant seals at Marion Island. University of Pretoria, Thesis

Hofmeyr G, Kirkman S, Pistorius P, Bester M (2012) Natal site fidelity by breeding female southern elephant seals in relation to their history of participation in the winter haulout. Afr J Mar Sci 34(3):373–382

Hooper AW, Berger RW, Rubin LS, McDonald BI, Crocker DE (2019) Maternal age influences offspring behaviour and growth efficiency during provisioning in northern elephant seals. Anim Behav 151:121–130

Horning M, Hill RD (2005) Designing an archival satellite transmitter for life-long deployments on oceanic vertebrates: the life history transmitter. IEEE J Oceanic Eng 30(4):807–817

Horning M, Mellish J-AE (2009) Spatially explicit detection of predation on individual pinnipeds from implanted post-mortem satellite data transmitters. Endanger Species Res 10:135–143

Horton TW, Hauser N, Zerbini AN, Francis MP, Domeier ML, Andriolo A, Costa DP, Robinson PW, Duffy CA, Nasby-Lucas N (2017) Route fidelity during marine megafauna migration. Front Mar Sci 4:422

Houser D, Costa D (2001) Protein catabolism in suckling and fasting northern elephant seal pups (*Mirounga angustirostris*). J Comp Physiol B 171(8):635–642

Houston AI, Carbone C (1992) The optimal allocation of time during the diving cycle. Behav Ecol 3(3):255–265

Huber H, Beckham C, Nisbet J (1991) Effects of the 1982–83 El Nino on northern elephant seals on the South Farallon Islands, California. In: Trillmich F, Ono KA (eds) Pinnipeds and El Niño. Springer, Berlin, Heidelberg, pp 219–233

Huber HR, Rovetta AC, Fry LA, Johnston S (1991) Age-specific natality of northern elephant seals at the South Farallon Islands, California. J Mammal 72(3):525–534

Hückstädt LA, Holser RR, Tift MS, Costa DP (2018) The extra burden of motherhood: reduced dive duration associated with pregnancy status in a deep-diving mammal, the northern elephant seal. Biol Let 14(2):20170722

Hückstädt LA, Koch PL, McDonald BI, Goebel ME, Crocker DE, Costa DP (2012) Stable isotope analyses reveal individual variability in the trophic ecology of a top marine predator, the southern elephant seal. Oecologia 169(2):395–406. https://doi.org/10.1007/s00442-011-2202-y

Huey LM (1924) Recent observations on the northern elephant seal. J Mammal 5(4):237–242

Huntley A, Costa D, Worthy G, Castellini M (1987) Approaches to marine mammal energetics. Spec. Publ. 1, Society for Marine Mammalogy. Allen Press, Lawrence, KS,

Huntley AC, LeBoeuf BJ (1989) Daily gross energy requirements of a female northern elephant seal *Mirounga angustirostris* at sea. 日本水産学会誌 55(12):2057–2063

Insley SJ (1992) Mother-offspring separation and acoustic stereotypy: a comparison of call morphology in two species of pinnipeds. Behaviour 120(1–2):103–122

Isaac JL (2005) Potential causes and life-history consequences of sexual size dimorphism in mammals. Mammal Rev 35(1):101–115

Jorgensen SJ, Anderson S, Ferretti F, Tietz JR, Chapple T, Kanive P, Bradley RW, Moxley JH, Block BA (2019) Killer whales redistribute white shark foraging pressure on seals. Sci Rep 9(1):6153

Jouma'a J, Le Bras Y, Richard G, Vacquié-Garcia J, Picard B, El Ksabi N, Guinet C (2016) Adjustment of diving behaviour with prey encounters and body condition in a deep diving predator: the southern elephant seal. Funct Ecol 30(4):636–648

Kienle SS (2019) Intraspecific variation and behavioral flexibility in the foraging strategies of seals. Dissertation, University of California Santa Cruz

Kirkman S, Bester M, Hofmeyr G, Jonker F, Pistorius P, Owen R, Strydom N (2004) Variation in the timing of the breeding haulout of female southern elephant seals at Marion Island. Aust J Zool 52(4):379–388

Kirkman S, Bester M, Pistorius P, Hofmeyr G, Jonker F, Owen R, Strydom N (2003) Variation in the timing of moult in southern elephant seals at Marion Island. S Afr J Wildlife Res 33(2):79–84

Kirkman S, Bester M, Pistorius P, Hofmeyr G, Owen R, Mecenero S (2001) Participation in the winter haulout by southern elephant seals (*Mirounga leonina*). Antarct Sci 13(4):380–384

Krüger O, Wolf JB, Jonker RM, Hoffman JI, Trillmich F (2014) Disentangling the contribution of sexual selection and ecology to the evolution of size dimorphism in pinnipeds. Evolution 68(5):1485–1496

Kuhn CE, Crocker DE, Tremblay Y, Costa DP (2009) Time to eat: measurements of feeding behaviour in a large marine predator, the northern elephant seal *Mirounga angustirostris*. J Anim Ecol 78(3):513–523. https://doi.org/10.1111/j.1365-2656.2008.01509.x

Labrousse S, Sallée J-B, Fraser AD, Massom RA, Reid P, Sumner M, Guinet C, Harcourt R, McMahon C, Bailleul F (2017) Under the sea ice: exploring the relationship between sea ice and the foraging behaviour of southern elephant seals in East Antarctica. Prog Oceanogr 156:17–40

Labrousse S, Vacquié-Garcia J, Heerah K, Guinet C, Sallée J-B, Authier M, Picard B, Roquet F, Bailleul F, Hindell M (2015) Winter use of sea ice and ocean water mass habitat by southern elephant seals: the length and breadth of the mystery. Prog Oceanogr 137:52–68

Labrousse S, Williams G, Tamura T, Bestley S, Sallée J-B, Fraser AD, Sumner M, Roquet F, Heerah K, Picard B (2018) Coastal polynyas: Winter oases for subadult southern elephant seals in East Antarctica. Sci Rep 8(1):3183

Laws R (1956) The elephant seal (*Mirounga leonina, Linn.*): III. The physiology of reproduction, vol 15. London, HMSO, p 66

Laws RM (1994) History and present status of southern elephant seal populations. In: Le Boeuf BJ, Laws RM (eds) Elephant seals: population ecology, behavior, and physiology. University of California Press, Berkeley, pp 49–65

Le Boeuf B, Condit R, Reiter J (2019) Lifetime reproductive success of northern elephant seals, Mirounga Angustirostris. Can J Zool 97(12):1203–1217

Le Boeuf B, Crocker D, Blackwell S, Morris P, Thorson P (1993) Sex differences in diving and foraging behaviour of northern elephant seals. Symp Zool Soc Lond 1993:149–178

Le Boeuf B, Reiter J (1988) Lifetime reproductive success in northern elephant seals. In: Clutton-Brock TH (ed) Reproductive success: studies of individual variation in contrasting breeding systems. University of Chicago Press, Chicago, pp 344–362

Le Boeuf B, Reiter J (1991) Biological effects associated with El Niño Southern Oscillation, 1982–83, on northern elephant seals breeding at Ano Nuevo, California. Trillmich F, Ono KA (eds) Pinnipeds and El Niño. Springer, Berlin, Heidelberg, pp 206–218

Le Boeuf BJ (1974) Male-male competition and reproductive success in elephant seals. Am Zool 14(1):163–176

Le Boeuf BJ (1991) Pinniped mating systems on land, ice and in the water: emphasis on the Phocidae. In: Renouf D (ed) The behaviour of pinnipeds. Springer, Dordrecht, pp 45–65

Le Boeuf BJ (2004) Hunting and migratory movements of white sharks in the eastern North Pacific. Mem Natl Inst Polar Res Spec Issue 58:89–100

Le Boeuf BJ, Briggs KT (1977) The cost of living in a seal harem. Mammalia 41(2):167–196

Le Boeuf BJ, Condit R, Morris PA, Reiter J (2011) The northern elephant seal (*Mirounga angustirostris*) rookery at Año Nuevo: a case study in colonization. Aquat Mamm 37(4):486

Le Boeuf B, Crocker DE, Costa DP, Blackwell SB, Webb PM, Houser DS (2000) Foraging ecology of northern elephant seals. Ecol Monogr 70(3):353–382

Le Boeuf BJ, Crocker D, Grayson J, Gedamke J, Webb PM, Blackwell SB, Costa DP (2000) Respiration and heart rate at the surface between dives in northern elephant seals. J Exp Biol 203(21):3265–3274

Le Boeuf BJ, Crocker DE (2005) Ocean climate and seal condition. BMC Biol 3(1):9

Le Boeuf BJ, Laws RM (1994) Elephant seals: population ecology, behavior, and physiology. University of California Press, Berkeley

Le Boeuf BJ, Ortiz CL (1977) Composition of elephant seal milk. J Mammal 683–685

Le Boeuf BJ, Peterson RS (1969) Social status and mating activity in elephant seals. Science 163(3862):91–93

Le Boeuf BJ, Whiting RJ, Gantt RF (1972) Perinatal behavior of northern elephant seal females and their young. Behaviour 43:121–156

LeBoeuf B, Costa D, Huntley A, Kooyman G, Davis R (1986) Pattern and depth of dives in northern elephant seals, *Mirounga angustirostris*. J Zool 208(1):1–7

LeBoeuf BJ, Crocker DE (1996) Diving behavior of elephant seals: implications for predator avoidance. In: Klimley AP, Ainley DG (eds) Great white sharks: the biology of *Carcharadon carcharias*. Academic Press, San Diego, pp 193–205

Le Bras Y, Jouma'a J, Guinet C, (2017) Three-dimensional space use during the bottom phase of southern elephant seal dives. Mov Ecol 5(1):18

Lewis R, O'Connell TC, Lewis M, Campagna C, Hoelzel AR (2006) Sex-specific foraging strategies and resource partitioning in the southern elephant seal (*Mirounga leonina*). Proc Biol Sci 273(1603):2901–2907. https://doi.org/10.1098/rspb.2006.3642

Ling J, Bryden M (1992) Mirounga leonina. Mamm Species 391:1–8

Ling J, Bryden M (1981) Southern elephant seal, *Mirounga leonina* Linnaeus, 1758. In: Ridgway S, Harrison R (eds) Handbook of marine mammals, vol 2. Seals. Academic Press, London, pp 297–327

Ling J, Thomas C (1967) The skin and hair of the southern elephant seal,(*Mirounga leonina* (L.) II. pre-natal and early post-natal development and moulting. Aust J Zool 15(2):349–365

Lloyd KJ, Oosthuizen WC, Bester MN, de Bruyn PN (2019) Trade-offs between age-related breeding improvement and survival senescence in highly polygynous elephant seals: Dominant males always do better. J Anim Ecol 89(3):897–909

Lowry MS, Condit R, Hatfield B, Allen SG, Berger R, Morris PA, Le Boeuf BJ, Reiter J (2014) Abundance, distribution, and population growth of the northern elephant seal (*Mirounga angustirostris*) in the United States from 1991 to 2010. Aquat Mamm 40(1):20–31

Lowther AD, Lydersen C, Kovacs KM (2016) The seasonal evolution of shelf water masses around Bouvetøya, a sub-Antarctic island in the mid-Atlantic sector of the Southern Ocean, determined from an instrumented southern elephant seal. Polar Res 35(1):28278

Lübcker N, Bloem LM, Du Toit T, Swart P, de Bruyn PN, Swart AC, Millar RP (2020) What's in a whisker? High-throughput analysis of twenty-eight C_{19} and C_{21} steroids in mammalian whiskers by ultra-performance convergence chromatography-tandem mass spectrometry. J Chromatogr B1141:122028

Lübcker N, Condit R, Beltran RS, de Bruyn PN, Bester MN (2016) Vibrissal growth parameters of southern elephant seals Mirounga leonina: obtaining fine-scale, time-based stable isotope data. Mar Ecol Prog Ser 559:243–255

Lübcker N, Reisinger RR, Oosthuizen WC, de Bruyn PN, van Tonder A, Pistorius PA, Bester MN (2017) Low trophic level diet of juvenile southern elephant seals Mirounga leonina from Marion Island: a stable isotope investigation using vibrissal regrowths. Mar Ecol Prog Ser 577:237–250

Malpress V, Bestley S, Corney S, Welsford D, Labrousse S, Sumner M, Hindell M (2017) Bio-physical characterisation of polynyas as a key foraging habitat for juvenile male southern elephant seals (*Mirounga leonina*) in Prydz Bay, East Antarctica. PloS One 12(9):e0184536

Maresh JL, Simmons SE, Crocker DE, McDonald BI, Williams TM, Costa DP (2014) Free-swimming northern elephant seals have low field metabolic rates that are sensitive to an increased cost of transport. J Exp Biol 217(9):1485–1495
Martin C, Bentaleb I, Steelandt S, Guinet C (2011) Stable carbon and nitrogen isotope variations in canine dentine growth layers of Kerguelen southern elephant seals. Mar Ecol Prog Ser 439:295–305
Mathevon N, Casey C, Reichmuth C, Charrier I (2017) Northern elephant seals memorize the rhythm and timbre of their rivals' voices. Current Biology 27(15):2352–2356. e2352
Maxwell SM, Frank JJ, Breed GA, Robinson PW, Simmons SE, Crocker DE, Gallo-Reynoso JP, Costa DP (2012) Benthic foraging on seamounts: a specialized foraging behavior in a deep-diving pinniped. Mar Mamm Sci 28(3):E333–E344
Maxwell SM, Hazen EL, Bograd SJ, Halpern BS, Breed GA, Nickel B, Teutschel NM, Crowder LB, Benson S, Dutton PH (2013) Cumulative human impacts on marine predators. Nat Commun 4:2688
McCann T (1982) Aggressive and maternal activities of female southern elephant seals (*Mirounga leonina*). Anim Behav 30(1):268–276
McConnell B, Fedak M, Burton HR, Engelhard G, Reijnders PJ (2002) Movements and foraging areas of naive, recently weaned southern elephant seal pups. J Anim Ecol 71(1):65–78
McDonald BI, Crocker DE (2006) Physiology and behavior influence lactation efficiency in northern elephant seals (*Mirounga angustirostris*). Physiol Biochem Zool 79(3):484–496
McIntyre T, Bester MN, Bornemann H, Tosh CA, de Bruyn PN (2017) Slow to change? Individual fidelity to three-dimensional foraging habitats in southern elephant seals, Mirounga leonina. Anim Behav 127:91–99
McIntyre T, De Bruyn P, Ansorge I, Bester M, Bornemann H, Plötz J, Tosh C (2010) A lifetime at depth: vertical distribution of southern elephant seals in the water column. Polar Biol 33(8):1037–1048
McMahon C, Burton HR, Bester MN (2000) Weaning mass and the future survival of juvenile southern elephant seals, *Mirounga leonina*, at Macquarie Island. Antarct Sci 12(2):149–153
McMahon CR, Bester MN, Burton HR, Hindell MA, Bradshaw CJ (2005) Population status, trends and a re-examination of the hypotheses explaining the recent declines of the southern elephant seal *Mirounga leonina*. Mammal Rev 35(1):82–100
McMahon CR, Bradshaw CJ (2004) Harem choice and breeding experience of female southern elephant seals influence offspring survival. Behav Ecol Sociobiol 55(4):349–362
McMahon CR, Burton HR (2005) Climate change and seal survival: evidence for environmentally mediated changes in elephant seal, *Mirounga leonina*, pup survival. Proceedings of the Royal Society b: Biological Sciences 272(1566):923–928
McMahon CR, Burton HR, Bester MN (2003) A demographic comparison of two southern elephant seal populations. J Anim Ecol 72(1):61–74
McMahon CR, Field IC, Bradshaw CJ, White GC, Hindell MA (2008) Tracking and data–logging devices attached to elephant seals do not affect individual mass gain or survival. J Exp Mar Biol Ecol 360(2):71–77
McMahon CR, Harcourt RG, Burton HR, Daniel O, Hindell MA (2017) Seal mothers expend more on offspring under favourable conditions and less when resources are limited. J Anim Ecol 86(2):359–370
McMahon CR, Hindell MA, Burton HR, Bester MN (2005) Comparison of southern elephant seal populations, and observations of a population on a demographic knife-edge. Mar Ecol Prog Ser 288:273–283
McMahon CR, Hindell MA, Charrassin J-B, Corney S, Guinet C, Harcourt R, Jonsen I, Trebilco R, Williams G, Bestley S (2019) Finding mesopelagic prey in a changing Southern Ocean. Sci Rep 9(1):1–11
McMahon CR, Roquet F, Baudel S, Belbeoch M, Bestley S, Blight C, Boehme L, Carse F, Costa DP, Fedak MA, Guinet C, Harcourt R, Heslop E, Hindell MA, Hoenner X, Holland K, Holland M, Jaine FRA, Jeanniard Du Dot T, Jonsen ID, Theresa RK, Kovacs KM, Labrousse S, Lovell P,

Lydersen C, March D, Mazloff M, McKinzie MK, Muelbert MMC, O'Brien KM, Phillips LR, Portela E, Pye J, Rintoul S, Sato K, Sequeira AMM, Simmons SE, Tsontos VM, Turpin V, van Wijk E, Vo D, Wege M, Whoriskey FG, Wilson K, Woodward BE (2021) Animal Borne Ocean Sensors – AniBOS – an essential component of the Global Ocean Observing System (GOOS) Frontiers in Marine Science 8. https://doi.org/10.3389/fmars.2021.751840

Meir JU, Ponganis PJ (2010) Blood temperature profiles of diving elephant seals. Physiol Biochem Zool 83(3):531–540

Meir JU, Robinson PW, Vilchis LI, Kooyman GL, Costa DP, Ponganis PJ (2013) Blood oxygen depletion is independent of dive function in a deep diving vertebrate, the northern elephant seal. PloS One 8(12):e83248

Miller PJOM, Shapiro AD, Deecke VB (2010) The diving behaviour of mammal-eating killer whales (*Orcinus orca*): variations with ecological not physiological factors. Can J Zool 88(11):1103–1112

Minias P (2020) Ecology and evolution of blood oxygen-carrying capacity in birds. Am Nat 195(5):000–000

Modig A, Engström H, Arnbom T (1997) Postweaning behaviour in pups of the southern elephant seal (Mirounga leonina) on South Georgia. Can J Zool 75(4):582–588

Naito Y, Bornemann H, Takahashi A, McIntyre T, Plötz J (2010) Fine-scale feeding behavior of Weddell seals revealed by a mandible accelerometer. Polar Sci 4(2):309–316

Naito Y, Costa DP, Adachi T, Robinson PW, Fowler M, Takahashi A (2013) Unravelling the mysteries of a mesopelagic diet: a large apex predator specializes on small prey. Funct Ecol 27(3):710–717

Naito Y, Costa DP, Adachi T, Robinson PW, Peterson SH, Mitani Y, Takahashi A (2017) Oxygen minimum zone: An important oceanographic habitat for deep-diving northern elephant seals, Mirounga Angustirostris. Ecol Evolut 7(16):6259–6270

Noren DP, Crocker DE, Williams TM, Costa DP (2003) Energy reserve utilization in northern elephant seal (*Mirounga angustirostris*) pups during the postweaning fast: size does matter. J Comp Physiol B 173(5):443–454. https://doi.org/10.1007/s00360-003-0353-9

Norris A, Houser D, Crocker DE (2010) Environment and activity affect skin temperature in breeding adult male elephant seals (*Mirounga angustirostris*). J Exp Biol 213(24):4205–4212

Oliver GW, Morris PA, Thorson PH, le Boeuf BJ (1998) Homing behavior of juvenile northern elephant seals. Mar Mamm Sci 14(2):245–256

Oosthuizen WC, Altwegg R, Nevoux M, Bester M, de Bruyn PN (2018) Phenotypic selection and covariation in the life-history traits of elephant seals: heavier offspring gain a double selective advantage. Oikos 127(6):875–889

Oosthuizen WC, Bester MN, Altwegg R, McIntyre T, de Bruyn PN (2015) Decomposing the variance in southern elephant seal weaning mass: partitioning environmental signals and maternal effects. Ecosphere 6(8):1–22

Oosthuizen WC, Postma M, Altwegg R, Nevoux M, Pradel R, Bester MN, de Bruyn PN (2019) Individual heterogeneity in life-history trade-offs with age at first reproduction in capital breeding elephant seals. Popul Ecol 61(4):421–435

Orgeret F, Cox SL, Weimerskirch H, Guinet C (2019) Body condition influences ontogeny of foraging behavior in juvenile southern elephant seals. Ecol Evol 9(1):223–236

Orr RT, Poulter TC, Hanna GD (1965) The pinniped population of Año Nuevo Island, California. California Academy of Sciences

Ortiz CL, Costa D, Le Boeuf BJ (1978) Water and energy flux in elephant seal pups fasting under natural conditions. Physiol Zool 51(2):166–178

Paterson JT, Rotella JJ, Link WA, Garrott R (2018) Variation in the vital rates of an Antarctic marine predator: the role of individual heterogeneity. Ecology 99(10):2385–2396

Paterson JT, Rotella JJ, Mannas JM, Garrott RA (2016) Patterns of age-related change in reproductive effort differ in the pre-natal and post-natal periods in a long-lived mammal. J Anim Ecol 85(6):1540–1551

Peterson SH, Ackerman JT, Costa DP (2015) Marine foraging ecology influences mercury bioaccumulation in deep-diving northern elephant seals. Proceed Royal Soc b Biological Sci 282(1810):20150710

Petrinovich L (1974) Individual recognition of pup vocalization by northern elephant seal mothers. Z Tierpsychol 34:308–312

Pirotta E, Schwarz LK, Costa DP, Robinson PW, New L (2019) Modeling the functional link between movement, feeding activity, and condition in a marine predator. Behav Ecol 30(2):434–445

Pistorius P, Bester M, Kirkman S (1999) Survivorship of a declining population of southern elephant seals, *Mirounga leonina*, in relation to age, sex and cohort. Oecologia 121(2):201–211

Pistorius P, Bester M, Kirkman S, Taylor F (2001) Temporal changes in fecundity and age at sexual maturity of southern elephant seals at Marion Island. Polar Biol 24(5):343–348

Pistorius PA, Bester MN, Hofmeyr GJ, Kirkman SP, Taylor FE (2008) Seasonal survival and the relative cost of first reproduction in adult female southern elephant seals. J Mammal 89(3):567–574

Pistorius PA, Bester MN, Kirkman SP, Taylor FE (2001) Pup mortality in southern elephant seals at Marion Island. Polar Biol 24(11):828–831

Pistorius PA, De Bruyn P, Bester MN (2011) Population dynamics of southern elephant seals: a synthesis of three decades of demographic research at Marion Island. Afr J Mar Sci 33(3):523–534

Postma M, Bester MN, De Bruyn PN (2013) Spatial variation in female southern elephant seal mass change assessed by an accurate non-invasive photogrammetry method. Antarct Sci 25(6):731–740

Queiroz N, Humphries NE, Couto A, Vedor M, Da Costa I, Sequeira AM, Mucientes G, Santos AM, Abascal FJ, Abercrombie DL (2019) Global spatial risk assessment of sharks under the footprint of fisheries. Nature 572(7770):461–466

Rasa OA (1971) Social interaction and object manipulation in weaned pups of the northern elephant seal, *Mirounga angustirostris*. Z Tierpsychol 29:82–102

Reiter J, Le Boeuf BJ (1991) Life history consequences of variation in age at primiparity in northern elephant seals. Behav Ecol Sociobiol 28(3):153–160

Reiter J, Panken KJ, Le Boeuf BJ (1981) Female competition and reproductive success in northern elephant seals. Anim Behav 29(3):670–687

Reiter J, Stinson NL, Le Boeuf BJ (1978) Northern elephant seal development: the transition from weaning to nutritional independence. Behav Ecol Sociobiol 3(4):337–367

Rick TC, DeLong RL, Erlandson JM, Braje TJ, Jones TL, Arnold JE, Des Lauriers MR, Hildebrandt WR, Kennett DJ, Vellanoweth RL (2011) Where were the northern elephant seals? Holocene archaeology and biogeography of *Mirounga angustirostris*. The Holocene 21(7):1159–1166

Riedman M, Ortiz CL (1979) Changes in milk composition during lactation in the northern elephant seal. Physiol Zool 52(2):240–249

Riedman ML, Le Boeuf BJ (1982) Mother-pup separation and adoption in northern elephant seals. Behav Ecol Sociobiol 11(3):203–215

Robinson PW (2009) Exploiting pelagic habitat: navigation, migration, and foraging success in the northern elephant seal (*Mirounga angustirostris*). Dissertation, University of California Santa Cruz,

Robinson PW, Costa DP, Crocker DE, Gallo-Reynoso JP, Champagne CD, Fowler MA, Goetsch C, Goetz KT, Hassrick JL, Hückstädt LA (2012) Foraging behavior and success of a mesopelagic predator in the northeast Pacific Ocean: insights from a data-rich species, the northern elephant seal. PloS One 7(5):e36728

Robinson PW, Simmons SE, Crocker DE, Costa DP (2010) Measurements of foraging success in a highly pelagic marine predator, the northern elephant seal. J Anim Ecol 79(6):1146–1156. https://doi.org/10.1111/j.1365-2656.2010.01735.x

Roquet F, Williams G, Hindell MA, Harcourt R, McMahon C, Guinet C, Charrassin J-B, Reverdin G, Boehme L, Lovell P (2014) A Southern Indian Ocean database of hydrographic profiles obtained with instrumented elephant seals. Sci Data 1:140028

Saijo D, Mitani Y, Abe T, Sasaki H, Goetsch C, Costa DP, Miyashita K (2017) Linking mesopelagic prey abundance and distribution to the foraging behavior of a deep-diving predator, the northern elephant seal. Deep Sea Res Part II 140:163–170

Salogni E, Galimberti F, Sanvito S, Miller E (2018) Male and female pups of the highly sexually dimorphic northern elephant seal (*Mirounga angustirostris*) differ slightly in body size. Can J Zool 97(3):241–250

Sandegren FE (1976) Agonistic behavior in the male northern elephant seal. Behaviour 57(1–2):136–157

Sanvito S, Galimberti F, Miller EH (2007) Observational evidences of vocal learning in southern elephant seals: a longitudinal study. Ethology 113(2):137–146

Schwarz L, Hindell M, McMahon C, Costa D (2012) The implications of assuming independent tag loss in southern elephant seals. Ecosphere 3(9):1–11

Schwarz LK, Villegas-Amtmann S, Beltran RS, Costa DP, Goetsch C, Hückstädt L, Maresh JL, Peterson SH (2015) Comparisons and uncertainty in fat and adipose tissue estimation techniques: the northern elephant seal as a case study. PloS One 10(6):e0131877

Sequeira A, Heupel M, Lea MA, Eguíluz V, Duarte CM, Meekan M, Thums M, Calich H, Carmichael R, Costa D (2019) The importance of sample size in marine megafauna tagging studies. Ecol Appl 29(6):e01947

Setsaas T, Bester M, Van Niekerk J, Roux J, Hofmeyr G (2008) Dispersion during the moult haulout of southern elephant seals at the Courbet Peninsula. Iles Kerguelen. Polar Biology 31(2):249

Simmons SE, Crocker DE, Hassrick JL, Kuhn CE, Robinson PW, Tremblay Y, Costa DP (2010) Climate-scale hydrographic features related to foraging success in a capital breeder, the northern elephant seal *Mirounga angustirostris*. Endanger Species Res 10:233–243

Simmons SE, Crocker DE, Kudela RM, Costa DP (2007) Linking foraging behaviour of the northern elephant seal with oceanography and bathymetry at mesoscales. Mar Ecol Prog Ser 346:265–275

Slip D, Burton H, Gales N (1992) Determining blubber mass in the southern elephant seal, *Mirounga leonina*, by ultrasonic and isotopic techniques. Aust J Zool 40(2):143–152

Slip DJ, Gales N, Burton HR (1992) Body-Mass loss, utilization of blubber and fat, and energetic requirements of male southern elephant seals, *Mirounga leonina*, during the molting fast. Aust J Zool 40(3):235–243

Spraker TR, Lyons ET, Kuzmina TA, Tift MS, Raverty S, Jaggi N, Crocker DE (2014) Causes of death in preweaned northern elephant seal pups (*Mirounga angustirostris*, Gill, 1866), Año Nuevo State Reserve, California, 2012. J Vet Diagn Invest 26(2):320–326

Spurlin SM, Peterson SH, Crocker DE, Costa DP (2019) Nitrogen and carbon stable-isotope ratios change in adult northern elephant seals (*Mirounga angustirostris*) during the breeding and molting fasts. Mar Mamm Sci 35(2):707–717

Stewart BS (1997) Ontogeny of differential migration and sexual segregation in northern elephant seals. J Mammal 78(4):1101–1116

Tarnawski BA, Cassini GH, Flores DA (2013) Skull allometry and sexual dimorphism in the ontogeny of the southern elephant seal (*Mirounga leonina*). Can J Zool 92(1):19–31

Thornton SJ, Spielman DM, Pelc NJ, Block WF, Crocker DE, Costa DP, LeBoeuf BJ, Hochachka PW (2001) Effects of forced diving on the spleen and hepatic sinus in northern elephant seal pups. Proc Natl Acad Sci 98(16):9413–9418

Thorson PH, Le Boeuf B (1994) Developmental aspects of diving in northern elephant seal pups. In: Le Boeuf BJ, Laws RM (eds) Elephant seals: population ecology, behavior, and physiology. University of California Press, Berkeley, pp 271–289

Thums M, Bradshaw CJ, Sumner MD, Horsburgh JM, Hindell MA (2013) Depletion of deep marine food patches forces divers to give up early. J Anim Ecol 82(1):72–83

Townsend CH (1912) The northern elephant seal. vol 2. New York Zoological Society

Trivers RL (1972) Parental investment and sexual selection. In: Campbell B (ed) Sexual selection and the descent of man 1871–1971. Alden Press, Chicago, pp 136–207

Van den Hoff J (2001) Dispersal of southern elephant seals (*Mirounga leonina L.*) marked at Macquarie Island. Wildl Res 28(4):413–418

Van den Hoff J, Burton H, Davies R (2003) Diet of male southern elephant seals (Mirounga leonina L.) hauled out at Vincennes Bay, East Antarctica. Polar Biol 26(1):27–31
Van den Hoff J, Kilpatrick R, Welsford D (2017) Southern elephant seals (*Mirounga leonina* Linn.) depredate toothfish longlines in the midnight zone. PloS One 12(2):e0172396
Van den Hoff J, McMahon CR, Simpkins GR, Hindell MA, Alderman R, Burton HR (2014) Bottom-up regulation of a pole-ward migratory predator population. Proceed Royal Soc B Biol Sci 281(1782):20132842
Van den Hoff J, Morrice MG (2008) Sleeper shark (*Somniosus antarcticus*) and other bite wounds observed on southern elephant seals (*Mirounga leonina*) at Macquarie Island. Mar Mamm Sci 24(1):239–247
Vázquez-Medina JP, Crocker DE, Forman HJ, Ortiz RM (2010) Prolonged fasting does not increase oxidative damage or inflammation in postweaned northern elephant seal pups. J Exp Biol 213(14):2524–2530
Walters A, Lea MA, van den Hoff J, Field IC, Virtue P, Sokolov S, Pinkerton MH, Hindell MA (2014) Spatially explicit estimates of prey consumption reveal a new krill predator in the Southern Ocean. PloS One 9 (1):e86452. https://doi.org/10.1371/journal.pone.0086452
White FN, Odell DK (1971) Thermoregulatory behavior of the northern elephant seal, Mirounga angustirostris. J Mammal 52(4):758–774
Williams TM (2001) Intermittent swimming by mammals: a strategy for increasing energetic efficiency during diving. Am Zool 41(2):166–176
Williams TM, Blackwell SB, Richter B, Sinding M-HS, Heide-Jørgensen MP (2017) Paradoxical escape responses by narwhals (*Monodon monoceros*). Science 358(6368):1328–1331
Williams TM, Davis R, Fuiman L, Francis J, Le B, Horning M, Calambokidis J, Croll D (2000) Sink or swim: strategies for cost-efficient diving by marine mammals. Science 288(5463):133–136
Wojczulanis-Jakubas K, Plenzler J, Jakubas D (2019) Indications of contagious behaviours in the southern elephant seal: an observational study. Behaviour 156(1):59–77
Worthy G, Morris P, Costa D, Le Boeuf B (1992) Moult energetics of the northern elephant seal (*Mirounga angustirostris*). J Zool 227(2):257–265
Yoshino K, Takahashi A, Adachi T, Costa DP, Robinson PW, Peterson SH, Hückstädt LA, Holser RR, Naito Y (2020) Acceleration-triggered animal-borne videos show a dominance of fish in the diet of female northern elephant seals. J Exp Biol 223:jeb212936

Chapter 12
Dive Types Matter: They Reveal the Foraging Ecology of Elephant Seals

B. J. Le Boeuf and Y. Naito

Contents

Abstract The different two-dimensional shapes of dives of reveal much about the foraging ecology of elephant seals (*Mirounga* spp.). We give a brief history of dive typing, instrument development that improved and validated the process, and address what dive types reveal about foraging strategies, foraging success, and foraging location. Dive types were initially classified by visual inspection and later with Principal Component Analysis and Clustering, or machine-learning algorithms. Time-depth recorders, with the additional measure of swim speed, yield four major dive types: pelagic foraging, benthic foraging, transit, and drift dives. Geolocation, 3D measures, and other sensors confirm the principal dive types and add additional relevant information about diving and the environment. Dive types have revealed that females forage pelagically over deep water in the open ocean on prey in the deep scattering layer, whereas males forage benthically near continental slopes. Transit dives are used to get to and from foraging sites. Passive drift dives may serve the funtions of digestion, rest, and sleep. The direction of drifting reflects buoyancy, which changes as the seal feeds, thus giving a measure of foraging success. Sex-specific differences

B. J. Le Boeuf (✉)
Department of Ecology and Evolutionary Biology and Institute of Marine Sciences, University of California, Santa Cruz, CA 95064, USA
e-mail: leboeuf@ucsc.edu

Y. Naito
National Institute of Polar Research, 10-3 Midorikawa, Tachikawa 190-8518, Tokyo, Japan

D. P. Costa and E. McHuron (eds.), *Ethology and Behavioral Ecology of Phocids*, Ethology and Behavioral Ecology of Marine Mammals,
https://doi.org/10.1007/978-3-030-88923-4_12

in dive types indicative of foraging begin to appear at two years of age. Dive typing has also been used to describe the foraging ecology of other pinnipeds as well as cetaceans.

12.1 Introduction

Marine mammals must dive to eat. The form and shape of dives are a guide to how the mammals forage. Dive types are simply dive shapes with similar characteristics that one can discern by inspection or with a classification program or computer-generated algorithm. In their simplest form, dive depth is depicted as a function of time. During the last three decades, dive type analysis has advanced understanding of the diving and foraging behavior of both species of elephant seals (*Mirounga angustirostris and M. leonina*) and is increasingly used to study the behavior of other phocids and cetaceans.

Our aim in this chapter is to provide a brief historical perspective of the dive types observed in the diving records of elephant seals and their function, as revealed by innovations in diving instruments, sensors, and computer programs. Dive typing became possible with the development of time-depth recorders. We address the role of dive types in understanding the diving behavior, foraging, and the migratory pattern, foraging success, and foraging location of elephant seals. Specifically, we emphasize (1) knowledge gained about the diving behavior of elephant seals from the dive types evident in diving records; (2) the influence of improved and new diving instruments and sensors for elucidating dive type function; (3) sex differences in diving and foraging revealed by dive types; (4) development of sex differences, as revealed by the appearance and frequency distributions of dive types in juveniles; (5) foraging behavior inferred from changes in body composition and buoyancy reflected by drifting behavior; and (6) dive types in other phocids and cetaceans. As dive types suggest dive function, their study helps us understand what the seal is doing during a single dive, as well as yielding insights into the dynamic pattern of multiple dive shapes during foraging/migration trips at sea lasting for months, foraging success, and the location of foraging.

12.2 History of Dive Typing in Elephant Seals

12.2.1 Early Dive Records and First Descriptions of Dive Types

In 1982, we knew the fundamentals of breeding and terrestrial behavior of northern elephant seals on land (e.g., Bartholomew 1952; Le Boeuf and Peterson 1969; Le Boeuf 1972; Reiter et al. 1978, 1981; Riedman and Le Boeuf 1982) but their lives at sea, where they spent most of the year, was a mystery. It was like they entered a black hole and disappeared. We knew when they left, how long they stayed, and that they fed and gained mass, but we did not know how and where they did this. All of this changed with the development of time-depth recorders (TDRs) that could be attached to elephant seals and recovered when they returned to the rookery months later.

The first TDRs were deployed on a variety of animals, such as Weddell seals (*Leptonychotes weddellii*), five species of fur seals, Hawaiian monk seals (*Neomonachus schauinslandi*), and Galapagos sea lions (*Zalophus wollebaeki*; Kooyman 1965, 1966, 1972, 1975; Kooyman et al. 1976, 1983; Feldkamp et al. 1983; Gentry and Kooyman 1985; Kooyman and Trillmich 1985). The study subjects were determined by the need to recover the archival instruments to retrieve the diving data. Weddell seals had to return to the same breathing hole where they entered the water. Nursing otariids had to return from brief foraging trips to feed their pups. When the animals returned from the water, the recorders containing the data were recovered. In these early days of biologging, elephant seals offered advantages because they were accessible, they could be immobilized for instrument attachment, the instruments were easily attached to the seals, their movements from land to sea (and vice versa) were known and predictable, and they were at sea foraging for months at a time.

The first diving record of an adult female northern elephant seal was obtained in 1983 as she went to sea to feed following breeding (Le Boeuf et al. 1986). The TDR was designed by G. Kooyman, R. Gentry, and J. Billups and fabricated by Billups at Meer Instruments (Kooyman et al. 1976, 1983; Le Boeuf et al. 1986). The recorder's outside dimension was about 5 cm, and the film strip was 32 mm wide (Figs. 12.1, 12.2; Table 12.1; Kooyman 1981; Kooyman et al. 1983). As the film spooled out over about 14 days, a light-emitting diode (LED) exposed a timing dot on the film every 12 min. Another LED stylus recorded depth (pressure) continuously via a Bourdon tube as the seal dived.

The dive record provided our first glimpse of the seal's behavior in the black hole. The performance was remarkable: continuous deep, long-duration dives for 11 days, the duration of the record. Dive shapes were too crude to suggest dive function. In a follow-up study in 1985 and 1986 (Le Boeuf et al. 1988), seven adult female northern elephant seals were instrumented with the same analog TDRs shortly after weaning their pups and before departing on their foraging trips. The attachment, a marine epoxy cradle glued to the pelage on the dorsum above the shoulders to which the TDR was attached with hose clamps (Fedak et al. 1983), remained securely in place until the females returned to the rookery, the seals molted, after which the instrument and epoxy cradle fell off, leaving the fur and skin below unblemished. The TDRs produced photographic records of dive duration and dive depth during the first two weeks at sea. The film was developed, and the print enlarged seven times and digitized.

From the diving records, we discerned six dive types and named them A through F. Despite the lack of resolution owing to the greatly enlarged diving records (Figs. 12.2 and 12.3), it was apparent that 65 to 77% of the dives were type D. These were deep dives with steady descent to depth (the dive bottom) where the animal remained for

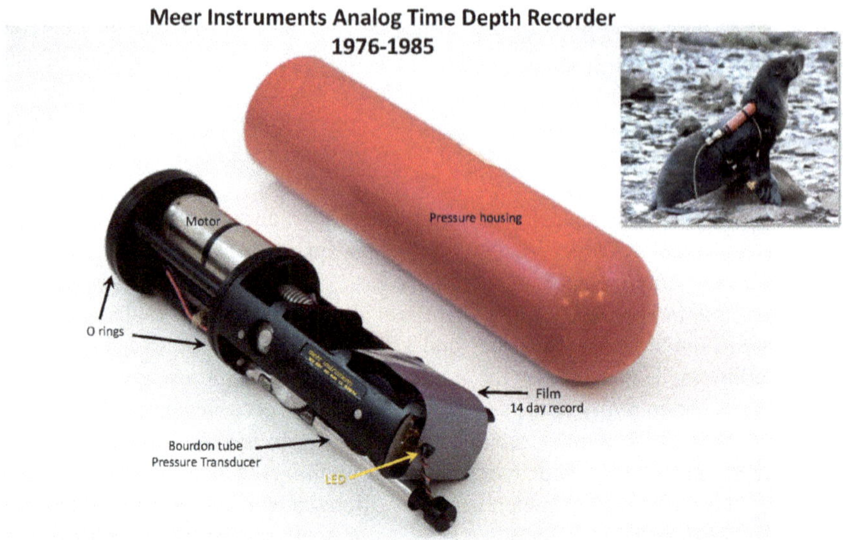

Fig. 12.1 The Meer Instrument analog TDR designed by G. Kooyman, R. Gentry, and J. Billups shown attached to an Antarctic fur seal (*Arctocephalus gazella*). Photograph by D. Costa

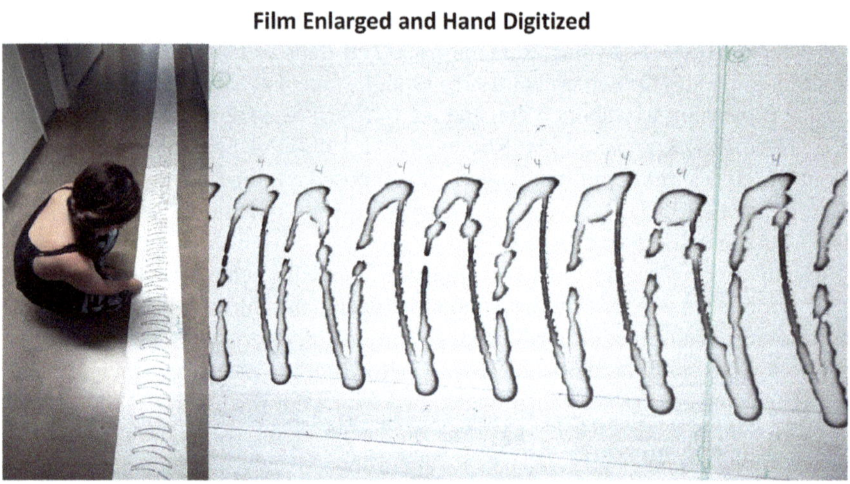

Fig. 12.2 The Meer Instrument TDR produced a photographic record of dive duration and dive depth in real time. The film was developed, the print enlarged by copy flow xerography, and hand digitized on a digitizing tablet and transferred to a 12 inch digital tape, at the Scripps Institution of Oceanography computer center. Photograph by D. Costa

Table 12.1 Examples of instruments that advanced understanding of elephant seal diving behavior using dive types

Instruments	Comments	Designer (D) Fabricator (F)	References
Kooyman / Gentry / Billups TDR	Analog, film, records 11–14 days	G. Kooyman (D) R. Gentry (D) J. Billups (D,F)	Kooyman et al. (1983)
Naito TDR	Analog, carbon-coated paper, records 135 days	Y. Naito (D) I. Suzuki (F)	Naito et al. (1989) Le Boeuf et al. (1989)
Swim speed-distance meter	Swim speed calculated from distance-time function	Y. Naito (D) I. Suzuki (F)	Le Boeuf et al. (1992) Asaga et al. (1994) Crocker et al. (1994) Hassrick et al. (2007)
Satellite relay data logger (SRDL)	Sensors, computers, transmit to Argos receiver when seal at surface	B. McConnell (D) M. Fedak (D) SMRU (F)	McConnell and Fedak (1996 Biuw et al. (2003, 2007) Bailleul et al. (2007a)
Swim-speed TDR with Tattletale Lite B-H datalogger and Logtron paddle wheel	Records depth and swim speed for 45–73 days	R. Andrews (D) C.A. Haverl (D) S.B. Blackwell (F)	Crocker et al. (1997) Webb et al. (1998) Hassrick et al. (2007)
Micro-processor TDRs and dive analysis software	A variety of TDRs with extensive memory and additional sensors	R. Hill (D) Wildlife Computers (F)	Numerous
Geolocation TDRs	Daylength and solar noon plus Argos satellite to determine location at sea	R. Hill (D) Wildlife Computers (F)	Numerous
3D video recorders	3D paths computed from depth, compass bearing and swim speed	R. Davis (D) W. Hagey (F)	Davis et al. (2001)
Stomach temperature recorder (HTR) and stomach temperature telemetry (STT)	Decrease in stomach temperature associated with dive types indicative of feeding	Numerous (D) Wildlife Computers (F)	Kuhn et al. (2009)
3D data loggers with swim speed, depth, temperature, tri-axial acceleration and magnetometry sensors	Elucidates body movements during drifting	Y. Naito (D) I. Suzuki (F)	Mitani et al. (2010)

(continued)

Table 12.1 (continued)

Instruments	Comments	Designer (D) Fabricator (F)	References
Jaw-motion recorder	Records jaw movements indication feeding	Y. Naito (D) I. Suzuki (F)	Naito et al. (2010)
Head-mounted camera	Confirms prey consumed by jaw motion events	Y. Naito (D) I. Suzuki (F)	Naito et al. (2010)

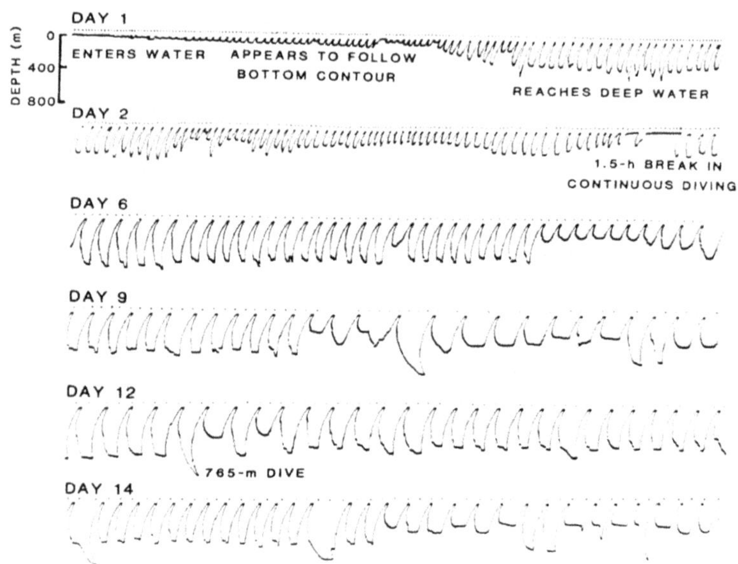

Fig. 12.3 Portions of the diving record of an adult female northern elephant seal during the post-breeding period at sea in 1985, showing the depth of dives as a function of time. The timing dots are 12 min apart. Traveling dives over the continental shelf (first half of Day 1) are excluded from analysis. Note the D dives (pelagic foraging dives) during the first half of Day 6. Adapted from Le Boeuf et al. (1988)

23 to 42% of the total dive duration before ascending directly to the surface. The bottom portion of these dives had frequent vertical changes in depth, i.e., a "jagged" or "wiggly" bottom. The dives occurred in long series at all times of the day or night. The dive depth and dive duration pattern suggested that the females were pursuing vertically migrating prey in the deep scattering layer, the mesopelagic zone between 200 and 1000 m that was nearer the surface at night (shorter and shallower dives) and deeper during the day (longer and deeper dives). We hypothesized that type D dives indicated pelagic foraging.

The function of the other dives was uncertain, but it was noted that A and B dives accounted for approximately 20% of the dives, C dives accounted for about 6% of the dives, and E and F dives were rare. It was suggested that these other dive types might represent feeding, transit, predator avoidance, rest, sleep, and conservation of energy. Here and for the remainder of this chapter, we do not address type F dives because they are restricted to traveling dives over the depth-limited continental shelf at the beginning and end of migrations (Fig. 12.3).

12.2.2 Swim Speed Adds a Dimension to Dive Typing

A significant advance in interpreting dive types was made from two instrument developments. The first was the Naito analog Pen TDR, which measured the depth and duration of dives for the entire post-breeding foraging trip of adult females that lasts approximately 70 days (Fig. 12.4; Table 12.1; Naito et al. 1989; Le Boeuf et al. 1989). The second development was linking the Pen TDR to a swim-speed-distance meter (SSDM; Le Boeuf et al. 1992; Asaga et al. 1994; Crocker et al. 1994), which measured distance traveled as a function of time. Swim speed was calculated from the distance-time function. The coupled instruments added an important second spatial dimension to depth, horizontal displacement, or distance traveled, from which descent and ascent angles could be estimated. Both instruments were attached to a lactating female at the Año Nuevo rookery in mid-February 1989, two days before she went to sea on a 2.5-month foraging trip.

From this deployment, five dive types were distinguished (Figs. 12.5 and 12.6). They were named by letter: A—direct descent to a sharp inflection point at the bottom of the dive, then direct ascent to the surface; B—direct descent to a rounded bottom of the dive followed by direct ascent to the surface. C—direct descent to a depth of about 200–400 m, at which point the descent rate decreased, suggested drifting to the bottom of the dive, followed by direct ascent to the surface; D—direct descent to a depth, at which point there occurred 2 to 12 up and down excursions, or "wiggles," followed by nearly vertical ascent to the surface; and E—direct, nearly

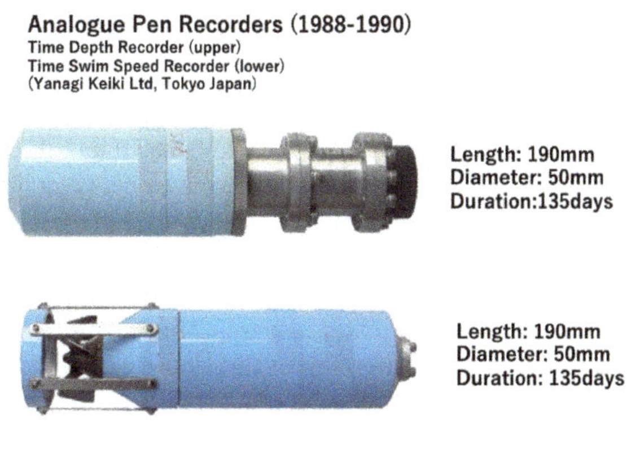

Fig. 12.4 The Naito analogue Pen TDR (top), which recorded depth and duration of dives for up to 135 days, was linked to the Naito swim-speed-distance recorder (SSDM) (bottom). Photograph by Y. Naito

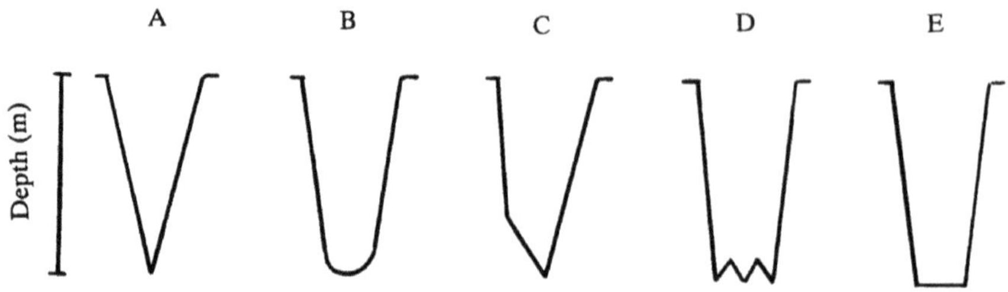

Fig. 12.5 Schematic representations of the five dive types in diving records. The putative functions of these dive types are: transit, travel or exploration for dives A and B; digesting food, rest, or sleep for dive C; pelagic foraging for dive D; and benthic foraging for Dive E. The dive types are not adjusted for depth or time. Adapted from Le Boeuf et al. (1992)

vertical descent to the bottom of the dive, which was flat, ending in a direct, nearly vertical ascent to the surface.

The diving record suggested that these different dive types served three main functions. The primary function of A and B dives, combined, was to serve as transit to foraging sites, and they were notable for the great horizontal distance covered, about 1.2 km per dive. They showed the shallowest angles of descent and ascent. During the first three weeks females were at sea, they occurred in long series and were interspersed with long bouts of D dives or putative pelagic foraging dives. It was argued that the long strings of AB dives, or "yo-yo diving," was an energy-efficient way of traveling (Weihs 1974). The strong symmetry in shape, great horizontal distance covered, and temporal patterning of these dives was consistent with this hypothesis (see Hindell 1990). It was hypothesized that the drift segment of C dives served as a processing function, such as digestion or clearing of metabolites or rest or sleep rather than external processes such as foraging. These dives showed the shortest horizontal distance covered. The marked reduction in swim speed in the second segment of descent indicated passive drifting rather than swimming. The drift phase occurred at the same depth throughout a series of C dives and accounted for almost half of the entire dive duration. Crocker et al. (1997) hypothesized that oxygen saved from reduced swim speed, or drifting, might be used in processing of food or metabolites incurred during diving. Oxygen savings could be used to perfuse organs that incurred restricted blood flow on other types of dives. Since feeding increases the metabolic rate by about 65% five hours after feeding (P. Thorson, pers. comm.; Alexis Barbour, unpublished data), the increase in energy expenditure due to digestion might affect the balance between the increased oxygen utilization required for active foraging and the hypoxia of diving (Thorson and Le Boeuf 1994). If the metabolic changes associated with food processing, however, were coupled temporally with decreased oxygen requirements due to decreased swimming effort, this impact could be lessened significantly. Thus, the overall decrease in swim speed and the temporal patterning of this dive type in records were consistent with these dives serving a physiological process such as digestion, sleep, or rest.

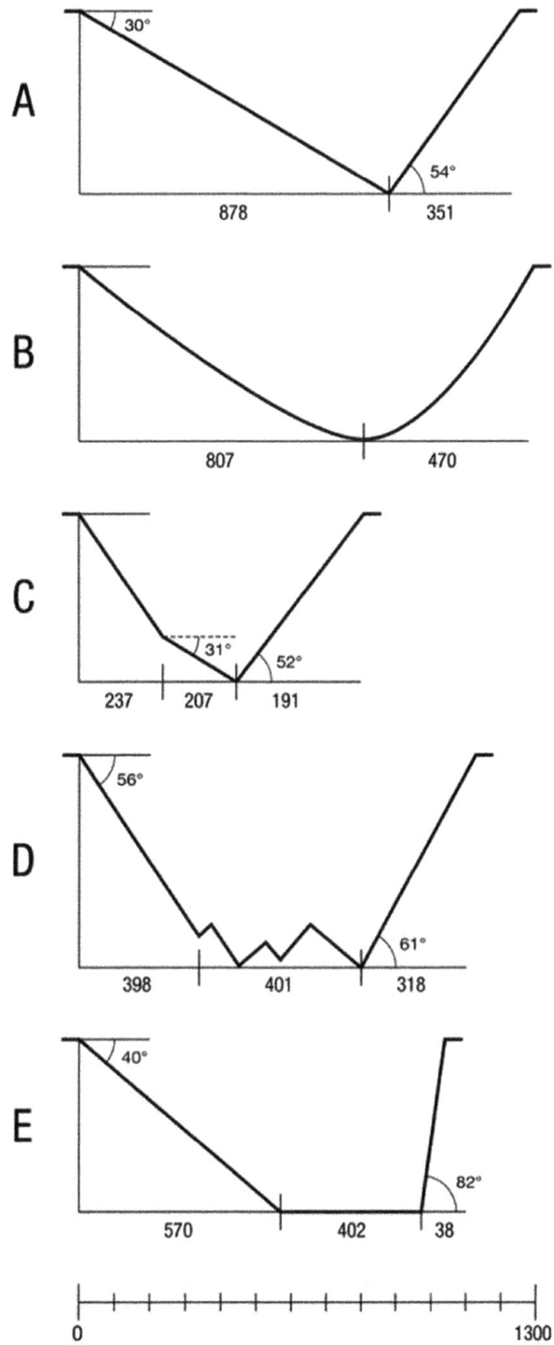

Fig. 12.6 Iconic representation of the dive types of an adult female northern elephant seal on a post-breeding foraging trip based on coupling a swim-speed recorder to a TDR. The mean angles of descent and ascent and the mean distance covered in meters are shown for entire dives and segments of dives. A and B dives serve a similar function and are often combined as AB dives. Adapted from Le Boeuf et al. (1992)

Type D and E dives had characteristics consistent with foraging. At the bottom of D dives, the vertical excursions, or wiggles, suggested searching, pursuing, and capturing prey in the water column. It was suggested that Type E dives, with their flat bottom shape, and uniform depth over time, represented benthic foraging. The shape of these dives indicated that the seal was either following the bottom topography of the ocean floor or sitting and waiting for the approach of prey. Moreover, these dives were associated with coastal regions and were rare in females because most females frequented the open ocean. At this time, the dive types described and their presumed function was based solely on the records of adult females.

12.2.3 Microprocessor TDRs Expand Diving Studies

In the early 1990s, microprocessors replaced analog TDRs as the diving instruments of choice. This was a significant advance because the instruments were small, there was ample and ever-increasing memory storage, sampling rates could be varied, additional sensors could be added, geolocation could be determined from light levels (Hill 1994), and 3D visualization of seal movements at sea could be obtained. The shape of dive types became more detailed, which facilitated typing the dive shapes by categories. Roger Hill started Wildlife Computers in 1984. The company provided numerous wildlife telemetry solutions to researchers over the years that continues today (see https://wildlifecomputers.com/#). An explosion of studies of the diving behavior of marine mammals followed. Indeed, the literature on elephant seal diving alone is so exhaustive that we are forced to be selective; we cannot review all relevant papers.

The first microprocessor TDRs ran for 60 days on average and up to a maximum of 95 days (Fig. 12.7). Additional channels on the diving instruments recorded ambient water temperature at l minute intervals, or less, and light levels to determine the seal's location (DeLong et al. 1992). These instruments not only sampled more variables, such as temperature and light, but they added precision to the interpretation of dive types, especially with sampling rates of dive depth at 30-s intervals or less. This enabled a more precise picture of the shape of a dive as sampling for depth increased from every 12 min to every 30 s to 5 s to 1 s to 0.1 s!

In the early 1990s, several research teams used various microprocessor-controlled recording units with add-on sensors to study elephant seal diving behavior from different breeding locations. In northern elephant seals, research was conducted at San Miguel Island in southern California (DeLong et al. 1992; Stewart and DeLong 1994) and at Año Nuevo in central California (Le Boeuf et al. 1986, 1988, 1989, 1992, 1993). Research on southern elephant seals on circumpolar Antarctic sites was conducted at Macquarie Island (Hindell 1991; Hindell et al. 1991; Slip et al. 1994), South Georgia (McConnell et al. 1992; McConnell and Fedak 1996), Marion Island (Jonker and Bester 1994, 1998; McIntyre et al. 2010, 2011, 2017), the Kerguelen Islands (Bailleul et al. 2007a, b; Dragon et al. 2012; Le Bras et al. 2016; Godard et al. 2020), and Península Valdés, Patagonia, Argentina (Campagna et al. 1995;

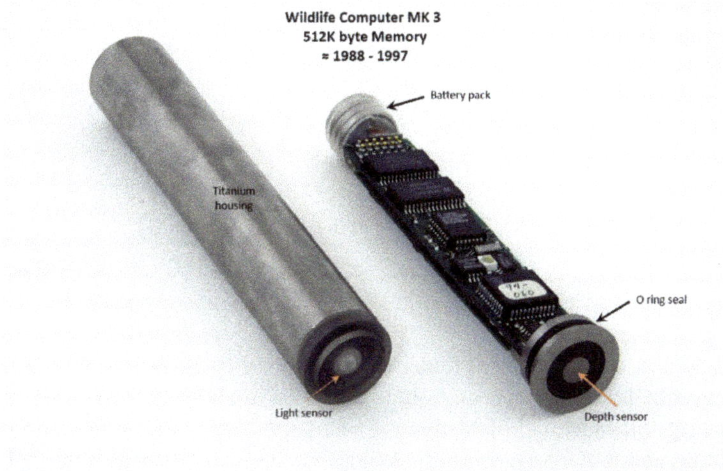

Fig. 12.7 Wildlife computer TDR Mk 3, which recorded depth and duration of dives for about 60 days, as well as light levels and time of day to provide location of the seal. Photograph by D. Costa

McGovern et al. 2019; Sala et al. 2011). In most of these studies, the classification of dive types was a key tool in studying various aspects of elephant seal foraging ecology. Dive-typing in other pinnipeds began at this time (Thompson et al. 1991; Bengston and Stewart 1992; Schreer and Testa 1996).

Several research groups studying elephant seals used their own distinctive nomenclature to classify dive types. For example, the same dive type was named a D dive by Le Boeuf et al. (1988), Type 1 by Hindell et al. (1991), Type 7 by Jonker and Bester (1994), Type W by McIntyre et al. (2011), and W-shape by Bailleul et al. (2007b). Multiple names for the same dives are unwieldy and confusing. To avoid overburdening the reader, we use the putative function of each dive type for the remainder of this chapter (pelagic foraging, benthic foraging, transit, drift/processing dives). We are aware that the precise function of some dives may be debated. Nevertheless, in the example above, the term pelagic foraging dives applies to D dives, Type 1 dives, Type 7 dives, Type W, and W-shape dives. Benthic foraging dives are dives with a flat-bottom performed over a shelf or close to shore. The dive depth of numerous dives is similar and may indicate the seabed depth. The primary function of transit dives is to travel to foraging areas, but these dives may also represent exploration and opportunistic feeding, roughly in that order of frequency of occurrence. A more appropriate term for them might be TEF dives, for transit, exploration, and feeding. That is, the seal, usually females, is in transit to a foraging site, and the dives are of a depth to cover a long distance efficiently. Still, the seals may explore and dive deeper from time to time to search for prey, and if prey are encountered, to consume them. Then, the female shifts to pelagic foraging dives. Drift or processing dives are characterized by three segments. In the first segment, the seal descends by stroking and gliding to a depth and stops swimming. The second segment, the drift segment, begins as the seal drifts passively up or down depending on its buoyancy. In the third segment, the seal strokes to the surface. These are the fundamental dive types; small variations on some of them occur in certain contexts.

12.2.4 Dive Typing in Southern Elephant Seals

12.2.4.1 Macquarie Island

Mark Hindell and colleagues (Hindell 1990; Hindell et al. 1991, 1992; Slip et al. 1994) described the dive types of adult female and adult male southern elephant seals that inhabit Macquarie Island in the Southern Ocean between the southern tip of Australia and Antarctica. Initially, Wildlife Computers devices with 512 Kbyte capacity were used to record depth every 30 s, and water temperature every 300 s. Changes in light levels provided an estimate of the seal's location.

Four principal dive types were distinguished, and they are analogous to those described for northern elephant seals. The expression of these dive types varied with the location of foraging, from near Macquarie Island, to the open waters to the south, and further south on or near the continental slope of Antarctica. The majority of dives (about 80%) were classified as pelagic and benthic foraging. The high frequency of pelagic and benthic foraging dives supports the interpretation that both species of elephant seals are foraging most of the time they are at sea, consistent with the need to build up fat reserves after fasting on land. Females never displayed benthic foraging dives, whereas males displayed both pelagic and benthic foraging dives, depending on the location of foraging. This contrasts with male northern elephant seals that rarely exhibit pelagic foraging dives and forage benthically along the continental slope (Le Boeuf et al. 2000; Kienle 2019). Processing dives, which Hindell and collaborators called resting dives, were observed about once a day, especially early in the morning, and accounted for 4–5% of total dives; this too was similar to the frequency of timing of these dives observed in northern elephant seals. For the seals at Macquarie Island, traveling or TEF dives accounted for only 1.3% of all dives and were restricted to near Macquarie Island. In contrast, transit dives were much more frequent in both male and female northern elephant seals, perhaps because the latter had to travel a greater distance to feed.

12.2.4.2 Marion Island

The findings of Hindell and colleagues and those in northern elephant seals (Le Boeuf et al 1988, 1992) were similar to those found for females at Marion Island (Jonker and Bester 1994, 1998) where microprocessor-controlled geolocation time-depth recorders (GLTDRs) were attached to post-breeding and post-molting females. Dive types were classified using the Wildlife Computer Dive Analysis.

The majority of dives were pelagic foraging (66.5%) and transit dives (26%). Drift dives accounted for only 6.6% of total dives. No benthic dives were observed in the records of females from Marion Island; the females foraged pelagically in the deep, open ocean with depth and duration of dives following a diel pattern, deeper and longer dives at mid-day, suggesting feeding pelagically on prey in the deep scattering layer.

McIntyre et al. (2011) classified the standard dive types exhibited by females and used them to examine water column use and to infer areas of increased foraging effort. The most common dives were transit (65%) and pelagic foraging dives (30.6%). The females did not forage near the continental shelf, and no benthic dives were recorded. There were few drift dives, which suggests that females from Marion Island did not forage as successfully as females from other colonies, i.e., needed less time to digest food than females from other colonies.

12.2.4.3 South Georgia

The picture from South Georgia was somewhat different. Researchers at this rookery addressed similar questions but with a different method and emphasis: the distribution of the seals in their oceanographic environment, their prey, and how, where and when they forage? Adult males and females were tracked with Satellite-Relay Data Loggers (SRDLs) and an Argos RF unit at South Georgia that provided 3D movements as the seals swam from South Georgia to distant areas of the Antarctic continental shelf (McConnell et al. 1992; McConnell and Fedak 1996; Bennett et al. 2001). The data collected consisted of behavioral and location information; dive types were addressed indirectly because the satellite tags lacked resolution for dive-typing. Pelagic or benthic dives were assumed depending on where the seals fed, i.e., in deep, oceanic waters or near the continental slope, respectively.

McConnell and Fedak (1996) showed that some females traveled eastward to the deep open ocean and foraged pelagically; others traveled southwest towards the Antarctic Peninsula and foraged benthically. The males dived and foraged close to South Georgia, spending prolonged periods diving benthically on the continental shelf or shelf edge around the island. All females traveled extensively. Evidently, the seals used different strategies to locate suitable foraging areas; similar strategies were observed in females from Marion Island (McIntyre et al. 2017).

Several studies exploited and extended the SLDR system and used dive types to study the relationship of diving seals to the environment (e.g., Bennett et al. 2001; Biuw et al. 2003, 2007; Bailleul et al. 2007a, b; McIntyre et al. 2010). For example, Biuw et al. (2010) examined temporal changes in the drift rates of processing dives of newly weaned elephant seals at South Georgia throughout their foraging trip to sea. They showed that this simple method could be used to estimate changes in lipid content of free-ranging seals at sea and in so doing provide a measure of body condition, foraging strategy and foraging success (see Chap. 6).

These studies revealed that post-breeding females travel long distances to forage benthically along the Antarctic continental shelf. They travel fast in transit to these areas and appear to forage pelagically and opportunistically along the way. It is notable that benthic diving was pronounced in females near the Antarctic continental shelf, but they also foraged pelagically in the open ocean; the foraging strategy depended on location and in some cases, sea-ice condition. This contrasts with southern elephant seals from Macquarie and Marion Island and northern elephant seal females where most females forage pelagically except for a few that forage near the coast (Le Boeuf et al. 2000). Kienle (2019) found that 6% of the females tracked spent some time foraging near the coast but they exhibited pelagic dives, not benthic dives. Perhaps females can and will forage benthically, depending on location, but these dives are rare. Males forage benthically almost exclusively, but in the southern hemisphere, they will forage pelagically in certain locations, as the studies of Hindell and his colleagues demonstrate.

12.2.4.4 Kerguelen Islands

At the Kerguelen Islands, seals traveled long distances to forage along the Antarctic continental shelf, with some pelagic foraging occurring opportunistically along the way (Bailleul et al. 2007a, b). Juvenile males and females foraged benthically while on the continental shelf; females switched to pelagic foraging further north in the marginal sea-ice zone, whereas males continued to forage benthically on the shelf. The dive shapes were consistent with traveling to foraging sites, benthic dives on the continental shelf, and pelagic foraging dives for females in the open sea away from the shelf.

12.2.4.5 Península Valdés, Argentina

Campagna et al. (1995) showed that females exhibited the same diving behavior as females from South Georgia and Macquarie Island. All females crossed the extensive 360 km continental shelf rapidly, in 2 ½ days, and did not forage until reaching deep water (the 200 m isobath). Pelagic foraging dives were most common, followed by transit dives. Prey were pursued in the deep scattering layer. Drift dives accounted for about 8% of the dives.

Sala et al. (2011) showed that juveniles, 2 to 3 years old, exhibited the same dive types displayed by adults. McGovern et al. (2019) classified the dives of free-ranging females based on the 3-dimensional movements and video-recorded observations. The three major dive types of females were identified as foraging, resting, and transit. Most prey captures occurred during foraging dives, and the primary prey were pencil smelt (*Nansenia* sp) and myctophids (*Lampadena* sp). Vertical head movements were associated with prey capture. There was great variation in pitch and roll during drift dives.

12.3 Sex Differences in Dive Types of Elephant Seals

12.3.1 Northern Elephant Seals

One expects sex differences in the diving behavior of elephant seals because males, being 3 to 10 times larger than females, must acquire more resources to attain great size as well as to maintain it (Le Boeuf et al. 1993). Moreover, males have less time in the year to forage than females, eight months of the year at sea for males compared to 10 months at sea for females. Does the diving pattern and dive types of the sexes differ, reflecting the need for males to acquire more resources or to acquire them more efficiently than females? Based on the size difference alone, males are expected to consume about three times more prey energy than females. They might do this by being more efficient at prey capture, consume larger prey, or pursue different prey with a higher energy density than the prey of females. The sexes might forage in different locations. For these reasons, it is expected that the diving pattern and dive types of the sexes would differ.

The foraging behavior of adult females received a lot of attention at Año Nuevo. Satellite tracks and dive records were obtained from over 400 females (e.g., Le Boeuf et al. 2000; Robinson et al. 2012). In contrast, few adult males were tracked because of the logistical challenges in handling the large males and recovering instruments.

Le Boeuf et al. (1993) attached Wildlife Computer TDRs, with geolocation capability from light level sensors, to free-ranging adult males, adult females, and juveniles. From strip-chart representations of the time-depth profile, two researchers classified dives "by inspection" independently into the four major dive types. There was a 91% agreement by classifiers on dive type categorization. This study was followed up with two subsequent studies: Le Boeuf et al. (2000) of 27 adult males and 20 adult females where dive types were identified by three individuals. Kienle (2019) tracked 128 adult females and 32 adult males, and dive types were classified using a hierarchical classification analysis based on diving variables, such as depth, shape, and duration.

Adult males migrated northward foraging along the continental margin at the distal point, from Oregon to the eastern Aleutian Islands. Their average travel speed was 90 km/day, faster than adult females (77 km/day) and juveniles (58 km/day). Upon reaching their destination, males exhibited repeated benthic diving in narrowly confined areas along the continental slope lasting from one to two months. These were called focal foraging areas (FFAs). In contrast, most adult females ranged widely in the open northeastern Pacific Ocean foraging pelagically as far as 56°N and 150°W. Females often remained in the same area for several days, suggesting concentrated foraging at numerous places along the migratory route. Similar sex-specific geographic separation in foraging areas was detected in more recent studies (Robinson et al. 2012; Kienle 2019).

There was a distinct sex difference in dive types (Table 12.2). Males preferred benthic foraging dives and exhibited few pelagic foraging dives. Females did the reverse, preferring pelagic foraging dives and exhibiting few benthic foraging dives,

Table 12.2 Dive types, their presumed function, and frequency in the diving records of male and female northern elephant seals. na = not available. Data from Le Boeuf et al. 2000

Dive type	Putative function	Location	Percentage of total dives	
			Adult male	Adult female
AB	Transit*	In transit On foraging area	77 20.8	48.3 17.1
C	Digest, sleep, rest	In transit On foraging area	na na	5 10.6
D	Pelagic foraging	In transit On foraging area	9.3 2.8	42.1 71.9
E	Benthic foraging	In transit On foraging area	9.4 73.2	2.7 0.1

*Transit, search, eat for females; transit only for males

most of the latter by a few females near the coastal zone. Both sexes exhibited an equal percentage of transit dives. There were no sex differences in the frequency of drift dives. Benthic diving in females needs more information, as it is not clear whether they are similar to the benthic dives on the seafloor of males or simply deep pelagic dives with a flat bottom.

To this point, sex-specific variation in foraging locations and dive types is consistent with the greater energy requirements of males compared to females. The shorter time foraging by males, compared to females, suggests that they feed on prey that are more energy-dense than the prey of females. In their transit to focal foraging areas, males prioritize direct, rapid, and concentrated transit dives until reaching a foraging area, which suggests that they do not feed *en route*. Once they reach their focal foraging area, they remain there for 1 to 2 months, gorge-feeding benthically around the clock. In contrast, females use TEF dives to transit to, search for, locate, and to consume prey opportunistically *en route*. When prey are located, females switch to pelagic foraging dives perhaps until sated or the prey diminish in number, and they move on to search for the next prey patch. Females do not appear to use transit dives just to travel; they are exploring and searching for prey along the way and consuming prey when located. Only then do they switch to pelagic foraging dives if the prey patch warrants the effort.

12.3.2 Southern Elephant Seals

The earliest studies of Mark Hindell on Macquarie Island (Hindell 1990; Hindell et al. 1991; Slip et al. 1994) showed that the sexes foraged in different locations (open waters for females and along continental slopes for males) and had preferred dive types (benthic for males and pelagic for females, although males foraged pelagically in certain locations), which suggested different foraging strategies and possibly different prey.

12.3.3 *Ontogeny of Diving Behavior*

Because adults of both sexes of elephant seals are prodigious deep and long-duration divers, the development of diving in young seals is of special interest. Study of the diving behavior in northern elephant seals was extended to juveniles during the first four foraging migrations (i.e., the first two years of life) by Le Boeuf et al. (1996). Thereafter, the land-sea schedule of males and females changes because some females give birth at age three, and in so doing, are defined as adults (Reiter and Le Boeuf 1991). Wildlife Computer Mk3 TDRs and geolocation GLTDRs were attached to 13 male and 11 female juveniles from the Año Nuevo and Piedras Blancas rookeries (Fig. 12.7). From strip charts of the time-depth profiles, dives were classified according to type.

Dive records of the youngest yearlings (first two trips to sea during the first year of life) revealed similar dive types. As the juveniles matured to 2-year-olds and became more experienced, females favored routes over the open ocean like adult females. In contrast, males favored routes closer to continental margins like adult males, confirming earlier observations (Le Boeuf et al. 1993; Le Boeuf 1994). That is, the sex differences in preferred dive types start early in life. Transit and foraging dives accounted for the majority of dives of both sexes. Females exhibited significantly more of the combined category of transit and pelagic foraging dives than males. Males displayed significantly more benthic foraging dives than females whereas females had only a few of these dives. Benthic foraging dives occurred along migratory paths proximal to the continental margin and not in locations more closely associated with the open ocean, where pelagic foraging dives predominated. Drift dives occurred with similar frequency in both sexes. A subsequent study by Zeno et al. (2007) detected alterations in the foraging behavior of female juveniles between the third and fourth foraging trips, supporting earlier findings that juveniles begin to emulate the diving pattern and location of diving of adults in their second year of life.

12.4 Validation of Dive Type Function

12.4.1 *Drift Dives*

Crocker et al. (1997) examined the drift dives of female northern elephant seals, concluding that they had an internal processing function. The sample consisted of 13 adult females recorded during the foraging trip after breeding and after molting during the last trimester of pregnancy. Mk 3 TDRs from Wildlife Computers were attached to all females, and in addition, one female carried a custom-made velocity–time-depth recorder (B-H Mk 1 Datagrabber) that recorded both depth and swim speed (Fig. 12.8). Mass and body composition were determined prior to attaching the instruments.

Fig. 12.8 B-H Mk 1 Datagrabber Swim speed TDR with Tattletale Fast Lite datalogger and Logtron paddle wheel designed and fabricated by S.B. Blackwell and C. Haverl. Photograph by S. B. Blackwell

Drift dives accounted for 6.3% of the dive types observed and occurred in bouts containing an average of 3.8 dives (range = 2–10); a bout of drift dives occurred at least once a day and peaked in the early morning just like those of southern elephant seals. The dives of post-breeding females lasted an average of 23 min, with over 50% of the dive spent in the drift segment (segment 2). Drift dives were the longest in females during the third trimester of pregnancy, lasting an average of 34 min. The depth at which the drift segment began was about 200 to 400 m and was highly consistent within a bout but varied considerably between bouts. The descent speed of the first segment of the dives was 1.45 m/s, similar to the descent swim speeds of TEF dives (1.58 m/s) and pelagic foraging dives (1.74 m/s); the drift segment of processing dives declined to 0.26 m/s, the stall speed of the recorder. Females did not swim during the drift segment but drifted passively. Body composition (buoyancy) determined the direction and rate of drifting and depth change.

Females that went to sea shortly after breeding drifted down while females in the third trimester of pregnancy drifted up, indicating that drift direction and speed were a function of buoyancy, i.e., the ratio of fat to lean body mass of the seal that was changing as the seal fed and increased its lipid composition. The association of drift dives with feeding was indicated by a positive correlation between their frequency and the frequency of pelagic foraging dives; in contrast, processing dive frequency was negatively correlated with TEF or transit dives. Processing dives were also closely associated with extended surface intervals (ESIs) that lasted 10 min or more. The ESIs preceded, followed, or occurred between processing dives, suggesting that if digestion could not be completed during drifting, which was presumably rare, it was completed at the surface. The duration of drift was similar to other dive types but covered only half the horizontal distance traveled.

Biuw et al. (2003) focused on the drift dives and changes in body composition of southern elephant seal pups from Macquarie Island during their first trip to sea. Initially, the seals drifted up because they were positively buoyant, but the descent rate decreased over the first 30–50 days after departure as they lost fat. Over the next 100 days, drift rates reversed and increased gradually as the seals fed and increased their lipid composition; during the last 20–45 days of the foraging trip, drift rates remained constant or decreased slightly. This suggested that the juveniles did not feed at first but fed during the second phase and fasted again as they returned to the rookery. Changes in the rate of drifting reflected changes in body condition and foraging success, as was found in northern elephant seals (Webb et al. 1998). This study was pivotal in showing that drift dives could be used to estimate the body composition of free-ranging elephant seals, as well as other marine mammals that exhibited drift dives.

Orgeret et al. (2018) also focused on drift dives and showed that juveniles from the Kerguelen Islands on their first foraging trip optimized swimming effort and dive duration during transit to benefit from buoyant forces. After eight months at sea, the dive depth and duration of juveniles was 50% and 70% that of adults, respectively. The diet of juveniles during their first year was predominantly krill (*Euphausia* sp). Mortality at sea was high during the first year, 45% of the sample, similar to the rate observed in northern elephant seals. Evidently, the swimming effort is adjusted early in life according to changes in buoyancy.

12.4.2 Sex Differences

Additional studies provided data consistent with the alleged functions of dive types and sex differences in dive types (Le Boeuf et al. 2000). Males displayed transit dives almost exclusively while moving directly to focal foraging areas (FFAs) along the continental shelf from central California to Oregon, Washington, Canada up to the western Aleutian Islands, where they remained foraging benthically for a month or more before starting the return trip to the rookery. Once on FFAs, transit dives changed to flat-bottomed benthic foraging dives (Table 12.2). Pelagic foraging dives were rare. Drift dives were not observed *en route;* on FFAs, drift could not be distinguished from benthic foraging dives. There was no diel pattern in the depth of male dives either in transit or on FFAs.

Females moved across a wider area of the northeastern Pacific than males, from 38° to 58° N and from near the coast at 125 ° W to as far west as 172.5 ° E. They appeared to alternate long strings of TEF dives with pelagic foraging dives that lasted about 3.5 days. That is, they appeared to forage at many sites *en route.* On FFAs, transit dives decreased by 65% and pelagic foraging dives increased by 71%. There were twice as many drift dives on FFAs as in transit, adding to the positive association of drift dives with foraging. Females exhibited a marked diel pattern in depth of dives during transit and on FFAs, deeper during the day than at night, suggesting that they were searching for prey while in transit and consuming prey when they encountered it. Females did not exhibit benthic foraging dives except when crossing the depth-limited continental shelf or when near the coast, usually during the early part of their foraging migration. The movements and diving patterns of females suggest that they feed predominantly on patchily distributed pelagic prey in the water column that migrates vertically by day with available light.

Sexual segregation in foraging behavior is characteristic of northern elephant seals. Males and females clearly pursue different foraging strategies as indicated by differences in where and how they forage and the dive types displayed.

12.4.3 Importance of Measuring Swim Speed

A study by Hassrick et al. (2007) augmented the earlier study that measured swim speed and the diving behavior of an adult female (Le Boeuf et al. 1992). This study addressed the function of dive types of adult males, adult females, and juveniles with respect to swimming speed, dive geometry, and swimming effort. Mk6 and Mk8 time-depth recorders (Wildlife Computers, Redmond, WA) and custom-built B-H Mk 1 Datagrabbers and swimming speed/time-depth recorders were deployed on ten juveniles of 1.4 and 1.8 years of age, and five adult female and five adult male elephant seals on foraging trips in 1995 and 1996 (Figs. 12.7 and 12.8). The adult seals also carried 0.5-W Argos-linked ST6 platform terminal transmitters (PTT) (Telonics, Mesa, AZ, USA). Body composition was determined using a combination of ultrasound and morphometric measurements (Gales and Burton 1987; Crocker 1995).

Sex differences were pronounced, especially with foraging dives (Table 12.2). Adult males exhibited transit dives until they reached their destination whereupon they switched to benthic foraging dives. Adult females switched from transit or TEF dives to pelagic foraging dives when they encountered a prey patch. Transit dives represented 35% of the dive record in juveniles, 36% of recorded dives in adult females, and 64% of recorded dives in adult males. Pelagic foraging dives represented 58% of the dive records of juveniles of both sexes, 57% for adult females, and 7% for adult males. There were wide oscillations in swimming speed at the bottom of these dives. Both sexes exhibited drift dives, but these dives were much more common in juveniles and females than in males (8, 6, and 1%, respectively). The low percentage of processing dives in males may have been due, in part, to the difficulty of differentiating them from benthic foraging dives.

Once on foraging grounds, 91% of male dives were the benthic foraging type. After reaching deep water, females and juveniles exhibited no benthic foraging dives. Two types of benthic foraging dives were distinguished in males: (1) flat-bottomed dives with a swim speed of 1.3 m/s (22% of the record) and (2) cessation of swimming (mean swim speed of 0.1 m/s) for the majority of the bottom time (6% of the record). It was suggested that the latter might serve the same function as drifting in drift dives.

No surface swimming was observed in any of the seals. Increased overall swimming speed was negatively correlated with dive duration. Overall swimming speed did not vary significantly between adult males, adult females, and juveniles. Buoyancy varied significantly between age and sex classes as they began a foraging trip: juveniles were most buoyant, followed by adult females and then adult males. Other studies showed that elephant seals employ a burst and glide form of locomotion influenced by buoyancy, which conserves oxygen reserves (Williams et al. 2000; but see Adachi et al. 2014).

Swimming speed on descent increased with depth. Ascent swim speed did not vary across classes. That is, all seals appeared to swim within a narrow range of speeds. For pelagic foraging dives of females and juveniles, swimming speed during the bottom phase was negatively correlated with bottom time. During transit, adult males traveled horizontally further than adult females. On some occasions, males cease swimming for the majority of the bottom segment of some benthic dives.

12.5 Dive Types, Foraging Tactics, and Foraging Success

12.5.1 3-D Movements, Body Position, and Activity

The early TDRs produced diving records that measured only one spatial dimension, depth. Coupling TDRs with measurement of swim speed (water speed) measured two spatial dimensions, depth and horizontal displacement, from which one could estimate distance traveled and descent and ascent angles. This considerably improved the interpretation of dives types. The third spatial dimension, dynamic orientation or pitch and yaw of the diver during a dive, remained to be measured to give a true multidimensional spatial analysis of behavior during a dive. Lacking the latter, the data supporting the putative function of dive types are important but circumstantial, not empirical. Nevertheless, in science, as opposed to a court of law, circumstances or context provides crucial information and is valid. In any case, two studies of northern elephant seals sought to address this third spatial dimension.

Davis et al. (2001) attached a video and data recorder to a 27-month-old juvenile northern elephant seal and recorded its behavior during transit dives on its return to the Año Nuevo rookery after being translocated beyond the continental shelf at the Año Nuevo rookery (Oliver et al. 1998). Three-dimensional dive paths were computed from depth, compass bearing, and swimming speed.

Transit dives were most common in the record because the seal was not foraging but simply returning "home." The seal returned to the rookery in a direct path, much like the direct paths taken by adult males in transit to their foraging destinations. Three modes of swimming were observed: continuous stroking, stroke-and-glide swimming, and prolonged gliding. The seal used continuous stroking from the surface to a mean depth of 20 m followed by stroke-and-glide swimming. Prolonged gliding started at a mean depth of 60 m and continued to the bottom of dives. For dives to depths of 300 m or more, 75% of the descent time was spent in prolonged gliding and 10% in stroke-and-glide swimming, amounting to about 6 to 10 min of passive descent per dive. Continuous stroking was restricted to the initial stage of descent and throughout most of the ascent to the surface. This study suggests that seals save substantial energy by gliding during transit dives, which has important implications for understanding diving energetics. Gliding reduces the cost of transport and enables the diver to increase dive duration.

Mitani et al. (2010) used the same homing paradigm to examine the drift dives of six juvenile seals translocated from the Año Nuevo rookery that were returning to the rookery. The aim was to elucidate the biomechanics of diving during the drift portion of processing dives. The seals were fitted with three-dimensional data loggers that recorded swim speed, depth, temperature, tri-axial acceleration, and magnetometry, as well as satellite transmitters. The emphasis was on examining the drift segment of dives that represented greater than 40% of the total dive duration, and a constant low vertical speed was less than 0.4 m/s.

Forty-five drift dives were identified. The seals descended almost vertically and stroked continuously at the beginning of dives. In most of them, the seal stopped stroking and began gliding before the second drift section began. This is similar to Davis et al. (2001), who found that seals stopped stroking long before reaching the bottom of transit dives. The drift segment started when the seals' pitch angle on descent changed from positive to negative, i.e., from nose down to nose up. At this time, and at a mean depth of about 220 m, the seals rolled over, changing their body orientation from the normal ventral-down position to a 'belly-up' position. This drifting phase lasted for half of the total dive duration. There were no flipper strokes during the belly-up phase, and the body wobbled slowly, pitching, and yawing. The reconstructed three-dimensional paths during the belly-up phase showed four different types of downward trajectories (the juveniles were negatively buoyant): circular, fluttering, a combination of circular and fluttering, and straight. Seventy-five percent of the trajectories were circular. When the belly-up phase ceased, seals started their ascent to the surface.

It was concluded that the drifting or the "falling leaf" phase was consistent with resting or sleep. It is not likely that drifting served food processing, as was argued for free-ranging females (Crocker et al. 1997), because these seals were merely returning to their resting place where they fast on land. This means that the drift dives of translocated seals may serve a different function than those of free-ranging, foraging seals. Diving to get "home" after translocation may require periodic rest or sleep, which may help to explain the high number of drift dives recorded.

It is not clear why the seals oriented belly-up and the function this served. It may be, as the authors suggest, that the blubber layer is thickest on the ventral surface, causing the body to roll belly-up during passive gliding. It was also suggested that the mode of descending in a circular fashion may reduce the speed of drifting down, effectively reducing the total depth of the dive, making it less energetically costly to power up during ascent.

This study sheds light on the biomechanics of drifting during this particular dive type as well as on diving energetics. Similar to the Davis et al. study, it does not, however, address diving and foraging behavior during migration.

Sala et al. (2011) deployed multi-sensor data loggers that registered bi-axial body position (pitch and roll) on 2–3-year old juveniles who restricted their diving to the Patagonian continental shelf. They recorded dives that were superficially similar to drift dives, although the descent phase was followed by a period on the seabed where the seals remained motionless, lying on either their right or left side, similar to some benthic dives. They were constrained by the shallowness of the seabed on the shelf.

12.5.2 Prey Density

Le Bras et al. (2016) used high-resolution accelerometers, magnetometers, pressure sensors, and GPS loggers to examine how southern elephant seals adjust their diving behavior and horizontal movements in response to prey encounter densities. These instruments were deployed on nine female southern elephant seal females from the Kerguelen Islands. The emphasis was on pelagic foraging dives of females. The detection of Prey Encounter Events (PEE) was followed Guinet et al. (2014) and Vacquié-Garcia et al. (2015). Prey Encounter Events were detected in 91% of non-drift dives; PEE occurred in 10.5% of descents, 78% in the bottom phase of dives, and 11.5% of ascents, i.e., most of the prey were encountered at the bottom, during the wiggle segment of pelagic foraging dives. The results suggest that the vertical aggregation of prey regulates prey density and influences foraging success of female elephant seals.

12.5.3 Body Condition, Foraging Success, and Cost of Transport

Several studies of both species followed up on the seminal studies of Crocker et al. (1997) and Biuw et al. (2003), showing that drift dives reflect changes in body condition, foraging success, and the location of foraging (Biuw et al. 2007; Thums et al. 2008, 2011; Robinson et al. 2010; Bradshaw and Hindell 2011; Dragon et al. 2012; Schick et al. 2013; Pirotta et al. 2018; Arce et al. 2019).

Arce et al. (2019) showed that quantifying buoyancy changes also provides a useful index for determining the location and amount of foraging resource that is acquired at sea. In the northeastern Pacific, reduced transit speed and increased drift dives are most prevalent in the North Pacific Transition Zone between the

Subarctic and Subtropical gyres. Clusters of intense foraging were characterized by slow transit or an elevated drift dives rate (Robinson et al. 2012). Southern elephant seals performed fewer drift dives than northern elephant seals and gained lipids at a lower rate (Schick et al. 2013). A southern elephant seal female that fed along the coast traveled a shorter distance and was estimated to feed more per day but accumulated smaller lipid reserves compared with females that fed further from the coast in the open ocean (Pirotta et al. 2018). Gordine et al. (2015), working with southern elephant seals from South Georgia, developed a model for detecting buoyancy changes in the records of any drift diving species.

Webb et al. (1998) showed that changes in body density led to changes in diving behavior. Aoki et al. (2011) showed that juvenile northern elephant seals were sensitive to their drift rate (buoyancy) and modulated stroking and gliding following density changes. The metabolic-cost models of Miller et al. (2012) show that one-way cost of transport is reduced when buoyancy forces assist movement, and cost of transport increases when buoyancy hinders movement. The models also show that the minimum round-trip cost of transport between the surface and a given depth is lowest at neutral buoyancy.

Since females exhibit few drift dives per day, Richard et al. (2014) reasoned that changes in active swimming effort and swim speed could be used as an alternative method of monitoring density variation throughout a foraging trip. They equipped six-post breeding females from Kerguelen Island with time-depth recorders and accelerometers. Both descent swimming speed and ascent swimming effort correlated positively with descent rate during drift dives, enabling the fine-scale monitoring of seal density change over the entire foraging trip. The implication is that this is a superior measure of foraging than tracking drift rate change. Moreover, a small change in seal density induces a 20% change in swimming effort, and this affects dive duration. Their data support the Miller et al. (2012) model stating that the cost of transport, or swimming effort, is minimal when seals are neutrally buoyant. Monitoring descent swimming speed and ascent swimming can be used on species with no drift dives.

Orgeret et al. (2018) showed that juveniles optimized swimming effort and dive duration during transit to benefit from buoyant forces. Similar to the findings of Richard et al. (2014), the drift rate was increasingly negative at first, but turned positive in the second phase. The seals showed gradual improvement in their foraging skills.

12.5.4 Feeding and Consumption of Prey

The putative function of dive types garnered from dive shapes, dive patterns, and dive location is useful information that reveals much about what the diver is doing. To fully understand foraging behavior, however, some argue that one must document prey consumption and measure it directly.

Stomach temperature provides a direct measure of the ingestion of prey (Gales and Renouf 1993; Bekkby and Bjørge 1998; Hedd et al. 1996). Since the seals feed on ectothermic prey, the prey is colder than the seals' core body temperature, resulting in a decrease in stomach temperature with consumption. Kuhn et al. (2009) measured feeding in 13 free-ranging adult female northern elephant seals with stomach temperature recorders, coupled with TDRs and satellite transmitters, as they began their post-breeding foraging trip to sea (Fig. 12.9). The stomach temperature records showed that feeding occurred with all dive types but with variable frequency. As expected, most feeding was associated with pelagic foraging dives (74.2%), followed by TEF dives (16.4%). Drift dives, benthic foraging dives, and shallow dives over the continental were associated with minimal estimates of feeding (2–4%). As for feeding during TEF dives, one expects the females to be opportunistic during transit and feed whenever they can, in single events as well as in long feeding bouts, depending on the circumstances. These results align with the idea that females' transit dives serve not only traveling but also exploration and searching for prey as they move through the environment. Of course, they consume prey if they find it.

The latency for the stomach telemetering device to register the arrival of a bolus of food is important. The size of the bolus may be especially important because females feed on small prey (Naito et al. 2013). The number of small myctophids that must be consumed to register as feeding is not clear. Biuw et al. (2003) list other reservations with this technique, such as short retention times of stomach tags and that estimates of meal size are unreliable.

Several studies, using different methods, have confirmed the consumption of prey. Jaw-motion recorders and head-mounted cameras were used to study the pursuit,

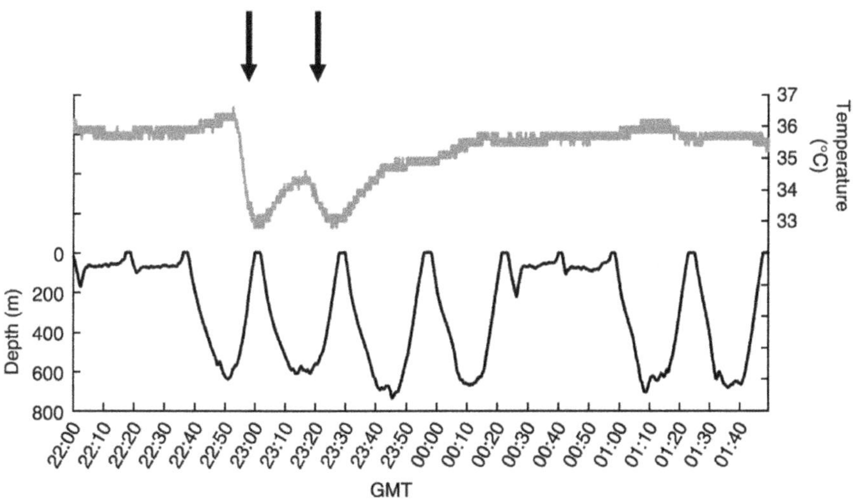

Fig. 12.9 A plot of stomach temperature (top, grey line) and dive depth (bottom, black line) against time for a free-ranging female northern elephant seal. Stomach temperature record shows two single feeding events (arrows) denoted by sharp declines in temperature. From Kuhn et al. (2009)

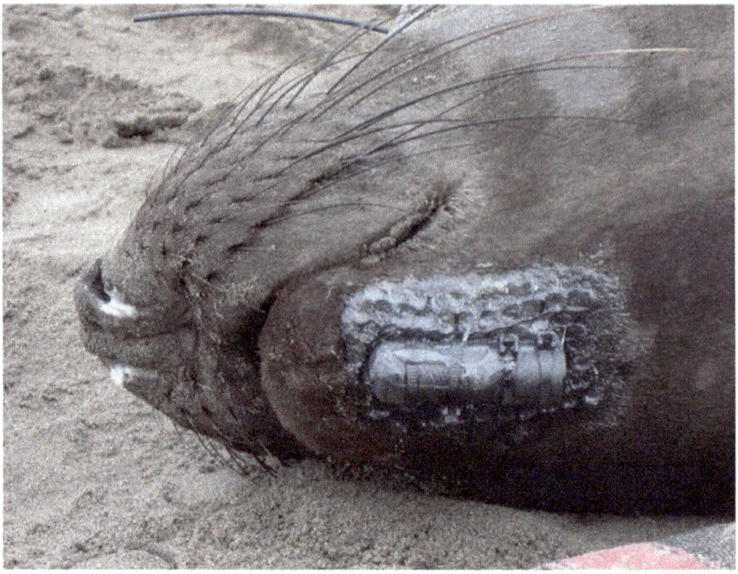

Fig. 12.10 The jaw-motion recorder attached to the lower jaw of an adult female northern elephant seal; a camera was mounted on her head to record prey capture. Photograph by D. Costa

capture, and identification of prey for northern elephant seals (Fig. 12.10; Naito et al. 2010, 2013). Attachment of the recorders to adult females on their post-breeding migration, along with satellite tags and TDRS, indicated that jaw motion events (JMEs) were recorded soon after departure, occurred every day, and lasted until the end of migration (Fig. 12.11). Dives were classified as JME dives and non-JME dives. Jaw motion events were associated primarily with the bottom segment of pelagic foraging dives, their jagged or wiggly part, as well as with the bottom parts of deep TEF or exploratory dives. The latter were interspersed between pelagic foraging dives. The head-mounted camera confirmed that JMEs were associated with consuming small prey by suction, such as myctophids or deep-sea smelts weighing 10 to 20 g and <12 cm in length. The frequency of JMEs was greatest in the gyre-gyre zone of the northeastern Pacific Ocean, where most females forage (Robinson et al. 2010, 2012). This study confirmed that the activity at the bottom of these dives is associated with the female opening and closing their jaws, which essentially confirms feeding.

Gallon et al. (2013) used acceleration loggers on adult females during a translocation study at Kerguelen Island. Rapid head movements were associated with prey encounter events during foraging dives. Le Bras et al. (2016), using high-resolution accelerometers, magnetometers, pressure sensors and GPS loggers, on females from Kerguelen Island detected prey encounter events (PEE) in 91% of non-drift dives. PEE were most frequent at the bottom of pelagic foraging dives (78%), following by 10.5% on descent, and 11.5% on ascent. McGovern et al. (2019), using a video-data

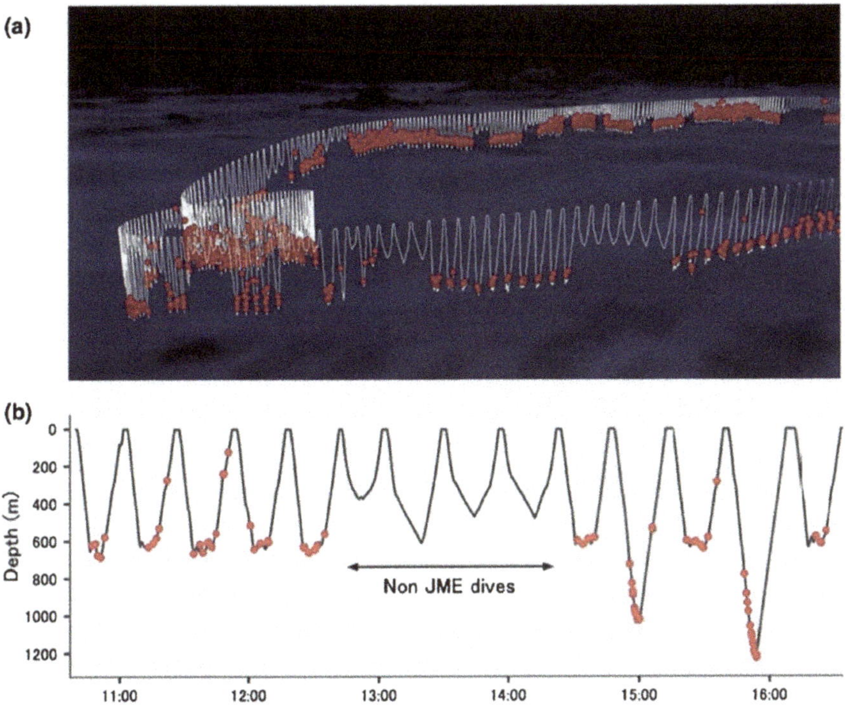

Fig. 12.11 The relationship between jaw-movement events (JMEs) and dive depth and dive type from the diving record of an adult female northern elephant seal: **a** The three-dimensional distribution of JMEs (red dots) for the dives (white lines) along the migratory path of the seal in the northeastern Pacific; **b** An excerpt from the diving record showing JME distribution. Dives 1 through 5 and 10, 12, and 14 are pelagic foraging dives; dives 7, 8 and 9 are drift dives (**c**); dives 11 and 13 are transit dives; dive 6 is not classified. From Naito et al. 2013

recorder, satellite-linked telemeters, and VHF radio transmitters, on females from Peninsula Valdes, Argentina, reported that 98% of prey captures occurred during pelagic foraging dives.

These different methods all report that females feed during pelagic foraging dives, especially at the dive bottom where the vertical up and down movements (the "wiggles") are observed.

12.5.5 Prey and Diet

Since one purpose of dive typing is to determine when the seals are feeding, it is necessary to address their prey. Early studies of the diet of northern elephant seals emphasized the study of dead animals' stomach contents (e.g., Antonellis and Fiscus 1980; Condit and Le Boeuf 1984; Antonelis et al. 1987, 1994; DeLong and Stewart

1991). This method reflected only prey consumed during the last few days of transit. Squid beaks and fish otoliths were prevalent, in part, because of slow passage through the seal's guts, which introduced bias. Little attempt was made to identify the prey of each sex.

Seals with spines of stingrays embedded in their mouths were noted. Some males were observed swallowing dogfish sharks, *Squalus acanthis*, whole (Albro 1980; see Fig. 1 in Condit and Le Boeuf 1984). Hacker (1986) reported that males fed on diverse prey such as skates, rats, ratfish, small sharks, and hagfish.

These studies showed that elephant seals fed on a variety of epi- and mesopelagic, bioluminescent cephalopods, teleosts such as Pacific hake (*Merluccius productus*), crustaceans, elasmobranchs, cyclostomes, and tunicates and the most frequent prey was cephalopods. The general conclusion was that squid was the most important prey (Antonelis et al. 1987, 1994).

Studies of southern elephant seals from the sub-Antarctic suggested a generalist diet (Hindell et al. 1991; Daneri et al. 2000; Piatkowski et al. 2002; Bradshaw et al. 2003; Field et al. 2005, 2007; Lewis et al. 2006; Campagna et al. 2007; Ducatez et al. 2008; Newland et al. 2009; Biuw et al. 2010; Eder et al. 2010). The diet was comprised primarily of fish and cephalopods (Green and Burton 1993; Daneri and Carlini 2002; Lewis et al. 2006; Newland et al. 2009; Eder et al. 2010). Females fed in deep waters on diel migrating mesopelagic prey, whereas males fed on demersal and benthic prey (Boyd and Arnbom 1991; Hindell et al. 1991; Campagna et al. 1998; Field et al. 2001; Eder et al. 2010; Hückstädt et al. 2012; McIntyre et al. 2011). During their first year at sea, the diet of juveniles was identified as predominantly krill (*Euphausia* sp.; Walters et al. 2014; Orgeret et al. 2018).

Using novel methods, recent studies shed new light on the prey of female elephant seals of both species. Naito et al. (2013) used a head-mounted camera and jaw-motion recorder attached to northern elephant seal females to determine that small mesopelagic prey (10–20 g), which included lanternfish (Myctophidae), were consumed during diel foraging dives. In a subsequent study (Naito et al. 2017), they identified other prey as subadult or adult ragfish (*Icosteus aenigmaticus*).

Goetsch et al. (2018) used quantitative fatty acid signature analysis and a library of prey fatty acid profiles from the mesopelagic eastern North Pacific, where female northern elephant seals forage, to estimate their diet composition. They concluded that mesopelagic fishes were the most important part of the diet of female northern elephant seals. The myctophids have high lipid content making them one of the most energetically-rich prey resources. They have an average energy content (gram/wet weight) up to an order of magnitude higher than that of squids. Viperfish and barracudina were also consumed, but they provide half as much energy as myctophids. Most of these species are bioluminescent and are visible to foraging elephant seals. The fishes are ingested whole by suction. Cues to the location of prey may be provided by visual sensing of bioluminescence or sensing by whiskers to detect the vortices created by swimming prey (Chap. 2).

Yoshino et al. (2020), using video cameras mounted on the head of female northern elephant seals, found that fish dominated the diet (78%) across all foraging locations, diving depths, and water temperatures, while squid comprised only 7% of the diet.

Fish included myctophids, *Merluccius* sp., and *Icosteus aenignaticus*, and squid such as *Histioteuthis* sp., *Octopoteuthis* sp., and *Taningia danae*. These results align with the fatty acid analysis study of Goetsch et al. (2018) but are in contrast with stomach lavage analysis that concluded that cephalopods were the most important component of the diet (Antonelis et al. 1987, 1994).

From a head-mounted video camera, McGovern et al. (2019) identified the prey of southern elephant seal females from Patagonia, Argentina, as lanternfishes (*Lampadena* sp.) from the family Myctophidae and pencil smelts (*Nansenia* sp.) from the Family Microstomatidae. Other prey identified was barbeled dragonfishes from the Family Stomiidae. The number of prey encountered was 3.4 per foraging dive.

The diet of male elephant seals of both species, who mostly feed benthically, is largely unknown and more study is needed. We know from observations in nature (Albro 1980; Condit and Le Boeuf 1984) and captivity (BL, personal observation) that male northern elephant seals consume dogfish sharks, *Squalus acanthis*. Hacker (1986) suggested that males prey on skates, rays, ratfish, small sharks, and hagfish, but the this was based on a small number of samples collected from stomach lavage of dead seals.

12.6 Dive Types in Other Pinnipeds and Whales

Do the dive types of elephant seals generalize to other pinnipeds and whales? A few generalizations apply. Each family, or species, has its own foraging strategy and dive types vary because they reflect different foraging strategies. For example, some species of toothed whales dive deeply and echolocate on descent to locate prey and then change to bursts as they feed at the bottom of dives. Balaenid mysticetes, the bowhead (*Balaena mysticetus*) and right whales (*Eubalaena* spp.), ram filter zooplankton continuously throughout the water column. Blue (*Balaenoptera musculus*) and fin whales (*B. physalus*) lunge feed, while gray whales (*Eschrichtius robustus*) feed on benthic invertebrates. Weddell seals are restricted to their breathing holes; harbor seals (*Phoca vitulina*) and gray seals (*Halichoerus grypus*) dive and forage in depth-limited areas near the coast (Chaps. 6, , 9, , 10). Comparisons of dive types across families and species are difficult, in part, because dive shape depends on the resolution, which varies with the depth sampling rates of the different methods of study, e.g., archival tags vs. satellite tags.

Dive types in other phocids are not easily compared with the dive shapes observed in elephant seals. For example, Davis et al. (2003) classified Weddell seal dives based on 3D movements and video-recorded observations. They described four dive types; dive type 1 were foraging dives with identified prey, whereas potential functions of the other types included hole-guarding and exploratory behavior. None of the dives bear a close resemblance to those observed in elephant seals; the foraging dives of Weddells most closely resemble the transit dives of elephant seals. Others have made similar categories of dive types for Weddell seals (Kooyman 1965, 1968; Shreer and Testa 1995; Plötz et al. 2001). For comparative purposes, the problem is that the seals are restricted to diving from an isolated breathing hole in the ice. The dives performed by free-ranging seals can be quite different with access to multiple holes and other seals.

Lesage et al. (1999) distinguished five dive types in harbor seals from the St. Lawrence estuary; four were U-shaped, and one was V-shaped. Feeding occurred during all dive types. Seals dove to the sea bottom (about 20 m) when performing U-shaped dives during the day; dives were shallower (8 m) at twilight and during the night. These dives are most like benthic foraging dives in elephant seals. Baechler et al. (2002) showed that U-shaped dives were associated with foraging and V-shaped dives were not. Eguchi and Harvey (2006) found that 80% of square dives (analogous to U-shaped dives) in Monterey Bay, California, were associated with foraging. Only 11% of the total dives were V-shaped.

Thompson et al. (1991) discerned three dive types in gray seals on the Farne Islands in England: V-shaped traveling dives, U-shaped, square-wave dives with slow swimming over short trips that served foraging, and shallow dives close to haul-out sites. The foraging dives resemble the benthic dives of elephant seals but there is activity at the bottom of the gray seal dives.

Little attention has been devoted to dive types in studies of otariids. For example, in a study of northern fur seals by Joy et al. (2015), archival dead-reckoner tags, Argos satellite platform transmitter terminals (Spot 5, Wildlife Computers), and a VHF transmitter tag were attached to female northern fur seals. Dive types are only mentioned indirectly; shallow, repeated dives with short surface time intervals were observed as well as deeper, repeated U-shaped dives.

The diving behavior of whales has been recorded with archival tags attached to the whales with suction cups and satellite tags attached to the whales with barbed hooks. The former records for a short time but gives a high-resolution picture of dive shape; the latter records for months but lacks the precision necessary for evaluating dive shapes for comparison with seal dives. Tyack et al. (2006) attached sound and orientation recording tags by suction cups to Cuvier's beaked whales, *Ziphius caviostris,* in the Ligurian Sea, Italy, and Blainville's beaked whales, *Mesoplodon densirostris* Blainville, in the Canary Islands. These two species dive deeper and longer than most whales and seals. The dive pattern consists of a deep dive to about 1800 m followed by several shallow dives to about 200 m, which last longer than the deep dive. There is a diel element to this pattern. Feeding occurs during deep dives, indicated by click vocalizations (searching for prey) on descent and buzzes (consuming prey) at the dive bottom (Madsen et al. 2005). Feeding dives resemble the pelagic foraging dives of female elephant seals because of the activity (buzzes) at the bottom of dives. The relatively shallow dives between deep dives resemble transit dives of elephant seals, but their function for the whales is unclear. No drift dives are reported. Using the same attachment method, similar findings were reported for both species of whales recorded in Hawaii (Baird et al. 2006; see also Aguilar de Soto et al. 2020).

Schorr et al. (2014) attached satellite tags to eight Cuvier's beaked whales in southern California and collected dive records for three months. The dive pattern was as described by Tyack et al. (2006) over a short duration. Dives were categorized as square, V-shaped, or U-shaped, but the sampling rate of satellite tags was too slow to provide a detailed picture of dive shapes. This was also the case for Shearer et al. (2019), who used satellite tags on Cuvier's beaked whales, with depth recorded only every 2.5 min. They noted that there were three times more shallow dives than deep dives and both types of dives took up an equal amount of time, 43% of the total dives recorded. The whales spent little time at the surface; the median duration was 2.2 min, but there were occasional extended surface intervals lasting over 30 min and one lasting 5.2 h. A figure depicting depth as a function of time shows a dive pattern of deep dives, most of them with flat bottoms, followed by 3 to 5 shallow dives. A study in the Bahamas (Joyce et al. 2017) showed that the deep dives of Cuvier's beaked whales were associated with foraging near the seafloor, similar to male elephant seals. Foraging occurred during flat-bottomed deep dives of low resolution (see Madsen et al. 2005). The whales were located in the Southern California Anti-submarine Warfare Range, where Naval sonar signals are produced that may have influenced the whales' diving behavior (Schorr et al. 2014).

Archival Dtag acoustic recording tags were attached with suction cups to sperm whales, *Physeter macrocephalus*, in the Atlantic Ocean, the Gulf of Mexico, and the Ligurian Sea (Watwood et al. 2006). On average, foraging behavior involved 45 min dives to depths of 400–1200 m with 9 min on the surface between dives. The whales echolocate on descent, searching for prey, and buzzes are recorded as they feed at the bottom of deep dives. Although the dive bottom may remain at a certain depth, such as 625 m, there are up and down movements, or wiggles, throughout the dive bottom as buzzes are heard and the whale feeds. The bottom portion of the dive takes up 62% of the total dive duration. In overall shape, the deep dives of sperm whales resemble the benthic foraging dives of male elephant seals; however, the up and down movements at the bottom of the dives resemble the wiggles observed in the pelagic foraging dives of female elephant seals. The whales make repeated deep foraging dives followed by shallow diving or occasional extended surface intervals. The dive shapes are of high resolution, but the tags were carried for only 3–6.5 h.

Narwhals, *Monodon monoceros*, dive to average depths between 200 and 800 m depth depending on location (Laidre et al. 2003; Ngô et al. 2019; Heide-Jørgensen et al. 2020). They echolocate throughout deep dives and emit buzzes indicative of feeding at the bottom of dives.

The continual lunge feeding of blue and fin whales occurs at about 140 m. Foraging dives resemble pelagic foraging dives of female elephant seals, with vertical excursions and feeding occurring at the bottom of dives. The foraging dives of North Atlantic right whales (*Eubalaena glacialis*) resembled the flat-bottomed benthic dives of male elephant seals, with little vertical activity at the bottom of dives (Baumgartner and Mate 2003). Travel and search dives resembled the transit dives of elephant seals.

While both archival and satellite tags provide useful information on the diving and foraging behavior of whales, the archival data provide sufficient resolution to allow comparisons with seal dives. Much of the foraging in the phocids and whales mentioned occurs at the bottom of dives. Drift dives were not observed in any of the seal or whale diving records. Toothed whales echolocate to find and consume prey; elephant seals and other pinnipeds do not echolocate.

12.7 Conclusions

In 1986 or 1987, when the diving records from our second deployment of TDRs on female elephant seals were inspected, we thought that the dives with the wiggly bottom were probably foraging dives. Over a decade later, in the early 2000s, YN showed BL a figure from the dive record of a female with a jaw motion sensor attached to her lower jaw. There were red dots in the wiggle portion of the putative pelagic foraging dives representing jaw-motion closures. The seal was feeding at the bottom of these dives. The dive shapes were no longer putative pelagic foraging dives; they were bonified pelagic foraging dives. This has since been confirmed by others. BL framed and hung the figure on the wall of his study. We had come full circle from speculation to confirmation.

We have learned much about the foraging ecology of elephant seals from the study of dive types, aided and advanced by the astounding development of biologging science. From its beginnings in the 1960s, the technological developments in biologging have been rapid and extraordinary in both hardware and software. So much so that we succumb to the temptation to paraphrase the last sentence in The Origin of Species (Darwin 1859): ".. from so simple a beginning endless... " technological developments "most beautiful and wonderful have been, and are being" developed.

Dive type classification started simply with trying to assess function from dive shapes presented in 2D, with depth as a function of time. With the addition of swim speed measurement and geolocation, we made great strides in understanding elephant seals' foraging ecology. From this alone, we discerned the four fundamental shapes of pelagic and benthic foraging dives, transit dives, and drift dives that we recognize today. We learned that the majority of foraging dives of females are pelagic and those of males are benthic. The sex differences are associated with location—the open ocean and continental slopes, respectively—and that these sex differences appear early in life. This relatively inexpensive method tells us much about the fundamental foraging ecology of these animals, i.e., when, where, and how each sex and the young forage and even their foraging success. Of course, improvements in TDRs, new sensors, 3D records, accelerometers, computer programs for data analysis and presentation of figures, and sophisticated algorithms for classifying and filtering dives, advanced the study of diving mammals to a new and much higher level.

We can now track the movement of seals in 3D, sense depth, speed, light level, conductivity, temperature, and sound (see McGovern et al. 2019). Video recordings show prey capture and the identity of prey. Location is provided from light levels or from satellite-linked telemetry when the animal surfaces. A saltwater switch determines when the animal is at the surface from which the beginning and end of dives is determined. All of this comes from relatively small devices attached to the free-ranging seals that provide continuous data recording for up to a year. Foraging dives are confirmed from video recordings of prey or the ingestion of prey or the crunching sound heard on the audio. Stroke frequency and gliding on descent and ascent reveal the energetic cost of transport. Algorithms predict prey encounters when the video is not used by using head movements and crunching sounds. The resulting representations give a precise and intimate picture of what the seal is doing and where it is foraging. Drift dives reflect changes in body condition, foraging success, and location of foraging during the course of the foraging trip. Swimming speed and diving effort measure the cost of transport. Notably, no differences are detected in the daily drift rates calculated from TDR-only data versus TDR-accelerometer data (Richard et al. 2014). Yesterday's inexpensive 2D technology works perfectly well on this particular task.

Besides advancing our understanding of the diving behavior of these animals, analysis of dive types, in 2D or 3D, provides vital information of the ecosystem the seals inhabit and exploit. The study of dive types has facilitated population studies and aids in conserving and managing the seals and the marine ecosystem.

Gaps remain in our knowledge of the diving behavior of many marine mammals. For elephant seals, we need more data on prey pursued, caught, and consumed. For example, when females are foraging pelagically and suddenly switch to very deep transit dives, are they pursuing a higher concentration of the same prey, or different prey that is more energy dense? Are they acting like the gold miners of the nineteenth century that left a productive claim to seek out an even more productive one? How precise is the link between what appears to be foraging activity in a dive and prey consumption as reflected by a decrease in stomach temperature? Is the latency in temperature decrease similar when ingesting small vs large prey? Are the infrequent benthic dives of females like males or are they flat-bottomed dives near the coast?

We know little about the foraging behavior of male elephant seals. What prey are they pursuing on benthic dives? Besides feeding, do males also rest or sleep during benthic dives? Do males digest food while inactive at the bottom of dives, as females appear to do in drift dives? What prey are targeted, and how are they pursued, caught, and eaten? Do males fast while transiting to focal foraging areas? A study using JMEs, head-mounted cameras, and swim speed recorders (Naito et al. 2013) or 3D studies, would reveal much about male foraging behavior.

The relationship between the diving behavior of translocated seals and free-ranging, migrating seals is not clear. Drift dives of migrating seals are associated with foraging, but this is unlikely in translocated seals that are not feeding. Do the drift dives in translocated seals indicate rest or sleep?

Many other marine mammals exhibit dive types similar to those described for elephant seals (e.g., Schreer and Testa 1996; Lesage et al. 1999; Page et al. 2003; Miller et al. 2008; Andersen et al. 2014; Watanabe and Takahasi 2013; Hindell et al. 2016). Comparison of dive types across marine mammal species is challenging but warrants attention to determine the commonalities or universals across species.

This review reveals that our understanding of the foraging ecology of elephant seals and marine mammals, in general, is aided by addressing dive types. Dive types matter, especially when coupled with increasingly sophisticated instruments and sensors, e.g., audio for echolocating whales. Indeed, the technological innovations we have witnessed in this discipline in the last few decades are analogous to great advances made in astronomy, most of which have come from new instruments such as novel scanners, laser technology and adaptive optics, and ever more powerful telescopes. We anticipate that advances in technology will continue to advance the study of marine mammals.

Acknowledgements We thank Susanna Blackwell, Claudio Campagna, Daniel Costa, Gerald Kooyman, and Elizabeth McHuron for reviewing drafts of the manuscript; Daniel Costa and Susanna Blackwell for photographs and figures; and Vicky Zavatieri and Larry Hogle for improving the figures.

References

Adachi T, Maresh JL, Robinson PW, Peterson SH, Costa DP, Naito Y, Watanabe Y, Takahashi V (2014) The foraging benefits of being fat in a highly migratory marine mammal. Proc Biol Sci 281

Aguilar de Soto N, Visser F, Tyack PL, Alcazar J, Ruxton G, Arranz P, Madsen PT, Johnson M (2020) Fear of killer whales drives extreme synchrony in deep diving beaked whales. Sci Rep 10:13

Albro RR (1980) Northern elephant seal vs. dogfish. Cetus J Whale Museum 21–2

Andersen JM, Stenson GG, Skern-Maurizen M, Wiersma YF, Rosing-Asvid A, Hammill MO, Boehme L (2014) Drift diving by hooded seals (Cystoophora cristata) in the North West Atlantic Ocean. PLoS ONE 9:3103072

Antonellis GA Jr., Fiscus CH (1980) The pinnipeds of the California Current. CalCOFI Rep XXI:68–78

Antonelis GA Jr, Lowry MS, DeMaster DP, Fiscus CH (1987) Assessing northern elephant seal feeding habits by stomach lavage. Mar Mamm Sc 3:308–322

Antonelis GA Jr, Fiscus CH, Stewart BS, Delong RL (1994) Diet of the northern elephant seal. In: Le Boeuf BJ, Laws RM (eds) Elephant seals: population ecology, behavior, and physiology. University of California Press, Berkeley, pp 211–223

Aoki K, Watanabe YY, Crocker DE, Robinson PW, Biuw M, Costa DP, Miyazaki N, Fedak MA, Miller PJ (2011) Northern elephant seals adjust gliding and stroking patterns with changes in buoyancy: validation of at-sea metrics of body density. J Exp Biol 214:2973–2987

Arce F, Bestley S, Hindell MA, McMahon CR, Wotherspoon S (2019) A quantitative, hioerarchical approach for detecting drift dives and tracking buoyancy changes in southern elephant seals. Sci Rep 9:8936

Asaga T, Naito Y, Le Boeuf BJ, Sakurai H (1994) Functional analysis of dive types of female northern elephant seals. In: Le Boeuf BJ, Laws RM (eds) Elephant seals: population ecology, behavior, and physiology. University of California Press, Berkeley, pp 310–327

Baechler J, Beck CA, Bowen WD (2002) Dive shapes reveal temporal changes in the foraging behavior of different age and sex classes of harbour seals (Phoca vitulina). Can J Zool 80:1569–1577

Bailleul F, Charrassin J-B, Monestiez P, Roquet F, Biuw M, Guinet C (2007) Successful foraging zones of southern elephant seals from the Kerguelen Islands in relation to oceanographic conditions. Philos Trans R Soc Lond Biol Sci 362(1487):2169–2181

Bailleul F, Charrassin J-B, Ezraty R, Girard-Ardhuin F, McMahon CR, Field IC, Guinet C (2007) Southern elephant seals from Kerguelen Islands confronted by Antarctic Sea ice. Changes in movement and in diving behavior. Deep-Sea Res 54:343–355

Baird RW, Webster DL, McSweeney DJ et al (2006) Diving behaviour of Cuvier's (Ziphius cavirostris) and Blainville's (Mesoplodon densirostris) beaked whales in Hawai'i. Can J Zool 84:1120–1128

Bartholomew GA (1952) Reproductive and social behavior of the northern elephant seal. Univ Calif Pub Zool 47:369–472
Baumgartner MF, Mate BR (2003) Summertime foraging ecology of North Atlantic right whales. Mar Ecol Prog Ser 264:123–135
Bekkby T, Bjørge A (1998) Variation in stomach temperature as indicator meal size in harbor seals, Phoca vitulina. Mar Mamm Sci 14:627–637
Bengston JL, Stewart BS (1992) Diving and haulout behavior of crabeater seals in the Weddell Sea, Antarctica, during March 1986. Polar Biol 12:635–644
Bennett KA, McConnell BJ, Fedak MA (2001) Diurnal and seasonal variations in the duration and depth of the longest dives in southern elephant seals (Mirounga leonina): possible physiological and behavioral constraints. J Exp Biol 204:649–642
Biuw M, McConnell B, Bradshaw CJA, Burton H, Fedak M (2003) Blubber and buoyancy: monitoring the body condition of free-ranging seals using simple dive characteristics. J Exp Biol 206:3405–3423
Biuw M, Boehme L, Guinet C et al (2007) Variations in behavior and condition of a Southern Ocean top predator in relation to in situ oceanographic conditions. Proceedings of the National Academy of Sciences of the United States of America 10413705–13710
Biuw M, Nøst OA, Stien A, Zhou Q, Lydersen C, Kovacs KM (2010) Effects of hydrographic variability on the spatial, seasonal and diel diving patterns of southern elephant seals in the eastern Weddell Sea. PLOS One 5:e13816
Boyd IL, Arnbom T (1991) Diving behaviour in relation to water temperature in the southern elephant seal: foraging implications. Polar Biol 11:259–266
Bradshaw CJ, Hindell MA (2011) In situ measures of foraging success and prey encounter reveal marine habitat-dependent search strategies. Ecology 92:1258–1270
Bradshaw C, Hindell M, Best N, Phillips K, Wilson G, Nichols P (2003) You are what you eat: describing the foraging ecology of southern elephant seals (Mirounga leonina) using blubber fatty acids. Proc R Soc (b Biol Sci) 270:1283–1292
Campagna C, Le Boeuf BJ, Blackwell SB, Crocker DE, Quintana F (1995) Diving behaviour and foraging location of female southern elephant seals from Patagonia. J Zool 236:55–71
Campagna C, Quintana F, Le Boeuf BJ, Blackwell SB, Crocker DE (1998) Diving behavior and foraging ecology of female southern elephant seals from Patagonia. Aquat Mamm 24:1–11
Campagna C, Piola AR, Marin MR, Lewis M, Zajaczkovski U, Fernández T (2007) Deep divers in shallow seas: southern elephant seals on the Patagonian shelf. Deep-Sea Res I Oceanogr Res Pap 54:1792–1814
Condit R, Le Boeuf BJ (1984) Feeding habits and feeding grounds of the northern elephant seal. J Mammal 65:281–290
Crocker DE (1995) Reproductive effort and fasting physiology of female northern elephant seals, Mirounga angustirostris. PhD dissertation, University of California
Crocker DE, Le Boeuf BJ, Naito Y, Asaga T, Costa DP (1994) Swim speed and dive function in a female northern elephant seal. In: Le Boeuf BJ, Laws RM (eds) Elephant seals: population ecology, behavior, and physiology. University of California Press, Berkeley, pp 328–342
Crocker DE, Le Boeuf BJ, Costa DP (1997) Drift diving in female northern elephant seals: implications for food processing. Can J Zool 75:27–39
Daneri G, Carlini A, Rodhouse P (2000) Cephalopod diet of the southern elephant seal, Mirounga leonina, at King George Island, South Shetland Islands. Antarct Sci 12:16–19
Daneri G, Carlini A (2002) Fish prey of southern elephant seals, Mirounga leonina, at King George Island. Polar Biol 25:739–743
Darwin C (1859) The origen of Species; by means of natural selection or the preservation of favored races in the struggle for life. John Murray, London
Davis RW, Fuiman LA, Williams TM, Le Boeuf BJ (2001) Three-dimensional movements and swimming activity of a northern elephant seal. Comp Biochem Physiol A Mol Integr Physiol 129:759–770

Davis RW, Fuiman LA, Williams TM, Horning M, an Hagey, W. (2003) Classification of Weddell seal dives based on 3 dimensional movements and video-recorded observations. Mar Ecol Prog Ser 264:109–122
DeLong RL, Stewart BS (1991) Diving patterns of northern elephant seal bulls. Mar Mamm Sci 7:369–384
DeLong RL, Stewart BS, Hill RD (1992) Documenting migrations of northern elephant seals using day length. Mar Mamm Sci 8:155–159
Dragon A-C, Bar-Hen A, Monestiez P, Guinet C (2012) Horizontal and vertical movements as predictors of foraging success in a marine predator. Mar Ecol Prog Ser 447:243–257
Ducatez S, Dalloyau S, Richard P, Guinet C, Cherel Y (2008) Stable isotopes document winter trophic ecology and maternal investment of adult female southern elephant seals (Mirounga leonina) breeding at the Kerguelen Islands. Mar Biol 155:41–420
Eder EB, Lewis MN, Campagna C, Koch PL (2010) Evidence of demersal foraging from stable isotope analysis of juvenile elephant seals from Patagonia. Mar Mamm Sci 26:430–442
Eguchi T, Harvey JT (2006) Diving behavior of the Pacific harbor seal (Phoca vitulina Richardii) in Monterey Bay, California. Mar Mamm Sci 21:283–295
Fedak MA, Anderson SS, Curry MG (1983) Attachment of a radio tag to the fur of seals. J Zool 46:298–300
Feldkamp SD, DeLong RL, Antonelis GA Jr (1983) Diving behavior of California sea lions (Zalophus californianus). In: Biennial conference on the biology of marine mammals No. 5. Nov. 27-Dec. 1, Boston, Massachusetts
Field I, Hindell MA, Slip DJ, Michael KJ (2001) Foraging strategies of southern elephant seals (Mirounga leonina) in relation to frontal zones and water masses. Antarct Sci 13:371–379
Field IC, Bradshaw CJA, Burton HR, Sumner MD, Hindell MA (2005) Resource partitioning through oceanic segregation of foraging juvenile southern elephant seals (Mirounga leonina). Oecologia 142:127–135
Field I, Bradshaw C, van den Hoff J, Burton H, Hindell M (2007) Age-related shifts in the diet composition of southern elephant seals expand overall foraging niche. Mar Biol 150:1441–1452
Gales NJ, Burton HR (1987) Ultrasonic measurement of blubber thickness of the southern elephant seal, Mirounga leonina (Linn.). Aust J Zool 35:207–217
Gales NJ, Renouf D (1993) Detecting and measuring food and water intake in captive seals using temperature telemetry. J Wild Man 87:514–519
Gallon S, Baillweul F, Charrassin J-Bk, Guinet C, Bost C-A, Handrick Y et al (2013) Identifying foraging events in deep diving southern elephant seals, Mirounga leonina, using acceleration data loggers. Deep Sea Res Part II Top Stud Oceanogr 88–89:14–22
Gentry RL, Kooyman GL 1985 (Eds) Fur seals: Maternal strategies on land and at sea. Princeton University Press, Princeton
Godard M, Manté C, Guinet C, Picard B, Nerini D (2020) Diving behavior of Mirounga leonina: a functional data analysis approach. Front Mar Sci 7:595
Goetsch C, Conners MG, Budge SM, Mitani Y, Walker WA, Bromaghin JF, Simmons SE, Reichmuth C, Costa DP (2018) Energy-rich mesopelagic fishes revealed as a critical prey resource for a deep-diving predator using quantitative fatty acid signature analysis. Front Mar Sci 5:1–19
Gordine SA, Fedak M, Boehme L (2015) Fishing for drifts: detecting buoyancy changes of a top marine predator using a step-wise filtering method. J Exp Biol 218:3816–3824
Green K, Burton H (1993) Comparison of the stomach contents of southern elephant seals, Mirounga leonina, at Macquarie and Heard Islands. Mar Mamm Sci 9:10–22
Guinet C, Vacquié-Garcia J, Picard B, Bessigneul G, Lebras Y, Dragon AC et al (2014) Southern elephant seal foraging success in relation to temperature and light conditions: insight into prey distribution. Mar Ecol Prog Ser 499:265–301
Hacker ES (1986) Stomach content analysis of short-finned pilot whales (Globicephala macrorhynchus) and northern elephant seals (Mirounga angustirostris) from the southern California Bight. Southwest Fisheries Center, La Jolla, Calif. Administrative Rep. NO. LJ-86–08C

Hassrick JL, Crocker DE, Zeno RL, Blackwell SB, Costa DP, Le Boeuf BJ (2007) Swimming speed and foraging strategies of northern elephant seals. Deep Sea Res Part II 54:369–383
Hedd A, Gales R, Renouf D (1996) Can stomach temperature telemetry be used to quantify prey consumption by seals? A re-examination. Polar Biol 16:261–370
Heide-Jørgensen M, Blackwell SB, Williams TM et al (2020) Some like it cold: temperature-dependent habitat selection by narwhals. Ecol Evol 10:8073–8090
Hill RD (1994) Theory of geolocation by light levels. In: Le Boeuf BJ, Laws RM (eds) Elephant seals: population ecology, behavior, and physiology. University of California Press, Berkeley, pp 227–236
Hindell MA (1990) Population dynamics and diving behaviour of a declining population of southern elephant seals. PhD thesis in the Department of Zoology, University of Queensland
Hindell MA, Slip DJ, Burton HR (1991) The diving behaviour of adult male and female southern elephant seals, Mirounga leonina (Pinnipedia: Phocidae). Aust J Zool 39:595–619
Hindell MA (1991) Some life-history parameters of a declining population of southern elephant seals, Mirounga leonine. J Anim Ecol 60:119–134
Hindell MA, Slip DJ, Burton HR, Bryden MM (1992) Physiological implications of continuous, prolonged, and deep dives of the southern elephant seal (Mirounga leonina). Can J Zool 70(2):370–379
Hindell MA, McMahon CR, Bester MN et al (2016) Circumpolar habitat use in the southern elephant seal: implications for foraging success and population trajectories. Ecosphere 7:1–27
Hückstädt LA, Koch PL, McDonald BI, Goebel ME, Crocker DE, Costa DP (2012) Stable isotope analyses reveal individual variability in the trophic ecology of a top marine predator, the southern elephant seal. Oecologia 169:395–406
Jonker FC, Bester MA (1994) The diving behaviour of adult southern elephant seals, Mirounga leonina, cows from Marion Island. S Afr J Antarct Res 24:75–93
Jonker FC, Bester MA (1998) Seasonal movements and foraging areas of adult southern elephant seals, Mirounga leonina, from Marion Island. Antarct Sci 10:21–30
Joy R, Dowd MG, Battaile BC, Lestenkof PM, Sterling JT, Trites AW, Routledge RD (2015) Linking northern fur seal dive behavior to environmental variables in the eastern Bering Sea. Ecosphere 6(5):75
Joyce TW, Durban JW, Claridge DE, Dunn CA, Fernbach H, Parsons KM, Andrews RD, Balance LT (2017) Physiological, morphological, and ecological tradeoffs nfluence vertical habitat use of dee-diving toothed-whales in the Bahamas. PLoS One 12:e018113
Kienle SS (2019) Intraspecific variation and behavioral flexibility in the foraging strategies of seals. PhD thesis, University of California Santa Cruz
Kooyman GL (1965) Techniques used in measuring diving capacities of Weddell seals. Polar Rec 12:391–394
Kooyman GL (1966) Maximum diving capacities of the Weddell seal, Leptonychotes weddelli. Science 151:1553–1554
Kooyman GL (1972) Deep diving behaviour and effects of pressure in reptiles, birds, and mammals. Symp SOCexp Biol 26: 295–31. Cambridge University Press, Cambridge
Kooyman GL (1975) A comparison between day and night diving in the Weddell seal. J Mammal 56:563–574
Kooyman GL (1981) Weddell seal: consummate diver. Cambridge University Press, Campbridge
Kooyman GL, Trillmich F (1985) Diving behavior of Galapagos sea lions. In: Gentry RL, Kooyman GL (eds) Fur seals: maternal behavior on land and at sea. Princeton University Press, Princeton, pp 209–219
Kooyman GL, Billups JO, Farwell WD (1983) Two recently developed recorders for monitoring diving activity of marine birds and mammals. In: Macdonald AG, Priede IG (eds) Experimental biology at sea. Academic Press, London, pp 197–214
Kooyman GL, Gentry RL, Urquhart DL (1976) Northern fur seal diving behavior: a new approach to its study. Science 193:411–412

Kuhn CE, Crocker DE, Tremblay Y, Costa DP (2009) Time to eat: measurements of feeding behaviour in a large marine predator, the northern elephant seal Mirounga angustirostris. J Anim Ecol 78:513–523

Laidre KL, Heide-Jørgensen M, Dietz R, Hobbs RC, Jørgensen OA (2003) Deep-diving by narwhals Monodon monoceros: differences in foraging behavior between wintering areas? Mar Ecol Prog Ser 261:269–281

Le Boeuf BJ (1972) Sexual behavior in the northern elephant seal, Mirounga angustirostris. Behaviour 41:1–26

Le Boeuf BJ (1994) Variation in the diving pattern of northern elephant seals with age, mass, sex, and reproductive condition. In: Le Boeuf BJ, Laws R (eds) Elephant seals: population ecology, behavior and physiology. University of California Press, Berkeley, pp 237–252

Le Boeuf BJ, Peterson RS (1969) Social status and mating activity in elephant seals. Science 163:91–93

Le Boeuf BJ, Costa DP, Huntley AC, Kooyman GL, Davis RW (1986) Pattern and depth of dives in northern elephant seals, Mirounga angustirostris. J Zoo Lond 208:1–7

Le Boeuf BJ, Costa DP, Huntley AC, Feldkamp SD (1988) Continuous, deep diving in female northern elephant seals, Mirounga angustirostris. Can J Zool 66:446–458

Le Boeuf BJ, Naito Y, Huntley AC, Asaga T (1989) Prolonged, continuous, deep diving by northern elephant seals. Can J Zool 67:2514–2519

Le Boeuf BJ, Naito Y, Asaga T, Crocker DE, Costa DP (1992) Swim speed in a female northern elephant seal: metabolic and foraging implications. Can J Zool 70:786–795

Le Boeuf BJ, Crocker DE, Blackwell SB, Morris PA, Thorson PH (1993) Sex differences in diving and foraging behavior of northern elephant seals. Symp Zool Soc Lond 66:149–178

Le Boeuf BJ, Morris PA, Blackwell SB, Crocker DE, Costa DP (1996) Diving behavior of juvenile northern elephant seals. Can J Zool 74:1632–1644

Le Boeuf BJ, Crocker DE, Costa DP, Blackwell SB, Webb PM, Houser DS (2000) Foraging ecology of northern elephant seals. Ecol Monogr 70:353–382

Le Bras Y, Jouma'a J, Picard B, Guinet C (2016) How elephant seals (Mirounga leonina) adjust their fine scale horizontal movement and diving behaviour in relation to prey encounter rate. Plos One 11(12):e0167226

Lesage V, Hammill MO, Kovacs KM (1999) Functional classification of harbor seal (Phoca vitulina) dives using depth profiles, swimming velocity, and an index of foraging success. Can J Zool 77:74–87

Lewis R, O'Connell TC, Lewis M, Campagna C, Hoelzel AR (2006) Sex-specific foraging strategies and resource partitioning in the southern elephant seal (Mirounga leonina). Proc R Soc B 273:2901–2907

Madsen PT, Johnson M, de Soto NA, Zimmer WMX, Tyack P (2005) Biosonar performance of foraging beaked whales (Mesoplodon densirostris). J Exp Biol 208:181–194

McIntyre PJ, de Bruyn N, Ansorge IJ, Bester MN, Bornemann H, Plöotz J, Tosh CA (2010) A lifetime at depth: vertical distribution of southern elephant seals in the water column. Polar Biol 33:1037–1048

McIntyre T, Bornemann H, Plötz J, Tosh CA, Bester MN (2011) Water column use and forage strategies of female southern elephant seals from Marion Island. Mar Biol 158:2125–2139

McIntyre T, Bester MN, Bornemann H, Tosh CA, Nico de Bruyn PJ (2017) Slow to change? Individual fidelity to three-dimensional foraging habitats in southern elephant seals, Mirounga leonina. Anim Behav 127:91–99

McConnell BJ, Fedak MA (1996) Movements of southern elephant seals. Can J Zool 74:1485–1496

McConnell BJ, Chambers C, Fedak MA (1992) Foraging ecology of southern elephant seals in relation to the bathymetry and productivity of the Southern Ocean. Antarct Sci 4(4):393–398

McGovern KA, Rodriguez DH, Lewis MN, Davis RW (2019) Classification and behavior of free-ranging female southern elephant seal dives based on three-dimensional movements and video0recorded observations. Mar Ecol Prog Ser 620:215–232

Miller PJO, Aoki K, Rendell LE, Armano M (2008) Stereotypical resting behaviour of the sperm whale. Curr Biol 18:21–23
Miller PJO, Biuw M, Watanabe YY, Thompson D, Fedak MA (2012) Sink fast and swim harder! Round-trip cost-of-transport for buoyant divers. J Exp Biol 215:3622–3630
Mitani Y, Andrews RD, Sato K, Kato A, Naito Y, Costa DP (2010) Three-dimensional resting behaviour of northern elephant seals: drifting like a falling leaf. Biol Lett 6(2):163–166
Naito Y, Le Boeuf BJ, Asaga T, Huntley AC (1989) Long-term diving records of an adult female northern elephant seal. Nankyoku Shiryo (polar Rec.) 33:1–9
Naito Y, Bornemann H, Takahashi A, McIntypre T, Plötz J (2010) Fine-scale feeding behavior of Weddell seals revealed by a mandible accelerometer. Polar Sci 4:309–316
Naito Y, Costa DP, Adachi T, Robinson PW, Fowler M, Takahashi A (2013) Unravelling the mysteries of a mesopelagic diet: a large apex predator specializes on small prey. Funct Ecol 27(3):710–717
Naito Y, Costa DP, Adachi T, Robinson PW, Peterson SH, Mitani Y, Takahashi A (2017) Oxygen minimum zone: an important oceanographic habitat for deep-diving northern elephant seals, Mirounga Angustirostris. Ecol Evol 7(16):6259–6270
Newland C, Field IC, Nichols PD, Bradshaw CJA, Hindell MA (2009) Blubber fatty acid profiles indicate dietary resource partitioning between adult and juvenile southern elephant seals. Mar Ecol Prog Ser 384:303–312
Ngô MC, Heide-Jorgensen MKP, Ditlevsen S (2019) Understanding narwhal diving behaviour using hidden markov models with dependent state distributions and long range dependence. PLoS Comput Biol 15:1–21
Oliver GW, Morris PA, Thorson PH, Le Boeuf BJ (1998) Homing behavior of juvenile northern elephant seals. Mar Mamm Sci 14:245–256
Orgeret F, Cox SL, Weimerskirch H, Guinet C (2018) Body condition influences ontogeny of foraging behavior in juvenile southern elephant seals. Ecol Evol 9:223–236
Page B, Welling A, Chambellant M, Goldsworthy SD, Dorr T, van Veen R (2003) Population status and breeding season chronology of Heard Island fur seals. Polar Biol 26:219–224
Piatkowski U, Vergani D, Stanganelli Z (2002) Changes in the cephalopod diet of southern elephant seal females at King George Island, during El Niño-La Niña events. J Mar Bio Assoc UK 82:913–916
Pirotta E, Schwarz LK, Costa DP, Robinson PW, New L (2018) Modeling the functional link between movement, feeding activity, and condition in a marine predator. Behav Ecol 30:1–12
Plötz J, Bornemann H, Knust R, Schröder A, Bester M (2001) Foraging behaviour of Weddell seals, and its ecological implications. Polar Biol 24:901–909
Reiter J, Stinson NL, Le Boeuf BJ (1978) Northern elephant seal development: the transition from weaning to nutritional independence. Behav Ecol Sociobiol 3:337–367
Reiter J, Panken KJ, Le Boeuf BJ (1981) Female competition and reproductive Success in northern elephant seals. Anim Behav 29:670–687
Reiter J, Le Boeuf BJ (1991) Life history consequences of variation in age at primiparity in northern elephant seals. Behav Ecol Sociobiol 28:153–160
Richard and Vacquié-Garcia1, J., Jouma'a1, J., Picard, B., Génin, A., Arnould, J.P.Y., Bailleul, F., and Guinet, C. , 2014.Richard G, Vacquié-Garcia1 J, Jouma'a1 J, Picard B, Génin A, Arnould JPY, Bailleul F, Guinet C (2014) Variation in body condition during the post-moult foraging trip of southern elephant seals and its consequences on diving behaviour. J Exp Biol 217:2609–2619
Riedman ML, Le Boeuf BJ (1982) Mother-pup separation and adoption in northern elephant seals. Behav Ecol Sociobiol 11:203–215
Robinson PW, Simmons SE, Crocker DE, Costa DP (2010) Measurements of foraging success in a highly pelagic marine predator, the northern elephant seal. J Anim Ecol 79(6):1146–1156
Robinson PW, Costa DP, Crocker DE et al (2012) Foraging behavior and success of a mesopelagic predator in the northeast Pacific Ocean: insights from a data-rich species, the northern elephant seal. PLoS ONE 7(5):e36728

Sala JE, Quintana F, Wilson RP, Dignani J, Lewis MN, Campagna C (2011) Pitching a new angle on elephant seal dive patterns. Polar Biol 34:1197–1209
Schick RS, New LF, Thomas L et al (2013) Estimating resource acquisition and at-sea body condition of a marine predator. J Anim Ecol 82:1300–1315
Schorr GS, Falcone EA, Moretti DJ, Andrews RD (2014) First long-term behavioral records from Cuvier's beaked whales (Ziphius cavirostris) reveal record-breaking dives. PLoS ONE 9:e92633
Shreer JF, Testa JW (1995) Statistical classification of diving bwhavior. Mar Mamm Sci 11:85–93
Schreer JF, Testa JW (1996) Classification of Weddell seal diving behavior. Mar Mamm Sci 12:227–250
Shearer JM, Quick NJ, Cioffi WR, Baird RW, Webster DL, Foley HJ, Swaim ZT, Waples DM, Bell JT, Read AJ (2019) Diving behaviour of Cuvier's beaked whales (Ziphius cavirostris) off Cape Hatteras, North Carolina. R Soc Open Sci 6:181728
Slip DJ, Hindell MA, Burton HR (1994) Diving behavior of southern elephant seals from Macquarie Island: an overview. In: Le Boeuf BJ, Laws R (eds) Elephant seals: population ecology, behavior and physiology. University of California Press, Berkeley, pp 253–270
Stewart BS, DeLong RL (1994) Postbreeding foraging migrations of northern elephant seals. In: Le Boeuf BJ, Laws R (eds) Elephant seals: population ecology, behavior and physiology. University of California Press, Berkeley, pp 290–309
Thompson D, Hammond PS, Nicholas KS, Fedak MA (1991) Movements, diving and foraging behaviour of grey seals (Halichoerus grypus). J Zool 224:223–232
Thorson PH, Le Boeuf BJ (1994) Developmental aspects of diving in northern elephant seal pups. In: Le Boeuf BJ, Laws R (eds) Elephant seals: population ecology, behavior and physiology. University of California Press, Berkeley, pp 271–289
Thums M, Bradshaw C, Hindell M (2008) Tracking changes in relative body composition of southern elephant seals using swim speed data. Mar Ecol Prog Ser 370:249–261
Thums M, Bradshaw C, Hindell M (2011) In situ measures of foraging success and prey encounter reveal marine habitat-dependent search strategies. Ecology 92:1258–1270
Tyack PL, Johnson M, Soto NA, Sturiese A, Madsen PT (2006) Extreme diving of beaked whales. J Exp Biol 209:4238–4253
Vacquié-Garcia J, Guinet C, Laurent C, Bailleul F (2015) Delineation of the southern elephant seal's main foragaing environments defined by temperature and light conditions. Deep Sea Res Part II Top Stud Oceanogr 113:145–153
Walters A, Lea MA, van den Hoff J, Field IC, Virtue P, Sokolov S, Pinkerton MH, Hindell MA (2014) Spatially explicit estimates of prey consumption reveal a New Krill Predator in the Southern Ocean. PLoS One 9(1):e86452
Watanabe YY, Takahasi A (2013) Linking animal-borne video to accelerometers reveals prey capture variability. PNAS 110:2199–2204
Watwood SL, Miller PJO, Johnson M, Madsen PT, Tyack PL (2006) Deep-diving foraging behavour of sperm whales (Physeter macrocephalus). J Anim Ecol 75:814–825
Webb PM, Crocker DE, Blackwell SB, Costa DP, Le Boeuf BJ (1998) Effects of buoyancy on the diving behavior of northern elephant seals. J Exper Biol 201:2349–2358
Weihs D (1974) Energetic advantages of burst swimming of fish. J Theoret Biol 48(1):215–229
Williams TM, Davis RW, Fuiman LA, Francis J, Le Boeuf BJ, Horning M, Calambokidis J, Croll DA (2000) Sink or swim: strategies for cost-efficient diving by marine mammals. Science 288:133–136
Yoshino K, Takahashi A, Adachi T, Costa DP, Robinson PW, Peterson SH, Hückstädt LA, Holser RR, Naito Y (2020) Acceleration-triggered animal-borne videos show a dominance of fish in the diet of female northern elephant seals. J Exper Biol 223:jeb212936
Zeno RL, Crocker DE, Hassrick JL, Allen SG, Costa DP (2007) Development of foraging behavior in juvenile northern elephant seals. J Zool 274:180–187

Chapter 13
The Weddell Seal: Eco-Physiological Adaptations to a High-Latitude Seasonal Environment

Michelle R. Shero and Jennifer M. Burns

Contents

Abstract As the world's southernmost living mammal, the Antarctic Weddell seal (*Leptonychotes weddellii*) experiences large shifts in light levels, temperature, ice extent, and prey availability. Understanding the extent that environmental constraints impact the seals' physiology and behavior has been supported by isolated hole diving experiments and technological advances in behavioral data loggers. A >50-year-long mark recapture program has revealed how organismal-level processes translate to demographic variation at the individual and population-scales. In combination, research has demonstrated that during the austral summer, Weddell seals utilize a mixed capital/income breeding strategy and have a short post-breeding recuperation period. This is shortly followed by the annual molt and onset of active gestation. Foraging effort is elevated during winter, particularly in pregnant females. While a suite of adaptations enable Weddell seals to thrive in highly seasonal environments, the same traits and specificity in timing of key life history events may make the species particularly vulnerable to climate and anthropogenic perturbations that impact habitat and prey accessibility. This chapter focuses on phenotypic and behavioral flexibility (and thresholds) to cope with environmental stressors, and knowledge gained from the unique attributes of the Weddell seal study system.

M. R. Shero (✉)
Biology Department, Woods Hole Oceanographic Institution, Woods Hole, MA 02543, USA
e-mail: mshero@whoi.edu

J. M. Burns
Department of Biological Sciences, Texas Tech University, Lubbock, TX 79409, USA

D. P. Costa and E. McHuron (eds.), *Ethology and Behavioral Ecology of Phocids*, Ethology and Behavioral Ecology of Marine Mammals,
https://doi.org/10.1007/978-3-030-88923-4_13

13.1 Introduction

The Weddell seal (*Leptonychotes weddellii*) is the world's southernmost living mammal, residing in the fast- and pack-ice covered waters surrounding the Antarctic continent year-round. This is an extremely seasonal environment, with light levels fluctuating from complete darkness to complete daylight, ambient temperatures varying from well below to just above freezing, and ice extent and thickness likewise showing extensive annual variation (Eicken 1993; Ackley and Sullivan 1994). These patterns drive seasonal variation in productivity and ease of access to prey resources, and influence the thermoregulatory costs for seals when in and out of the water (Hindle et al. 2015; Walcott et al. 2020). Thus, Weddell seals have evolved a suite of traits that make them uniquely suited for enduring and thriving in a harsh and highly seasonal environment. While these traits have permitted the species to thrive in an environment inhabited by few other large vertebrate predators, their specializations may leave Weddell seals particularly vulnerable in the face of ongoing shifts in climate and increased fishing pressures.

Changes in climate that influence sea ice dynamics and productivity are having an impact on marine mammal populations at both poles (Haas 2003; Siniff et al. 2008; Kovacs et al. 2011). An understanding of how any species will respond to alterations in habitat and prey availability is critical to predicting population vital rates, but such understanding is difficult for many marine mammal species. This is particularly true for those in the Arctic, where the transient pack ice habitat and long history of interactions with terrestrial predators makes species particularly wary of humans and difficult to relocate reliably for long-term monitoring. In contrast, Antarctic species are not disturbed by human approach, and not exposed to terrestrial predators.

Of all high-latitude pinnipeds, Weddell seals are the most accessible for detailed study, due to their high-affinity for fast-ice, and close proximity of colonies to research stations. Individuals show a high degree of site fidelity, allowing the same seals to be observed repeatedly and serve as subjects in research studies that range from days to generations. This accessibility has revealed that Weddell seal behavior and life history parameters vary in response to annual variation in sea ice (Proffitt et al. 2007a; Paterson et al. 2015), interannual shifts in oceanographic conditions due to the Southern Oscillation Index (SOI) and the El-Niño Southern Oscillation (ENSO; Testa et al. 1991), and unusual events such as the presence of the B15 iceberg (which led to a 4-year hiatus in the annual breakout of sea ice from the Ross Sea and McMurdo Sound area Chambert et al. 2012, 2013). In combination, these qualities make the Weddell seal an exceptionally tractable study model that can shed light on how other polar species may respond to ongoing environmental perturbations. This chapter focuses on features that make this species best able to exploit high-latitude niches, phenotypic and behavioral flexibility (and constraints) to cope with environmental stressors, and the knowledge-set gained from the unique attributes of the Weddell seal study system.

13.2 Population Size and Distribution

The first Weddell seal specimen was collected from the South Orkney Islands in 1823 on a voyage led by Captain James Weddell. The first description of the species (*weddelli*) was made by Weddell (1825) and refined by Lesson (1826). The initial assignment of genus *Leptonyx* by Gray (1837) was later replaced by Gill (1872) as *Leptonychotes*. Adult Weddell seals typically reach sizes of 2.5–3.5 m in length and 350–550 kg in mass, with females achieving a slightly larger body size than males. On average, Weddell seals live to 18 years (Cameron and Siniff 2004); however, there are documented cases of individuals living to be >30 years old (J. Rotella *pers. comm*). The Weddell seal has a circumpolar distribution centered in the fast-ice around the Antarctic continent, but extending to the pack ice edge in the Antarctic convergence and ice-free islands of the Western Antarctic Peninsula (WAP). Individuals have also been observed at most of the subantarctic islands (Fig. 13.1; Bertram 1940; Mansfield 1958).

Weddell seals were never commercially exploited, but were hunted by early Antarctic explorers for food for men and dogs. Weddell seals gained some protections after the Antarctic Treaty was signed in 1961, and additional protections under the Antarctic Conservation Act (ACA) of 1978 and through regulations imposed by the Convention on the Conservation of Antarctic Marine Living Resources (CCAMLR) established in 1982, as well as other relevant national laws (e.g., the U.S. Marine Mammal Protection Act, MMPA, enacted in 1972). Harvests continued to support dog teams at some research stations into the 1980s (e.g., Erebus Bay from 1966–1984; Stirling 1971), and recovery from those harvests has likely occurred only recently (D. Ainley, *pers comm*). Currently, human caused mortality is estimated at <10 individuals/yr and is no longer a threat to any local colony or population. With the exception of permitted scientific research activities, all Antarctic seals are protected from

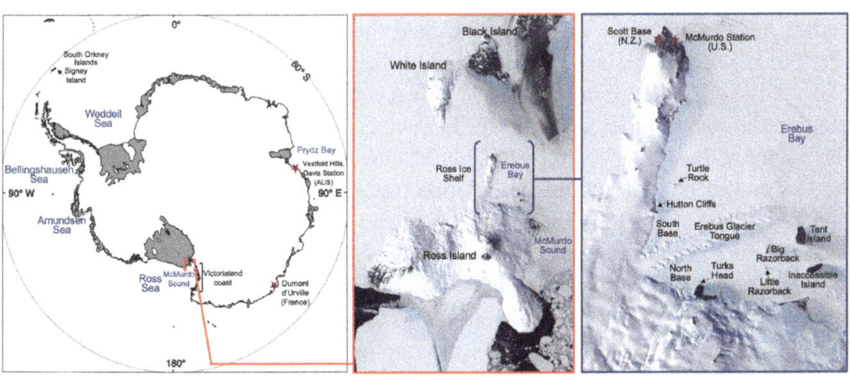

Fig. 13.1 The Weddell seal resides around the Antarctic continent, and all colonies and sites referred to throughout this chapter are labeled. Much research on the species has taken place around Ross Island (red inset) and the Erebus Bay breeding colonies (blue inset). Satellite images from USGS Landsat Image Mosaic of Antarctica

human disturbances. However, there is growing awareness that increased tourism in the Antarctic may pose threats due to inadvertent disturbances, and introduced diseases or pathogens (Frenot et al. 2005).

Circumpolar abundance estimates determined from shipboard surveys, which are highly uncertain, range from approximately 200,000 to 1 million individuals (averaging 600,000–800,000; Erickson and Hanson 1990; Southwell et al. 2012). The Antarctic Pack Ice Survey conducted in 2002 generated regional estimates of 331,000 (144,000–759,000) Weddell seals in the Ross and Amundsen Seas (150°E–100°W) and 302,000 (77,000–576,000) seals in the Bellingshausen and Weddell Seas, and along the WAP (90°W–30°W; Southwell et al. 2012). Most recently, high resolution satellite imagery suggests a much smaller global population count of just 202,000 female Weddell seals (85,345–523,140; 95% confidence interval; LaRue et al. 2021). While there are no apparent barriers to animal movements among regions, tagging studies and genetic sequence differences in serum transferrins suggest that immigration and emigration rates between regions are low (Shaughnessay 1969; Stirling 1969a). However, as new habitat in the Ross Sea became available ~50,000 years ago through glacial retreat, seal populations expanded rapidly into the region (Zappes et al. 2017). Similarly, open water access to White Island in the southern Ross sea led to the recent establishment of a small population there (see Sect. 13.7). These retrospectives indicate that under the right conditions, dispersal rates are sufficient to permit range expansions through the establishment of new breeding colonies.

13.3 Demography and Vital Rates

High site fidelity of mature Weddell seals has provided an unparalleled opportunity for demographic studies that have improved our understanding of how intrinsic and extrinsic factors influence vital rates in this high-latitude predator. Mark-recapture studies have taken place at colonies around the Antarctic continent including the Vestfold Hills, Signy Island, DuMont D'Urville, Prydz Bay, and in the Weddell Sea (Fig. 13.1; Siniff 1981; Siniff et al. 1977; Testa et al. 1990; Lake 1997). The Weddell seal population in Erebus Bay, McMurdo Sound (77.8°S latitude) has received particular attention due to its proximity to United States and New Zealand research stations. Studies on this relatively stable population (total abundance ~1800 seals; Chambert et al. 2012) began in 1969 (Siniff et al. 1977) and have been on-going since, resulting in one of the longest, continuous demographic studies conducted on any mammalian species—terrestrial or marine. Over the past half-century, individuals have been tracked through the use of identification tags applied to the hind-flipper interdigital webbing; currently, all newborn pups and immigrants into the breeding colonies are outfitted with two Dalton Superflexi tags (Fig. 13.2). These tags have high retention rates (Testa and Rothery 1992), and because damaged, faded, or lost tags are replaced as necessary, >80% of animals within the Erebus Bay population are known individuals. Multiple censuses across the summer ensure that all animals that return to the region are sighted each year, and yield lifetime records of colony presence

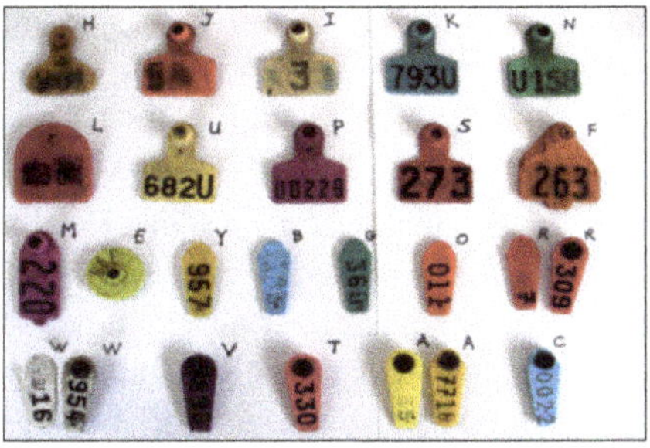

Tag Code	Gross Color	Color	Common Name by Study Team	Make	Model	Size	First Year Deployed
A	Yellow	Amarillo	Amarillo	Dalton	Supertag	1-Small	1997
B	Blue	Blue	Blue	Dalton	Rototag	1-Small	1985
E	Yellow	Yellow	EID Tag	Dalton	FXD RFID	0-Round	1999
F	Orange	Orange	Fearing	Fearing	Duflex	2-Medium	1988
G	Green	Green	Green	Dalton	Rototag	1-Small	1989
H	Yellow	Old Yellow	H-Tag	Allflex	Duplex	2-Medium	1971
I	White	White	I-Tag	Allflex	Duplex	2-Medium	1974
J	Orange	Orange	J-Tag	Allflex	Duplex	2-Medium	1973
K	Blue	Blue	K-Tag	Allflex	Duplex	2-Medium	1976
L	Red	Red	L-Tag	Dalton	Goliath	2.5-Medium	1979
M	Red	Magenta	Magenta	Fearing	Sheep/Goat	1.5-Small	1987
N	Green	Green	N-Tag	Allflex	Duplex	2-Medium	1977
O	Orange	Orange	Orange	Dalton	Rototag	1-Small	1990
P	Purple	Purple	Purple	Allflex	Duplex	2-Medium	1981
Q	Metal	Metal	Metal	Monel	Metal	1-Small	1969
R	Red	Red	Red	Dalton	Supertag	1-Small	1992
S	Red	Scarlet	Scarlet	Allflex	Duplex	2-Medium	1984
T	Pink	Hot Pink	Hot Pink	Dalton	Supertag	1-Small	1996
U	Yellow	Yellow	U-Tag	Allflex	Duplex	2-Medium	1980
V	Purple	Violet	Violet	Dalton	Supertag	1-Small	1994
W	White	White	White	Dalton	Supertag	1-Small	1992
X	White	White	X-Tag	Allflex	Duplex	3-Large	1971
Y	Yellow	Yellow	Yellow	Dalton	Rototag	1-Small	1983

Fig. 13.2 A wide array of identification tags have been applied to the Weddell seal's hind-limb interdigital webbing to maintain records of unique individuals in Erebus Bay. Courtesy of Jay Rotella

and reproductive events. This work has provided a more nuanced understanding of demographic responses to baseline, long-term, and acute environmental shifts than has been possible in most other marine mammal species.

Impacts of environmental conditions are apparent in the Weddell seal lifecycle. In the southernmost Erebus Bay population, Weddell seal females recruit into the breeding population at 7–8 years of age on average (range 4–>11), which is significantly older than that at Signey Island (60°S; 4–5 years; Hadley et al. 2006). This suggests that the shorter summer season, more severe winter, and more constrained foraging environment in Erebus Bay slows individual growth and maturity rates. In addition, fewer than 75% of reproductively mature females in Erebus Bay give birth each year (Testa et al. 1990; Hadley et al. 2007a; Chambert et al. 2012) and, on average, females fail to produce a pup one out of every three years. As a result, Weddell seals in Erebus Bay have an annual reproductive rate that is notably lower than that seen in phocids from more benign environments (Hadley et al. 2007a). That these differences are tied to local conditions is further suggested by the fact that in years of relatively low summer sea ice extent and high local productivity (and positive phase of SOI), birth rates are higher (Paterson et al. 2015), females attain greater body mass, and thus pups are weaned larger (Proffitt et al. 2007a) compared to years of high summer sea ice extent.

The extensive lifetime records of individual Weddell seal reproductive success have also revealed that female behavior does not fully align with that predicted for long-lived, slow growing K-selected species. While the Prudent Parent Hypothesis (Lack 1968; Goodman 1974) suggests that females should conserve maternal resources to not incur survival costs in association with reproduction, the survival rates of females from the Erebus Bay population are slightly reduced in the year following a reproductive event. For example, while adult annual survival rates are quite high on average (94%), survival declines to 91% after a reproductive year (Hadley et al. 2007a), likely in response to energetic constraints imposed by lactation (see Sect. 13.4). While the reduction in survival probability appears small, the cumulative effect across an individual's lifetime would likely be substantial. In addition, females are ~10% less likely to reproduce the next year if they gave birth in the current year, as compared to those that did not (67 vs. 76% probability of pupping the next year; Chambert et al. 2013) again reflective of the high energetic costs associated with reproduction. There is also an impact of age, as the probability of failing to reproduce in the subsequent year is ~30% greater for young primiparous mothers than older, more experienced ones (Chambert et al. 2013; Hadley et al. 2007a). Finally, pre-weaning pup mortality rates are nearly double for primiparous vs. experienced (7.1 vs. 3.2%) females (Hastings and Testa 1998). The age-effect observed across almost all aspects of reproductive potential in Weddell seals is likely a function of female experience and maternal attendance behavior, and also the amount of energy females can afford to allocate to offspring rearing.

Female body mass and pup birth mass both increase with maternal age until females reach 17 years old, after which there is a senescent decline. Older females have lower chances of surviving to reproduce the following year, and increasing resource allocation toward the current year's offspring may be the best strategy to enhancing the female's fitness (Terminal Investment Strategy). Indeed, pup birth mass and maternal mass at parturition increase from age of first reproduction to prime reproductive age (~10–20 years old), suggesting that younger, smaller females face constraints in how much energy can be allocated to offspring. Pre- versus postnatal energy allocation from female to pup exhibits additional shifts with age, and older females tend to give birth to smaller pups yet devote a greater percentage of their energy stores to the pup postnatally, likely incurring a greater cost to her own survival (Mannas 2011; Paterson et al. 2016). Ultimately, pups born to older, more experienced females are also more likely to survive and be recruited back into the population (Hastings and Testa 1998; Paterson et al. 2016). The high energetic costs of reproduction are also suggested by the fact that pups born after a one-year hiatus tend to be larger (and born earlier in the season) than sequential births (Paterson et al. 2016; Rotella et al. 2016). Thus, opting not to produce a pup in one year appears to allow the females to recuperate more mass and energy than is possible for gestating females or if females incur a net energetic deficit due to nursing pups year-after-year (Paterson et al. 2016).

There is an age-related decline in pupping probability in old females, with reproductive senescence emerging in females older than 27 years old (Paterson et al. 2016). Senescent declines have been attributed to reduced foraging success due to changes in muscle architecture as well as dental wear (Stirling 1969b; Hindle et al. 2009). Reduced reproductive success at both ends of the lifespan likely stem from energy constraints, which are more severe for young, still growing females, and those that are less capable at recovering the mass and energy loss associated with the long lactation period. Environmental perturbations that reduce prey availability or increase reproductive costs are likely to have larger effects on these segments of the population than on 'prime-age' females (Hadley et al. 2007a, b; Proffitt et al. 2007b; Chambert et al. 2012). When the B15 and C16 icebergs (11,000 km^2 and 650 km^2, respectively) blocked McMurdo Sound, there were significant increases in local ice thickness and reductions in primary productivity (Fig. 13.3). As a result, many demographic parameters reached record lows. While colony attendance, reproductive rates, and first year pup survival all declined, impacts were less evident in prime

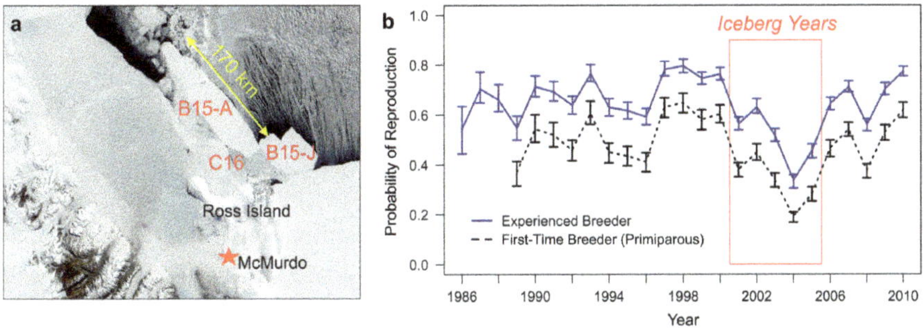

Fig. 13.3 **a** The B15 and C16 icebergs caused greater ice extent in McMurdo Sound, satellite image from NASA Earth Observatory Nov. 9, 2004. **b** Reproductive rates of first-time and experienced breeding female Weddell seals in Erebus Bay declined significantly during the 'iceberg years' (red box), from Chambert et al. (2012)

age females. For example, colony attendance was reduced by 42% for 10-year-old females (prime age) but by 59% for younger (5-year-old) females. Similarly, reproductive rates decreased by 27% for experienced breeders but by 37% for first-time breeders (Chambert et al. 2012).

While the cumulative effects of lowered survival probabilities following offspring rearing across an animal's lifetime are expected to have substantial impacts on female lifespan and fitness (Hadley et al. 2007a), there is substantial intraspecific heterogeneity in the population's reproductive rates. Thus, 'higher-quality' females produce up to 9 pups per decade, whereas 'lower-quality' females produce as few as 4 pups in 10 years (Chambert et al. 2013, 2014). In addition, females that have a history of producing pups more frequently also give birth to larger pups, which would tend to have higher postweaning survival rates (Paterson et al. 2016). This individual heterogeneity in reproductive fitness appears fixed, as the offspring of 'high-quality' females also produce significantly more pups once they are recruited into the population (Chambert et al. 2014). Individual quality accounts for as much variation in population reproductive rates as does study year. Even under harsh conditions such as occurred during the 'B-15 iceberg years' (Fig. 13.3), it was the same inherently 'higher-quality' individuals that were most likely to produce a pup (Chambert et al. 2013). That a select few 'high-quality' individuals contribute unequally to the population gene pool and population maintenance has been shown in other mammals (Wilson and Nussey 2010; Bergeron et al. 2011), pinnipeds (Le Boeuf et al. 2019), and long-lived seabirds (Lescroël et al. 2009; Jenouvrier et al. 2015; Tuck et al. 2015). This may make Weddell seal populations particularly vulnerable to environmental shifts that lower survival or hinder reproductive success of these relatively few 'high-quality' animals.

13.4 The Annual Cycle—Temporal Alignment with Environmental Constraints

13.4.1 Female Pupping and Lactation

The dramatic seasonality in the Weddell seal's high-latitude environment places tight temporal constraints on energetically costly life history events and exerts strong control over the annual cycle of behavior and condition (Fig. 13.4). At the start of the austral summer, Weddell seals return to the colonies in peak body condition with large lipid reserves. Female Weddell seals give birth at the beginning of the austral summer to a single pup per year; there are very few records of possible twinning (<0.1% of births genetically determined to be twins; Stirling 1969a; Gelatt et al. 2001). Females will rarely 'adopt' and nurse nearby abandoned pups (Tedman and Bryden 1979). In cases where the female suckles multiple pups, it is most likely that neither pup is successfully weaned due to the inability of a single female to provide sufficient calories to both pups (Gelatt et al. 2001). Across the species'

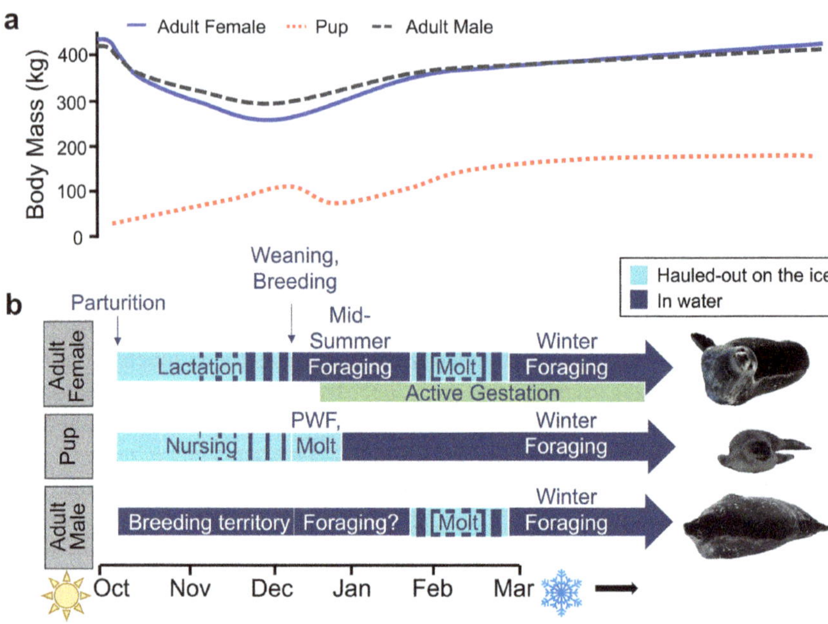

Fig. 13.4 **a** Annual changes in Weddell seal body mass associated with **b** life history events for adult female-pup pairs and adult males. Females give birth and nurse their pups at the start of the austral summer (October in McMurdo), forage to recuperate some mass, before undergoing the annual molt and then starting the long winter foraging period while pregnant (active gestation) [mass dynamics adapted from Salas et al. 2017]. After weaning, pups undergo a postweaning fast (PWF) and molt prior to foraging. Meanwhile, males establish breeding territories at the start of the summer before the molt and subsequent winter foraging

range, birth timing varies in response to both environmental and female conditions. The peak pupping period occurs on September 7 on Signy Island (60°S latitude) and September 20 at Graham Land (65–68°S), while at the southernmost extent of the Weddell seal's range in Erebus Bay (77°S) females give birth 1–2 months later (mid-October through mid-November; Stirling 1969a; Testa et al. 1990). Females typically arrive at breeding colonies just a few (1–2) days before giving birth (Tedman and Bryden 1979); however, parturition timing may be delayed in the event of unfavorable weather conditions. There are very few births during the harsh storms that still occur at the start of the summer season. Yet, on the next benign day (warm, low wind) following a storm, pupping rates are often high, suggesting a mechanism exists to delay parturition by a short duration (Liggins and Creasy 1980).

Parturition of Weddell seals occurs relatively synchronously within populations to ensure that females give birth to pups when ambient conditions are conducive to pup survival, similar to other species of high-latitude environments. The Erebus Bay breeding colonies are surveyed every 1–3 days during the pupping season, and precise pup birth dates are recorded. There are strong maternal effects on parturition dates, which range across 49 days within Erebus Bay (Rotella et al. 2016). Females give birth progressively earlier until they reach 16 years old, and then proceed to give birth progressively later into the pupping season (Rotella et al. 2016). Females that reach primiparity at a relatively young age also consistently give birth earlier than females that were older at their first year of reproduction (Rotella et al. 2016). While the pup sex ratio is approximately 1F:1M in Erebus Bay, male pups are born on average 2 days earlier than female pups (Rotella et al. 2016). Further, females with higher indices of reproductive success produce more male offspring throughout their lifetime, suggesting that Weddell seals conform to the Trivers-Willard sex allocation model (Proffitt et al. 2008b). This model hypothesizes that higher-quality females should produce more and/or devote more resources to male offspring that are more 'profitable' with higher lifetime reproductive pay-offs. Indeed, maternal energy allocation has a greater impact on male than female pup survival (Proffitt et al. 2008a). First-year survival is related to birth dates, as pups born relatively late in the season typically endure better ambient conditions (i.e., avoid early summer storms) and have higher pre-weaning survival rates. However, pups born earlier in the season may be weaned in better condition and/or have greater foraging experience, conferring higher postweaning survival rates (Proffitt et al. 2008a; Rotella et al. 2016).

Without the evolutionary constraints of land predators or ephemeral pack ice to select for a short lactation period, female Weddell seals nurse their pups for the longest duration of any phocid (except Mediterranean monk seals, *Monachus monachus*; Chap. 16): 5–7 weeks on average (Chaps. 7 and 8). The Weddell seal's fast-ice breeding colonies overlap spatially with the availability of prey, which allows for supplemental foraging during lactation (Testa et al. 1989; Sato et al. 2002). The ability to employ a mixed capital and income strategy, where the costs of lactation are supported by energy reserves and foraging (Chap. 8), is crucial to sustaining their protracted lactation period. However, there is variation among individuals with some female Weddell seals using a nearly pure 'capital' investment strategy, while others supplement their reserves by foraging extensively during the lactation period.

The probability of relying on income obtained through foraging may be higher in younger females (Hill 1987), is more common for females in worse condition (Sato et al. 2002), and increases later in lactation (Hindell et al. 2002; Sato et al. 2002). Any time spent foraging must be balanced against time spent nursing the pup, as females remain hauled-out with their pup for the first two weeks post-partum before they re-commence foraging. Diving during late-lactation is highly variable among individuals and ranges from 1.7 to 17.8 h per day. Some seals rarely dive deeper than 50 m, while others spend up to 9.3 h per day making deep dives (from >50–487 m; Sato et al. 2002) and travel further (~1100 m) away from the breeding colony (Hindell et al. 2002). Because a large number of females are actively foraging while concentrated near the breeding colonies, prey depletion occurs in the local area across the austral summer (Testa et al. 1985; Hindell et al. 2002). As females spend more time foraging, the frequency and duration of nursing decreases across lactation from approximately 15 suckling events, totaling 5 h per day during the first week of lactation, to only half that by the sixth week of lactation (Tedman and Bryden 1979).

Despite supplemental caloric intake from foraging during lactation, females nevertheless lose 2–6 kg per day, most of which comes from their large lipid reserves (22–45% of total mass at the start of lactation; Wheatley et al. 2006; Shero et al. 2015c). However, protein reserves also contribute a substantial proportion of the energy transferred, particularly in smaller females, which may help explain the higher survival costs associated with reproducing at a younger age (Wheatley et al. 2006). In addition to considerable heterogeneity in resource allocation, there is a wide range in lactation duration (from 28–54 days). In particular, smaller females often nurse their pups for significantly shorter periods than larger females (Stirling 1969a; Hill 1987; Wheatley et al. 2006).

By the end of lactation, individual variation in lactation length, fuel use, and foraging success have compounded, and females have the lowest energy reserves for the year. Reproductive females have lost between 18 and 55% of their total body mass and 13–47% of their protein reserves (Fig. 13.5). Younger and smaller females that started lactation with smaller energetic reserves (absolute and as a proportion of body

Fig. 13.5 Female-pup pair at **a** start of lactation and **b** end of lactation. Females lose >30% of their body mass while pups triple or quadruple in size. NMFS 87-1851-04. *Photo credits* **a** Michelle Shero, **b** Daniel Costa

mass) may be particularly vulnerable to changes in environmental conditions that reduce early-to-mid-summer foraging success. Any such impacts would ultimately translate to pup weaning success. Pups assimilate 25–46% of the mass lost by females, gaining 0.5–2.5 kg per day and almost quadrupling in size by the time they are weaned (from ~25 kg at birth to ~100 kg at weaning; Fig. 13.5; Wheatley et al. 2006). At the time of weaning or shortly after that, female serum progesterone concentrations increase associated with ovulation, and the breeding season begins (Hill 1987).

13.4.2 Breeding Behavior

As females aggregate in dense numbers at the colonies, males establish underwater mating territories along breathing holes and tide cracks in the sea ice in preparation for the breeding season, characterized as an exploded lek, moderately polygynous mating system. In an exploded lek mating system, males may display for female attention outside of visual range but within auditory range of other males. In contrast to terrestrial breeding pinnipeds, Weddell seal males are slightly smaller than females, presumably because of the selective pressures of maintaining agility in the water to defend territories (Croxall and Hiby 1983; Cameron and Siniff 2004; Harcourt et al. 2007a, b). Further, the ability of territorial males to forage during the reproductive periods (Siniff et al. 1977; Harcourt et al. 2007b) likely reduces the advantage that larger body size and fuel reserves provide to males in terrestrial breeding species as they must fast throughout the season (Chaps. 4, 7, and 8).

It is not clear how body size at the start of the breeding season is related to variation in territory size (3.0–6.5 km^2), in part because there appears to be spatial overlap among territories and shared use of breathing holes (Harcourt et al. 2007b). There is some advantage to increased body size and experience (Harcourt et al. 2007a), as males younger than five are not present at the breeding colonies, and it does not appear that males are successful at competing for territory until they are 8–9 years old (Bartsh et al. 1991; Testa 1997). Territorial males exhibit a rise in serum testosterone concentrations late October and proceed to spend the most time in the water (17–>23 h per day) in the weeks preceding the female's postweaning annual estrus from early November through mid-December (Bartsh et al. 1991). Conversely, smaller males presumed to have lost territory, are seen hauled-out on the ice for prolonged durations at this time, often with injuries to the eyes, flippers, and penile opening (Fig. 13.6; Kaufman et al. 1975; Bartsh et al. 1991).

Weddell seal mating occurs underwater, but there has only been one observation of copulation. The copulatory event lasted ~5 min and occurred at shallow depths (1–3 m), and the male oriented himself dorsal to the female and clasped his fore-flippers around the female's body (Cline et al. 1971). This event occurred prior to the female weaning her pup, but was quite late into lactation (43 days post-partum). An experiment conducted by Hill (1987) supported this initial observation of the timing of mating, in which colored grease applied to the male's penile opening was transferred to females during copulation, on average 35 days post-partum. Females

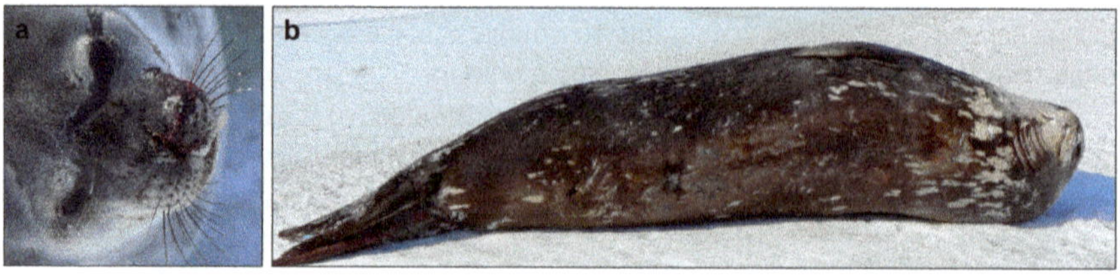

Fig. 13.6 **a** Male Weddell seals are often seen with injuries to the face and **b** scratches all along their ventral side, penile opening, and flippers during the breeding season. *Photo credits* Henry Kaiser, NMFS 15,748

were only sighted with one color grease marking, suggesting that copulating with multiple males likely occurs infrequently (Hill 1987).

A singular male is often responsible for the vast majority of the breeding efforts at a given colony (Hill 1987). During the breeding season at McMurdo, the female-to-male ratio of animals present at the colonies is ~4:1. However, the functional (i.e., reproductively-active) female-to-male ratio may be as high as 16:1, as both direct monitoring of male mating behaviors (Hill 1987) and genetic studies (Zappes et al. 2017) suggest that only a subset of observed males breed each year. Paternity DNA testing has shown that half the resident males at the Turtle Rock colony sired no pups while other individuals fathered up to five pups (Harcourt et al. 2007a). Therefore, similar to a few remarkable 'high-quality' females that contribute disproportionately to the population's gene pool, only a small number of males ultimately sire pups in the population each year. Less is known regarding multiyear dynamics and lifetime siring success in males than in females.

As the breeding season comes to a close, males have lost up to 24% of their body mass and sex hormones, spermatogenesis, and testis size all quickly decline by mid-December (Smith 1966; Bartsh et al. 1991; Harcourt et al. 2007a). At more northerly latitudes, the timing of testicular regression occurs 4–6 weeks earlier (Smith 1966) and mirrors the forward shift observed in the females' arrival at birthing colonies compared to the southernmost extent of the range. This seasonal recrudescence in males followed by reduced mating competence may prevent males from mating with females that wean their pups and/or ovulate very late in the season. It would be disadvantageous for either sex to expend energy on reproduction outside this optimal window as it would lead to a temporal mismatch in the next year's pupping and breeding season and reduce fitness.

13.4.3 Vocal Behavior

Weddell seals have diverse vocalizations (Chap. 3), but are relatively silent for most of the year, and most vocalizations are associated with breeding activities. While females remain hauled-out on the ice with their newborn pups for the first few days post-partum, playback experiments show that females are attentive and raise their heads, check on their pup, or call after hearing a pup vocalization. During this time, females do not respond differently to their own pup as compared with pups at the same or even from different breeding colonies (Van Opzeeland et al. 2012), and there is likely a combination of vocal and tactile, olfactory, or spatial cues that contribute to the female-pup bond (Tedman and Bryden 1979). After the pups reach ~10 days of age, they develop more individually distinctive vocalizations (Collins et al. 2006). This age coincides with the time that pups begin entering the water with females (see Sect. 13.4.4; Tedman and Bryden 1979; Sato et al. 2003).

Territorial males have an extensive repertoire of complex vocalizations that they direct toward challengers during underwater disputes. The vocalizations can be heard through 4 m of ice to reach seals hauled-out on the ice as well as in the water (Thomas 1982). Overall, males are much more vocal than females, and many vocalization types are exclusively made by males, including long-duration calls (averaging 5–42 s) of descending frequency with harmonics (Thomas 1982). Weddell seal repertoires also exhibit geographic variation, and animals make shorter and lower frequency calls at the Antarctic Peninsula compared to McMurdo (Thomas and Stirling 1983).

13.4.4 Pup Dive and Post-Weaning Development

Breeding colonies are located along tidal cracks that provide access to the water for both mothers and pups (Hill 1987), and Weddell seals are one of the few phocids where pups enter the water during the nursing period (along with bearded, *Erignathus barbatus*; ringed, *Pusa hispida*; and harbor seals, *Phoca vitulina*; see also Chaps. 10 and 15). At these colonies, mothers ream the ice with their teeth to modify ice cracks for young pups to enter and exit the water (Fig. 13.7). For the first several days to weeks of life, pups rarely (if ever) enter the water. When pups are a few weeks old, females begin to coax them into the water by sitting in the water and vocalizing (M. Shero and J. Burns, *pers. obs*). Pups often require assistance in hauling back out of the water, and females will use their noses or chests to push the pup onto the ice. Approximately one-half of the pups that die prior to weaning are never seen on the ice surface (Schreer et al. 1996), suggesting that early aquatic activities are risky. Within the Erebus Bay region, breeding colonies located in areas with broad, shallow, spreading cracks are occupied by more mature and experienced females and tend to have higher rates of pup survival as compared to those with steep and/or narrow cracks (Hastings and Testa 1998). Kaufman et al. (1975) suggested that female-pup pairs compete for space along suitable tidal cracks because of these early dives.

Once in the water, mother–pup pairs dive and swim together at the surface (Sato et al. 2003), but aquatic foraging by pups likely doesn't begin for several weeks after these initial 'swimming lessons' (Burns and Castellini 1996; Burns et al. 1999). When diving with their pups, females make shorter, shallower (5.5 vs. 202 m mean depth), and slower (0.7 vs. 1.3 m per second) dives than when diving by themselves, and active foraging during dives has not been observed in video records (Sato et al.

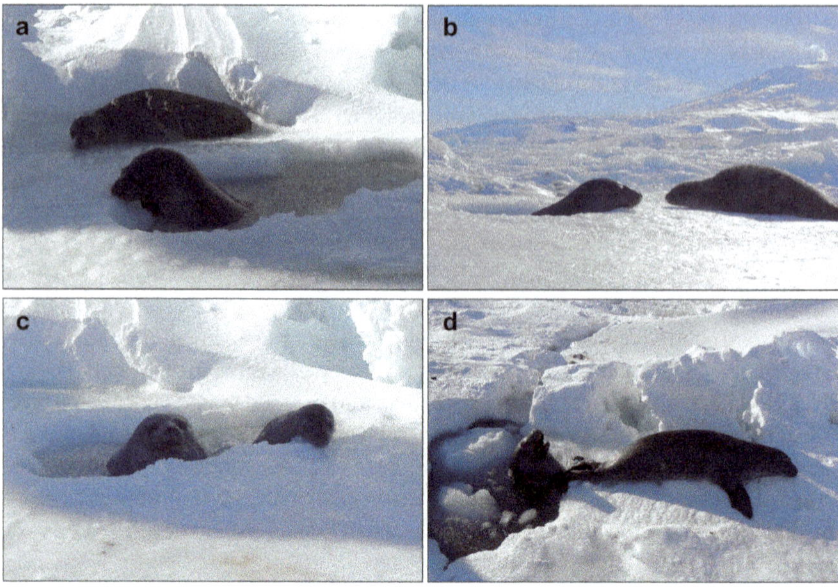

Fig. 13.7 **a** Female reaming ice with her teeth to create a shallow ledge for her pup, **b** coaxing pup into the water, **c** swimming with her pup, and **d** helping her pup back out of water. NMFS 87-1851-04. *Photo credit* Michelle Shero

2003). Since this early aquatic activity carries both energetic (increased activity and thermoregulatory costs; Pearson et al. 2014) and survival risk (Schreer et al. 1996), early diving must also provide some net benefit—perhaps by training the pups in how to navigate under the ice and locate breathing holes, or providing the hypoxic stimuli needed to trigger maturation of diving oxygen (O_2) stores (De Miranda et al. 2012; Geiseler et al. 2013) or heart rate regulation (Greaves et al. 2005).

As in other pinniped species, substantial physiological and behavioral post-weaning development is required for neonates and young-of-the-year to attain adult dive and foraging capabilities (Burns and Castellini 1996; Burns 1999; Burns et al. 2007). Both dive duration and depth increase rapidly in Weddell seal pups from the time they enter the water at 2–3 weeks through the time they are weaned at 5–7 weeks of age (Burns 1999). The diving and foraging activities of newly weaned pups are not sufficient to meet their energy demands, and pups go through a short period of mass loss during mid-summer. This period of mass loss appears to end later in the summer at a time when mean dive depths (but not duration) plateau at values relatively similar to yearlings and adults (weaned pups: 105.8 m, yearlings: 118.5 m, adults: 163.1 m; Burns 1999). This suggests that during late summer, all age classes are targeting the same prey species (although perhaps of different sizes; see Sect. 13.6; Fig. 13.8a). Weaning may be timed to coincide with a summer shallowing of the prey field, which would improve access to prey for the physiologically immature pups.

Because of their smaller O_2 stores and higher mass-specific diving metabolic rates (DMRs), the weaned pups' mean dive durations remain significantly below

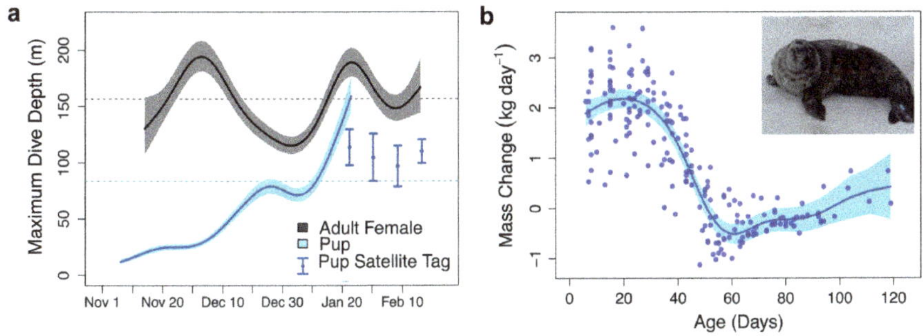

Fig. 13.8 **a** Both adults and pups make shallower dives mid-summer coincident with the phytoplankton bloom, and **b** it is not until late January or ~90 days of age that pups begin gaining mass after they've been weaned (GAMM, shaded region shows 95% CI). Courtesy of Jennifer Burns

adult values (weaned pups: 5.82 min, yearlings: 8.72 min, adults: 13.14 min; Burns 1999). Perhaps to compensate, young pups make far more dives per day than adults and a higher proportion of the dives they make reach or exceed their calculated aerobic dive limit (Burns and Castellini 1996; Burns 1999). While this may improve caloric intake rates, it is not until pups are close to three months old that they begin to gain mass again (Fig. 13.8b). By this point, many of the pups have left the local area and traveled north into the coastal regions of the greater Ross Sea (Burns 1999; Burns et al. 1999). Overwinter foraging patterns and success are not known for this segment of the population, as very few individuals return to Erebus Bay the next year as yearlings. Those that return are of similar mass but slightly longer than weaned pups, suggesting that overwinter foraging is challenging (Burns 1999). Estimated first-year survival rates vary from ~15–65%, averaging ~50% (Hastings and Testa 1998). Variation in first-year survival has been linked to maternal size and colony, as well as mass and condition at weaning. Few pups weaned at <75 kg are seen again (J. W. Testa *pers. comm.*), suggesting that relatively large reserves are needed for pups to survive the transition from nursing pup to independent forager.

13.4.5 Late Summer Activities

By the end of the lactation and breeding season, reproductive adults have lost significant amounts of mass from lipid and lean reserves. The late summer foraging period (January/February) is therefore a critical window for recovery before the onset of molt and winter darkness. In the high Antarctic, mid-summer is characterized by near 24 h daylight and significant increases in primary productivity (Smith and Nelson 1986; Jones and Smith 2017). Zooplankton and fishes are more abundant in surface waters than later in the year (Cisewski et al. 2010), and foraging success of many top predators is higher in summer than winter (Curtice et al. 2015; O'Toole et al. 2015). For adult Weddell seals in Erebus Bay, mid-summer foraging allows rapid recovery

(>1 kg per day) of a significant proportion of the mass lost (~15%, A. Kirkham and J. Burns *pers. comm.*), while making relatively short and shallow dives due to the high summer productivity (Beltran 2018). While seals spend significant time foraging during mid-summer, this varies by age and past reproductive effort. For example, weaned pups and juveniles rarely haul out and instead spend a considerable amount of time foraging outside the colonies (Burns et al. 1999). Post-reproductive females spend more time in the water than females that did not give birth earlier in the season (13.6 vs. 10.4 h per day; Walcott 2019). However, given the short period between the end of breeding and the beginning of molt, even such intensive foraging does not allow females (particularly those that wean pups late) or males to fully recoup stores before molting or the onset of winter foraging.

13.4.6 The Annual Molt

The annual molt in most pinnipeds occurs several months after the breeding season ends. This allows for post-reproductive individuals to perform their post-breeding foraging to recoup mass and condition (as discussed above) before initiating the energetically costly molt (Carlini et al. 2005). In pinnipeds, the embryo generally remains in a state of diapause (or delayed implantation) throughout the post-breeding recuperation period and the molt. The gap between the end of the lactation period and onset of visible hair shedding is exceptionally short for Weddell seals. Recent work (A. Kirkham *pers. comm.*) shows that Weddell seal hair follicles begin to progress into the molt before the end of lactation, which is much earlier than is known for any other pinniped species. Moreover, embryo implantation occurs shortly after breeding and well before completion of the annual molt (Fig. 13.9; Shero et al. 2015a). As a result, females simultaneously support the minimal energetic costs associated with the earliest stages of pregnancy and higher costs of hair regrowth. As in other species, molting occurs earlier for non-reproductive and juvenile animals compared to reproductive males and females (Beltran et al. 2019). These patterns are likely due to the energetic costs associated with molting, and the highly seasonal environment in which Weddell seals live.

The Weddell seal's molt is characterized as a brief period when old fur is shed, the dermal layer maintained, and the new pelage emerges. It consists of two portions: the preparatory period during which the hair follicles are activated, begin active cell division, and the new hair shaft begins to grow subdermally; and the period of visible shedding, as the new hair emerges beyond the skin surface and the old fur is lost. In Weddell seals, the preparatory period begins during the breeding season (November), but visible shedding occurs several weeks to months later (Fig. 13.10; Beltran et al. 2019). Changes in a suite of hormones likely initiate molt, several of which also regulate the reproductive and annual cycle (Paus and Cotsarelis 1999; Conrad and Paus 2004; Craven et al. 2006). The molt is also accompanied by increased perfusion to the dermal papilla and follicles that increases the conductivity of the dermal layer and thermoregulatory costs by ~8% (Walcott 2019).

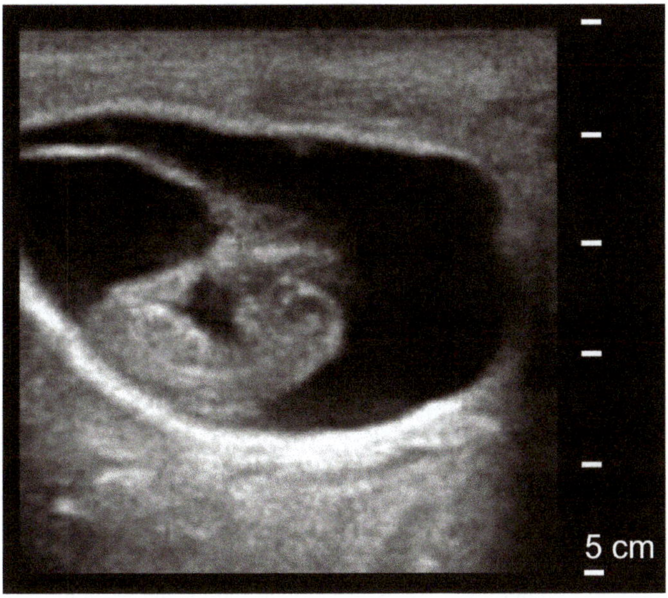

Fig. 13.9 Ultrasound image of Weddell seal embryo in January/February; detectably pregnant females were still in the molting process. Scale bars show 1 cm increments. Adapted from Shero et al. (2015a)

Fig. 13.10 Molting Weddell seals, transitioning from **a** entirely old, dull, and brown fur, **b** growth of new black fur on the dorsum and head, to **c** an entirely new black-silver pelage. NMFS 87-1851-04 and 17411. *Photo credits* Michelle Shero, Roxanne Beltran, Daniel Costa

Several behavioral shifts accompany molting: during the preparatory phase and following pup weaning, territorial males and post-parturient female disperse from the breeding colonies and begin active foraging (more below). However, the seals remain within the local area, as they shortly return to remnant fast-ice as the visible shedding begins. Seal density during the molt can be quite high, but males no longer defend ice cracks, nor do they react to other animals hauled-out nearby (Smith 1965, 1966; M. Shero and J. Burns *pers. obs.*). At this point, molting Weddell seals spend (~12%) less time in the cold water, particularly earlier in the visible shedding cycle, as this is when perfusion rates are likely highest (Beltran et al. 2017; Walcott 2019). Haulout activity is centered in the early afternoon when ambient conditions are warmest (Lake et al. 1997). Reduced aquatic activity slows heat loss to the environment, but also reduces the time available for foraging activities during a period when protein demands (for new hair synthesis) are high. At the end of the molting period, individual seals are still less massive (on both lean and lipid basis) than when they return to the colonies in late spring (at the start of their annual cycle; Shero et al. 2014, 2015c).

While pups molt their lanugo pelage during the nursing period or just thereafter, all other age classes molt after breeding (DeMaster 1979; Beltran et al. 2019). Juvenile and non-reproductive individuals can molt earlier in the summer when ambient conditions are more benign because they did not experience the physiological costs associated with reproductive activities. By molting when it is warmer, they may incur reduced thermoregulatory costs during this period. Conversely, individuals that molt late in the season may experience higher thermoregulatory costs due to colder ambient conditions by February/March. Animals that molt later would also have lower accessibility to their preferred stable and largely predator-free fast-ice substrate due to the increased prevalence of storms breaking fast-ice away from the coastal zone (Walcott 2019). Thus, selective pressures may favor molting before the onset of fall storms and poor weather, even though this reduces the time that post-reproductive individuals have to forage during the summer window when prey are potentially more accessible and shallower in the water column (Beltran 2018). The spread of molt dates among individual Weddell seals in the Erebus Bay population (visible shedding initiated as early as December and as late as March) reflects the balancing of these constraints by individuals with different past histories. Changes in climate patterns and/or summer prey availability would therefore likely differentially impact various population segments.

13.4.7 The Austral Winter

Weddell seals begin intensive foraging at the onset of the austral winter. During this period, females regain the remainder of the mass lost during lactation and also support the costs of late-stage pregnancies. To do so, adults increase time spent in the water from 40–45% late summer (January/February) to 50–65% of the day diving from March to October (Shero et al. 2018). Despite significantly increasing foraging effort over the austral winter, female Weddell seals gain considerably less body mass and lipid stores in preparation for the next year's lactation relative to other phocid species. Female Weddell seals only gain 10–15% of their body mass, whereas other phocids such as gray (*Halichoerus grypus*) and northern elephant seals (*Mirounga angustirostris*) increase their body mass by 55–>70% (Shero et al. 2015c. The lower amount of mass recovered/accumulated over winter may be due to difficulties finding and exploiting prey at depth and in complete darkness, particularly in areas with heavy ice cover and low winter productivity rates.

Weddell seals move away from their breeding colonies to improve foraging success once they are no longer constrained to be within habitats suitable for young pups. Across their range, Weddell seals appear to have a great affinity for areas with modified circumpolar deep water (MCDW) being advected onto the shelf and increased mixed layer depths (MLD). These are regions associated with higher productivity and prey availability, particularly around bathymetric troughs and banks (Goetz 2015). The distance animals need to travel to reach prime foraging areas has shifted through time and differs across populations. For example, of 15 female Weddell seals from Erebus Bay tracked overwinter in 1990–1991, all but two animals stayed within 50 km of Ross Island (Testa 1994). In contrast, the vast majority of seals tagged in Erebus Bay and the Victorialand coastline from 2010 to 2012 utilized the entirety of the Western Ross Sea. The Ross Sea Weddell seals preferred habitat with lower ice cover (<60%) and often traveled >500 km from their tagging sites overwinter to reach MCDW (Fig. 13.11, Goetz 2015).

In contrast to the above, Weddell seal breeding colonies in East Antarctica at Dumont D'Urville in Adelie Land (66.4°S, 140°E) and Davis Station (68.6°S, 78°E) are close to MCDW and animals do not need to expend as much time and energy transiting to productive waters. Instead, these seals remained within 34 km and 88 km of their tagging sites, respectively, while preferring to forage in areas on the continental shelf with little bathymetric variability and locations with higher ice cover (>80%; Heerah et al. 2013, 2016). Weddell seals at Dumont D'Urville maintained small home ranges while hunting within a 5–6 km radius. In contrast to the Ross Sea seals that travel long distances overwinter, seals in East Antarctica only appeared to travel as far as they could with one breath to remain close to breathing holes (Lake et al. 2006). Weddell seals actively maintain breathing holes during heavy ice cover conditions by reaming the ice with their teeth. Thus, re-use of the same hole for longer periods would reduce wear on the teeth as seals age, which is believed to contribute to adult mortalities (Stirling 1969b).

Weddell seal dives become progressively longer and deeper from the summer to the winter months (Testa 1994; Heerah et al. 2013; Goetz 2015; Shero et al.

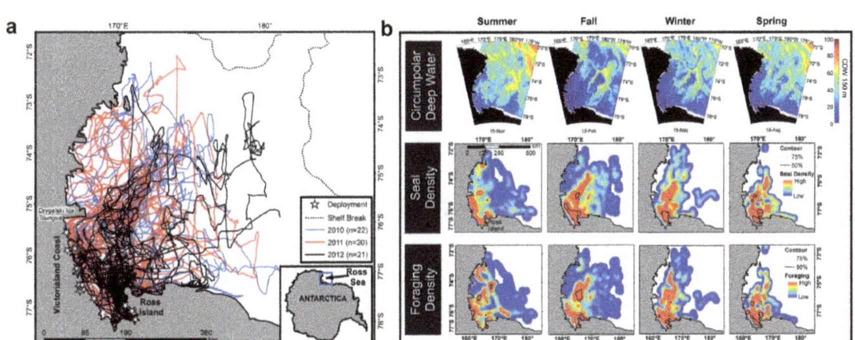

Fig. 13.11 **a** Weddell seal track lines throughout McMurdo Sound, Ross Sea and **b** seasonal shifts in seal foraging patterns associated with oceanographic features. (*Top*) A representative day from each season showing areas with Circumpolar Deep Water (CDW) relative to seal location (red points show individual seals for the representative day in year 2010). CDW is an important factor associated with seasonal shifts in seal density and foraging density (determined from first-passage time and area-restricted search analyses). Seals were originally instrumented at Ross Island and the Victorialand coastline. From Goetz (2015)

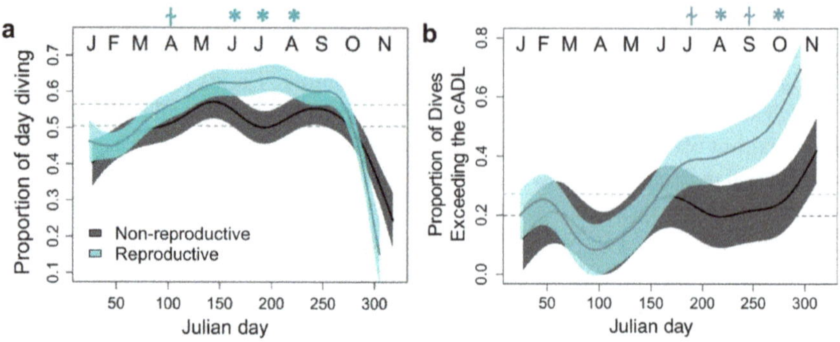

Fig. 13.12 Female Weddell seals that successfully give birth to a pup after the long overwinter foraging period tended to **a** spend more time diving per day and **b** exceeded their calculated aerobic dive limits more often during late winter (and gestation) when compared to non-reproductive females (†: $P < 0.1$ and *: $P < 0.05$, GAMM, shaded region shows 95% CI). From Shero et al. (2018)

2018), increasing from an average of 8.2 min in summer to 12.3 min in winter, and from depths of 114 m in summer to 223 m by late winter and spring (Goetz 2015). Moreover, the 8-month overwinter foraging period coincides with active gestation (Smith 1966; Shero et al. 2018), and females that successfully give birth to a pup the next year tend to dive longer, deeper, have longer foraging bouts, and spend more time in the water (by 1.1 h per day) than females that fail to produce offspring (Fig. 13.12; Shero et al. 2018). To support the additional energetic costs of pregnancy females likely need to push their physiological dive limits. Indeed, females that produced offspring after the winter foraging period exceeded their calculated aerobic dive limits (cADL's) more often on dives than females that did not give birth (Shero et al. 2015b, 2018). These differences in dive behavior between females that successfully produced a pup versus those that did not become most apparent in late winter, when the energetic costs of maintaining a third-trimester pregnancy is highest.

13.5 Physiological Underpinnings of Dive Behavior

Because Weddell seals are large, easily accessible, and tolerant of repeated human handling, they have frequently served as the test case for new approaches in the study of diving behavior, and the physiological adaptations that permit marine mammals to exploit underwater prey resources. Weddell seals are also remarkable divers among the Phocidae, second only to elephant seals. While on average dives typically last ~12–15 min, the longest-ever recorded dive by a Weddell seal is 96 min (Heerah et al. 2013). The ability to make such long dives is ultimately due to large endogenous oxygen stores (hemoglobin, myoglobin) and low diving metabolic rates (Kooyman 1989; Ponganis et al. 2011), allowing seals to maximize their breath-hold abilities and reach extraordinary depths of up to 904 m (Heerah et al. 2013).

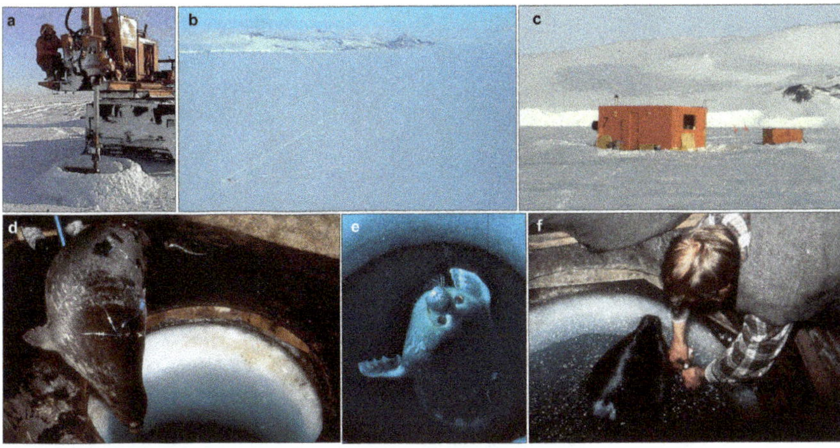

Fig. 13.13 **a–c** Drilling a hole in the sea ice isolated from all other cracks in the area, for a research camp. **d–f** Weddell seals are brought to the research camp for isolated hole experiments where they can dive freely but must return to the single breathing hole. Physiologic samples can be collected upon resurfacing. *Photo credits* **a–e** Randall Davis, **f** Randall Davis/Michael Castellini

The Weddell seal's fast-ice habitat in Erebus Bay made it possible to conduct isolated hole experiments, which were crucial to unraveling the series of physiologic events in the 'dive response'. These experiments were carried out from a human-made dive hole drilled in the ice at a location devoid of nearby tidal cracks (for ~3–6 km). Thus, seals could dive freely but were constrained to return to the single, isolated breathing hole (Kooyman 1981). Upon resurfacing, scientists were able to collect blood samples and other measurements to gain a detailed understanding of the physiological mechanisms that permit Weddell seals to attain great dive depths and durations (Fig. 13.13). In the study that jump-started the field of diving physiology, Gerald Kooyman and his team used this approach to investigate the aerobic dive limits of seals. They demonstrated that the dive reflex is not an all-or-nothing response but instead is graded on voluntary dives. Following dives shorter than ~20 min, there were only low (background) concentrations of lactate in the Weddell seal's blood, but an exponential increase in anaerobic byproducts once dives exceeded this duration (Kooyman et al. 1980). Behaviorally, high lactate results in exponentially longer post-dive recovery times when animals must remain at the water's surface (Kooyman et al. 1980) or limits seals to shorter dives (Castellini et al. 1988). This provides a physiologic threshold (the 'aerobic dive limit', Chap. 4) with significant ecological implications (Chap. 8). As evidenced by the use of the earliest time-depth recorders (TDRs) and film-based Billups dive recorders (Fig. 13.14a, b), the vast majority (~95%) of Weddell seal dives remain aerobic in nature (Kooyman et al. 1980). A myriad of subsequent studies combined the isolated hole approach with new technologies to obtain some of the first records of diving heart rate, diving heat flux, flipper stroke frequency, oxygen consumption, and diving metabolic rate (DMR). In combination, these approaches led to a mechanistic understanding of the

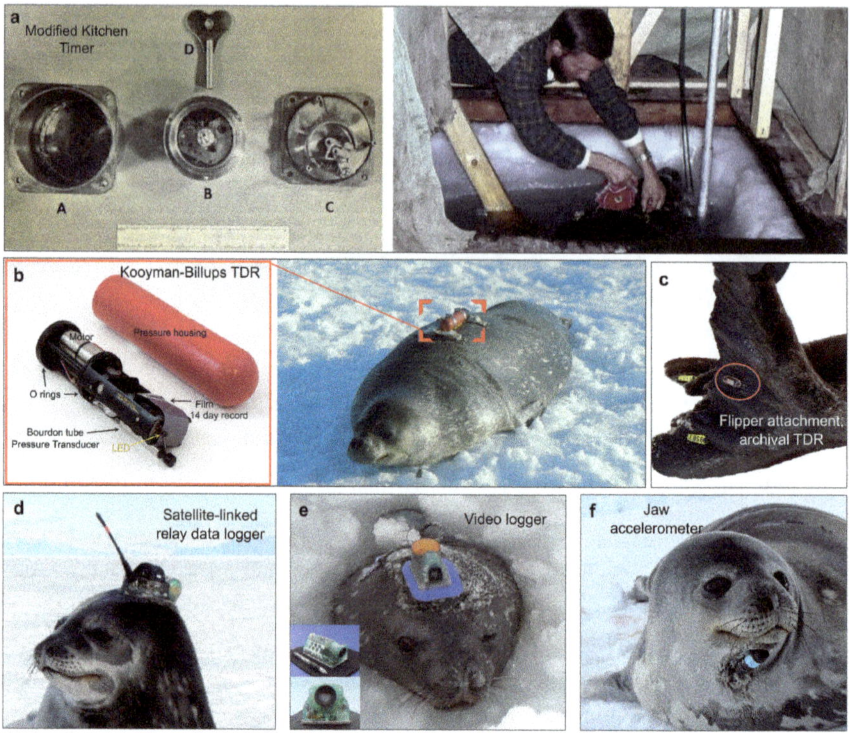

Fig. 13.14 **a** The earliest time-depth recorder (TDR) was a modified kitchen timer which recorded a dive trace (from Kooyman 1965, 1966), followed by the **b** Kooyman-Billups TDR that was capable of recording dive behavior for 14 days, with a pressure transducer moving an LED and exposing film, **c** a contemporary TDR capable of recording dive behavior for > 1 year (LOTEK Lat series), **d** contemporary relay logger that transmits dive behavioral data via satellite so that instruments do not need to be physically recovered (Sea Mammal Research Unit), **e** contemporary video logger capable of recording 27 h, **f** contemporary jaw accelerometer used to detect prey capture events (Little Leonardo co.). *Photo credits* **a** Gerald Kooyman, **b**, **d** Daniel Costa, NMFS 87-1851-04, **c**, **f** Roxanne Beltran, NMFS 17,411, **e** Randall Davis, NMFS 984-1814

behavioral and physiological adaptations that allow Weddell seals to sustain aerobic metabolic processes during most dives.

Over the subsequent 50 years, film time-depth records quickly gave way to digital recorders, and advances in technology improved both the temporal resolution and duration over which behavior could be recorded, while reducing the size, weight, and potential biological impact of instrumentation (Fig. 13.14c–f). For example, electrocardiograms and the development of microprocessor 'backpacks' for controlled blood sampling during dives have shown that the onset of Weddell seal dives are marked by profound bradycardia (heart rate <10 bpm). This further revealed that lung collapse occurs at 25–50 m depths to prevent gas exchange and avoid 'the bends' (i.e., nitrogen narcosis; Falke et al. 1985; Chap. 4). Without an influx from the lungs, oxygen is a finite resource that is partitioned to maximize dive durations.

Near-infrared spectrophotometers implanted in-vivo revealed that myoglobin-bound oxygen stores are consumed by the working muscles. Vasoconstriction prevents muscles from being replenished with blood-O_2 stores on dives, and oxygenated blood is shunted to the vital tissues (heart, brain, and lungs) instead (Guyton et al. 1995). Post-dive oxygen consumption also increases proportionately with the number of strokes an animal makes (net cost of a stroke: 0.044 mL O_2 per kg), portraying the additional energetic costs of attaining greater dive depths. However, swimming stroke costs are lower than the energetic expenditures of strides in terrestrial mammals (Williams et al. 2004). Overall post-dive oxygen consumption is 1.3–3.5 times greater than surface resting metabolic rates. Also, post-dive oxygen consumption increases due to prey capture success (by 45%, for up to 5-h) because of the warming of ingested fish and heat increment of feeding (Williams et al. 2004). As a result, we now know that Weddell seals have large tissue O_2 stores, exquisite regulation of diving blood flow patterns, and low locomotory costs and diving metabolic rates, allowing for aerobic metabolism to persist for 10–20 min into their dives (Fig. 13.15; Kooyman et al. 1980; Kooyman 1989; Williams et al. 2015).

Advances in animal-borne imaging cameras and jaw accelerometers (Fig. 13.14e, f) have also provided new insights into the linkages between prey abundance and dive behavior, and how well actual foraging behavior reflects theoretical predictions of Optimal Foraging Theory (Chap. 6). In the vicinity of the Erebus Bay breeding colonies, maximum prey density is located in the water column at 250–350 m;

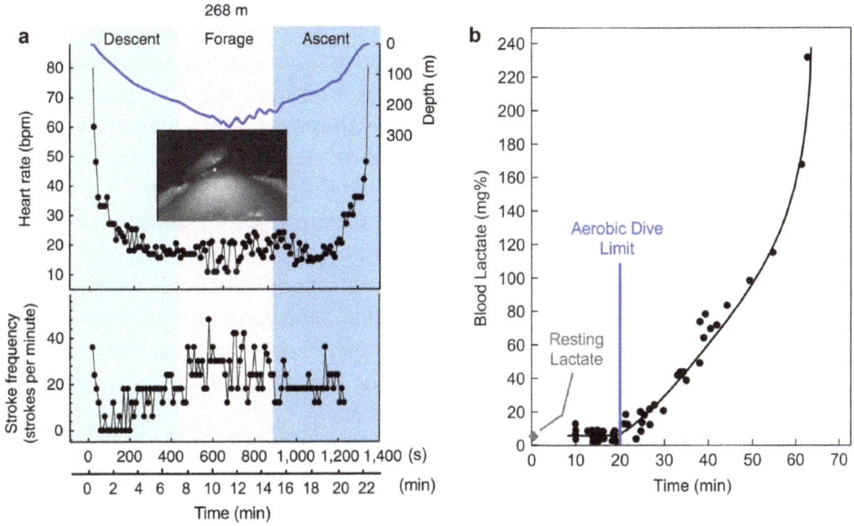

Fig. 13.15 **a** Weddell seal diving heart rate is affected by dive duration, depth, and stroke frequency (from Williams et al. 2015) and animals significantly lower their heart rate (bradycardia) on dives. **b** As a result, Weddell seals prolong the duration that they can rely on aerobic metabolism for their dives (the aerobic dive limit) as shown by blood lactate concentrations only beginning to increase above background levels after 20 + minute dives (from Kooyman et al. 1980)

however, seals focus their dive efforts at the shallow extent of this range (250–300 m) despite the ability to easily achieve 350 m depths (Sato et al. 2002; Watanabe et al. 2003). Because the costs of locomotion increase linearly on a dive, the animals should only travel as deep in the water column as is necessary to reach relatively high prey density to maximize net energetic gains and forage optimally (Watanabe et al. 2003; Mitani et al. 2004). It may be beneficial for animals to exceed their aerobic dive limits to exploit rich prey patches if the energetic gains from these resources outweigh the metabolic costs and extra time needed to breakdown lactate byproducts after surfacing (Houston and Carbone 1992). Indeed, Weddell seals diving longer than 20 min (and thus presumed to rely more on anaerobic glycolysis) encountered more prey per unit area, determined from cameras deployed on the animals (Mori et al. 2005). Understanding the interplay between oxygen stores, dive duration, and foraging optimality and success has supported modeling efforts helping to elucidate the Weddell seal's year-round energy budgets (Beltran et al. 2017; Salas et al. 2017).

13.6 Diet

At-sea habitat use and vertical dive behavior through the water column are driven by availability and predictability of prey resources—both on a daily and seasonal scale. In McMurdo Sound, scat and stable isotope analyses indicate that the majority of the Weddell seal's diet is comprised of mature (~14 cm) Antarctic silverfish (*Pleuragramma antarcticum*), and supplemented by other small notothenioid fishes such as *Pagothenia borchgrevinki* and *Trematomus* species (Burns et al. 1998; Goetz et al. 2017). As *P. antarcticum* accounts for >90% of the fish biomass in the McMurdo Sound ecosystem, it is not surprising that they are the primary prey item of adult seals (Eastman 1985). Video cameras deployed on Weddell seals revealed that silverfish are frequently captured at the midwater depths (~200 m, preferred by adult fish), with more than 50% of dives involving multiple prey capture events (Fuiman et al. 2007). In contrast, seals employ a bubble blowing strategy to flush the smaller *P. borchgrevinki* from their shelter in near-surface platelet ice (Davis et al. 1999; Fuiman et al. 2007) and also forage in nearshore benthic habitats on *Trematomus* species (Goetz et al. 2017; Brault et al. 2019). Invertebrate species (squid, octopus, krill, amphipods, and other crustaceans) occur in the diet of Weddell seals, with their frequency differing by age and location. For example, cephalopod beaks, primarily from *Brachioteuthiid picta*, occur more commonly in scat from juvenile Weddell seals (Burns et al. 1998), whereas crustacean prey are more common in the diet of Weddell seals from along the Western Antarctic Peninsula and the Davis Sound region than in McMurdo Sound (Green and Burton 1987). Seasonal differences in diet are poorly known, although studies in the Weddell sea suggest that seals forage mainly on *P. antarcticum* in the summer but concentrate on other species in the spring (Plötz 1986; Green and Burton 1987; Plötz et al. 1991; Castellini et al. 1992).

While most prey items available to Weddell seals are small, the Antarctic toothfish (*Dissostichus mawsoni*) is a large (max length >1.7 m), energy-dense (Lenky et al.

2012) prey species. It is likely to be a critically important component of the seals' diet, particularly during the mid-summer mass recovery period (Salas et al. 2017). Video recordings have documented seals stalking large toothfish by backlighting fish from below and, when encountered, large toothfish were generally at shallower depths as compared with silverfish. However, toothfish encounters were relatively rare overall (8% of dives; Fuiman et al. 2007). This fits with other dietary studies, as neither scat nor stable isotope analyses suggest that toothfish form a significant portion of the diet. Weddell seals do not ingest many of the bones or the large head (Ainley and Siniff 2009), thus toothfish could potentially be underrepresented in diet based on the recovery of hard parts. However, such bias does not occur with isotopic studies, and stable isotope analysis suggests that the average contribution of toothfish to the Weddell seal's diet is low (1.6%). There is significant individual variability in the estimated contribution, with toothfish making up a greater portion of the diet in older, larger seals (Fig. 13.16, range: 0.8–19.3%; Goetz et al. 2017). Indeed, seals may focus more heavily on toothfish when it is in high abundance. For example, during an isolated hole experiment, three Weddell seals consumed 28 toothfish across 37 days, and one seal even caught and consumed three toothfish in a single day (Ponganis and Stockard 2007). Such observations raised concern regarding commercial fishing of toothfish within the Ross Sea, helping to lead to the establishment of the world's largest Marine Protected Area (1.5 million km^2) in 2016 (Fig. 13.17).

There is ongoing debate regarding the importance of toothfish to the Weddell seal's diet (Ainley and Siniff 2009; Goetz et al. 2017), and the impact of increased fishing pressure on toothfish stocks, as well as the associated uncertainty about how that pressure is impacting the availability of toothfish and silverfish to Weddell seal populations in the Ross Sea area. These issues are compounded with the potential impacts of climate change on the timing and availability of sea ice, fast-ice, and prey distributions (Saenz et al. 2020), suggesting that a more nuanced understanding of

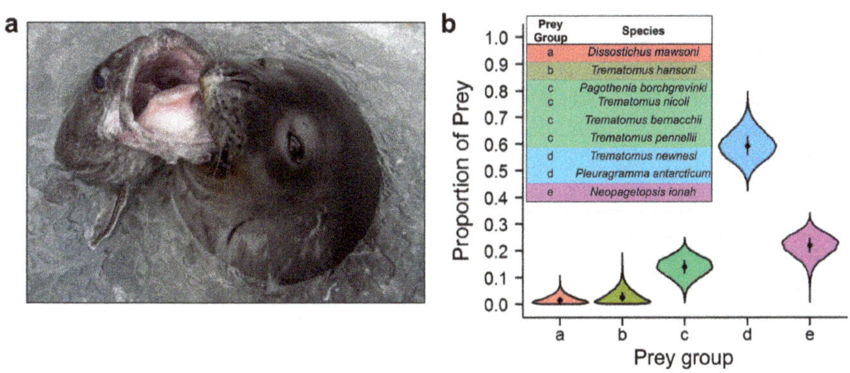

Fig. 13.16 **a** A Weddell seal bringing a toothfish, *Dissostichus mawsoni*, to the surface for consumption. *Photo credit* Jessica Meir. **b** Proportion of the Weddell seal's diet comprised of fish species groups, derived from stable isotope analysis of red blood cells in Goetz et al. (2017)

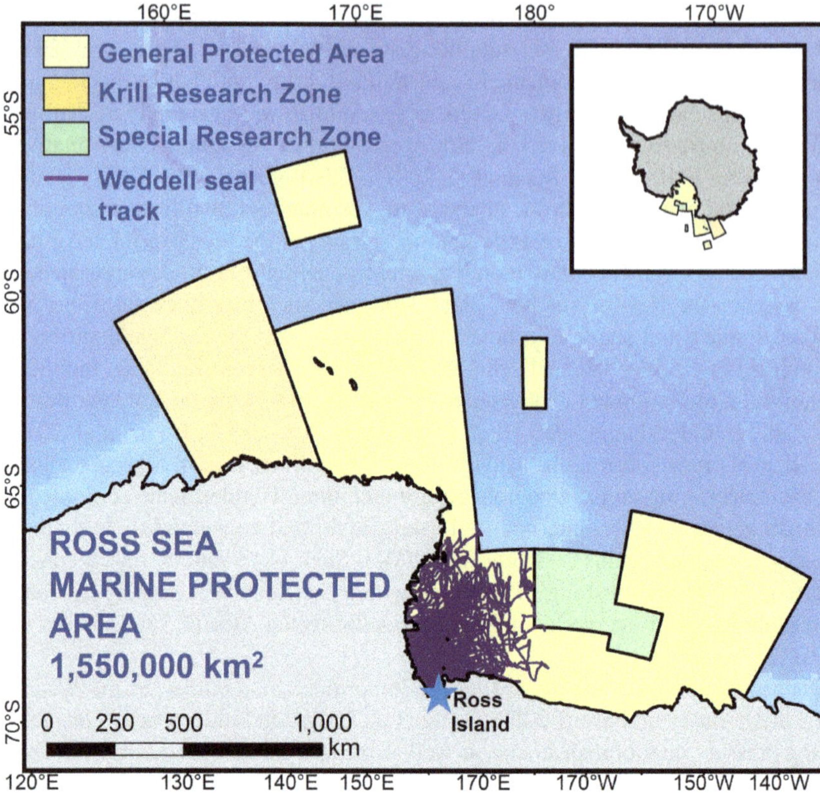

Fig. 13.17 The Ross Sea Marine Protected Area, which has substantial spatial overlap with habitat utilized by Weddell seals in McMurdo Sound (animal track lines; seals were originally instrumented at Ross Island and along the Victorialand coastline). Courtesy of Kimberly Goetz

the factors influencing trophic structure in the Ross Sea is needed. For example, when Hückstädt et al. (2017) compared stable isotope values in modern-day fur samples to samples recovered from Weddell seal pelts left in Robert Falcon Scott's Discovery Hut between 1901 and 1904, bulk isotope analysis suggested a change in the seal's trophic position. However, compound-specific stable isotope analysis of individual amino acids revealed that this was due to a shift in isotopic values at the base of the food chain, potentially due to climate regime shifts associated with the end of the Little Ice Age (Hückstädt et al. 2017; Brault et al. 2019). When this was factored into the analysis, Hückstädt et al. (2017) concluded that contemporary Weddell seals' trophic position was consistent with that from 100 years ago, despite the recent fishing pressure on toothfish. Nevertheless, apparent declines in the abundance of Weddell seals in the greater Ross Sea (Ainley et al. 2015) suggest that a precautionary approach to setting fishing limits needs to be undertaken.

13.7 A Population at the Edge

Most Weddell seal populations are found in coastal areas with fast-ice and ready access to open ocean habitat, but the population at White Island is completely enclosed and geographically isolated. As such, these seals have been used as a natural model for conservation demography to better understand the fate of exceedingly small populations (Testa and Scotton 1999; Gelatt et al. 2010). It is believed that the population was founded in the mid-1950s when a short-term break in the ice shelf provided seals from Erebus Bay with access to White Island (77° 10'S, 167° 20'E), which is more than 20 km to the south. Two seals were first observed at White Island in 1958 (Heine 1960), and genetic analysis suggests that the founding group of seals consisted of only three females and two males. Since the founding event, the seals have been trapped by surrounding thick (15 m) ice and the Ross and McMurdo Ice Shelves, approximately 20 km from cracks or any open water (Castellini et al. 1984; Gelatt et al. 2010). Periodic tagging and census work indicate that the population reached 11 seals in the 1960s, and increased to the current size of ~26 seals by 1994 (Stirling 1972; Gelatt et al. 2010). Area-wide censuses have provided no evidence of migration between the two regions (Stirling 1972; Testa and Scotton 1999). However, in 2015 a carcass of a juvenile male tagged at White Island was found in the underwater habitat near Turtle Rock in Erebus Bay (R. Robbins, J. Rotella *pers. comm.*). It is unclear how this White Island seal made it to Erebus Bay—perhaps, it died at White Island and was carried by underwater currents into Erebus Bay.

The White Island Weddell seals live under the most extreme environmental conditions of any Weddell seal population. The Ross ice shelf is extremely thick, even along the shores of White Island, making entry to and egress from the water challenging for seals. In addition, light levels under the shelf are much lower than under the annual sea ice. Local prey resources may be less abundant due to the greater distance from the ice edge and sources of primary productivity. There are several notable differences in demographic parameters between the White Island and Erebus Bay populations that may result from these environmental factors: female reproductive rates are much lower (33% at White Island vs. 50–75% in Erebus Bay), and the sex ratio is highly skewed (2.2F:1M vs. 1F:1M). In addition, pup survival rates are quite low: of the 64 pups born 1990–2000, only seven (6F:1M) were subsequently observed (Gelatt et al. 2010). Genetic analyses indicate that inbreeding coefficients are high (0.094–0.313) with all pups sired by just three males, and several pups the result of mother-son matings. Pups that survive their first year have lower inbreeding coefficients (0.071) than pups that do not (0.141), and skeletal deformities occur at higher frequency in White Island mortalities (Testa and Scotton 1999). Population projections suggest that in the absence of new immigrants, inbreeding coefficients for all individuals will soon exceed that of non-surviving offspring (Gelatt et al. 2010).

There is one additional difference between seals from the two locations: seals at White Island are notably larger and fatter than their Erebus Bay ancestors (Stirling 1972). White Island seals forage on prey underneath the ice shelf and do not face competition for that prey from penguins or whales. In addition, the small seal population likely does not deplete the local prey resources, which may result in high foraging success (Castellini et al. 1984). The summer pulse of regional productivity may occur later further to the south under the shelf, which would account for the fact that females give birth approximately three weeks later in the summer at White Island as compared to Erebus Bay (Castellini et al. 1984). Nevertheless, the beneficial foraging environment and freedom from predation by killer whales seem unlikely to offset potential negative impacts of continued inbreeding, and this population may not be viable if it remains isolated over the long run.

13.8 Concluding Remarks

The Weddell seal has been an invaluable model species for pioneering research in the fields of marine mammal demography, life history, and physiology. From the beginning, it has been evident that all aspects of a Weddell seal's life—diving patterns, foraging success, behavior, energetics, physiology, reproduction, and annual cycle—are shaped by the marked seasonal changes in their highly polar environment. Adaptations that permit Weddell seals to exploit seasonally abundant food resources and the (relatively) more benign ambient summer conditions opened a unique niche for this species. Physiological and behavioral specializations allow Weddell seals to thrive under conditions that might otherwise be considered the world's harshest and most inhospitable environment for a warm-blooded, air-breathing, semi-aquatic species. However, these same adaptations to life at high latitudes may make them more susceptible to unexpected perturbations or climate regime shifts than temperate phocids. Individuals may be particularly vulnerable to changes in prey abundance or predictability if shifts in resource availability occur during periods when Weddell seals are already approaching their physiologic limits. For example, Weddell seals would be particulately affected if resource predictability decreased during lactation or the breeding season when animals experience peak energetic demands. Weddell seals would be similarly impacted if prey availability declined at times when animals are pushing their aerobic dive limits and foraging capacities (winter for adults—especially pregnant females; during late summer for newly weaned pups). The recognition that not all segments of the population may be similarly vulnerable to anthropogenic or climate-related shifts in habitat suitability highlights the critical need to continue monitoring efforts even for species that are often thought to be well-studied.

References

Ackley SF, Sullivan CW (1994) Physical controls on the development and characteristics of Antarctic sea ice biological communities—a review and synthesis. Deep Sea Res 41:1583–1604

Ainley DG, LaRue MA, Stirling I, Stammerjohn S, Siniff DB (2015) An apparent population decrease, or change in distribution, of Weddell seals along the Victoria Land coast. Mar Mamm Sci 31(4):1338–1361

Ainley DG, Siniff DB (2009) The importance of Antarctic toothfish as prey of Weddell seals in the Ross Sea. Antarct Sci 21(4):317–327

Antarctic Conservation Act, 16 U.S.C §§ 2401–2413, *et seq.* (1978)

Bartsh SS, Johnston SD, Siniff DB (1991) Territorial behavior and breeding frequency of male Weddell seals (*Leptonychotes weddelli*) in relation to age, size, and concentrations of serum testosterone and cortisol. Can J Zool 70:680–692

Beltran RS (2018) Bridging the gap between pupping and molting phenology: behavioral and ecological drivers in Weddell seals. Dissertation, University of Alaska Fairbanks

Beltran RS, Kirkham AL, Breed GA, Testa JW, Burns JM (2019) Reproductive success delays moult phenology in a polar mammal. Sci Rep 9(1):5221. https://doi.org/10.1038/s41598-019-41635-x

Beltran RS, Testa JW, Burns JM (2017) An agent-based bioenergetics model for predicting impacts of environmental change on a top marine predator, the Weddell seal. Ecol Modell 351:36–50. https://doi.org/10.1016/j.ecolmodel.2017.02.002

Bergeron P, Baeta R, Pelletier F, Réale D, Garant D (2011) Individual quality: tautology or biological reality? J Anim Ecol 80(2):361–364. https://doi.org/10.1111/j.1365-2656.2010.01770.x

Bertram GCL (1940) The biology of the Weddell and crabeater seals. Br Graham Land Exped Sci Rep 1:1–139

Brault EK, Koch PL, Costa DP, McCarthy MD, Hückstädt LA, Goetz KT, McMahon KW, Goebel ME, Karlsson O, Teilmann J, Harkonen T, Harding KC (2019) Trophic position and foraging ecology of Ross, Weddell, and crabeater seals revealed by compound-specific isotope analysis. Mar Ecol Prog Ser 611:1–18

Burns JM (1999) The development of diving behavior in juvenile Weddell seals: pushing physiological limits in order to survive. Can J Zool 77:737–747

Burns JM, Castellini MA (1996) Physiological and behavioral determinants of the aerobic dive limit in Weddell seal (*Leptonychotes weddellii*) pups. J Comp Physiol B 166:473–483

Burns JM, Castellini MA, Testa JW (1999) Movements and diving behavior of weaned Weddell seal (*Leptonychotes weddellii*) pups. Polar Biol 21:23–36

Burns JM, Lestyk K, Folkow LP, Hammill MO, Blix AS (2007) Size and distribution of oxygen stores in harp and hooded seals from birth to maturity. J Comp Physiol B 177:687–700

Burns JM, Trumble SJ, Castellini MA, Testa JW (1998) The diet of Weddell seals in McMurdo Sound, Antarctica as determined from scat collections and stable isotope analysis. Polar Biol 19:272–282

Cameron MF, Siniff DB (2004) Age-specific survival, abundance, and immigration rates of a Weddell seal (*Leptonychotes weddellii*) population in McMurdo Sound, Antarctica. Can J Zool 82:601–615

Carlini AR, Daneri GA, Márquez MEI, Bornemann H, Panarello H, Casaux R, Ramdohr S, Plötz J (2005) Food consumption estimates of southern elephant seal females during their post-breeding aquatic phase at King George Island. Polar Biol 28(10):769–775. https://doi.org/10.1007/s00300-005-0004-6

Castellini MA, Davis RW, Davis M, Horning M (1984) Antarctic marine life under the McMurdo ice shelf at White Island: a link between nutrient influx and seal population. Polar Biol 2:229–231

Castellini MA, Davis RW, Kooyman GL (1988) Blood chemistry regulation during repetitive diving in Weddell seals. Physiol Zool 61(5):379–386

Castellini MA, Davis RW, Kooyman GL (1992) Annual cycles of diving behavior and ecology of the Weddell seal. Bull Scripps Inst Oceanogr 28:1–54

Chambert T, Rotella JJ, Garrott RA (2012) Environmental extremes versus ecological extremes: impact of a massive iceberg on the population dynamics of a high-level Antarctic marine predator. Proc R Soc B 279:4532–4541

Chambert T, Rotella JJ, Garrott RA (2014) An evolutionary perspective on reproductive individual heterogeneity in a marine vertebrate. J Anim Ecol 83(5):1158–1168. https://doi.org/10.1111/1365-2656.12211

Chambert T, Rotella JJ, Higgs MD, Garrott RA (2013) Individual heterogeneity in reproductive rates and cost of reproduction in a long-lived vertebrate. Ecol Evol 3(7):2047–2060

Cisewski B, Strass VH, Rhein M, Krägefsky S (2010) Seasonal variation of diel vertical migration of zooplankton from ADCP backscatter time series data in the Lazarev Sea, Antarctica. Deep Sea Res Part I 57(1):78–94. https://doi.org/10.1016/j.dsr.2009.10.005

Cline DR, Siniff DB, Erickson AW (1971) Underwater copulation of the Weddell seal. J Mammal 52(1):216–218. https://doi.org/10.2307/1378453

Collins KT, Terhune JM, Rogers TL, Wheatley KE, Harcourt RG (2006) Vocal individuality of in-air Weddell seal (*Leptonychotes weddellii*) pup "primary" calls. Mar Mamm Sci 22(4):933–951. https://doi.org/10.1111/j.1748-7692.2006.00074.x

Conrad F, Paus R (2004) Estrogens and the hair follicle. J Ger Soc Dermatol 2(6):412–413

Craven AJ, Nixon AJ, Ashby MG, Ormandy CJ, Blazek K, Wilkins RJ, Pearson AJ (2006) Prolactin delays hair regrowth in mice. J Endocrinol 191:415–425

Croxall JP, Hiby L (1983) Fecundity, survival and site fidelity in Weddell seals, *Leptonychotes weddellii*. J Appl Ecol 20:19–32

Curtice C, Johnston DW, Ducklow H, Gales N, Halpin PN, Friedlaender AS (2015) Modeling the spatial and temporal dynamics of foraging movements of humpback whales (*Megaptera novaeangliae*) in the Western Antarctic Peninsula. Mov Ecol 3(1):13. https://doi.org/10.1186/s40462-015-0041-x

Davis RW, Fuiman LA, Williams TM, Collier SO, Hagey WP, Kanatous SB, Kohin S, Horning M (1999) Hunting behavior of a marine mammal beneath the Antarctic fast ice. Science 283:993–996

De Miranda MA, Schlater AE, Green TL, Kanatous SB (2012) In the face of hypoxia: myoglobin increases in response to hypoxic conditions and lipid supplementation in cultured Weddell seal skeletal muscle cells. J Exp Biol 215(5):806–813. https://doi.org/10.1242/jeb.060681

DeMaster D (1979) Weddell seal. In: Mammals in the seas, vol 2. FAO fisheries series, p 5

Eastman JT (1985) *Pleuragramma antarcticum* (Pisces, Nototheniidae) as food for other fishes in McMurdo Sound, Antarctica. Polar Biol 4:155–160

Eicken H (1993) The role of sea ice in structuring Antarctic ecosystems. In: Hempel G (ed) Weddell sea ecology. Springer, Berlin, p 3–13

Erickson AW, Hanson MB (1990) Continental estimates and population trends of Antarctic ice seals. In: Kerry KR, Hempel G (eds) Antarctic ecosystems. Springer Verlag, Berlin, pp 253–264

Falke KJ, Hill RD, Qvist J, Schneider RC, Guppy M, Liggins GC, Hochachka PW, Elliott RE, Zapol WM (1985) Seal lungs collapse during free diving: evidence from arterial nitrogen tensions. Science 229:556–558

Frenot Y, Chown SL, Whinam J, Selkirk PM, Convey P, Skotnicki M, Bergstrom DM (2005) Biological invasions in the Antarctic: extent, impacts and implications. Biol Rev 80(1):45–72. https://doi.org/10.1017/s1464793104006542

Fuiman LA, Madden KM, Williams TM, Davis RW (2007) Structure of foraging dives by Weddell seals at an offshore isolated hole in the Antarctic fast ice environment. Deep Sea Res II 54:270–289

Geiseler SJ, Blix AS, Burns JM, Folkow LP (2013) Rapid postnatal development of myoglobin from large liver iron stores in hooded seals. J Exp Biol 216(10):1793–1798. https://doi.org/10.1242/jeb.082099

Gelatt TS, Davis CS, Siniff DB, Strobeck C (2001) Molecular evidence for twinning in Weddell Seals (*Leptonychotes weddellii*). J Mammal 82(2):491–499

Gelatt TS, Davis CS, Stirling I, Siniff DB, Strobeck C, Delisle I (2010) History and fate of a small isolated population of Weddell seals at White Island, Antarctica. Conserv Genet 11(3):721–735. https://doi.org/10.1007/s10592-009-9856-6

Gill T (1872) Arrangement of the families of mammals; with analytical tables. Smithson Misc Coll No 230(11):6–98
Goetz KT (2015) Movement, habitat, and foraging behavior of Weddell seals (*Leptonychotes weddellii*) in the western Ross Sea, Antarctica. Dissertation, University of California Santa Cruz
Goetz KT, Burns JM, Hückstädt LA, Shero MR, Costa DP (2017) Temporal variation in isotopic composition and diet of Weddell seals in the western Ross Sea. Deep sea research part II: topical studies in oceanography 140 (Supplement C), pp 36–44. https://doi.org/10.1016/j.dsr2.2016.05.017
Goodman D (1974) Natural selection and a cost ceiling on reproductive effort. Am Nat 108(961):247–268. https://doi.org/10.1086/282906
Gray JE (1837) Description of some new or little known mammalia, principally in the British Museum collection. Mag Nat Hist 1:577–587
Greaves DK, Schreer JF, Hammill MO, Burns JM (2005) Diving heart rate development in postnatal harbour seals, *Phoca vitulina*. Physiol Biochem Zool 78:9–17
Green K, Burton HR (1987) Seasonal and geographical variation in the food of the weddell seals, *Leptonychotes weddellii*, in Antarctica. Aust Wildl Res 14:475–489
Guyton GP, Stanek KS, Schneider RC, Hochachka PW, Hurford WE, Zapol DG, Liggins GC, Zapol WM (1995) Myoglobin saturation in free-diving Weddell seals. J Appl Physiol 79(4):1148–1155
Haas C (2003) Sea ice: a critical habitat for polar marine mammals and birds. In: Thomas DN, Dieckmann GS (eds) Sea ice: an introduction to its physics, chemistry, biology, and geology. Blackwell, Oxford, p 240–266. https://doi.org/10.1002/9780470757161.ch8
Hadley GL, Rotella JJ, Garrott RA (2007a) Evaluation of reproductive costs for Weddell seals in Erebus Bay, Antarctica. J Anim Ecol 76:448–458
Hadley GL, Rotella JJ, Garrott RA (2007b) Influence of maternal characteristics and oceanographic conditions on survival and recruitment probabilities of Weddell seals. Oikos 116(4):601–613. https://doi.org/10.1111/j.0030-1299.2007.15528.x
Hadley GL, Rotella JJ, Garrott RA, Nichols JD (2006) Variation in probability of first reproduction of Weddell seals. J Anim Ecol 75(5):1058–1070. https://doi.org/10.1111/j.1365-2656.2006.01118.x
Harcourt RG, Kingston JJ, Cameron MF, Waas JR, Hindell MA (2007a) Paternity analysis shows experience, not age, enhances mating success in an aquatically mating pinniped, the Weddell seal (*Leptonychotes weddellii*). Behav Ecol Sociobiol 61:643–652
Harcourt RG, Kingston JJ, Waas JR, Hindell MA (2007b) Foraging while breeding: alternative mating strategies by male Weddell seals? Aquat Conserv 17(S1):S68–S78. https://doi.org/10.1002/aqc.915
Hastings KK, Testa JW (1998) Maternal and birth colony effects on survival of Weddell seal offspring from McMurdo Sound, Antarctica. J Anim Ecol 67:722–740
Heerah K, Andrews-Goff V, Williams G, Sultan E, Hindell M, Patterson T, Charrassin J-B (2013) Ecology of Weddell seals during winter: Influence of environmental parameters on their foraging behaviour. Deep Sea Res Part II 88–89:23–33. https://doi.org/10.1016/j.dsr2.2012.08.025
Heerah K, Hindell M, Andrew-Goff V, Field I, McMahon CR, Charrassin J-B (2016) Contrasting behavior between two populations of an ice-obligate predator in East Antarctica. Ecol Evol 7(2):606–618. https://doi.org/10.1002/ece3.2652
Heine AJ (1960) Seals at White Island, Antarctica. Antarctica 2:272–273
Hill SE (1987) Reproductive ecology of Weddell seals (*Leptonychotes weddellii*) in McMurdo Sound. Thesis, University of Minnesota, Antarctica
Hindell MA, Harcourt R, Waas JR, Thompson D (2002) Fine-scale three-dimensional spatial use by diving, lactating female Weddell seals *Leptonychotes weddellii*. Mar Ecol Prog Ser 242:275–284
Hindle AG, Horning M, Mellish JE (2015) Estimating total body heat dissipation in air and water from skin surface heat flux telemetry in Weddell seals. Anim Biotel 3:50
Hindle AG, Horning M, Mellish JE, Lawler JM (2009) Diving into old age: muscular senescence in a large-bodied, long-lived mammal, the Weddell seal (*Leptonychotes weddellii*). J Exp Biol 212:790–796

Houston AI, Carbone C (1992) The optimal allocation of time during the diving cycle. Behav Ecol 3(3):255–265

Hückstädt LA, McCarthy MD, Koch PL, Costa DP (2017) What difference does a century make? Shifts in the ecosystem structure of the Ross Sea, Antarctica, as evidenced from a sentinel species, the Weddell seal. Proc Royal Soc B Biol Sci 284(1861):20170927. https://doi.org/10.1098/rspb.2017.0927

Jenouvrier S, Péron C, Weimerskirch H (2015) Extreme climate events and individual heterogeneity shape life-history traits and population dynamics. Ecol Monogr 85(4):605–624. https://doi.org/10.1890/14-1834.1

Jones RM, Smith WO (2017) The influence of short-term events on the hydrographic and biological structure of the southwestern Ross Sea. J Mar Syst 166:184–195. https://doi.org/10.1016/j.jmarsys.2016.09.006

Kaufman GWLC, Dubuque I (Dept. of Biology, USA), Siniff DB, Reichle R (1975) Colony behavior of Weddell seals, *Leptonychotes weddelli*, at Hutton Cliffs, Antarctica, vol 169

Kooyman GL (1981) Weddell seal: consummate diver. Cambridge University Press, Cambridge

Kooyman GL (1989) Diverse divers: Physiology and behavior. Springer-Verlag, Berlin

Kooyman GL, Wahrenbrock EA, Castellini MA, Davis RW, Sinnett EE (1980) Aerobic and anaerobic metabolism during voluntary diving in Weddell seals: evidence of preferred pathways from blood chemistry and behavior. J Comp Physiol 138:335–346

Kovacs KM, Lydersen C, Overland JE, Moore SE (2011) Impacts of changing sea-ice conditions on Arctic marine mammals. Mar Biodivers 41(1):181–194. https://doi.org/10.1007/s12526-010-0061-0

Lack D (1968) Ecological adaptations for breeding in birds. Methuen, London

Lake SE (1997) Aspects of the ecology of Weddell seals at the Vestfold Hills, Prydz Bay. Thesis, University of Tasmania, East Antarctica

Lake SE, Burton HR, Hindell MA (1997) Influence of time of day and month on Weddell seal haul-out patterns at the Vestfold Hills, Antarctica. Polar Biol 18:319–324

Lake S, Burton H, Wotherspoon S (2006) Movements of adult female Weddell seals during the winter months. Polar Bio 29(4) 270–279.https://doi.org/10.1007/s00300-005-0050-0

LaRue M, Salas L, Nur N, Ainley D, Stammerjohn S, Pennycook J, Dozier M, Saints J, Stamatiou K, Barrington L, Rotella J (2021) Insights from the first global population estimate of Weddell seals in Antarctica. Sci Adv 7(39):https://doi.org/10.1126/sciadv.abh3674

Le Boeuf B, Condit R, Reiter J (2019) Lifetime reproductive success of northern elephant seals (*Mirounga angustirostris*). Can J Zool 97(12):1203–1217. https://doi.org/10.1139/cjz-2019-0104

Lenky C, Eisert R, Oftedal OT, Metcalf V (2012) Proximate composition and energy density of nototheniid and myctophid fish in McMurdo Sound and the Ross Sea, Antarctica. Polar Biol 35(5):717–724. https://doi.org/10.1007/s00300-011-1116-9

Lescroël A, Dugger KM, Ballard G, Ainley DG (2009) Effects of individual quality, reproductive success and environmental variability on survival of a long-lived seabird. J Anim Ecol 78(4):798–806. https://doi.org/10.1111/j.1365-2656.2009.01542.x

Lesson RP (1826) Sur le phogue léopard de mer (Sea Leopard) des Orcades australes; par James Weddell. Reviewed by Lesson. [Ferussac's] Bulletin des Sciences Naturelles et de Géologie 7:437–438

Liggins GC, Creasy RK (1980) Experimental studies in the fetal seal. In: Nathanielsz PW (ed) Animal models in fetal medicine. Elsevier/North Holland Biomedical Press, pp 235–243

Mannas JM (2011) Assessing Weddell seal maternal investment in offspring. Montana State University, Thesis

Mansfield AW (1958) The breeding behaviour and reproductive cycle of the Weddell seal (*Leptonychotes weddellii* Lesson). Falkland Islands Dependencies Survey, Scientific Reports

Marine Mammal Protection Act, 16 U.S.C. §§ 1361 et seq. (1972)

Mitani Y, Watanabe Y, Sato K, Cameron MF, Naito Y (2004) 3D diving behavior of Weddell seals with respect to prey accessibility and abundance. Mar Ecol Prog Ser 281:275–281

Mori Y, Watanabe Y, Mitani Y, Sato K, Cameron MF, Naito Y (2005) A comparison of prey richness estimates for Weddell seals using diving profiles and image data. Mar Ecol Prog Ser 295:257–263

O'Toole MD, Lea M-A, Guinet C, Schick R, Hindell MA (2015) Foraging strategy switch of a top marine predator according to seasonal resource differences. Front Mar Sci 2(21). https://doi.org/10.3389/fmars.2015.00021

Paterson JT, Rotella JJ, Arrigo KR, Garrott RA (2015) Tight coupling of primary production and marine mammal reproduction in the Southern Ocean. Proc R Soc B 282:20143137

Paterson JT, Rotella JJ, Mannas JM, Garrott RA (2016) Patterns of age-related change in reproductive effort differ in the pre-natal and post-natal periods in a long-lived mammal. J Anim Ecol 85(6):1540–1551. https://doi.org/10.1111/1365-2656.12577

Paus R, Cotsarelis G (1999) The biology of hair follicles. N Engl J Med 341:491–497

Pearson LE, Liwanag HEM, Hammill MO, Burns JM (2014) Thermoregulatory strategy varies among neonatal polar phocids. Comp Biochem Physiol A Mol Integr Physiol 178:59–67

Plötz J (1986) Summer diet of Weddell Seals (*Leptonychotes weddelli*) in the eastern and southern Weddell Sea, Antarctica. Polar Biol 6(2):97–102. https://doi.org/10.1007/BF00258259

Plötz J, Ekau W, Reijnders PJH (1991) Diet of Weddell seals *Leptonychotes weddellii* at Vestkapp, Eastern Weddell Sea (Antarctica), in relation to local food supply. Mar Mamm Sci 7(2):136–144. https://doi.org/10.1111/j.1748-7692.1991.tb00560.x

Ponganis PJ, Meir JU, Williams CL (2011) In pursuit of Irving and Scholander: a review of oxygen store management in seals and penguins. J Exp Biol 214:3325–3339

Ponganis PJ, Stockard TK (2007) The Antarctic toothfish: how common a prey for Weddell seals? Antarct Sci 19(4):441–442

Proffitt KM, Garrott RA, Rotella JJ (2008a) Long-term evaluation of body mass at weaning and postweaning survival rates of Weddell seals in Erebus Bay, Antarctica. Mar Mamm Sci 24(3):677–689. https://doi.org/10.1111/j.1748-7692.2008.00207.x

Proffitt KM, Garrott RA, Rotella JJ (2008b) Variation in offspring sex ratio among individual Weddell seal (*Leptonychotes weddellii*) females of different quality. Behav Ecol Sociobiol 62(10):1679–1687. https://doi.org/10.1007/s00265-008-0596-2

Proffitt KM, Garrott RA, Rotella JJ, Siniff DB, Testa J (2007a) Exploring linkages between abiotic oceanographic processes and a top-trophic predator in an Antarctic ecosystem. Ecosystems 10(1):119–126

Proffitt KM, Garrott RA, Rotella JJ, Wheatley KE (2007b) Environmental and senescent related variations in Weddell seal body mass: implications for age-specific reproductive performance. Oikos 116(10):1683–1690

Rotella JJ, Paterson JT, Garrott RA (2016) Birth dates vary with fixed and dynamic maternal features, offspring sex, and extreme climatic events in a high-latitude marine mammal. Ecol Evol 6(7):1930–1941. https://doi.org/10.1002/ece3.1985

Saenz BT, Ainley DG, Daly KL, Ballard G, Conlisk E, Elrod ML, Kim SL (2020) Drivers of concentrated predation in an Antarctic marginal-ice-zone food web. Sci Rep 10:7282. https://doi.org/10.1038/s41598-020-63875-y

Salas L, Nur N, Ainley D, Burns J, Rotella J, Ballard G (2017) Coping with loss of large, energy-dense prey: a potential energy bottleneck for Weddell seals in the Southern Ross Sea. Ecol Appl 27:10–25

Sato K, Mitani Y, Cameron MF, Siniff DB, Watanabe Y, Naito Y (2002) Deep foraging dives in relation to the energy depletion of Weddell seal (*Leptonychotes weddellii*) mothers during lactation. Polar Biol 25:696–702

Sato K, Mitani Y, Naito Y, Kusagaya H (2003) Synchronous shallow dives by Weddell seal mother-pup pairs during lactation. Mar Mamm Sci 19(2):384–395. https://doi.org/10.1111/j.1748-7692.2003.tb01116.x

Schreer JF, Hastings KK, Testa JW (1996) Preweaning mortality of Weddell seal pups. Can J Zool 74:1775–1778

Shaughnessay P (1969) Transferrin polymorphism and population structure of the Weddell seal *Leptonychotes weddelli* (Lesson). Aust J Biol Sci 22(6):1581–1584. https://doi.org/10.1071/BI9691581

Shero MR, Adams GP, Burns JM (2015a) Field use of ultrasonography to characterize the reproductive tract and early pregnancy in a phocid, the Weddell seal (*Leptonychotes weddellii*). Anat Rec 298(12):1970–1977

Shero MR, Costa DP, Burns JM (2015b) Scaling matters: Incorporating body composition into Weddell seal seasonal oxygen store comparisons reveals maintenance of aerobic capacities. J Comp Physiol B 185(7):811–824

Shero MR, Goetz KT, Costa DP, Burns JM (2018) Temporal changes in Weddell seal dive behavior over winter: are females increasing foraging effort to support gestation? Ecol Evol 8(23):11857–11874. https://doi.org/10.1002/ece3.4643

Shero MR, Krotz RT, Costa DP, Avery JP, Burns JM (2015c) How do overwinter changes in body condition and hormone profiles influence Weddell seal reproductive success? Funct Ecol 29(10):1278–1291

Shero MR, Pearson LE, Costa DP, Burns JM (2014) Improving the precision of our ecosystem calipers: a modified morphometric technique for estimating marine mammal mass and body composition. PLoS One 9 (3):e91233. https://doi.org/10.1371/journal.pone.0091233

Siniff D, Garrott RA, Rotella JJ, Fraser WR, Ainley D (2008) Projecting the effects of environmental change on Antarctic seals. Antarct Sci 20(5):425–435

Siniff DB (1981) Seal population dynamics and ecology. J R Soc N Z 11:317–327

Siniff DB, DeMaster DP, Hofman RJ, Eberhardt LL (1977) Analysis of the dynamics of a Weddell seal population. Ecol Monogr 47:319–335

Smith MSR (1965) Seasonal movements of the Weddell seal in McMurdo Sound, Antarctica. J Wildl Manag 29:464–470

Smith MSR (1966) Studies on the Weddell seal (*Leptonychotes weddelli* lesson) in McMurdo Sound Antarctica. Dissertation, University of Canterbury

Smith WO, Nelson DM (1986) Importance of ice edge phytoplankton production in the southern ocean. Bioscience 36(4):251–257. https://doi.org/10.2307/1310215

Southwell C, Bengston J, Bester M, Blix AS, Bornemann H, Boveng P, Cameron M, Forcada J, Laake J, Nordoy E, Plötz J, Rogers T, Southwell D, Steinhage D, Stewart BS, Trathan P (2012) A review of data on abundance, trends in abundance, habitat use and diet and ice-breeding seals in the Southern Ocean. CCAMLR Sci 19:49–74

Stirling I (1969a) Ecology of the Weddell seal in McMurdo Sound Antarctica. Ecology 50(4):573–586

Stirling I (1969b) Tooth wear as a mortality factor in the Weddell seal, *Leptonychotes weddelli*. J Mammal 50(3):559–565. https://doi.org/10.2307/1378783

Stirling I (1971) Population dynamics of the Weddell seal (*Leptonychotes weddelli*) in McMurdo Sound, Antarctica, 1966–1968. In: Burt WH (ed) Antarctic pinnipedia, vol 18. American Geophysical Union, Washington DC, p 141–161. https://doi.org/10.1029/AR018p0141

Stirling I (1972) Regulation of numbers of an apparently isolated population of Weddell seals (*Leptonychotes weddelli*). J Mammal 53:107–115

Tedman RA, Bryden MM (1979) Cow-pup behaviour of the Weddell seal, *Leptonychotes weddellii*, in McMurdo Sound, Antarctica. Aust Wildl Res 6:19–37

Testa JW (1994) Over-winter movements and diving behavior of female Weddell seals (*Leptonychotes weddellii*) in the southwestern Ross Sea, Antarctica. Can J Zool 72:1700–1710

Testa JW (1997) A comment on the age at onset of breeding of male Weddell seals (*Leptoychoes weddellii*). Can J Zool 74:156–157

Testa JW, Hill SEB, Siniff DB (1989) Diving behavior and maternal investment in Weddell seals (*Leptonychotes weddellii*). Mar Mamm Sci 5:399–405

Testa JW, Oehlert G, Ainley DG, Bengtson JL, Siniff DB, Laws RM, Rounsevell D (1991) Temporal variability in Antarctic marine ecosystems: Periodic fluctuations in the phocid seals. Can J Fish Aquat Sci 48(4):631–639. https://doi.org/10.1139/f91-081

Testa JW, Rothery P (1992) Effectiveness of various cattle ear tags as markers for Weddell seals. Mar Mamm Sci 8:344–353

Testa JW, Scotton BD (1999) Dynamics of an isolated population of Weddell seals (*Leptonychotes weddelli)* at White Islana Antarctica. J Mammal 80(1):82–90

Testa JW, Siniff DB, Croxall JP, Burton HR (1990) A comparison of reproductive parameters among three populations of Weddell seals (*Leptonychotes weddellii*). J Anim Ecol 59:1165–1175

Testa JW, Siniff DB, Ross MJ, Winter JD (1985) Weddell seal—Antarctic cod interactions in McMurdo Sound, Antarctica. In: Siegfried WR, Condy PR, Laws RM (eds) Antarctic nutrient cycles and food webs. Springer-Verlag, Berlin Heidelberg, pp 561–565

Thomas JA (1982) Quantitative analysis of Weddell seal (*Leptonychotes weddelli*) underwater vocalizations at McMurdo Sound, Antarctica. J Acoust Soc Am 72:1730–1738

Thomas JA, Stirling I (1983) Geographic variation in the underwater vocalizations of Weddell seals (*Leptonychotes weddelli*) from Palmer Peninsula and McMurdo Sound, Antarctica. Can J Zool 61(10):2203–2212. https://doi.org/10.1139/z83-291

Tuck GN, Thomson RB, Barbraud C, Delord K, Louzao M, Herrera M, Weimerskirch H (2015) An integrated assessment model of seabird population dynamics: can individual heterogeneity in susceptibility to fishing explain abundance trends in Crozet wandering albatross? J Appl Ecol 52(4):950–959. https://doi.org/10.1111/1365-2664.12462

Van Opzeeland IC, Van Parijs SM, Frickenhaus S, Kreiss CM, Boebel O (2012) Individual variation in pup vocalizations and absence of behavioral signs of maternal vocal recognition in Weddell seals (*Leptonychotes weddellii*). Mar Mamm Sci 28(2):E158–E172. https://doi.org/10.1111/j.1748-7692.2011.00505.x

Walcott SM (2019) Evaluating the dynamics of physiological, environmental and behavioral parameters to the cost of the annual pelage molt in a polar pinniped: the Weddell seal (*Leptonychotes weddellii*). University of Alaska Anchorage, Thesis

Walcott SM, Kirkham AL, Burns JM (2020) Thermoregulatory costs in molting Antarctic Weddell seals: impacts of physiological and environmental conditions: themed issue article: conservation of Southern Hemisphere mammals in a changing world. Conserv Physiol 8(1). https://doi.org/10.1093/conphys/coaa022

Watanabe Y, Mitani Y, Sato K, Cameron MF, Naito Y (2003) Dive depths of Weddell seals in relation to vertical prey distribution as estimated by image data. Mar Ecol Prog Ser 252:283–288

Weddell J (1825) A voyage towards the South Pole, performed in the years 1822–24: containing an examination of the Antarctic Sea to the seventy-fourth degree of latitude: and a visit to Tierra del Fuego, with a particular account of the inhabitants, to which is added, much useful information on the coasting navigation of Cape Horn and the adjacent lands. Longman, Rees, Orme, Brown, and Green, London

Wheatley KE, Bradshaw CJ, Davis LS, Harcourt RG, Hindell MA (2006) Influence of maternal mass and condition on energy transfer in Weddell seals. J Anim Ecol 75(3):724–733. https://doi.org/10.1111/j.1365-2656.2006.01093.x

Williams TM, Fuiman LA, Horning M, Davis RW (2004) The cost of foraging by a marine predator, the Weddell seal *Leptonychotes weddellii*: pricing by the stroke. J Exp Biol 207:973–982

Williams TM, Fuiman LA, Kendall T, Berry P, Richter B, Noren SR, Thometz N, Shattock MJ, Farrell E, Stamper AM, Davis RW (2015) Exercise at depth alters bradycardia and incidence of cardiac anomalies in deep-diving marine mammals. Nat Commun 6(1):6055. https://doi.org/10.1038/ncomms7055

Wilson AJ, Nussey DH (2010) What is individual quality? An evolutionary perspective. Trends Ecol Evol 25(4):207–214. https://doi.org/10.1016/j.tree.2009.10.002

Zappes IA, Fabiani A, Sbordoni V, Rakaj A, Palozzi R, Allegrucci G (2017) New data on Weddell seal (*Leptonychotes weddellii*) colonies: a genetic analysis of a top predator from the Ross Sea, Antarctica. PLoS One 12(8):e0182922. https://doi.org/10.1371/journal.pone.0182922

Chapter 14
The Harp Seal: Adapting Behavioral Ecology to a Pack-Ice Environment

Mike O. Hammill and Garry B. Stenson

Contents

Abstract The harp seal (*Pagophilus groenlandicus*) is the most abundant pinniped in the northern hemisphere, with an estimated total of 9.5 million animals. Commercially exploited since the eighteenth century, there is a large historical body of ecological knowledge that has provided insights into environmental factors that affect this species' behavioral dynamics. Often referred to as the ice-loving seal from Greenland, harp seals breed and rest in spring on the drifting pack ice at the southern limits of their range, then migrate northwards to summer at the edge of the Arctic polar ice pack. Harp seals are gregarious during the breeding season. Ice-based research opportunities have provided insights into how harp seals locate conspecifics, care for their young, and how young transition from a 'terrestrial' to a marine environment. Fine-scale observations of animals outside of the breeding season have been more limited as animals disperse over hundreds of kilometers to the north of the breeding areas to molt and then feed. Nonetheless, the deployment of biologgers, working with seal hunters, and multidisciplinary studies have provided insights into factors affecting productivity and how environmental factors such as climate change may impact harp seals in the longer term.

M. O. Hammill (✉)
Department of Fisheries and Oceans, Maurice Lamontagne Institute, PO Box 1000, Mont-Joli, QC G5H3Z4, Canada
e-mail: mike.hammill@dfo-mpo.gc.ca

G. B. Stenson
Department of Fisheries and Oceans, Northwest Atlantic Fisheries Centre, PO Box 5667, St. John's, NL A1C 5X1, Canada

D. P. Costa and E. McHuron (eds.), *Ethology and Behavioral Ecology of Phocids*, Ethology and Behavioral Ecology of Marine Mammals,
https://doi.org/10.1007/978-3-030-88923-4_14

Keywords *Pagophilus groenlandicus* · Ice seal · Loup marin · Pack ice · Capelin · Lactation · Diving · Behavior · Harp seal · Saddleback seal

14.1 Introduction

Behavior represents individuals' integrated response to the environmental and biological factors surrounding them to maximize the probability of survival. Among pinnipeds, some of these strategies are guided by phylogeny. Pinnipeds have evolved in response to tradeoffs between conflicting needs for obtaining energy while at sea and for a solid substrate for parturition and molting (Bartholomew 1970; Stirling 1983; Costa 1993; Chaps. 7 and 8).

The harp seal (*Pagophilus groenlandicus*) occurs throughout many northern parts of the North Atlantic and the Arctic Ocean. Adults are characterized by a silver-colored body, a black face, and two large black side saddles joined anteriorly, which in the past someone thought resembled a harp (Fig. 14.1). In Europe, they are also referred to as the 'saddleback' seal.

The harp seal is the most abundant pinniped in the northern hemisphere, with a total of 9.5 million (±62, 1000) animals (ICES 2019; DFO 2020). Three populations are recognized. There are two populations in the Northeast Atlantic, the White Sea/Barents Sea population that pups in the White Sea and the Greenland Sea population that pups on the pack ice near Jan Mayen Island. The third population occurs in the Northwest Atlantic (Fig. 14.2; Nansen 1925; Sergeant 1991). The Northwest Atlantic population pups in two primary locations, one off southern Labrador and northeast Newfoundland (Nfld; an area referred to as the 'Front'), while the second group pups on ice floes in the Gulf of St. Lawrence (referred to as the 'Gulf'; Sergeant 1991). The Northwest Atlantic population is the largest at an estimated 7.6 million animals, followed by the White Sea/Barents Sea (1.5 million) and the Greenland Sea (400,000; Fig. 14.3; ICES 2019; Hammill et al. 2021). The question of whether there is exchange between the whelping (pupping) areas or whether different whelping

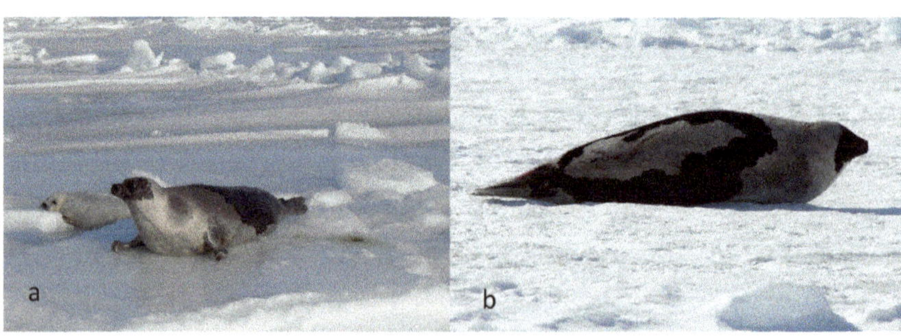

Fig. 14.1 **a** Female harp seal with her whitecoat pup. Note the dark face and saddle on her side **b** A dorsal view of a male harp seal showing the harp on the back. The blood and scratches are from males fighting underwater. *Photo credits* MO Hammill (**a**), GB Stenson (**b**)

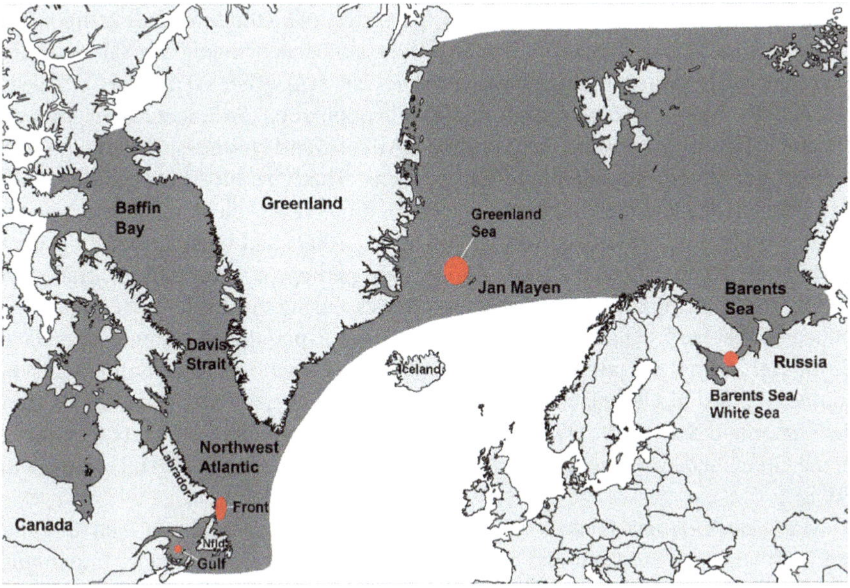

Fig. 14.2 Range (gray) and location of whelping concentrations (red) of harp seals in the North Atlantic

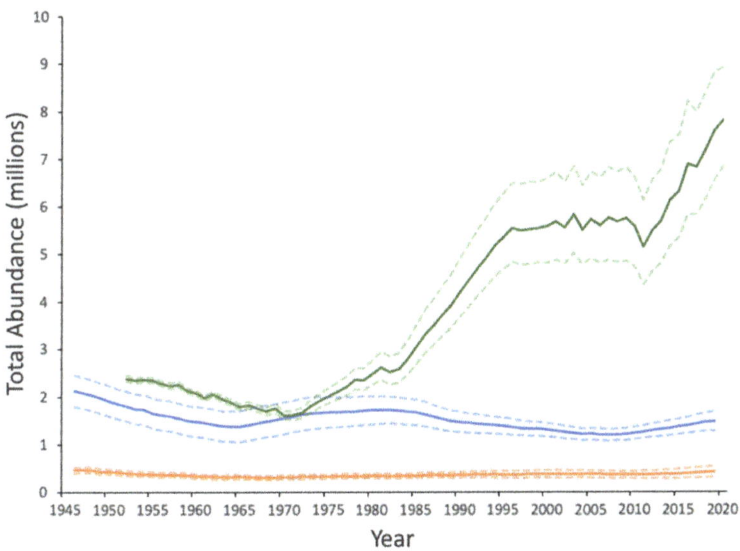

Fig. 14.3 Estimated abundance of Greenland Sea (orange), the White Sea/Barents Sea (blue), and Northwest Atlantic (green) harp seals and 95% CI. Data from ICES (2019) and Hammill et al. (2021)

groups are distinct has been examined using a variety of methods, including flipper tag recoveries, skull morphometrics, vocalizations, and genetic methods (Yablokov and Sergeant 1963; Sergeant 1965; Terhune 1994; Carr et al. 2008, 2015; Van Opzeeland et al. 2009). Front and Gulf seals form a single population. Still, there are significant genetic differences between the Northwest Atlantic and Northeast Atlantic populations of the Greenland and White/Barents Seas. The separation between the White Sea/Barents Sea and the Greenland Sea populations is less clear, with some genetic studies finding no difference (e.g., Perry et al. 2000) and others separating them (Carr et al. 2015). These disparate findings are perhaps not surprising. Large scale deployments of flipper tags and their recoveries during the 1960, 1970, and 1980s indicated that harp seals from the Northwest Atlantic population summered in Davis Strait, Baffin Bay, and portions of the Canadian Arctic, with some seals moving to the southeast coast of Greenland. There is considerable overlap between animals from the Greenland Sea and the White Sea/Barents Sea populations, which both summer in the Greenland and Barents Seas (Sergeant 1991; Folkow et al. 2004; Nordøy et al. 2008).

Many phocids use land as a platform for resting, reproduction, and molting, but harp seals are rarely found on land. In fact, their scientific name, *Pagophilus groenlandicus*, means ice-loving seal from Greenland. In one sense, this may be a misnomer because harp seals do spend large parts of the year away from ice, in what some consider a continuous migration outside of the breeding and molting periods (Chapskii 1961; Folkow et al. 2004; Nordøy et al. 2008). However, in general, the seasonal distribution of harp seals reflects the seasonal changes in the distribution of this habitat, with movement to the north limited by the edge of the polar pack ice. Their southward migration is limited by the southernmost extent of the winter pack ice. The most significant migration is by the Northwest Atlantic population, which summers as far north as 74°N in Baffin Bay and migrates south during the fall to feed and pup off southern Labrador, northeast Newfoundland (~50°N), or the Gulf of St Lawrence (~465°N), a distance of almost 3,500 km (Fig. 14.2).

Whereas land is a fixed feature, the moving pack ice is ephemeral, forming and expanding throughout the harp seal range in winter as temperatures drop. The seals' drifting ice platforms disintegrate in spring as temperatures rise, making way for the spring bloom in productivity that recedes northwards. Overall, this vibrant, dynamic habitat has been a significant driver in harp seals' life history and behavioral ecology. Although there is variability in the seasonal occurrence of ice, there is a certain amount of predictability in the general location where it forms each winter and where pupping occurs. This facilitates the need for conspecifics to find each other during the breeding season and has been beneficial to predators, including humans, who have learned where to search for the whelping herds of harp seals in traditional areas for economic benefit (e.g., hunting, tourism).

The ice used by harp seals is always drifting, moved by currents and wind often at rates of tens of kilometers in a day with leads between the pans of ice opening up or closing, resulting in the ice being compressed and crushed together, or separating and spreading out widely (Dorofeev 1939; Sergeant 1991; Bajzak et al. 2011). In all three populations, pupping occurs on ice that tends to be retained by local currents and physical features. These features delay the drift of ice to open water for 30–40 days until the time when pups begin to leave the ice and enter the water to dive and forage on their own (Fig. 14.4; Sergeant 1991; Wilkinson and Wadhams 2005; Bajzak et al. 2011; Stenson and Hammill 2014). In the Greenland Sea, the Jan Mayen current creates a gyre that forms and retains a tongue-shaped ice feature, known as the Odden, in the vicinity of Jan Mayen during the spring (Wilkinson and Wadhams 2005). In recent years, the ice edge has shifted westward towards the Greenland coast and the Odden feature has not developed. This has resulted in a shift in pupping closer to the Greenland coast, where it falls under the influence of the southward moving

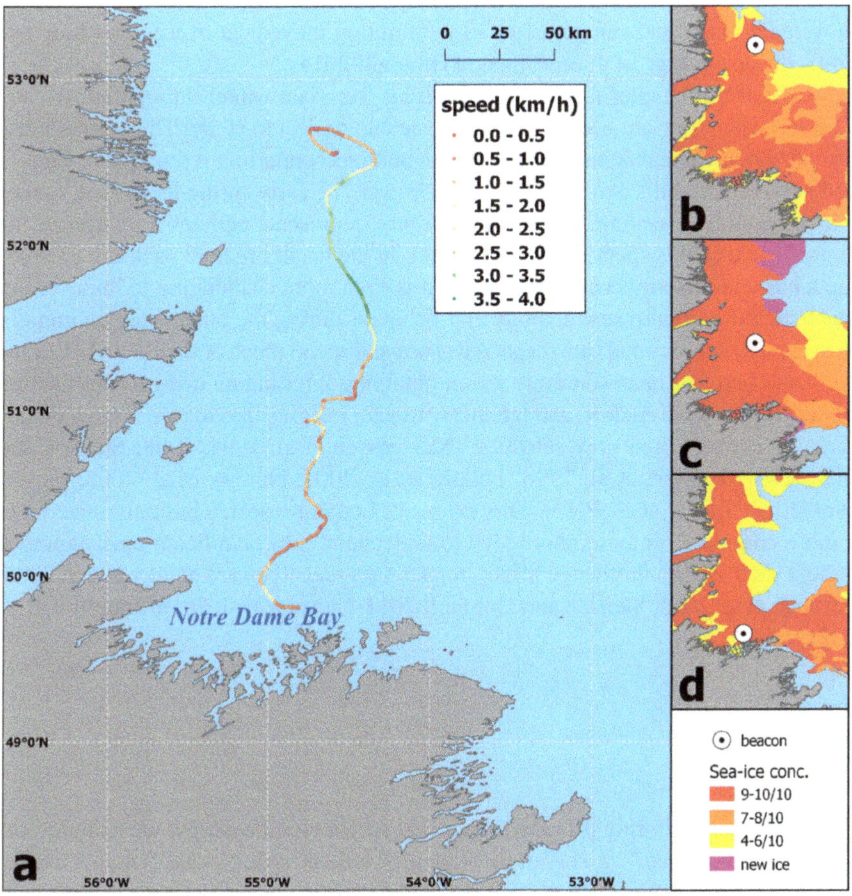

Fig. 14.4 Movement of satellite-linked GPS transmitters deployed on the ice in whelping concentrations at the Front to monitor ice movement during the March 2017 harp seal survey. The colored line shows drift speed (**a**). The panels on the right **b**–**d** show the ice concentration at different points along this drift trajectory (Extracted from Canadian Ice Service: https://www.canada.ca/en/environment-climate-change/services/ice-forecasts-observations/latest-conditions/archive-overview.html)

Greenland current. This rapidly drifting ice may be less stable and more vulnerable to destruction, negatively impacting young of the year survival (Stenson et al. 2020c). In the White Sea, pupping occurs on the pack ice near the southern opening of a narrow strait called the 'Gorlo', meaning throat, which slows the drift of the ice into the northern part of the White Sea referred to as the 'Voronka' or funnel, and delays the drift into the Barents Sea (Fig. 14.2; Sergeant 1991; Potelov et al. 2003). In the Gulf, the herd forms next to the Magdalen Islands, whose presence forces the ice southwards and slows movement out through Cabot Strait (Bajzak et al. 2011; Stenson and Hammill 2014). At the Front, the seals tend to haul out on the drifting ice at the southeastern portion of Labrador; the southward drifting Labrador current transports the ice to Notre Dame Bay, which acts as a barrier to the southern movement of ice and can generate a gyre limiting movement of ice eastwards and down the coast (Fig. 14.4; Stenson and Hammill 2014).

Much of this chapter focuses on harp seals from late winter through late spring, a period when they are accessible on the ice and easiest to study. This research has provided many insights into behavioral adaptations supporting reproductive success and survival of adults and their young. However, the early spring ice period, during which pupping, breeding, and molting occurs, represents perhaps a quarter of the annual cycle of this species. While research in the harsh offshore pack-ice environment has undoubtedly been challenging, it is even more challenging to improve our understanding of harp seal ecology and behavior during the other three quarters of the year when the animals are dispersed at sea and spend much of their time below the surface. There has been some success in obtaining information using a combination of field and captive studies, and telemetry to gain insights into seasonal movements, foraging ecology, and prey selection (Mårtensson et al. 1994, 1998; Stenson and Sjare 1997; Lawson et al. 1998; Folkow et al. 2004; Nordøy et al. 2008; Øigård et al. 2013; Tucker et al. 2013). This extensive body of research has culminated into a more complete synthesis that seeks to understand how significant environmental factors such as climate change might impact harp seals (Kovacs and Lydersen 2008; Peacock et al. 2013; Stenson and Hammill 2014; Stenson et al. 2016; Hammill et al. 2021).

14.2 History

Before further exploring this ice-loving seal's behavioral ecology, we examine the long history of man's association with this unique species. Like many marine mammals, harp seals have been exploited for food by subsistence hunters for millennia (Meldgaard 2004; Renouf and Bell 2008). However, the pack-ice habitat limited access to animals for subsistence hunters to points where animals passed near shore, and the movement of small vessels among the leads in spring as animals moved from their southern wintering to summering areas in the Arctic (Ryan 1994; Castonguay 2003; Renouf and Bell 2008). Commercial exploitation for oil and furs began early in the eighteenth century, refining the local subsistence hunters' techniques to develop extensive net fisheries. By the end of the century, schooners were used to enter the ice where seal herds were located. European whalers found they could make a sealing trip to the offshore icefields near Jan Mayen before heading to the Arctic for the summer commercial bowhead whale (*Balaena mysticetus*) harvest (Ryan 1994; Sergeant 1991). The most extensive period of harvesting occurred during the age of sail. The Greenland Sea and the Northwest Atlantic populations were already significantly depleted by the late 1800s (Ryan 1994; Hammill et al. 2011). In the Northwest Atlantic, the number of vessels, participants, and total catches peaked at 400 sailing ships, 13,000 hunters, and almost 700,000 animals by the middle of the 1800s before declining (Fig. 14.5; Colman 1937; Ryan 1994; Hammill et al. 2011). Commercial and subsistence harvesting of animals from this population continues to this day. In 2019, estimated removals included a reported 32,000 animals taken in the Canadian commercial harvest, and subsistence hunts estimated to be 1,000 animals in the Canadian Arctic and 59,000 in Greenland (ICES 2019; Stenson and Upward 2020). Intensive harvesting of the White Sea/Barents Sea population began somewhat later during the beginning of the twentieth century. However, as with other populations, there was a rapid decline in seal abundance (Skaug et al. 2007). In all three areas, stocks declined significantly following the World Wars and reached population minima during the early 1970s. At that time, significant management measures were put in place to protect breeding females and reduce catches. These measures contributed to some recovery, but only the Northwest Atlantic population has recovered to levels last seen during the nineteenth century despite continued high levels of commercial harvesting (Figs. 14.3 and 14.5; Hammill et al. 2011,

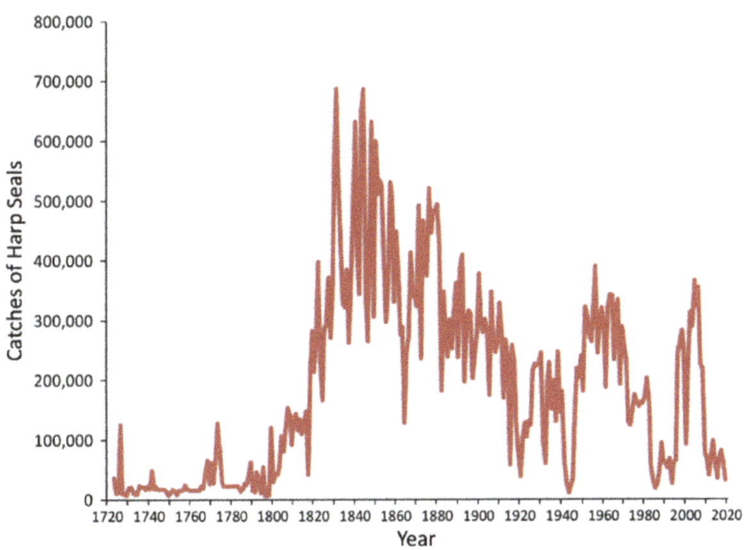

Fig. 14.5 Commercial catches of Northwest Atlantic harp seals (1720–2019) in Canadian waters

2021; Skaug et al. 2007; ICES 2019). Harvest levels in the Greenland Sea and the White Sea areas have remained very low, but after showing some small degree of recovery, these populations have remained very low or continued to decline, possibly due to ecological changes in their habitat (Øigård et al. 2013; ICES 2019; Stenson et al. 2020c).

The intense economic interest and logistical challenges associated with accessing the harsh offshore environment limited the study of harp seals to accounts of the general seasonal distribution and migration observed mainly in the southern parts of their range during the winter, timing, location, and ice conditions during whelping, and behavioral anecdotes (Robinson 1897; Nansen 1925). In the early twentieth century, biological research was conducted by scientists onboard hunting vessels, which provided access to animals in the breeding patches, but also limited research to avoid slowing down or interfering with the hunt. Observations at this time focused primarily on biological samples of vital rates as well as the types of ice selected by seals, seasonal changes in distribution, diet, pup growth and development, and the timing of molt among adults (Dorofeev 1939; Sivertsen 1941; Chapskii 1961, 1964; Surkov 1957). Interestingly, the first studies to show the high-fat content in harp seal milk were reported from this time in Sivertsen (1941). The discovery by Laws (1952) that seals could be aged by counting growth layers in their canine tooth provided an opportunity to obtain more quantitative demographic information related to harp seal productivity (Fisher 1954, 1955).

Increased research efforts continued during the 1960 and 1970s as herds were declining. The research was directed towards understanding interdependence among herds through the recovery of tags from marked animals (Sergeant 1965; Popov 1970; Øien and Øritsland 1995) and variation in cranial measures (Yablokov and Sergeant 1963), and increasing understanding of demographic rates (Nazarenko 1975; Sergeant 1966, 1973; Bowen et al. 1981; Bowen and Sergeant 1983; Kjellqwist et al. 1995; Frie et al. 2003; Sjare and Stenson 2010). Considerable effort was also directed towards improvements in methodologies to estimate pup production using mark-recapture, aerial surveys, and population modeling, which has continued to the present (Sergeant and Fisher 1960; Lavigne 1976; Roff and Bowen 1983; Stenson et al. 1993; Øigård et al. 2014; Hammill et al. 2015).

Since the 1970s, the adoption of new technologies has opened up a whole new realm for research on pack-ice seals. As small helicopters became affordable, researchers were able to operate independently of the sealing fleets since they could now look for groups of seals and land on the ice away from hunters. This new tool made possible quantitative research into harp seal communication, behavioral ecology, and ecophysiology during the breeding season. Insights into maternal investment, reproductive effort, and lactation strategies were gained using a combination of longitudinal and cross-sectional methods to study energy transfer from females to their pups, and changes in body composition of developing neonates, as well as research on mother–pup behavior (Stewart and Lavigne 1980, 1984; Lavigne et al. 1982a; Stewart 1986; Kovacs and Lavigne 1985; Kovacs 1987; Oftedal et al. 1989; Kovacs et al. 1991, Chaps. 7 and 8). Research on wild animals, coupled with studies of animals in captivity, improved our understanding of the physiological mechanisms driving behavior related to communication (Terhune and Ronald 1971, 1972; MØhl et al. 1975; Terhune 1976; Merdsoy et al. 1978; Terhune et al. 1979; Serrano 2001; Moors and Terhune 2003; Chap. 3), and the transition from being essentially a terrestrial animal to a fully functioning diving mammal (Worthy and Lavigne 1982, 1983; Worthy 1987, 1991; Burns et al. 2007, 2015; Pearson et al. 2014b, 2019). The integration of new technologies such as doubly-labeled water techniques and deployment of recoverable time-depth recorders also allowed us to further extend our understanding of maternal investment and lactation strategies (Oftedal et al. 1987, 1996; Lydersen and Kovacs 1993, 1996, 1999; Chap. 8). Long-term studies have provided opportunities to examine in finer detail the relationships between density-dependent and density-independent factors and their impacts on the behavioral ecology of harp seals, and the fine-scale drivers of habitat use, such as changes in ice conditions and factors affecting food resources (Hammill et al. 1995; Nilssen et al. 1995a, b; Chabot and Stenson 2002; Buren et al. 2014; Stenson and Hammill 2014; Stenson et al. 2016; Hammill and Sauvé 2017; Stenson et al. 2020a).

14.3 Breeding

The production and survival of viable young represent the successful integration of the previous year's foraging, combined with the transfer of energy to offspring under conditions favoring offspring survival. Animals returning from the Arctic in the fall show improved energy reserves, but deposition differs between the Northwest Atlantic and the Northeast Atlantic populations. The Barents Sea harp seals, including both the Greenland Sea and the White Sea animals, appear to buildup their energy reserves in the form of blubber during summer. They must then maintain stored energy levels throughout the fall until pupping (Nilssen et al. 1997). Northwest Atlantic harp seals, on the other hand, show only a small improvement in condition over the summer and instead acquire a significant proportion of their energy during their last trimester of pregnancy (November–March) while on the feeding grounds of the southeast Labrador-northeast Newfoundland area and the Gulf of St. Lawrence (Beck et al. 1993b; Chabot and Stenson 2002; Stenson et al. 2016).

Pupping occurs during spring in all three populations, but there are slight differences in timing between them. In the Northwest Atlantic population, pupping occurs at the time of peak ice formation, when the strength of the sun during the day tends to offset any ice formation. In the Gulf of St Lawrence, pupping begins in the last week of February and can continue until mid-March, although most pups are born by March 3rd. At the Front, pupping begins about five days later in early March and can continue until mid-month, with the majority of pups born by March 10th (Stenson et al. 1993, 2002, 2003, 2020a; Sergeant 1991; Bajzak et al. 2011; Stenson and Hammill 2014). Pupping in the White Sea is similar to that in the Gulf, with births beginning in late February and continuing into early March, while pupping in the Greenland Sea occurs the latest, with pupping not beginning until the third week of March (Sivertsen 1941; Popov 1966; Sergeant 1991; Potelov et al. 2003; Øigård et al. 2014; ICES 2019).

Pupping in all populations occurs in areas where oceanographic or geographical features result in a slowing of ice drift and retention of ice in areas where the integrity of the pans persists until pups enter the water. This provides a level of predictability of suitable ice when adults return to the same area year after year, but with ice extending over many kilometers, seals require additional mechanisms to find each other. The most efficient way to achieve this for marine mammals is to use sound for long-distance communication (Terhune and Ronald 1986). Under the ice, sound-level measurements of 140–180 dB re 1 μPa emitted by individuals are likely to be detected by other seals at distances of at least 2 km and possibly as far away as 15 km or more, depending on model assumptions and underwater conditions (Rossong and Terhune 2009). In addition to high sound levels, calls end abruptly, and there is a very high degree of repetitive calling, with calls often being repeated more than 10–20 times. These call characteristics provide a means of signaling the presence of the herd at a considerable distance against the random nature of background noise (MØhl et al. 1975; Watkins and Schevill 1979; Terhune and Ronald 1986; Serrano and Terhune 2002; Rossong and Terhune 2009; Chap. 3). The calls of nearby animals generate a positive feedback loop. As the number of animals in the area increases, the size of the acoustic herd also increases. Thus, the herd's audible presence could extend over several tens of kilometers, which facilitates identifying the herd's location by other animals arriving at the whelping area. Once females haul out and give birth, their movements become more restricted because of the pup. This results in an unusual situation where the female, relatively speaking, does not venture far from her pup. However, the mother–pup pair's actual geographical location may change by tens of kilometers each day as the ice drifts (Fig. 14.4).

The birth of the pup appears to be similar in all respects to those of other pack-ice seals (Stewart et al. 1981). The female hauls out on the ice shortly prior to parturition, which occurs rapidly. Presentation is caudal, and the fetal membranes tear during birth; the placenta may or may not be expelled immediately (Sergeant 1991). The female does not lick the pup to remove the fetal membranes, nor does she consume the placenta, which is left on the ice to be scavenged by gulls and the occasional fox (Lavigne and Kovacs 1988; Sergeant 1991). Immediately after birth and during the first two days, the female remains almost continually with her pup; she spends more time checking for her pup and engages in nose-to-nose behavior more frequently than during the remainder of the nursing period (Kovacs 1987). Owing to their lack of coordination, the pups are most vulnerable at this time. However, there are few apparent predators other than gulls who scavenge the placenta and any dead pups.

Pup attendance is much more variable for harp seals than other phocids, such as gray seals (*Halichoerus grypus*) or hooded seals (*Cystophora cristata*) where the female normally remains very close (>80% of the time) to her pup throughout the nursing period (Bowen et al. 1985; Kovacs and Lavigne 1992; Haller et al. 1996). After the first two days, females in the Northwest Atlantic only spend about 15% of their time beside their pup, with peak periods occurring in the morning (08h00–09h00) and late afternoon (16h00–17h00) based on on-ice observations (Kovacs

1987; Perry et al. 2017). In the Gulf, time-depth recorder deployments on females have shown that throughout lactation females only spend on average 29% (±13) of their time hauled out on the ice; the rest of their time is spent in the water at the surface or diving (Lydersen and Kovacs 1996). A similar pattern has been reported for lactating females equipped with satellite transmitters in the White Sea (Nordøy et al. 2008). Females initiate nursing by hauling out on the ice and approaching the pup, initiating contact by nosing behavior. The female then presents herself for the pup to suckle. Nursing bouts are often broken by the female moving a few meters away from the pup, then presenting again. The pup usually follows and resumes nursing (Kovacs 1987). Females appear to spend more time in attendance at the Front and in the Greenland Sea, with females in attendance 85.1% (±28.7) and 52.2% (±38.7) of the time, respectively (Van Opzeeland et al. 2009; Perry et al. 2017). The reasons for these differences between populations are not clear, but pan size tends to be smaller and drift rates more rapid at the Front and in the Greenland Sea compared to the Gulf.

During lactation, females haul out significantly more often during the day (07h00–19h00), than at night (Lydersen and Kovacs 1996), which is in agreement with results from acoustic studies that found harp seals were vocally active all night and least active at daybreak and in the early afternoon (Terhune and Ronald 1986). At the Front, shipboard studies observed that the proportion of females hauled out increased until 09h00 in the morning, then declined during the rest of the morning, before rising to a peak in mid-afternoon. However, the proportion of females hauled out was affected by environmental conditions. Perry et al. (2017) found that females reduced attendance during high winds and/or low temperatures, while the likelihood of attending females nursing during these poor weather conditions was greater than when conditions were better. That is, females were less likely to be present when weather conditions were poor, but when present, they were more likely to be provisioning their pups. The authors hypothesized that this strategy could help females defray the thermoregulatory demands during the nursing period while ensuring that their young attain weights that are likely to increase postweaning survival.

Occasionally pups fall off the ice pan on their own. If this occurs, they use the claws of their fore flippers to pull themselves out of the water onto the ice. If the pans are unstable and breaking or being pushed together and rafted up, the female may lead the pup to another ice pan, and we have observed them crossing leads of 15 m or more. If the pup doesn't follow close enough, the female returns and bites the pup at the neck to briefly pull it forward, similar to how harbor seal (*Phoca vitulina*) and ringed seal (*Pusa hispida*) females grab their pups and pull them quickly through the water (MO Hammill and GB Stenson personal observations).

For harp seals and other phocids, where the female frequently returns to the water during lactation, relocation of the pup represents a significant challenge, particularly in the pack ice, which is mobile and subject to structural change. While in the water, the female must remain in close proximity to the ice flow where the pup is located. Females are often seen in the water along the ice edge where they appear to be 'monitoring' their pup, particularly at the Front where the pans are smaller and ice drift is rapid (Fig. 14.6). To add to this challenge, the herd is often aggregated,

Fig. 14.6 Harp seal female in the water keeping an eye on her pup. *Photo credit* GB Stenson

although not to the same extent of crowding observed among many land breeding seals, requiring females to be close enough to identify their own pup. Spatial, visual, auditory, and olfactory cues are likely factors involved in facilitating female-pup reunions (Terhune et al. 1979; Van Opzeeland et al. 2004; Chap. 3). However, in most cases, harp seal mother–pup reunions are usually simple, with the majority of reunions occurring where females utilize an access point on the ice close to their pups. This access point may be shared with several females. If the pup changes its location, the female will adjust her behavior to use another access point closer to her pup. Once hauled out, the female heads directly to the last location of her pup, indicating a strong sense of spatial orientation (Popov 1966; Kovacs 1987, 1995). Individual variability in calls emitted by pups suggests that acoustic cues are also used (Van Opzeeland et al. 2004). However, acoustic and visual cues are considered of secondary importance, with final confirmation by nose-to-nose behavior (Kovacs 1987, 1995). The relocation mechanism appears to be quite effective, as the mortality rate due to starvation during the first few weeks of life is extremely low (Kovacs et al. 1985). Few abandoned animals are ever spotted, particularly when compared with observations from other species such as gray seals (MO Hammill and GB Stenson personal observations).

14.3.1 Lactation

Lactation represents a major proportion of maternal reproductive effort (Lavigne et al. 1982b; Hammill et al. 2008; Stenson et al. 2016, 2020a). Since pinnipeds use a solid substrate for nursing, this results in the bulk of the female's reproductive energy expenditure occurring on the ice (or land) in an environment in which she cannot feed, which has resulted in a wide range of behavioral, physiological, and anatomical adaptations (Oftedal et al. 1987; Chap. 8). Generally, phocids have a capital breeding system where they accumulate energy and nutrient reserves throughout the year to support some, or all, of the lactation costs. By averaging their energy buildup over many months, they need only to increase their daily food intake by about 18% (Perez and Mooney 1986; Hammill et al. 2008). In contrast, otariids have an income breeding system where they rely on food resources from foraging trips to meet the female's daily energy requirements and the costs of producing milk to feed her offspring. The phocid strategy is considered more economical. For example, lactating northern fur seals (*Callorhinus ursinus*) consume 80% more food than non-lactating fur seals across the 4-month lactation period (Costa 1993). During lactation, the typical phocid females fasts while she remains close to her pup and produces milk with a high fat content. These factors, together with the ability of the pup to deposit much of the energy as fat, facilitate rapid energy transfer. The subsequently shortened lactation duration reduces her energetic overhead, and the time she must remain hauled out, where she is vulnerable to surface predators (Oftedal et al. 1987; Costa 1993; Boness and Bowen 1996). However, not all phocids exhibit a simple capital breeding system. The ability to store energy prior to parturition is tied to body mass and phocid females (mean mass = 229 kg), generally being larger than otariid females (mean mass = 80 kg), are capable of storing more energy (Costa 1993). There is a gradation among phocids, with the smallest species adopting a more otariid strategy of a relatively long lactation period and continued feeding owing to their relatively small size, which limits how much energy can be stored (e.g., ringed seal, mass = 81 kg, lactation length = 42 d; Hammill et al. 1991).

Adult harp seal females weigh on average 138 ± 67 kg (range 117–151 kg) at the start of lactation, but actual mass varies with age, ecosystem conditions, and population density (Stewart 1986; Kovacs et al. 1991; Lydersen and Kovacs 1991; Hammill et al. 2008; Hammill and Sauvé 2017). At the time of parturition, females comprised of approximately 33% fat, 16% protein, 49% water, and 2% ash (Lydersen and Kovacs 1996). Female harp seals lose 2.9–3.2 kg per day during a lactation period of approximately 10–13 days. Approximately 50% of mass loss is fat, and 11% is protein, with much of the loss by females occurring from the blubber layer (Lydersen and Kovacs 1996; Stewart and Lavigne 1984; Stewart 1986). While females are losing mass, pups more than triple their birth weights, increasing from approximately 10 kg at birth to 35 kg at weaning (Oftedal et al. 1989; Sergeant 1991; Stewart and Lavigne 1984; Stewart 1986; Kovacs 1987; Kovacs et al. 1991; Lydersen and Kovacs 1991; Oftedal et al. 1996). This rapid pup growth/female mass loss results in an even greater efficiency of mass transfer than the 63% transfer efficiency observed in hooded seals, which also pups in the pack ice. However, hooded seal adults are more than double the size of harp seals, and lactation lasts only 4 days (Bowen et al. 1985). Harp seals are among the smaller phocids, and their lactation period is still quite short, suggesting they may have limited capacity to wean a pup using stored energy reserves alone. Harp seals spend considerable time in the water during lactation (Terhune et al. 1979;

Kovacs 1987; Beck et al. 1993b; Nordøy et al. 2008; Lydersen and Kovacs 1996; Van Opzeeland et al. 2009; Perry et al. 2017). Much of this time in the water is spent diving, although mean dive depths of 30 m and mean dive durations of 3.2 ± 2.4 min suggest that females are not moving far from the ice pans where their pups are located (Lydersen and Kovacs 1993; Nordøy et al. 2008). However, some animals dove for as long as 13 min and to depths of 90 m, and evidence from other studies shows that some animals are feeding, mostly on invertebrates to varying degrees, with some females consuming up to 4.8 kg of food per day (Stewart and Murie 1986; Beck et al. 1993b; Lydersen and Kovacs 1993, 1996; Nilssen et al. 2000; Potelov et al. 2000). For harp seals, birth weights and growth rates are similar for female and male pups. The amount of energy invested in successfully weaned pups is relatively constant regardless of female age. If the energy investment in pups is relatively constant then the relative costs for females that are smaller because they are younger, or their condition is poorer due to environmental or density-dependent factors, will be greater (Stewart 1986; Hammill and Sauvé 2017). Under these conditions, smaller animals might forage more extensively than would be the case for larger, older individuals. Similarly, when population densities are high, or in years where environmental conditions for foraging have been less favorable, females may compensate by increasing foraging efforts during lactation, but this has yet to be examined.

14.3.2 Males

Comparatively less is known about males than females because, unlike females who are tending to their pup, they are much more difficult to recapture. Instead, we must rely on cross-sectional studies of mass changes to provide insights into the costs of reproduction for males. During the breeding season, males lose about 39 kg, which represents 26% of their initial mass with most (56%) of the mass loss occurring from the muscle or core area and the remainder of the mass loss occurring from the surrounding blubber layer (Chabot and Stenson 2002). The overall mass loss by mature males and females is similar, indicating that both sexes invest a similar amount of energy in reproduction, females during lactation and males probably during underwater copulation and competition with other males for access to females.

14.3.3 Pup Development

At birth, pups weigh 9–11 kg; they are lean, with fat comprising approximately 7% of total body mass (Lydersen and Kovacs 1996). During the first several days, there is very little growth as pups develop some coordination, seek out the female's teats, and learn to suckle. Initially, they only gain 1.5–1.8 kg per day, but this rate increases to as high as 2.5–3.5 kg per day (Popov 1966; Kovacs and Lydersen 1996; Oftedal et al. 1996). Popov (1966) suggested that mass gain slowed just before weaning. However, Lydersen and Kovacs (1996) and Oftedal et al. (1996) did not observe the same changes. Overall, the pups consume approximately 3.65 ± 0.24 kg of milk per day, increasing from 38.0 ± 8.2% fat at the start of lactation to 58.5 ± 0.6% fat by day 11 (Lydersen and Kovacs 1996; Oftedal et al. 1996). Seventy-one percent of the mass gain is in fat, 5% in protein, and 25% in water. After the first week, the mass gain is primarily in water (46%), followed by fat (37%) and protein (15%). Weaning occurs abruptly, with the female entering the water after a nursing bout and never returning (Stewart and Lavigne 1980; Stewart 1986; Oftedal et al. 1989; Kovacs et al. 1991; Lydersen and Kovacs 1996).

The pup passes through several developmental stages during nursing (Fig. 14.7). Newborn pups, wet from blood and amniotic fluid ejected as part of the birth process, are very lean and lack coordination. This stage generally lasts for less than one day. Pups vocalize often soon after birth, but only vocalize when hungry as they get older (Kovacs 1987). A few hours after birth, depending on conditions, the fur dries with

Fig. 14.7 Developmental stages of harp seal pups **a** Newborn, **b** Yellowcoat, **c** Thin Whitecoat, **d** Fat Whitecoat, **e** Raggedy jacket, and **f** Beater. *Photo credit* GB Stenson

a yellow hue from the amniotic fluid that can persist for one to two days. During this time, the pups are referred to as 'yellowcoats'. After their first day, the pups have become more coordinated and begin gaining weight. They are sufficiently strong and coordinated to hold their heads up by their second day and locate and move between the two teats quickly. At this point, pups begin to gain mass more quickly (Lydersen and Kovacs 1996). By this time, the sun bleaches out the yellow stain and, as the pups grow, they are referred to as thin, and eventually fat, 'whitecoats' by the end of their first week (Stewart and Lavigne 1980). The pups can move like adults by the thin whitecoat stage, using the front flippers simultaneously or alternatively to pull the body forward. Young pups rarely engage in agonistic behavior, which does not appear until the pups have reached the fat whitecoat stage; at about 5 days old, they may emit a growl or lunge their upper bodies toward each other (Kovacs 1987). By their tenth day, the underlying juvenile pelt becomes apparent through the white fur, giving the fur a grayish tinge and they are sometimes referred to as graycoats because of this greyish hue. About 50% of the pups are weaned at this stage when they have a mass of around 32–38 kg. No differences are observed between male and female pups in growth rates or weaning mass, nor in female behavior towards male or female pups (Stewart and Lavigne 1980; Kovacs 1987; Oftedal et al. 1989; Lydersen and Kovacs 1996). As the white fur begins to shed by about 10–13 days, patches of the juvenile fur begin to appear (referred to as a 'raggedy jacket'), and most pups have weaned, although about 10% may still be suckling (Stewart and Lavigne 1980; Kovacs 1987). This molt lasts for about a week, and by about the age of 18 days, the animals have only the juvenile fur, which they retain for their first year. They are known as 'beaters' because they swim on the water surface by 'beating' their foreflippers (Sivertsen 1941; Stewart and Lavigne 1980; Sergeant 1991).

14.3.4 Pup Thermoregulation

Heat loss for adult marine mammals is minimized using a variety of mechanisms, including a reduced surface area to volume ratio compared to terrestrial animals of similar size, the development of circulatory mechanisms, and effective insulation in the form of a blubber layer, which also functions as an organ for storing energy (Oftedal et al. 1989; Pearson et al. 2014a; Chap. 4). These mechanisms are poorly developed among phocid neonates, and animals are faced with significant challenges for thermoregulation in an environment characterized by below zero temperatures. Harp, Weddell (*Leptonychotes weddellii*), and hooded seal pups are all similarly insulated at birth. Still, there are significant differences among the three species in how this is achieved. Harp and Weddell seals rely initially on lanugo for insulation, while hooded seals are born with a subcutaneous layer of blubber, approximately 1.5 cm thick (Oftedal et al. 1989). At birth, harp seal pups are wet and have a relatively high surface to volume ratio leading to high heat loss. While the fur is wet, it has almost no insulative value and pups initially rely on shivering for thermogenesis. Muscle glycogen stores that fuel the shivering response are exhausted within hours, and pups turn to non-shivering thermogenesis by burning brown adipose tissue (Blix et al. 1979; Pearson et al. 2014b). As the fur dries, it becomes a much more effective insulator than blubber of a similar thickness. It is light, so small pups can carry a thicker lanugo than they could carry blubber (Ryg et al. 1993). For the neonate, the lanugo accounts for 84% of the total insulation. However, as the pups deposit fat the blubber layer plays an increasingly important role as an insulator, increasing from around 16% of the total insulation for a dry neonate to 60% of the total insulation by day 8, and 78% of the overall insulation by weaning. At weaning, the blubber is approximately 4 cm thick, giving it insulative qualities that are similar to the lanugo, but more importantly, blubber is more effective as an insulator in water (Pearson et al. 2014b). At this point, the insulative quality of the lanugo deteriorates as pups begin to molt (Øritsland and Ronald 1973; Stewart and Lavigne 1980; Pearson et al. 2014b).

14.3.5 Postweaning Transition

For most mammals, the offspring transition from milk to solid foods occurs while still with their mothers. Among pinnipeds, particularly the phocids, this does not appear to be the case (Oftedal et al. 1989). For phocids, there are two main developmental pathways (Boness and Bowen 1996): long lactation periods with early aquatic activity of the pups and short (or no) aquatic postweaning fasts (e.g., bearded *Erignathus barbatus*, ringed, Weddell, and harbor seals), or brief lactation periods followed by longer terrestrial (or ice-based) postweaning fasts with little aquatic activity during lactation (e.g., elephant *Mirounga* spp., gray, harp, and hooded seals) (Burns et al. 2007). Despite these different strategies, it is rare for pups to begin independently foraging before one month of age. This period does not appear to differ if the pups enter the water during the nursing period or wait until the end of the postweaning fast. Even after foraging begins, it may be days or weeks before pups forage well enough to meet their energy demands, making it difficult to pinpoint when pups become nutritionally independent (Muelbert and Bowen 1993).

The harp seal pup nurses for a relatively short time followed by a relatively long postweaning fast. This minimizes the time females spend tied to a single location and tied to the pup, which decreases her vulnerability to predators while still transferring the energy needed for pups to survive until foraging begins. Once weaned, the animals remain on the ice for several weeks; they may form small groups (MO Hammill and GB Stenson personal observation), but interactions among animals have not been studied.

The function of the subsequent postweaning fast is not clear. On the one hand, initial growth is so rapid, with much of that growth deposited in the form of blubber, such that the fat content of newly weaned pups is approximately 58% (Lydersen and Kovacs 1996). This leaves animals with positive buoyancy, limiting their ability for sustained diving activity until some of those reserves can be metabolized. The fast may also provide the additional time needed to develop the physiological mechanisms necessary for efficient foraging (Oftedal et al. 1989).

During the postweaning fast, the pup's mass declines from 32–38 kg at weaning to approximately 24 kg after one month of fasting (Stewart and Lavigne 1980; Oftedal et al. 1989; Worthy and Lavigne 1983). In a captive study, animals initially lost about 0.43 kg per day during the first week of the postweaning fast. However, the rate of weight loss declined to 0.31 kg per day after one month (Worthy and Lavigne 1983). During the fast, animals rely on energy from lipids and protein from the core and the blubber to meet their daily energy requirements, with mass loss initially occurring more rapidly from the core (Worthy and Lavigne 1983; Oftedal et al. 1989). However, after the first two weeks, mass loss from the blubber tends to increase. While the total mass has declined by the end of the one-month fast, the relative importance of the blubber, skin (sculp), and core as a proportion of overall body mass have remained the same (Stewart and Lavigne 1980; Worthy 1987; Worthy and Lavigne 1987; Oftedal et al. 1989). This pattern of mass loss appears to differ when compared to species that fast on land. In a controlled study comparing the mass loss of harp seals and gray seals fasting on land or in the water, no differences were observed in the overall relative rates of mass loss between species nor between treatments. Still, the pattern of loss differed between the two species (Worthy and Lavigne 1987). While harp seals lost mass equally between their blubber and the core regardless of where the fasting occurred (land or in the water), more than 70% of the mass loss in gray seals occurred from the blubber. This suggests that there may be a selection for preserving the insulative blubber layer in harp seals that spend more time in the water and would be forced to cope with water's higher thermal conductivity during the postweaning fast than gray seals.

One possible reason for the postweaning fast might be the pup's need to undergo further physiological development necessary for diving (Oftedal et al. 1989; Burns et al. 2007). In this situation, the buildup of blubber is necessary for the pup to survive until they have physiologically matured. Neonates have significantly lower total body oxygen stores than adults when examined on a mass-specific basis. During lactation, there is an absolute increase in these stores due to an increase in blood oxygen storage. However, the overall mass-specific oxygen storage capacity actually declines from the dilution effect of the massive gain in lipid reserves as well as a delay due to the time required for the production of new blood cells (Burns et al. 2005; Clark et al. 2006). At weaning, pups also have lower muscle oxygen stores on a mass-specific basis owing to low myoglobin, acid buffering capacity, and lower aerobic and anaerobic enzyme activity, indicating that pup muscles are less able to sustain aerobic and anaerobic energy production during diving than adults (Burns et al. 2005, 2007, 2010). Pups lose mass during the postweaning fast, enabling them to increase their mass-specific total body oxygen stores (i.e., to reverse the dilution effect). It has been suggested that pups remain in this terrestrial phase until some rebalancing of total body oxygen stores occurs between the blood and muscle (Burns et al. 2005, 2007). This pattern echoes that seen in other terrestrial and marine species, suggesting that it reflects phylogenetic constraints on early muscle development (Burns et al. 2015).

By the end of the fast, however, total body mass-specific oxygen stores are only about 2/3 those of adults, primarily due to limited oxygen storage capacity in the muscle, which in turn is tied in part to low muscle myoglobin levels (Burns et al. 2007). Thus, in addition to naivety in searching, identifying, and capturing prey, pups entering the water are also physiologically limited in their capacity to forage compared with adults. Perhaps as a result of these limitations, young of the year initially take invertebrates such as amphipods (*Parathemisto libellula*) euphausiids (*Thyssanoessa* spp), and small fishes such as capelin (*Mallotus villosus*) and Arctic cod (*Boreogadus saida*), which tend to aggregate and may be easier to learn to capture (Sergeant 1973; Haug et al. 2000).

14.3.6 Behavioral Responses to Danger

Animals exposed to danger usually respond by escaping or defending themselves. A few species use a third strategy referred to as freezing or paralysis, where animals remain motionless while the threat is present. When harp seal pups are approached or handled by humans, and adult harp seals are captured, they often show paralysis (Ronald et al. 1970; Lydersen and Kovacs 1995). This behavior is unique among pinnipeds. Paralysis is characterized by animals becoming rigid, bringing their fore-flippers into the sides of the body, pressing their hindflippers closely together, closing their eyes, and retracting their head into the fat layer circling the neck. The animals always urinate and often defecate (Ronald et al. 1970; Lydersen and Kovacs 1995; MO Hammill and GB Stenson personal observation). In the case of pups, they can be picked up or turned over without them moving. Paralysis is more frequent among pups than adults; it is observed more often among older pups, occurring spontaneously in 61% of graycoats, compared to only 25% of the time among yellowcoats, while some pups never show paralysis (20% yellowcoats; 6% graycoats). The usual behavior among adults is for animals to leave the ice or show aggression when approached by humans. However, some will exhibit paralysis spontaneously when approached (7% females, 0% males) or live-captured (46% females, 80% males; Lydersen and Kovacs 1995). Physiologically, paralysis is characterized by two phases, apnea accompanied by a heart rate that declines to about 30 beats per minute for approximately 100 s, followed by hyperventilation with heart rates increasing to as much as an average of 171 beats per minute (Lydersen and Kovacs 1995).

The paralysis behavior in harp seals appears to be an evolutionary response to reduce predation mortality resulting from a complex suite of lactation-related adaptations and ecological circumstances. Harp seals have adapted to the presence of polar bear predation in the Arctic and on the whelping aggregations (Peacock et al. 2013; Smith and Stirling 2019). Compared with Arctic fast-ice breeding seals, harp seals breed in dense aggregations on unstable ice in areas occurring at the limits of the polar bear range. These aggregations of animals would reduce the likelihood of a bear encountering an individual, and by freezing, the seal may avoid detection, or the predator may lose interest in a non-moving prey (Lydersen and Kovacs 1995).

14.3.7 Mating Behavior

Harp seals show little sexual dimorphism in body size, age of maturity, and sex ratio. Males are only slightly larger than females (asymptotic length M:169–173 cm, F:165–169 cm) (Hammill et al. 1995; Chabot and Stenson 2002; Frie et al. 2003; Hammill and Sauvé 2017). The age of sexual maturity for males and females ranges between 4 and 8 years of age. It varies considerably between populations, likely reflecting differences in how the populations are responding to density-dependent and density-independent factors acting on individuals in each herd (Nazarenko 1975; Sergeant 1966, 1991; Kjellqwist et al. 1995; Frie et al. 2003; Sjare and Stenson 2010; Øigård et al. 2013; Stenson et al. 2016).

The dispersed nature of females on the ice and their frequent diving during lactation limits male opportunities to restrict access to females by other males. Thus, males appear unable to guard or otherwise monopolize access to females during the lactation period (Kovacs 1995). At the start of pupping, the males are seen in clusters on pans around the periphery of what will later develop into the whelping or pupping patch (Fig. 14.8). The males haul out at the ice edge in the sun within 1–2 body lengths of each other. There is little interaction among individuals, and if disturbed, animals will enter the water and disperse individually or as a group to another area. This differs from the female pattern (Fig. 14.8), where most animals lie about 5–10 m apart and within a similar distance to open water, a lead or access hole to the water, and whose movements are limited by the need to attend her pup (Kovacs 1995). After the first week, individual or small groups of males are seen swimming in the leads between the floes where the females and their pups are hauled out. Most of the time, the animals swim on their backs, which, owing to the shape of their skulls and placement of their eyes, provides them with a better view of conspecifics, prey, and predators below them (Figs. 14.1 and 14.6). As animals move through the leads, they lift their heads to take a breath, then return to monitor any activity below (Merdsoy et al. 1978; GB Stenson personal observation). Some males also haul out within the female herd, but do so individually and with females dispersed around them. Generally, there is little interaction between males and females on the ice. However, there is a considerable increase in snorting and bubble blowing behavior by males in the access holes and leads near females (Kovacs 1987). Occasionally males approach females who respond aggressively (MO Hammill and GB Stenson personal observation). As the pups approach the graycoat stage, there is an increase in male–male and male–female interactions. It has been suggested that these on-ice interactions are frequent in the White Sea, ending in copulation, and thus may act to terminate lactation (Popov 1966).

However, while these behaviors and on-ice copulation have also been observed in other areas, they are infrequent (Kovacs 1987; Lavigne and Kovacs 1988). Instead, much of the courting and mating activity occurs under the ice. As pups are weaned, there is an increase in the number of fresh wounds around the hind flippers and blood stains of males hauled out on the ice (Fig. 14.9), indicating that a considerable amount of male–male fighting involves underwater biting activity (Kovacs

Fig. 14.8 Differences in distribution between male (**a**) and female (**b**) harp seals on the whelping ice. *Photo credit* GB Stenson

1995). This, combined with the extensive acoustic activity, suggests that male agility, underwater calls, and acrobatic and display behavior are the fundamental components affecting male courtship and mating success (MØhl et al. 1975; Terhune and Ronald 1976; Merdsoy et al. 1978; Kovacs 1995). As pups are weaned and females leave, 'pairs' of adults on the ice outside of the whelping concentrations are sometimes seen (GB Stenson personal observation). These pairs are spread around outside of the whelping area until late March. It is not clear if they are male–female pairs. Still, if they are, it suggests that males and females may form temporary (transitory) pairs and that they may spend at least some time together before dispersing.

The males produce a strong odor that is associated with the rut (Sergeant 1965). This odor may be associated with sebum secretions from highly active sebaceous glands concentrated in the muzzle, as observed in ringed seals and to a lesser extent among gray seals (Hardy et al. 1991). The role of such an odor for harp seals is not clear since much of the male reproductive activity occurs in the water. The increase in adult male underwater display activity when pups reach the graycoat stage overlaps with the peak in testicular size in March (Fisher 1954; Miller et al. 1998).

Fig. 14.9 A group of males hauled out at the Front. The wounding results from fighting among males under the ice. *Photo credit* GB Stenson

Females return to the water to feed after breeding, replacing some of the mass lost during lactation between the end of breeding and the start of molt in late April (Sergeant 1991; Beck et al. 1993b; Nordøy et al. 2008). At this time, the ice is breaking up, providing access to energy-rich capelin that are still aggregated in relatively dense schools under the ice (Winters 1970; Buren et al. 2014), as well as other important prey such as small forage fish, including Atlantic herring (*Clupea harengus*), sandeels (*Ammodytes* spp), and invertebrates (Beck et al. 1993a; Lawson et al. 1995; Nilssen et al. 1995a, b; Lawson and Stenson 1997).

14.4 Molt

Phocid seals molt their fur and skin once per year. Harp seals form large aggregations on suitable ice to shed their old fur during this annual molt (Fig. 14.10), often in the same areas as the whelping concentrations or slightly further north. After spending the night in the water, seals begin to haul out on the ice each morning, and by mid-morning large numbers of seals can be seen basking in the sun. The sun's heat is thought to stimulate the development of new follicles, speeding the molt that usually takes several weeks for an individual (Ashwell-Erickson et al. 1986). Harp seals feed very little during the molting period and rely upon their energy stores during this period of the year.

Fig. 14.10 Molting harp seals. *Photo credit* D McKinnon

There is age and sex segregation in the timing of molt among harp seals. Generally, immature seals and mature males begin the molting process first, with mature females following (Sergeant 1965; Nazarenko and Timoshenko 1974; Chabot and Stenson 2002). Immature seals are usually the first to arrive on the molting ice, usually beginning in early-mid April in the Northwest Atlantic and later in the Greenland Sea. Unlike the adult females that spend some time feeding following breeding, adult males appear to remain in the area of the whelping patch until all of the pups are weaned and the females bred before moving to nearby ice to molt (GB Stenson and MO Hammill unpublished data; Chabot and Stenson 2002). Few adult females are observed in the molting patch until late April, and most of the seals have completed their molt by mid-May.

Males do not show any recovery in body mass post-breeding and continue to lose mass throughout the spring. Throughout April, approximately 60% of the mass loss occurs from the muscle or core, but by the end of April, there is a shift to preserving the core, and 75% of loss now occurs from the outer blubber (Beck et al. 1993b; Chabot and Stenson 2002). At the end of the molt in May, males and females are at their minimum mass for the year (Sergeant 1965, 1991; Nilssen et al. 1997; Chabot and Stenson 2002).

14.5 Post-Molt

Much of our information on seasonal movements (and diets) outside of winter when harp seals are in southern waters is based on returns of flipper tags and biological samples obtained from animals harvested by Arctic subsistence hunters (e.g., Sergeant 1965; Kapel 2000). Biologgers have been deployed to study movements and diving behavior; their use has been more limited than with some other species because harp seals are difficult to capture outside of the pupping period. Unfortunately, biologgers deployed during pupping (March) would only provide information for a short time until lost during the molt (April–May). Capturing molting harp seals requires access to the ice where animals are hauled out (usually using an ice-reinforced vessel) and once captured, the seals must then be kept until they finish the molt before the biologgers can be glued to their fur and the seal released (Stenson and Sjare 1997; Folkow et al. 2004).

After the molt, all animals are at their minimum blubber thickness in the annual cycle (Beck et al. 1993b; Nilssen et al. 1995a; Chabot and Stenson 2002). Seals from all populations disperse to feed in more polar regions. Animals from the White Sea/Barents Sea population follow the receding ice edge, gradually moving northwards and north-eastwards to feed around Svalbard and in the northern Barents Sea throughout the summer (Nordøy et al. 2008). They overlap considerably with Greenland Sea harp seals at this time of year, who also summer in the Barents Sea, feeding in the waters around Spitzbergen and northwards towards the polar ice pack (Haug et al. 1994; Folkow et al. 2004). Diving behavior is similar between the two populations, but harp seals from the Greenland Sea tend to dive to deeper depths than harp seals from the Barents Sea/White Sea population (Nordøy et al. 2008). Harp seals are capable of diving to depths of 400 m, but the majority of dives are to depths of less than 100 m (Nordøy et al. 2008). In both populations, animals tend to dive deeper during the day, females tend to dive deeper than males, and dives are shallower during summer than during the winter months just prior to pupping (Nordøy et al. 2008).

In the Northwest Atlantic population, the northward dispersal of seals extends over a greater distance. Most animals remain offshore on the Newfoundland and Labrador shelves out to the 200 m contour (Stenson and Sjare 1997; Lacoste and Stenson 2000; GB Stenson personal observation). From there, they move north during June and July along both the Greenland and Canadian continental shelves. The majority of seals move northwards along the Canadian shelf, with some dispersing into Hudson Bay, Foxe Basin, and further north into Lancaster Sound and Baffin Bay where they remain until the fall (Sergeant 1991; Stenson and Sjare 1997). The majority of animals that move to Greenland follow the west coast northwards to Baffin Bay, while some may move along the southeast coast of Greenland towards the north. As in the Northeast Atlantic, most dives are relatively shallow (50–100 m), although a dive deeper than 700 m has been recorded (GB Stenson unpublished data).

As the ice disintegrates in July, harp seals penetrate into the fjords and channels of the eastern Arctic archipelago (Finley et al. 1990). In the high Arctic, seals are seen traveling in groups of 5–50 animals or more, often porpoising when traveling

Fig. 14.11 Imagery of harp seals on the Grand Banks, 160 km off the east coast of Newfoundland, showing how large harp seal groups can be when viewed from the air. *Photo credit* Dept of Fisheries and Oceans Canada

through the water. Feeding aggregations of several hundred are not uncommon and are usually associated with schools of Arctic cod. At times, groups of harp seals can be seen pushing schools of Arctic cod towards the shore where the fish may be captured more easily (Finley et al. 1990). This suggests some cooperative strategy, but the level of communication and coordination among individuals is not known. Adults migrate in groups, whereas the young of the year travel individually and do not aggregate in groups of more than several animals before the end of their first year (Meldgaard 2004). This development of gregarious behavior in adult harp seals may be an adaptive feeding strategy among adults that parallels a shift in diet from feeding on pelagic crustaceans to small schooling fish (Sergeant 1973; Finley et al. 1990). The majority of groups reported by Lacoste and Stenson (2000) contained fewer than 10 seals, although a few groups of more than 100 seals were seen. Groups of several hundred to a few thousand seals have been observed on the ice or swimming on their backs during the fall and winter off the Grand Banks (Fig. 14.11; Sergeant 1991; GB Stenson personal observation). Unfortunately, little is known of the duration or structure of these groups (Finley et al. 1990; Sergeant 1991).

14.6 Fall

In the Northeast Atlantic, the significant improvement in body condition observed among seals over the summer months is maintained throughout the fall and winter until pupping. In November–December, the White Sea/Barents Sea population returns southward towards the breeding areas, overwintering in the southeastern Barents Sea and White Sea, whereas Greenland Sea harp seals winter off southeast Greenland in the Denmark Strait (Folkow et al. 2004; Nordøy et al. 2008).

The fall migration timing is similar in the Northwest Atlantic, with harp seals from this population returning to southern Labrador in November. The majority of the population feeds on the Newfoundland Shelf and northern Grand Bank during the winter, but some animals may move into the Gulf of St Lawrence by December (Sergeant 1991). In both areas, animals continue feeding until moving to the whelping ice in late February (Stenson and Sjare 1997; Lacoste and Stenson, 2000; GB Stenson unpublished data).

In contrast to the Northeast Atlantic harp seals, the condition of the Northwest Atlantic animals does not plateau from summer until pupping. Instead, the most significant increase in weight occurs when they return to the southern feeding areas off the southern Labrador coast or in the Gulf of St Lawrence. At this time, just prior to pupping, the condition of seals in the Northwestern Atlantic population reaches a maximum (Beck et al. 1993b; Nilssen et al. 1995b; Hammill et al. 2008; Chabot and Stenson 2002).

The buildup of energy stores by female harp seals is necessary for the successful birth of the pup (Stenson et al. 2016, 2020a). Control of pregnancy can occur at various stages of the reproductive cycle, beginning with the decision to mate in March. Boyd (2000) proposed that the decision to reproduce in some pinnipeds may occur at the time of implantation, which in the case of harp seals occurs in late July or early August (Chabot and Stenson 2000). A minimal level of energy stores is likely necessary before a female can proceed with implantation and early gestation. The energetic cost of supporting a fetus is relatively low for the first two-thirds of gestation but increases significantly during the last trimester (Stewart et al. 1989; Chabot and Stenson 2000; Yunker et al. 2005). However, the most energetically expensive phase involves caring for the pup after birth (see Sect. 14.3.1). Overall, the physical condition of females and reproductive rates of harp seals in the Northwest Atlantic population has declined, likely due to density-dependent factors. However, there is considerable inter-annual variability in pregnancy rates that appear to be influenced by capelin biomass and ice conditions (as a proxy for ecological conditions) acting through changes in body condition (Fig. 14.12; Stenson et al. 2016, 2020a).

Stenson et al. (2016, 2020a) found that high rates of late-term abortion were associated with low capelin biomass and poor winter ice conditions. Since gestation is relatively inexpensive compared to lactation, this suggests that a final critical control point relies on winter foraging conditions. If the female can attain a minimum condition level to support lactation, then the pregnancy is maintained. However, if this threshold is not attained, the seal may abort rather than giving birth to a pup that the female has little chance of successfully rearing, or which may also jeopardize her survival (Stenson et al. 2016, 2020a).

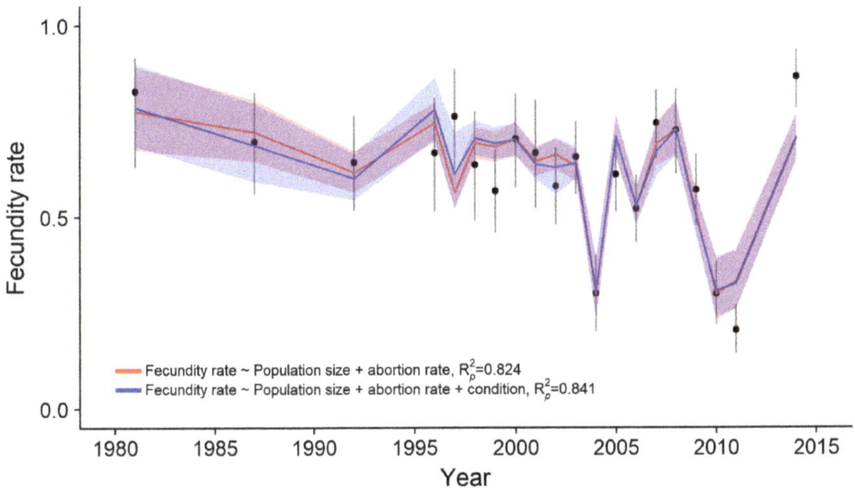

Fig. 14.12 Generalized additive model fit to describe the fecundity rate using relative body condition as an explanatory parameter (Stenson et al. 2020a)

14.7 Diet

Throughout its range, the diet of harp seals varies seasonally and geographically. However, it tends to be dominated by relatively few species, in particular, capelin, Arctic cod, Atlantic herring, euphausiids, and amphipods, *Themisto* spp (Sergeant 1973; Kapel 2000; Murie and Lavigne 1991; Wallace and Lavigne 1992; Beck et al. 1993a; Lawson and Stenson, 1997; Lawson et al. 1998; Nilssen et al. 2000; Lindstrøm et al. 2013; Hammill et al. 2005). Arctic cod and crustaceans appear to be of particular importance during summer and autumn feeding in the northern parts of their range (Sergeant 1973; Kapel 2000; Finley et al. 1990; Lydersen et al. 1991; Nilssen et al. 1995a, b). As ice cover expands southwards in late autumn and winter, the diet of southward migrating seals switches from mostly crustaceans to mainly fish, particularly Arctic cod, capelin, and herring (Nilssen et al. 1995b; Lindstrøm et al. 1998). The factors affecting prey selection are complex but may involve interactions between animal age, prey availability, energy density, and handling costs (Lawson et al. 1998; Lindstrøm et al. 1998, 2013).

14.8 Conclusions and Future Directions

Throughout this chapter, we have identified the mechanisms that have affected the behavioral ecology of harp seals, and in particular, how their association with ice has influenced them. Despite the challenges of working on a pack-ice loving species, a considerable amount of research has focused on the challenges of this transient environment and the behavioral-ecological mechanisms that have evolved for harp seals to adapt to these conditions. Much of this research has concentrated around the breeding season when harp seals are most accessible. Much more needs to be done during the open water period when seals are distributed throughout the offshore and northern portions of their range. The advent of new technologies is helping to fill this data gap. Throughout its range, the harp seal's behavior is very closely associated with the pack ice and the distribution of capelin, although not entirely. Within the pack ice, they may reduce exposure to aquatic predators such as killer whales (*Orcinus orca*), but animals sleeping on the ice are vulnerable to predation from polar bears (Fabricius 1790; Peacock et al. 2013; Smith and Stirling 2019). While ice is essential to harp seals as a platform, ice cover and ice-breakup are also crucial to the dynamics of their prey (e.g., bottom-up control of capelin; Buren et al. 2014; Lewis et al. 2019).

In a warming North Atlantic, there has been a decline in the extent of the seasonal pack ice. In the Northeast Atlantic, an increase in the inflow of warmer Atlantic water has contributed to warmer water temperatures. This has led to changes in the distribution of whelping ice in the Greenland Sea and increased frequency of poor ice years in the White Sea, both of which are likely to increase pup mortality. An increase in the flow of Atlantic water has also led to an increase in the abundance of boreal fish species. This, combined with very high Atlantic cod stocks, may have led to increased competitive interactions between seals and cod, to the detriment of harp seal productivity (Stenson et al. 2020c). In the Northwest Atlantic, the cold Labrador current that brings Arctic water southward remains dominant. However, a decline in the southern extension of the seasonal pack ice is predicted, with the southeastern Labrador Sea being virtually ice-free by 2100 (Han et al. 2019). There has already been an increase in the frequency of poor ice conditions in the Gulf of St. Lawrence, at the southern limit of their range (Johnston et al. 2005; Stenson and Hammill 2014). Harp seals have not responded to this deterioration in breeding habitat by moving onshore to raise their pups, as has been observed among gray seals in the southern Gulf of St. Lawrence (Stenson and Hammill 2014; Hammill et al. 2017). Instead, in years where some ice forms but is not extensive, pup mortality appears to increase, whereas in years where very little ice forms, animals appear to shift from the Gulf to the Front, or further north along the Labrador coast (Stenson and Hammill 2014, Stenson et al. 2020b). This suggests that over the longer term, harp seals will respond to changes in the extent of the seasonal pack ice by shifting their summer and breeding distribution northwards with the resultant loss of pupping in the southern portions of their range. As such, harp seals are an excellent indicator of the changes in their pack-ice environment.

Acknowledgements The authors thank all of the people who have helped improve our understanding of harp seals over the past five decades, including W. Penney, D. Wakeham, D. McKinnon, A. Buren, J. Lawson, T. Haug, J-F Gosselin, S. Turgeon, P. Carter, P. Goulet, L. Shepherd, B. Stockwood, W. Hoek, for help in the field, and P. Goulet and Dr. J. Lawson for help with the figures. Support was provided by the Department of Fisheries and Oceans.

References

Ashwell-Ericksson S, Fay FH, Elsner R, Wartzok D (1986) Metabolic and hormonal correlates of molting and regeneration of pelage in Alaskan harbor and spotted seals (*Phoca vitulina* and *Phoca largha*). Can J Zool 64:1086–1094

Bajzak C, Hammill MO, Stenson GB, Prinsenberg S (2011) Drifting away: implications of changes in ice conditions for a pack-ice-breeding phocid, the harp seal (*Pagophilus groenlandicus).* Can J Zool 89:1050–1062

Bartholomew GA (1970) A model for the evolution of pinniped polygyny. Evolution 24:546–559

Beck GG, Hammill MO, Smith TG (1993a) Seasonal variation in the diet of harp seals (*Phoca groenlandica*) from the Gulf of St Lawrence and western Hudson Strait. Can J Fish Aquat Sci 50:1363–1371

Beck GG, Smith TG, Hammill MO (1993b) Evaluation of body condition in the Northwest Atlantic harp seal (*Phoca groenlandica*). Can J Fish Aquat Sci 50:1372–1381

Blix AS, Grav HJ, Ronald K (1979) Some aspects of temperature regulation in newborn harp seal pups. Am J Physiol Regul Integr Comp Physiol 236:188–197

Boness D, Bowen WD (1996) The evolution of maternal care in pinnipeds. Bioscience 46:645–654

Bowen WD, Sergeant DE (1983) Mark-recapture estimates of harp seal pup, *Pagophilus groenlandicus*, production in the northwest Atlantic. Can J Fish Aquat Sci 40:728–743

Bowen WD, Capstick CK, Sergeant DE (1981) Temporal changes in the reproductive potential of female harp seals, *Pagophilus groenlandicus*. Can J Fish Aquat Sci 38:495–503

Bowen WD, Oftedal OT, Boness DJ (1985) Birth to weaning in 4 days: remarkable growth in the hooded seals, *Cystophora cristata*. Can J Zool 63:2841–2846

Boyd IL (2000) State-dependent fertility in pinnipeds: Contrasting capital and income breeders. Funct Ecol 14:623–630

Buren, AD, Koen-Alonso M, Pepin P, et al (2014) Bottom-up regulation of capelin, a keystone forage species. PLoS One 9:e87589

Burns JM, Costa DP, Frost K, Harvey JT (2005) Physiological development in juvenile harbor seals. Physiol Biochem Zool 78:1057–1068

Burns JM, Lestyk K, Folkow LP, Hammil MO, Blix AS (2007) Size and distribution of oxygen stores in harp and hooded seals from birth to maturity. J Comp Physiol B 177:687–700

Burns JM, Skomp N, Bishop N, Lestyk K, Hammill M (2010) Development of aerobic and anaerobic metabolism in cardiac and skeletal muscles from harp and hooded seals. J Expt Biol 213:740–748

Burns JM, Lestyk K, Freistroffer D, Hammill MO (2015) Preparing muscles for diving: age-related changes in muscle metabolic profiles in harp (*Pagophilus groenlandicus*) and hooded (*Cystophora cristata*) seals. Physiol Biochem Zool 88:167–182

Carr SM, Marchall HD, Duggan AT, Flynn SMC, Johnstone KA, Pope AM, Wilkerson CD (2008) Phylogeographic genomics of mitochondrial DNA: highly resolved patterns of intraspecific evolution and a multi-species, microarray-based DNA sequency strategy for biodiversity studies. Comp Biochem Physiol Part D 3:1–11

Carr SM, Duggan AT, Stenson GB, and Marshall HD (2015) Quantitative phylogenomics of within-species mitogenome variation: monte carlo and non-parametric analysis of hylogeographic structure among discrete transatlantic breeding areas of harp seals (*Pagophilus groenlandicus*). PLoS One 10(8):e0134207. https://doi.org/10.1371/journal.pone.0134207

Castonguay D (2003) L'exploitation du loup-marin et son incidence sur l'occupation de la Côte par les Montagnais de la traite de Tadoussac, au XVIII siècle Recherches Amérindiens au Québec 33:61–72

Chabot D, Stenson GB (2000) Implantation date, growth rate and allometric relationships in foetal Northwest Atlantic harp seals (*Phoca groenlandica*). Can J Zool 78:501–505

Chabot D, Stenson GB (2002) Growth and seasonal fluctuations in size and condition of male northwest Atlantic harp seals *Phoca groenlandica*: an analysis using sequential growth curves. Mar Ecol Prog Ser 227:25–42

Chapskii KK (1961) Nekotorye ekologicheskie obosnovaniy sezonnoi dinamiki areala belomorskoi populyatsii grenlandskogo tyulenya (*Pagophoca gorenlandica*). [Some ecological factors determining seasonal changes in distribution of the White Sea harp seal population (*Pagophoca groenlandica*)] Tr Soveshch Ikhtiol Kom Akad Nauk SSSR 12:150–163 Transl from Russian by Fish Res Board Transl ser No 380

Chapskii KK (1964) Nachal'nyi period postnatal' nogo rosta belomorskogo lysuna (*Pagophoca groenlandica*). [The initial period of postnatal growth in the harp seal (*Pagophoca groenlandica*) Morskie mlekopitayuschie Akademiya Nauk SSS,R Moscow: 138–157 Transl from Russian By the Dept of the Secretary of State of Canada no 4173

Clark CA, Burns JM, Schreer JF, Hammill MO (2006) Erythropoietin concentration in developing harbor seals (*Phoca vitulina*). Gen Comp Endocrinol 147:262–267

Colman JS (1937) The present state of the Newfoundland seal fishery. J of Anim Ecol 6:145–159

Costa DP (1993) The relationship between reproductive and foraging energetics and the evolution of Pinnipedia. Symp Zool Soc Lond 66:293–314

Dorofeev SV (1939) Vliyanie ledovykh uslovii na povedenie grenlanskogo tyulenya. [The influence of ice conditions on the behavior of harp seals] Zool Zh 18:748–761 Transl from Russian by Secretary of State Multilingual Services division transl No 1155996

DFO (2020) 2019 Status of Northwest Atlantic harp seals, *Pagophilus groenlandicus*. DFO. Can Sci Advis Sec Sci Advis Rep 2020/020

Fabricius O (1790) The seal of Greenland cited in: Kapel, F (2005) Otto Fabricius and the seals of Greenland. Biosci 55:50. (Monographs on Greenland subseries)

Finley KJ, Bradstreet MSW, Miller GW (1990) Summer feeding ecology of harp seals (*Phoca groenlandica*) in relation to Arctic cod (*Boreogadus saida*) in the Canadian High Arctic. Polar Biol 10:609–618

Fisher HD (1954) Studies on reproduction in the harp seal, *Phoca groenlandica*, Erxleben, in the Northwest Atlantic. Fish Res Board Can MS Rep 588:109

Fisher HD (1955) Utilization of Atlantic harp seal populations. Trans 20th N Amer Wildl Conf 507–518

Folkow LP, NordØy ES, Blix AS (2004) Distribution and diving behavior of harp seals (*Pagophilus groenlandicus*) from the Greenland Sea stock. Polar Biol 27:281–298

Frie AK, Potelov VA, Kingsley MCS, Haug T (2003) Trends in age-at maturity and growth parameters of female Northeast Atlantic harp seals, *Pagophilus groenlandicus* (Erxleben, 1777). ICES J Mar Sci 60:1018–1032

Haller MA, Kovacs KM, Hammill MO (1996) Maternal behavior and energy investment by grey seals (*Halichoerus grypus*) breeding on land-fast ice. Can J Zool 74:1531–1541

Hammill MO, Sauvé C (2017) Growth and condition in harp seals: evidence of density-dependent and density-independent influences. ICES J Mar Sci 74:1395–1407

Hammill MO, Lesage V, Carter P (2005) Problems associated with determining diet composition in harp seals by reconstruction of stomach contents or faecal samples. Can J Zool 83:1365–1372

Hammill, MO, Ryg M, Chabot D (2008) Seasonal changes in energy requirements of harp seals. J Northw Atl Fish Sci 42:135–152

Hammill MO, Kingsley MCS, Beck G, Smith TG (1995) Growth and condition in the Northwest Atlantic harp seal. Can J Fish Aquat Sci 52:478–488

Hammill MO, Stenson GB, Kingsley MCS (2011) Historical abundance of Northwest Atlantic harp seals (*Pagophilus groenlandicus*): influence of harvesting and climate. DFO Can Sci Advis Sec Res Doc 2011 100(iv):16

Hammill MO, Lydersen C, Ryg M, Smith TG (1991) Lactation in the ringed seal (*Phoca hispida*). Can J Fish Aquat Sci 48:2471–2476

Hammill MO, Stenson GB, Doniol-Valcroze T, Mosnier A (2015) Conservation of northwest Atlantic harp seals: past success, future uncertainty? Biol Cons 192:181–191

Hammill MO, den Heyer CE, Bowen WD, Lang SLC (2017) Grey seal population trends in Canadian Waters, 1960–2016 and harvest advice. DFO Can. Sci. Advis. Sec. Res. Doc. 2017/052. v + 30 p

Hammill MO, Stenson GB, Mosnier A, Doniol-Valcroze T (2021) Trends in harp seal, *Pagophilus groenlandicus*, abundance in the Northwest Atlantic, 1952–2019. DFO Can Sci Advis Sec Res Doc 2021 006(iv):29

Han G, Ma Z, Long Z, Perrie W, Chasse J (2019) Ocean and sea-ice model under an A1B forcing scenario 2011–2069. Atmos Ocean 57:3–17

Hardy MH, Roff E, Smith TG, Ryg M (1991) Facial skin glands of ringed and grey seals, and their possible function as odoriferous organs. Can J Zool 69:189–200

Haug T, Nilssen KT, Lindblom L (2000) First independent feeding of harp seal (*Phoca groenlandica*) and hooded seal (*Cystophora cristata*) pups in the Greenland Sea. NAMMCO Sci Publ 2:29–39

Haug T, Nilssen KT, Oien N, Potelov V (1994) Seasonal distribution of harp seals (*Phoca groenlandica*) in the Barents Sea. Pol Res 13:163–172

ICES (2019) ICES/NAFO/NAMMCO Working group on harp and hooded seals (WGHARP). ICES Sci Rep 1:72–193. https://doi.org/10.17895/ices.pub.5617

Johnston DW, Friedlaender AS, Torres LG, Lavigne DM (2005) Variation in sea ice cover on the east coast of Canada from 1969 to 2002: climate variability and implications for harp and hooded seals. Clim Res 29:209–222

Kapel FO (2000) Feeding habits of harp and hooded seals in Greenland waters. NAMMCO Sci Publ 2:50–64

Kjellqwist SA, Haug T, Øritsland T (1995) Trends in age composition, growth and reproductive parameters of Barents Sea harp seals *Phoca groenlandica.* ICES J Mar Sci 52:197–208

Kovacs KM (1987) Maternal behavior and early behavioral ontogeny of harp seals, *Phoca groenlandica.* Anim Behav 35:844–855

Kovacs KM, Lavigne DM (1992) Mass-transfer efficiency between hooded seal (*Cystophora cristata*) mothers and their pups in the Gulf of St. Lawrence. Can J Zool 70:1315–1320

Kovacs KM (1995) Harp and hooded seals-a case study in the determinant of mating systems in pinnipeds. In: Blix AS, WallØe L, Ultang Ø (eds) Whales, seals, fish and man. Elsevier, Amsterdam, p 329–335

Kovacs KM, Lavigne DM (1985) Neonatal growth and organ allometry of northwest Atlantic harp seals (*Phoca groenlandica*). Can J Zool 63:2793–2799

Kovacs KM, Lavigne DM, Innes S (1991) Mass transfer efficiency between harp seal (*Phoca groenlandica*) mothers and their pups during lactation. J Zool Lond 223:213–221

Kovacs KM, Lavigne DM, Stewart REA (1985) Early postnatal mortality in northwest Atlantic harp seals, *Phoca groenlandica.* J Mammal 66:556–558

Kovacs KM, Lydersen C (2008) Climate change impacts on seals and whales in the north Atlantic Arctic and adjoining waters. Sci Progr 91:117–150

Lacoste KN, Stenson GB (2000) Winter distribution of harp seals (*Phoca groenlandica*) off eastern Newfoundland and southern Labrador. Polar Biol 23:805–811

Lavigne DM, Kovacs KM (1988) Harps and hoods: ice-breeding seals of the northwest Atlantic. University of Waterloo Press, Waterloo

Lavigne DM (1976) Counting harp seals with ultra-violet photography. Polar Rec 18:269–277

Lavigne DM, Stewart REA, Fletcher F (1982a) Changes in the composition and energy content of harp seal milk during lactation. Physiol Zool 55:1–9

Lavigne DM, Barchard WW, Innes S, Øritsland NA (1982b) Pinnipeds bioenergetics. In: Gordon J, Goodman J (eds) Mammals in the seas: small Cetaceans, seals, sirenians and otters, vol IV. FAO Fisheries Ser 5, Rome, p 191–235

Laws RM (1952) A new method of age determination for mammals. Nature 169:972

Lawson JW, Stenson GB (1997) Diet of northwest Atlantic harp seals (*Phoca groenlandica*) in offshore areas. Can J Zool 75:2095–2106

Lawson JW, Anderson JT, Dalley EL, Stenson GB (1998) Selective foraging by harp seals, *Phoca groenlandica* in nearshore and offshore waters of Newfoundland, 1993 and 1994. Mar Ecol Progr Ser 163:1–10

Lawson JW, Stenson GB, McKinnon D (1995) Nearshore diet of harp seals (*Phoca groenlandica*) in the northwest Atlanditc during 1990–1993. Can J Zool 75:1805–1818

Lewis KP, Buren AD, Regular PM, Mowbray FK, Murphy HM (2019) Forecasting capelin *Mallotus villosus* biomass on the Newfoundland shelf. Mar Ecol Prog Ser 616:171–183

Lindstrøm U, Harbitz A, Haug T, Nilssen KT (1998) Do harp seals *Phoca groenlandica* exhibit particular prey preferences? ICES J Mar Sci 55:941–953

Lindstrøm U, Nilssen KT, Pettersen LMS, Haug T (2013) Harp seal foraging behaviour during summer around Svalbard in the northern Barents Sea: diet composition and the selection of prey. Polar Biol 36:305–320. https://doi.org/10.1007/s00300-012-1260-x

Lydersen C, Angantyr LA, Wiig Ø, Øritsland T, (1991) Feeding habits of Northeast Atlantic harp seals (*Phoca groenlandica*) along the summer ice edge of the Barents Sea. Can Fish Aquat Sci 48:2180–2183

Lydersen C, Kovacs KM (1993) Diving behavior of lactating harp seal, *Phoca groenlandica*, females from the Gulf of St. Lawrence Canada. Anim Behav 46:1213–1221

Lydersen C, Kovacs KM (1996) Energetics of lactation in harp seals (*Phoca groenlandica*) from the Gulf of St. Lawrence Canada. J Comp Physiol B 166:295–304

Lydersen C, Kovacs KM (1999) Behavior and energetics of ice-breeding, north Atlantic phocid seals during the lactation period. Mar Ecol Prog Ser 187:265–281

Mårtensson P-E, NordØy ES, Blix AS (1994) Digestibility of crustaceans and capelin in harp seals (*Phoca groenlandica*). Mar Mamm Sci 10:325–331

Mårtensson P-E, NordØy ES, Messelt EB, Blix AS (1998) Gut length, food transit time and diving habit in phocid seals. Polar Biol 20:213–217

Meldgaard M (2004) Ancient harp seal hunters of Disko Bay: subsistence and settlement at the Saqqaq culture site Qeqertasussuk (2400–1400 BC), West Greenland. Meddelelser om Grønland I, Man & Society, vol 30. Copenhagen, Danish Polar Center

Merdsoy BR, Curtsinger WR, Renouf D (1978) Preliminary underwater observations of the breeding behavior of the harp seal (*Pagophilus groenlandicus*). J Mamm 59:180–185

Miller EH, Stewart ARJ, Stenson GB (1998) Bacular and testicular growth, allometry, and variation in the harp seal (*Pagophilus groenlandicus*). J Mamm 79:502–513

MØhl B, Terhune JM, Ronald K (1975) Underwater calls of the harp seal. Rapp P-V Reun Cons Int Explor Mer 196:533–543

Moors HB, Terhune JM (2003) Repetition patterns within harp seal (*Pagophilus groenlandicus*) underwater calls. Aquat Mamm 29:278–288

Muelbert MMC, Bowen WD (1993) Duration of lactation and postweaning changes in mass and body composition of harbour seal, *Phoca vitulina*, pups. Can J Zool 71:1405–1414

Murie DJ, Lavigne DM (1991) Food consumption of wintering harp seals, *Phoca groenlandica*, in the St. Lawrence estuary. Can J Zool 69:1289–1296

Nansen F (1925) Hunting and adventure in the Arctic. Duffield and Co, New York

Nazarenko YI (1975) Sexual maturity, reproductive rate and missed pregnancy in female harp seals. Rapp P-V Reun Cons Int Explor Mer 169:413–415

Nazarenko YI, Timoshenko YK (1974) Age structure and sex ratio in the White Sea population of *Pagophoca groenlandica* as an index of efficiency of protective measures. Zoologischeskii Zhumal 53:256–262

Nilssen KT, Haug T, Potelov V, Stasenkov VA, Timoshenko YK (1995a) Food habits of harp seals (*Phoca groenlandica*) during lactation and molt in March-May in the southern Barents Sea and White Sea. ICES J Mar Sci 52:33–41

Nilssen KT, Haug T, Potelov V, Timoshenko Y (1995b) Food habits of harp seals (*Phoca groenlandica*) during early summer and autumn in the northern Barents Sea. Polar Biol 15:485–493

Nilssen KT, Haug T, Grotnes PE, Potelov V (1997) Seasonal variation in body condition of adult Barents Sea harp seals (*Phoca groenlandica*). J Northwest Atl Fish Sci 22:17–25

Nilssen KT, Pedersen O-P, Folkow LP, Haug T (2000) Food consumption estimates of Barents Sea harp seals. NAMMCO Sci Publ 2:9–27

Nordøy ES, Folkow LP, Potelov V, Prischemikhin V, Blix AS (2008) Seasonal distribution and dive behavior of harp seals (*Pagophilus groenlandicus*) of the White Sea-Barents Sea stock. Polar Biol 31:1119–1135
Oftedal OT, Bowen WD, Boness DJ (1996) Lactation performance and nutrient deposition in pups of the harp seal, *Phoca groenlandica*, on ice floes off southeast Labrador. Physiol Zool 69:635–657
Oftedal OT, Boness DJ, Tedman RA (1987) The behavior, physiology and anatomy of lactation in pinnipedia. Curr Mamma 1:175–245
Oftedal OT, Bowen WD, Widdowson EM, Boness DJ (1989) Effects of suckling and postsuckling fast on weights of the body and internal organs of harp and hooded seal pups. Biol Neonate 56:283–300
Øien N, Øritsland T (1995) Use of mark-recapture experiments to monitor seal populations subject to catching. In: Blix SD, Walløe L, Ulltang Ø (eds) Whales, seals, fish and man. Elsevier, Amsterdam, pp 35–45
Øigård TA, Haug T, Nilssen KT (2014) From pup production to quotas: current status of harp seals in the Greenland Sea. ICES J Mar Sci 71:537–545. https://doi.org/10.1093/icesjms/fst155
Øigård TA, LindstrØm U, Haug T, Nilssen KT, Smout S, (2013) Functional relationship between harp seal body condition and available prey in the Barents Sea. Mar Ecol Progr Ser 484:287–301
Øritsland NA, Ronald K (1973) Effects of solar radiation and windchill on skin temperature of the harp seal, *Pagophilus groenlandicus*, (Erxleben, 1777). Comp Biochem Physiol A 44:519–525
Peacock E, Taylor M, Laake J, Stirling I (2013) Population ecology of polar bears in Davis Strait, Canada and Greenland. J Wildl Mgt 77:463–476
Pearson LE, Liwanag HEM, Hammill MO, Burns JM (2014a) To each its own: thermoregulatory strategy varies among neonatal polar phocids. Comp Biochem Physiol A 178:59–67
Pearson LE, Liwanag HEM, Hammill MO, Burns JM (2014b) Shifts in thermoregulatory strategy during ontogeny in harp seals (*Pagophilus groenlandicus*). J Therm Biol 44:93–102
Pearson LE, Weitzner EL, Burns JM, Hammill MO, Liwanag HEM (2019) From ice to ocean: changes in the thermal function of harp seal pelt with ontogeny. J Comp Physiol B 189:501–511
Perez MA, Mooney EE (1986) Increased food and energy-consumption of lactating northern fur seals, *Callorhinus ursinus*. Fish Bull 84:371–381
Perry EA, Stenson GB, Buren AD (2017) Attendance and nursing patterns of harp seals in the harsh environment of the northwest Atlantic. Polar Biol 40:151–160
Perry EA, Stenson GB, Bartlett SE, Davidson WS, Carr SM (2000) DNA sequence analysis identifies genetically distinguishable populations of harp seals (*Pagophilus groenlandicus*) in the Northwest and Northeast Atlantic. Mar Biol 137:53–58
Popov LA (1966) Na l'dine s tyulenyami Ledovyi dreif biologov v Belom more Priroda 9 93-101 Preliminary translation by DE Sergeant Fish Res Board Can 11. (Translation series no 814)
Popov LA (1970) Soviet tagging of harp and hooded seals in the north Atlantic. Fiskdir skrServ HavUnders 16:1–9
Potelov VA, Golikov AP, Bondarev VA (2003) Estimated pup production of harp seals *Pagophilus groenlandicus* in the White Sea, Russia, in 2000. ICES J Mar Sci 60:1012–1017
Potelov V, Nilssen KT, Svetochev V, Haug T (2000) Feeding habits of harp seals (*Phoca groenlandica*) and hooded seals (*Cystophora cristata*) during late winter, spring and early summer. NAMMCO Sci Pub 2:40–49
Renouf MAP, Bell T (2008) Dorset paleoeskimo skin processing at Phillip's garden, Port au Choix, Northwestern Newfoundland. Arctic 61:35–47
Robinson G (1897) Ice-riding pinnipeds. JD Potter, London
Roff DA, Bowen WD (1983) Population dynamics and management of the northwest Atlantic harp seal (*Phoca groenlandica*). Can J Fish Aquat Sci 40:914–932
Ronald K, Johnson E, Foster M, Vander Pol D (1970) The harp seal, *Pagophilus groenlandicus*, (Erxleben 1777). I. Methods of handling, molt, and diseases in captivity. Can J Zool 48:1035–1040
Rossong MA, Terhune JM (2009) Source levels and communication-range models for harp seal (*Pagophilus groenlandicus*) underwater calls in the Gulf of St. Lawrence, Canada. Can J Zool 87:609–617

Ryan S (1994) The ice hunters: a history of Newfoundland sealing to 1914. Newfoundland History Series 8 Breakwater Press, St John's NL, Canada

Ryg et al., 1993. Ryg M, Lydersen C, Knutsen LO, Bjorge A, Smith TG, Øritsland NA (1993) Scaling of insulation in seals and whales. J Zool 230:193–206

Sergeant DEA (1965) Migration of harp seals *Pagophilus groenlandicus* (Erxleben) in the northwest Atlantic. J Fish Res Bd Can 22:433–464

Sergeant DE (1966) Reproductive rates of harp seals *Pagophilus groenlandicus* (Erxleben). J Fish Res Bd Canada 23:757–766

Sergeant DE (1973) Feeding, growth and productivity of northwest Atlantic harp seals (*Pagophilus groenlandicus*). J Fish Res Board Can 30:17–29

Sergeant DE (1991) Harp seals, man and ice. Can Spec Publ Fish Aquat Sci 114:153

Sergeant DE, Fisher HD (1960) Harp seal populations in the western north Atlantic from 1950 to 1960. Fish Res Board Can Arctic. (Unit circular no 5)

Serrano A (2001) New underwater and aerial vocalizations of captive harp seals (*Pagophilus groenlandicus*). Can J Zool 79:75–81

Serrano A, Terhune JM (2002) Antimasking aspects of harp seal (*Pagophilus groenandicus*) underwater vocalizations. J Acoust Soc Am 112:3083–3090

Sivertsen E (1941) On the biology of the harp seal, *Phoca groenlandica* Erxl. Investigations carried out in the White Sea 1925–1937. Hvalradets Skr 26:1–166

Sjare B, Stenson GB (2010) Changes in the reproductive parameters of female harp seals (*Pagophilus groenlandicus*) in the Northwest Atlantic. ICES J Mar Sci 67:304–315

Skaug HJ, Frimannslund L, Øien NI (2007) Historical population assessment of Barents Sea harp seals (*Pagophilus groenlandicus*). ICES J Mar Sci 64:1356–1365

Smith TG, Stirling I (2019) Predation of harp seals, *Pagophilus groenlandicus*, by polar bears, *Ursus maritimus*, in Svalbard. Arctic 72:197–202

Stenson GB, Hammill MO (2014) Can ice breeding seals adapt to habitat loss in a time of climate change? ICES J Mar Sci 71:1977–1986

Stenson GB, Sjare B (1997) Seasonal distribution of harp seals, (*Phoca groenlandica*) in the northwest Atlantic. ICES CM 1997/CC 10:23

Stenson GB, Buren AD, Koen-Alonso M (2016) The impact of changing climate ad abundance on reproduction in an ice-dependent species, the northwest Atlantic harp seal, *Pagophilus groenlandicus*. ICES J Mar Sci 73:250–262

Stenson GB, Buren AD, Sheppard GL (2020a) Updated estimates of reproductive rates in Northwest Atlantic harp seals and the influence of body condition. DFO Can Sci Advis Sec Res Doc 2020. (in press)

Stenson GB, Gosselin J-F, Lawson JW, Buren A, Goulet P, Lang SLC, Nilssen K, Hammill MO (2020b) Estimating pup production of Northwest Atlantic harp seals, *Pagophilus groenlandicus*, in 2017. DFO Can Sci Advis Sec Res Doc 2020. (in press)

Stenson GB, Haug T, Hammill MO (2020c) Harp seals: monitors of change in differing ecosystems. Front Mar Sci 7:569258. https://doi.org/10.3389/fmars.2020.569258

Stenson GB, Hammill MO, Kingsley MC, Sjare S, Warren WG, Myers RA (2002) Is there evidence of increased pup production in northwest Atlantic harp seals, *Pagophilus groenlandicus*? ICES J Mar Sci 59:81–92

Stenson GB, Rivest L-P, Hammill MO, Gosselin J-F, Sjare B (2003) Estimating pup production of harp seals, *Phoca groenlandica*, in the Northwest Atlantic. Mar Mamm Sci 19:141–160

Stenson GB, Myers RA, Hammill MO, Ni I-H, Warren WG, Kingsley MCS (1993) Pup production of harp seals, *Phoca groenlandica*, in the northwest Atlantic. Can J Fish Aquat Sci 50:2429–2439

Stenson GB, Upward P (2020) Updated estimates of harp seal bycatch and total removals in the Northwest Atlantic. DFO Can Sci Advis Sec Res Doc 2020/014

Stewart REA (1986) Energetics of age-specific reproductive effort in female harp seals *Phoca groenlandica*. J Zool Lond 208:503–517

Stewart REA, Lavigne DM (1980) Neonatal growth of northwest Atlantic harp seals *Pagophilus groenlandicus*. J Mamm 61:670–680

Stewart REA, Lavigne DM (1984) Energy transfer and female condition in nursing harp seals *Phoca groenlandica*. Hol Ecol 7:182–194

Stewart REA, Murie DJ (1986) Food habits of lactating harp seals (*Phoca groenlandica*) in the Gulf of St. Lawrence in March. J Mammal 67:186–188

Stewart REA, Lightfoot N, Innes S (1981) Parturition in harp seals. J Mammal 67:186–188

Stewart REA, Stewart BE, Lavigne DM, Miller GW (1989) Fetal growth of northwest Atlantic harp seals, *Phoca groenlandica*. Can J Zool 67:2147–2157

Stirling I (1983) The evolution of mating systems in pinnipeds. Spec Publs Am Soc Mammal No 7:489–527

Surkov SS (1957) Raspredelenie I Zapasy Lysuna v Belom More [Distribution and stocks of the harp seal in the White Sea]. PINRO 59pp transl from Russian Fish Res Board Can Transl Ser No 202

Terhune JM (1976) Examining harp seal behavioral patterns via their underwater vocalizations. Appl Anim Ethol 2:261–264

Terhune JM (1994) Geographical variation of harp seal underwater vocalizations. Can J Zool 72:892–897

Terhune JM, Ronald K (1971) The harp seal, *Pagophilus groenlandicus* (Erxleben, 1777). X. The air audiogram. Can J Zool 49:385–390

Terhune JM, Ronald K (1972) The harp seal, *Pagophilus groenlandicus* (Erxleben, 1777). lll. The underwater audiogram. Can J Zool 50:565–569

Terhune JM, Ronald K (1976) Examining harp seal behavioural patterns via their underwater vocalizations. Appl. Anim. Ethol. 2:261–264

Terhune JM, Ronald K (1986) Distant and near-range functions of harp seal underwater calls. Can J Zool 64:1065–1070

Terhune JM, Terhune ME, Ronald K (1979) Location and recognition of pups by adult female harp seals. Appl Anim Ethol 5:375–380

Tucker S, Stenson GB, Bowen WD, Iverson SJ (2013) Fueling phocids: divergent exploitation of primary energy sources and parallel ontogenetic diet switches among three species of subarctic seals. Mar Mamm Sci 29:E428–E447

Van Opzeeland IC, Parijs SM (2004) Individuality in harp seal, *Phoca groenlandica*, pup vocalizations. Anim Behav 68:1115–1123

Van Opzeeland IC, Corkeron PJ, Risch D, Stenson G, Van Parijs SM (2009) Geographic variation in vocalizations of pups and mother-pup behavior of harp seals *Pagophilus groenlandicus*. Aquat Biol 6:109–120

Wallace SD, Lavigne DM (1992) A review of stomach contents of harp seals (*Phoca groenlandica*) from the Northwest Atlantic. International Marine Mammal Association Inc. (Technical report no 92-03)

Watkins WA, Schevill WE (1979) Distinctive characteristics of underwater calls of the harp seal, *Phoca groenlandica*, during the breeding season. J Acoust Soc Am 66:983–988

Wilkinson JP, Wadhams P (2005) A method of detecting change in the ice conditions of the central Greenland Sea by the whelping locations of harp seals. J Clim 18:1216–1226

Winters GH (1970) Biological changes in coastal capelin from the over-wintering to the spawning condition. J Fish Res Board Can 27:2215–2224

Worthy GAJ (1991) Insulation and thermal balance of fasting harp and grey seal pups. Comp Biochem Physiol A 100:845–8511

Worthy GAJ (1987) Metabolism and growth of young harp and grey seals. Can J Zool 65:1377–1382

Worthy GAJ, Lavigne DM (1982) Changes in blood properties of fasting and feeding harp seal pups, *Phoca groenlandica*, after weaning. Can J Zool 60:586–592

Worthy GAJ, Lavigne DM (1983) Energetics of fasting and subsequent growth in weaned harp seal pups, *Phoca groenlandica*. Can J Zool 61:447–456

Worthy GAJ, Lavigne DM (1987) Mass loss, metabolic rate and energy utilization by harp and grey seals pups during the postweaning fast. Physiol Zool 60:352–364

Yablokov AV, Sergeant DE (1963) Izmenchivost kraniologicheskikh priznakov grenlandskogo tiulenia (*Pagophilus groenlandicus* Erxleben 1777) Zool Zh 42:1857–1865 Transl from Russian Fish Res Bd Can Translation ser No 485

Yunker GB, Hammill MO, Gosselin J-F, Dion DM, Schreer JF (2005) Foetal growth in north-west Atlantic grey seals (*Halichoerus grypus*). J Zool Lond 265:411–419

Chapter 15
The Ringed Seal: Behavioral Adaptations to Seasonal Ice and Snow Cover

Brendan P. Kelly

Contents

Abstract Behavioral adaptations for maintaining breathing holes in sea ice and lairs in the overlying snow cover allow ringed seals to occupy sea ice environments from which other marine mammals are excluded for much of the year. The broad distribution of ringed seals in the ice-covered seas of the northern hemisphere created a niche for an apex predator, and polar bears have evolved as specialists preying on ringed seals. The earliest occupation of the Arctic by people, likewise, depended heavily on the year-round availability of ringed seals. Reliance on breathing holes limits ringed seal movements for as much as 9 months of the year, with implications for foraging and reproductive behaviors as well as for predator avoidance. Ringed seals forage throughout the year with the greatest intake during a brief foraging period between the breakup and subsequent formation of sea ice. Predator avoidance appears to explain the allocation of time better than optimal foraging, at least during ice-bound periods. The under-ice behavior of breeding adults is more consistent with mate or resource guarding than with territoriality as a mating strategy. Feeding through lactation requires a trade-off between time spent foraging and attendance of pups vulnerable to predation. Ringed seals are becoming more vulnerable to predators in the water and on the ice as refuge on ice and under snow diminishes in a warming Arctic.

B. P. Kelly (✉)
International Arctic Research Center, University of Alaska Fairbanks, Fairbanks, AK 99775, USA
e-mail: bpkelly@alaska.edu

D. P. Costa and E. McHuron (eds.), *Ethology and Behavioral Ecology of Phocids*,
Ethology and Behavioral Ecology of Marine Mammals,
https://doi.org/10.1007/978-3-030-88923-4_15

Keywords *Pusa hispida* · Sea ice · Snow cover · Site fidelity · Mate guarding · Navigation · Foraging · Predator avoidance · Trade-offs

15.1 Introduction

Ringed, Baikal, and Caspian seals inhabit ice in the Arctic Ocean and adjacent seas, Lake Baikal, and the Caspian Sea, respectively (Fig. 15.1). They are recognized as primitive and distinct among the Phocinae and comprise *Pusa*, considered a subgenus of *Phoca* (Árnason et al. 2006; Burns and Fay 1970; Higdon et al. 2007; McLaren 1960) or a distinct genus (Allen 1880; Demere et al. 2003; Rice 1998; Scheffer 1958). The Taxonomy Committee of the Society for Marine Mammalogy follows Rice (1998) in classifying *Pusa* as a genus. The freshwater Baikal seal (*Pusa sibirica*) and Caspian seal (*Pusa caspica*) diverged from a common ancestor of Arctic ringed seals (*Pusa hispida hispida*; Fig. 15.2a) 2–3 million years ago. Still, their route to their respective basins is unclear (Palo and Väinölä 2006). More recent vicariance events associated with expansion and contraction of sea ice and ice sheets left refugial populations of Arctic ringed seals in the Baltic Sea (*P. h. botnica*) and the Sea of

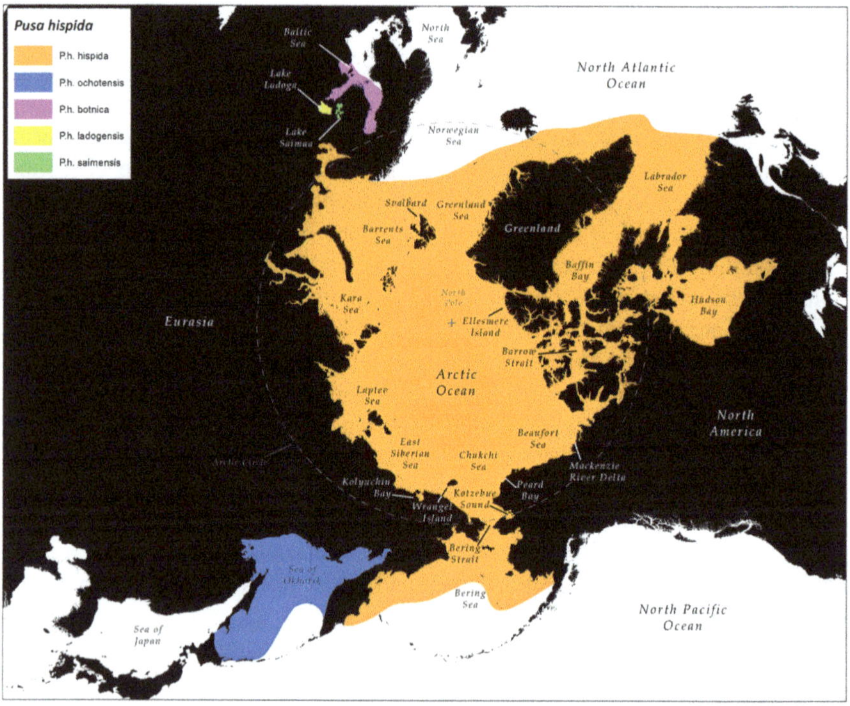

Fig. 15.1 Distribution of five subspecies of ringed seals (*Pusa hispida*). (Modified after Kelly et al. 2010c)

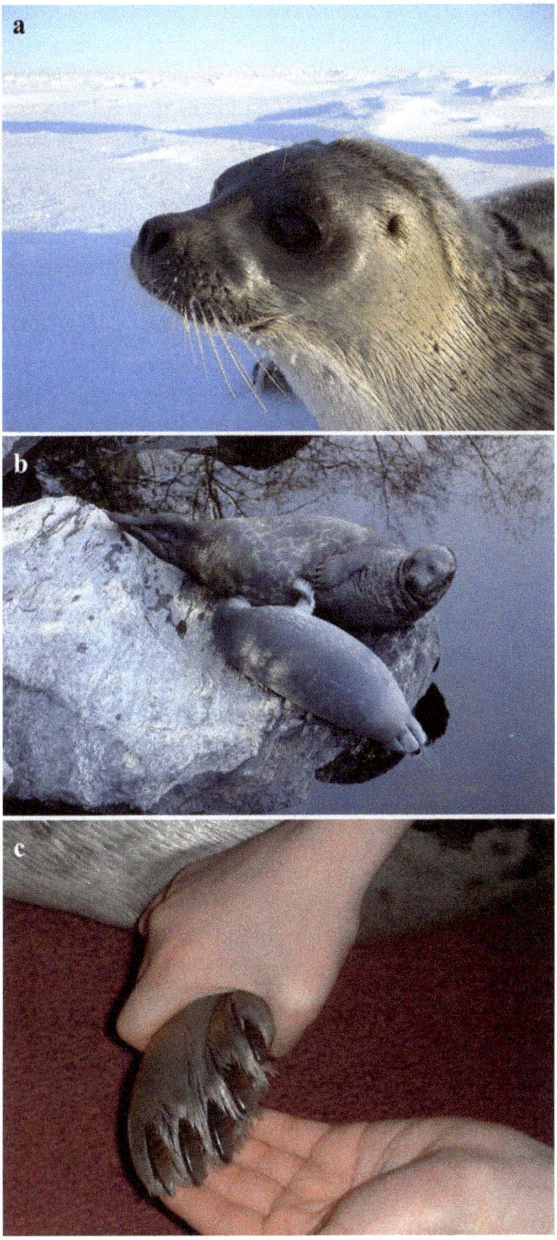

Fig. 15.2 **a** Arctic ringed seal on shore-fast sea ice in the Chukchi Sea. (Photo by John Moran). **b** Saimaa ringed seal and her pup on the shoreline of Lake Saimaa, Finland. (Photo by Mervi Kunnasranta). **c** Claws of an Arctic ringed seal (Photo by Oriana Badajos)

Okhotsk (*P. h. ochotensis*) about 11,000 years ago and in Lake Ladoga, Russia (*P. h. ladogensis*) and Lake Saimaa, Finland (*P. h. saimensis*; Fig. 15.2b) 9,000–8,000 years ago (Forstén and Alhonen 1975; Heaton and Grady 2003; Hyvärinen and Nieminen 1990; Müller-Wille 1969; Nyman et al. 2014; Reeves 1998).

Among phocid seals, ringed seal behavior is among the least known to science primarily because this Arctic species spends approximately nine months of the year hidden under snow and ice (Allen 1880; McLaren 1958). The *Sikumiut*, the people who use sea ice (Wenzel 1984), and other Indigenous hunters developed considerable insights into ringed seal behavior over thousands of years (Wenzel 1984; Murray 2005). While other marine mammal species mostly retreat from the High Arctic in winter, ringed seals maintain breathing holes, allowing them to remain even in areas of continuous ice cover throughout the year. The availability of ringed seals, even in winter, made possible the colonization of the High Arctic by Inuit (Wenzel 1984; Murray 2008), and researchers have benefitted a great deal from the accumulated Indigenous Knowledge of ringed seals and other pinnipeds (e.g., McLaren 1958; Fay 1982; Freeman 1984; Furgal et al. 2002; Smith 2009; Ferguson et al. 2012; Moore and Hauser 2019). In recent decades, telemetry has extended the scientific study of ringed seal behavior, especially in diving and movement patterns. Social behavior, however, remains poorly known from meager observations and suppositions.

Excavating and maintaining breathing holes with the claws of their fore flippers allows ringed seals to exploit sea ice that excludes many of their competitors for much of the year. Other than exceptionally robust claws (Fig. 15.2c), ringed seals are morphologically and physiologically little differentiated from seals of the genus *Phoca* (King 1964; Smodlaka and Henry 2014; Smodlaka et al. 2006). Their adaptations to occupying sea ice year-round are largely behavioral and reflect selection pressures from the extreme thermal environment (Kelly 2001; Smith et al. 1991) and predation by polar bears (*Ursus maritimus*; Hall 1865; Smith 1980; Hammill and Smith 1991) and other carnivores as well as large birds (Lukin 1980; Lydersen and Smith 1989). During the subnivean period—the months when ringed seals surface to breathe in holes in the ice and rest, whelp, and nurse young in lairs excavated in the snow—their behavior is primarily driven by predation avoidance and maintenance of thermal neutrality (Smith et al. 1991; Lydersen 1995; Kelly 2001; Kelly et al. 2010b).

The same ice cover that conceals ringed seals for most of the year also provides a platform for research activities, allowing us to accumulate behavioral observations over long periods. As a consequence, we know more about the species' behavior during that period than in the few months when seals forage in areas of reduced or no ice cover. Foraging rates are highest during this latter period when seals range widely (Lowry et al. 1980; Ryg et al. 1990; Weslawski et al. 1994; Oksanen et al. 2015). In this chapter, I focus on *P. hispida*, the subspecies with which I am most familiar. I briefly summarize the history of ringed seal studies, review what we know about behavior during three periods of the year, and conclude with a perspective on future research needs.

I draw on Indigenous Knowledge and scientific studies—including my own observations during thousands of hours in the field—that inform our knowledge of ringed seal behaviors. I start with an overview of ringed seal behaviors important to occupying sea ice and snow. With that background, I consider movements; navigation; diving; foraging; social behavior; and predator avoidance in the *foraging*, *subnivean*, and *basking periods*, the latter being when epidermal tissue is shed and regenerated.

15.2 History

Ringed seals were exploited by the earliest human inhabitants of the Arctic some 4,500 years ago and subsequently "are ubiquitous in archaeological deposits, regardless of cultural affiliation, region, or time period" (Murray 2008). The archaeological data indicate specialized hunting methods in winter originated early (Murray 2005). European explorers frequently referred to ringed seals in print in the late nineteenth century (e.g., Allen 1880; Hall 1865; Nordqvist 1883; Murdoch 1885), and biological investigations followed in Alaska, Russia, and Canada (Bailey and Hendee 1926; Chapskii 1940; McLaren 1958).

Canadian researchers, in particular, partnered with Inuit hunters and obtained demographic data and estimates of productivity via dentinal annuli and reproductive organs, as well as data on diet composition from stomach content analyses (McLaren 1958; Johnson et al. 1966; Mansfield 1967; Smith 1973, 1987; Lowry et al. 1980). Researchers adopted the Inuit method of locating the subnivean breathing holes and lairs of ringed seals with dogs, a particularly effective method that was essential to studying the species during the majority of the year (Smith and Stirling 1975; Lukin 1980; Kelly and Quakenbush 1990; Kelly 2009). Ringed seals—and their predators—leave considerable evidence of their behavior in the snow, and on-ice observations during ecological and population studies hinted at some aspects of social behaviors and predation. For example, McLaren (1958) inferred competition for breeding sites based on increasing age of seals with distance from the seaward edge of shore-fast ice and that maternal care extended for 2 months based on the dates when hunters harvested lactating females. Others inferred a polygynous territorial breeding system based on scent marking by adult males, disparate sex ratios in some breeding areas, and intra-specific wounds and increased underwater vocalizations in the breeding season (Smith and Hammill 1981; Stirling et al. 1983; Smith 1987; Hardy et al. 1991; Krafft et al. 2007; Rautio et al. 2009). Careful observations of subnivean lairs located by trained dogs revealed which were used as birth and nursing sites and incidences of predation attempts by Arctic foxes (*Alopex lagopus*) and polar bears (Smith and Stirling 1975, 1978; Smith and Hammill 1981; Kelly et al. 1988).

More direct evidence of seal behavior came with the application of radio telemetry (Kelly 1988; Kelly et al. 1986, 1988; Kelly and Quakenbush 1990; Hyvärinen et al. 1995; Kunnasranta et al. 2002; Carlens et al. 2006; Niemi et al. 2019), acoustic and thermal monitoring (Lydersen 1991; Kelly and Wartzok 1996; Kelly 2009), time-depth recorders (Lydersen and Hammill 1993), and satellite-linked transmitters and dive recorders (Teilmann et al. 1999; Gjertz et al. 2000; Born et al. 2004; Freitas et al. 2008a; Härkönen et al. 2008; Kelly et al. 2010b; Martinez-Bakker et al. 2013; Citta et al. 2018; Crawford et al. 2019; Von Duyke et al. 2020).

The generous sharing of knowledge and methods by Inuit hunters has largely continued among ringed seal researchers. Thomas Smith (then at the Department of Fisheries and Oceans in Canada) and the late Jimmy Memorana (Ulukhaktok, Northwest Territories, Canada) taught Mike Hammill (Department of Fisheries and Oceans in Canada) and me how to train dogs to find ringed seal breathing holes and lairs. I shared the method with colleagues in Alaska as well as Canadian and Finnish researchers. Smith's generosity extended beyond dog training. He shared his extensive knowledge of ringed seal biology with U.S., Canadian, and Norwegian researchers. Christian Lydersen and Kit Kovacs have been extremely productive studying ringed seals and other marine mammals in Norway. They have trained many students, as have Steve Ferguson in Canada and Mervi Kunnasranta and colleagues studying ringed seals in Finland and Russia.

My own studies benefitted from learning skills for living and working on the ice from Jimmy Memorana, Tom Smith, and Inupiaq and Yupik hunters in Alaska. Over four decades, I lived the equivalent of 5 years on the ice in the Bering, Chukchi, Beaufort, and Baltic seas. Most of that time was on shore-fast ice in the months spanning the breeding and molting seasons of ringed seals. That concentrated time on the ice yielded admonishments from my dermatologist but also observations of behaviors typically hidden by snow and ice cover.

15.3 Behavioral Adaptation to Ice and Snow

The reproductive success of pinnipeds depends on strategies for coupling aquatic foraging with habitat suitable for resting, whelping, nursing young, and molting (Costa 1991; Chaps. 7 and 8). Among extant pinnipeds, ice is the most common out-of-water habitat, in part because it provides expansive areas of refuge from surface predators (Kelly 2001). In polar seas and certain lakes, the behavior of seals reflects seasonally quite different habitats: one constrained by ice cover and another offering greater access to air. The restrictions imposed by ice cover may be compensated by reduced predator access (Kelly 1996a, 2001). Nonetheless, they shape the seals' foraging, social, and predator avoidance behavior. Ringed seals have evolved behaviors for excavating and maintaining breathing holes through ice cover and, thereby, have access to under-ice prey and on-ice resting and nursing sites over nearly the entire Arctic Ocean, seasonally ice-covered seas, and some freshwater lakes of the northern hemisphere. After spring mating, Arctic ringed seals molt and regenerate epidermal tissues while resting on the decaying ice, while lake-inhabiting seals complete their molt on offshore rocks or shorelines. After the molt, ringed seals forage in waters with discontinuous or no ice cover until the subsequent freeze-up.

Ringed seals have the broadest distribution of the Arctic pinnipeds primarily due to behavioral adaptations for excavating ice and snow (Allen 1880; Scheffer 1958; Kelly 1988). Some of the associated behaviors have been discerned from physical evidence left in the snow and ice, remotely monitoring the sounds of seals at their breathing holes, radio telemetry, and direct visual observations, but many behaviors remain unknown.

As the sea surface freezes in the fall, ringed seals concentrate in the cracks and leads that are last to freeze (Hammill 1987) and where they push their heads through the thin ice to breathe (Lukin and Potelov 1978). As the ice surrounding the incipient breathing hole thickens, the seals maintain access to the air by abrading the ice with the claws on their front flippers (Bailey and Hendee 1926; Smith and Stirling 1975; Lukin and Potelov 1978). In much of the ringed seal's habitat, these holes are the only access to air for several months. The seals actively maintain them by planting the claws of one fore flipper into the ice to anchor themselves while raking the ice with the claws of the other flipper. Periodically repositioning the anchoring claw, the seal slowly spins around in the hole, incising a spiral pattern in the ice. Before the ice melts in the following summer, the breathing holes may extend through ice more than 2 m thick. The holes typically are 35–40 cm in diameter at the ice's upper surface and twice that diameter at the undersurface (Kelly 1996b).

Early in the subnivean period, snow accumulates over the breathing holes, insulating them from the atmosphere and slowing the rate at which they re-freeze. The seals still must abrade the ice frequently through the coldest months of the winter. They also use their claws to excavate a cavity for their head in the snow immediately above the breathing hole. At some breathing holes with sufficient snow cover, males and females expand that cavity to form a lair in which they rest, whelp, and nurse their young (Fig. 15.3; Chapskii 1940; McLaren 1958). Sufficient air penetrates the snow cover to permit breathing.

Excavation of subnivean lairs requires snow depths $\geq$30 cm. In marine environments, sufficient snow accumulates mainly in drifts that form next to hummocks and pressure ridges in the ice. In Lake Saimaa, however, snowdrifts sufficiently deep for lair excavation are primarily limited to areas along the shorelines (Kunnasranta et al. 2021). In these areas, lairs are concentrated in drifts emanating from the shore. Seals accessing those lairs must do so in very shallow water (Niemi et al. 2019). Most lairs in Lake Ladoga also have been found along shorelines, although some also have been located in drifts associated with pressure ridges in the ice (Sipilä et al. 1996; Kunnasranta et al. 2001). Even in the Arctic, some lairs form in drifts along the shoreline. For example, at Nuvuk (Point Barrow), Alaska, where sea ice is often pushed up onto the shore, I have found lairs that extended up onto the gravel beach in the associated snowdrifts.

The internal height of lairs can be as much as 65 cm and the length from 30 cm to 5 m (Smith and Stirling 1975; Kelly and Quakenbush 1990; Kelly 2009). Most lairs have a single, more-or-less oval chamber. Others have multiple convoluted chambers and were recognized by Inuit hunters as being used by more than one adult seal (Smith and Stirling 1975), a finding confirmed by tracking Arctic ringed seals telemetrically (Kelly 1988). Exceptionally large, multi-chambered lairs in Lake Ladoga also were thought to have accommodated more than one seal (Kunnasranta et al. 2001).

Individual ringed seals use multiple breathing holes and lairs in the ice-bound portion of the year. Recording lairs occupied by radio-tagged seals, we showed that

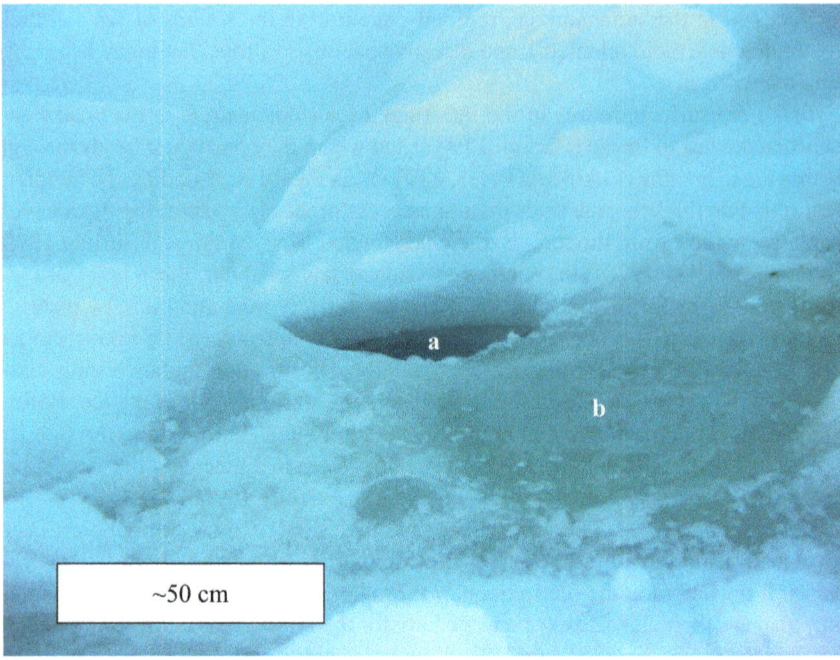

Fig. 15.3 Interior of a ringed seal's subnivean lair. The breathing hole **a** allows passage in and out of the lair. A seal's body melted a slight depression in the floor **b**. (Photo by Brendan Kelly)

the number of lairs used reached an asymptote of 4.5 lairs/seal (Kelly et al. 2010b). Movements of Saimaa ringed seals in the subnivean period are quite similar to those of Arctic ringed seals, and individual Saimaa seals also used an average of 4.5 lairs during those months (Liukkonen et al. 2018).

In the Arctic and the Sea of Okhotsk, ringed seals give birth in subnivean lairs in March and April; in the White and Baltic seas and lakes Ladoga and Saimaa, pups are born in lairs in February and March (Kelly et al. 2010c; Trukhanova et al. 2017). The pups are nursed for as little as 3 weeks in pack ice environments (Fedoseev 1975), 6–8 weeks in shore-fast sea ice (Lydersen and Hammill 1993), and 7–12 weeks in Lake Saimaa (Sipilä and Hyvärinen 1998; Niemi 2013). Their mothers alternate daily nursing bouts with foraging under the ice, and pups are left in the lair unattended and vulnerable to predation by polar bears and Arctic foxes (Smith 1976; Hammill and Smith 1991; Pilfold et al. 2012). The risk to the pups is lessened—but not eliminated—in birth lairs with deeper snow.

Breathing holes, including those providing access to lairs, begin to freeze if left unattended, which often happens if a predator disturbs the site. If a seal returns to a hole that has frozen over, it may abrade the ice until its nares can reach the surface. Typically, a seal can reopen a frozen hole in a single visit if the ice layer is 10 cm or less. If the ice is thicker, the seal may depart to another breathing hole for air before returning to finish the excavation. Our acoustic monitoring and visual observations at breathing holes revealed specialized behaviors when seals reopen frozen breathing holes. When excavating through thick ice, the seal's claws penetrate the ice layer before it can fit its face into the developing hole. In those instances, ringed seals sometimes rise and lower repeatedly in the breathing hole using their bodies like pistons that cause the water in the hole to surge up and down. When the water surface is drawn below the ice being excavated, the seal abruptly thrusts itself upward and rapidly inhales from the momentarily available air pocket.

When ringed seals surface in fully open breathing holes, the piston effect pushes water above the upper surface of the ice, where it freezes around the perimeter of the hole. Eventually, a dome of ice, 3–4 cm thick, forms under the snow and over the breathing hole (Fig. 15.4). Air does not penetrate the ice domes, but the domes are perforated by a characteristic hole about 3 cm in diameter and situated to one side of the dome's apex. As no claw marks surrounded the ports, it was not obvious how the domes were formed until we monitored breathing holes by way of microphones planted in the snow and that transmitted to our huts. Visits to breathing holes sometimes were nearly silent, with only the opening of the nares being faintly audible (see 15.8). Such cautious surfacing behaviors were typical of the first in a bout of surface visits. Ringed seals appear to become more confident after they

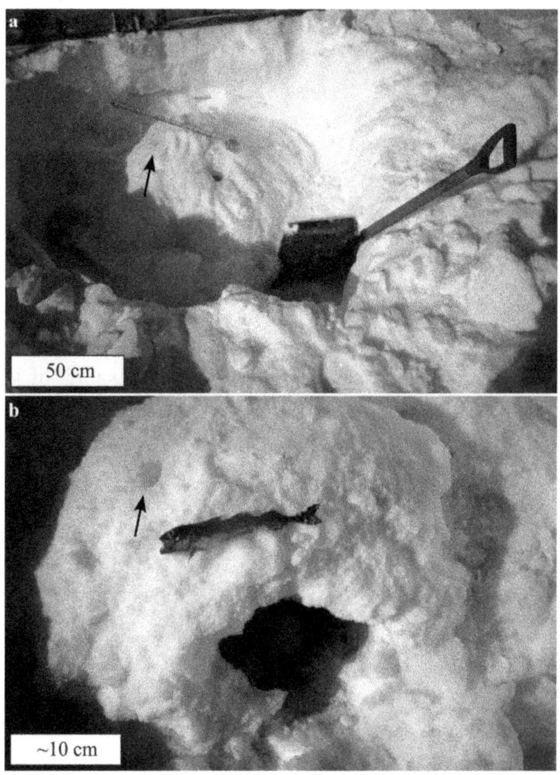

Fig. 15.4 Domes of ice formed above breathing holes from water pushed ahead of surfacing ringed seals: **a** an entire dome cleared of overlying snow cover, arrow points to hole maintained in the dome by seal exhalations, the larger hole in foreground was chiseled by an investigator; **b** close up of a dome, arrow points to hole maintained in the dome by seal exhalations, the larger hole in foreground was chiseled by an investigator, the Arctic cod was found floating dead in the breathing hole. (Photos by Brendan Kelly)

have repeatedly visited the same hole. Between dives, they surface with readily audible surges of water, and their inhalations and exhalations also are easily heard. At holes with well-established ice domes, however, seals occasionally interrupt a typical breathing session with a series of forceful exhalations aimed at the port in the dome. Those forceful exhalations keep the ports open, thereby allowing air to penetrate the domes. The location of the ports correlates with the orientation of the nares of a seal resting at the surface.

When temperatures warm sufficiently, typically in early summer, the snow over ringed seal breathing holes and lairs begins to soften and eventually fails. Ringed seals then begin basking in the open, either next to breathing holes that lacked a lair or at lairs that collapsed or from which the seals tunneled to the surface.

After abandoning their lairs and before the ice break up, radio-tagged seals increased time out of the water to one-half or more of each day (Kelly et al. 2010b). Emergence from the lairs was nearly synchronous—happening over a few days in some years—and distributed over two or more weeks in other years. Whereas the seals mostly came out of the water at night during the subnivean period, during the basking period, they shifted to being on the ice in the late afternoon. The variance in time on the ice during the basking period was substantial, but the probability of a seal being out of the water often exceeded 0.50, especially in the afternoon hours. After completing the molt in June or early July, Arctic ringed seals move more widely than during the subnivean and basking periods.

15.4 Movements

Descriptions by Indigenous hunters of annual, directional movements of ringed seals, including through the Bering Strait and along the coasts of the Chukchi and Beaufort seas, suggested that the species made regular, seasonal migrations (Bailey and Hendee 1926; Burns 1970; Smith 1973; Smith and Stirling 1978; Kapel et al. 1998). Recoveries of a few tagged immature ringed seals showed that some had made extensive movements of hundreds or thousands of kilometers (Smith 1976; Smith and Stirling 1978; Kelly and Wartzok 1996; Kapel et al. 1998). The movements of individuals tracked telemetrically, however, indicate seasonal patterns in movements of ringed seals.

15.4.1 Foraging Period

While ringed seals forage throughout the year, most foraging occurs between ice break up and the subsequent freeze-up. During that foraging period, movements of telemetrically tracked ringed seals are broadly bimodal, with some staying within a few hundred kilometers of their breeding sites and others making excursions of over 2,500 km to feed at the summer ice edge, glacier fronts, or other productive locations

(Gjertz et al. 2000; Freitas et al. 2008a; Kelly et al. 2010b; Harwood et al. 2012; Crawford et al. 2012, 2019; Yurkowski et al. 2016; Martinez-Bakker et al. 2013; Von Duyke et al. 2020). Distances traveled during the foraging period ranged from 2 to 75 km/day with evidence of regional variation (Yurkowski et al. 2016).

In northern Svalbard, Norway, ringed seals tagged in July showed no significant difference in habitat selection by age class (Frietas et al. 2008a). Adult seals tagged in Kotzebue Sound (Chukchi Sea) in October stayed close to their capture sites during the foraging period, while 14 juveniles tagged in the same location moved south into the Bering Sea, possibly to take advantage of areas of higher productivity (Crawford et al. 2012, 2019).

Most movement studies of ringed seals have relied on capturing seals in open-water and gluing instruments to their hair after the summer molt (Citta et al. 2018). Consequently, the tags are shed after less than one year, and the relationship between the seals' capture sites and their previous breeding locations was not known. To tag and track ringed seals between successive breeding sites, we used trained dogs to locate—and specially designed nets to capture—ringed seals in their breathing holes during the breeding season (Fig. 15.5; Kelly 1996b, 2009). We attached satellite-linked transmitters to the seals' hind flippers so they would not shed the tags during the annual molt. In that way, we were able to track 27 ringed seals tagged in their breeding sites in the Canadian Beaufort Sea, Peard Bay (approximately 100 km south of Nuvuk, Alaska), and Kotzebue Sound for up to 14 months (Kelly et al. 2010b; Martinez-Bakker et al. 2013). Twenty-three of those seals were tracked for more than one month, and several moved extensively during the foraging period. Fourteen of them foraged within 400 km of the breeding sites at which they were tagged, but nine ranged up to 2,600 km to the Arctic Ocean's ice edge during the foraging period (Fig. 15.6).

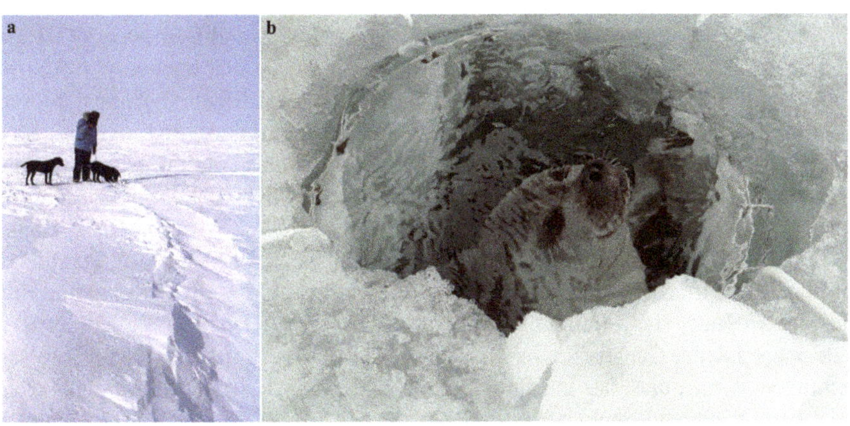

Fig. 15.5 **a** Trained dogs locate a ringed seal breathing hole under the snow (Photo by William Stortz); **b** a ringed seal surfaces after a net closed off the underside of its breathing hole (Photo by Daniella Swenton)

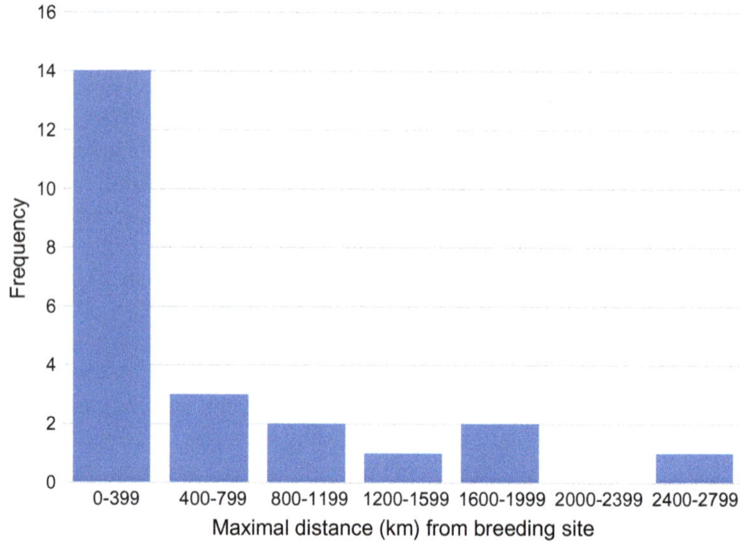

Fig. 15.6 Maximal distance 23 ringed seals moved from their breeding sites. Each seal was tracked for 2–14 months

Regardless of the distance traveled during the foraging period, ringed seals tended to return to the region in which they bred. For example, an adult male seal captured in Peard Bay's shore-fast ice in May 2005 was tracked for 13 months (Fig. 15.7). He stayed within 100 km of Point Barrow throughout the following year except for two long-distance excursions. In July 2005, he moved east to the MacKenzie River plume offshore the Canadian Beaufort Sea, just under 1,000 km. In this region, copepods concentrate at the interface of the plume and marine waters, attracting bowhead whales (*Balaena mysticetus*) in summer (Ainley and Demaster 2009) and aggregations of hundreds of ringed seals on the surface at that boundary in August (personal observation). The seal returned to the Pt. Barrow region in August, moved southwest over 700 km across the Chukchi Sea to a point north of Kolyuchin Bay, Siberia in November, and then returned in December to his Peard Bay breeding site where he remained through June of 2006. The waters off Kolyuchin Bay concentrate plankton and are essential to feeding sea birds and whales as well as seals (Banks et al. 2001; Khim et al. 2003). An adult female, also satellite-tagged in Peard Bay in May 2005, transited the Chukchi Sea to the same region north of Kolyuchin Bay in May. In June, she swam 2,591 km across the Arctic Ocean to the ice north of Ellesmere Island, Canada. In July, she was back in the Peard Bay area where she remained at least until her transmitter stopped in August. Similar movements were recorded for an adult male tagged in Peard Bay in May 2006 (Fig. 15.8). He swam northeast over 1,700 km to the ice in the central Arctic Ocean in July; returned to the Peard Bay region in August; made a northwesterly excursion of nearly 700 km in September to an area north of Wrangel Island, Chukchi Sea; and was back in Peard Bay in October through January 2007 when his transmitter failed.

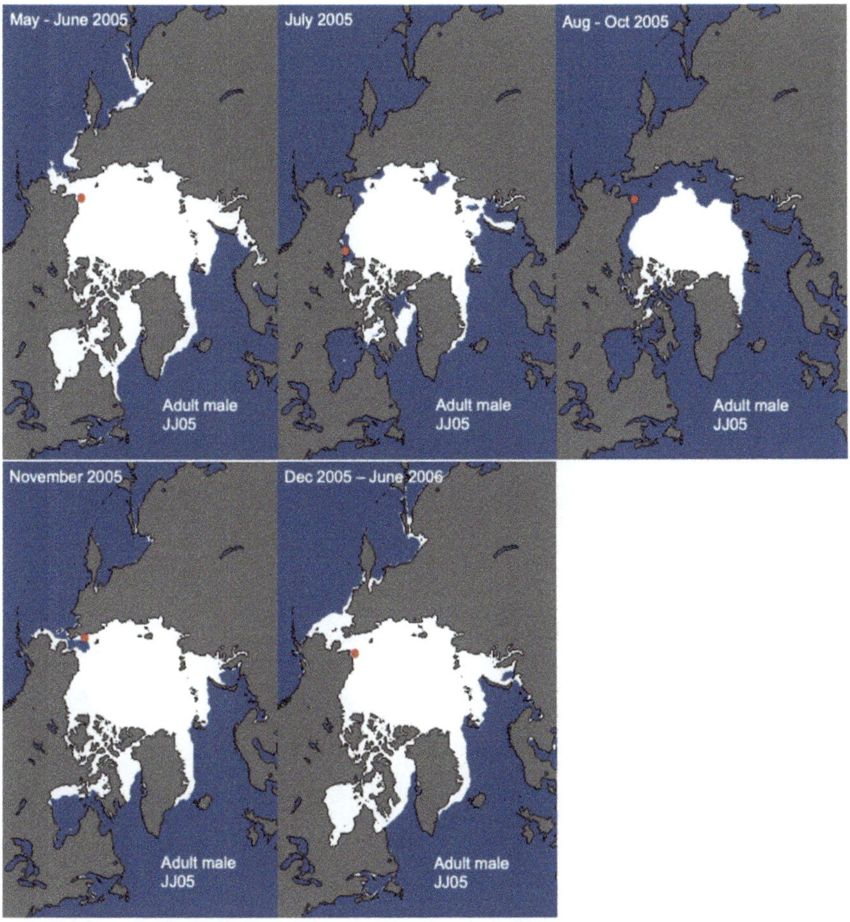

Fig. 15.7 On-ice locations (red dots) of an adult male ringed seal (JJ05) tracked by satellite between 25 May 2005 and 21 June 2006. The seal was tagged at his breeding site in Peard Bay, Alaska and remained in the vicinity for most of the year except for excursions to the Tuktoyuktuk Peninsula, Beaufort Sea in July 2005 and to the coast of Siberia, Russia in November 2005. Sea ice extent (area in white) from Fetterer et al. (2017)

None of the seals we tagged in their Chukchi or Beaufort sea breeding locations moved into the Bering Sea, in stark contrast to the seals tagged after the summer molt by Harwood et al. (2012), Crawford et al. (2012, 2019), and Von Duyke et al. (2020). Given the extensive movements northward toward the retreating ice edge in early summer observed among seals tagged in their Chukchi and Beaufort sea breeding sites, many of the seals others captured in the region after they molted may have bred further south in the Bering Sea. Their subsequent movements in winter to the Bering Sea would then be consistent with our observations of ranging widely in summer and fall and then returning to sites occupied in the previous breeding season.

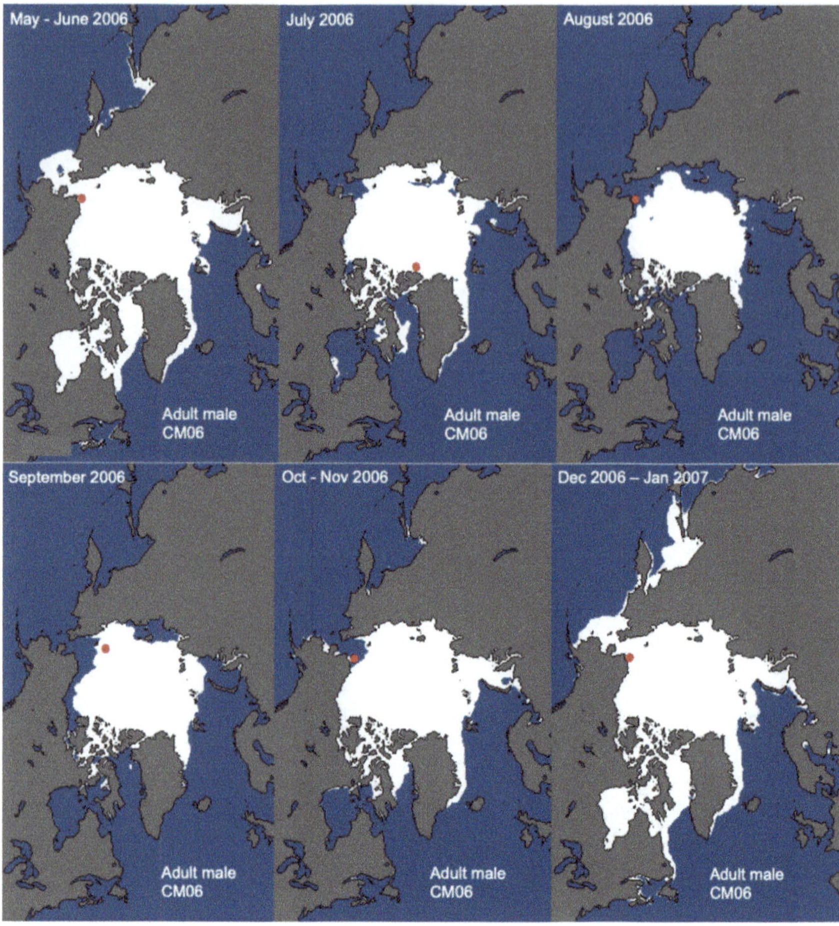

Fig. 15.8 On-ice locations (red dots) of an adult male ringed seal (CM06) between 14 May 2006 and 17 January 2007. The seal was tagged at his breeding site in Peard Bay, Alaska and remained in the vicinity for most of the year except for excursions to the Elsmere Island in July 2006 and to the ice north of Wrangell Island in September 2006. Sea ice extent (area in white) from Fetterer et al. (2017)

Arctic ringed seals rarely come ashore, but ringed seals in the White Sea, Sea of Okhotsk, Baltic Sea, and lakes Ladoga and Saimaa lack access to sea ice in the foraging period. Consequently, they rest on island shores, offshore reefs, and rocks (Ognev 1935; Härkönen et al. 1998; Trukhin 2000; Kunnasranta 2001; Lukin et al. 2006).

15.4.2 Subnivean Period

By the start of the fall freeze-up, adult ringed seals return to the same sites at which they bred the previous year. Reports from Indigenous hunters and biologists suggested that individual seals occupy the same breeding sites over successive years (Smith and Hammill 1981; Huntington 2000; Krafft et al. 2007; Kelly et al. 2010b). We tested this by tracking 12 ringed seals from one breeding season to the next (Kelly et al. 2010a) and found that each seal bred within 750–2000 m of their previous breeding site (Fig. 15.9).

As seals occupy breeding sites throughout the winter and spring, especially among those individuals that inhabit shore-fast ice, their movements are limited (Kelly et al. 2010b; Niemi et al. 2019). Fifty-eight seals that had ranged hundreds or thousands of kilometers in summer and early fall spent the late fall, winter, and spring in the shore-fast ice along the Beaufort Sea coast of Alaska (Kelly et al. 2010b). They typically moved between a few lairs within 2 km of one another but, in rare cases, almost 10 km apart.

Three of the radio-tracked seals and nine tracked by satellite ranged up to 2,600 km from their winter/spring home ranges in summer, but they returned to the same small (1–2 km^2) areas during the following year's ice-bound months. The restricted

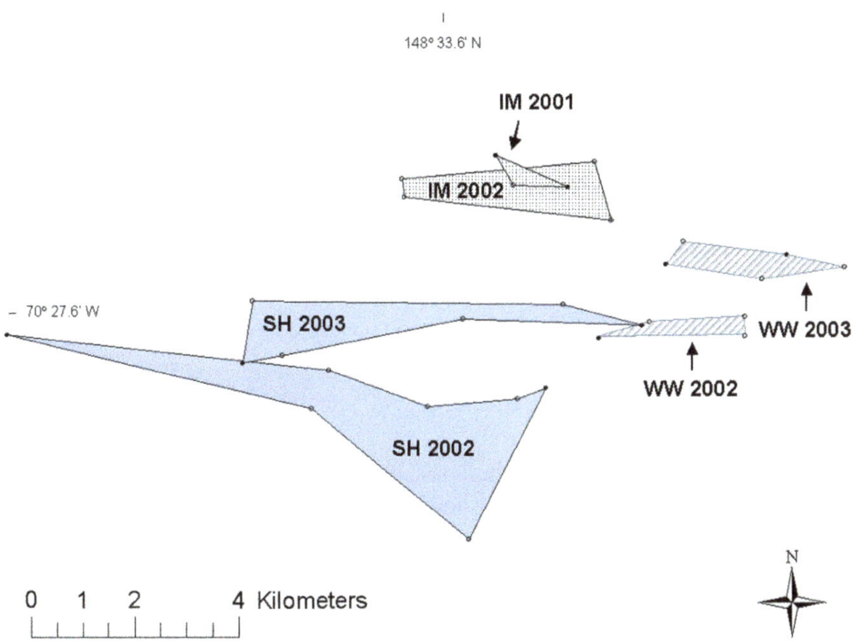

Fig. 15.9 Home ranges (minimum convex polygons) of two adult male (IM and WW) and an adult female (SH) ringed seals tracked during successive breeding seasons. (After Kelly et al. 2010b)

movements of ringed seals during the ice-bound season—including the breeding season—limits their foraging activities for most of the year.

15.4.3 *Basking Period*

Ringed seal pups shed their natal pelage just as they are weaned (Kelly 1988; Lydersen and Hammill 1993), typically in mid-April to mid-June in the Arctic Ocean and Sea of Okhotsk and in March and April in the Baltic Sea and lakes Saimaa and Ladoga (Kelly et al. 2010c; Kunnasranta et al. 2021). Subsequent annual molts take place when the seals emerge from their subnivean lairs in mid-May to mid-July (Arctic Ocean and Sea of Okhotsk) and late March to mid-June in the Baltic region. Ringed seals forage little if at all during that time of year, and their metabolism is believed to decline substantially as in other molting phocids (Ashwell-Erickson et al. 1986). During the molt, ringed seal behavior shifts in several ways; the duration of periods lying on the surface of the ice increases, the timing of those episodes shifts to mid-day, they increasingly move beyond the breathing holes used during the subnivean period, and they often aggregate around a single breathing hole in contrast to their more solitary behavior in the subnivean period (Smith 1973; Smith and Hammill 1981; Kelly and Quakenbush 1990; Kelly et al. 2010b; Martinez-Bakker et al. 2013).

15.5 Navigation

In summer and early fall, ringed seals navigate long distances across the Arctic basin and surrounding seas and return to previous breeding sites. There, they spend as much as 9 months—including the dark winter—navigating between breathing holes under ice and snow cover.

15.5.1 *Foraging Period*

Ringed seals that remain along the coast during the foraging period may navigate using visual cues, similar to northern elephant seals (*Mirounga angustirostris*; Matsumura et al. 2011; Chap. 2). Those that move long distances to the ice edge are navigating away from coastal landmarks and cannot rely on spatial memory, as the ice edge is dynamic within and between years. Freitas et al. (2008a) suggested that ringed seals may find their way to offshore foraging areas by following decreasing water temperatures and increasing concentrations of ice floes. Other modalities, such as celestial (Mauck et al. 2008) or magnetic cues (Matsumura et al. 2011) might also be involved (Chap. 2).

15.5.2 Subnivean Period

During the dark winter and early spring months, light is further reduced by snow and ice cover as ringed seals navigate between breathing holes and under-ice foraging locations. Moreover, blind ringed seals (and other pinnipeds) appear to otherwise be healthy in the wild (Poulter 1963; Hyvärinen 1989), and it has been suggested that, in the near or total absence of vision, seals navigate by echolocation (Poulter 1963; Renouf and Davis 1982), vibrissal sensation (Stephens et al. 1973; Hyvärinen 1989; Dehnhardt et al. 1998), or acoustic cues (Elsner et al. 1989; Wartzok et al. 1992a). Evidence for echolocation by pinnipeds has been unconvincing (Wartzok et al. 1984; Kunnasranta et al. 1996; Chap. 2), however, experiments with ringed and Weddell seals (*Leptonychotes weddellii*) diving under ice in captivity and in the wild (Elsner et al. 1989; Wartzok et al. 1992a) yielded evidence for navigation relying on visual, tactile, and passive listening (Schusterman et al. 2000).

Ringed seals tracked acoustically under the ice of a frozen freshwater pond relied primarily on vision for navigation (Wartzok et al. 1992a, b). When exploring under the ice, they were attracted to bright locations associated with holes in the ice or reduced snow cover on top of the ice. If no hole was found, the seals would scratch at the underside of ice where the most light penetrated. When we shoveled snow in a path leading away from a release hole, the seals would follow that path when exploring under the ice. That seals oriented to bright spots under the ice in the experiments was consistent with behavior in the natural habitat. Breathing holes excavated and maintained by seals in Arctic sea ice frequently are concentrated in refrozen cracks, which transmit more light than does the surrounding, thicker ice (Kelly and Quakenbush 1990; Kelly 2009).

We tested the abilities of an adult and a juvenile male ringed seal to navigate in the dark by tracking them blindfolded under the pond's ice (Wartzok et al. 1992a). Blindfolded, both were reluctant to venture away from the release hole and did so only when the hole was blocked from above. Their subsequent movements were unusually convoluted with more frequent turns than when swimming without blindfolds. Neither located a previously unvisited breathing hole while blindfolded until we provided an acoustic cue by tapping a steel rod against the ice at the new hole. The adult and the juvenile responded immediately and similarly to the acoustic stimulus on the first presentations. At the onset of the acoustic stimulus, each swam directly toward the source (75 m distant) for 15–30 m, turned and swam for approximately 40 m at a right angle to the initial bearing, and then turned back to the direct bearing for the remaining distance before surfacing in the new hole (Fig. 15.10). Neither seal had been exposed to the tapping stimulus when it was first presented, yet both immediately responded in a similar manner. That both swam directly toward the sound during the first and final legs of the navigation indicated good directional hearing. Swimming on a different bearing in the intermediate leg likely provided the seals with information about the distance to the source and, hence, if it was within swimming range.

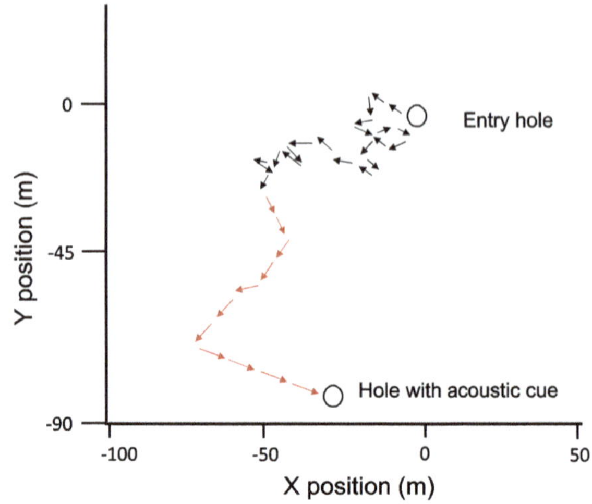

Fig. 15.10 Under-ice track of blindfolded juvenile ringed seal from an entry hole to a novel hole. Black arrows depict path prior to onset of an acoustic cue; red arrows depict path while investigators tapped with a steel rod at a previously undiscovered hole. (After Wartzok et al. 1992b)

That both seals responded immediately and similarly to a novel acoustic cue was consistent with behavior in the wild. Inuit hunters are aware of and exploit ringed seals' sensitivity to sound at the surface. In Barrow Strait (Northwest Territories, Canada), I observed an Inuk hunting ringed seals in April 1997. He paralleled a narrow crack in the ice on his snowmobile until he encountered a seal's breathing hole. He dismounted and stood quietly above the hole while directing me to continue along the crack on my snowmobile. In less than 2 min, a seal surfaced at the hole where the hunter was waiting. The hunter explained that he could discern—by the appearance of the snow and ice around the breathing hole—if a seal had recently visited the site. If so, he said, the seal would likely be nearby and inclined to return and surface upon hearing whatever approached the hole depart. I observed that using this method, the hunter was successful on most attempts. Ringed seals in subnivean lairs also are sensitive to sounds and escape into the water in response to human footfalls as far away as 600 m (Kelly et al. 1988).

Phocid seals have highly innervated vibrissae (Dykes 1975; Hyvärinen and Katajisto 1984; Chap. 2). Hyvärinen (1989) proposed that vibration of their vibrissae may inform ringed seals about their spatial orientation as well as swimming speed and direction. Dehnhardt et al. (2001) showed that blindfolded harbor seals (*Phoca vitulina*) can track and capture fish by sensing hydrodynamic wakes with their vibrissae. Blindfolded ringed seals surfacing in breathing holes during the navigation experiments sometimes seemed to use their vibrissae to center themselves within the hole, but did not seem to rely on their vibrissae while searching for breathing holes even when blindfolded (Elsner et al. 1989; Wartzok et al. 1992a). It seems likely that ringed seals track prey in dark waters through vibrissal vibration, as demonstrated in harbor seals, although it is not clear what other information they gain from the vibrissae.

Our navigation experiments revealed no evidence that ringed seals used echolocation. Instead, we found that vision was the primary sense used in navigation, but that ringed seals also relied on directional hearing and a ranging behavior to navigate under the ice. Those conclusions were reinforced when we released the adult male ringed seal used in the experiments described above. We released him in the Beaufort Sea at a hole we cut in the ice near his original capture site of a year earlier. We positioned the release hole 74 m from a breathing hole that showed signs of once having had a lair above it but was then maintained only as a breathing site. Our intent was to ensure that the released seal had at least one alternative breathing site. We monitored the seal's movements by way of a light tether line for a series of initial dives from the release hole. He did not discover the alternative breathing hole until we provided the tapping cue. After he successfully navigated three times between the breathing holes, we removed the tether.

One week later, we visited the release hole and saw the seal breathing there. Nearly two months later, a radio signal led us to the seal in a subnivean lair 1.2 km beyond the release site. When the snow began to warm a few weeks later, it became apparent that the entrance to that lair was excavated through the ice of a refrozen crack that also contained the breathing hole to which we had acoustically led our seal. On top of the ice, those cracks were only visible to us when the snow warmed and slumped into the refrozen crack, the surface of which is typically 4–5 cm below that of the surrounding ice (Fig. 15.11). To seals under the ice, however, those cracks were visible as bright linear features and indicated ice thin enough to readily allow the excavation of new breathing holes. The released ringed seal used acoustic and visual cues under sea ice just as it had in the experimental pond. In the natural environment, acoustic cues may be provided by sounds generated in the ice. Thermal and other stresses in sea ice generate "broadband impulses, frequency modulated (FM) tones, and high-frequency broadband noise" (Kinda et al. 2015), and it may be that ringed seals gain information—including the locations of holes—from the interactions of those sounds with the architecture of the ice.

15.5.3 Basking Period

Navigational needs are minimal in late spring–early summer when molting ringed seals mainly rest on the ice. During that period, they spend as much as 60% of their time lying next to breathing holes (Martinez-Bakker et al. 2013), typically in an area they occupied for the previous several months (Kelly and Quakenbush 1990; Kelly et al. 2010b). Food intake also is minimal at that time, further diminishing navigational needs.

Fig. 15.11 A ringed seal breathing hole in a crack refrozen in the shore-fast ice of the Beaufort Sea. A melted depression where a seal laid is in the foreground. A gray area extending away from the breathing hole delineates where warming snow has collapsed onto the surface of the crack, which is a few centimeters lower than the surrounding ice. (Photo by Brendan Kelly)

15.6 Diving and Foraging

Pinniped diving is shaped mainly by the three-dimensional distribution of prey (Kienle and Berta 2018) and, to some extent, by predator avoidance (Kelly 1996a; Crocker et al. 1997; Mitani et al. 2010) and the need to rest or sleep (Thompson et al. 1991; Le Boeuf et al. 1992; Crocker et al. 1997). Ringed seal diving and foraging additionally reflect adaptations to occupying seasonal sea ice. While they acquire most of their energy in a few months, they are not strictly capital breeders but forage throughout the year, including during breeding and lactation (Lydersen and Hammill 1993; Kelly and Wartzok 1996; Lydersen and Kovacs 1999). They do most of their feeding in the summer and fall when their movements are not constrained by ice cover, mating, nursing young, or molting (McLaren 1958; Fedoseev 1965; Lowry et al. 1980; Smith 1987; Ryg et al. 1990).

Direct observations of ringed seals feeding are quite rare, but intensive foraging in the summer and fall—when ice cover is reduced—is inferred from substantial body mass gained during those months (Smith 1987; Ryg et al. 1990), samples of stomach contents (Johnson et al. 1966; Lowry et al. 1980), and movements to areas known to be productive (Born et al. 2004; Freitas et al. 2008b; Crawford et al. 2012; Von Duyke et al. 2020). Arctic ringed seals eat a wide variety of mostly invertebrates and fish that form dense aggregations (Kovacs 2007). There is little evidence of diet variation between sexes (Johnson et al. 1966; Lowry et al. 1980; Holst et al. 2001), but stomach contents suggest that juvenile ringed seals consume more crustaceans and fewer fish than do adults (Fedoseev 1965; Lowry et al. 1980; Smith 1987). Regional variation in water depth, productivity, and prey availability likely account for much of the observed variation in diet (Kelly et al. 2010c). The diet of Saimaa ringed seals of all ages is made up almost entirely of fish with invertebrates rarely ingested, possibly incidental to consuming fish (Kunnasranta et al. 1999; Auttila et al. 2015).

Ringed seals consume live prey by suction in captivity and in the wild (Smith 1987; Kelly 1988; Kienle et al. 2018). A seal carrying a video camera was observed ingesting by suction an amphipod—possibly *Onisimus litoralis* (Carey 1992; Lowry and Stoddart 1993)—along the underside of the ice within 5 s of diving from a breathing hole (Kelly 2008). The profile of ringed seal teeth has led some researchers to suggest that crustaceans and other small prey may be consumed by filtering (McLaren 1958; Fedoseev 1965).

15.6.1 Foraging Period

Unfortunately, we know little about ringed seal behavior in the period of the year when foraging is most intensive. In those months, ringed seals range widely, and our knowledge of feeding is mainly inferred from seasonal changes in body condition (McLaren 1958; Fedoseev 1965; Smith 1987; Ryg et al. 1990; Härkönen et al. 2008; Young and Ferguson 2013), locations (Gjertz et al. 2000; Born et al. 2004; Freitas et al. 2008a, b), analyses of stomach contents and stable isotopes (Dehn et al. 2007; Young and Ferguson 2013), and dive profiles obtained through satellite telemetry (Teilmann et al. 1999; Gjertz et al. 2000; Born et al. 2004; Freitas et al. 2008a, b; Härkönen et al. 2008; Kelly et al. 2010b; Crawford et al. 2012, 2019; Harwood et al. 2012, 2015; Martinez-Bakker et al. 2013; Von Duyke et al. 2020).

Arctic ringed seals' distribution spans a wide latitudinal gradient with substantial variation in the extent and duration of ice cover. In higher latitudes, ringed seals traveled more in their comparatively brief foraging period than did ringed seals in lower latitudes characterized by longer foraging periods (Yurkowski et al. 2016). Ringed seals of all ages spent comparable times in the resident state with no evidence of adults competitively excluding juveniles from foraging patches.

The depth and duration of dives vary with body size, season, and location. Maximal dive durations are proportional to body mass in ringed seals (Kelly and Wartzok 1996; Teilmann et al. 1999; Kunnasranta 2001), with a large adult male in North Baffin Bay diving to 500 m (Born et al. 2004). Most dives by adults and juveniles in that area were to <100 m. The longest of nearly 10,000 dives by 14 ringed seals monitored in Barrow Strait, Nunavut, Canada was 26 min (Kelly and Wartzok 1996).

Broad patterns of habitat use and diversity hotspots were examined in two recent syntheses that compiled marine mammal tracking data from the Arctic. Citta et al. (2018) summarized marine mammal movements in the western Arctic from many telemetry studies. They attempted to identify core use areas based on locations in two broadly defined seasons of the year, but noted the data were unlikely to reflect the population level distribution for ringed seals and other ice-associated pinnipeds given the influence of location and tagging season. In many of the ringed seal studies, captures were made just after the molt, and whether the animals were moving toward or away from areas occupied in the subnivean period was unknown. Yurkowski et al. (2016) determined biological hotspots based on telemetric tracking of ringed seals and 20 other species of marine mammals and seabirds in the Arctic. Their analyses showed hotspots concentrated along the continental shelf and slope in summer and fall and in offshore pack ice in winter and spring.

Gray seals (*Halichoerus grypus*) and elephant seals are known to rest and sleep underwater (Thompson et al. 1991; Le Boeuf et al. 1992; Crocker et al. 1997), and dives lasting over 10 min by Saimaa ringed seals were attributed to resting and possibly sleeping (Hyvärinen et al. 1995; Kunnasranta et al. 2001). Analysis of three-dimensional movements of ringed seals diving under Arctic sea ice showed no evidence of resting while diving (Simpkins et al. 2001a), although resting in breathing holes and subnivean lairs is common (Kelly and Quakenbush 1990).

15.6.2 Subnivean Period

The behavior of ringed seals diving under shore-fast ice was recorded via acoustic tags in the subnivean period. The tags' pulse rates were modulated by the depth and recorded by way of sub-ice hydrophones. Lydersen (1991) used this approach to record an adult female ringed seal diving under the ice in a Svalbard fjord. The seal was monitored for 7 days in May when she spent 55% of the time in the water. Dives lasted up to 17 min (mean = 2.7, SD = 2.7 min) and reached a maximal depth of 40 m (mean = 10.6, SD = 9.0 m).

We acoustically monitored almost 10,000 dives by 14 ringed seals during April, May, and June in Barrow Strait, Nunavut, Canada (Kelly 1996a; Kelly and Wartzok 1996). The seals in our study spent substantially more time in the water (70–93%) than did the seal monitored by Lydersen (1991), probably because that seal was monitored later in the spring. We recorded shallow (<25 m) transit dives in which ascents immediately followed descents to another breathing hole. Occasionally, transit dives occurred in a series of two or three. In contrast, foraging dives generally occurred in a series of 10 or more. They included a period of two or more minutes at a static depth, typically the maximal depth for the dive. During a foraging bout, seals typically returned to the same depth on successive dives consistent with observations of zooplankton and fish concentrating in a few distinct layers under the ice of Barrow Strait (Crawford and Jorgenson 1990). The maximal dive depth was to the bottom (222 m), and maximal dive durations were strongly correlated with body mass (r^2 =

Table 15.1 Dive depths and durations of three adult male and three adult female ringed seals diving under shore fast sea ice in Barrow Strait, Nunavut, Canada. (After Kelly and Wartzok 1996)

Seal	Sex	N	Max. depth (m)	Median depth (m)	Max. duration (min)	Median duration (min)
HO91	M	210	94	28	22	2
TR91	M	1953	153	35	23	3
UM92	M	532	149	48	13	6
ME91	F	804	186	89	26	7
HE91	F	278	153	106	23	6
LF92	F	688	168	137	22	8

0.94, $F = 142.52$, $p < 0.0001$) and less so with age ($r^2 = 0.17$, $F = 2.86$, $p = 0.13$). Maximal dive depths and durations were similar for adult males and adult females (Table 15.1). Dives by adult males were shallower and shorter than dives by adult females. Depths and durations of dives by three juvenile males were similar to those of the adult females.

Female ringed seals divided their time at depth and the surface according to their reproductive status. Before giving birth, two females tracked in Kotzebue Sound (Chukchi Sea), each averaged less than 1 h/day on the ice in subnivean lairs (Kelly and Quakenbush 1990). After giving birth, both occupied lairs daily for up to 16 h with means of 4.2 (SD = 4.2) and 8.4 (SD = 17.6) hours/day. One occupied her birth lair daily from 24 March until 23 April 1984 when it was flooded by overflow from a nearby crack in the ice. The other female also occupied her birth lair daily on 24 March but ceased doing so between 31 March and 11 April, apparently having lost her pup. Once they ceased daily visits to the birth lairs, the average duration of bouts in lairs dropped to 1.2 (SD = 16.9) and 1.4 (SD = 8.6) hour/day.

Dive depths and durations for three adult female ringed seals tracked in Barrow Strait also reflected reproductive status (Kelly 1996a). Two seals, nursing pups in subnivean lairs, divided their time between foraging at depths below 100 m and attending pups at the surface. They spent 73 and 59% of their respective time within 25 m of the surface. A third female in the same study area lost her pup; she was within 25 m of the surface for only 34% of the time, and spent 51% of her time foraging deeper than 100 m.

In Barrow Strait, ringed seals prey primarily on Arctic cod (*Boreogadus saida*), pelagic amphipods, and several species of amphipods associated with the benthos or the undersurface of the sea ice (Welch 1992). Remains of Arctic cod and amphipods were common in the regurgitation and feces of seals during our study. The depth distribution of Arctic cod is bimodal with peaks reflecting partial refuges from predation (Bradstreet et al. 1986; Crawford and Jorgenson 1990; Welch et al. 1993), and ringed seal foraging dives in Barrow Strait reflected that bimodal distribution. Adult females without pups and juvenile ringed seals in Barrow Strait allocated considerable time to foraging below 100 m. Because those depths required extended travel time, they afforded the seals minimal patch time and, therefore, provided a partial refuge from predation for Arctic cod. Cod also found refuge in the frazil (soft, amorphous) ice and cracks in the underside of the ice, and yearling ringed seals concentrated their foraging on the underside of the ice, apparently because their small size made foraging at depth inefficient (Kelly and Wartzok 1996).

The necessity of returning to the surface after foraging dives is analogous to central place foraging (Houston and McNamara 1985; Chap. 6). The added need to return to one of a few breathing holes at the end of each dive suggests that the relationship between surface intervals and dive durations might follow the marginal value theorem (Kramer 1988; Houston and Carbone 1992; Boyd et al. 1995), but the sub-ice diving behavior of ringed seals did not fit the predictions. Because oxygen is acquired at a diminishing rate, foraging seals would optimize patch time by spending less time at the surface before shallow dives. I tested the prediction by analyzing the allocation of time to the descent, foraging, ascent, and surface phases of 9,778 under-ice dives by 13 ringed seals (Kelly 1996a). Their surface intervals did not correlate consistently with the depths of the following dives; slopes of regressions of surface intervals versus dive depths were significantly positive for 2 seals, significantly negative for 3 seals, and not significantly different from 0 for 8 seals. When slopes were significant, dive depth accounted for 12% or less of the variance in surface intervals. I further tested the predicted relationship between surface interval and dive depth with the data restricted to surface intervals $\leq$2 min, sufficient time for ringed seal tissues to be saturated with oxygen (Thompson and Fedak 1993). The number of seals showing significant positive slopes of surface intervals on dive depth increased from 2 to 8 for the 39% of foraging dives that followed surface intervals of $\leq$2 min. Nevertheless, the slopes remained very shallow with r^2 values ranging from 0.10 to 0.36. The proportion of time spent at the surface decreased, and the proportion in foraging patches increased with increasing dive depth, contrary to the predictions based on the marginal value theorem suggesting that ringed seals optimize a currency other than patch time (see sect. 15.8).

The assumption that the bottom phase of ringed seal dives represents patch time was challenged when we analyzed the three-dimensional movements of ringed seals diving under the ice. Those analyses indicated that foraging was not restricted entirely to the putative patch time at the bottom phase of dives (Simpkins et al. 2001b). The foraging seals actively pursued prey horizontally during the bottom phase of dives (mostly in the water column) but also occasionally during descent or ascent. They rarely employed sit-and-wait foraging, but strong currents in the Barrow Strait study area complicated the analysis. On the few occasions in which seals seemed to be stationary, it was not clear if they were swimming into a current. On the other hand, we observed a seal, oriented vertically in the water column, passively drift under a hole in the ice. The seal was looking upward and could have been searching for prey backlit against the underside of the ice.

Fewer than 10% of the recorded dives involved traveling from one breathing hole to another, with most dives appearing to include searching for prey. Convoluted movements consistent with foraging occurred in U-shaped and V-shaped dives underscoring that time-depth profiles are not reliable for inferring functions of dives in this species (Lesage et al. 1999; Harcourt et al. 2000; Simpkins et al. 2001b). The three-dimensional analyses of ringed seals diving under sea ice showed that bottom phases included the search for prey within patches and transit between patches (Simpkins et al. 2001b).

Ringed seals appear more responsive to episodic changes in prey availability during the subnivean period than during the open-water period when they can search for prey over more extensive areas. In Barrow Strait, we saw evidence of dramatic changes in prey availability between and within years. Amphipods predominated in fecal remains in 1990, 1992, 1996, and 1997, but Arctic cod otoliths dominated in 1991. In 1991, we frequently encountered dead Arctic cod on the surface of seal breathing holes (Fig. 15.4b) but saw no such instances in any other year. The spines of the dead cod were broken immediately posterior to the opercula, a condition I recognized from herring fed ad libitum to captive ringed seals. When the captive seals were sated, they stopped accepting fish and typically regurgitated several minutes after they swallowed the last fish. The spines of regurgitated herring were broken just behind the opercula. It would seem that Arctic cod were sufficiently abundant in Barrow Strait in 1991 that ringed seals frequently returned sated to their breathing holes where they regurgitated the last ingested fish. While dead cod were not seen in the breathing holes in 1996, we observed dense aggregations of them in echograms on two occasions that year. The aggregations abruptly appeared and filled the bottom 100 m of the echosounder beam for 30–120 min. A Canadian fisheries biologist recorded a school of Arctic cod one summer that stretched for many kilometers along the coast near Barrow Strait, attracting large numbers of foraging seals and belugas (Buster Welch, pers. com.).

15.6.3 Basking Period

Ringed seals appear to have evolved flexible foraging behaviors in response to the ephemeral abundance of prey. Long periods tied to a few breathing holes restrict movements and require feeding throughout the breeding season. Ringed seal foraging is reduced only during the basking period, a period that also sees an increased opportunity for social interactions.

15.7 Social and Mating Behavior

15.7.1 Foraging Period

Social behavior in the foraging period is mostly unknown for Arctic ringed seals. Sightings of ringed seals at sea are primarily of individuals. Even when aggregated around abundant prey, for example, at the edge of the McKenzie River plume in the

Beaufort Sea, they seem to maintain a distance of several body lengths. Similarly, when resting on ice in the foraging period, ringed seals typically are alone.

Saimaa ringed seals also are described as solitary, but Ladoga ringed seals come ashore in herds as large as 50 individuals during summer (Kunnasranta et al. 1996; Sipilä et al. 1996). Agonistic behaviors in those herds include splashing, scratching, biting, growling, and open-mouthed threats associated with competition for favored resting sites (Kunnasranta et al. 1996). The vocal repertoire of Ladoga ringed seals is richer than in the other subspecies, and in-air vocalizations were described by Kunnasranta et al. (1996) as comprising clicks, burst pulses, knocks, chirps, yelps, and low-frequency vocalizations. Underwater vocalizations included barks, growls, and roars and were most frequent when seals submerged immediately after agonistic interactions ashore. Ladoga seals swimming at the surface splash the surface with their fore flippers when passing other seals resting on the shore (Kunnasranta et al. 1996).

15.7.2 Subnivean Period

Ringed seals are mostly solitary in the subnivean period, and mating has not been observed but is presumed to occur under the ice (Stirling 1983). In the absence of direct observations, the mating system has been inferred from indirect evidence as territorial or based on males guarding access to estrous females. A polygynous mating system was inferred with males maintaining large under-ice territories containing multiple female territories. In the breeding season, adult male ringed seals emit a strong odor that persists in the snow where they have breathed, and those males are known as *tigak* to the Inuit (Smith 1987; Hardy et al. 1991; Ryg et al. 1992). Breathing holes imbued with the odor of *tigak* were thought to be used exclusively by breeding males, while "complexes" of pupping lairs and nearby breathing holes were thought to be used by a single female and her pup (Smith and Stirling 1975; Smith and Hammill 1981; Smith et al. 1991). Smith and Stirling (1975) noted, however, that they were not able "to obtain direct evidence of movement by a seal between the lairs in a complex."

Movements of seals between breathing holes and lairs documented in the Beaufort Sea north of Alaska and Barrow Strait, Nunavut, Canada, provided more direct evidence of social interactions (Kelly et al. 2010b). Detailed mapping of lairs used by 55 ringed seals during the breeding season revealed substantial overlap in areas used by seals of both sexes (Fig. 15.12). Home ranges—based on surfacing locations—of adult males (median = $0.62\ km^2$) were not significantly different than those of adult females (median = $0.65\ km^2$), and breathing holes often were shared among and between sexes. Observations of hole sharing included capturing and tracking multiple seals to the same breathing hole; nets set in 42 breathing holes captured 58 seals.

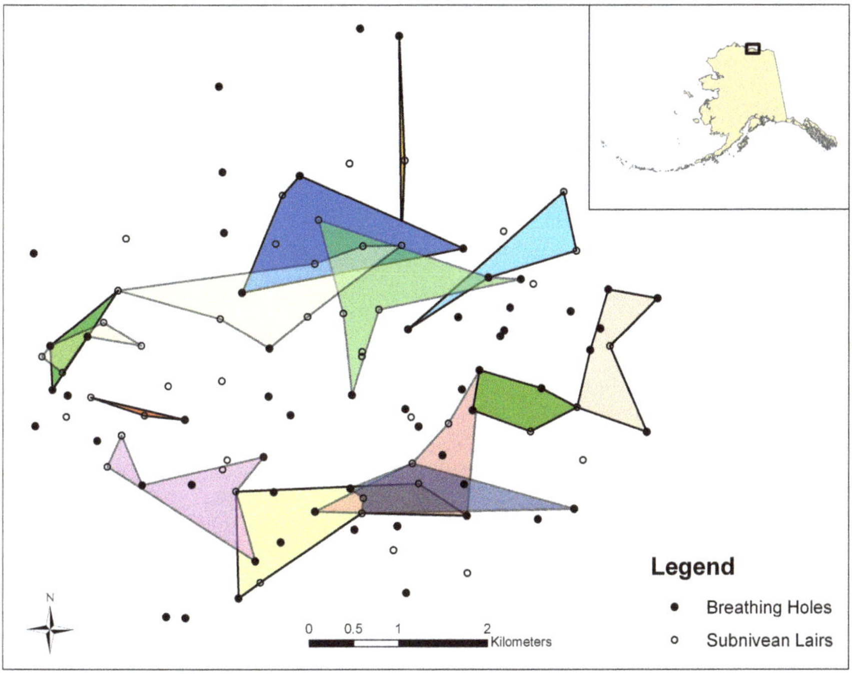

Fig. 15.12 Home ranges of 14 ringed seals occupying shore-fast sea ice along the Beaufort Sea coast of Alaska during the subnivean period in the year 2003. (After Kelly et al. 2010b)

The three-dimensional under-ice movements of ringed seals in the breeding season were inconsistent with the territorial hypothesis, which supposed large male territories containing multiple female territories. Three-dimensional ranges of adult female ringed seals were twice as large as those of juvenile and adult males (Fig. 15.13). Considerable spatial overlap was also observed in the under-ice ranges of pairs of seals that shared the use of one or more breathing holes and were tracked in the same volume of water. Under-ice volumes overlapped 56% for a yearling female and an adult male, 40% for a juvenile male and an adult male, 46% for two adult females (Fig. 15.13c, d), and 55% for a juvenile male and an adult female (Kelly and Wartzok 1996; Kelly et al. 2010b). We have few data on male-male interactions, but we observed substantial overlap in home ranges of males during the breeding season (Kelly et al. 2010b). Moreover, males remained near the underside of the ice in the breeding season, while females made deep foraging dives. Guarding access to the breathing hole at a pupping lair—to which a female is sure to return—would likely be a more successful male strategy than attempting to patrol the large three-dimensional area used by adult females. Mate or resource guarding suggests a minimal potential for polygyny (Bertram 1940; Stirling 1975, 1983; Chap. 8) and is similar to the breeding strategy reported for Weddell seals, the only other seal that breeds primarily in fast-ice (Chap. 13).

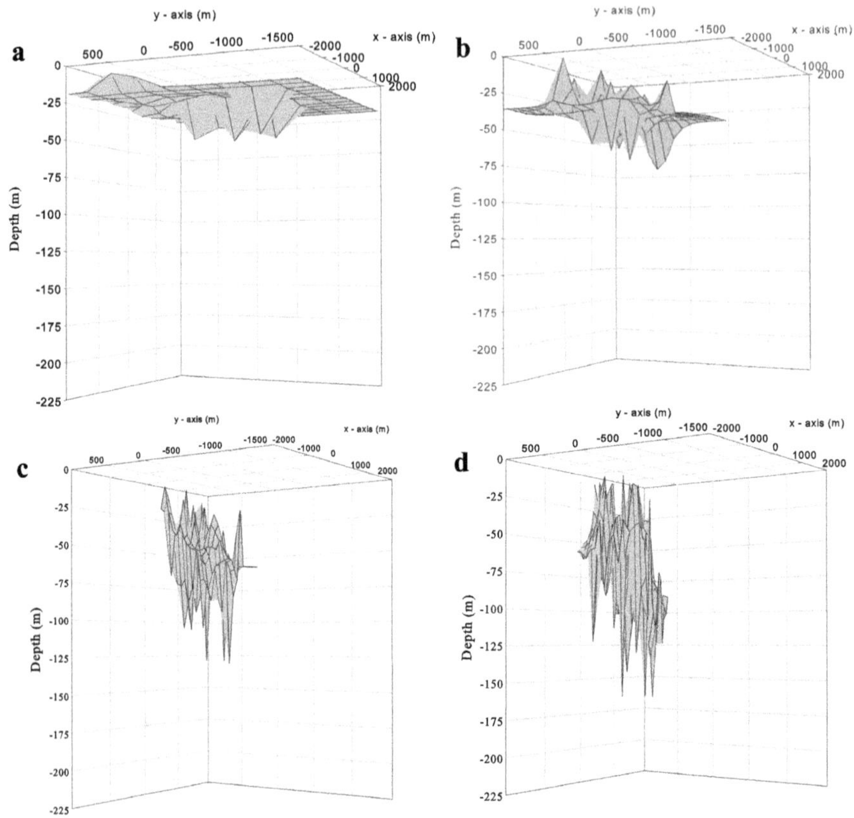

Fig. 15.13 Sub-ice home ranges of adult ringed seals in Resolute Passage, Northwest Territories, Canada: **a** and **b** adult males in 1990; **c** and **d** adult females tracked in 1991. Grids in **c** and **d** span the same volumes, and the home ranges of these two females overlapped by 46%. (After Kelly et al. 2010b)

Elsewhere, the local distribution of the sexes during the breeding has been interpreted as evidence of territoriality. For example, areas used by female Saimaa ringed seals tracked in the breeding season did not overlap (Niemi et al. 2019). In a Svalbard fjord, some adult males were proximal to concentrations of adult females in the breeding season, while other males occupied areas with lower densities of females (Krafft et al. 2007).

Bite wounds to male ringed seals (Smith and Hammill 1981; Smith 1987) and increased underwater vocalizations (Stirling et al. 1983; Rautio et al. 2009) during the breeding season have also been cited as support for the territorial hypothesis. It was only assumed, however, that the bites were inflicted by other males, and wounds from other seals are consistent with mate or resource guarding as well as territoriality. Nor were the individuals vocalizing or the function of their calls identified in field recordings (Stirling 1973; Richardson et al. 1995). Behavioral observations coupled with recordings in captivity do suggest, however, that low amplitude vocalizations—especially knocks and snorts lasting as long as 4 s—are used by males courting females near breathing holes (Mizuguchi et al. 2016).

The size of adult males captured in the Chukchi and Beaufort seas during the breeding season did not correlate with age ($r^2 = 0.052$, $p = 0.08$). Among 60 males >5 years of age, the mean weight was 55.3 (SD = 12.7) kg, but 10 of those seals weighed less than 45 kg, with some as little as 24 kg, and they showed the black facial pigmentation and odor of rut. Conceivably, small body size among some breeding males may reflect an alternative "sneaker" mating strategy that minimizes competition with larger breeding males.

A skewed age distribution of hunted ringed seals suggested that adults exclude juveniles from breeding areas (McLaren 1958; Smith 1973; Krafft et al. 2007). Juveniles were common, however, among ringed seals we captured within shore-fast ice in Kotzebue Sound, the Beaufort Sea, and Barrow Strait. Agonistic behavior of yearlings was consistently observed in the more than 100 seals that we handled in those areas. Almost without exception, yearlings hissed, scratched, and attempted to bite when captured (Kelly 1996b). Only one seal older than one year was similarly aggressive when captured; the rest were quite placid when captured.

The behavior of a yearling male (20 kg) tracked in Barrow Strait in 1997 suggested that the agonistic behavior of yearlings is directed at conspecifics as well as biologists. That yearling (TL97) repeatedly surfaced and rested in a hole cut in the ice under our hut despite the presence of people and dogs. We captured him in that hole on 8 May and observed fresh, conspecific bite marks on his hind flippers and torso. We attached an acoustic tag to his hair and released him on the floor of the hut. Although he had free access to the hole, he spent the night resting in the hut and only returned to the water at 0800 the next day. For the next 17 days, TL97 surfaced only at three sites: the hut (where he frequently rested on top of the ice), an enlarged transducer hole 5 m away, and a lair (designated 047) 800 m from the hut. When he visited the hut, he ignored our presence unless one of us or a dog passed too closely. In those instances, he would snarl and lunge at the intruder or splash them by smacking the water surface with a fore flipper. TL97 also occasionally rested in lair 047 and, on two occasions, was in that lair when a second seal entered. Both seals could be heard breathing in the lair, and TL97 emitted a near-constant crying sound until the second seal left a few hours later.

Although he was not the only seal to breathe at the hole under our hut, TL97 likely had less competition for that hole than for natural seal holes. Similarly, the transducer hole that he enlarged likely also provided some refuge from other seals. Once he adopted those, the only other hole he used was the access hole to lair 047, but—judging from his vocalizations—he was distressed on the two occasions when another seal joined him in that lair. The wounds suffered by TL97 and his persistent antagonistic behaviors toward people, dogs, and other seals are consistent with our observations that yearlings, in the wild and captivity, are especially aggressive and the suggestion by several authors that adults minimally tolerate juvenile ringed seals.

15.7.3 *Basking Period*

Social encounters appear to increase during the basking period when seals rest without snow cover at holes away from hummocks or pressure ridges. Those deformities in the ice often are responsible for snowdrifts in which lairs are formed, so many seals move at least a short distance to bask (Smith and Hammill 1981; Kelly et al. 2010b). As many as eight or more seals may congregate around some basking holes, and vocal and physical aggression (water splashing, biting, scratching) are frequent (Smith and Hammill 1981).

15.8 Predator Avoidance

Ringed seal behavior has mostly been shaped by predators (Stirling 1975; Kelly 2001) and—as the species responds to rapidly diminishing snow and ice cover (Kelly et al. 2010c; Ferguson et al. 2017)—predation may well exert new selection pressures (Ferguson et al. 2010, 2012; Hezel et al. 2012; Auttila et al. 2014; Auttila 2015). Extant predators of ringed seals include at least nine species of mammals, two species of birds, and one shark species (summarized in Kelly et al. 2010c). While some predators rarely take ringed seals, they have exerted significant selection pressure on their behavior. Other predators, such as polar bears and Arctic foxes, continue to put substantial pressure on ringed seal populations (Smith 1976; Pilfold et al. 2012).

15.8.1 *Foraging Period*

During the foraging period, ringed seals are vulnerable in the water to Greenland sharks (*Somniosus microcephalus*; Heptner et al. 1976; Leclerc et al. 2012) and killer whales (*Orcinus orca*; Melnikov and Zagrebin 2005; Ferguson et al. 2010, 2012) and on ice to polar bears (Smith 1980; Hammill and Smith 1991; Pilfold et al. 2012), walruses (*Odobenus rosmarus*; Fay 1960), and killer whales (Ferguson et al. 2010; Higdon et al. 2012).

Arctic ringed seals spend 90% of their time in the water during the foraging period (Martinez-Bakker et al. 2013). Inuit report that seals attempt to escape killer whales by moving into shallow water (Ferguson et al. 2012; Higdon et al. 2012). Predation by Greenland sharks has been confirmed from analysis of shark stomach contents. Shark attacks have not been observed, however, Leclerc et al. (2012) suggested that ringed seals are taken while resting or sleeping.

When ringed seals rest on the ice during the foraging period, they typically lie with their heads at the edge of drifting floes. That behavior favors quick escape from a predator approaching on the floe but leaves them vulnerable to polar bears, walruses, or killer whales approaching from the water. Polar bears stalk resting seals by approaching on the ice or from the water (Stirling 1974; Stirling and van Meurs 2015). On the other hand, walruses approach from the water and use their tusks to attack seals resting on the ice (Fay 1960; Lowry and Fay 1984). Inuit describe killer whales dislodging ringed seals from the ice by tipping floes either by directly pushing them from underneath or by creating waves that wash the seal into the water (Ferguson et al. 2012), similar to killer whale predation behavior observed in the Southern Ocean (Smith et al. 1981; Visser et al. 2008). Ringed seals frequently scan for predators while resting on the ice in the basking period and, presumably, do so when on the ice in the foraging period as well.

15.8.2 Subnivean Period

Excavating and occupying subnivean lairs protects ringed seals from extreme cold and many potential predators (Lukin 1980; Taugbøl 1982; Kelly 1988; Hammill and Smith 1991; Smith et al. 1991). Gulls (Laridae), ravens (*Corvus corax*), and possibly snowy owls (*Bubo scandiacus*) prey on newborn ringed seals in the absence of subnivean lairs. The protective snow cover likely sets a southern limit to the ringed seals' distribution (Kumlien 1879; Gjertz and Lydersen 1983; Lydersen and Smith 1989; Lydersen and Ryg 1991). Thus, avian predation might be a ghost of predation past (Silliman et al. 2018) that may also figure prominently in the ringed seals' future (Hezel et al. 2012). Ringed seals comprise 90% of polar bear prey in the Beaufort Sea, and female bears with cubs focus their hunting efforts on ringed seals during the subnivean period (Pilfold et al. 2012). Adult and newborn ringed seals each make up close to 40% of the kills, and the sex ratio of seals killed is equal.

While avian predation is thwarted by the ringed seals' use of subnivean lairs, that strategy is only partially effective against polar bears and other mammalian predators able to excavate the lairs (Stirling 1975; Kunnasranta 2001; Sipilä 2003). A behavior unique to polar bears involves throwing their entire weight on to their front paws to collapse a lair before its occupant(s) can escape (Stirling and Latour 1978; Smith 1980; Preuß et al. 2009; Kelly et al. 2010a). The deeper snow cover of pupping lairs hinders but does not prevent polar bear predation. Natural cavities in pressure ridges provide better protection from polar bears but are rarely used as lairs, possibly because they are less stable and less well insulated than lairs excavated in snow. As snow cover has declined on Lake Saimaa, the pups of Saimaa ringed seals are increasingly vulnerable to predation by red foxes (*Vulpes vulpes*), invasive raccoon dogs (*Nyctereutes procyonoides*), gulls, and ravens (Auttila et al. 2014; Auttila 2015; Kunnasranta et al. 2021).

Ringed seals are vulnerable to polar bears when they surface in breathing holes and rest on the ice. Seals in lairs are alert to sounds and dive from lairs at the approach of potential predators (Kelly et al. 1988). Our observations of the shared use of breathing holes and lairs indicate that ringed seals have multiple escape options during the subnivean period. Nonetheless, the pups remain vulnerable to Arctic foxes as well as bears when alone in lairs while their mothers are away foraging (Smith

1976, 1980). Birth lairs regularly are excavated in snow >90 cm deep, whereas non-birth lairs are more typically in snow depths of 50–75 cm (Kelly et al. 1986). The deeper snow thwarts polar bears from collapsing the lair roofs by pouncing on them. Bears and Arctic foxes must tunnel into lairs in deep snow, giving the seals more time to escape. If a pup's mother detects a predator excavating a lair, she will take the pup to another lair (Taugbøl 1982; Smith et al. 1991). Ringed seal pups often excavate one or more tunnels as extensions of the main lair. Those tunnels are too small to accommodate an adult seal, can reach 2 or more meters in length, and may help conceal the pups from predators.

Protecting pups from predation may influence the depths at which female ringed seals forage. Female ringed seals decrease foraging depths during lactation allowing them to more frequently attend pups in lairs, thereby increasing the probability of detecting predators and evacuating their pups to safety (Kelly 1996a).

Similarly, the allocation of time breathing at the surface and foraging beneath the ice among ringed seals that we acoustically tracked in Barrow Strait suggested a trade-off between optimal foraging and predator avoidance. The proportion of time those seals spent at the surface was not scaled to the depths at which they foraged (Kelly 1996a). Instead, the duration of their stays at the surface between foraging dives was consistently sufficient to saturate their tissues with oxygen. Thus, ringed seals were not optimizing time in their sub-ice foraging patch, perhaps, in favor of optimizing predator avoidance. Saturating before every dive may allow ringed seals to return to the surface with sufficient reserves to reject surfacing at a hole that shows evidence of a waiting predator and, instead, extends the dive long enough to surface at a safe breathing hole.

The most common polar bear hunting strategy involves lying or sitting above a breathing hole, waiting for a seal to surface, and then grabbing the seal with its teeth (Stirling 1974; Stirling and Latour 1978). The behavior of surfacing ringed seals reflects this danger. Ringed seals are extremely sensitive to the amount of light penetrating the ice at and around their breathing holes. More light than usual could signal that a bear had excavated the snow above and less light that a bear is lying above, a tactic common when bears have excavated a deep lair. In those instances, bears will block the light in the entry tunnel with their body and position themselves above the breathing hole. The likelihood of seals surfacing at breathing holes where we set nets decreased if we did not restore the overlying snow such that light penetration was substantially unaltered (Kelly 1996b, 2009).

Ringed seals warily approach breathing holes that they have not recently used. First visits often are preceded by one or more instances of water surging in the breathing hole as the seal swims past the underside of the hole, possibly looking for signs of a predator. Such swim-bys are frequently followed, after several minutes, by the seal releasing a train of bubbles into the breathing hole. Bubbles might induce a startle response and, thereby, give away a polar bear waiting above (Smith and Stirling 1975), but bears likely would learn to tame such a response. Alternatively, the bubbles may serve to momentarily clear snow and ice floating in the breathing hole, providing a clear view. When we set nets, an unobstructed view down the breathing hole required clearing ice and snow floating on the surface, but snow settled continuously onto the surface. Blowing a few times on the surface, however, pushed the snow to the periphery of the breathing hole and momentarily afforded a clear view down the breathing hole.

A seal confident that a predator is not waiting will rise in the hole and breathe with its entire head out of the water. A wary seal will either depart for another breathing hole or surface very quietly. In the latter case, only the nares break the surface, and the breaths are nearly inaudible. Lying 20–30 cm directly above seals breathing under our hut, I could faintly hear their nares open with each breath, but I could not hear the inhalations or exhalations.

The sensitivity of ringed seals to dangers above the breathing hole can be seen in their response to objects placed in the breathing holes. In developing our *aglu* traps, we experimented with various mechanisms for supporting snares, nets, and triggering devices and noted considerable tolerance for objects in the breathing hole as long as they were at least a few centimeters below the surface. Objects close to or above the surface, however, were not tolerated by the seals. Similarly, when removing seals trapped in the breathing holes, we found that any movements close to the surface immediately caused the seal to dive to the bottom of the net. Movements below the surface, however, drew little or no response. The nets closed off the bottom of the breathing hole and allowed the seal to surface and breathe at will. When we arrived at a capture site, the seal typically dove to the bottom of the net and surfaced after a few minutes, breathing with only the nares exposed. Snow and bits of ice on the surface blocked the seal's view of what was above. If we brushed the seal's head as we reached into the hole, it would abruptly dive, but if we slid a hand down the side of the breathing hole, carefully avoiding the seal's head, we could then reach over and touch the seal's torso repeatedly with no response. We blindly felt our way around the seal's torso until locating by feel a fore flipper, its claws inevitably pressed into the ice. We then grasped the flipper, still with no response, finally pulling the seal from the hole.

15.8.3 Basking Period

Ringed seals spend 37% of their time on the surface—more than in any other periods—during the basking period (Kelly et al. 2010b), a period when polar bears are feeding most intensively (Stirling and Latour 1978). The seals rest for several hours each day next to breathing holes that, for the most part, are on flat ice with minimal visual obstructions. They repeatedly lift their heads and survey their surroundings. The frequency of surveying seems not to be influenced by the number of seals around a hole (Smith and Hammill 1981). Still, the grouping behavior may decrease the probability of predation by increasing the likelihood of detecting an approach and/or through a selfish herd effect (Hamilton 1971).

The increased probability of detecting a predator likely reaches an asymptote, after which additional seals around a breathing hole are a net deficit to the group. When fleeing the surface, only one seal can pass down the breathing hole at one time, but two or more panicked ringed seals often enter the breathing hole simultaneously and become stuck while only partially submerged. For several seconds, the entire group's exposure is increased before the breathing hole is cleared. Eight or more ringed seals may bask at a single breathing hole, but the optimal group size—in terms of joint surveillance—may be smaller depending on the cost of repelling new arrivals (Caraco and Wolf 1975).

15.9 Future Directions

Despite recent advances, scientific knowledge of ringed seal behavior remains paltry, principally because of the challenges of observing a species that spends most of the year under snow and ice (Kelly 2001). At the same time, snow and ice cover are diminishing rapidly in the warming Arctic (Hezel et al. 2012; Haine and Martin 2017; Landrum and Holland 2020), resulting in large scale loss of habitat (Stirling and Smith 2004; Kelly et al. 2010c; Hezel et al. 2012; Auttila et al. 2014; Yurkowski et al. 2016; Ferguson et al. 2017). Future studies of ringed seal behavior need to be undertaken in the context of diminishing habitat, as well as to advance understanding of the evolution of pinniped behaviors.

Ringed seals currently are listed as threatened (marine subspecies) or endangered (freshwater subspecies) under the U.S. Endangered Species Act primarily because of habitat loss resulting from climate change (https://www.fisheries.noaa.gov/action/listing-4-subspecies-ringed-seals-arctic-okhotsk-baltic-and-ladoga-under-esa). Thus, conservation efforts would benefit from an improved understanding of behaviors involved in habitat selection. Studies carefully designed to test specific hypotheses could improve inferences based on comparisons of habitat selection across species. For example, an assessment of threats based on a multi-species comparison concluded that ringed seals should be the least sensitive of Arctic marine mammals to climate-induced habitat change (Laidre et al. 2008), but the scoring matrix did not include snow cover on ice. Snow on sea ice is not an important feature for most Arctic marine mammals, but its critical importance in ringed seal habitat selection has been documented across the Arctic (Kumlien 1879; McLaren 1958; Lukin and Potelov 1978; Lukin 1980; Smith and Hammill 1981; Gjertz and Lydersen 1983; Helle et al. 1984; Lydersen and Gjertz 1987, Lydersen et al. 1987; Lydersen and Smith 1989; Lydersen and Ryg 1990; Lydersen 1998; Ryg and Øristland 1991; Kelly 2001; Kunnasranta et al. 2001; Stirling and Smith 2004; Ferguson et al. 2005; Kelly et al. 2006; Lukin et al. 2006; Hezel et al. 2012; Auttila et al. 2014). The species' sensitivity to rapidly diminishing snow cover could be assessed more directly, however, through on-ice behavioral observations of lair formation and abandonment (e.g., Stirling and Smith 2004) coupled with pan-Arctic quantification of snow melts using satellite-borne backscatter data (Belchansky et al. 2004; Kelly et al. 2006).

The importance of sea ice as ringed seal habitat is well established through observations of Indigenous hunters (Shapiro et al. 1979; Noongwook 2000; Pungowiyi 2000) and scientific investigations (Vibe 1950, 1967; Fay 1974; Burns et al. 1981;

Braham et al. 1981; Stirling 1997; Tynan and DeMaster 1997; Harwood et al. 2000; Smith and Harwood 2001; Kelly et al. 2010c). Nonetheless, a multi-species synthesis of movements by Arctic marine mammals was silent on the role of sea ice. Despite a stated aim of "understanding the potential impacts of changes that are underway," including diminishing sea ice, Citta et al. (2018) concluded that the effects of sea ice were outside the scope of their multi-species synthesis. Hypothesis-driven behavioral studies, on the other hand, can elucidate the impacts of sea ice cover on foraging and energy expended in travel (e.g., Yurkowski et al. 2016).

Biotic features of ringed seal habitat also are changing (Wassmann et al. 2011; McMeans et al. 2013), and responses should be predicted with models of habitat selection that include water masses, location, dive activity, bathymetry, predation, and sea ice data (Freitas et al. 2008b; McClintock et al. 2017). Predictions concerning changing prey availability could be tested with the behavior of seals carrying appropriate instruments (Gryba et al. 2019). It will be more challenging, but at least as important, to assess the likely impacts of changing availability of ringed seals to predators. How will the seals respond to increased vulnerability to killer whale and polar bear predation as refuge on ice and under snow diminishes?

From an evolutionary perspective, it remains of interest to better understand the sensory basis of ringed seal navigation over long distances between foraging areas and breeding sites and more locally when foraging under ice and at depth in the dark of winter. The fidelity to breeding sites among adult ringed seals has ecological and evolutionary implications. Long-term behavioral observations are needed to ascertain whether and to what degree juvenile seals are faithful to specific sites.

A more detailed understanding of the mating behavior of ringed seals and other ice-associated pinnipeds will be necessary to fully understand the evolution of pinniped mating behavior. The number of pinniped species breeding in sea ice environments is greater than the numbers breeding in mid and lower latitudes (Kelly 2001), but the mating behavior of the latter is much better known. Does the dynamic nature of sea ice habitats lead to more intra-specific variation in mating strategies? Does substantial variation in body size of adult ringed seals reflect differences in shore-fast versus pack ice environments as suggested by Fedoseev (1975), distinct stocks as suggested by Indigenous observations (the late Chester Noongwook reported that the St. Lawrence Island Yupik recognize two species of ringed seals; one in which adults are small and one in which adults are large), or alternative mating strategies in which small adult males avoid competing with larger males for mating opportunities?

Of course, it remains the case that many aspects of ringed seal behavior will remain challenging to observe. Inferences from telemetry and extrapolation from other species may yield further insights, especially if tempered and extended by the in-depth knowledge that comes from many hours of direct observation (Stirling et al. 2016). A wealth of such knowledge among Indigenous hunters has benefitted—and could further expand—scientific understanding.

References

Ainley DG, Demaster DP (2009) The upper trophic levels in polar marine ecosystems. In: Smith W (ed) Polar Oceanography. Part B. Chemistry, Biology and Geology. Academic Press, San Diego, pp 599–630

Allen JA (1880). History of North American pinnipeds. A monograph of the walruses, sea-lions, sea-bears, and seals of North America. Washington, DC: U.S. Geol. Geogr. Surv. Terr. Misc. Publ. 12. U.S. Print. Off

Árnason U, Gullberg A, Janke A, Kullberg M, Lehman N, Petrov EA, Väinölä R (2006) Pinniped phylogeny and a new hypothesis for their origin and dispersal. Mol Phylogenet Evol 41:345–354

Ashwell-Erickson S, Fay FH, Elsner R, Wartzok D (1986) Metabolic and hormonal correlates of molting and regeneration of pelage in Alaskan harbor and spotted seals (*Phoca vitulina* and *Phoca largha*). Can J Zool 64:1086–1094

Auttila M, Niemi M, Skrzypczak T, Viljanen M, Kunnasranta M (2014) Estimating and mitigating perinatal mortality in the endangered Saimaa ringed seal (*Phoca hispida saimensis*) in a changing climate. Ann Zool Fenn 51(6):526–534

Auttila M, Sinisalo T, Valtonen M, Niemi M, Viljanen M, Kurkilahti M, Kunnasranta M (2015) Diet composition and seasonal feeding patterns of a freshwater ringed seal (*Pusa hispida saimensis*). Mar Mammal Sci 31:45–65

Auttila M (2015). The endangered Saimaa ringed seal in a changing climate–challenges for conservation and monitoring. Dissertations in Forestry and Natural Sciences No 194, University of Eastern Finland

Bailey AM, Hendee RW (1926) Notes on the mammals of northwestern Alaska. J Mammal 7:9–28

Banks D, Williams M, Pearce J, Springer A, Hagenstein R, Olson D (2001) Ecoregion-based conservation in the Bering Sea: identifying important areas for biodiversity conservation. World Wildlife Fund and The Nature Conservancy of Alasak, Washington, D.C

Belchansky GI, Douglas DC, Mordvintsev IN, Platonov NG (2004) Estimating the time of melt onset and freeze onset over Arctic sea-ice area using active and passive microwave data. Remote Sens Environ 92(1):21–39

Bertram GLC (1940) The biology of the Weddell and crabeater seals: with a study of the comparative behaviour of the pinnipedia. Br Mus Nat Hist Sci Rep 1

Born EW, Teilmann J, Acquarone M, Riget FF (2004) Habitat use of ringed seals (*Phoca hispida*) in the North Water area (North Baffin Bay). Arctic 57:129–142

Boyd IL, Reid K, Bevan RM (1995) Swimming speed and the allocation of time during the dive cycle of Antarctic fur seals. Anim Behav 50:769–784

Bradstreet MS, Finley KJ, Sekerak AD, Griffiths WB, Evans CR, Fabijan MN, Stallard HE (1986) Aspects of the biology of Arctic cod (*Boreogadus saida*) and its importance in the Arctic marine food chains. Department of Fisheries and Oceans, Central and Arctic Region

Braham HW, Burns JJ, Fedoseev GA, Krogman BD (1981) Distribution and density of ice- associated pinnipeds in the Bering Sea. NOAA, National Marine Fisheries Service, Northwest and Alaska Fisheries Center, National Marine Mammal Laboratory. p 27

Burns JJ (1970) Remarks on the distribution and natural History of pagophilic pinnipeds in the Bering and Chukchi Seas. J Mammal 51:445

Burns JJ, Fay FH (1970) Comparative morphology of the skull of the ribbon seal, *Histriophoca fasciata*, with remarks on systematics of Phocidae. J Zool 161:363–394

Burns JJ, Shapiro LH, Fay FH (1981) Ice as marine mammal habitat in the Bering Sea. In: Hood DW, Calder JA (eds) The Eastern Bering Sea shelf: oceanography and resources, vol. 2. U.S. Department of Commerce, NOAA, U.S. Department of Interior, Office of Marine Pollution Assessment, and Bureau of Land Management, Washington, DC

Caraco T, Wolf LL (1975) Ecological determinants of group sizes of foraging lions. Am Nat 109:343–352

Carey AG (1992) The ice fauna in the shallow southwestern Beaufort Sea, Arctic Ocean. J Mar Syst 3:225–236

Carlens H, Lydersen C, Krafft BA, Kovacs KM (2006) Spring haul-out behavior of ringed seals (*Pusa hispida*) in Kongsfjorden. Svalbard. Mar. Mammal Sci. 22:379–393
Chapskii KK (1940) The ringed seal of western seas of the Soviet Arctic. Tr. Vses. Arckt. Inst. 145:1–72
Citta JJ, Lowry LF, Quakenbush LT, Kelly BP, Fischbach AS, London JM, Jay CV, Frost KJ, Crowe GOC, Crawford JA et al (2018) A multi-species synthesis of satellite telemetry data in the Pacific Arctic (1987–2015): overlap of marine mammal distributions and core use areas. Deep. Res. Part II Top. Stud. Oceanogr. 152:132–153
Costa DP (1991) Reproductive and foraging energetics of pinnipeds: implications for life history patterns. In: Renouf D (ed) The behaviour of pinnipeds. Springer, Dordrecht, pp 300–344
Crawford R, Jorgenson J (1990) Density distribution of fish in the presence of whales at the Admirality Inlet landfast ice edge. Arctic 43:215–222
Crawford JA, Frost KJ, Quakenbush LT, Whiting A (2012) Different habitat use strategies by subadult and adult ringed seals (*Phoca hispida*) in the Bering and Chukchi seas. Polar Biol 35:241–255
Crawford JA, Frost KJ, Quakenbush LT, Whiting A (2019) Seasonal and diel differences in dive and haul-out behavior of adult and subadult ringed seals (*Pusa hispida*) in the Bering and Chukchi seas. Polar Biol 42:65–80
Crocker DE, Boeuf BJL, Costa DP (1997) Drift diving in female northern elephant seals: implications for food processing. Can J Zool 75:27–39
Dehn LA, Sheffield GG, Follmann EH, Duffy LK, Thomas DL, O'Hara TM (2007) Feeding ecology of phocid seals and some walrus in the Alaskan and Canadian Arctic as determined by stomach contents and stable isotope analysis. Polar Biol 30:167–181
Dehnhardt G, Mauck B, Bleckmann H (1998) Seal whiskers detect water movements. Nature 394:235–236. https://doi.org/10.1038/28303
Dehnhardt G, Mauck B, Hanke W, Bleckmann H (2001) Hydrodynamic trail-following in harbor seals (*Phoca vitulina*). Science 293:102–104
Demere TA, Berta A, Adam PJ (2003) Pinnipedimorph evolutionary biogeography. Bull. Am. Museum Nat. Hist. 279:32–76
Dykes RW (1975) Afferent fibers from vibrissae of cats and seals. J Neurophysiol 38:650–662
Elsner R, Wartzo D, Sonafrank NB, Kelly BP (1989) Behavioral and physiological reactions of arctic seals during under-ice pilotage. Can J Zool 67:2506–2513
Fay FH (1960) Carnivorous Walrus and Some Arctic Zoonoses. Arctic 13:111–122
Fay FH (1982) Ecology and biology of the Pacific walrus, *Odobenus rosmarus divergens* Illiger. North American Fauna 74:1–279
Fay FH (1974) The role of ice in the ecology of marine mammals of the Bering Sea. In: Hood DW, Kelly EJ (eds) Oceanography of the Bering Sea. Institute of Marine Science, Hakodate, Japan, pp. 389–399
Fedoseev GA (1965) Food of the ringed seal. In: Poppino ML (ed)Annual Reports for Principal Investigators, Translated from Russian. U.S. National Oceanic and Atmospheric Administration: Boulder, CO, pp 373–383
Fedoseev GA (1975) Ecotypes of the ringed seal (*Pusa hispida* Schreber, 1777) and their reproductive capabilities. In: Biology of the Seal. Proceedings of a Symposium Held in Guelph 14-17 August 1972, Conseil International pour l'Éxploration de la Mer 169, pp 156–160
Ferguson SH, Stirling I, McLoughlin P (2005) Climate change and ringed seal (*Phoca hispida*) recruitment in western Hudson Bay. Mar Mamm Sci 21:121–135
Ferguson SH, Higdon JW, Chmelnitsky EG (2010) The rise of killer whales as a major Arctic predator. In: Ferguson SH, Loseto LL, Mallory ML (eds) In: A little less Arctic. Springer, Dordrecht, pp 117–136
Ferguson SH, Higdon JW, Westdal KH (2012) Prey items and predation behavior of killer whales (*Orcinus orca*) in Nunavut, Canada based on Inuit hunter interviews. Aquat Biosyst 8:3

Ferguson SH, Young BG, Yurkowski DJ, Anderson R, Willing C, Nielsen O (2017) Demographic, ecological, and physiological responses of ringed seals to an abrupt decline in sea ice availability. PeerJ 5:e2957

Fetterer F, Knowles K, Meier WN, Savoie M, Windnagel AK (2017) *Sea Ice Index, Version 3.* [Sea Ice Extent]. NSIDC: National Snow and Ice Data Center, Boulder, Colorado USA. https://doi.org/10.7265/N5K072F8 Accessed on June 2020

Forstén A, Alhonen P (1975) The subfossil seals of Finland and their relation to the history of the Baltic Sea. Boreas 4:143–155

Freeman MMR (1984) Contemporary Inuit exploitation of the sea-ice environment. Sikumiut: The people who use the Sea Ice. Canadian Arctic Resources Committee, Ottawa, pp 73–96

Freitas C, Kovacs KM, Ims RA, Fedak MA, Lydersen C (2008a) Ringed seal post-moulting movement tactics and habitat selection. Oecologia 155:193–204

Freitas C, Kovacs KM, Lydersen C, Ims RA (2008b) A novel method for quantifying habitat selection and predicting habitat use. J Appl Ecol 45:1213–1220

Furgal CM, Innes S, Kovacs KM (2002) Inuit spring hunting techniques and local knowledge of the ringed seal in Arctic Bay (Ikpiarjuk). Nunavut Polar Res 21:1–16

Gjertz I, Lydersen C (1983) Ringed seal *Phoca hispida* pupping in the Svalbard area. Fauna 36:65–66

Gjertz I, Kovacs KM, Lydersen C, Wiig Ã (2000) Movements and diving of adult ringed seals (*Phoca hispida*) in Svalbard. Polar Biol 23:651–656

Gryba RD, Wiese FK, Kelly BP, Von Duyke AL, Pickart RS, Stockwell DA (2019) Inferring foraging locations and water masses preferred by spotted seals *Phoca largha* and bearded seals *Erignathus barbatus*. Mar Ecol Prog Ser 631:209–224. https://doi.org/10.3354/meps13145

Haine TW, Martin T (2017) The Arctic-Subarctic sea ice system is entering a seasonal regime: implications for future Arctic amplification. Sci Rep 7:4618

Hall CF (1865) Arctic researches, and life among the Esquimaux: being the narrative of an expedition in search of Sir John Franklin, in the years 1860, 1861, and 1862. Harper and Brothers, New York, p 595

Hamilton WD (1971) Geometry for the selfish herd. J Theor Biol 31:295–311

Hammill MO, Smith TG (1991) The role of predation in the ecology of the ringed Sseal in Barrow Strait, Northwest Territories Canada. Mar Mamm Sci 7:123–135

Hammill MO (1987) Ecology of the ringed seal (*Phoca hispida* Schreber) in the fast-ice of Barrow Strait, Northwest Territories. PhD Dissertation, Macdonald College of McGill University, Montreal, Quebec, Canada, p 108

Harcourt RG, Hindell MA, Bell DG, Waas JR (2000) Three-dimensional dive profiles of free-ranging Weddell seals. Polar Biol 23:479–487

Hardy MS, Roff E, Smith TG, Ryg M (1991) Facial skin glands of ringed and grey seals, and their possible function as odoriferous organs. Can J Zool 69:189–200

Härkönen T, Stenman O, Jüssi M, Jüssi I, Sagitov R, Verevkin M (1998) Population size and distribution of the Baltic ringed seal (*Phoca hispida botnica*). In: Heide-Jørgensen MP, Lydersen C (eds) Ringed seals in the North Atlantic. NAMMCO Scientific Publications, vol. 1. Tromsø, Norway, pp. 167–180

Härkönen T, Jüssi M, Jüssi I, Verevkin M, Dmitrieva L, Helle E, Sagitov R, Harding KC (2008) Seasonal activity budget of adult Baltic ringed seals. PLoS One 3:e2006

Harwood LA, Smith TG, Melling H (2000) Variation in reproduction and body condition of the ringed seal (*Phoca hispida*) in Western Prince Albert Sound, NT, Canada, as assessed through a harvest-based sampling program. Arctic 53:422–431

Harwood LA, Smith TG, Auld JC (2012) Fall migration of ringed seals (*Phoca hispida*) through the Beaufort and Chukchi seas, 2001–02. Arctic 65:35–44

Harwood LA, Smith TG, Auld JC, Melling H, Yurkowski DJ (2015) Seasonal movements and diving of ringed seals, *Pusa hispida*, in the western Canadian Arctic, 1999–2001 and 2010–11. Arctic 68:193–209

Heaton TH, Grady F (2003) The late Wisconsin vertebrate history of Prince of Wales Island, southeast Alaska. In: Schubert BW, Mead JI, Graham RW (eds) Ice age cave faunas of North America. Indiana University Press, Bloomington, pp 17–53
Helle E, Hyvärinen H, Sipilä T (1984) Breeding habitat and lair structure of the Saimaa ringed seal *Phoca hispida saimensis* Nordq. in Finland. Acta Zoologica Fennica 172:125–127
Heptner LVG, Chapskii KK, Arsen'ev VA, Sokolov VA (1976) Ringed seal. *Phoca* (*Pusa*) *hispida* Schreber, 1775. In: Heptner LVG, Naumov NP, Mead J (eds) Mammals of the Soviet Union. Volume II, Part 3–Pinnipeds and Toothed Whales, Pinnipedia and Odontoceti. Vysshaya Shkola Publishers, Moscow, pp 218–260
Hezel PJ, Zhang X, Bitz CM, Kelly BP, Massonnet F (2012) Projected decline in spring snow depth on Arctic sea ice caused by progressively later autumn open ocean freeze-up this century. Geophys Res Lett 39:6–11
Higdon JW, Bininda-Emonds ORP, Beck RMD, Ferguson SH (2007) Phylogeny and divergence of the pinnipeds (Carnivora: Mammalia) assessed using a multigene dataset. BMC Evol Biol 7:216
Higdon JW, Hauser DDW, Ferguson SH (2012) Killer whales (*Orcinus orca*) in the Canadian Arctic: distribution, prey items, group sizes, and seasonality. Mar Mammal Sci 28:E93–E109
Holst M, Stirling I, Hobson KA (2001) Diet of ringed seals (*Phoca hispida*) on the east and west sides of the North Water Polynya, northern Baffin Bay. Mar Mammal Sci 17:888–908
Houston AI, Carbone C (1992) The optimal allocation of time during the diving cycle. Behav Ecol 3:255–265
Houston AI, McNamara JM (1985) A general theory of central place foraging for single-prey loaders. Theor Popul Biol 28:233–262
Huntington HP (2000) Traditional ecological knowledge of seals in Norton Bay, Alaska. Report submitted to the Elim-Shaktoolik-Koyuk Marine Mammal Commission and the National Marine Fisheries Service
Hyvärinen H (1989) Diving in darkness: whiskers as sense organs of the ringed seal *Phoca hispida saimensis*. J Zool 218:663–678
Hyvärinen H, Katajisto H (1984) Functional structure of the vibrissae of the ringed seal (*Phoca hispida* Schr.). Acta Zool. Fenn. 171:27–30
Hyvärinen H, Nieminen M (1990) Differentiation of the ringed seal in the Baltic Sea, Lake Ladoga and Lake Saimaa. Finnish Game Res 47:21–27
Hyvärinen H, Hämäläinen E, Kunnasranta M (1995) Diving behavior of the Saimaa ringed seal (*Phoca hispida saimensis* Nordq.). Mar Mammal Sci 11:324–334
Johnson ML, Fiscus CH, Ostenson BT, Barbour ML (1966). Marine mammals. In: Wilimovsky NJ, Wolfe JN (eds) Environment of the Cape Thompson Region, Alaska, U.S. Atomic Energy Commission, Oak Ridge, TN, pp 877–924
Kapel FO, Christiansen J, Heide-Jørgensen MP, Härkönen T, Born EW, Knutsen LØ, Riget F, Teilmann J (1998) Netting and conventional tagging used to study movements of ringed seals (*Phoca hispida*) in Greenland. NAMMCO Sci Publ 1:211
Kelly BP (1988) Ringed Seal. In: Lentfer JW (ed) Selected marine mammals of Alaska: species accounts with research and management recommendations. Marine Mammal Commision, Washington, DC, pp 57–75
Kelly BP (2009) Studying seals in their sea ice habitat. In: Eicken H, Gradinger R, Salganek M, Shirasawa K, Perovich DK, Lepparanta M (eds) Field techniques for sea-ice research. University of Alaska Press, Fairbanks, pp 301–344
Kelly BP, Quakenbush LT (1990) Spatiotemporal use of lairs by ringed seals (*Phoca hispida*). Can J Zool 68:2503–2512
Kelly BP, Wartzok D (1996) Ringed seal diving behavior in the breeding season. Can J Zool 74:1547–1555
Kelly BP, Badajos OH, Kunnasranta M, Moran JR, Martinez-Bakker M, Wartzok D, Boveng P (2010a) Seasonal home ranges and fidelity to breeding sites among ringed seals. Polar Biol 33:1095–1109

Kelly B, Whiteley A, Tallmon D (2010c) The Arctic melting pot: hybridization in polar species could hit biodiversity hard. Nature 468:891

Kelly BP, Quakenbush LT, Rose JR (1986) Ringed seal winter ecology and effects of noise disturbance. Outer Continental Shelf Environmental Assessment pp. 447–536

Kelly BP, Burns JJ, Quakenbush LT (1988) Responses of ringed seals (*Phoca hispida*) to noise disturbance. In: Port and Ocean Engineering Under Arctic Conditions, vol. 2, Symposium on Noise and Marine Mammals, pp 27–38

Kelly BP, Badajos OH, Kunnasranta M, Moran J (2006) Timing and re-interpretation of ringed seal surveys. Final Report to the Coastal Marine Institute, University of Alaska Fairbanks. p 60

Kelly BP, Bengston JL, Boveng PL, Cameronm MF, Dahle SP, Jansen JK, Logerwell EA, Overland JE, Sabine LC, Wilder JM (2010b) Status review of the ringed seal. NOAA Tech. Memo. NMFS-AFSC-212

Kelly BP (1996a) Behavior of ringed seals diving under shore-fast sea ice. PhD Dissertation, Purdue University, West Lafayette, Indiana, p 91

Kelly BP (1996b) Live capture of ringed seals in ice-covered waters. J Wildl Manage 60:678–684

Kelly BP (2001) Climate change and ice breeding pinnipeds. In: "Fingerprints" of Climate Change, pp 43–55

Kelly BP (2008) Ringed seals: an instrumentation challenge. In: Marshall G (ed) Proceedings of the 2007 animal-borne imaging symposium. National Geographic Society, Washington, DC, pp 145–148

Khim BK, Krantz DE, Cooper LW, Grebmeier JM (2003) Seasonal discharge of estuarine freshwater to the western Chukchi Sea shelf identified in stable isotope profiles of mollusk shells. J Geophys Res C Ocean 108:C9

Kienle SS, Berta A (2018) The evolution of feeding strategies in phocid seals (Pinnipedia, Phocidae). J Vertebr Paleontol 38:e1559172

Kienle SS, Hermann-Sorensen H, Costa DP, Reichmuth C, Mehta RS (2018) Comparative feeding strategies and kinematics in phocid seals: Suction without specialized skull morphology. J Exp Biol 221

Kinda GB, Simard Y, Gervaise C, Mars JI, Fortier L (2015) Arctic underwater noise transients from sea ice deformation: characteristics, annual time series, and forcing in Beaufort Sea. J Acoust Soc Am 138:2034–2045

King JE (1964) Seals of the World. British Museum (Natural History), London

Kovacs KM (2007) Background document for development of a circumpolar ringed seal (*Phoca hispida*) monitoring plan. Work to Dev Monit Plans Arct Mar Mamm 45

Krafft BA, Kovacs KM, Lydersen C (2007) Distribution of sex and age groups of ringed seals *Pusa hispida* in the fast-ice breeding habitat of Kongsfjorden. Svalbard Mar Ecol Prog Ser 335:199–206

Kramer DL (1988) The behavioral ecology of air breathing by aquatic animals. Can J Zool 66:89–94

Kumlien L (1879) Mammals. Contributions to the natural history of Arctic America, made in connection with the Howgate Polar Expedition 1877–78. Government Printing Office, Washington, D.C., pp 55–61

Kunnasranta M, Hyvärinen H, Sipilä T, Koskela JT (1999) The diet of the Saimaa ringed seal *Phoca hispida saimensis*. Acta Theriol (warsz) 44:443–450

Kunnasranta M, Hyvärinen H, Sipilä T, Medvedev N (2001) Breeding habitat and lair structure of the ringed seal (*Phoca hispida ladogensis*) in northern Lake Ladoga in Russia. Polar Biol 24:171–174

Kunnasranta M, Hyvärinen H, Häkkinen J, Koskela JT (2002) Dive types and circadian behaviour patterns of Saimaa ringed seals *Phoca hispida saimensis* during the open-water season. Acta Theriol (warsz) 47:63–72

Kunnasranta M, Hyvärinen H, Sorjonen J (1996) Underwater vocalizations of Ladoga ringed seals (*Phoca hispida ladogensis* Nordq.) in summertime. 12:611–618

Kunnasranta M, Niemi M, Auttila M, Valtonen M, Kammonen J, Nyman T (2021) Sealed in a lake—biology and conservation of the endangered Saimaa ringed seal: a review. Biol Cons 253 https://doi.org/10.1016/j.biocon.2020.108908

Kunnasranta M (2001) Behavioural biology of two ringed seal (*Phoca hispida*) subspecies in the large European lakes Saimaa and Ladoga. University of Joensuu
Laidre KL, Stirling I, Lowry LF, Wiig Ø, Heide-Jørgensen MP, Ferguson SH (2008) Quantifying the sensitivity of arctic marine mammals to climate-induced habitat change. Ecol Appl 18:S97–S125. https://doi.org/10.1890/06-0546.1
Landrum L, Holland MM (2020) Extremes become routine in an emerging new Arctic. Nat Clim Chang 10:1108–1115
Le Boeuf BJ, Naito Y, Asaga T, Crocker D, Costa DP (1992) Swim speed in a female northern elephant seal: metabolic and foraging implications. Can J Zool 70:786–795
Leclerc LME, Lydersen C, Haug T, Bachmann L, Fisk AT, Kovacs KM (2012) A missing piece in the Arctic food web puzzle? Stomach contents of Greenland sharks sampled in Svalbard Norway. Polar Biol 35:1197–1208
Lesage V, Hammill MO, Kovacs KM (1999) Functional classification of harbor seal (*Phoca vitulina*) dives using depth profiles, swimming velocity, and an index of foraging success. Can J Zool 77:74–87
Liukkonen L, Ayllón D, Kunnasranta M, Niemi M, Nabe-Nielsen J, Grimm V, Nyman AM (2018) Modelling movements of Saimaa ringed seals using an individual-based approach. Ecol Modell 368:321–335
Lowry LF, Fay FH (1984) Seal eating by walruses in the Bering and Chukchi Seas. Polar Biol 3:11–18
Lowry JK, Stoddart HE (1993) The Onisimus problem (Amphipoda, Lysianassoidea, Uristidae). Zool Scr 22:167–181
Lowry LF, Frost KJ, Burns JJ (1980) Variability in the diet of ringed seals, *Phoca hispida*. Alaska Can J Fish Aquat Sci 37:2254–2261
Lukin LR (1980) The habitat of the White Sea ringed seal during early postnatal development. Mar Biol 5:33–37
Lukin LR, Potelov VA (1978) Living conditions and distribution of ringed seal in the White Sea in winter. Biol Morya 3:62–69
Lukin LP, Ognetov GN, Boiko NS (2006) Ecology of the ringed seal in the White Sea Uro RAN. Ekaterinburg, Russia
Lydersen C (1991) Monitoring ringed seal (*Phoca hispida*) activity by means of acoustic telemetry. Can J Zool 69:1178–1182
Lydersen C (1995) Energetics of pregnancy, lactation and neonatal development in ringed seals (*Phoca hispida*). Dev Mar Biol 4:319–327
Lydersen C, Gjertz I (1987) Population parameters of ringed seals (*Phoca hispida* Schreber, 1775) in the Svalbard area. Can J Zool 65:1021–1027
Lydersen C, Hammill MO (1993) Activity, milk intake and energy consumption in free-living ringed seal (*Phoca hispida*) pups. J Comp Physiol B 163:433–438
Lydersen C, Kovacs KM (1999) Behaviour and energetics of ice-breeding, North Atlantic phocid seals during the lactation period. Mar Ecol Prog Ser 187:265–281
Lydersen C, Smith TG (1989) Avian predation on ringed seal *Phoca hispida* pups. Polar Biol 9:489–490
Lydersen C, Jensen PM, Lydersen E (1987) Studies of the ringed seal *(Phoca hispida)* population in the Van Mijen fiord, Svalbard, in the breeding period 1986. Norsk Polarinstitutt Rapportserie No 34:89–112
Lydersen C, Ryg MS (1990) An evaluation of Tempelfjorden and Sassenfjorden as breeding habitat for ringed seals *Phoca hispida*. In: Severinsen T, Hansson R (eds). Environmental Atlas Gipsdalen, Svalbard. vol. 3, Reports on the Fauna of Gipsdalen. Norsk Polarinstitutt Rapportserie, No.66, pp 33–40
Lydersen C, Ryg M (1991) Evaluating breeding habitat and populations of ringed seals *Phoca hispida* in Svalbard fjords. Polar Rec 27:223–228

Lydersen C (1998) Status and biology of ringed seals (*Phoca hispida*) in Svalbard. In: Heide-Jørgensen MP, Lydersen C (eds) Ringed seals in the North Atlantic. NAMMCO Scientific Publications, vol. 1. Tromsø, Norway, pp 46–62

Mansfield AW (1967) Seals of arctic and eastern Canada. Bull Fish Res Board Canada 137:1–35

Martinez-Bakker ME, Sell SK, Swanson BJ, Kelly BP, Tallmon DA (2013) Combined genetic and telemetry data reveal high rates of gene flow, migration, and long-distance dispersal potential in Arctic ringed seals (*Pusa hispida*). PLoS One 8:e77125

Matsumura M, Watanabe YY, Robinson PW, Miller PJO, Costa DP, Miyazaki N (2011) Underwater and surface behavior of homing juvenile northern elephant seals. J Exp Biol 214:629–636

Mauck B, Gläser N, Schlosser W, Dehnhardt G (2008) Harbour seals (*Phoca vitulina*) can steer by the stars. Anim Cogn 11:715–718

McClintock BT, London JM, Cameron MF, Boveng PL (2017) Bridging the gaps in animal movement: hidden behaviors and ecological relationships revealed by integrated data streams. Ecosphere 8(3):e01751

McLaren IA (1960) On the origin of the Caspian and Baikal seals and the paleoclimatological implications. Am J Sci 258:47–65

McLaren IA (1958) The biology of the ringed seal (*Phoca hispida* Schreber) in the eastern Canadian Arctic. Fisheries Research Board of Canada, Montreal

McMeans BC, Rooney N, Arts MT, Fisk AT (2013) Food web structure of a coastal Arctic marine ecosystem and implications for stability. Mar Ecol Prog Ser 482:17–28. https://doi.org/10.3354/meps10278

Melnikov VV, Zagrebin IA (2005) Killer whale predation in coastal waters of the Chukotka Peninsula. Mar Mammal Sci 21:550–556

Mitani Y, Andrews RD, Sato K, Kato A, Naito Y, Costa DP (2010) Three-dimensional resting behaviour of northern elephant seals: drifting like a falling leaf. Biol Lett 6:163–166

Mizuguchi D, Tsunokawa M, Kawamoto M, Kohshima S (2016) Underwater vocalizations and associated behavior in captive ringed seals (*Pusa hispida*). Polar Biol 39:659–669. https://doi.org/10.1007/s00300-015-1821-x

Moore S, Hauser D (2019) Marine mammal ecology and health: finding common ground between conventional science and indigenous knowledge to track arctic ecosystem variability. Environ Res Lett 14:075001. https://doi.org/10.1088/1748-9326/ab20d8

Müller-Wille LL (1969) Biometrical comparison of four populations of *Phoca hispida* Schreb. in the Baltic and White Seas and Lakes Ladoga and Saimaa. Comment Biol Soc Sci Fenn 31:1–12

Murdoch J (1885) Mammals, part IV, division 1. In: Ray PH (ed) Report of the International polar expedition to point barrow. In: 48th Congress, 2nd Session, House Executive Document 44, Alaska. Washington, DC, pp 92–103

Murray MS (2005) Prehistoric use of ringed seals: a zooarchaeological study from Arctic Canada. Environ Archaeol 10:19–38

Murray MS (2008) Zooarchaeology and arctic marine mammal biogeography, conservation, and management. Ecol Appl 18:S41–S55

Niemi M, Liukkonen L, Koivuniemi M, Auttila M, Rautio A, Kunnasranta M (2019) Winter behavior of Saimaa ringed seals: Non-overlapping core areas as indicators of avoidance in breeding females. PLoS One 14:e0210266

Niemi M (2013) Behavioural ecology of the Saimaa ringed seal–implications for conservation. PhD thesis, University of Eastern Finland, Joensuu, Finland

Noongwook G (2000) Native observations of local climate change around St. Lawrence Island. In: Huntington HP (ed) Impacts of changes in sea ice and other environmental parameters in the Arctic. Marine Mammal Commission, Bethesda, MD, pp 21–24

Nordqvist O (1883) Anteckningar och studier till Sibiriska Ishafskustens däggdjursfauna. In: Nordenskiöld AE (ed) Vega-Expeditionens Vetenskapliga Isakttagelser, vol. 2. F. & G. Beijers, Stockholm, pp 61–117

Nyman T, Valtonen M, Aspi J, Ruokonen M, Kunnasranta M, Palo JU (2014) Demographic histories and genetic diversities of Fennoscandian marine and landlocked ringed seal subspecies. Ecol Evol 4:3420–3434
Ognev SI (1935). Mammals of the U.S.S.R. and adjacent countries, vol. 1. Carnivora Glavpushnina NKVT, Moscow
Oksanen SM, Niemi M, Ahola MP, Kunnasranta M (2015) Identifying foraging habitats of Baltic ringed seals using movement data. Mov Ecol 3:33. https://doi.org/10.1186/s40462-015-0058-1
Palo JU, Väinölä R (2006) The enigma of the landlocked Baikal and Caspian seals addressed through phylogeny of phocine mitochondrial sequences. Biol J Linn Soc 88:61–72
Pilfold NW, Derocher AE, Stirling I, Richardson E, Andriashek D (2012) Age and sex composition of seals killed by Polar bears in the Eastern Beaufort sea. PLoS One 7:e41429
Poulter TC (1963) Sonar signals of the sea lion. Science 139:753–755
Preuß A, Gansloßer U, Purschke G, Magiera U (2009) Bear-hybrids: behaviour and phenotype. Zoologische 78:204–220
Pungowiyi C (2000) Native observations of change in the marine environment of the Bering Strait Region. In: Impacts of changes in sea ice and other environmental parameters in the Arctic. Final report of the Marine Mammal Commission Workshop Girdwood, Alaska, p 98
Rautio A, Niemi M, Kunnasranta M, Holopainen IJ, Hyvärinen H (2009) Vocal repertoire of the Saimaa ringed seal (*Phoca hispida saimensis*) during the breeding season. Mar Mammal Sci 25:920–930
Reeves RR (1998) Distribution, abundance and biology of ringed seals (*Phoca hispida*): an overview. In: Heide-Jorgensen MP, Lydersen C (eds) Ringed seals in the North Atlantic. NAMMCO Scientific Publications, Tromsø, Norway, pp 9–45
Renouf D, Davis MB (1982) Evidence that seals may use echolocation. Nature 300:635–637
Rice DW (1998) Marine mammals of the world: systematics and distribution. Society for Marine Mammology, Lawrence, KS
Richardson WJ, Greene CR Jr, Malme CI, Thomson DH (1995) Marine mammals and noise. Academic Press, San Diego
Ryg M, Smith TG, Oritsland NA (1990) Seasonal changes in body mass and body composition of ringed seals (*Phoca hispida*) on Svalbard. Can J Zool 68:470–475
Ryg M, Solberg Y, Lydersen C, Smith TG (1992) The scent of rutting male ringed seals (*Phoca hispida*). J Zool 226:681–689
Ryg M, Øritsland NA (1991) Estimates of energy expenditure and energy consumption of ringed seals (Phoca hispida) throughout year. In: Sakshaug E, Hopkins CCE, Øritsland NA (eds.) Proceedings of the pro mare symposium on polar marine ecology, Trondheim, 12–16. Polar Research 10(2):595
Scheffer VB (1958) Seals, sea lions and walruses: a review of the Pinnipedia. Stanford University Press, Stanford, California
Schusterman RJ, Kastak D, Levenson DH, Reichmuth CJ, Southall BL (2000) Why pinnipeds don't echolocate. J Acoust Soc Am 107:2256–2264
Shapiro LH, Metzner RE, Toovak K (1979) Historical references to ice conditions along the Beaufort Sea coast of Alaska. In: Environmental Assessment of the Alaskan Continental Shelf. Annual Reports of Principal Invesitoators for the year ending March 1979. Volume 9. Hazards. Boulder, CO, U.S. Department of Commerce, NOAA and U.S. Department of Interior, Bureau of Land Management, pp 632-682
Silliman BR, Hughes BB, Gaskins LC, He Q, Tinker MT, Read A, Nifong J, Stepp R (2018) Are the ghosts of nature's past haunting ecology today? Curr Biol 28:R532–R537
Simpkins MA, Kelly BP, Wartzok D (2001a) Three-dimensional diving behaviors of ringed seals (*Phoca hispida*). Mar Mammal Sci 17:909–925
Simpkins MA, Kelly BP, Wartzok D (2001b) Three-dimensional analysis of search behaviour by ringed seals. Anim Behav 62:67–72

Sipilä T, Medvedev NV, Hyvärinen H (1996) The Ladoga seal (Phoca hispida ladogensis Nordq.). In: Simola H, Viljanen M, Slepukhina T, Murthy R (Eds) The first international Lake Ladoga symposium. Dordrecht, Kluwer Academic Publishers, 193–198

Sipilä T, Hyvärinen H (1998) Status and biology of Saimaa (*Phoca hispida saimensis*) and Ladoga (*Phoca hispida ladogensis*) ringed seals. NAMMCO Sci Publ 1:83–99

Sipilä T (2003). Conservation biology of Saimaa ringed seal (*Phoca hispida saimensis*) with reference to other European seal populations. Dissertation, University of Helsinki

Smith TG (1973) Population dynamics of the ringed seal in the Canadian eastern Arctic. Bull Fish Res Board Canada 181:55

Smith TG (1976) Predation of ringed seal pups (*Phoca hispida*) by the arctic fox (*Alopex lagopus*). Can J Zool 54:1610–1616

Smith TG (1980) Polar bear predation of ringed and bearded seals in the land-fast sea ice habitat. Can J Zool 58:2201–2209

Smith TG (1987) The ringed seals, *Phoca hispida*, of the Canadian western arctic. Can Bull Fish Aquat Sci 216:1–81

Smith TG (2009) Jimmy Memorana (1919–2009). Arctic 62:489–490

Smith TG, Hammill MO (1981) Ecology of the ringed seal, *Phoca hispida*, in its fast ice breeding habitat. Can J Zool 59:966–981

Smith TG, Harwood LA (2001) Observations of neonate ringed seals, *Phoca hispida*, after early break-up of the sea ice in Prince Albert Sound, Northwest Territories, Canada, spring 1998. Polar Biol 24:215–219

Smith TG, Stirling I (1975) The breeding habitat of the ringed seal (*Phoca hispida*). The birth lair and associated structures. Can J Zool 53:1297–1305

Smith TG, Stirling I (1978) Variation in the density of ringed seal (*Phoca hispida*) birth lairs in the Amundsen Gulf, Northwest Territories. Can J Zool 56:1066–1070

Smith TG, Siniff DB, Reichle R, Stone S (1981) Coordinated behavior of killer whales, *Orcinus orca*, hunting a crabeater seal, *Lobodon carcinophagus*. Can J Zool 59:1185–1189

Smith TG, Hammill MO, Taugbol G (1991) A review of the developmental, behavioural and physiological adaptations of the ringed seal, *Phoca hispida*, to life in the Arctic winter. Arctic 44:124–131

Smodlaka H, Henry RW (2014) Gross anatomy of the ringed seal (*Pusa hispida*) gastro-intestinal tract. J Vet Med Ser C Anat Histol Embryol 43:230–238

Smodlaka H, Reed RB, Henry RW (2006) Microscopic anatomy of the ringed seal (*Phoca hispida*) lower respiratory tract. J Vet Med Ser C Anat Histol Embryol 35:35–41

Stephens RJ, Beebe IJ, Poulter TC (1973) Innervation of the vibrissae of the California sea lion, *Zalophus californianus*. Anat Rec 176:421–441

Stirling I (1973) Vocalization in the ringed seal (*Phoca hispida*). J Fish Res Board Can 30:1592–1594

Stirling I (1974) Midsummer observations on the behavior of wild polar bears (*Ursus maritimus*). Can J Zool 52:1191–1198

Stirling I (1975) Factors affecting the evolution of social behaviour in the pinnipedia. J. Int. Counc. Explor. Sea 169:205–212

Stirling I (1997) The importance of polynyas, ice edges, and leads to marine mammals and birds. J Mar Syst 10:9–21

Stirling I, Latour PB (1978) Comparative hunting abilities of polar bear cubs of different ages. Can J Zool 56:1768–1772

Stirling I, Smith TG (2004) Implications of warm temperatures and an unusual rain event for the survival of ringed seals on the coast of Southeastern Baffin Island. Arctic 57(1):59–67

Stirling I, van Meurs R (2015) Longest recorded underwater dive by a polar bear. Polar Biol 38:1301–1304

Stirling I, Calvert W, Cleator H (1983) Underwater vocalizations as a tool for studying the distribution and relative abundance of wintering pinnipeds in the High Arctic. Arctic 36:262–274

Stirling I, Spencer C, Andriashek D (2016) Behavior and activity budgets of wild breeding polar bears (*Ursus maritimus*). Mar Mammal Sci 32:13–37

Stirling I (1983) The evolution of mating systems in pinnipeds. In: Eisenberg JF, Kleiman DG (eds) Advances in the study of mammalian behavior. Am Soc Mammalog 7:489–527

Taugbøl G (1982). Ringed seal thermoregulation, energy balance and development in early life, a study on *Pusa hispida* in Kongsfj. University of Oslo, Svalbard.

Teilmann J, Born EW, Acquarone M (1999) Behaviour of ringed seals tagged with satellite transmitters in the North Water polynya during fast-ice formation. Can J Zool 77:1934–1946

Thompson D, Fedak MA (1993) Cardiac responses of grey seals during diving at sea. J Exp Biol 174:139–154

Thompson D, Hammond PS, Nicholas KS, Fedak MA (1991) Movements, diving and foraging behaviour of grey seals (*Halichoerus grypus*). J Zool 224:223–232

Trukhanova IS, Grachev AI, Somov AG, Burkanov VN, Laidre KL, Boveng PL (2017) The commercial harvest of ice-associated seals in the Sea of Okhotsk, 1972–1994. PLoS One 12:e0182725

Trukhin AM (2000). Ringed seal on the eastern coast of Sakhalin Island. In: Marine Mammals of the Holarctic. Archangel, Russia, Pravda Severa, p 4

Tynan CT, DeMaster DP (1997) Observations and predictions of Arctic climatic change: potential effects on marine mammals. Arctic 50:308–322

Vibe C (1950) The marine mammals and the marine fauna in the Thule District (Northwest Greenland) with observations on ice conditions in 1939-41. In: Meddelelser om Grønland, Bind 150. Copenhagen, Denmark, C.A. Reitzels Forlag, pp 1–115

Vibe C (1967) Arctic animals in relation to climate fluctuations, the Danish zoogeographical investigations in Greenland. Bergen, Norway, Meddelelser om Grønland

Visser IN, Smith TG, Bullock ID, Green GD, Carlsson OGL, Imberti S (2008) Antarctic peninsula killer whales (*Orcinus orca*) hunt seals and a penguin on floating ice. Mar Mamm Sci 24:225–234

Von Duyke AL, Douglas DC, Herreman JK, Crawford JA (2020) Ringed seal (*Pusa hispida*) seasonal movements, diving, and haul-out behavior in the Beaufort, Chukchi, and Bering Seas (2011–2017). Ecol Evol 10(12):5595–5616

Wartzok D, Shusterman RJ, Gailey-Phipps J (1984) Seal echolocation? Nature 308:1753

Wartzok D, Elsner R, Stone H, Kelly BP, Davis RW (1992a) Under-ice movements and the sensory basis of hole finding by ringed and Weddell seals. Can J Zool 70:1712–1722

Wartzok D, Saygegh S, Stone H, Barchak J, Barnes W (1992b) Acoustic tracking system for monitoring under-ice movements of polar seals. J Acoust Soc Am 92:682–687

Wassmann P, Duarte CM, Agustí S, Sejr MK (2011) Footprints of climate change in the Arctic marine ecosystem. Glob Change Biol 17:1235–1249. https://doi.org/10.1111/j.1365-2486.2010.02311.x

Welch HE (1992) Energy flow through the marine ecosystem of the Lancaster Sound region, Arctic Canada. Arctic 45:343–357

Welch HE, Crawford RE, Hop H (1993) Occurrence of Arctic cod (*Boreogadus saida*) schools and their vulnerability to predation in the Canadian High Arctic. Arctic 46:331–339

Wenzel G (1984) Archaeological evidence of prehistoric Inuit use of the sea-ice environment. In: Cooke A, Alstine EV (eds) Sikumiut: The people who use the sea ice. Canadian Arctic Resources Committee, Ottowa, pp 41–59

Weslawski JM, Ryg M, Smith TG, Oritsland NA (1994) Diet of ringed seals (*Phoca hispida*) in a fjord of west Svalbard. Arctic 47:109–114

Young BG, Ferguson SH (2013) Seasons of the ringed seal: Pelagic open-water hyperphagy, benthic feeding over winter and spring fasting during molt. Wildl Res 40:52–60

Yurkowski DJ, Semeniuk CAD, Harwood LA, Rosing-Asvid A, Dietz R, Brown TM, Clackett S, Grgicak-Mannion A, Fisk AT, Ferguson SH (2016) Influence of sea ice phenology on the movement ecology of ringed seals across their latitudinal range. Mar Ecol Prog Ser 562:237–250

Chapter 16
The Hawaiian Monk Seal: Ethology Applied to Endangered Species Conservation and Recovery

Stacie Robinson, Michelle Barbieri, and Thea Johanos

Contents

Abstract The Hawaiian monk seal (*Neomonachus schauinslandi*) is among the world's rarest pinnipeds, with a population abundance of around 1,400 individuals. This species fills a unique ecological niche as the only tropical phocid. Since their listing as depleted under the Marine Mammal Protection Act and endangered under the Endangered Species Act, there has been substantial effort to recover the species. Ethological research has played an essential role in the conservation efforts for Hawaiian monk seals, contributing to our understanding of the seals' interactions with their environments and the threats they face. In some cases, behavioral patterns have increased monk seals' risk of harm, such as foraging in nearshore waters, increasing the potential for interactions with fishing gear. In other instances, management agencies have been able to exploit behavioral characteristics to design successful conservation interventions. For example, given their limited tendency for dispersal, Hawaiian monk seal pups can be safely translocated to areas that increase their chance of survival. This chapter reviews over four decades of research dedicated to Hawaiian monk seals, with particular attention to the ethological insights revealed in areas of reproductive behavior, social behavior, foraging behavior, and

S. Robinson (✉) · M. Barbieri · T. Johanos
Pacific Islands Fisheries Science Center, National Oceanographic and Atmospheric Administration, National Marine Fisheries Service, Hawaiian Monk Seal Research Program, Honolulu, HI, USA
e-mail: stacie.robinson@noaa.gov

D. P. Costa and E. McHuron (eds.), *Ethology and Behavioral Ecology of Phocids*, Ethology and Behavioral Ecology of Marine Mammals,
https://doi.org/10.1007/978-3-030-88923-4_16

dispersal behavior. We highlight the conservation efforts ranging from animal rescues to translocations that have relied upon this information to work toward the recovery of this endangered seal.

Keywords Hawaiian monk seal · Conservation · Reproductive behavior · Social behavior · Foraging behavior · Dispersal behavior

16.1 Overview and Conservation Status

Hawaiian monk seals (*Neomonachus schauinslandi*) are part of an ancient phocid lineage that includes the Mediterranean monk seal (*Monachus monachus*; endangered) and the Caribbean monk seal (*Neomonachus tropicalis;* extinct, last seen in 1952). It is estimated that the Hawaiian monk seal diverged from Caribbean seals between 3–11 million years ago after crossing over the submerged Isthmus of Panama into the Pacific Ocean (Fyler et al. 2005; LeDuc 2009). While the bulk of this chapter will focus on Hawaiian monk seals, we have included information about the Mediterranean monk seal in Appendix 16.7.

Hawaiian monk seals range throughout the Hawaiian Archipelago and are characterized as a metapopulation with semi-isolated subpopulations distributed among a number of islands and atolls (Antonelis et al. 2006). Across this >2,500 km archipelago, there is considerable variation in breeding sites inhabited by monk seals (Fig. 16.1). Six islands/atolls (Kure Atoll to French Frigate Shoals) in the Northwestern Hawaiian Islands (NWHI) comprise the most-studied breeding sites for Hawaiian monk seals. The largest of the NWHI landmasses, Laysan Island, has only 8 km of shoreline, whereas many of the atolls host several small islets totaling less than 1 km^2 of land area. While the land area is small, these sites are surrounded by large lagoons and fringing reefs, and a series of submerged banks and seamounts extends the underwater habitat for monk seals. At the southeastern end of the NWHI lie Mokumanamana (Necker) Island and Nihoa Island, volcanic remnants with steep cliffs and small beach areas consisting of rock shelves and minimal sand beaches. These islands form something of a nexus between the primary breeding sites in the NWHI and the main Hawaiian Islands. The main Hawaiian Islands (MHI), the eight islands populated by humans, are large islands (50–430 km shoreline) with mountainous terrestrial terrain, beaches of sand and rock, and considerable human density (just under 1,000,000 people on O'ahu; 2010 US Census).

Monk seals likely inhabited the entire Hawaiian Archipelago prior to human settlement, with the population in the MHI being extirpated shortly after the arrival of Polynesians (~800–1200 CE). Archaeological evidence indicates that monk seals were present in the MHI prior to European contact in 1778, though their limited role in Hawaiian culture or legends suggests they were rare in the populated islands (Rosendahl 1994). Newspapers reference occasional monk seal sightings on the populated islands throughout the 1800–1900s (Watson et al. 2011). Commercial sealing expeditions reduced the remainder of the Hawaiian monk seal population in

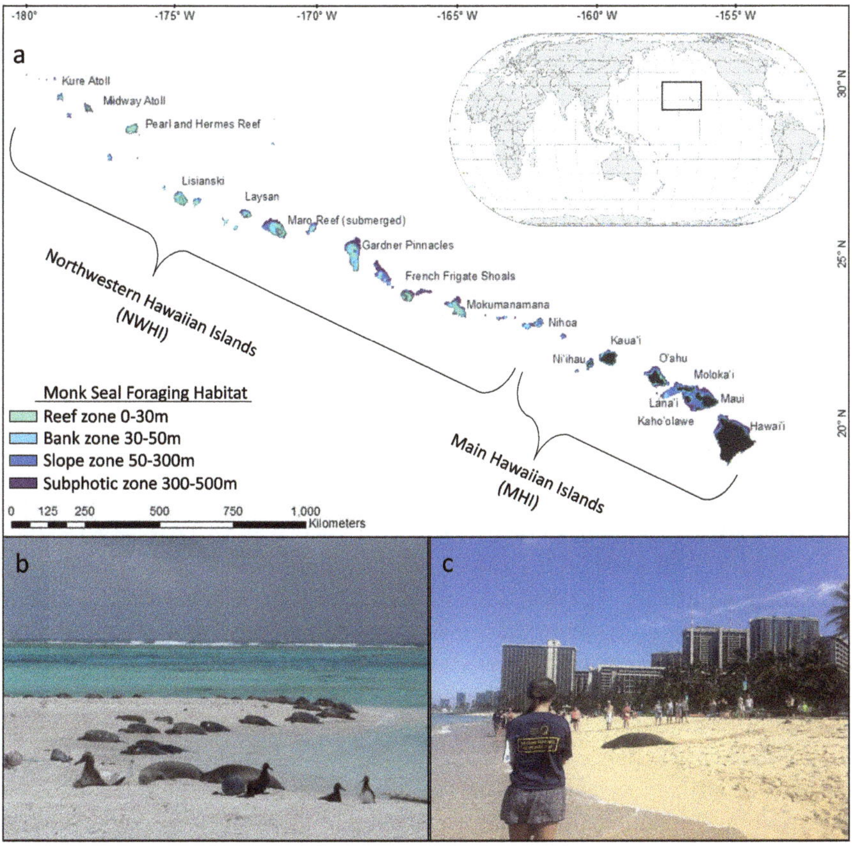

Fig. 16.1 The Hawaiian Archipelago constituting the range of the Hawaiian monk seal. **a** Colors indicate habitat in depth zones utilized by foraging monk seals. Photographs illustrate the contrasting haulout habitat in the Northwestern (**a**) versus Main Hawaiian Islands (**c**)

the NWHI to near extinction in the mid-1800s, and guano miners, bird hunters, and whalers further depleted the population by the early 1900s (Cobb 1902; Kenyon and Rice 1959).

16.1.1 Conservation Status and Threats

The first formal (aerial) surveys of the Hawaiian monk seal population counted a minimum of 1,013 seals on land in the summer of 1957 (Kenyon and Rice 1959); this likely represented a population of closer to 3,000 seals (accounting for haulout correction factors later developed; Harting et al. 2017). Follow-up surveys in the 1960s and 1970s suggested a 50% decline in abundance since the 1950s (Kenyon

1973; DeLong et al. 1976; Johnson et al. 1982). This decline led Hawaiian monk seals to be the first species designated *depleted* under the Marine Mammal Protection Act in 1976, as well as *endangered* under the Endangered Species Act. In more recent years, monk seal surveys have been standardized and conducted annually, revealing subtler variation in abundance trends over time. Population counts in the NWHI declined 4.2% per year from 1985–1993; this decline slowed to 1.9% per year from 1993–2003 (Carretta et al. 2004).

Meanwhile, monk seal sightings in the MHI remained relatively rare until the mid-1990s, but rebounded considerably in the early 2000s (Baker and Johanos 2004). By 2013, the population could be estimated across the entire archipelago, and was showing signs of a stabilizing trend (Baker et al. 2016b). Recent data have shown promising signs of positive trends, with the range-wide population increasing at a rate of 2% per year from 2013–2019 (NMFS unpublished). In light of this positive outlook and decisive management actions, Hawaiian monk seals were moved from *critically endangered* to *endangered* on the IUCN Red List in 2015 (IUCN 2015). The population in 2019 was just over 1,400 individuals with about 1,100 seals residing in the remote NWHI and about 300 in the MHI (Hawaiian Monk Seal Research Program 2019). Within the overall positive trend, local demographics and age-specific survival rates continue to vary between sites, with both juvenile survival and population growth rates generally lower in the NWHI than MHI (Baker et al. 2011b).

As with localized population trends, particular threats vary for monk seals across their range. Poor juvenile survival associated with prey limitation has been suggested as the primary proximate cause for the decline of monk seals in the NWHI in recent decades (Craig and Ragen 1999). Other significant threats in the NWHI include entanglement in marine debris (Henderson 2001), male aggression leading to female injury and death (particularly at Laysan Island; Hiruki et al. 1993b, Johanos et al. 2010), shark predation on seal pups (particularly at French Frigate Shoals; Gobush and Farry 2012), and the threat of island disappearance with sea-level rise (Baker et al. 2006). Meanwhile, threats in the MHI tend to be more anthropogenic in nature, with intentional seal killings, direct interactions with fisheries (hook ingestion or net entanglement), and disease (particularly infection with the parasite *Toxoplasma gondii,* spread by cats) comprising the leading causes of death in the MHI (Harting et al. 2020). Human disturbance and pollution also pose threats in the MHI (Baker et al. 2011b; Lopez et al. 2014). The current Recovery Plan for Hawaiian Monk Seals (a conservation planning document required for all ESA-listed species; National Marine Fisheries Service 2007) prioritizes activities to address threats throughout the monk seal range: (1) improve survivorship of females of all ages, particularly juveniles and yearlings in the NWHI, (2) maintain or expand existing field efforts, (3) ensure natural recovery of the Hawaiian monk seal in the MHI, and (4) reduce the probability of inadvertent introduction of infectious diseases into the Hawaiian monk seal population.

16.1.2 Modes of Research and Ethological Observation

Monk seals spend approximately one-third of their time hauled out on land for parturition, nursing, molting, and resting (Cahoon 2011; Harting et al. 2017, Wilson et al. 2017a). This time on land offers valuable opportunities for the observation that forms the backbone of population monitoring and ethological data (Fig. 16.2). The National Marine Fisheries Service Hawaiian Monk Seal Research Program (NMFS, HMSRP) began consistently tagging weaned monk seal pups in the 1980s (Antonelis et al. 2006). These tagged pups are resighted throughout their lifetimes using the applied flipper tags as well as natural markings or other identifiable characteristics documented in a digital photography database (Harting et al. 2004). Resighting efforts and methods vary across the monk seal range. In the remote NWHI, seals are surveyed over the course of 2–5-month field camps (established nearly every year since 1982; Baker and Johanos 2004; Johanos et al. 2014). Mokumanamana and Nihoa Islands are typically surveyed 1–4 times per year in conjunction with research missions to the NWHI. Unlike in the NWHI, where a relatively large number of seals are concentrated on small islands, in the MHI a small number of seals are distributed over extensive (and often inaccessible) coastlines. Thus, seals in the MHI are monitored by networks of volunteers and public seal watchers as well as biologists, yielding year-round coverage that provides a dataset sufficient to determine a minimum abundance in this region. Data from all sites and site-specific survey methods are combined for estimating population abundance and vital rates throughout the full Hawaiian monk seal range (Baker et al. 2016b).

In addition to tracking population trends, field observations are a primary means of collecting ethological data on monk seals. The timing of field camps is aligned to maximize the observation of pupping and breeding patterns to understand reproductive biology (Johanos et al. 1994). Throughout field camps in the NWHI, and some focused studies in the MHI, animal behavior and social interactions are observed at haulout areas (Johanos et al. 2010; Robinson et al. 2018). Many aspects of seal behavior require observation at sea. To understand diving and foraging behavior, we have employed a number of different technologies such as time-depth-recorders (DeLong et al. 1984), satellite and GPS tags (Stewart et al. 2006; Cahoon 2011), and seal-borne video cameras (Parrish and Littnan 2007; Wilson et al. 2017a).

Fig. 16.2 A biologist sneaks around a tiny islet at French Frigate Shoals in the Northwestern Hawaiian Islands, carefully collecting survey data without disturbing the resting seals

An understanding of Hawaiian monk seals' behavior in the context of their ecology and survival threats has allowed the NMFS HMSRP to design conservation interventions aimed at enhancing the survival of individuals and ongoing recovery of the species. The cumulative impact of these intervention efforts has been one of the cornerstones of progress in Hawaiian monk seal recovery, with up to 32% of the population comprised of seals benefitting from such interventions (Harting et al. 2014). Throughout this chapter, we review the science related to Hawaiian monk seal behavior, with a particular focus on highlighting how insights from ethological research have informed conservation actions that benefit monk seals.

16.2 Reproductive Behavior—Influences on Reproduction and Survival

16.2.1 Reproductive Biology

Hawaiian monk seals typically give birth on beaches near shallow water that offers protection from rough seas and sharks. Females may pup at sites different from their most-used haulout locations, and typically show up to their pupping site immediately prior to pupping (Johanos et al. 1994). Monk seals exhibit asynchronous mating (Kenyon and Rice 1959; Johanos et al. 1994), a likely result of their tropical-subtropical environment lessening some of the pressures selecting for synchronous breeding in many other pinniped species. Resources are consistently available to support gestating females or newly weaned seals throughout the year, and seasonal thermal gradients are less extreme than in higher latitudes. While births have been observed throughout all months of the year, there is a (protracted) seasonality to monk seal breeding (Kenyon and Rice 1959; Johanos et al. 1994). The pupping season is concentrated from late February through early June in the NWHI (Johanos et al. 1994), and April–July in the MHI (Robinson et al. 2020). A similarly protracted breeding season occurs in the most southern population of Mediterranean monk seals (Appendix 16.7).

As capital breeders (Chap. 8), female Hawaiian monk seals remain hauled out or near shore with the pup and fast throughout the ~ 40 day nursing period, providing nutrients for pups to gain 50–80 kg (approximately tripling their birth mass). This is a contrast to Mediterranean monk seals, with females nursing for longer duration and leaving periodically to feed (Appendix 16.7). When the Hawaiian monk seal female's resources are depleted, weaning occurs abruptly as the female leaves the pup, swimming offshore to forage. Females in the MHI exhibit longer nursing periods than those in the NWHI (43.7 vs. 39.0 days in the NWHI, 12% longer nursing for MHI pups; Johanos et al. 1994; Robinson et al. 2020). Despite this longer nursing period, MHI females regain body condition more quickly and are able to undergo resource-intensive molting sooner after weaning than their NWHI counterparts (54 days post-weaning /98 days post-pupping vs. 66/105 days for northwestern females; Robinson et al. 2020). This increase in nursing period likely provides a significant nutritional benefit to MHI pups, consistent with reports of larger size-at-weaning and accelerated post-weaning growth in MHI versus NWHI pups (Baker and Johanos 2004; Baker et al. 2014).

Critical factors of the reproductive cycle, such as estrus and gestation, are less flexible than nursing and molting intervals. The mean inter-birth interval of 381–382 days has been recorded with remarkable consistency across studies spanning great geographic and temporal range (NWHI; Wirtz 1968, Johanos et al. 1994, MHI; Robinson et al. 2020). The consistency of inter-birth intervals suggests that polyestrous cycling described in a single captive monk seal (Pietraszek and Atkinson 1994) is not typical in wild seals. Estrus is believed to occur after weaning a pup and before molting (or ~ 1.5 month before molting for females without a pup in a given year). Because mating activity occurs offshore and is rarely observed, the timing of mating injuries (detailed in Sect. 16.2.3) is a critical factor in inferring time of estrus. Females hauled out on shore are typically attended by a male during the interval between weaning a pup and molting (Johanos et al. 1994). Injuries were observed an average of 26 days after the end of maternal investment, with the rate of injury decreasing as females approached molt, suggesting a consistent relationship with termination of nursing and onset of the next estrous cycle (Johanos et al. 1994).

Reproductive output varies across breeding sites and among individuals. Site-specific reproductive rates closely follow body condition and trends in growth, with age at primiparity (first pupping) aligning with the age at which females reach adult size (Harting et al. 2007). Females in the MHI tend to have robust body condition and pup at a younger age than NWHI females (Baker and Johanos 2004). The reproductive rate for MHI seals exceeds 0.70 for all ages 8–18 years (after accounting for unobserved pups; Robinson et al. 2020). Among closely studied NWHI sites, Laysan Island has the highest reproductive rate (0.73) and earliest mean age at primiparity among NWHI sites (7.4 years), whereas French Frigate Shoals females produce their first pups at an older age (~10 years) and exhibit lower reproductive rates (0.61). Prime reproductive years occur from ages 10–15, with reproductive senescence becoming evident around 15–20 years of age (Harting et al. 2007). On an individual level, some females consistently maintain either higher or lower reproductive rates throughout their lives, but we have found no relationship between reproductive investment and female survival (Harting et al. 2007).

16.2.2 Mom and Pup Interactions

Throughout the nursing period, mothers and pups spend time nursing and resting while hauled out and swim together in shallow waters near the birth site (Fig. 16.3). Swimming forays typically do not range far or deep, and there is no evidence of mother–pup pairs foraging together. Mothers and their offspring do not appear to maintain close interactions after weaning (NMFS HMSRP unpublished database).

Fig. 16.3 On a popular pupping beach in the Northwestern Hawaiian Islands, several mothers (larger grey seals)–pup (smaller black seals) pairs are nursing, resting, or swimming nearshore

Nursing female monk seals tend to be particularly asocial. Female monk seals occur in low density compared to most land-breeding seals, with densities as low as 1.5 females per 1000 m^2 on East Island, a small sand islet within French Frigate Shoals, and just 0.5 females per 1000 m^2 on Laysan, the largest of the NWHI sites (Boness et al. 1998). Even at such low densities, animals still showed avoidance of neighbors with daily nearest neighbor distances (mean 27.0 m; range 14.2–52.6 m) greater than expected by chance (Boness et al. 1998). Nursing females are intolerant when approached by other females sharing a nursing beach or males seeking attention (Johanos et al. 1994). When aggressive interactions occur between neighboring nursing mothers, it is common for the pups to be mixed up in the melee, often leading to switches in the mother–pup pairs (Boness 1990).

Fostering is common in Hawaiian monk seals and is described in a number of studies (Alcorn and Henderson 1984; Boness 1990; Johanos et al. 1994; Boness et al. 1998). While females begin nursing their biological pup after birth, many mothers foster one or more pups (in sequence, not simultaneously) and may not reunite and resume nursing their biological pup. An early study found fostering rates as high as 53% (Laysan Island) to 90% (East Island at French Frigate Shoals) (Boness et al. 1998). The average female spent 34% (range 1–91%) of her nursing period fostering an average of 2.3 pups (max observed 5; Boness 1990). While the most common cause of fostering is pup-switching between females, pups may also seek out a foster mother to nurse within a few days after being weaned. One extreme example of this is a pup, who after being weaned by its mother, proceeded to nurse from a foster mother that had lost her pup early, finally weaning after a total of 68 days of nursing from two mothers (Alcorn and Henderson 1984). Fostering is well documented in some pinniped species including Mediterranean monk seals (Appendix 16.7) and more distantly related species (elephant seals; Riedman and Le Boeuf 1982, harbor seals; Schaeff et al. 1999).

The frequency of pup-switching and fostering in Hawaiian monk seals calls into question the monk seal's ability to recognize individuals or distinguish close kin such as mothers or pups. Research on vocal recognition is inconclusive. An early study found that, despite significant differences in vocal attributes among pups, discriminant function analysis failed to distinguish individuals based on vocalizations (Job et al. 1995). Playback experiments in the field also indicated that females did not discriminate between the voices of their own pup versus another, leading to the conclusion that females were unable to recognize their pups by voice (Job et al. 1995). An ongoing study has found stronger distinction between vocal attributes, especially the "bah" and "mah" calls used by pups (Chaudun 2018). This study found that their discriminant function correctly classified 54% of individual calls, suggesting a better-than-random chance of vocally distinguishing individuals (Chaudun 2018). Further research may determine whether this discriminant ability is realized in the wild.

Because there is a strong relationship between girth at weaning and survival (Baker 2008), there can be winners and losers in a pup-switch. If pups of substantially different ages switch, the older pup may get an extended nursing period from the mother with more days left to nurse, while the younger pup might be weaned early by the mother whose resources are more depleted. A prematurely weaned pup that does not attain sufficient girth stands a poor chance of survival (Baker 2008). Given the importance of females in maximizing the reproductive potential of the population, it has been the policy of the HMSRP to intervene in cases where a female pup is disadvantaged by a male pup in a pup-switch (i.e., the female pup ends up with the mother with less milk left to give). Field biologists remove the switched pups from their respective foster mothers and place them back next to their original mothers. There may also be cases in which a mother and pup become separated, but no other nursing mother fosters the pup. In these cases, field biologists locate the mother and place the pup in her proximity. In many cases nursing resumes and the pup weans normally, but in some cases a mother may repeatedly reject her pup.

16.2.3 Male and Female Interactions

Hawaiian monk seals are polygamous, and mating occurs throughout the year. As a result, the breeding and pupping aggregations common in other pinniped species are not seen in monk seals. Males do not maintain specific breeding territories nor monopolize particular females over the long term, but they may haulout next to available females (presumably to attain a beneficial position to follow her to sea for mating), and spar with other males that approach. Mating occurs at sea and is rarely observed, but occasional observations have documented males holding a female with teeth and foreflippers in an attempt to mount her (Wirtz 1968; Hiruki et al. 1993a; Johanos et al. 1994). In some cases, termed multiple male aggression, more than one male may attempt to mount a single female (or an immature seal of either sex; Fig. 16.4a). Multiple male aggression events observed in nearshore waters have involved from 2–32 males, and have lasted from moments to >8 h (Johanos et al. 2010). Mating typically leaves females with injuries that can range from minor scratches or punctures to subcutaneous hemorrhage, or severe gaping wounds covering much of the dorsum (Fig. 16.4b), with the more severe injuries being associated with multiple male aggression incidents (Hiruki et al. 1993a). Wounds associated with mating can be an indicator of breeding activity, or an essential indicator of imbalances in social structure that pose severe risks to female survival.

Multiple male aggression reached particularly dangerous levels on Laysan Island in the 1980s. Approximately 70% of adult females on Laysan exhibited scars associated with injuries from multiple male aggression, and, over the same time, 45%

Fig. 16.4 Several males try to mount a single female in incidents of multiple male aggression. **a** These events can leave females with serious injuries **b** that can lead to death

of adult females at French Frigate Shoals had such scars (Hiruki et al. 1993a). The seasonal timing when females appear with severe wounds is similar to when females appear with more minor mating injuries (e.g., scratches), suggesting that both are associated with higher male attraction during estrus (Johanos et al. 1994). A study of hormone levels in captive male monk seals suggests that testosterone varies seasonally (peaking in summer, and declining when animals approach molt in the fall), which would support heightened male aggression during seasonal breeding (Atkinson and Gilmarti 1992). While seeking mating opportunities appears to be the impetus of multiple male aggression, animals other than adult females can become targets. During the peak years of aggression at Laysan, while 70% of observed mounting injuries were incurred by adult females, the other 30% were incurred by males or immature females for which estrus could not have been a factor (Johanos et al. 1994).

Injurious mating aggression has been documented in other phocids (Le Boeuf and Mesnick 1991; Mesnick and Le Boeuf 1991; Rose et al. 1991), but it seems to be particularly prevalent and problematic in Hawaiian monk seals. This could be due to differences in mating systems and female availability, or impacts may be exacerbated by small population sizes. The severity and prevalence of injuries from multiple male aggression has the potential to substantially impact the monk seal population by removing breeding females. During a peak period of aggressive activity on Laysan (1983–89), 87% of adult females that died had been seriously injured (Hiruki et al. 1993b). Post mortem examinations confirmed that the majority of females dying after male-inflicted injury were in estrus (Atkinson et al. 1994). Injuries that occur shortly prior to pupping or during nursing also have the potential to impact survival of the female's pup (Hiruki et al. 1993b). Females injured early in their breeding cycle typically had sufficient time to heal and regain body condition prior to pupping so that, provided mortality was avoided, wounding had little impact on female reproductive rate, nursing period, or pup survival (Hiruki et al. 1993b).

While monk seals do not have a rigid dominance hierarchy, males observed in aggressive interactions were likely to exhibit more subdominant behaviors. They were less likely to be sighted regularly hauled out, less likely to be sighted hauled out in proximity to a female, more likely to be sighted cruising (swimming along the shoreline, presumably looking for a female), and more likely to be sighted throughout many island sectors (wider/more varied space use). Only about 10–20% of males were classified as dominant (spending more time hauled out on land, and in proximity to adult females) in a given year. Given the association of multiple male aggression with male-biased sex ratios, it is likely that dominant males become overwhelmed by the more numerous competitors. During the peak period of aggression on Laysan Island, male:female ratios were ~2:1, but could be locally elevated due to animal haulout preferences. Space use differed between dominant and subdominant males, with dominant males hauling out primarily on the northeastern side of Laysan where adult females concentrated. Meanwhile, nondominant males had a more generalized distribution spanning northeast to southwest. Thus, the southwestern side of the island developed both an extreme male-biased sex ratio (as much as 25:1), and a high concentration of subdominant males. Females hauling out on these southwestern beaches were most likely to be observed with severe injuries (Johanos et al. 2010).

16.2.4 Conservation Interventions to Mitigate Threats of Male Aggression

Long-lasting population impacts of losing breeding females provided a strong impetus for human managers to mitigate the threat of female mortality from mounting injuries. One early avenue that was explored was the treatment of males with gonadotrophin-releasing hormone agonist to lower testosterone levels, and thus aggression. Atkinson et al. (1993) showed that testosterone was suppressed for 7–8 weeks after treatment, and recommended treatment at 2-month intervals to control male aggression, which was unrealistic in the field setting for wild populations. The association of multiple male aggression incidence with male-biased sex ratios and over-abundance of subdominant males led to the suggestion that adjustment of sex ratios to re-balance social structures within the population could provide a solution. Simulation models that evaluated different management scenarios according to outcomes for overall population size, number of breeding females, and sex ratio supported the removal of males to balance sex ratios (Starfield et al. 1995). Behavioral profiles were developed for males on Laysan Island to help identify males for removal that would (1) maximize the benefit of mitigating harmful injuries to females, and (2) be less disruptive to the social structure on the island (Johanos et al. 2010). From 1984–1994, 37 males were removed from Laysan Island (translocated either to Johnston Atoll or the main Hawaiian Islands), reducing the male sex ratio from 2.1 to 0.9. Following these translocations, female mortality decreased from 4.1% to 0.3% and non-lethal injury rates decreased from 10.9% to 2.3% (non-lethal injury rates went from 10.9%–2.3%; Johanos et al. 2010).

16.3 Social Behavior—Influences on Disease Spread and Mitigations

Hawaiian monk seals are not highly social (Kenyon and Rice 1959). Seals are often solitary, both when foraging and hauled out, which may in part be due to the fact that they naturally occur at low densities. An island or atoll in the NWHI may be shared by only 50–250 animals, even at the most populous breeding sites. At Laysan Island, the breeding population of ~250 seals is spread across just 8 km of coastline, resulting in densities of ~25 seals/km (NMFS HMSRP unpublished database). In the MHI, the population of ~300 seals is spread over >1500 km of coastline on eight islands, or just ~0.2 seals/km. A notable aggregation might be tens of animals hauled out on one beach in the NWHI, while in the MHI, well-known haulout "hotspots" might only host 3–6 seals (NMFS HMSRP unpublished database). While seals sometimes aggregate around a favorable haulout location, they appear to neither seek out interaction nor exhibit territorial behaviors (Kenyon and Rice 1959; though see reproductive behavior described in Sect. 15.2).

Observations of at-sea interactions are rare, but seal-mounted camera studies have recorded occasional interactions between seals underwater. Most of these interactions have been curious and brief, not demonstrating either agonistic or cooperative behavior (Parrish and Littnan 2007; Wilson et al. 2017a). Some video-sound recordings have detected underwater vocalizations, but it is unknown if the purpose of these vocalizations was to seek out nearby animals or deter them from a possible territory (CritterCam video footage; NMFS HMSRP unpublished instrument data). A number of telemetry studies have shown a considerable overlap in space use at sea in both the NWHI (Parrish and Abernathy 2006; Stewart et al. 2006; Curtice et al. 2011) and MHI (Littnan et al. 2006; Cahoon 2011, Wilson et al. 2017b), indicating the potential for interaction to occur.

Interactions among animals while hauled out have been more closely studied. Since the early 1980s, seals' behavior on Laysan Island and other sites has been recorded on breeding, aggression, and dominance interactions (Johanos et al. 1994; 2010). These data were later analyzed in a social network context to estimate the contact rates between seals (Baker et al. 2016a). Observations of seals on O'ahu were made to document types of contact and interaction rates in a lower-density population (Robinson et al. 2018). These studies revealed the subtle patterns of interactions within monk seals' low degree of sociality (Fig. 16.5).

Most associations between monk seals are short-lived. On Laysan Island, most seal pairings (82%) occurred only once over a 5-month study period, and two seals were seldomly recorded together five or more times (2%, primarily mother–pup pairs; Baker et al. 2016a). The most common type of association between seals involves resting nearby (Fig. 16.5a); of the interactions observed on O'ahu, 78% of associations were based solely on haulout proximity (Robinson et al. 2018). Direct interactions were twice as likely to be casual sniffing or nudging (14.5% of total interactions), with aggressive interactions (sparring or biting) making up only 7.5% of the total (Fig. 16.5b; Robinson et al. 2018).

Seals do not mix randomly, but rather show preferential interaction among age and sex classes. Mothers and pups have a highly elevated rate of interaction throughout the nursing period (Johanos et al. 1994), but nursing females tend to avoid each other (Boness et al. 1998, Baker et al. 2016a). Males show more social interaction than females. For example, males exhibit a "cruising" behavior that involves swimming near the shoreline, apparently searching for females, and frequently come to shore and have brief interactions with other seals hauled out, which can raise their contact rate with numerous animals (Baker et al. 2016a). Males are likely to spar with one another, especially to attend a female; males may fight to displace one another or maintain their position near females (Johanos et al. 2010). On Laysan Island, subadult males were the most interactive, with interaction rates approaching mother–pup pairs

Fig. 16.5 Contact between Hawaiian monk seals may vary from merely lying in close proximity on a beach (**a**) to intensive interactions during play or aggressive incidents (**b**). A contact network of seals using one O'ahu beach shows the varying degree of contact between individuals, with male seals (blue) having the highest contact levels (line weight indicates proximity vs. intensive contact, line distance indicates relative frequency)

(Baker et al. 2016a). On O'ahu, where there is a smaller population and fewer subadult males, adult males had the highest rate of interactions, and specifically aggressive interactions (Robinson et al. 2018). This difference is likely due to differences in sex ratio and the age structure of the given population. At the time of observation, Laysan Island had a highly male-biased sex ratio with numerous subadult and adult males, heightening the level of competitive behavior among (likely subdominant) subadult males. Meanwhile, on O'ahu, with few subadults in the local population, it appears that adult males showed more competitive interactions among themselves.

16.3.1 Implications for Disease Spread and Vaccination

For endangered species existing in small populations, such as Hawaiian monk seals, a disease outbreak could have disastrous consequences. Understanding social behavior in the context of potentially infectious contact rates is particularly important for assessing disease risk. While they do not show social aggregations common in some pinniped species, monk seals encounter and interact with enough seals to effectively spread disease through a small population. Hawaiian monk seal researchers have used behavioral observations and social network analysis to calculate contact rates, model disease spread, and design risk mitigations (Baker et al. 2016a; Baker et al. 2017; Robinson et al. 2018). Morbillivirus has been the focus of this work because viruses in this family, specifically phocine distemper virus (PDV) and canine distemper virus (CDV), have caused mass die-offs of other phocids (Grachev et al. 1989; Heide-Jørgensen et al. 1992; Kennedy et al. 2000; Jensen et al. 2002). Previous epidemiological screening has shown that Hawaiian monk seals were naïve to morbilliviruses (including canine distemper virus, phocine distemper virus, dolphin morbillivirus, porpoise morbillivirus) in both the MHI (Littnan et al. 2006) and NWHI (Aguirre et al. 2007). Due to their isolation in the Pacific and depauperate genetic diversity (Schultz et al. 2009), there is concern that Hawaiian monk seals could be highly susceptible if exposed to morbillivirus. The epidemiological models constructed for morbillivirus are likely to apply to other diseases directly spread between monk seals in close proximity (as by respiratory/aerosol substance). Monk seals face several disease threats other than morbillivirus (Barbieri et al. 2016; Harting et al. 2020) and some, like toxoplasmosis (transmitted by oocysts contaminating the environment via cat feces), are not related to direct contact among seals and cannot be prevented through vaccination.

Contact networks based on observed monk seal interactions indicated that networks were not subdivided into cliques despite heterogeneous contact structure. That is, all seals using an island were connected in a single component indicating pathogens could spread from one seal to another (Baker et al. 2016a; Robinson et al. 2018). Social connectivity was higher for seals in denser populations in the NWHI compared to those in the MHI, though perhaps not as different as expected given the disparity in seal density. Seals on Laysan had an average of 15 connections in the network of 250 seals using this island with 8 km of shoreline (Baker et al. 2016a), while seals on O'ahu had a mean of 10 contacts in a network of just 45 seals using 365 km of coastline (Robinson et al. 2018). The slight relationship between population density and contact is likely indicative of (a) numerous animals frequenting a few favored haulout locations, and (b) some animals such as subadult/adult males cruising the shoreline seeking out other seals, thus increasing connectivity throughout the population.

While connectivity was high among seals using a single island, the population's spatial subdivision works to limit contact among islands. Outbreaks simulated in a single admixed population grew in 91% of simulations, whereas in spatially subdivided populations (based on the MHI, with relatively frequent movement between nearby islands), the outbreak only grew in 38% of simulations (Baker et al. 2017). A recognition of the population's accurate spatial structure also helps in assessing the percentage of the population that needs to be vaccinated. Achieving herd immunity (in 95% of simulated scenarios) would require vaccinating 86% of the admixed population, but could be achieved after vaccinating just 60% of the spatially structured population (Baker et al. 2017).

16.3.2 Vaccination as a Conservation Intervention

Morbillivirus is one of the few pathogens for which vaccination is a management option. The American Association of Zoological Veterinarians recommended a recombinant vaccine to CDV (Purevax, Merial), which has been used extensively in zoological collections (Bronson et al. 2007) and has been tested for safety in Hawaiian monk seals (Yochem unpublished). Epidemic simulations showed that vaccination in response to an outbreak would have little efficacy in stemming the extent of the outbreak, primarily due to pathogens spreading faster than vaccines could be administered and confer immunity (Baker et al. 2017). Thus, in 2016, NOAA initiated a prophylactic vaccination program to protect Hawaiian monk seals against the possibility of morbillivirus exposure. During the pilot effort, 21 monk seals on O'ahu were successfully vaccinated (2 injections, 3–5 weeks apart), and showed no ill effects (Fig. 16.6; Robinson et al. 2018). Since that time, vaccination efforts have been expanded to all segments of the population in the NHI and NWHI, with work continuing to reach vaccination numbers sufficient for herd immunity. The Hawaiian monk seal system is particularly well-suited to protection through vaccination; nearly all pups are handled each year for flipper tagging, thus there is a good opportunity to vaccinate the majority of individuals in each new cohort of seals. After the 2019 field season, enough seals had been vaccinated to support herd immunity goals in 60–80% of simulations in the NWHI given the spatial structure of the population. This level of population protection was possible only after understanding social interactions and behavior relative to disease transmission and vaccination.

Fig. 16.6 A biologist uses a pole syringe to inject a sleeping Hawaiian monk seal with Morbillivirus vaccine

16.4 Foraging Behavior—Influences on Survival and Exposure to Risk of Fisheries Interactions

16.4.1 Diet

Hawaiian monk seals consume a wide variety of prey species of more than 40 families of fish, cephalopods, and crustaceans, many of which have a cryptic, benthic lifestyle. Prey include diurnal and nocturnal species that inhabit a range of habitats from the shallow reefs to subphotic depths (Goodman-Lowe 1998; Longenecker 2010; Iverson et al. 2011). Within this diversity of prey, there is considerable individual variation, with individual seals showing specialization on certain prey species or foraging areas (Parrish et al. 2000; Longenecker 2010; Iverson et al. 2011). Similarly, Mediterranean monk seals also consume a variety of predominantly benthic prey (Appendix 16.7).

Diet composition varies considerably depending on the method of diet analysis (Fig. 16.7). For example, fatty acid analysis (Iverson et al. 2011) indicates that snapper (Lutjanidae), octopus and squid (varied cephalopods) are significant parts of the monk seal diet; however, fecal remains analysis (Goodman-Lowe 1998; Cahoon et al. 2013) found these to be minor components, with triggerfishes (Balistidae) and crustaceans (varied) being the among the most frequent diet items. Both methods found tang and surgeonfishes (Acanthuridae) to be important diet items. Stable isotope analysis

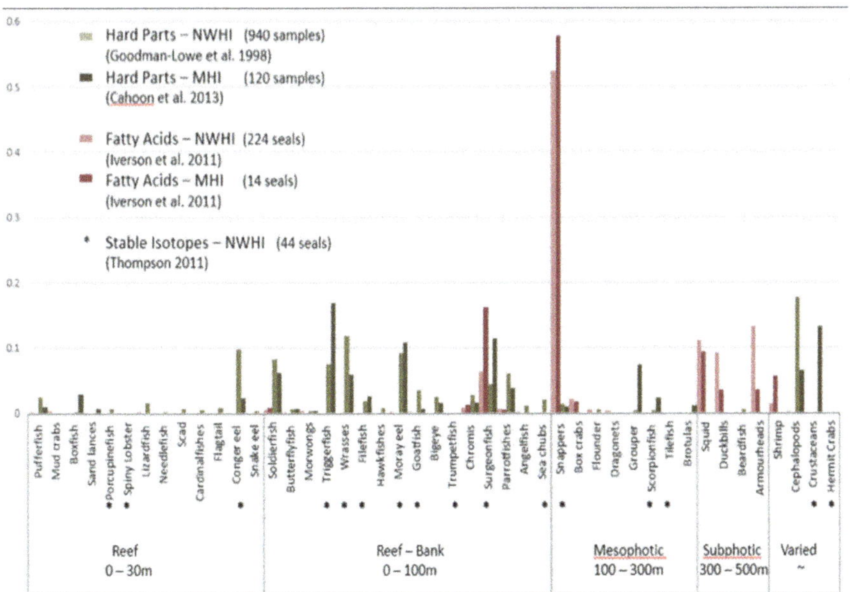

Fig. 16.7 The Hawaiian monk seal diet is diverse, focusing on benthic fishes, cephalopods, and crustaceans. The precise composition differs considerably according to the analytical method used (fecal hard part analysis, fatty acid analysis, and stable isotope analysis (presence only))

indicates that a subset of these prey families are in their diet (Thompson 2011). No one method is likely to provide a precise or complete picture of the monk seal diet, as each method is subject to its own biases. Fecal analysis is highly dependent on digestion rates and only provides a snapshot of the most recently consumed prey, whereas fatty acid and stable isotope analyses can be influenced by model assumptions and the prey libraries used to parameterize them.

Comparative studies have found differences in the diet composition of monk seals in the MHI versus NWHI. Based on fecal remains analysis, MHI seals consumed more fish and fewer cephalopods or crustaceans than NWHI seals (97.9% vs. 78.6% fish, 1.2% vs. 5.7% crustaceans, 0.9% vs. 15.7% cephalopods Cahoon et al. 2013). Further, the MHI monk seals' diet represented a subset (20 families; Cahoon et al. 2013) of the prey taxa included in the NWHI diet (31 families; Goodman-Lowe 1998). Fatty acid analysis found that the proportion of boarfish (Pentacerotidae), duckbill (Percophidae), squid (Ommastrephidae), and squirrelfish snapper (*Etelis carbunculus*) steadily decreased from the NWHI to the MHI, while the proportions of flower snapper (*Pristipomoides zonatus*) and surgeonfish (Acanthuridae) steadily increased (Iverson et al. 2011). Dietary differences have also been detected at a finer scale, among NWHI sites. For example, one study detected a greater proportion of octopi in the seals' diet at French Frigate Shoals, suggesting lesser fish availability, which could be linked to monk seals' especially poor condition and juvenile survival at this site (Goodman-Lowe 1998).

Ontogenetic shifts in monk seal diet show that foraging preferences or strategies evolve as animals mature, though there may not be consistent directions of change (Goodman-Lowe 1998; Iverson et al. 2011). When individuals were sampled as juveniles and adults using fatty acid analysis, individuals' diet compositions tended to change as they matured, but the prey species involved in the shifts were highly individual (Iverson et al. 2011). Analysis of fecal remains (but interestingly not fatty acids; Iverson et al. 2011), has consistently showed eel (Muraenidae and Ophichthidae) as an important diet item (Kenyon and Rice 1959; DeLong et al. 1984; Goodman-Lowe 1998; Longenecker et al. 2006; Longenecker 2010). Goodman-Lowe (1998) found that juveniles and subadults tended to have a higher percentage of nocturnal prey and eel in their diet, while adults consumed more wrasse (Labridae) and triggerfish. This has been suggested as a potential link between environment conditions and juvenile foraging success (Longenecker 2010), as both eel recruitment (Miller and McCleave 1994) and juvenile body condition appear to improve in response to oceanographic phenomena such as the periodic warming of the tropical Pacific, termed El Niño (Antonelis et al. 2003).

16.4.2 Habitat Use—Horizontal

Hawaiian monk seals appear to orient around submarine ridge systems, often following ridges from their home island/atoll to other foraging grounds (Abernathy 1999; Stewart et al. 2006). Areas of highest use tend to be concentrated within the

200 m bathymetric contour (Abernathy 1999; Littnan et al. 2006; Cahoon 2011). Home ranges of monk seals are typically on the order of 100 s of km^2 in the MHI (Wilson et al. 2017b) to 1000 s of km^2 in the NWHI (Curtice et al. 2011). This variation in home range size tends to reflect local topography and proximity of banks and seamounts used for foraging, but also may be indicative of the availability of prey in areas near the islands used for hauling out.

Monk seal home ranges often overlap, areas used by instrumented seals tend to overlap partially or entirely with others (Abernathy 1999), and seal-mounted video studies have shown tolerant interactions between animals underwater (Parrish et al. 2000). Seals from different sites often overlap at popular foraging locations. Seals from Midway and Kure Atolls use the Nero Seamount between the two atolls, seals from Laysan and Lisianski Islands overlap at the Hampton Seamounts between these sites, and animals from several islands in the MHI utilize Penguin Banks off the shore of Moloka'i (Stewart et al. 2006; Curtice et al. 2011; Cahoon et al. 2013, Wilson et al. 2017b). Individual seals may have favored foraging locations that they frequently visit, but the favored locations, and thus typical travel distances, are highly variable among individuals (Abernathy 1999, Wilson et al. 2017b). In addition to what might be the typical range for a given seal, most seals also show occasional exploratory movements well beyond their primary utilization distribution (Wilson et al. 2017b).

Differences in distance and duration of foraging trips highlight the extra effort dedicated to foraging in the NWHI, supporting the hypothesis that resources are more readily available for monk seals in the MHI. In the NWHI, seals typically use multiple foraging sites outside their home atoll or island (4 on average; Stewart et al. 2006), traveling 24–322 km to reach foraging sites on trips lasting an average of 17–20 days (Abernathy 1999). A notable exception in previous NWHI studies was Pearl and Hermes Reef, where animals seldom left the atoll to forage (throughout a study lasting one season), making home ranges and trip distances smaller for animals at this site (Stewart et al. 2006; Cahoon 2011). In the MHI, seals readily move between islands separated by as little as 15–100 km (~35–37% of individuals; Littnan et al. 2006; Wilson et al. 2017b), favoring foraging locations close to the island on which they commonly haulout. For example, seals from Kaua'i typically traveled between Kaua'i and neighboring Ni'ihau, while seals from Moloka'i often foraged at a submerged bank extending offshore (Cahoon 2011, Wilson et al. 2017b). Foraging trips in the MHI are typically much shorter than those in the NWHI, extending 10–50 km and lasting just 0.5–3.8 days (Cahoon 2011; Gobush et al. 2017).

In most cases, adult animals utilize larger areas and make longer foraging trips than juveniles. In the NWHI, younger animals were less likely than adults to leave their home atoll to forage at distant banks (Abernathy 1999; Stewart et al. 2006; Norris et al. 2017). Adults had more extensive home ranges than weaned pups and juveniles at all sites except Lisianski Island, where juveniles maintained the largest home ranges, followed by weaned pups, with adults having the smallest ranges (Curtice et al. 2011). In the MHI, like most of the NWHI, weaned pups used the smallest areas, followed by juveniles, with adults ranging the farthest.

16.4.3 Habitat Use—Vertical

Monk seals primarily dive to shallow depths (<50 m) for durations of 4–8 min (DeLong et al. 1984; Abernathy 1999; Stewart et al. 2006; Cahoon 2011, Wilson et al. 2017b). While the longest dive on record is ~20 min, dives lasting more than 10 min are rare (<5%; Abernathy 1999). Individual animals might diurnally shift their activity, showing a preference either for day or night foraging (Abernathy 1999), but there are not consistent or significant diurnal patterns in monk seal foraging activity (Parrish et al. 2000, Wilson et al. 2017a). Monk seals have also been recorded sleeping (34% of the time) and socializing with other seals (9% of the time) in waters as deep as 80 m (Parrish et al. 2000).

In the MHI, dives tend to be shallow, with a mean dive depth of 17.3 ± 16.8 m (Wilson et al. 2017a and 90% of dives <20 m (Cahoon 2011). Prey capture activity may be concentrated deeper than the average dive depth; in the NWHI, Parrish et al. (2000) found that prey searching and capture activity was concentrated at 50–60 m, while in the MHI, it occurred at a mean of 25.3 ± 16.2 m (Wilson et al. 2017a). Several studies have shown rare dives over 100 m and even >300 m, indicating what might be occasional explorations or directed deepwater foraging (DeLong et al. 1984; Parrish et al. 2002, Wilson et al. 2017a) as well as a few individuals with a focus on diving deeper banks (Parrish et al. 2000; Norris et al. 2017).

16.4.4 Foraging Behaviors

The Hawaiian monk seals' foraging strategy of focusing on cryptic prey leads to continuous foraging as seals move between small patches of habitat harboring individual prey (Wilson et al. 2017b). Seal-mounted cameras have been essential in revealing the ways that monk seals interact with their foraging habitat. Seals may swim in the upper water column (typically making shallow dives of 15–20 m) while transiting to an offshore bank, but most of their swimming time is focused near the seafloor where they can find demersal and benthic prey (Fig. 16.8; Parrish et al. 2000). Seals show the greatest foraging effort in benthic areas with moderate complexity, such as sand beds with occasional rocks or coral heads (Parrish et al. 2000, Wilson et al. 2017a). Uniform smooth or hard substrates likely offer fewer hiding spots to harbor the monk seals' prey, whereas highly complex areas such as extensive coral reefs likely make maneuvering and prey capture less efficient (Parrish et al. 2000).

When monk seals find a prey item, they typically invert their body, using their head to dig down into the sand or turn over rock or coral fragments to catch prey (Parrish et al. 2000, Wilson et al. 2017a). This characteristic and abrupt change in body position has made it possible to develop models that can successfully use sensors, such as triaxial accelerometers, to detect prey capture attempts from biologging instruments (Wilson et al. 2017a). These models have shown that the probability of

Fig. 16.8 Images from seal-mounted cameras show the seal-eyed view of benthic habitat and reveal hunting behaviors to enhance understanding of foraging ecology

prey capture attempts increases on long, deep dives with more time at the bottom and with increased body motion (Wilson et al. 2017a).

Monk seals often target small prey that are quickly consumed while continuing to swim and forage (80% of prey items <20 cm; Parrish et al. 2000). Large prey require substantially more handling time and are often consumed at the surface (Parrish et al. 2000). Handling time may also be impacted by the type of prey targeted; on average, seals spent three times longer digging out and eating burrowing prey vs. prey items camouflaged on the surface of the sand floor (Parrish et al. 2005). Individual seals can specialize in employing these different hunting tactics, with some focusing on sand fields where they adeptly dig burrowing prey from the sand, while others routinely search under large loose rocks for hiding prey (Parrish and Littnan 2007).

There is evidence that juvenile seals adjust their foraging habits over time, either developing physical strength, physiological capacity, or increasing local knowledge. Translocations (of animals after rehabilitation) have helped illustrate the learning process. Juvenile seals that were removed from the local environment for a period of captive rehabilitation showed wider exploratory movements upon release, whereas those that remained in their natural habitat had ranges more focused on particular foraging areas (Norris et al. 2011). Over time, the foraging range and diving habits of the post-rehabilitation seals converged with the controls as they learned the local foraging environment (Norris et al. 2011). Video-based studies at French Frigate Shoals found that juvenile seals used sand fields more extensively and targeted smaller prey (<10 cm) than adults (Parrish et al. 2005). Age-specific activity budgets (from video and triaxial accelerometry data; NMFS HMSRP unpublished instrument data) have shown that juvenile animals make more exploratory movements, use more energetically costly foraging techniques, and spend less time resting at sea than adults. This likely occurs because adults are more adept at finding profitable foraging areas, spending less time searching before making capture attempts, and using more energy-efficient prey capture techniques. These differences in foraging behavior illustrate the learning curve that young animals must overcome and provide clues to the mechanisms that may drive differential survival across age classes.

Interspecific competition is thought to be one factor underlying the nutritional limitation and poor body condition that is commonly seen in juvenile seals in the NWHI (National Marine Fisheries Service 2007). Seal-mounted videos have shown that predatory fish (most commonly jacks, *Caranx* spp., and sharks, *Carcharhinus* spp.) travel with foraging seals to exploit the seals' superior ability to flush cryptic prey from benthic habitats (Parrish et al. 2008). While escorts by such competitors were common in the NWHI (predatory fish followed seals for 17% of their foraging time; Parrish et al. 2008), they are much rarer in the MHI (few escorts observed in video footage; Wilson et al. 2017a) where predatory fish populations have been decreased by human fishing pressure. This difference in competitive pressure may be one factor leading to more robust nutritional status and improved survival of seals in the MHI (Baker et al. 2011b; Wilson et al. 2017a).

16.4.5 Interventions to Mitigate Food Limitation

Nutritional limitation leading to poor body condition and starvation takes a substantial toll on juvenile monk seals in the NWHI, with survival rates dipping as low as 0.20 at some sites (Baker and Thompson 2007). Ecological factors might impact nutritional condition including intra- and inter-specific competition (Parrish et al. 2008, Baker et al. 2011b), differences in prey type, abundance, or quality (Craig and Ragen 1999; Friedlander and Demartini 2002; Parrish et al. 2005), and accessibility and effort required to acquire food (Parrish et al. 2005; Cahoon 2011; Curtice et al. 2011). Because the juvenile survival bottleneck has lasting impacts on population dynamics and recovery potential, the HMSRP designates enhancing juvenile survival as a top priority (National Marine Fisheries Service 2007). While nutritional limitation may stem from ecosystem-level issues, interventions at the individual seal level have proven effective in boosting survival (Harting et al. 2014). Two key intervention strategies are used to mitigate nutritional limitation: translocating weaned pups from sites with the lowest survival rates to those with higher survival (see Sect. 16.5), and rehabilitation of young monk seals in poor body condition.

Pups tend to have insufficient maternal investment if they wean with an axillary girth <90 cm; these pups are typically found in the NWHI, and they have a little chance of first-year survival (Baker 2008). These undersized pups are prioritized for rehabilitation, and female pups in particular because they are the species' reproductive future. Juveniles (and one subadult) in poor body condition have also been rehabilitated. Early rehabilitation efforts included supplemental feeding in beach pens in the remote NWHI and rehabilitation at NMFS facilities and other facilities. While > 50 seals were successfully released back into the wild in the NWHI, both approaches had mixed results, with some animals dying or being deemed non-releasable (Ragen and Lavigne 1999; Norris et al. 2011). In 2014, The Marine Mammal Center opened a dedicated monk seal rehabilitation facility, Ke Kai Ola, in Kailua-Kona. Since this hospital's establishment, >30 NWHI seals have been rehabilitated, with all being released back into the wild. Young seals released from rehabilitation have survived at rates greater than would be expected without rehabilitation, given their poor body condition at the time of stranding (NMFS HMSRP unpublished database).

16.4.6 Interventions to Mitigate Fisheries Interactions

The foraging habits of monk seals in the MHI can bring seals into contact with anthropogenic threats. Because foraging in the MHI tends to focus mainly on shallower nearshore areas, much of the seals' space use overlaps with areas humans use for recreation and fishing. Accidental interactions with fishing gear, including ingestion or external lodging of fish hooks ("hookings") and entanglement in lay nets, can pose a significant risk to monk seals. Net entanglements and hookings pose one of the greatest threats to seal survival in the MHI, depressing potential population growth by as much as 50% (Harting et al. 2020).

NOAA documented 232 monk seal–fisheries interactions in the MHI between 1976 and 2019 (Gobush et al. 2017, NMFS HMSRP unpublished database). These included 16 deaths linked to net drownings and 7 from ingested hooks (Harting et al. 2020). While net entanglements pose a greater immediate mortality risk, they are more cryptic than hookings, which are often more visible and, therefore, readily detected and mitigated. Between 1988 and 2014, approximately one quarter (26%) of monk seals in the MHI had at least one documented hooking (Gobush et al. 2017). Detected interactions likely represent only a portion of the complete number as detection is imperfect, and seals vary in their detectability based on their haulout areas' accessibility. Fisheries interactions have been the highest on the islands of Kaua'i and O'ahu where there is a high intersection of seal haulout use and human fishing effort in addition to much public surveillance (Gobush et al. 2017). The rates of seal–fisheries interactions have increased as the MHI seal population has grown (Gobush et al. 2017).

Mitigating threats from fisheries interactions involves a multifaceted approach, working with state agencies, local fishing communities, and the visiting and resident public to encourage wildlife awareness and responsible fishing, reporting, and finally, intervening to remove dangerous fishing gear from seals. HMSRP has successfully removed hooks from ~ 50% of seals that have been detected hooked, with 5% of these cases requiring surgery (NMFS HMSRP unpublished database).

16.5 Dispersal Behavior and Site Fidelity—Influences on the Success of Translocation as a Conservation Tool

The nature of the Hawaiian Archipelago, a string of atolls and islands separated by 10–100 s of kms, leads to fragmentation of island populations. Across the archipelago, subpopulations of Hawaiian monk seals at different sites (individual islands or atolls) are impacted by localized threats and exhibit variation in demographic rates (Baker and Thompson 2007). Despite local effects, Hawaiian monk seals move widely throughout the archipelago (Stewart et al. 2006; Johanos et al. 2014), and genetic analysis suggests that gene flow is sufficient to avoid genetic isolation (Schultz et al. 2011). An understanding of the degree of dispersal and population connectivity can be vital for informing conservation actions. Here we review research on monk seal dispersal and site fidelity and consider implications for translocation as a conservation tool.

16.5.1 Dispersal Patterns and Site Fidelity

The foundational knowledge regarding monk seal dispersal and site fidelity has come from the long history of visual monitoring as well as tracking through instrumentation. A 30-year retrospective study resighted 4,438 seals (4,320 NWHI and 118 MHI from 1981–2011), detecting 1,161 movements between sites (different islands or atolls) by 373 individuals (Johanos et al. 2014). Telemetry studies have documented many movements among sites and can capture more short-term movements (such as foraging trips) in addition to longer-term dispersals (Stewart et al. 2006).

The probability of a monk seal dispersing away from its natal site increases with age. Weaned pups spend much of their time near the beach where they weaned, and gradually increase their range in the months after weaning (Henderson and Johanos 1988; Norris et al. 2017). By the age of 10, 14% of seals had dispersed to a site other than their natal site, with an additional 28% making some brief forays before returning to their natal site (Johanos et al. 2014). Likewise, telemetry studies suggest that older animals are increasingly likely to move away from their home sites for longer foraging trips, and show that proportions of animals traveling between sites are likely even higher than what is recorded through visual observations (Stewart et al. 2006). Dispersal does not appear to be sex-biased, as both males and females appear equally likely to disperse from their natal sites (Johanos et al. 2014). Once females pup, they show high fidelity to their pupping site, with 95% of subsequent pups born at the same island or atoll as the female's first pup (Johanos et al. 2014).

Distance between sites influences connectivity and movement patterns, with closer sites exchanging more seals than those farther apart. Johanos et al. (2014) observed relatively high movement rates between sites separated by 100 km or less and very little between locations separated by >400 km (Johanos et al. 2014). In the MHI, where islands are separated by only 10 s of kms, over 30% of seals use multiple islands (Littnan et al. 2006; Wilson et al. 2017b). While less common, longer distance dispersals occur, including trips from the NWHI to the MHI, which would have been important in the early recovery of the MHI monk seal population (Johanos et al. 2014).

16.5.2 Translocation as a Conservation Intervention

Knowledge of monk seal movement patterns has informed (and been informed by) the use of translocation as a conservation intervention. For over 35 years, NMFS's HMSRP has been successfully utilizing translocation as a means to enhance the survival of Hawaiian monk seals. The primary purpose of Hawaiian monk seal translocations has been to improve the survival probability of young seals in the NWHI by moving weaned pups away from islets with especially heavy shark predation or by moving pups and rehabilitated yearlings from sites with lower survival to sites with higher survival and presumed favorable habitat (Fig. 16.9; Baker et al. 2011a, Baker et al. 2020). Translocation has also been used to mitigate various other survival threats. As discussed earlier, adult male monk seals were moved from Laysan Island to Johnston Atoll and the MHI to decrease the male-biased sex ratio to reduce the incidence of male aggression that led to high female mortality (Johanos et al. 2010, see Sect. 16.2). On occasion, seals have also been translocated within the MHI, or from the MHI to the NWHI to (a) move seals that have shown dangerous signs of habituation to human interaction, or (b) preempt habituation by moving weaned pups born on beaches with high human use to more remote locations (Baker et al. 2011a).

Translocating Hawaiian monk seals has proven safe overall, and seals have shown survival rates matching those of seals native to the destination sites (Baker et al. 2011a; Baker et al. 2020). However, the effectiveness of translocations has varied depending on the conservation goal and the target animal (Baker et al. 2011a). Upon reviewing 247 translocations taking place from 1984 through 2009, Baker et al. (2011a) found that weaned pups were the most amenable to translocation, while older animals were more likely to roam farther from their destination sites.

Because Hawaiian monk seal pups remain relatively sedentary during a post-weaning fast, they typically develop little foraging experience or strong ties to their home site prior to translocation (Henderson and Johanos 1988). Thus, weaned pups adapt remarkably well to their translocation site, typically remaining near their release beach for >40 days (Baker et al. 2011a). This has led to success in translocating

Fig. 16.9 Two biologists carry a weaned Hawaiian monk seal pup to translocate it from one islet with heavy shark predation to another, just a few kilometers across the French Frigate Shoals atoll, where the probability of survival will be substantially higher

weaned pups even small distances; for instance, at French Frigate Shoals, moving a pup from the islets with the heaviest shark predation to an islet just a few kms away can dramatically improve early survival (Baker et al. 2011a). When moved from one island/atoll to another, weaned pups typically remained at their new site until maturity (Baker et al. 2020). When tracked along with control animals at release sites, translocated pups developed similar foraging movements and maintained ranges similar to resident weaned pups (Norris et al. 2017).

In contrast, older seals (juveniles, subadults, adults) have already developed extensive home ranges (see Sect. 16.4) and have proven less likely to remain close to their site of translocation. Of the adult males translocated from the NWHI to the MHI, most resighted seals were seen on multiple islands within the MHI (Baker et al. 2011a) and one of these males made two round trips between the MHI and Nihoa (Johanos et al. 2014). One or more of the males translocated to Johnston Atoll remained there at least 2 years after release (though note low monitoring effort at this location; NMFS HMSRP unpublished database). While these males did not show strong fidelity to their new site, none are known to have returned to the NWHI. Thus, these translocations successfully mitigated a threat to the females at the origin site of Laysan Island.

Translocations of older seals showing signs of human habituation have been less successful (Baker et al. 2011a). In cases where seals were moved within the MHI, their range of normal foraging movements allowed them to come in contact with humans again quickly, and dangerous behaviors (interacting with swimmers or fishers) often resumed. In the few cases where MHI seals were moved to the NWHI to achieve a greater distance from human interaction, the seals failed to adapt to their new sites and disappeared soon after translocation, sometimes after making long exploratory movements (NMFS HMSRP unpublished instrument data). It is unknown whether this may be caused by a drive to return to the home site in the MHI, or lack of familiarity with the increased competition for resources with increased numbers of seals and other predators in the NWHI.

16.6 Conclusions and Future Directions

Behavioral ecology and ethological research have been essential in informing and evaluating conservation interventions, which are a crucial part of the effort to recover the Hawaiian monk seal. An integral part of the success of the HMSRP has been the extensive database collected from decades of fieldwork, which includes behavioral observations. This database has facilitated the ability to link monk seal ecology and behavior to survival threats and allowed successful interventions to help recover the population. In addition to collecting critical data, researchers' time in the field directly correlates with the ability to detect seals in need and conduct life-saving interventions (Harting et al. 2014).

Technology has dramatically enhanced the ability of researchers to study the underwater behavior of marine mammals. As technology continues to improve, future research directions will benefit from improvements such as smaller instruments, improved modes of data transmission, and longer battery life/solar charging capabilities. It is a consistent goal of NMFS HMSRP to collect data from animals with the least possible disturbance or impact on natural behaviors. In recent years, smaller instrument sizes have allowed us to expand our instrumentation of smaller seals, advancing our understanding of juvenile foraging ecology and the survival threats facing this vulnerable demographic class.

The recolonization of the MHI has provided a unique opportunity to study the influence of environment on monk seal behavior, reproductive success, and population dynamics. In light of population declines in the NWHI and expansion in the MHI, substantial research effort has focused on understanding the drivers underlying regional differences in population dynamics. Studies of foraging behavior have offered valuable clues regarding differences in activity budgets and interactions with competitors (Wilson et al. 2017a). Future research will explore how seals in the MHI versus NWHI may interact differently with their environments. While ecological conditions appear favorable in the MHI, concern has grown about a number of anthropogenic threats (Harting et al. 2020), and ethological research will take on new importance in the effort to understand how monk seal space use, diet, and behavior may impact individual exposure to risk factors.

The Hawaiian Archipelago, shaped dynamically over the past 30 + million years, has undoubtedly undergone substantial changes since monk seals first inhabited the Hawaiian Islands. With climate change altering ocean conditions, coral reef ecosystems, and coastal lands, monk seals are likely to experience more changes in the near future. In particular, severe weather events and sea-level rise have the potential to impact the low-lying sandy islands of the NWHI where the majority of seals reside. We have already seen shifts in favored pupping beaches as islets have shifted or disappeared at some atolls (Baker et al. 2006), and it will be essential to continue to assess monk seals' behavioral capacity to adapt to the changing environment. Further, safeguarding the MHI population in light of anthropogenic threats will become increasingly important to ensure safe refuge in the higher elevation islands.

Acknowledgements We are indebted to the entire Hawaiian Monk Seal Research Program, to the generations of biologists who dedicated their time in the remotest reaches of the Hawaiian Archipelago to build the strong data foundation upon which species recovery efforts have been built. We are likewise grateful to the staff of partner agencies, volunteers, and engaged members of the public that make our research and recovery activities possible. Many thanks to valuable comments in the early review from Jason Baker and Charles Littnan.

16.7 Appendix

Mediterranean Monk Seal *Monachus Monachus*

By Joan Gonzalvo and Giuseppe Notarbartolo di Sciara; Tethys Research Institute, Milan, Italy.

Mediterranean monk seal, swimming (left) and hauled out on rocks in a sea cave (right). Photo credit: Joan Gonzalvo; Tethys Research Institute.

Mediterranean monk seals (*Monachus monachus*) were once widely and continuously distributed in the Mediterranean and Black Seas, and in North Atlantic waters from Morocco to Mauritania, including Cape Verde and the Canary Islands, Madeira, and the Azores (Johnson et al. 2006). Today fewer than 700 individuals are thought to survive in isolated subpopulations in the eastern Mediterranean, the archipelago of

Madeira, and the Cabo Blanco area in the north-eastern Atlantic Ocean (Karamanlidis et al. 2015). The largest aggregations of Mediterranean monk seals are found near Cabo Blanco (González and Fernandez de Larrinoa 2012, Martínez-Jauregui et al. 2012). Principal sites in the Mediterranean are located in the Ionian and Aegean seas, including the National Marine Park of Alonissos (Trivourea et al. 2011) and the Gyaros Marine Protected Area (Dendrinos et al. 2008), both in Greece.

Mediterranean monk seals have teetered on the brink of extinction for about one-half century (Notarbartolo di Sciara and Kotomatas 2016). After being classified as Critically Endangered for almost two decades, their status was reassessed as Endangered on the IUCN's Red List (Karamanlidis and Dendrinos 2015). Reassessment was based on the notion that monk seal numbers in specific locations (i.e., the Madeira archipelago, Western Sahara-Mauritania, and selected localities in the Aegean Sea) have been stable or may even be increasing. Moreover, there are recent indications that seals might be frequenting areas in the Mediterranean Basin within their historical range where they had been extirpated in previous decades (Bundone et al. 2019).

Historical evidence suggests that Mediterranean monk seals commonly hauled out on open beaches (Johnson and Lavigne 1999, González 2015). In more recent times—probably as an adaptation to increased human disturbance—they generally seek refuge in remote marine caves. These natural rocky shelters share common morphological characteristics, including one or more entrances above or below water level, an entrance corridor, an internal pool, and a beach that provides a dry haulout area (Dendrinos et al. 2007).

The monk seal populations at Cabo Blanco in the Atlantic and at Gyaros Island in the eastern Mediterranean are the only large extant aggregations of the species that still preserve the structure of a colony. The remaining subpopulations in the eastern Mediterranean are usually small, fragmented groups of <20 individuals (Karamanlidis et al. 2015). Most of our knowledge about Mediterranean monk seals' mating system comes from observations made in the Cabo Blanco area. Mating has only been observed to occur in the water, where males, who are markedly larger than females (Samaranch and González 2000), patrol and defend clearly delineated long-lasting aquatic territories. Aggressive interactions with intruders are characterized by loud puffs (i.e., short, explosive burst of breath), chases, and subaquatic fights. This aquatic breeding strategy results in very low levels of polygyny, as evidenced by the observed low or null relatedness among pups belonging to the same cohort (Pastor et al. 2011).

Pupping periods show differences between locations; in the colony of Cabo Blanco, births are recorded from April to November (González et al. 2002), while elsewhere, births are concentrated between October and November (Littnan et al. 2018). For the first week after giving birth, mothers spend much time on land lying close to their newborns. After this time and throughout the nursing period, females leave their pups unattended for extended periods, lasting up to several hours to forage. As lactation progresses, pups begin to swim and engage in short, increasingly independent trips between caves. During mother–pup encounters, mothers use nuzzling and vocalization to recognize their pups (Aguilar et al. 2007, Gazo and Aguilar 2005).

Fostering and milk stealing are common behavior patterns for lactating females and pups, respectively (Pires 2004, Aguilar et al. 2007). Weaning of pups gradually occurs at four to five months of age (Aguilar et al. 2007, Karamanlidis et al. 2015), when pups begin to forage on their own (Pastor and Aguilar 2003).

Mediterranean monk seals have a varied diet, consisting mainly of demersal fishes, cephalopods (the common octopus, *Octopus vulgaris*, being the most frequent prey item), and crustaceans (Salman et al. 2001, Karamanlidis et al. 2014, Pierce et al. 2011, Pinela et al. 2010, Kiraç and Ok 2019). Body parts of green turtles (*Chelonia mydas*) were also recently found in the stomach of an adult seal stranded in Turkey (Tonay et al. 2016). When foraging, monk seal pups typically dive to depths of a few tens of meters (mean depth = 11.6 ± 9.5 m), and their dive performance increases with age, in time spent diving and in dive parameters (Gazo et al. 2006). In Greece, a tagged rehabilitated monk seal performed 5–7 min dives with a mean maximum depth of 41 m and an overall maximum dive depth of 123 m (Dendrinos et al. 2007). Tagged seals have descended to the euphotic zone (Littnan et al. 2018) on dives lasting up to 18 min (Kiraç et al. 2002).

While at sea, Mediterranean monk seals have been reported sleeping, either at the surface floating (vertically or horizontally) with eyes closed, or resting underwater on the seafloor or over seagrass beds with eyes and nostrils shut (Karamanlidis et al. 2017, Mpougas et al. 2019). We have also observed similar episodes in the Ionian Sea (J. Gonzalvo, unpublished). On all occasions, seals woke up when approached by humans.

The main threats faced by Mediterranean monk seals include: (a) deliberate killings, mostly by artisanal fishers retaliating against net depredation and damage; (b) critical habitat deterioration, destruction, and fragmentation; (c) disturbance caused by tourists entering breeding caves during the reproductive season, as well as seal–boat interactions; and (d) bycatch in fishing gear, mainly of young inexperienced individuals (Güçlüsoy et al. 2004, Karamanlidis et al. 2020, Karamanlidis and Dendrinos 2015, Mpougas et al. 2019, Notarbartolo di Sciara and Kotomatas 2016). Other threats, such as disease outbreaks (e.g., morbillivirus), toxic algal blooms, prey depletion, and contamination by pollutants and oil spills in critical habitat, also have the potential of reducing monk seal survival (Karamanlidis et al. 2015). Inbreeding also raises concern, since the species appears to be "one of the most genetically depauperate mammals on Earth" (Karamanlidis et al. 2016), a condition believed to substantially increase the probability of extinction.

The Mediterranean monk seal is listed in Appendices I and II of the Convention on the Conservation of Migratory Species of Wild Animals (CMS), in Appendix I of the Convention on International Trade of Endangered Species (CITES), in Annex II of the Barcelona Convention Protocol on Specially Protected Areas and Biodiversity in the Mediterranean, and is a priority species of Community Interest, listed in Annexes II and IV of the EU Habitats Directive. The species is protected throughout its range through national laws. Parties to the Barcelona Convention have adopted a "Regional strategy for the conservation of Mediterranean monk seal" in 2013, revised and updated in 2019.

References

Aguilar A, Cappozzo LH, Gazo M, Pastor T, Forcada J, Grau E (2007) Lactation and mother–pup behaviour in the Mediterranean monk seal *Monachus monachus*: an unusual pattern for a phocid. J Mar Biol Assoc UK 87:93–99.

Bundone L, Panou A, Molinaroli E (2019) On sightings of (vagrant?) monk seals, *Monachus monachus*, in the Mediterranean Basin and their importance for the conservation of the species. Aquat Conserv Mar Freshw Ecosyst 29:554–563.

Dendrinos P, Karamanlidis AA, Kotomatas S, Legakis A, Tounta E, Matthiopoulos J (2007) Pupping habitat use in the Mediterranean monk seal: a long-term study. Mar Mamm Sci 23:615–628.

Dendrinos P, Karamanlidis AA, Kotomatas S, Paravas V, Adamantopoulou S (2008) Report of a new Mediterranean monk seal (*Monachus monachus*) breeding colony in the Aegean Sea, Greece. Aquat Mamm 34:355–361.

Gazo M, Aguilar A (2005) Maternal attendance and diving behavior of a lactating Mediterranean monk seal. Mar Mamm Sci 21:340–345.

Gazo M, Lydersen CV, Aguilar A (2006) Diving behaviour of Mediterranean monk seal pups during lactation and post weaning. Mar Ecol Progress Ser 308:303–309.

González LM, Cedenilla MA, Fernández de Larrinoa P, Layna JF, Aparicio F (2002) Changes in the breeding variables of the Mediterranean monk seal (*Monachus monachus*) colony of Cabo Blanco Peninsula after a mass mortality episode. Mammalia 6:173–182.

González LM, Fernandez de Larrinoa P (2012) Mediterranean monk seal *Monachus monachus* distribution and fisheries interactions in the Atlantic Sahara during the second half of the 20th century. Mammalia 77:41–49.

González LM (2015) Prehistoric and historic distributions of the critically endangered Mediterranean monk seal (*Monachus monachus*) in the eastern Atlantic. Mar Mamm Sci 31:1168–1192.

Güçlüsoy H, Kýraç CO, Veryeri NO, Savas Y (2004). Status of the Mediterranean monk seal, *Monachus monachus* (Hermann, 1779) in the coastal waters of Turkey. E.U. J Fish Aquat Sci 21(3–4):201–210.

Johnson WM, Karamanlidis AA, Dendrinos P, Fernández de Larrinoa P, Gazo M, González LM, Güçlüsoy H, Pires R, Schnellmann M (2006) Monk seal fact files. Biology, behaviour, status and conservation of the Mediterranean monk seal, *Monachus monachus*. The Monachus Guardian. http://www.monachus-guardian.org/factfiles/medit01.htm

Johnson WM, Lavigne DM (1999) Monk seals in antiquity. The Mediterranean monk seal (*Monachus monachus*) in ancient history and literature. Mededelingen 35:1–101.

Karamanlidis AA, Adamantopoulou S, Kallianiotis AA, Tounta E, Dendrinos P (2020) An interview-based approach assessing interactions between seals and small-scale fisheries informs the conservation strategy of the endangered Mediterranean monk seal. Aquat Conserv Mar Freshw Ecosyst 3:928–936.

Karamanlidis AA, Curtis PJ, Hirons AC, Psaradellis M, Dendrinos P, Hopkins III JB (2014) Stable isotopes confirm a coastal diet for critically endangered Mediterranean monk seals. Isot Environ Health Stud 50:1–11.

Karamanlidis AA, Dendrinos P (2015) *Monachus monachus* (errata version published in 2017). The IUCN Red List of Threatened Species 2015: e.T13653A117647375. https://dx.doi.org/10.2305/IUCN.UK.2015-4.RLTS.T13653A45227543.en. Downloaded on 09 June 2020.

Karamanlidis AA, Dendrinos P, Fernández de Larrinoa P, Gücü AC, Johnson WM, Kiraç CO, Pires R (2015) The Mediterranean monk seal *Monachus monachus*: status, biology, threats, and conservation priorities. Mamm Rev 46:92–105.

Karamanlidis AA, Gaughran S, Aguilar A, Dendrinos P, Huber D, Pires R, Schultz J, Skrbinsek T, Amato G (2016) Shaping species conservation strategies using mtDNA analysis: the case of the elusive Mediterranean monk seal (*Monachus monachus*). Biol Conserv 193:71–79.

Karamanlidis AA, Lyamin O, Adamantopoulou S, Dendrinos P (2017) First observations of aquatic sleep in the Mediterranean monk seal (*Monachus monachus*). Aquat Mamm 43(1):82–86. https://doi.org/10.1578/AM.43.1.2017.82.

Kiraç CO, Savas Y, Güçlüsoy H, Veryeri NO (2002) Observations on diving behaviour of free ranging monk seals *Monachus monachus* on the Turkish coasts. Monachus Guard. https://www.monachus-guardian.org/mguard09/09scien2.htm.

Kiraç OC, Ok M (2019) Diet of a Mediterranean monk seal *Monachus monachus* in a transitional post-weaning phase and its implications for the conservation of the species. Endanger Species Res 39:315–320.

Littnan C, Karamanlidis AA, Dendrinos P (2018) Monk seals. In: Würsig B, Thewissen JGM, Kovacs K (eds) Encyclopedia of marine mammals, 3rd edn. Academic Press, pp 622–653

Martínez-Jauregui M, Tavecchia G, Cedenilla MA, Coulson T, Fernández de Larrinoa P, Muñoz M, González LM (2012) Population resilience of the Mediterranean monk seal *Monachus monachus* at Cabo Blanco peninsula. Mar Ecol Progress Ser 461:273–281.

Mpougas E, Waggitt JJ, Dendrinos P, Adamantopoulou S, Karamanlidis AA (2019) Mediterranean Monk Seal (*Monachus monachus*) behavior at sea and interactions with boat traffic: implications for the conservation of the species in Greece. Aquat Mamm 45:419–424.

Notarbartolo di Sciara G, Kotomatas S (2016) Are Mediterranean monk seals, *Monachus monachus*, being left to save themselves from extinction? Adv Mar Biol 75:259–296.

Pastor T, Aguilar A (2003). Reproductive cycle of the female Mediterranean monk seal in the western Sahara. Mar Mamm Sci 19:318–330.

Pastor T, Cappozzo HL, Grau E, Amos W, Aguilar A (2011) The mating system of the Mediterranean monk seal in the Western Sahara. Mar Mamm Sci 27(4):E302–E320.

Pierce GJ, Hernandez-Milian G, Santos MB, Dendrinos P, Psaradellis M, Tounta E, Androukaki E, Edridge A (2011) Diet of the monk seal (*Monachus monachus*) in Greek waters. Aquat Mamm 37: 284–297.

Pinela AM, Borrell A, Cardona L, Aguilar A (2010) Stable isotope analysis reveals habitat partitioning among marine mammals off the NW African coast and unique trophic niches for two globally threatened species. Mar Ecol Progress Ser 416:295–306.

Pires R (2004) One pup –three mothers. Monachus Guard 7:33–34.

Salman A, Bilecenoglu M, Güçlüsoy H (2001) Stomach contents of two Mediterranean monk seals (*Monachus monachus*) from the Aegean Sea, Turkey. J Mar Biol Assoc UK 81:719–720.

Samaranch R, González LM (2000) Changes in morphology with age in Mediterranean monk seals (*Monachus monachus*). Mar Mamm Sci 16:141–157.

Tonay AM, Danyer E, Dede A, Öztürk B, Öztürk AA (2016) The stomach content of a Mediterranean Monk Seal (*Monachus monachus*): finding of Green Turtle (*Chelonia mydas*) remains. Zool Middle East 62:1–5.

Trivourea MN, Karamanlidis AA, Tounta E, Dendrinos P, Kotomatas S (2011) People and the Mediterranean monk seal (*Monachus monachus*): a study of the socioeconomic impacts of the National Marine Park of Alonissos, Northern Sporades, Greece. Aquat Mamm 37:305–318.

References

Abernathy KJ (1999) Foraging ecology of Hawaiian monk seals at French Frigate Shoals. Thesis, University of Minnesota Hawaii

Aguirre AA, Keefe TJ, Reif JS, Kashinsky L, Yochem PK, Saliki JT, Stott JL, Goldstein T, Dubey J, Braun R (2007) Infectious disease monitoring of the endangered Hawaiian monk seal. J Wildl Dis 43:229–241

Alcorn D, Henderson J (1984) Resumption of nursing in "weaned" Hawaiian monk seal pups. Elepaio 45:11–12

Antonelis GA, Baker JD, Johanos TC, Braun R, Harting A (2006) Hawaiian monk seal (*Monachus schauinslandi*): status and conservation issues. Atoll Res Bull 543:75–101

Antonelis GA, Baker JD, Polovina JJ (2003) Improved body condition of weaned Hawaiian monk seal pups associated with El Nino events: potential benefits to an endangered species. Mar Mamm Sci 19:590–598

Atkinson S, Becker BL, Johanos TC, Pietraszek JR, Kuhn BC (1994) Reproductive morphology and status of female Hawaiian monk seals (*Monachus schauinslandi*) fatally injured by adult male seals. J Reprod Fertil 100:225–230

Atkinson S, Gilmarti W (1992) Seasonal testosterone pattern in Hawaiian monk seals (*Monachus schauinslandi*). J Reprod Fertil 96:35–39

Atkinson S, Gilmartin W, Lasley B (1993) Testosterone response to a gonadotrophin-releasing hormone agonist in Hawaiian monk seals (*Monachus schauinslandi*). J Reprod Fertil 97:35–38

Baker J, Becker B, Wurth T, Johanos T, Littnan C, Henderson J (2011a) Translocation as a tool for conservation of the Hawaiian monk seal. Biol Cons 144:2692–2701

Baker JD, Harting AL, Wurth TA, Johanos TC (2011b) Dramatic shifts in Hawaiian monk seal distribution predicted from divergent regional trends. Mar Mamm Sci 27:78–93

Baker JD (2008) Variation in the relationship between offspring size and survival provides insight into causes of mortality in Hawaiian monk seals. Endanger Species Res 5:55–64

Baker JD, Barbieri MM, Johanos TC, Littnan CL, Bohlander JL, Kaufman AC, Harting AL, Farry SC, Yoshinaga CH (2020) Conservation translocations of Hawaiian monk seals: accounting for variability in body condition improves evaluation of translocation efficacy. Marine Mammal Science submitted

Baker JD, Harting AL, Barbieri MM, Johanos TC, Robinson SJ, Littnan CL (2016a) Estimating contact rates of Hawaiian monk seals (*Neomonachus schauinslandi*) using social network analysis. J Wildl Dis 52:533–543

Baker JD, Harting AL, Johanos TC, Littnan CL (2016b) Estimating Hawaiian monk seal range-wide abundance and associated uncertainty. Endanger Species Res 31:317–324

Baker JD, Harting AL, Barbieri MM, Robinson SJ, Gulland FMD, Littnan CL (2017) Modeling a morbillivirus outbreak in Hawaiian Monk Seals to aid in the design of mitigation programs. J Wildl Dis 53:736–748

Baker JD, Johanos TC (2004) Abundance of the Hawaiian monk seal in the main Hawaiian Islands. Biol Cons 116:103–110

Baker JD, Johanos TC, Wurth TA, Littnan CL (2014) Body growth in Hawaiian monk seals. Mar Mamm Sci 30:259–271

Baker JD, Littnan CL, Johnston DW (2006) Potential effects of sea level rise on the terrestrial habitats of endangered and endemic megafauna in the Northwestern Hawaiian Islands. Endanger Species Res 4:1–10

Baker JD, Thompson PM (2007) Temporal and spatial variation in age-specific survival rates of a long-lived mammal, the Hawaiian monk seal. Proc R Soc B Biol Sci 274:407–415

Barbieri MM, Kashinsky L, Rotstein DS, Colegrove KM, Haman KH, Magargal SL, Sweeny AR, Kaufman AC, Grigg ME, Littnan CL (2016) Protozoal-related mortalities in endangered Hawaiian monk seals *Neomonachus schauinslandi*. Dis Aquat Org 121:85–95

Boness DJ (1990) Fostering behavior in Hawaiian monk seals: is there a reproductive cost? Behav Ecol Sociobiol 27:113–122

Boness DJ, Craig MP, Honigman L, Austin S (1998) Fostering behavior and the effect of female density in Hawaiian monk seals, *Monachus schauinslandi*. J Mammal 79:1060–1069

Bronson E, Deem SL, Sanchez C, Murray S (2007) Serologic response to a canarypox-vectored canine distemper virus vaccine in the giant panda (*Ailuropoda melanoleuca*). 38:363–366

Cahoon M (2011) The foraging ecology of monk seals in the main Hawaiian Islands. Thesis, University of Hawaii

Cahoon M, Littnan C, Longenecker K, Carpenter J (2013) Dietary comparison of two Hawaiian monk seal populations: the role of diet as a driver of divergent population trends. Endanger Species Res 20:137–146

Carretta JV, Forney KA, Muto MM, Barlow J, Baker J, Lowry M (2004) U. S. Pacific Marine Mammal Stock Assessments: 2003. NOAA-TM-NMFS-SWFSC-358291

Chaudun L (2018) Description du répertoire vocal aérien du phoque moine d'Hawaii (*Neomonachus schauinslandi*) et recherche de signatures vocales individuelles. Institut des Neurosciences Paris-Saclay, CNRS-UMR 9197.

Cobb JN (1902) Commercial fisheries of the Hawaiian Islands (extracted from the US Fish Commission Report for 1901). US Government Printing Office, Washington, DC, pp 353–499, Plates 21–27

Craig MP, Ragen TJ (1999) Body size, survival, and decline of juvenile Hawaiian monk seals, *Monachus schauinslandi*. Mar Mamm Sci 15:786–809

Curtice C, Schick RS, Dunn DC, Halpin PNJAM (2011) Home range analysis of Hawaiian Monk Seals (*Monachus schauinslandi*) based on colony, age, and sex. Aquat Mamm 37:360–371

DeLong R, Kooyman G, Gilmartin W, Loughlin T (1984) Hawaiian monk seal diving behavior. Acta Zoologica Fennica 172:129–131

DeLong RL, Fiscus CH, Kenyon KW (1976) Survey of monk seals (*Monachus schauinslandi*) populations of the Northwestern (Leeward) Hawaiian Islands (Internal report). NMFS, Seattle, WA

Friedlander AM, Demartini EE (2002) Contrasts in density, size and biomass of reef fishes: APEX predators. Mar Ecol Prog Ser 230:253–264

Fyler CA, Reeder TW, Berta A, Antonelis G, Aguilar A, Androukaki E (2005) Historical biogeography and phylogeny of monachine seals (Pinnipedia: Phocidae) based on mitochondrial and nuclear DNA data. J Biogeogr 32:1267–1279

Gobush K, Wurth T, Henderson J, Becker B, Littnan C (2017) Prevalence of interactions between Hawaiian monk seals (*Nemonachus schauinslandi*) and nearshore fisheries in the main Hawaiian Islands. Pac Conserv Biol 23:25–31

Gobush KS, Farry SC (2012) Non-lethal efforts to deter shark predation of Hawaiian monk seal pups. Aquat Conserv Mar Freshwat Ecosyst 22:751–761

Goodman-Lowe G (1998) Diet of the Hawaiian monk seal (*Monachus schauinslandi*) from the Northwestern Hawaiian islands during 1991 to 1994. Mar Biol 132:535–546

Grachev M, Kumarev V, Mamaev L et al (1989) Distemper virus in Baikal seals. Nature 338:209

Harting A, Baker J, Becker B (2004) Non-metrical digital photoidentification system for the Hawaiian monk seal. Mar Mamm Sci 20:886–895

Harting AL, Baker JD, Johanos TC (2007) Reproductive patterns of the Hawaiian monk seal. Mar Mamm Sci 23:553–573

Harting AL, Baker JD, Johanos TC (2017) Estimating population size for Hawaiian monk seals using haulout data. J Wildl Manag 81:1202–1209

Harting AL, Barbieri MM, Baker JD, Mercer T, Johanos TC, Robinson SJ, Littnan CL, Colegrove KM, Rotstein DS (2020) Population-level impacts of natural and anthropogenic causes-of-death for Hawaiian Monk Seals in the Main Hawaiian Islands. Ecological Applications submitted

Harting AL, Johanos TC, Littnan CL (2014) Benefits derived from opportunistic survival-enhancing interventions for the Hawaiian Monk Seal: the silver BB paradigm. Endanger Species Res 25:89–96

Hawaiian Monk Seal Research Program (2019) Population summary for Hawaiian Monk seals in 2019. Internal Report IR 20-001; Pacific Island Fisheries Science Center, Honolulu

Heide-Jørgensen MP, Härkönen T, Åberg PJA (1992) Long-term effects of epizootic in harbor seals in the Kattegat-Skagerak and adjacent areas. Ambio 21:511–516

Henderson JR (2001) A pre-and post-MARPOL Annex V summary of Hawaiian monk seal entanglements and marine debris accumulation in the Northwestern Hawaiian Islands, 1982–1998. Mar Pollut Bull 42:584–589

Henderson JR, Johanos TC (1988) Effects of tagging on weaned Hawaiian monk seal pups. Wildl Soc Bull 16:312–317

Hiruki LM, Gilmartin WG, Becker BL, Stirling I (1993) Wounding in Hawaiian monk seals (*Monachus schauinslandi*). Can J Zool 71:458–468

Hiruki LM, Stirling I, Gilmartin WG, Johanos TC, Becker BL (1993) Significance of wounding to female reproductive success in Hawaiian monk seals (*Monachus schauinslandi*) at Laysan Island. Can J Zool 71:469–474

Iverson S, Piché J, Blanchard W (2011) Hawaiian monk seals and their prey: assessing characteristics of prey species fatty acid signatures and consequences for estimating monk seal diets using Quantitative Fatty Acid Signature Analysis. NOAA Technical Memorandum NMFS-PIFSC-23

Jensen T, van de Bildt M, Dietz HH, Andersen TH, Hammer AS, Kuiken T, Osterhaus AJS (2002) Another phocine distemper outbreak in Europe. Science 297:209

Job DA, Boness DJ, Francis JM (1995) Individual variation in nursing vocalizations of Hawaiian monk seal pups, *Monachus schauinslandi* (Phocidae, Pinnipedia), and lack of maternal recognition. Can J Zool 73:975–983

Johanos TC, Becker BL, Baker JD, Ragen TJ, Gilmartin WG, Gerrodette T (2010) Impacts of sex ratio reduction on male aggression in the critically endangered Hawaiian monk seal *Monachus schauinslandi*. Endanger Species Res 11:123–132

Johanos TC, Becker BL, Ragen TJ (1994) Annual reproductive cycle of the female Hawaiian monk seal (*Monachus schauinslandi*). Mar Mamm Sci 10:13–30

Johanos TC, Harting AL, Wurth TA, Baker JD (2014) Range-wide movement patterns of Hawaiian monk seals. Mar Mamm Sci 30:165–1174

Johnson A, Delong RL, Fiscus CH, Kenyon KW (1982) Population status of the Hawaiian monk seal (*Monachus schauinslandi*), 1978. J Mammal 63:415–421

Kennedy S, Kuiken T, Jepson PD, Deaville R, Forsyth M, Barrett T, Van de Bildt M, Osterhaus A, Eybatov T, Duck C, Kydyrmanov A (2000) Mass die-Off of Caspian seals caused by canine distemper virus. Emerg Infect Dis 6:637

Kenyon KW (1973) The Hawaiian monk seal. International Union for the Conservation of Nature and Natural Resources, Publications, New Series, Supplemental Paper 39: 88–97

Kenyon KW, Rice DW (1959) Life history of the Hawaiian monk seal. Pac Sci 13:215–252

Le Boeuf BJ, Mesnick S (1991) Sexual behavior of male northern elephant seals: I lethal injuries to adult females. Behaviour 116:143–162

LeDuc R (2009) Biogeography. In: Perrin WF, Würsig B, Thewissen JGM (eds) Encyclopedia of marine mammals. Academic Press, San Diego, pp 112–115

Littnan CL, Stewart BS, Yochem PK, Braun R (2006) Survey for selected pathogens and evaluation of disease risk factors for endangered Hawaiian monk seals in the main Hawaiian Islands. EcoHealth 3:232–244

Longenecker K (2010) Fishes in the Hawaiian monk seal diet, based on regurgitate samples collected in the Northwestern Hawaiian Islands. Mar Mamm Sci 26:420–429

Longenecker K, Dollar RA, Cahoon M (2006) Increasing taxonomic resolution in dietary analysis of the Hawaiian monk seal. Atoll Res Bull 543:101–113

Lopez J, Hyrenbach KD, Littnan C, Ylitalo GM (2014) Geographic variation of persistent organic pollutants in Hawaiian monk seals *Monachus schauinslandi* in the main Hawaiian Islands. Endanger Species Res 24:249–262

Mesnick SL, Le Boeuf BJ (1991) Sexual Behavior of male northern elephant seals: II. Female response to potentially injurious encounters. Behaviour 117:262–280

Miller MJ, McCleave JD (1994) Species assemblages of leptocephali in the Subtropical Convergence Zone of the Sargasso Sea. J Mar Res 52:743–772

National Marine Fisheries Service (2007) Recovery Plan for the Hawaiian Monk Seal (Monachus schauinslandi): Revision

NMFS HMSRP. unpublished database. Hawaiian monk seal sightings database. https://inport.nmfs.noaa.gov

NMFS HMSRP. unpublished instrument data. Instrument deployments on Hawaiian monk seals.

Norris TA, Littnan CL, Gulland FMD (2011) Evaluation of the captive care and post-release behavior and survival of seven juvenile female Hawaiian monk seals (*Monachus schauinslandi*). Aquat Mamm 37:342–353

Norris TA, Littnan CL, Gulland FM, Baker JD, Harvey JT (2017) An integrated approach for assessing translocation as an effective conservation tool for Hawaiian monk seals. Endanger Species Res 32:103–115

Parrish FA, Abernathy K (2006) Movements of monk seals relative to ecological depth zones in the lower Northwestern Hawaiian Islands. Atoll Res Bull 543:115–130

Parrish FA, Abernathy K, Marshall GJ, Buhleier BM (2002) Hawaiian monk seals (*Monachus schauinslandi*) foraging in deep-water coral beds. Mar Mamm Sci 18:244–258

Parrish FA, Craig MP, Ragen TJ, Marshall GJ, Buhleier BM (2000) Identifying diurnal foraging habitat of endangered Hawaiian monk seals using a seal-mounted video camera. Mar Mamm Sci 16:392–412

Parrish FA, Littnan CL (2007) Changing perspectives in Hawaiian monk seal research using animal-borne imaging. Mar Technol Soc J 41:30–34

Parrish FA, Marshall GJ, Buhleier B, Antonelis GA (2008) Foraging interaction between monk seals and large predatory fish in the Northwestern Hawaiian Islands. Endanger Species Res 4:299

Parrish FA, Marshall GJ, Littnan C, Heithaus M, Canja S, Becker B, Braun R, Antoneijs GA (2005) Foraging of juvenile monk seals at French Frigate Shoals, Hawaii. Mar Mamm Sci 21:93–107

Pietraszek J, Atkinson S (1994) Concentrations of estrone sulfate and progesterone in plasma and saliva, vaginal cytology, and bioelectric impedance during the estrous cycle of the Hawaiian monk seal (*Monachus schaunslandi*). Mar Mamm Sci 10:430–441

Ragen TJ, Lavigne DM (1999) The Hawaiian monk seal: biology of an endangered species. In: Twiss JR, Reeves RR (eds) Conservation and management of marine mammals. Smithsonian Institution Press, Washington, DC, pp 224–245
Riedman ML, Le Boeuf BJ (1982) Mother-pup separation and adoption in northern elephant seals. Behav Ecol Sociobiol 11:203–215
Robinson S, Harting A, Mercer T, Johanos T, Baker J, Littnan C (2020) Sightings patterns reveal cryptic pupping events to revise reproductive rate estimates for Hawaiian monk seals in the Main Hawaiian Islands. Mar Mamm Sci **submitted**
Robinson SJ, Barbieri MM, Murphy S, Baker JD, Harting AL, Craft ME, Littnan CL (2018) Model recommendations meet management reality: implementation and evaluation of a network-informed vaccination effort for endangered Hawaiian monk seals. Proc R Soc B 285: 20171899
Rose NA, Le Boeuf BJ, Deutsch CJ (1991) Sexual behavior of male northern elephant seals: III. The mounting of weaned pups. Behaviour 119:171–192
Rosendahl PH (1994) Aboriginal Hawaiian structural remains and settlement patterns in the upland archeological zone at Lapakahi, Island of Hawaii. J Hawaii Archeol 3:15–70
Schaeff CM, Boness DJ, Bowen WD (1999) Female distribution, genetic relatedness, and fostering behaviour in harbour seals, Phoca vitulina. Anim Behav 427–434
Schultz JK, Baker JD, Toonen RJ, Bowen BW (2009) Extremely low genetic diversity in the endangered Hawaiian monk seal (*Monachus schauinslandi*). J Hered 100:25–33
Schultz JK, Baker JD, Toonen RJ, Harting AL, Bowen BW (2011) Range-wide genetic connectivity of the Hawaiian monk seal and implications for translocation. Conserv Biol 25:124–132
Starfield AM, Roth JD, Ralls K (1995) "Mobbing" in Hawaiian monk seals (*Monachus schauinslani*): The value of simulation modeling in the absence of apparently crucial data. Conserv Biol 9:166–174
Stewart BS, Antonelis GA, Baker JD, Yochem PK (2006) Foraging biogeography of Hawaiian monk seals in the northwestern Hawaiian Islands. Atoll Res Bull 543:1–145
Thompson NM (2011) Changes in Northwest Hawaiian Island monk seal (*Monachus schauinslandi*) populations as evidenced by stable isotope ratios. Nova Southeastern University, Thesis
Watson TK, Kittinger JN, Walters JS, Schofield TD (2011) Culture, conservation, and conflict: assessing the human dimensions of Hawaiian monk seal recovery. Aquat Mamm 37:386–396
Wilson K, Littnan C, Halpin P, Read A (2017a) Integrating multiple technologies to understand the foraging behaviour of Hawaiian monk seals. R Soc Open Sci 4:160703
Wilson K, Littnan C, Read A (2017b) Movements and home ranges of monk seals in the main Hawaiian Islands. Mar Mamm Sci 33:1080–1096
Wirtz WO (1968) Reproduction, growth and development, and juvenile mortality in the Hawaiian monk seal. J Mammal 49:229–238
Yochem PK (unpublished) Vaccination of Hawaiian monk seals with canarypox recombinant canine distemper vaccine (Merial Purevax Ferret Distemper Vaccine). Internal Report to Pacific Island Fisheries Science Center, Honolulu.

Ethology and Behavioral Ecology of Phocids: Concluding Remarks

Daniel P. Costa and Elizabeth A. McHuron

A consistent theme in this book is that phocid seals have evolved a broad range of behavioral, physiological, and anatomical adaptations that allowed them to invade diverse habitats from the tropics to poles. They are more derived than their close relatives, the sea lions, fur seals, and walruses, but still must return to a solid substrate to give birth after foraging at sea. This book describes the reproductive, foraging, and movement behaviors of phocid seals, and the sophisticated and extreme sensory systems and physiological adaptations that have enabled seals to invade such diverse habitats. Our understanding of phocid evolution, physiology, ecology, and behavior has expanded dramatically from the early work predominately carried out as part of commercial seal harvests (Laws 1953) on or near breeding colonies. Some ideas and concepts have been reinforced, while others have not been supported. Recent findings provide the basis for a more complete understanding and remind us that it is best to not hold too tightly to established notions of phocid behavior.

Phocid evolutionary history provides one example of how our perceptions have changed through time. It was historically thought that phocids and otariids arose independently from different terrestrial ancestors (diphyletic), but modern molecular tools support the hypothesis that pinnipeds (phocids, otariids, and odobenids) share a common terrestrial ancestor related to modern bears (monophyletic). More recently, the belief that monachine seals originated in the northern hemisphere (and later dispersed to the southern hemisphere) has been challenged by the discovery of an extinct monk seal in New Zealand (Rule et al. 2020). Fossil monk seals had previously only been found in the northern hemisphere, leading to the conclusion that they were exclusively a northern tribe. Rule and colleagues suggest that this discovery supports

D. P. Costa (✉)
Center for Ocean Health, University of California, Santa Cruz, Santa Cruz, CA, USA

E. A. McHuron
University of Washington, Seattle, WA, USA

D. P. Costa and E. McHuron (eds.), *Ethology and Behavioral Ecology of Phocids*, Ethology and Behavioral Ecology of Marine Mammals,
https://doi.org/10.1007/978-3-030-88923-4

the hypothesis that monachines largely evolved in the southern hemisphere, with multiple equatorial crossings in their evolutionary history.

At first glance a chapter on evolution may seem out of place in an ethology book, but the fossil record provides a window into the evolution of life-history patterns and behavior. For example, large body size and sexual dimorphism appeared quite early in the fossil record, suggesting that at least some ancestral phocids exhibited a polygynous breeding system and perhaps used a capital-breeding strategy to some degree. Examinations of bony orbit size indicate that vision was the primary means of locating food, supporting the hypothesis that ancestral phocids were shallow divers. Other traits, such as the teeth or skull morphology, remain as critical pieces to investigate the origin and evolution of foraging patterns of both extant and extinct species.

Behavior is an expression of how individuals interact with their environment. As such, it is worthwhile to first consider their "sensory ecology", how seals perceive and interpret information from the environment. Like many other aspects of their life history, phocid sensory systems are specialized to accommodate their semi-aquatic lifestyle. Much of these adaptations revolve around the need to "see" in an environment with very little light, such as large eyes, highly innervated vibrissae that allow them to sense water movements created by fish movement and respiration, and a hearing range that spans nearly eight octaves. Adaptations to their visual and auditory systems enable phocids to successfully transition between sea and land, despite the vastly different physical properties of water and air. The ability to hear and see well on land is crucial during the breeding season, even for aquatic-mating species, since sound and vision are important in mother–pup recognition.

Similar to the sensory challenges posed by their semi-aquatic lifestyle, seals face conflicting physiological demands that constrain behavior while on land and at sea. Foraging at depth coupled with the need to return to the surface to breath requires extreme physiological adaptations aimed at increasing onboard oxygen stores, reducing oxygen requirements, and coping with extreme pressures at depth. These adaptations include large muscle myoglobin stores, increased blood volume, high hematocrit, and modified airway structures, along with a strong dive response characterized by extreme bradycardia. There is considerable intraspecific variation in diving behavior that corresponds well with the extent of these adaptations, with some species foraging at relatively shallow depths and others reaching depths on par with the deepest diving mammals, the beaked (family Ziphiidae) and sperm whales (*Physeter macrocephalus*). The need to reproduce on a stable substrate requires seals to fast (to some degree) during energetically expensive periods. This separation is facilitated by a large body size, and alterations in metabolism and endocrine regulation compared with terrestrial mammals. Behavioral endocrinology studies are largely lacking for phocids. Still, existing research indicates that hormones play a central role in the expression of behavior. This is apparent during the breeding season, where testosterone (in males) and oxytocin (in females) levels affect territorial behavior and pup attentiveness, respectively.

The interaction between physiology and the environment has shaped the breeding systems of phocid seals, which range from serial monogamy to extreme polygyny. This interplay was described in 1970 by George Bartholomew based on observations of just a few species. It is a concept that has survived the test of time. This paper was instrumental in directing my (Dan's) own research into pinniped reproductive patterns and energetics. Since the 1970s, thousands of hours of observations by countless researchers have supported and enriched Bartholomew's concepts. Highly polygynous breeding systems, such as those typified by elephant (*Mirounga* spp.) and gray seals (*Halichoerus grypus*), are associated with stable substrates with limited space where females aggregate in time and space. Sexual size dimorphism (males larger than females) is common in polygynous breeders, which facilitates males' ability to hold territory while fasting. Weddell (*Leptonychotes weddellii*) seals exhibit reverse sexual size dimorphism and are an exception to this pattern, likely because males hold aquatic territories where large body size is not advantageous. Unstable substrates where females are dispersed seem to favor serial monogamy, likely because males cannot monopolize access to multiple females simultaneously. The breeding systems of many ice-breeding phocids are unknown, but we can hypothesize about what they might be from these seemingly general patterns. The ability to do so is vital in identifying fundamental shifts in breeding behavior that may occur as sea ice disappears.

While some aspects of phocid breeding behavior have been well-described for decades, some relatively new and exciting avenues are being explored. Much of this revolves around studies of individual behaviors not attributable to the more commonly cited factors, such as the physical environment, sex, age, or body size. For example, Robinson and colleagues found that plasma oxytocin levels influenced maternal behavior, such that gray seal mothers with high oxytocin levels stayed closer to their pups, reducing mother–pup separations and increasing pup survival (Robinson et al. 2015). Pups of moms with high oxytocin levels had high oxytocin levels themselves; they exhibited higher growth rates without any additional energetic investment, perhaps due to behavioral differences that reduced pup energy expenditure (Robinson et al. 2019). Working in the same system, Twiss and colleagues have documented consistent individual differences in alertness and space use of tenured males across multiple breeding seasons, and vigilance behavior of breeding females (see Chap. 9 for references). This individual behavioral variation is akin to personalities. Any researcher that has spent time observing and working with phocids likely has anecdotal observations of personality differences among individual seals. For example, several adult female elephant seals at the Año Nuevo breeding colony are notorious for their excessive aggressive behavior towards researchers. In gray seals, there appear to be differences in reproductive performance associated with individual behavioral variation, highlighting the importance of further research into seal personalities.

The first studies of seal behavior were undertaken on land for the simple reason that it was primarily impossible to study behavior at sea. Today, we can characterize the movement patterns, foraging behavior, and habitat utilization of seals across a range of temporal and spatial scales, as well as the underlying mechanisms driving these behaviors. Like most other marine predators, their movements are primarily driven by the distribution of their prey, which range from krill, to fish and cephalopods, to other marine mammals. Ultimately prey distribution is dictated by the physical environment. As a result, seals are often associated with oceanographic (e.g., frontal zones, eddies, polynyas) and bathymetric (e.g., the shelf break, seamounts) features that tend to predictably concentrate and retain prey. Near the poles, seal distribution is also influenced by sea ice, with preferences in ice association varying within and among species. Given the close association between seals and dynamic ocean features, it is not surprising that foraging behavior and success vary regionally, intra- and interannually, and in response to large-scale climatic events, such as the El Niño Southern Oscillation.

Extrinsic factors are not the sole drivers of variation in foraging behavior, as sex, age, and body size and condition also play roles in shaping behavior. Once weaned, phocid pups generally exhibit rapid development in foraging behavior, with increases in dive duration and depth in the weeks and months following weaning. Adult foraging patterns develop within the first several months to years of life, but much remains to be learned about the ontogeny of foraging behavior. Body size, which is often inherently linked with sex, affects diving ability, fasting capacity, energetic needs, and reproductive strategies, all of which in turn affect foraging behavior. There is often vertical and spatial segregation in foraging in species with sexual size dimorphism, such as gray and elephant seals. In contrast, there is little to no differentiation in species where males and females are similarly sized. Sex-specific differences appear to develop early in life, often before size dimorphism becomes evident.

The book's final Chaps. 9–16 describe what is known about the best-studied seal species and provide species-specific examples of the earlier described concepts. A surprising amount of information has been acquired on elephant, gray, harbor (*Phoca vitulina*), Weddell, harp (*Pagophilus groenlandicus*), ringed (*Pusa hispida*), and Hawaiian monk seals (*Neomonachus schauinslandi*), despite that several of these species inhabit some of the most remote regions of the planet. We know the most about elephant, Weddell, and gray seals. They are the most accessible to researchers, form breeding colonies on a stable substrate (land or fast ice), and tolerate human activities well.

Gray and harbor seals are common fixtures throughout the northern hemisphere, with distributions covering almost the complete range of habitats used by all phocids. The incredible flexibility in habitat type and the range of substrates they use for breeding and resting provides an excellent opportunity for comparative studies. In Chap. 9, Twiss and colleagues highlight how variations in the physical environment influence gray seal site fidelity, time-activity budgets, breeding and lactation strategies, and conspecific interactions. Gray seals have and continue to be at the forefront of behavioral and other biological research; they were the focus of some of the early studies on diving physiology and reproductive energetics, and now are leading the marine mammal field in studies on animal personality. Unfortunately, comparative studies on harbor seals are more difficult due to their skittish nature and aquatic breeding strategy. It is clear that local environmental conditions can influence many harbor seal behaviors, such as haulout behavior, territorial defense, and diet. Still, there is currently little evidence that a harbor seal's behavior or life-history strategy in the Arctic is fundamentally different than one from a lower latitude.

There is a long history of research on elephant and Weddell seals, with long-term monitoring programs that have been in place since the 1970s and 1960s, respectively (Chaps. 11, 12 and 13). Both species more or less lack terrestrial predators, which has facilitated research since they are easily approached and tolerant of human handling (Weddell seals more so than elephant seals) compared with other species. Perhaps more than any other phocid, research on these two species has provided incredible insight into how physiology constrains the expression of behavior.

Elephant seals are the epitome of an extreme seal; they are the largest and deepest diving pinniped, migrate thousands of kilometers across ocean basins, exhibit extreme sexual size dimorphism and a highly polygynous breeding system, and rely entirely on body stores to support energetically costly life-history events. Size dimorphism has resulted in sex-specific variation in foraging habitat and presumably diet, which likely reflects the different energy needs and growth strategies of males and females. Their large body size enables elephant seals to utilize distant and deep prey patches at sea, while facilitating their extreme fasting ability on land. Technological developments, particularly accelerometers, video monitors, and ocean-sensing tags, have allowed us to view their behavior hundreds of meters below the surface and link such behavior with oceanographic features. If it wasn't for such advancements, we might still believe that elephant seals were a relatively coastal species, as was surmised early on from shipboard surveys of northern elephant seals (Fig. 1).

Weddell seals have provided incredible insight into the diving physiology of marine mammals, and are second only to elephant seals in their diving capabilities. Studies on their diving heart rate, metabolism, and heat flux have provided an unapparelled look at the mammalian "dive response" and revealed how marine mammals maximize their time at depth spent foraging. At the same time, long-term demographic studies have been instrumental in revealing how intrinsic and extrinsic factors influence reproductive behavior and vital rates. From this research, we have learned that there is a survival cost to reproducing at an early age, and that a few "high-quality" individuals contribute disproportionately to the population gene pool. This has also recently been noted for female northern elephant seals (Le Boeuf et al. 2019), raising the question of whether the disproportionate contribution of a few individuals to the gene pool is a broader trend among phocids. Weddell seals also serve as a good reminder that while large body size is a pre-requisite for being a capital breeder, it is not the sole factor that dictates lactation strategy. Body size, condition, and age all play roles in determining where individual seals fall along the capital-income spectrum; seals that are younger, smaller, and in poorer body condition rely more on supplemental foraging.

Harp and ringed seals live in the Arctic pack and fast ice, respectively, the characteristics of which have indelibly left their mark on all aspects of their behavior. Of all the species in Part II, these two are the most challenging to access, primarily due to the nature of the environments they inhabit. Nevertheless, there is still a significant amount of information on their reproductive energetics, breeding behavior, and

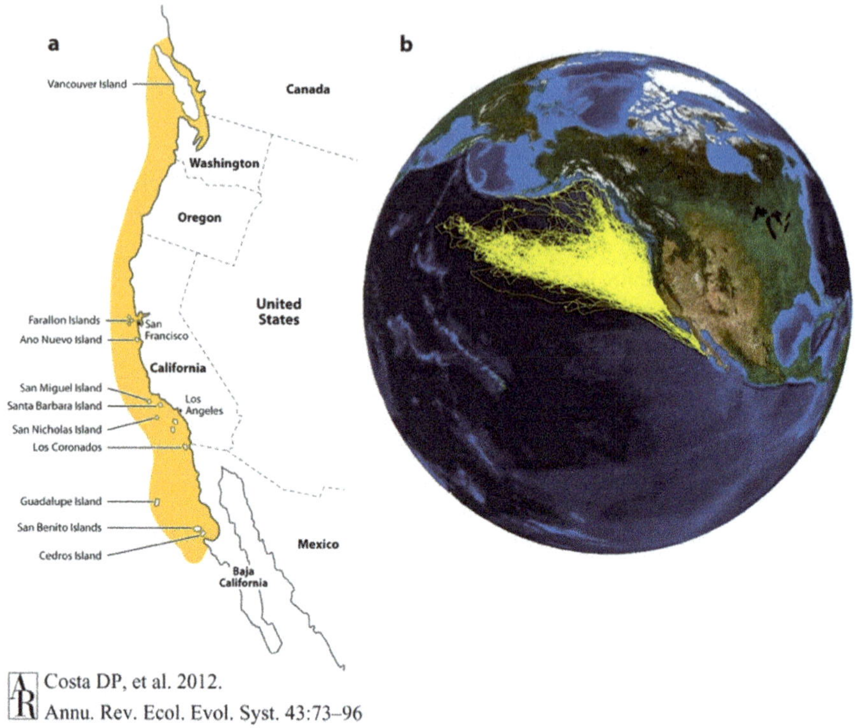

Fig. 1 Tagging technology radically changed our perception of the at-sea distribution of northern elephant seals. The distribution of northern elephant seals as determined from boat- and plane-based surveys is shown in **a** (from Riedman 1990). In **b**, the actual distribution is determined with satellite telemetry. Figure from Costa et al. (2012)

foraging patterns. It is a testament to those who work on them that such excellent data have been acquired. Behavioral studies on these two species have benefited from human exploitation, albeit in very different ways, with early studies facilitated by commercial (harp) and subsistence (ringed) harvests. Traditional Knowledge from Inuit hunters has provided valuable insight into ringed seal behavior, and will likely be particularly important in identifying behavioral responses to a rapidly changing Arctic. Interestingly, harp and ringed seals have vastly different behavioral responses to a shared predator, the polar bear (*Ursus maritimus*), with harp seals exhibiting paralysis and ringed seals the more traditional flight response. Hammill and Stenson discuss that the relatively unique behavioral response of paralysis may have evolved because harp seals breed in dense aggregations (which itself is driven by the instability of the pack ice), highlighting the role of extrinsic factors in modulating the evolution of behavioral responses to predators.

As members of the only tropical phocid genus, Hawaiian monk seals provide a unique glimpse into what it takes to live in a low-latitude, oligotrophic environment. The disappearance and subsequent recolonization of monk seals in the main Hawaiian

Islands has provided a natural experiment of sorts on the environment's influence (mainly food availability) on behavior. Such studies have revealed that seals in the main Hawaiian Islands forage closer to haulout sites, have shorter trip durations, and have a less diverse diet than seals in the Northwest Hawaiian Islands. Research and conservation efforts have been effective in protecting Hawaiian monk seals, the best-studied monk seal. They appear to be making a slow but steady recovery, with conservation interventions benefitting from ethological research. Observations of social interactions have been used to model disease spread and inform large-scale vaccination efforts, critical to preventing a large-scale morbillivirus outbreak. Knowledge of site fidelity, movements, and dispersal strategies have helped refine translocation efforts, revealing that translocation is an effective conservation tool more so for pups than adults because pups remain close to their translocation site.

It is essential that we see that our efforts to protect endangered species can succeed. History shows us that it is possible to bring a species back from the brink of extinction. Elephant seals are such a success story; they were hunted for their oil, and by 1896 the northern elephant seal was reduced to something around 25 individuals. Today, there are over 220,000 individuals that have largely recolonized their historic range. The more recent success with Hawaiian monk seals gives us hope for the future. The fate of Mediterranean monk seals (*Monachus monachus*) hangs in the balance, with fewer than 700 individuals remaining in the population. They face numerous threats, much of them of human origin. Still, there is some cause for hope because they appear to be recolonizing areas in the Mediterranean basin within their historical range.

While a wealth of information exists on the ethology of phocids, much remains to be learned. Most of our understanding has come from the work carried out on the eight species described above. We know very little about the breeding biology, reproductive behavior, and movement patterns of most ice-breeding seals (e.g., crabeater, leopard, ribbon, bearded, and Ross seals), making it difficult to predict how environmental perturbations will alter their behavior, ecology, and ultimately population dynamics. We have just begun to obtain sufficient data to examine the importance of variation in individual characteristics, such as age, sex, experience, and personality. From these studies, it is clear that a seal is not just a seal (as stated by Twiss et al. in Chap. 9), whether we consider two individuals within the same or different populations. Studies on all demographic groups are critical, not only because we know comparatively little about the ontogeny of behavior, but also because different ages, sex, and size classes may respond differently to anthropogenic or climate-related stressors. Similar to behavioral research, much of what we have learned about sensory abilities has been gleaned from just a few species and many questions remain. Comparative studies between shallow- and deep-diving species would help elucidate the role that phylogeny and environment have played in the evolution of sensory adaptations. How the different sensory modalities interact is unknown, which is essential to understand the impact of an environment altered by humans on phocid sensory perception, communication, and behavior.

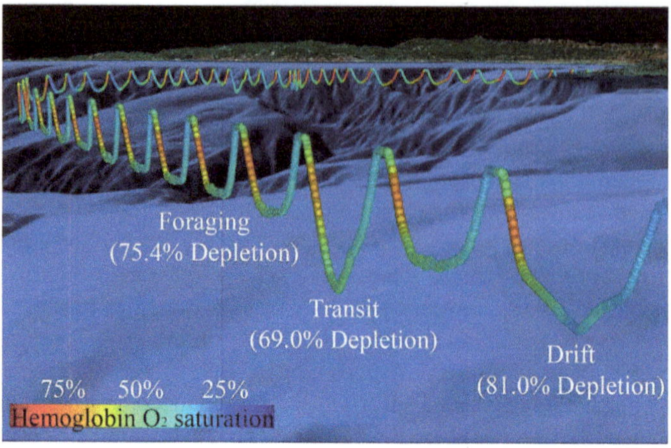

Fig. 2 Sophisticated sensors can be deployed on animals and the resulting data visualized in 3D. This dive profile is from a northern elephant seal crossing Monterey Bay, CA. The percent O_2 saturation of hemoglobin (Hb) is shown during each dive. Warmer colors represent greater % Hb O_2 saturation and cooler colors a lower Hb O_2 saturation. Blood O_2 saturation decreased throughout the dive, with a temporary increase in blood O_2 saturation early in the dive, indicating that gas exchange continued early in the dive. Interestingly, the pattern of O_2 depletion is similar across all dive types. Figure from Meir et al. (2013)

Technology has played a critical role in our study of phocid ethology and behavior, from early studies with simple time-depth recorders to today where we can simultaneously measure 3D movements, feeding events, and the external environment (Fig. 2). Technological advances continue to develop, enabling predictions of movement and a greater understanding of the ecological context and physiological mechanisms that couple behavior to demography. Other emerging tools include genomics, which makes possible significant advances in understanding physiological mechanisms, phylogeny, evolution, diet, population structure, and individual relatedness. Finally, we must interpret data in the context of how known and hypothesized processes and mechanisms affect populations. Studies of this kind can only be accomplished with long-term data sets where appropriate parameters are measured and related to how they impact reproductive success and survival.

A key driver of much ongoing behavioral research is to understand how individuals and populations will respond to a rapidly changing environment and one that is increasingly dominated by human activities. Throughout this book, there are numerous examples of how the physical environment has shaped the evolution and expression of phocid behavior, leaving little doubt that further change is coming. The extent of that change and the ability of phocids to adapt remain to be seen.

References

Bartholomew GA (1970) A model for the evolution of pinniped polygyny. Evolution 24:546–559.

Costa DP, Breed GA, Robinson PW (2012) New insights into pelagic migrations: implications for ecology and conservation. Annu Rev Ecol Evol Syst 43:73–96.

Laws RM (1953) The elephant seal (*Mirounga leonina* Linn.) I. Growth and Age. FIDS Sci Reports 8:1–67.

Le Boeuf B, Condit R, Reiter J (2019) Lifetime reproductive success of northern elephant seals (*Mirounga angustirostris*). Can J Zool 97:1203–1217.

Meir JU, Robinson PW, Vilchis LK, Kooyman GL, Costa DP, Ponganis PG (2013) Blood oxygen depletion is independent of dive function in a deep diving vertebrate, the northern elephant seal. PLoS One 8(12):e83248.

Riedman M (1990) The pinnipeds: seals, sea lions, and walruses. University of California Press, Berkeley, p 439.

Robinson KJ, Twiss SD, Hazon N, Pomeroy PP (2015) Maternal oxytocin is linked to close mother-infant proximity in gray seals (*Halichoerus grypus*). PloS One 10:e0144577.

Robinson KJ, Hazon N, Twiss SD, Pomeroy PP (2019) High oxytocin infants gain more mass with no additional maternal energetic costs in wild gray seals (*Halichoerus grypus*). Psychoneuroendocrinolo 110:104423.

Rule JP, Adams JW, Marx FG, Evans AR, Tennyson AJD, Scofield RP, Fitzgerald EMG (2020) First monk seal from the Southern Hemisphere rewrites the evolutionary history of true seals. Proc Royal Soc B 287:20202318.

GPSR Compliance
The European Union's (EU) General Product Safety Regulation (GPSR) is a set of rules that requires consumer products to be safe and our obligations to ensure this.

If you have any concerns about our products, you can contact us on

ProductSafety@springernature.com

In case Publisher is established outside the EU, the EU authorized representative is:

Springer Nature Customer Service Center GmbH
Europaplatz 3
69115 Heidelberg, Germany

www.ingramcontent.com/pod-product-compliance
Ingram Content Group UK Ltd.
Pitfield, Milton Keynes, MK11 3LW, UK
UKHW022253091026
15624UKWH00016B/359
* 9 7 8 3 0 3 0 8 8 9 2 2 7 *